Combinatorial Algorithms for Integrated Circuit Layout

Applicable Theory in Computer Science *

A Wiley – Teubner Series in Computer Science

*Previously known as the Wiley – Teubner Series in Computer Science

Combinatorial Algorithms
for Integrated Circuit Layout

Thomas Lengauer
University of Paderborn
Paderborn, West Germany

B. G. TEUBNER
Stuttgart

JOHN WILEY & SONS
Chichester · New York · Brisbane · Toronto · Singapore

Copyright © 1990 by John Wiley & Sons Ltd
Softcover reprint of the hardcover 1st edition 1990
    Baffins Lane, Chichester
    West Sussex PO19 1UD, England
  &B. G. Teubner
    Industriestrasse 15
    70565 Stuttgart
    Germany

Reprinted February 1994

Other Wiley Editorial Offices

John Wiley & Sons, Inc., 605 Third Avenue,
New York, NY 10158–0012, USA

Jacaranda Wiley Ltd, G.P.O. Box 859, Brisbane,
Queensland 4001, Australia

John Wiley & Sons (Canada) Ltd, 22 Worcester Road,
Rexdale, Ontario M9W 1L1, Canada

John Wiley & Sons (SEA) Pte Ltd, 37 Jalan Pemimpin #05-04,
Block B, Union Industrial Building, Singapore 2057

CIP-Titelaufnahme der Deutschen Bibliothek:

Lengauer, Thomas:
Combinatorial algorithms for integrated circuit layout / Thomas
Lengauer - Stuttgart ; Teubner - Chichester : Wiley, 1990
  ISBN 978-3-322-92108-6   ISBN 978-3-322-92106-2 (eBook)
  DOI 10.1007/978-3-322-92106-2

Library of Congress Cataloging-in-Publication Data:

Lengauer, T. (Thomas)
  Combinatorial algorithms for integrated circuit layout / Thomas
Lengauer.
   p. cm.
  Includes bibliographical references (p.) and indexes.
  ISBN 978-3-322-92108-6
  1. Integrated circuits—Very large scale integration—Design and
construction. 2. Algorithms. I. Title.
TK7874.L36 1990
621.39′5—dc20   90-12489
                  CIP

British Library Cataloguing in Publication Data:

Lengauer, Thomas
  Combinatorial algorithms for integrated circuit layout.
  1. Electronic equipment. Circuits. Design. Applications of
computer systems.
  I. Title II. Series
  621.38153

ISBN 978-3-322-92108-6

To Sybille and Sara

Contents

Foreword

The last decade has brought explosive growth in the technology for manufacturing integrated circuits. Integrated circuits with several hundred thousand transistors are now commonplace. This manufacturing capability, combined with the economic benefits of large electronic systems, is forcing a revolution in the design of these systems and providing a challenge to those people interested in integrated system design. Modern circuits are too complex for an individual to comprehend completely. Managing tremendous complexity and automating the design process have become crucial issues.

Two groups are interested in dealing with complexity and in developing algorithms to automate the design process. One group is composed of practitioners in computer-aided design (CAD) who develop computer programs to aid the circuit-design process. The second group is made up of computer scientists and mathematicians who are interested in the design and analysis of efficient combinatorial algorithms. These two groups have developed separate bodies of literature and, until recently, have had relatively little interaction. An obstacle to bringing these two groups together is the lack of books that discuss issues of importance to both groups in the same context. There are many instances when a familiarity with the literature of the other group would be beneficial. Some practitioners could use known theoretical results to improve their "cut and try" heuristics. In other cases, theoreticians have published impractical or highly abstracted toy formulations, thinking that the latter are important for circuit layout.

With *Combinatorial Algorithms for Integrated Circuit Layout*, Dr. Lengauer has taken a major step in unifying the work of practitioners and theoreticians interested in circuit layout. This unification provides benefits to both groups. The CAD practitioners benefit by seeing the most up-to-date results, with practical formulations, from the combinatorial and optimization community presented in the context of integrated circuit design. Layout algorithms are placed on a firm, theoretical foundation and are discussed in rigorous terms. This foundation often provides insights into the combinatorial structure of layout problems. The theoreticians benefit from an understanding of the important theoretical issues in circuit layout presented within the context of very large-scale integrated circuit technology. The book points the way to innovations needed in algorithm design. Practical formulations of layout problems allow the reader to evaluate the relevance of algorithm design to circuit layout.

This book concentrates on the constructive aspects of integrated-circuit layout. Since most of these problems are NP-hard, or harder, approximations and heuristics are important. In contrast to other authors who deal with this degree of difficulty in a less systematic fashion, Dr. Lengauer places approximations and heuristics on a sound, formal basis.

The coherent presentation of important combinatorial problems in the context of integrated-circuit design allows this book to be used as either a reference or as a text. Researchers, CAD tool builders, and people interested in algorithm design and analysis will all find this book to be a valuable resource as they jointly explore the frontiers of circuit-layout algorithms.

Palo Alto, January 1990   Bryan Preas

Preface

The layout of integrated circuits on chips and boards is a complex task. A major portion of the research in the area of design automation has been devoted to the development of efficient and easy-to-use systems that support circuit layout. There are two aspects of building a layout system that are difficult. One is the combinatorial aspect and the other is the systems aspect. The combinatorial aspect is that most of the optimization problems that have to be solved during integrated-circuit layout are intractable. More specifically, they are usually NP-hard (a notion that will be defined rigorously in Section 2.2) or harder. This observation implies that finding the optimal solution for each given problem instance in reasonable time is out of the question. We have to find other ways of solving the problem adequately. The systems aspect deals with all problems that remain, once the combinatorial problems have been solved. Among those are the maintenance of the consistency in the design database in the presence of simultaneous changes of the same design object, the development of an ergonomic and robust human interface, and the appropriate guidance through a partially interactive design process.

Both aspects of complexity are equally important, and they strongly influence each other. As an example, an efficient algorithm that optimally solves a combinatorial layout problem such as block placement obviates the necessity of providing designer guidance through an interactive process of iteratively improving a heuristically found placement. On the other hand, a certain model of the design process leads to specific combinatorial problems that do not have to be solved in other design scenarios. For instance, in highly interactive design systems, dynamic versions of combinatorial problems prevail.

Goal

This book is concerned with only the combinatorial aspect of circuit layout. Its purpose is to give an overview of what are currently the most important combinatorial problems in circuit layout, and to describe their solutions. We discuss algorithms used in layout systems today, as well as new algorithmic developments that as of now only exist as theoretical proposals but that bear some promise of stimulating the improvement of layout systems in the future.

The combinatorial aspect of circuit layout has been under intensive investigation in the past decade. Contributions have been made both by the CAD community and by mathematicians and computer scientists that specialize in the design and analysis of efficient algorithms for combinatorial problems. Initially, these two communities were quite disjoint, but recently they have been growing together to form a group of scientists that employs advanced techniques in algorithm and system design to solve the circuit-layout problem.

The overall goal of this book is to accelerate this process of unification. To this end, the book aims to provide experts from both communities with a solid basis for advanced research in the area. Therefore, the book caters to several groups of readers. On the one hand, it aims at the CAD practitioner who is interested in the basics of layout algorithms and in new algorithmic developments in the area of circuit layout. On the other hand, the book provides experts in the area of algorithm design and analysis with information on where the current hot spots are in layout-algorithm design.

Organization of the Book

The organization of the book is tailored to this purpose. Specifically, Part I of the book is a detailed introductory section that spans over 200 pages. The material presented in this section provides the foundations for the discussions of the layout problems investigated in Part II of the book.

Part I, Chapters 1 through 4, contains material from both layout practice and the mathematical disciplines contributing methods to layout optimization. Chapter 1 gives a general introduction into layout practice. The chapter does not dwell on details of specific fabrication and design technologies. Rather, the goal is to give a broad overview over layout strategies, so that we can understand and evaluate different mathematical methods in terms of their appropriateness for circuit-layout optimization.

Chapters 2 to 4 present the mathematical foundations of circuit layout. These chapters provide the essential material that the CAD expert needs in order to understand and develop circuit-layout methods. Much of this material has entered the design-automation community in bits and pieces through research papers. In this book, it has been collected and specially adapted for the needs of the CAD expert for the first time. The inclusion of these chapters makes the book self-contained and frees the reader from having to consult much more extensive mathematical texts, many of which are aimed exclusively at mathematicians or computer scientists.

Chapter 2 discusses the foundations of algorithm theory and introduces the important notions of asymptotic analysis and NP-hardness. Chapter 3 gives an extensive account of methods from graph theory that are of use in circuit layout. In particular, this chapter focusses on the methods of path searching (Section 3.8) and network design (Sections 3.9 and 3.10) that are central to wire routing. In addition, other classical material from graph theory is provided.

At the end of Chapter 3, an introduction into the space- and time-efficient methods of hierarchical graph processing is presented (Section 3.13). These methods gain special importance because of the tremendous increase in the volume of design data that have to be processed during layout optimization.

Chapter 4 gives an introduction to the results from combinatorial optimization that are of importance in layout. A particular focus of the chapter is efficient methods for solving integer linear programs. General techniques such as branch-and-bound (Section 4.6.2) and dynamic programming (Section 4.7) are discussed, and newer techniques such as Lagrangian relaxation (Section 4.6.3) are surveyed, also. In addition, this chapter discusses nonlinear optimization methods (Section 4.8) and statistical methods, especially simulated annealing (Section 4.3). The general material on simulated annealing is complemented with sections on the specific applications of simulated annealing to layout problems in Part II of the book.

Part II, Chapters 5 through 10, discusses the actual layout optimization problems. Chapter 5 provides the theoretical background by presenting a general layout technique that has a provable worst-case behavior that is quite good, theoretically. The method is not competitive, in practice, but it implies essential structural insights into circuit layout—especially as it relates to circuit partitioning.

Chapter 6 gives a taxonomy of the wide variety of partitioning problems. Furthermore, the chapter discusses various methods for circuit partitioning. On the one hand, the chapter presents iterative methods that are widely used in practice (Section 6.2). On the other hand, more recent methods based on network flow (Section 6.5) and nonlinear optimization (Section 6.6) are presented. Among special cases of the partitioning problem that are discussed in the chapter are clustering—that is, partitioning a circuit into many small parts (Section 6.7)—and partitioning planar graphs (Section 6.4). The latter problem is important in divide-and-conquer algorithms for layout. Here, we partition not circuits, but rather planar graphs representing the routing region on the substrate.

Chapter 7 discusses placement and floorplanning—that is, the task of assigning (approximate) coordinates to the components of the circuit. This problem is an optimization problem that has to deal with two cost functions. One cost function is an estimate of the area taken up by the subsequent wire-routing phase. The other cost function estimates the area consumed by the packing of the two-dimensional arrangement of the circuit components, disregarding the wiring. Both cost functions and methods for their optimization are discussed in turn. Section 7.1 discusses the cost of wiring. Section 7.2 discusses the packing aspect and, eventually, both cost functions in conjunction. At the center of the discussion is a robust method for placement and floorplanning that is based on circuit partitioning. The chapter also discusses different—for instance, nonlinear—methods of placement. Additional combinatorial problems occurring in this phase of layout are discussed in Section 7.3.

Chapter 8 discusses global routing—that is, the phase of layout that determines the approximate course of the wires. Global routing is formulated as an integer program. The classical maze-running and line-searching techniques are discussed as special ways of solving the integer program (Section 8.4). Much of this discussion refers back to the graph algorithms presented in Chapter 3. In addition to presenting classical routing methods, Chapter 8 introduces advanced methods for global routing that are based on more appropriate integer-programming techniques. In particular, two such methods are discussed. In hierarchical routing, the integer program is decomposed into small pieces that are solved exactly (Section 8.5). In global routing by relaxation, linear-programming techniques are employed, followed by a random or deterministic correction phase (Section 8.6). Chapter 8 also surveys newer research on integrating the floorplanning with the global-routing phase (Section 8.8).

Chapter 9 is the most extensive chapter of the book. It contains a taxonomy of detailed-routing problems and the methods for their solution. In detailed routing, the approximate wire routes determined during global routing are made exact. The chapter is so extensive because there is a wealth of different combinatorial problems and results in this area. We perform the classification of these results using two criteria. The first is the shape of the detailed-routing region. The second is the detailed-routing model—that is, the rules that have to be obeyed during routing. These rules are the formal counterpart of the technological boundary conditions. The chapter spans a wide spectrum of results, from the widely applied results on Manhattan channel routing (Section 9.6.1.3) to advanced theories of detailed routing (Section 9.2).

Chapter 10 discusses compaction. During compaction, the design rules of the fabrication process enter the layout optimization process. The emphasis of the chapter is the most popular approach to compaction—namely, graph-based one-dimensional compaction (Section 10.2.2). The alternative compression-ridge approach is also discussed (Section 10.2.1). On the foundation of one-dimensional compaction, extensions are discussed, such as two-dimensional compaction (Section 10.3) and hierarchical compaction (Section 10.4). At the end of the chapter, a survey of the new field of topological compaction, compaction before wire routing, is given (Section 10.5).

Layout verification and cell synthesis are not discussed in the book. We decided to omit layout verification because we want to emphasize, from a design-methodology standpoint, the more advanced, constructive aspects of circuit layout. Furthermore, combinatorial aspects of layout verification are well surveyed in existing textbooks on circuit layout and on computational geometry. Cell synthesis is still in a very early stage, combinatorially. We decided that the field is not mature enough to merit the inclusion of a special chapter on cell synthesis at this time. We expect a subsequent edition of the book to include such a chapter.

How to Use this Book

The book fulfills the dual purpose of a textbook for a graduate-level course and a reference for experts in CAD and in optimization. A basic understanding of algorithms, and of mathematical prerequisites such as linear algebra and linear programming, is required. Knowledge of VLSI design, such as communicated by Mead and Conway [313] is helpful.

The book can be used as a basis or as an accompanying text for courses in computer science, electrical engineering, mathematics, and operations research. Depending on the students' backgrounds, the courses can stress applied issues of layout (then the book would be an accompanying text), focus on layout algorithms (with the book as a basis for the course), or even focus on optimization, in general, and use circuit layout as a practical example. The role of Part I of the book depends on the background of the course participants and has to be decided by the lecturer in each case. Part II of the book relies heavily on the notation introduced in Part I. Thus, even if the course participants know the material in Part I, they will have to refer back to Part I in order to understand the notation. Therefore, Part I should always be background reading for any course using the book. On the other hand, Part I should be easy to read for somebody with knowledge of the corresponding material.

The book contains over 150 exercises. Many exercises answer questions raised in the text, or extend the discussion of the text. Exercises that are open research problems are marked with a star (*). Some of the other exercises are also quite complex. For most of the more difficult exercises, hints are provided to guide the solution. Some exercises contain references to the literature that contains the solution. For other exercises, the references to the literature are given at the text location that cites the exercise. No complete solutions to exercises are given in the book itself.

In addition to serving as a textbook, the book is designed as a reference for the field. To this end, Part II contains several survey sections that describe material presented in the literature but that do not go into technical detail. These survey sections can serve as background reading in a course based on the book. In addition, they can be used as a guideline for selecting topics for an advanced seminar on the subject. The survey sections are marked with an asterisk (*).

The book has an extensive list of references to the literature that contains papers that appeared prior to the summer of 1989. The reference list is not exhaustive, but it does contain most major papers on the discussed topics.

The book has an extensive hierarchical subject index, to help the reader to locate points where key terms are discussed. If a term is discussed over a section spanning several pages, only the beginning page of this section is listed. If there is a primary point of discussion of the term—for instance its definition—the corresponding page number is written in bold. However, some terms do not have a primary reference in the text.

An author index lists the points where the names of authors are mentioned in the book. Only the authors mentioned by name in the text are listed. To find the discussion of a paper, you should look up the name of the first author. Some papers are mentioned in the text without explicit reference to their authors. In this case, no listing will appear in the author index.

Paderborn, January 1990   Thomas Lengauer

Acknowledgments

A book is never the product of a single person: the author of a book always relies on the cooperation of many helpful people. In the case of the present book, both people and institutions have been critical to the completion of the project.

First, I wish to thank XEROX PARC. The backbone of the book was written during a 7-month period, while I stayed at XEROX PARC in summer 1988. The unique research environment at the Computer Systems Laboratory at XEROX PARC provided both the technical stimulation and the required flexibility to seclude myself, whenever I needed either. At XEROX PARC, Bryan Preas deserves my special thanks. Bryan was much more than a host for the time of my stay at PARC. He scrutinized my early drafts and provided me with much information on layout practice that helped me to put into perspective the results discussed in the book. Furthermore, Bryan suggested that I apply for a grant with the Special Interest Group on Design Automation (SIGDA) of the Association of Computing Machinery (ACM) to help produce the book. This grant was awarded, and it saved about a year of production time and substantially increased the production quality of the book. Thanks also to Waldo Magnuson, who handled the grant flexibly and with much engagement.

Many people have supported the book at XEROX PARC. Marshall Bern was a valuable discussion partner on path and Steiner-tree problems; he reviewed drafts of several chapters of the book. John Gilbert provided valuable feedback on graph partitioning. Dan Greene helped me prepare my stay at XEROX PARC and triggered many contacts with other researchers. Lissy Bland, Pavel Curtis, Carl Hauser, Willie-Sue Orr, and Michael Plass provided substantial support with the use of the Cedar programming environment of the Computer Systems Laboratory and with LaTeX. They had to put up with my frequently dropping into their offices unannounced, ripping them out of their work context, and asking for immediate response on detailed technical requests—and they always offered their help willingly and remained friendly. Lisa Alfke from the XEROX PARC library showed great identification with the project and provided many pieces of literature amazingly quickly. A large part of the reference list has been compiled with her help.

Many other people provided input for the book, among them Greg Cooper,

Wei-Ming Dai, Susanne Hambrusch, Kye Hedlund, David Johnson, Norbert Köckler, Kurt Mehlhorn, Rolf Möhring, Manfred Nowak, Ralph Otten, Franz Rendl, Majid Sarrafzadeh, Hans-Ulrich Simon, Bob Tarjan, Ioannis Tollis, Frank and Dorothea Wagner, and Andy Yao. Ulrich Lauther gave me extensive and very valuable comments on the first draft of the book. Wei-Ming Dai, Majid Sarrafzadeh, and Rolf Möhring used parts of the book as material for courses on the subject.

At the end of my stay at XEROX PARC the book was far from finished. The people in my research group at Paderborn helped me to set up the production environment in Germany and provided a smooth transition for further work on the book. Friedhelm Wegener, Michael Schmidt, and Michael Utermöhle provided me with reliably running hardware and software. Michael Janich contributed invaluable help with digging into the bowels of TeX during the final editorial work. Bertil Munde gave much appreciated support with the integration of the text and the figures.

In my research group at Paderborn, Jürgen Doenhardt, Jörg Heistermann, Franz Josef Höfting, Rolf Müller, Egon Wanke, and Charlotte Wieners-Lummer provided valuable feedback. They also did the proof-reading, and helped to organize a student seminar on the basis of the first draft. Above all, they quietly accepted the substantial lack of my availability during the last stages of the production. My students Anja Feldmann, Albert Specovius, Dirk Theune, and Christoph Wincheringer provided valuable comments on selected parts of the book.

I want to extend special thanks to Carmen Buschmeyer, who spent an extraordinary amount of time and energy on typesetting the book and drawing all figures. Figure drawing was the bottleneck for large intervals during the preparation of the book, and her quick grasp of my chaotic sketches of often quite intricate graphical material was instrumental in producing all figures in time.

The linguistic quality of a book contributes a great deal to its readability. In Lyn Dupré, I was very fortunate to have a copy editor who showed great identification with the project and zeroed in on the strange mistakes in the English language that only a German can produce. I would like to thank her for providing effective help over a large distance—and for encouraging me to adjust the names of classical mathematical problems. Lyn changed the way I perceive the English language.

In Bill Clancey and Greg Cooper I have very special friends who provided a lot of support going far beyond technical matters.

Finally, I want to extend special thanks to my wonderful wife Sybille and to my sweet little daughter Sara. They gave the most for receiving the least.

T.L.

Part I

Background

Chapter 1

Introduction to Circuit Layout

1.1 Combinatorial Aspects of Circuit Layout

In the combinatorial sense, the layout problem is a constrained optimization problem. We are given a description of a circuit—most often as a netlist, which is a description of switching elements and their connecting wires. We are looking for an assignment of geometric coordinates of the circuit components—in the plane or in one of a few planar layers—that satisfies the requirements of the fabrication technology (sufficient spacing between wires, restricted number of wiring layers, and so on) and that minimizes certain cost criteria (the area of the smallest circumscribing rectangle, the length of the longest wire, and so on). Practically all versions of the layout problem as a whole are intractable; that is, they are NP-hard. Thus, we have to resort to heuristic methods. One of these methods is to break up the problem into subproblems, which are then solved one after the other. Almost always, these subproblems are NP-hard as well, but they are more amenable to heuristic solution than is the layout problem itself. Each one of the layout subproblems is decomposed in an analogous fashion. In this way, we proceed to break up the optimization problems until we reach primitive subproblems. These subproblems are not decomposed further, but rather are solved directly, either optimally—if an efficient optimization algorithm exists—or approximately. The most common way of breaking up the layout problem into subproblems is first to do component *placement*, and then to determine the approximate course of the wires in a *global-routing* phase. This phase may be followed by a *topological compaction* that reduces the area requirement of the layout, after which a *detailed-routing* phase determines the exact course of the wires without changing the layout area. After detailed routing, a *geometric-compaction* phase may further reduce the area requirement

3

of the layout. This whole procedure may be done hierarchically, starting with large blocks as circuit components, which are themselves laid out recursively in the same manner. This recursive process may be controlled by algorithms and heuristics that allow for choosing among layout alternatives for the blocks such that the layout area of the circuit is minimized. If cells are *variable* in this sense, the placement phase is called *floorplanning*. Exactly how a given version of the layout problem is broken up into subproblems depends on both the design and the fabrication technology. For instance, in standard-cell design the detailed-routing phase essentially reduces to channel routing. In gate-array layout, the placement phase incorporates an assignment of functional circuit components to cells on the master.

The sequence of phases for solving the layout problem that we have described is the result of long-term experience with heuristic methods for laying out circuits. A necessary consequence of this approach is that the resulting layout is not optimal—it is only very nearly (we hope) optimal. Deviations from the optimum enter in two places. On the one hand, a heuristic or approximate solution of a primitive layout subproblem implies that any layout problem to which it contributes is also solved only approximately. On the other hand, and not quite as trivially, the different phases of the layout process optimize (or try to optimize) cost functions that do not quite match each other. As an example, when doing placement, we already have in mind the goal of minimizing wiring area—or of maximizing wirability. Therefore, the cost function to be minimized in placement correlates with the wiring area. However, since wiring is done after placement, the wiring area can enter the cost function during placement only as a rough estimate. We often choose the half-perimeter of the smallest rectangle surrounding all terminals of a wire as an estimate for the length of this wire. This estimate is accurate only, if the wire makes no detours, which it may have to do if there are obstructions in all its direct paths. Thus, in general, the estimate is only a lower bound on the real wire length, and placement intrinsically optimizes the "wrong" cost function for many problem instances.

Mismatch between cost functions is a central problem in circuit layout and, for that matter, in combinatorial optimization in general. Not only do we deviate from the optimal solution, but also it is difficult to bound the size of the deviation. The algorithm becomes *heuristic* in the true sense of the word. We find a solution, but we have no concrete knowledge as to how good it is. A consequence of this fact is that, as of today, the performance of a layout system can be evaluated *only relative to that of another*. To evaluate a layout system we expose it to a set of supposedly representative or especially difficult test cases, the *benchmarks*. The system's performance is compared to that of competing layout systems, and an assessment of the quality of each of the layout systems is made on this basis. (In the area of circuit layout, the situation is aggravated by the fact that no set of benchmarks exists that is universally agreed on. Thus, each person chooses whatever benchmarks are convenient for him, and it is difficult to evaluate any experimental data on layout systems. Fortunately, the process of defining a comprehensive and representative set of benchmarks is underway, the Microelectronics Center of North Carolina is coordinating the creation of a reliable set of benchmarks.)

For certain subproblems of the layout problem there are excellent heuristics, and almost-matching lower bounds prove that the heuristics find almost optimal solutions on many problem instances. An example is the channel-routing problem (in the Manhattan model, see Section 9.6.1). There are channel routers that route to within one or two of the smallest number of tracks achievable on most examples occurring in practice. However, this solution quality is lost as the solutions of the subproblems are combined to solve the overall layout problem. As a result, no absolute statements can be made as to the quality of the resulting layout.

Many investigations by algorithm designers are aimed at remedying this defect of heuristic layout optimization. They are looking for algorithms that are efficient, yet that yield a layout whose cost is guaranteed to deviate from the cost of an optimal layout by at most a small margin, say 5 percent. It is difficult to obtain such algorithms, even for simple subproblems of the layout problem, and the first such algorithms often achieve deviations that reach several *hundred* percent. Of course, such a statement about the performance of an algorithm is of no practical value. The worth of such an algorithm is that it presents a starting point for a heuristic refinement, at whose end we may obtain algorithms that are both provably good in the worst case *and* so efficient on typical examples that they can be used in practice.

This scenario is what we hope to bring about. In some cases, however, provably good algorithms use methods that are simply of no use for practical purposes. In other words, their margins are high and cannot be decreased if we stick to the method. Even those algorithms can provide valuable insights into the combinatorial structure of layout problems. For instance, their analysis can reveal that certain structural parameters, and only those parameters, determine the quality of a layout. There are several such examples in the layout literature; for instance, the connection between graph separation and layout established by Bhatt and Leighton [34], and the connection between channel width in the Manhattan model and the parameters density and flux, established by Baker, Bhatt, and Leighton [18]. Knowledge of this sort represents fundamental insight into the combinatorial structure of a layout problem. Such insight can be expected to be a prerequisite for a thorough understanding of the problem that can lead to efficient and reliable optimization algorithms.

One way in which we try to improve layout heuristics is to soften the boundaries between layout phases. In other words, rather than just estimating the effect of a succeeding layout phase, we integrate that phase into the current layout phase. An example is the integration between floorplanning and global routing that is being attempted by several researchers. Another example is the development of topological compacters that effectively integrate compaction with detailed routing.

In this section, we have identified the central problems in optimizing circuit layouts. The next section gives a more detailed account of the kinds of strategies used to lay out circuits.

1.2   Layout Methodologies

1.2.1   Semicustom Versus Full-Custom Layout

On a high level, we distinguish between two different kinds of circuit layout—semicustom layout and full-custom layout. In *full-custom layout*, we start with an empty piece of silicon (the circuit-layout specialists call this *real estate*), and we are given wide-ranging freedom regarding how to lay out the circuit components, how to place them, and how to connect them with wires. The utmost full-custom layout style is design by hand; here, the layout designer has full control over her product. In *semicustom layout*, this freedom is more severely restricted by rules that are imposed either by the design technology or by the fabrication technology. Examples of semicustom layout are layout styles that start on prefabricated silicon that already contains all the switching elements (*gate arrays*, see Section 1.2.4.3), or layout styles that involve geometrically restricted libraries of basic circuit components, and essentially do one-dimensional block placement (*standard cells*, see Section 1.2.4.2).

Traditionally, full-custom layout was supported by hardly any tools. The only things available were an interactive editor for hand-designing the layout, on the mask level (see Section 1.2.3.1), tools for verifying the correctness of the hand-layout, and capabilities for assemblying pieces of layout in very straight-forward ways. In contrast, semicustom layout styles were supported by layout tools that did the layout with little or no manual intervention. As the layout tools mature, the distinction between semicustom and full-custom layout becomes fuzzier, because full-custom layout can be supported with tools now, as well. Today, whether a layout style is called full-custom or semicustom is to some extent a matter of choice; commercial aspects are involved as well, since the notion of full-custom layout suggests a higher quality of the produced layouts. As a rule, full-custom design technologies allow for arbitrary two-dimensional placement of rectangular block layouts on initially empty silicon, perhaps involving floorplanning. Figure 1.1 shows a layout that is composed of several blocks that are arranged on the chip surface and connected by wires through intervening routing channels. The part of the layout process that places and routes the rectangular blocks is also called *general-cell placement*. Furthermore, a full-custom layout procedure provides a large variety of geometries for block layouts. In many cases, the layouts of the blocks are designed by hand, but the use of automatic *cell generators*, which are a combination of hand design and automated assembly, increases. Full-custom layout still

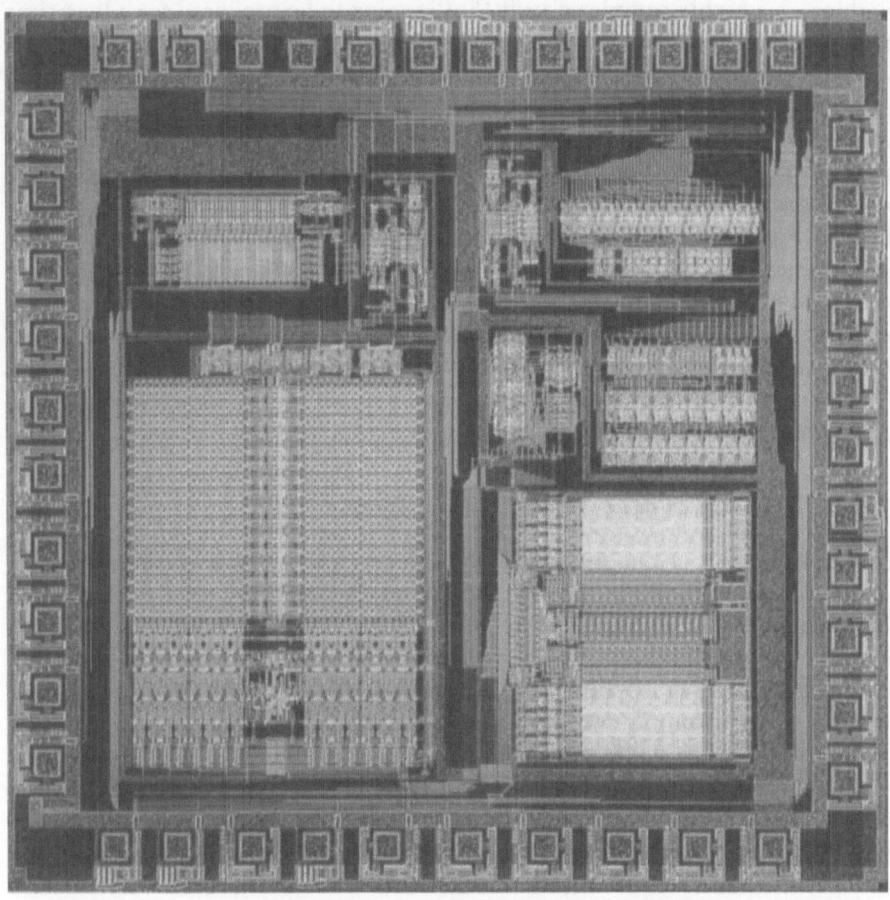

Figure 1.1: *A layout of a chip that is obtained by general-cell place-ment followed by wire routing (courtesy* SIEMENS AG*).*

requires a substantial amount of manual intervention, and the time is far off that completely automatic tools will produce layouts that are comparable in quality with hand-designed layouts. Nevertheless, the areas of semicustom and full-custom layout are starting to move toward each other. Recent progress in the fabrication technology has brought forth gate arrays on which general-cell placement is possible. Layout based on these *sea-of-gates* technologies (see Section 1.2.4.4) is located on the boundary between semicustom and full-custom layout.

When should a layout designer choose a semicustom and when a full-custom layout style? The answer depends on the type of product he is designing. If he is working on a highly complex circuit that is going to be produced for a long time in large quantities, then it pays to go the rockier full-custom route and to spend a lot of time producing a highly optimized layout. If, however, the product is an *application-specific integrated circuit* (ASIC), such as go into products such as appliances or telephones, a more highly automated semicustom layout style may be preferable. ASICs are usually produced in smaller quantities, their performance demands are not as high, and they tend to be updated more often. Thus, for an ASIC, a short design time has priority over the degree to which the product is optimized. As layout tools are improved, the more automated semicustom layout styles will approach the full-custom layout styles in quality; or, seen from the other side, tool support will make full-custom layout styles so economical that we can afford to design ASICs in this way.

Having given a high-level classification of layout styles, let us now examine a generic scenario for circuit layout.

1.2.2   Input to the Layout Problem

The input to the layout process is some functional or structural description of the circuit. More specifically, the circuit is described as a collection of components that are connected by wires. Formally, the circuit is a weighted hypergraph (see Section 3.1). The components can be transistors, gates, or more complicated subcircuits (*blocks*) that may be described recursively using the same mechanism. In the latter case, the circuit description is called *hierarchical*, and the subcircuits are called *blocks* or *cells*. We will use the word *cell* to denote a subcircuit, in general, or its layout on the chip surface, in particular. If it is important to distinguish between the subcircuit and its layout, the subcircuit will be called a *block* and the layout a *cell* (see Chapter 7).

1.2.2.1   Hierarchical Circuit Description

Hierarchical circuit descriptions are very powerful, since they allow for replicating complicated cells without repeating the full description of the cell for each copy. Rather, a copy of the cell (also called an *instance* of the block) consists of only the *interface* of the component, which is a description of the *pins* at which wires connect to it, a name identifying the type of the cell, and a name identifying the cell instance. Thus, the length of a hierarchical description of a circuit may be much smaller than is the size of the circuit. This data compression is what makes possible the economical design of circuits with many hundreds of thousand transistors. Figure 1.2(a) shows a diagram form of the hierarchical description of a 32-bit adder. We call this succinct hierarchical description of a circuit the *folded* hierarchy. It has the shape of a directed acyclic graph. The folded hierarchy is implemented as a pointer data structure that describes the circuit. Nodes in the folded hierarchy represent cells, and edges represent instances of cells.

The most efficient way to process a hierarchical circuit description is to work on the folded hierarchy directly. An algorithm working on the folded hierarchy handles all instances of a cell in a unified manner. Although it may provide different solutions for different instances, it formulates all such solutions in a closed form, and this closed-form description is small and can be computed quickly. Such hierarchical algorithms exist, for instance, for electric-verification problems and for compaction problems.

In the area of circuit layout, we often have to solve problems for which the solution of a cell instance depends so heavily on the neighborhood in which the instance is placed that processing the folded hierarchy is not possible. In this case, we expand the hierarchy selectively by giving a new type to subsets of cell instances of the same type that have similar neighborhoods. This approach expands the directed acyclic graph that describes the folded hierarchy, and makes it more treelike. The corresponding modification of the directed acyclic graph describing the hierarchical circuit operation duplicates a node that has more than one incoming edge into two nodes, and distributes the incoming edges between the two nodes. This operation is called *unfolding*. In the extreme case, when each cell type has only one instance, we speak of the *segmented* hierarchy. The segmented hierarchy of the adder in Figure 1.2(a) is shown in Figure 1.2(b). The directed acyclic graph describing a segmented hierarchy is a rooted tree. The segmented hierarchy represents a data structure in which each instance of a cell is described fully. This form of a hierarchical description is called *segmented* because it suggests segmenting the storage image of the circuit description such that each storage segment holds the information pertaining to a single node in the segmented hierarchy—that is, a single cell instance. The segmented hierarchy uses no less storage space than does the complete nonhierarchical circuit description, but structures storage in a way that makes processing the design more efficient. Indeed, great efforts are underway to make computer-aided design (CAD) tools *hierarchical* in the sense that the tools process the circuit description in some traversal of the segmented hierarchy during which edges of the hierarchy are traversed as seldom as possible. Using hierarchy we reduce the amount of local storage required to whatever is necessary to store small neighborhoods of a cell instance, and it keeps low the overhead connected with the swapping of storage segments. The complexity of such a hierarchical solution of a CAD problem is generally much higher than what we could achieve if we could work on the folded hierarchy, but, for many problems, this is the best we can do. In some cases, a partial expansion of the folded hierarchy that does not go all the way to the segmented hierarchy will do the trick.

Hierarchical concepts are central to circuit layout and to CAD in general. Historically, they have entered the field of circuit design on the descriptive level first, giving the circuit designer the power to express large designs succinctly. They are now changing and improving analysis and optimization procedures. The reader will find the issue of hierarchical layout design addressed through-

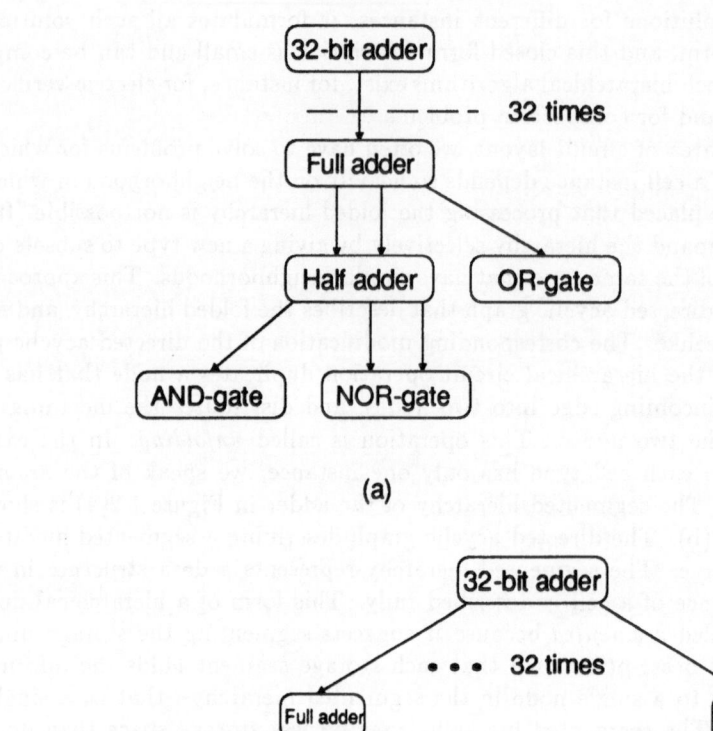

(a)

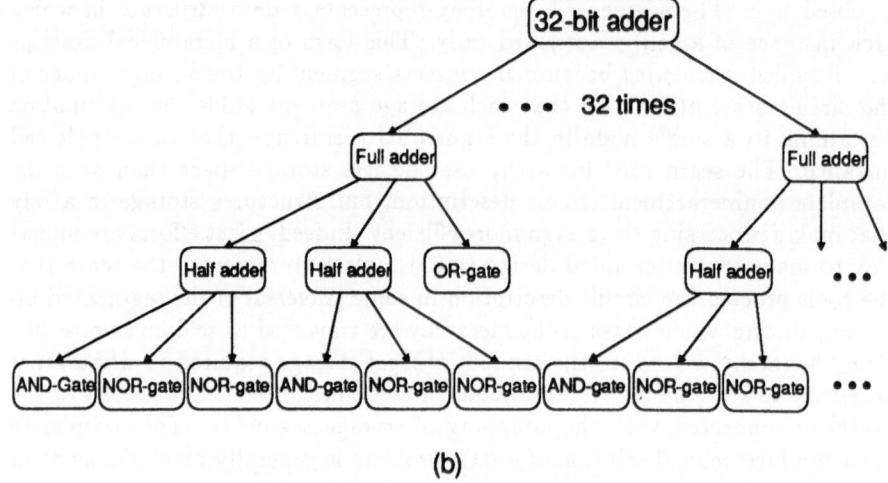

(b)

Figure 1.2: *A hierarchical circuit description of a 32-bit adder. (a) Folded hierarchy, (b) segmented hierarchy.*

out this book. Sections 3.13, 7.2, and 8.5 are worth mentioning specifically here. In Section 3.13, we discuss hierarchical processing of graphs in a general context. In Section 7.2, we present hierarchical floorplanning methods. Finally, in Section 8.5, we use the notion of hierarchy in a different sense; there, the hierarchy used is segmented. However, that hierarchy does not come out of the circuit description, but rather is artificially imposed by the algorithm. Section 10.4 discusses hierarchical compaction.

1.2.2.2 Complex Cells and Cell Generators

Often, the primitive components of a circuit description—that is, the leaves of the tree representing the segmented hierarchy—are not the switching elements or transistors, but rather are more complex circuits, the *leaf cells*. Leaf cells are traditionally designed by hand. In more advanced design technologies, leaf cells are generated by the computer. The specification of a leaf cell is here not a netlist, but rather lies on a higher level of abstraction. For instance, a circuit performing arithmetic and logic operations (an *arithmetic–logic unit*, ALU) may be a leaf cell that is described on a behavioral level, say, by a list of the operations it has to perform. For *random logic*—that is, for the type of unstructured logic that, for instance, generates the control signals for processors—the description may be given by a set of Boolean functions defining the control signals, or by a description of a finite automaton, if internal states are involved. In such cases, the description of the leaf cell must be converted to a layout by a *cell generator*. For instance, a behavioral description of an ALU could be converted into silicon by a *data-path generator*. For random logic, we might use a programmable-logic-array (PLA) *generator*. A cell generator implements a certain design philosophy for generating the layout. For instance, the PLA generator creates a *programmable logic array*, a matrixlike layout structure that is quite simple to generate but that consumes a lot of area and is slow.

Figure 1.3 gives a schematic version of a cell generator specifying a family of carry look-ahead adders. The cell generator takes three parameters:

n	an integer determining the length of the numbers to be added
overflow	a Boolean value determining whether an overflow flag should be generated
delay	a real value giving an upper bound on the circuit delay

If $n = 1$, a specific cell generator for one-bit adders is used. This cell generator is not detailed here. If $n > 1$, then the generator for carry look-ahead adders is called recursively twice to generate two adders of about one-half of the required size. The resulting rectangular cells are abutted in the x direction and are placed below a cell generated with the procedure carrylogic, which is a generator for an appropriate component of the carry chain. The required delay is split among the three components.

```
(1)        cell adder(n : integer, overflow : Boolean, delay : real);
(2)        begin
(3)           if n = 1 then
(4)              return(onebitadder(overflow, delay))
(5)           else
(6)              return(carrylogic(n, overflow, delay/3)
(7)                        above (adder(⌊n/2⌋, false, delay/3)
(8)                        leftof adder(⌈n/2⌉, false, delay/3)))
(9)        end;
```

Figure 1.3: *A cell generator specifying a family of recursively designed adders.*

Figure 1.4(a) depicts the block layout plan of a four-bit adder generated with this cell generator; and Figure 1.4(b) depicts the corresponding folded hierarchy. Note that different assignments to the parameters yield different hierarchies. The number of possible parameter settings is infinite.

Because of the multitude of designs generated by a cell generator, the analysis of cell generators is a difficult algorithmic problem. We discuss methods for processing cell generators based on hierarchical circuit designs in Section 3.13.

1.2.3 Output of the Layout Process

Now that we have described the structure of the input to the layout problem, let us discuss what the layout process is supposed to produce.

1.2.3.1 Geometric Layouts, Mask Data

On the lowest level, the layout is a set of geometric masks that guides the fabrication process of the circuit. The masks can be thought of as frames of photographic film with black (opaque) and white (transparent) regions on them. In many fabrication processes, the masks actually are turned into celluloid, but there are other fabrication methods in which a digital encoding of the information on the masks directly drives the machines that fabricate the circuit. Each mask describes the structural pattern of some material deposited on the silicon during the fabrication of the circuit. For instance, one mask contains the locations of the wires run on top of the circuit in metal to make connections between transistors. Another mask contains the areas in which ions have to be diffused into the monocrystalline silicon substrate to create the doped regions that are necessary to form transistors. A typical fabrication process is guided

carrylogic(4, **true**, *d* /3)			
carrylogic(2, **false**, *d* /9)		carrylogic(2, **false**, *d* /9)	
onebitadder(**false**, *d*/9)	onebitadder(**false**, *d*/9)	onebitadder(**false**, *d*/9)	onebitadder(**false**, *d*/9)

(a)

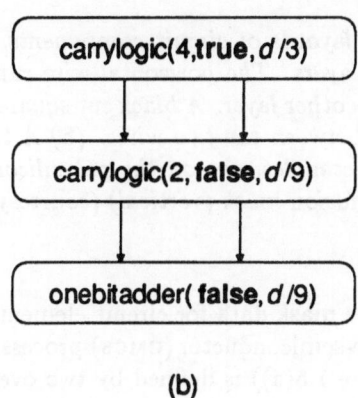

(b)

Figure 1.4: *The four-bit adder generated by the cell generator depicted in Figure 1.3. (a) Structure of the adder generated with the statement* adder(4, **true**, *d*); *(b) folded hierarchy of this adder.*

by from six to well over a dozen masks. We will not describe in detail how the masks guide the fabrication process. For more information on this topic, the reader is referred to introductory textbooks on design of very large-scale integrated (VLSI) circuits, such as [313, 334, 460].

To produce a technically functioning circuit, we must ensure that the images on each mask meet certain requirements, and that the different masks are consistent with respect to one another. The rules that determine what is allowed here are called the *design rules*, and they are particular to the given fabrication process. In their structure, however, the different sets of design rules are much alike. All of them contain rules that require sufficient *spacing* between different opaque and transparent regions on the same mask or on two different masks, and sufficient *width* of a single opaque (or transparent) region. Furthermore, there are rules that require that the layout structures for circuit elements have predescribed shapes.

(a)   (b)

Figure 1.5: *Mask layouts of circuit components. (a) A contact be-tween two wiring layers. The horizontal wire runs on one layer, the vertical wire on the other layer. A black cut square ensures that a con-tact is fabricated between the two wires. (b) A transistor formed by two overlapping rectangles, one on the polysilicon mask (horizontal) and one on the diffusion mask (vertical) (courtesy* XEROX PARC*).*

Figure 1.5 shows the mask data for circuit elements in a typical bulk com-plementary metal-oxide-semiconductor (CMOS) process. A contact between two wiring layers (see Figure 1.5(a)) is defined by two overlapping rectagonal fea-tures, one on the mask for each of the two wiring layers. Another rectangle on a specific *cut* mask (black) is used to create a hole connecting the two wiring layers. A transistor (see Figure 1.5(b)) is defined by two crossing rectangles, one on the mask for the gate layer (polysilicon, horizontal wire) and one on the mask for the channel layer (diffusion, vertical wire). The type of the transistor is defined using a special *implant* mask that determines the type of implant (n-type or p-type, not shown in the figure) to be deposited in the channel.

Figure 1.6 gives a pictorial representation of several design rules. Figure 1.7 presents the mask layout of an inverter using this process.

The reason why all design rules have a similar structure lies in the charac-teristics of the photolithographic fabrication process. Because of physical and technological limitations, the structure on the masks can be transferred to the silicon only within certain tolerance levels. Possible sources of error are mis-alignments of the different masks, the limited resolution of the photographic process dependent on the wavelength of the light used, and the fact that, in the later steps of the fabrication, the surface of the chip becomes increasingly nonplanar, which reduces the accuracy of subsequent steps of depositing and etching away material on the chip surface. The spacing rules prevent the inad-vertent fusion of regions on the silicon, the minimum-width rules ensure that no region on the chip surface that should be connected is broken during the fabrication process, and the shape rules ensure that the geometry of a circuit

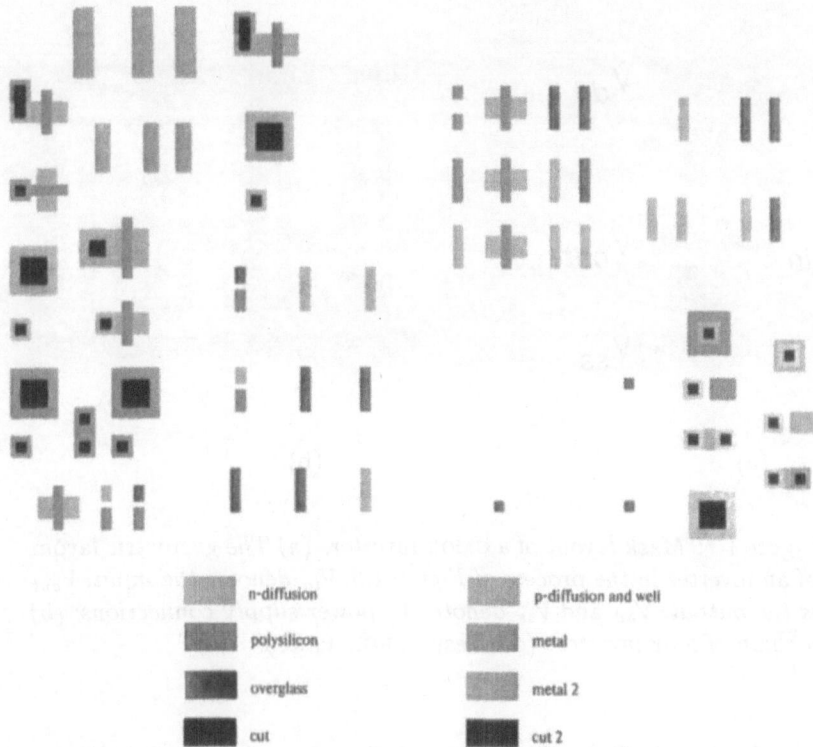

Figure 1.6: *Examples of design rules for a typical bulk* CMOS *process. The diagram shows the minimum feature sizes and the minimum separations between different features such as wires on different wiring layers, transistors, and contacts. The legend is added in order to list all mask layers. The reader is not expected to be able to discern the masks from their shading in the figure. The originals of this mask layout image and the mask layouts in Figures 1.5, 1.7, 1.10, and 1.11 are in color (courtesy* XEROX PARC*).*

element is transferred correctly from the masks onto the silicon. In addition, the design rules ensure sufficient electrical stability of the circuit. More details on design rules can be found in Section 10.1. Here, it shall suffice to mention that the great majority of the geometric structures on masks are *iso-oriented* rectangles; that is, rectangles that are aligned with the Cartesian coordinate axes. In mask data for leaf cells, rotated rectangular features or, more seldom, polygonal features may occur. Circular features occur in the design of layout for only analog circuitry.

Traditionally, the mask domain was the domain in which full-custom layout

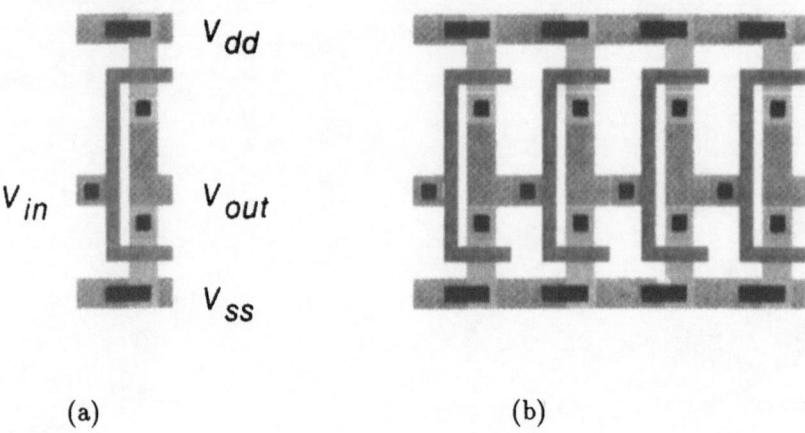

(a)   (b)

Figure 1.7: *Mask layout of a* CMOS *inverter. (a) The geometric layout
of an inverter in the process of Figure 1.6, V_{in} denotes the input, V_{out}
is the output, V_{dd} and V_{ss} denote the power supply connections; (b)
a chain of four inverters (courtesy* XEROX PARC*).*

was done (by hand). Today, however, even though every layout process eventually has to turn out mask data, an increasingly smaller part of the layout process
is concerned directly with the mask data. Rather, in the early stages of layout,
we work on higher levels of abstraction. The reason for doing so is that the
mask data themselves constitute a rather complicated domain that is awkward
for the designer to handle. In some sense, the mask domain is the machine-
programming level of hardware design. By using higher levels of abstraction,
we gain the same advantages as we would by using high-level languages in
software design—and we do so at the same cost. The advantages are that we
become independent of specific fabrication processes, and the process of layout
design becomes more human-oriented and less machine-oriented. The cost is a
decreased efficiency of the generated code. In the context of hardware design,
this means that the mask data compiled from high-level descriptions tend to
be less efficient in terms of area and performance than are hand-designed mask
layouts. Since real estate is still quite a valuable resource—much as storage
was in the late fifties—automatic generation procedures of mask data have to
perform competitively with hand design in order to be used in industrial applications. However, we expect a similar development in hardware design as has
already occurred in software design. An increasing supply of real estate will
essentially eliminate the need of the designer to handle most data directly.

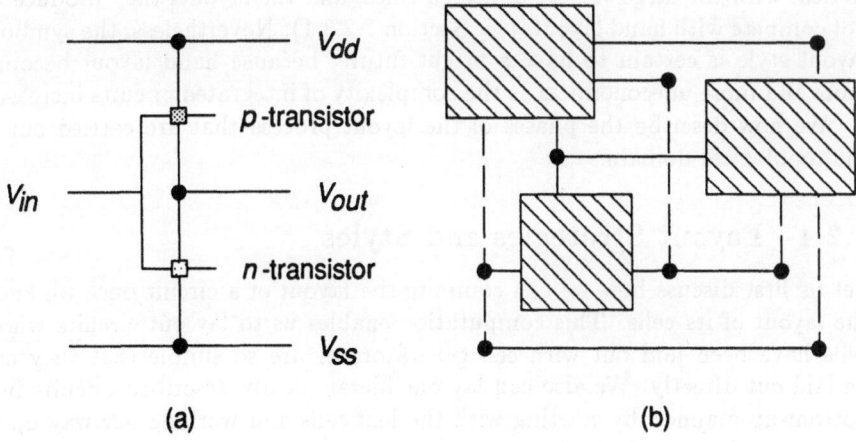

Figure 1.8: *Topological layouts. (a) Stick diagram of the inverter in Figure 1.7; (b) topological block layout.*

1.2.3.2 Topological Layouts

The first step in supporting full-custom layout design with tools was to free the designer from having to use the mask data level. Today, innovative interactive layout editors represent layouts on a topological level. Such layout editors are called *sticks editors*; this name is derived from the appearance of a topological layout (see Figure 1.8(a)). For leaf cells the corresponding form of layout is also called *symbolic layout*. In the topological domain, the basic structure on which layouts are composed is a grid. Wires are paths running along edges of the grid, and components are rectangular or connected rectagonal regions of the grid (see Figure 1.8). Today, all of the placement and routing is usually done in this *topological* domain. The topological domain still exhibits some dependence on the fabrication process. For instance, the number of wiring layers is a parameter of the fabrication process that carries over into the topological domain. Other parameters are preferred directions for wires on certain layers or special provisions for routing power and ground wires. By and large, however, the geometric parameters of the fabrication process have been eliminated. This abstraction eases the construction of layouts considerably.

The transformation from the topological domain into the mask domain is a translation task. It is performed by geometric compacters, which, in addition, do a process-dependent area optimization on the topological layout. Compaction is discussed in detail in Chapter 10. Despite its advantages, symbolic layout and compaction have not yet pervaded the practice of industrial layout systems. They have not done so because computers are not yet robust enough to deal with the large variety of design rules, and the layouts they produce do not compete with hand layouts (see Section 1.2.3.1). Nevertheless, the symbolic layout style is certain to have a bright future, because hand layout becomes more and more uneconomical as the complexity of integrated circuits increases.

We now describe the phases of the layout process that are carried out in the topological domain.

1.2.4   Layout Strategies and Styles

Let us first discuss how we can compute the layout of a circuit once we know the layout of its cells. This computation enables us to lay out circuits whose cells have been laid out with cell generators or are so simple that they can be laid out directly. We also can lay out hierarchically described circuits in a bottom-up manner, by starting with the leaf cells and working our way up to more complicated subcircuits.

1.2.4.1   Placement and Routing of General Cells

Today, doing all of the layout at once is still outside the range of practicality. Even so, striking theoretical results exist on this approach. They are discussed in Chapter 5. They show that, at the heart of the layout problem, is the problem of cutting the circuit into small pieces in a balanced manner without severing many wires. This is a theoretical justification of many layout heuristics that work with this approach (see Sections 7.1.3 and 7.2.1).

In practice, the layout process is subdivided into a placement and a routing phase. During the (general-cell) placement phase, the cells are placed onto the substrate surface such as to minimize certain cost criteria that, it is to be hoped, are correlated with the area necessary for wiring. In the subsequent routing phase, the course of the wires is determined. Most often, the routing phase is itself split into two phases—global routing and detailed routing. The global-routing phase determines how a wire maneuvers around and through the cells. The detailed-routing phase does not change the result of the global-routing phase, but simply computes the exact paths the wires take through the grid. (This is sometimes called *homotopic* routing, a term that comes from the area in topology that is concerned with this kind of problem.)

For a restricted set of detailed-routing problems, the *channel-routing* problems (see Chapter 9), there are ways of minimizing the routing area in the topological domain. The idea is to reduce the size of the routing area while preserving routability. Recently, this topological style of compaction *before* detailed routing has been extended to more general detailed-routing problems. The basis of these methods is progress in the theory of homotopic routing that has been made in several routing models. So far, the results have not been practically applied, but there is hope that the enhancement of the methods with new heuristics will extend them to other routing models and will make them feasible in practice. We discuss this new style of compaction—*topological compaction*—in Section 10.5.

Actually, with channel routing, we can go even one step further. By integrating the compaction phase with the routing phase, we can produce a compacted routing directly *on the mask data level.* Such channel routers are called *gridless.* They can achieve smaller routing area because the topological domain abstracts from geometric dimensions of wires and components, and thus the area measure in the topological domain may differ from the area measure in the mask domain, which is the one in which we are really interested.

In some cases, the detailed-routing phase does not determine which layer should be used for which wire. In this case, it has to be succeeded by a layer-assignment phase, discussed in Chapter 9.5.3. In general, however, it is a better idea to do the layer assignment at the same time as the detailed routing.

Often, this idealized acyclic version of the layout process is modified in practice, to allow for repeating some phases with improved layout data. For instance, the placement phase may use estimates of wiring area in the cost function it minimizes. At first, the estimates are rough, but once global routing has been done, they can be refined. Thus, it may pay to reiterate the placement phase with the new estimates on wiring area. The resulting placement may change the global routing, and so on.

The intermediate data structures through which the different layout phases communicate with one another vary in different layout methods. Figure 1.8(b) illustrates the intermediate result of a layout process that places cells directly on a grid with prescribed size. Such an approach to layout is quite rare. More often, the result of the placement phase is not an assignment of cells to locations on a fixed grid, but rather is a *floorplan*—that is, a tiling of a rectangle representing the circuit. Each tile represents a cell. The global-routing phase then assigns to the wires paths in this floorplan graph or in a variant of it called the *routing graph.* The detailed-routing phase expands each edge of the routing graph to a routing channel that contains the wires that had been assigned to the edge by the global-routing phase (see Figure 1.9).

In different layout styles, this overall structure of the layout process varies. Three restrictive layout styles that deserve special mention here are standard cells, gate arrays, and sea of gates.

1.2.4.2 Standard Cells

In standard-cell layout, the cells (standard cells) are small and rectangular; often, all cells have the same height; and the cells have fixed connections on the left and right side (clocks and/or power) that abut with each other. The placement phase places the standard cells in horizontal rows. The global-routing phase determines where the wires switch between the rows of standard cells. These locations are called *feedthroughs.* The detailed-routing phase amounts to a set of channel-routing problems, one for each routing channel between two

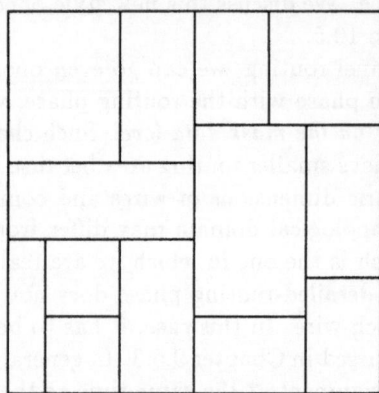

Figure 1.9: *A floorplan; wires are paths along the edges of the tiling.*

adjacent rows.

Figure 1.10 shows five different standard cells. Note that all cells have the same height and can be abutted horizontally. However, the widths of the cells are different. Figure 1.11 shows a small standard-cell layout.

Note that standard-cell layout is an intrinsically nonhierarchical layout methodology. If given a hierarchical circuit, it has to be expanded before the layout design can begin. However, we can combine standard-cell layout hierarchically with placement and routing of general cells, as Figure 1.12 shows.

If a standard cell provides the same electrical signal on different pins at different sides of the cell, there is a question as to what pin ought to be used in the layout. It is even possible to use both pins and to create a feedthrough at no extra cost. Here, we want to minimize routing area by choosing the right pins. Such *pin assignment* problems are generally difficult optimization problems (see Section 7.3.2).

The standard-cell methodology is actually a leftover from the days of small-scale integration. Back then, we assembled transistor–transistor logic (TTL) circuits (a kind of bipolar technology) on boards. The libraries of TTL components were extensively used, and many designs of circuits existed as TTL netlists. When large-scale integration came around, there was the need to upgrade the old TTL designs economically to the new technology. This upgrade was accomplished most easily by converting the TTL components into standard cells and writing programs for laying them out and wiring them up. The one-dimensional layout style was chosen such that the corresponding layout problems would be more easily solvable by heuristics. Such a layout style was possible, because the TTL components were small enough to translate into small

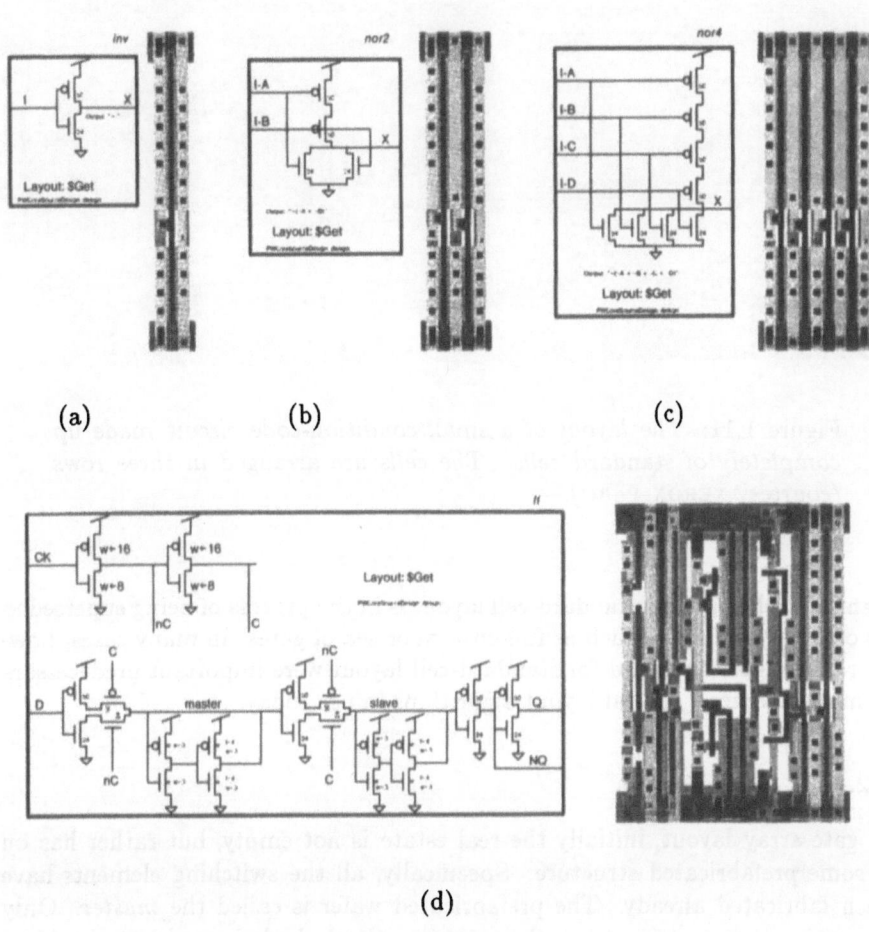

(a)   (b)   (c)

(d)

Figure 1.10: *Five standard cells, represented as logic diagrams and by their mask layouts. (a) Inverter, (b) two-input* NOR*-gate, (c) four-input* NOR*-gate, (d) master-slave flipflop (courtesy* XEROX PARC*).*

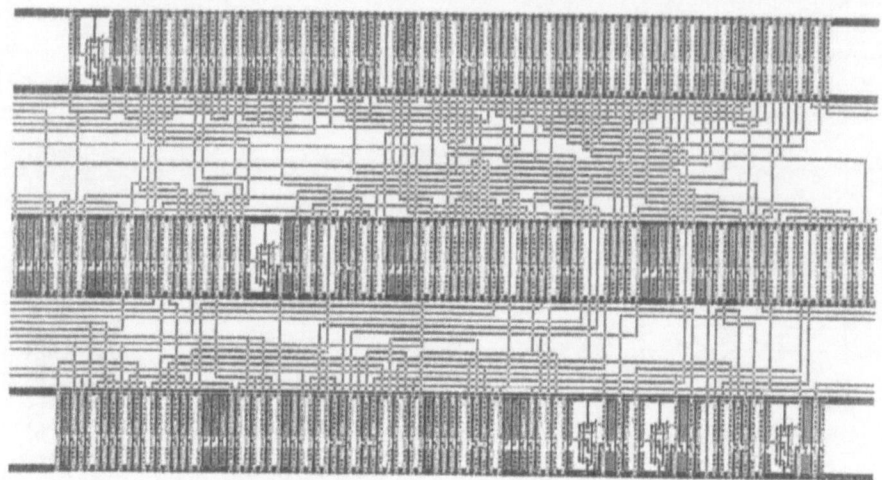

Figure 1.11: *The layout of a small condition-code circuit made up completely of standard cells. The cells are arranged in three rows (courtesy* XEROX PARC*).*

standard cells. Today, standard-cell layout is in the process of being superseded by other layout styles, such as full-custom or sea of gates. In many cases, however, the algorithms used for standard-cell layout were important predecessors of more advanced general layout algorithms in use today.

1.2.4.3   Gate Arrays

In gate-array layout, initially the real estate is not empty, but rather has on it some prefabricated structure. Specifically, all the switching elements have been fabricated already. The prefabricated wafer is called the *master*. Only the wiring that custom-tailors the master to the desired circuit has to be done, using the top two or more layers of metallization. This reduces the set of masks that describe the layout by a great number. Now one mask per wiring layer and masks describing contacts between the different wiring layers suffice. Not only the layout data are reduced. The fabrication has been made simpler, too— only the last few steps of the fabrication process have to be custom-tailored to the circuit. All other steps are the same for all masters. Thus, gate arrays are much less expensive than are custom chips, and their fabrication takes less time. However, because of the large amount of rigidity imposed both by the design technology and by the prefabrication of the master, gate arrays do not achieve the same level of performance and amount of density as do full-

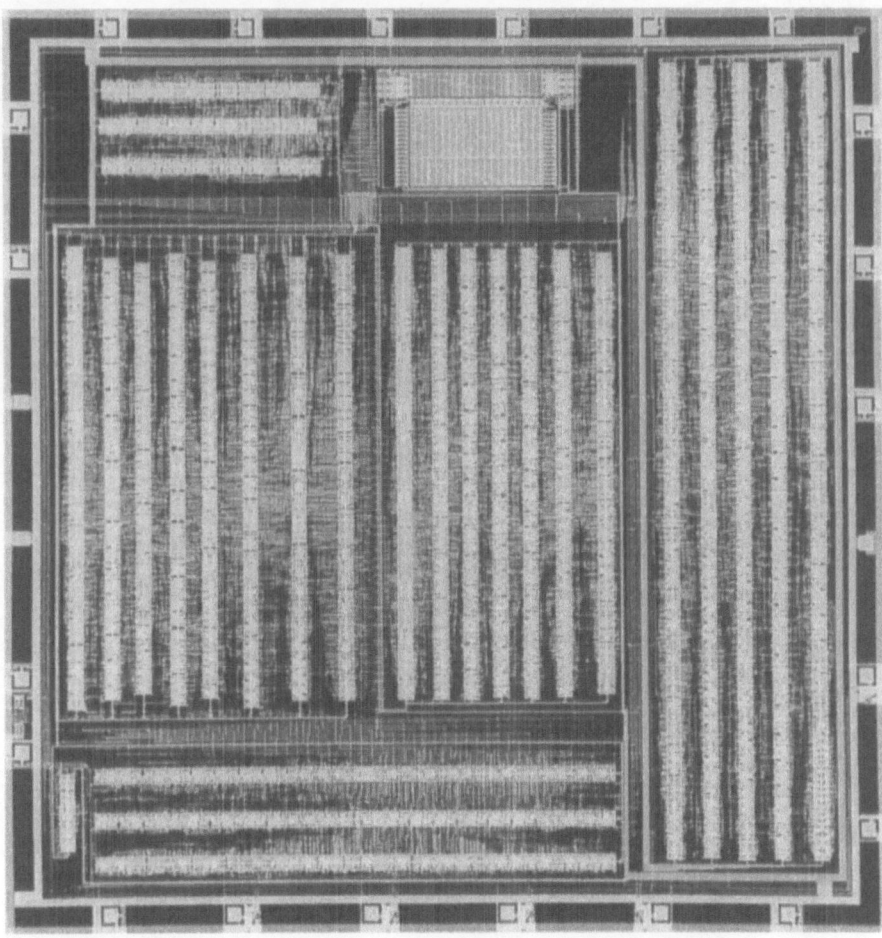

Figure 1.12: *A chip made of of general rectangular cells five of which are blocks of standard cells (courtesy* SIEMENS AG).

custom chips. Figure 1.13 shows a gate-array master and its customization. The prefabricated layout structures on the master are called *cells* or *gates*. On the master depicted in Figure 1.13 the cells are arranged in *rows*. Other masters show a *matrix pattern* of cells, separated by horizontal and vertical channels (see also Figure 7.1). In such cases the cells are also called *islands*. Switching elements, such as different types of Boolean gates or flip-flops, can be fabricated by appropriate metallization over the cells on the master. These switching elements can be connected by wires that are run through the channels. Both types of metallization structures together make up the customization of the master to form a specific circuit. In gate-array layout, the placement problem

Figure 1.13: *A customized gate array (courtesy* SIEMENS AG*).*

is actually an assignment problem. We have to assign to each gate in the netlist of the circuit a set of cells on the master that will implement this gate. Depending on how the master looks, we may not be able to rotate the layout for a gate, or we may even have to place the gate into a specific area on the master that contains the required cells. The routing problem is characterized by the fact that the routing channels on the master have a fixed width. This is in contrast to standard-cell and full-custom layout, where the channel width can be chosen by the layout procedure. Gate-array layout is a nonhierarchical layout style as well, partly because the master is so inhomogeneous, and partly because the level of intergration is too small to warrant hierarchical approaches.

1.2.4.4 Sea of Gates

The most recent generations of gate-array masters have a structure that is fundamentally different from the one shown in Figure 1.13. On these *sea-of-gates* masters, we can no longer discern routing channels. Rather, the whole real estate is filled with transistors. This makes it possible to put many more transistors onto a master. However, we can use only a certain fraction of these transistors. The others are unusuable because the area where their connections would have to be placed is occupied by wires. The other distinguishing feature of sea-of-gates technologies is an increased number of wiring layers. Sea-of-gates technologies are a hybrid between semicustom and full-custom layout. On the one hand, they involve prefabricated masters, as do the traditional gate-array technologies. On the other hand, the concept of a routing channel has been pushed from the hardware into the software domain. The layout procedure can decide whether it wants to route in channels and what the channels should look like. The success of sea-of-gates technologies, especially for CMOS circuits, has been promoted by the fact, that even in bulk CMOS, the design of leaf cells for full-custom circuits follows a certain semicustomlike pattern. Thus, for CMOS, the difference between sea-of-gates and full-custom layout is especially small. In addition, sea-of-gates masters allow a customization to form dense logic, such as random-access memories, directly on the master.

1.2.4.5 Printed Circuit Boards

So far, we have been concerned with laying out structures on the silicon surface only of a chip. Actually, many of the combinatorial layout problems have a similar flavor when we consider the placement and wiring of chips on *printed circuit boards* (PCBs) either of the standard or of the more advanced packing technologies. Of course, the variety of packing technologies is even larger than that of silicon technologies, and one packing technology can differ greatly from the other. For printed circuit boards, as for chips, the layout is defined by a set of masks, essentially one mask for each wiring layer. The fabrication processes for printed circuit boards differ from those for chips, in that they are not completely photolithographic, but also involve steps in which layers of structured planar material are sandwiched on top of each other. This allows for many more wiring layers. In fact, standard technologies today have up to eight wiring layers, and the number of wiring layers in advanced technologies reaches 30. Different layers may serve different purposes—such as power and ground wiring or signal wiring—and have different preferred directions in which the wires should be run.

On printed circuit boards, contact cuts—which are also called *vias* in this context—are created by drilling holes. This can be done such that contact cuts do not connect all but a subset of adjacent wiring layers (*partial via*). One can even connect different sets of wiring layers above the same point on the printed circuit board (*stacked vias*). The rules pertaining to vias can be quite involved. Figure 1.14 shows the kind of layouts produced for printed circuit boards.

Despite these differences, there are enough similarities between PCB layout and chip layout, that we can draw from the same set of layout algorithms. In PCB layout, as in chip layout, placement precedes routing. The distinction between global routing and detailed routing is usually not made in PCB layout, since this distinction depends on a planar routing concept, which gets increasingly blurry as the number of wiring layers increases. Geometric compaction has no place in PCB layout, since compaction is based on the design rules that are particular to silicon technologies. In fact, the topological gridlike domain for representing layout information is the only one used in PCB layout. Local deviations from the grid layout are introduced by postprocessors at the end of the layout process. When we describe layout algorithms, we will note whether they have applications in chip layout, in PCB layout, or in both.

1.2.4.6   Floorplanning

In this section, we discuss how the layout process we have described can be adapted to exploit hierarchical circuit descriptions more fully. We use segmented hierarchies. The central point is that we cannot expect to know what a cell looks like at the time that we want to place it. If we do know, we have a purely bottom-up layout construction. But then a cell instance is laid out with no concern as to its neighborhood. Thus, a pin of the cell may be placed on the opposite side of the block where we need it, or a block may get an aspect ratio that is completely unfitting to where the block has to go. Clearly, what we need is some information to propagate top down in the hierarchy. The solution to this problem is to replace the placement process by a floorplanning process. In floorplanning, we actually consider several layout alternatives for each block. We can determine the set of feasible alternatives for a block initially by statistical means; that is, by estimating the expected area requirement of the cell on the basis of easily accessible parameters, such as transistor count, external and internal wires, and so on. Alternatively, we can get some estimate by a bottom-up pass over the hierarchy that accumulates the relevant information. Or we use a mixture of these two strategies. One way of representing the space of layout alternatives for a block is by a sizing function that, given a width, returns the minimum height that a cell of that width must have to accommodate the block. Figure 1.15 shows several alternatives for sizing functions. Floorplanning now chooses a good layout alternative for each block and pieces together a good placement for the resulting cells. Like placement, floorplanning partly relies on inaccurate data. In floorplanning, in addition to the inaccuracy of the cost functions that we optimize, the area requirements for the cells may be inaccurate. As we move down the hierarchy, it may turn out that some cells need more area than we allocated for them, and that others need less. This mismatch calls for a reiteration of the floorplanning process

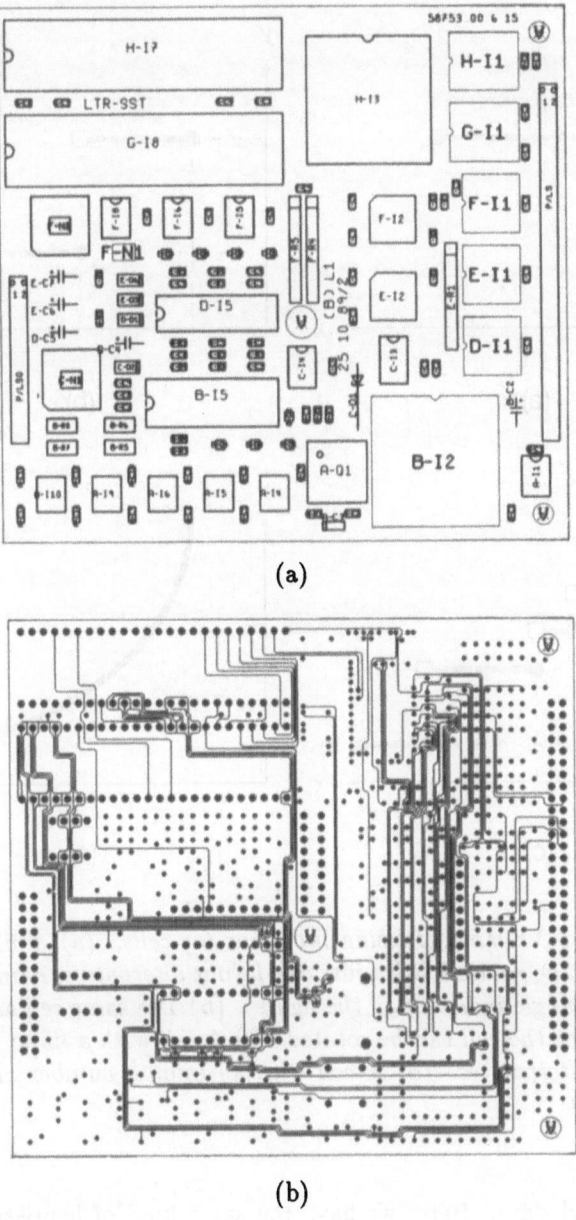

(a)

(b)

Figure 1.14: Layout masks for a printed circuit board. (a) Diagram
of the placement of the components on the board; (b) mask for one
of six different wiring layers. The dots at which wires end are vias.
The routing on each layer is planar, that is, no wires cross (courtesy
Nixdorf Computer AG).

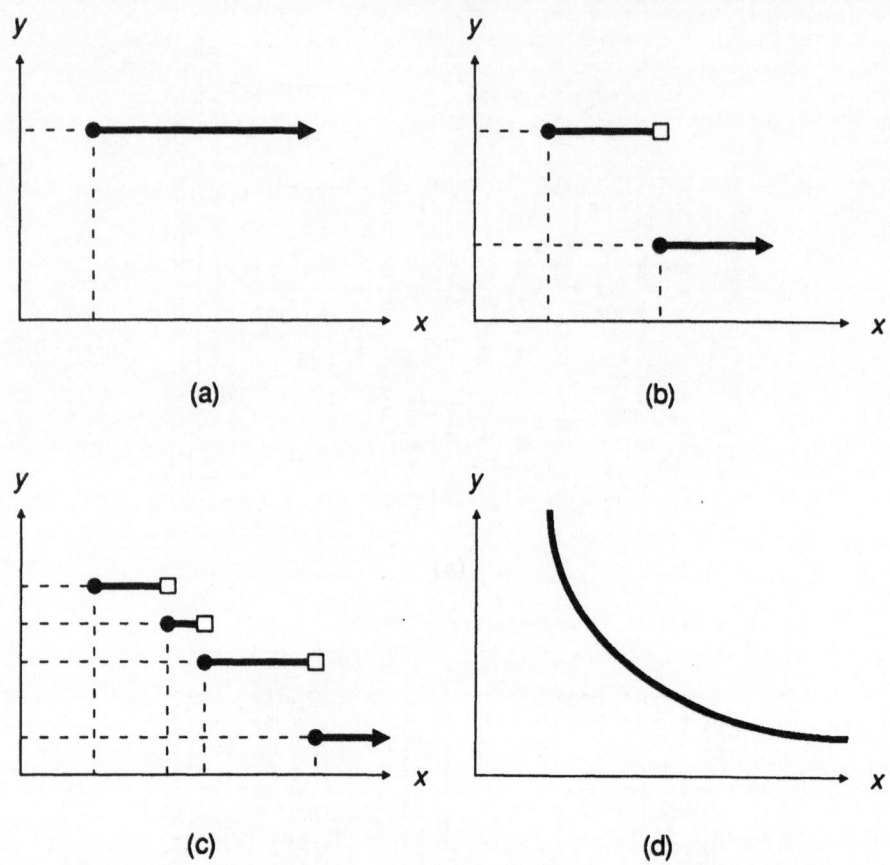

Figure 1.15: *Variants of sizing functions for cells. (a) A fixed cell that cannot be rotated (with just one layout alternative; namely, the dashed rectangle indicated in the figure). (b) The same cell as in (a), but this time the cell can be rotated. (c) A cell with a finite number of layout alternatives. (d) A cell with an infinite number of layout alternatives.*

using the revised data. Here, we have the same kind of heuristic loop that we encountered between placement and global routing in Section 1.2.4.1. In fact, to a large extent, the area estimate for a cell depends on how much area we allocate for wiring. There are methods under development for integrating floorplanning and global wiring to cut down on the leakage that occurs between these two phases as we go around the loop. We regard these developments as so important that we dedicate a special section (Section 8.8) to them.

1.3 Summary

We have briefly described the design technologies used in circuit layout on chips as well as on printed circuit boards. By doing so, we have provided the application framework for the identification and analysis of combinatorial problems occurring in circuit layout. Before we can start delving into this discussion, we have to review the methods from theoretical computer science and from applied mathematics that we use for the analysis of such combinatorial problems. We shall carry out this review in Chapters 2, 3, and 4.

Chapter 2

Optimization Problems

2.1 Basic Definitions

In this chapter, we will lay the formal framework for discussing the algorithmic problems arising in circuit layout. We will be rather formal right from the beginning, so as to be able to describe clearly the algorithmic structures with which we will be working. The first definition we will give is that of an algorithmic problem.

2.1.1 Algorithmic Problems

Definition 2.1 (Algorithmic Problem) *An* **algorithmic problem** Π *is a mapping* $\Pi : I \to 2^S$, *where* I *is the set of* **problem instances** *and* S *is the set of* **configurations**. *For a problem instance* $p \in I$, *the configurations in* $\Pi(p)$ *are called* **solutions** *of* p *or* **legal configurations** *for* p. *Both* I *and* S *may be infinite sets. If* S *is a singleton set, then* Π *is called a* **decision problem**.

For instance, several network problems have as the set I of instances the set of connected undirected graphs; and, as the set of solutions, they have the set of undirected trees. (For basic graph definitions, see Section 3.1.) The mapping Π assigns to a graph $G \in I$ the set of G's spanning trees; that is, those trees that are subgraphs of G and contains all vertices of G. Each spanning tree of G is a solution to the instance G of the problem Π.

If S is a singleton set, say the set {yes}, then 2^S can be identified with {0, 1}. The set {yes} corresponds to 1, and the empty set corresponds to 0. The binary range of Π is the basis for the notion "decision problem". If Π is a decision problem, then an instance $p \in I$ such that $\Pi(p) = 0$ is called a 0-instance; when $\Pi(p) = 1$, it is called a 1-instance.

For the purpose of solving algorithmic problems with algorithms, we have to specify in more detail what the sets I and S look like. We will assume that

31

both of them are sets of strings that suitably encode the problem instances
and solutions. The alphabet Σ over which these strings are formed has a
suitable finite set of characters, including binary digits, parentheses, and certain
delimiting characters. Encodings can be defined in a straightforward manner
for the structures we will be discussing. We can think of the representation of
integers as being in binary. We form lists of objects by separating the objects
with delimiters and parenthesizing the list. Matrices are represented as a list of
their rows. Graphs are represented by their adjacency matrix or their adjacency
lists, and so on. The exact definition of the encoding is tedious and, in most
cases, is not of importance for the algorithms we will be discussing. If the exact
encoding plays a major role, it will be specified.

2.1.2   Deterministic Algorithms

We will mainly consider deterministic sequential algorithms for solving algorithmic problems. The main performance measures with which we will be
concerned are *run time* and *storage space*. A suitable model of computation
is the random-access machine with unit-cost measure. In the unit-cost measure, we measure time by counting the number of executed instructions, and
we measure space by counting the number of used memory cells. Each memory
cell can hold an arbitrarily large number. In certain cases, if the size of the
numbers involved is important, the logarithmic-cost measure is appropriate. In
the logarithmic-cost measure, each bit needed for storing a number in memory
is counted, and the time an instruction takes is linear in the length of the numbers involved. We will not go into more detail about the model of computation
(see, for instance, [5]). In the following, we will always use the random access
machine model with the unit-cost measure, unless otherwise specified.

Definition 2.2 (Complexities of Algorithms) *An algorithm A* solves *the
algorithmic problem* Π *if, given an input that is a suitable encoding for some
instance* $p \in I$, *the algorithm computes the encoding of some solution* $s \in \Pi(p)$.
The run time of algorithm A on some instance $p \in I$ *is denoted by* $T(A, p)$; *the
storage space is denoted by* $S(A, p)$.

Usually, it is very hard or impossible to determine $T(A, p)$ and $S(A, p)$ exactly.
Thus, we blur the performance measures to make them easier to estimate. To
this end, we associate with each instance $p \in I$ a number $\ell(p)$, the *length*
of the instance. This number is usually identical or related to the length of
the encoding of the instance in bits (in the logarithmic cost measure) or the
number of integers occurring in the encoding of the instance (in the unit-cost
measure). We group together all instances of the same length, and compute
one performance value for them. The rule by which this value is computed
determines the complexity measure we use. The following rule defines the
worst-case complexity of algorithm A:

$$T_{\text{wc}}(A, n) \quad := \quad \max\{T(A, p) \mid p \in I, \ell(p) = n\}$$

$$S_{wc}(A, n) := \max\{S(A, p) \mid p \in I, \ell(p) = n\}$$

In worst-case complexity, the performance of the algorithm on instances of length n is determined by the hardest instance. This complexity measure is very conservative. Even if there is only one instance of a given length on which the algorithm has a poor performance, it will dominate the better performance of the algorithm on all other instances of that length. This one hard instance may never come up in practice, so the complexity of the algorithm does not necessarily give a good indication of how the algorithm performs in practice. There are two reasons why this complexity measure is so prevalent in research. The first is that it is the measure we need in scenarios where even one hard instance of the problem can hurt us severely. The second reason, and definitely the more important one, is that this complexity measure is by far the easiest to analyze. An upper bound on the performance of the algorithm can be found by proving the bound for all instances, and a lower bound can be found by exhibiting one instance on which the algorithm's performance exceeds the lower bound. The great majority of results with which we are concerned is about worst-case complexity.

Most often, however, we are more interested in the behavior of the algorithm on typical problem instances. One way of being precise about what we mean by "typical" here is to use average-case complexity. The prerequisite for an average-case analysis is that we know with what likelihood a particular instance occurs. For an instance $p \in I$, with $\ell(p) = n$, let $\pi(p)$ be the probability that p occurs among all instances of length n. The *average-case complexity* of algorithm A is then defined as follows:

$$T_{ave}(A, n) := \sum_{p, \ell(p) = n} \pi(p) \cdot T(A, p)$$

$$S_{ave}(A, n) := \sum_{p, \ell(p) = n} \pi(p) \cdot S(A, p)$$

If there are just a few hard instances of a given length, and these instances are unlikely to occur, they will not appreciably affect the average-case complexity of the algorithm. Thus, if we know the probability distribution of the problem instances, average-case complexity is a more realistic complexity measure than is worst-case complexity in many situations. The average-case complexity measure has two disadvantages. The first is that it is quite difficult to determine the average-case complexity of an algorithm, even for simple probability distributions. Average-case results exist for only the simplest algorithms, and even their proofs are much more difficult than are the proofs of the corresponding worst-case complexity results. The second reason is that it is difficult to ascertain the probability distribution for the instances of the problem in a given application. Indeed, the distribution may vary with time. Furthermore, if we happen to hit a hard instance, the average-case complexity will be a gross underestimate of the real performance of the algorithm. This is actually a deficiency of the algorithm: It will behave badly on hard instances, and there is nothing we can do about it. Average-case complexity gives the illusion of good performance by randomizing over the problem instances.

2.1.3   Randomized Algorithms

To overcome the fact that some instances are intrinsically and inevitably hard for the algorithm, we can introduce randomization internally to the algorithm. We allow the random-access machine to flip an unbiased coin and to proceed according to the outcome. Say, the machine stores a 0 or a 1 into a specific register, depending on how the coin falls. Now the randomization is internal to the algorithm, and, if the algorithm is lucky throwing coins, every instance may have a chance of being solved efficiently. Formally, a randomized algorithm A is a deterministic algorithm having two inputs (p, σ), one being the problem instance p and one being an infinite binary string σ containing the outcomes of all coin tosses. The outcomes are read from σ in left to right order. The algorithm A must solve the problem instance p for each coin-tossing string σ. Because the coin is fair, for each k, each binary string of length k is equally likely to occur as a prefix of σ. The corresponding complexity measure is the *randomized* complexity or, more precisely, the *Las Vegas* complexity. It is defined as follows:

$$T_{\text{LV}}(A, p) \; := \; \lim_{k \to \infty} 2^{-k} \sum_{\sigma \in \{0,1\}^k} T(A, (p, \sigma))$$

$$S_{\text{LV}}(A, p) \; := \; \lim_{k \to \infty} 2^{-k} \sum_{\sigma \in \{0,1\}^k} S(A, (p, \sigma))$$

$$T_{\text{LV}}(A, n) \; := \; \max_{\ell(p)=n} T_{LV}(A, p)$$

$$S_{\text{LV}}(A, n) \; := \; \max_{\ell(p)=n} S_{LV}(A, p)$$

Here $T(A, (p, \sigma))$ is the run time of the randomized algorithm A on instance p and with coin-tossing string σ. Note that the performance estimates in the Las Vegas complexity measure are expected values, but this time the randomization is internal to the algorithm and happens on every problem instance. Thus, the algorithm has the chance of being lucky on every problem instance. Modulo the internal averaging process, the Las Vegas complexity measure is a kind of worst-case complexity.

2.1.4   Asymptotic Analysis

It is still very hard to determine exactly the performance measures introduced in the previous section. Furthermore, the exact value of the performance measure depends on many details of the implementation, such as the instruction set of the underlying machine model, programming habits, and compiler characteristics (if the machine program is obtained by compilation from a high-level programming language). In the area of algorithm analysis, however, we are not interested in evaluating the merits of a specific implementation. Rather, we want to capture the complexities that are intrinsic to the *algorithm*, independent of the details of its implementation. As a consequence, we do not want to determine the performance measures exactly, but just to make statements as to how they increase with the length of the problem instance. Such statements are obtained by *asymptotic analysis*. For asymptotic analysis, we commonly use the following notation, introduced by Knuth [237]:

$$
\begin{aligned}
f(n) = \quad & O(g(n)) && \text{if there are constants } n_0 \in \mathbb{N} \text{ and } c > 0 \\
& && \text{such that, for all } n > n_0, \text{ we have } f(n) \leq c \cdot g(n) \\
f(n) = \quad & \Omega(g(n)) && \text{if there are constants } n_0 \in \mathbb{N} \text{ and } \varepsilon > 0 \\
& && \text{such that, for all } n > n_0, \text{ we have } f(n) \geq \varepsilon \cdot g(n) \\
f(n) = \quad & \Theta(g(n)) && \text{if } f(n) = O(g(n)) \text{ and } f(n) = \Omega(g(n))
\end{aligned}
$$

$f(n) = O(g(n))$ means that, as n increases, $f(n)$ always stays below $g(n)$ multiplied by some suitable constant. $f(n) = \Omega(g(n))$ means that, as n increases, $f(n)$ always stays above $g(n)$ multiplied by another suitable constant. If $f(n) = \Theta(g(n))$, then $f(n)$ can be bounded from below and above by $g(n)$ multiplied by suitable constants. In this case, we say that $f(n)$ has the same asymptotic growths $g(n)$. The asymptotic growth of $T_{wc}(A, n)$ (or of $S_{wc}(A, n)$) is called the *asymptotic worst-case time* (or *space*) *complexity* of algorithm A, and is a characteristic only of the algorithm and not of its implementation. Similar statements hold for the other complexity measures. We will focus mainly on time complexity, and the remainder of this chapter will also consider only time complexity. Some of the definitions carry over to space complexity.

The *worst-case complexity of an algorithmic problem* is obtained by minimizing over the complexities of all algorithms for the problem:

$$
T_{wc}(\Pi, n) := \min_{A \text{ solves } \Pi} T_{wc}(A, n)
$$

Analogous definitions hold for the average-case and the Las Vegas complexity of Π.

The advantage of being independent from the details of the implementation has been obtained at a cost. Since no further restrictions are placed on the constants c and ε, they can take on values that are outside the range of practical interest. As an example, the function $f(n) = 10^4 n$ grows more slowly asymptotically than does the function $g(n) = 10^{-4} n^2$, in the sense that $f(n) = O(g(n))$ but not $g(n) = O(f(n))$. However, $f(n)$ is larger than $g(n)$ for all $n < 10^8$. Thus, the range where $g(n)$ exceeds $f(n)$ is probably outside the range of practical interest. Since many asymptotic analyses are such that the constant factors c and ε are not determined explicitly, it is easy to come up with results whose relevance is difficult to evaluate. Therefore, it is very important to keep track of these constants. Even if their size can be estimated, they often take on impractical values. This does not mean that the corresponding asymptotic analysis is worthless, but it does qualify the result as being of purely theoretical interest. In many cases, both the algorithms and their analysis can be refined such as to bring the results into the range of practical interest.

In summary, asymptotic analysis is an essential tool for estimating the complexity of algorithms and of algorithmic problems. Depending on the size of the constants involved, the results may be of purely theoretical interest, or they may lead to practically interesting algorithms and bounds. If we aim to develop practically interesting algorithms, we also must refine their asymptotic analysis, such that we obtain practical constant factors that prove the algorithm's efficiency.

2.1.5   An Example: Quicksort

Let us now give an example that demonstrates the definitions given earlier.

Consider the following algorithm, called *quicksort* [192, 314, 410]. The algorithm is given a sequence $p_1, \ldots, p_n$ of n numbers. It picks the ith number p_i as a *partitioning number*. It constructs two subsequences of numbers of p: the sequence $p^<$ containing all numbers less than p_i, and the sequence $p^>$ containing all numbers greater than p_i. The sequences $p^<$ and $p^>$ are sorted recursively. Finally, the results are concatenated with p_i to yield the sorted output sequence.

It is not hard to see that quicksort takes time $O(n^2)$ in the worst case. The algorithm makes $O(n^2)$ comparisons, and the time for the choice and evaluation of each comparison is bounded by constant (in the unit-cost measure on a random-access machine).

If i is chosen deterministically, there are input sequences on which quicksort needs $\Omega(n^2)$ comparisons (Exercise 2.4). If all permutations of numbers are equally likely in the input, however, then quicksort takes only time $O(n \log n)$ on the average. This is little consolation, however, if we happen to hit a bad sequence that causes quicksort to compare many numbers. If the partitioning index i is chosen randomly, with each number between 1 and n being equally likely, then there are no longer any bad inputs. On each input, we have the chance of sorting in time $O(n \log n)$. In fact, we will sort this quickly even in the expected case.

2.1.6   Optimization Problems

We will be especially concerned with a certain class of algorithmic problems, the *optimization problems*.

Definition 2.3 (Optimization Problem) *A* **minimization problem** *is an algorithmic problem for which there exists a* **cost** function $c : S \to \mathbf{N}$. *An* *algorithm A that, on input $p \in I$, computes a legal configuration s such that $c(s)$ is minimum is said to* minimize Π. *The corresponding solution is called an* optimal *solution.*

A similar definition can be made for *maximization problems*—that is, for problems that maximize the cost function. Minimization and maximization problems together form the class of *optimization problems*. We will use a certain notation for defining optimization problems. This notation is exemplified with the well-known TRAVELING SALESPERSON problem.

TRAVELING SALESPERSON

Instance: An $n \times n$ matrix $D = ((d_{ij}))$ of positive integers representing a complete weighted directed graph

Configurations: All permutations π of the numbers $1, \ldots, n$

Solutions: All configurations

Minimize: $c(\pi) := \sum_{i=1}^{n} d_{\pi_i, \pi_{(i \bmod n)+1}}$

In the TRAVELING SALESPERSON problem, we are given a matrix of distances between n points. We are supposed to find a tour that goes through each point once and has minimum total length. This problem is one of the main problems in operations research and has been studied extensively [263]. It is discussed in more detail in Chapter 4.

2.2 NP-Hardness

2.2.1 Feasible and Infeasible Algorithms

In general, it is very difficult to determine the asymptotic complexity of an algorithmic problem. Even rough statements on how hard a problem is are difficult to make. If the length of the problem instance is the length of its encoding, and all of the input must be read to solve the problem, then a linear-time algorithm will be asymptotically optimal. This is a trivial statement. The disconcerting fact is that there is no algorithmic problem for which we know, say, an algorithm with run time $O(n^{\alpha})$ for some $\alpha > 1$ *and* can also show that there is no linear-time algorithm (on random-access machines). Such *nonlinear lower bounds* on the run time are known only for problems that require *exponential* run time—that is, for which all algorithms have run times $\Omega(2^{n^{\alpha}})$ for some $\alpha > 0$. Solving such problems exactly is infeasible for all but the smallest instances. Thus, we regard such problems as infeasible.

Definition 2.4 (Feasible Problem) *A* feasible *(or* polynomial-time*) algorithm A is an algorithm with $T_{wc}(A, n) = O(n^{\alpha})$ for some $\alpha \geq 0$. An al-*

gorithmic problem Π *is called* feasible *or* polynomial time *if there is an algorithm A solving* Π *that is feasible. The class of feasible decision problems is denoted by* P.

This notion of feasibility is somewhat problematic, for the same reasons as discussed previously in the context of asymptotic analysis. For a feasible algorithm, neither the constant factor nor the exponent of its run time are restricted in any way. Thus, an algorithm with a run time of n^{1000} is feasible, whereas an algorithm with a run time of $2^{n/1000}$ is infeasible. Even so, the second algorithm is faster than the first for $n \leq 2.4 \cdot 10^7$. Nevertheless, researchers usually agree that exponential growth is fundamentally worse than polynomial growth, such that this very rough classification can be made. Furthermore, most interesting polynomial-time algorithms have exponents that are quite small (less than 3 or 4). We should be careful, however, when resting our judgment about the feasibility of an algorithm purely on the fact that that algorithm is polynomial. Recent results in the area of graph algorithms have shown the existence of quadratic ($O(n^2)$ time) or cubic ($O(n^3)$ time) algorithms for several graph problems that were not previously known to be in P. However, the constants of these algorithms are huge [112, 379, 380]. As the number of such examples increases, basing the notion of feasibility on polynomial run time becomes more doubtful.

For the sake of our discussion, we will base our notion of feasibility on Definition 2.4, and we will make sure to estimate the constants involved.

As we noted, there are a few problems for which infeasibility can be shown [5], but the list is not extensive, and none of the problems is of interest to us. In general, it is very difficult to show infeasibility. But there is a large class of problems for which infeasibility is extremely likely. These are the NP-*hard* problems.

2.2.2   Nondeterministic Algorithms

To define the notion of NP-hardness, we need to introduce the concept of nondeterminism. This is an algorithmic concept that is defined only for algorithms solving decision problems.

Definition 2.5 (Nondeterministic Algorithm) *Let* Π *be a decision problem. A* nondeterministic *algorithm A for* Π *is an algorithm that, at each step, is allowed to make a binary guess as to how to proceed. The algorithm is said to* solve *the problem, if for each input* $p \in I$, *the following holds:*

- *If* $\Pi(p) = 1$, *then there is a way for A to guess such that 1 is computed. In this case, we define* $T_{\text{ndet}}(A, p)$ *to be the run time of the* shortest *such computation.*

- *If* $\Pi(p) = 0$, *then there is no way for A to guess such that a 1 is computed. In this case, we define* $T_{\text{ndet}}(A, p) := 1$.

Definition 2.6 (NP) *The class of all decision problems that have feasible nondeterministic algorithms is denoted by* NP.

At first, the definition of a nondeterministic algorithm seems nonintuitive, and it is certainly not very practical. A nondeterministic algorithm is similar to a randomized algorithm that has an optimally biased coin. If it is given a 1-instance, it always chooses the shortest computation to produce the 1. But, in contrast to a randomized algorithm, a nondeterministic algorithm is not at all concerned with 0-instances. If it is given a 0-instance, it rejects that instance immediately, without doing any computation. Thus, a nondeterministic algorithm is both clairvoyant and partial to the 1-instances—a rather unrealistic model of computation. But nondeterministic algorithms are not supposed to run on real machines. Rather, the central idea behind a nondeterministic algorithm is the following: *Given a 1-instance, a nondeterministic algorithm produces a short (namely, polynomial-time) proof of the fact that the instance is a 1-instance.* Indeed, if we have proved that the nondeterministic algorithm A solves problem Π, then any computation of A that yields a 1 on instance p is a proof that $\Pi(p) = 1$. This interpretation is also the reason why we restrict ourselves to decision problems here. A nondeterministic computation that, say, optimizes an optimization problem, does by itself not constitute a proof that the solution it computes is optimal.

Thus, if a problem is easy nondeterministically, there are short proofs for instances being 1-instances of the problem. Of course, by design, the nondeterministic algorithm avoids having to find such a proof—a task that any realistic algorithm will have to perform. Therefore, we expect the nondeterministic complexities of problems to be lower than are the problems' deterministic complexities in general. This is indeed the case, and the difference can amount to as much as an exponential.

To apply the concept of nondeterminism to optimization problems, we will associate with an optimization problem a naturally correlated decision problem.

Definition 2.7 (Threshold Problem) *Let* $\Pi : I \to 2^S$ *be a minimization problem with cost function c. The* **threshold problem** $\Pi_{TH} : I \times \mathbb{N} \to \{0, 1\}$ *is the decision problem that, given an instance* $(p, k) \in I \times \mathbb{N}$, *yields a 1 if there is a configuration* $s \in \Pi(p)$ *such that* $c(s) \leq k$.

As an example, the threshold problem TRAVELING SALESPERSON$_{TH}$ that corresponds to the TRAVELING SALESPERSON problem asks whether there is a tour with cost at most k. There is a simple nondeterministic algorithm for solving this problem. The algorithm can guess any tour, compute its length, and check whether that length stays below k. The algorithm can be made to run in $O(n \log n)$ time; thus, TRAVELING SALESPERSON$_{TH}$ is in NP. On the other hand, no deterministic polynomial-time algorithm for TRAVELING SALESPERSON$_{TH}$ has been found yet, and it is very unlikely that one exists.

Our intuition tells us that finding a proof must be much harder than just checking a proof that has been given to us. Nevertheless, the question of whether P = NP persistently resists being answered, and can be regarded as the main open problem in theoretical computer science today. However, the evidence that P $\neq$ NP grows overwhelmingly, as more and more problems like the TRAVELING SALESPERSON problem are found. These problems share the property with the TRAVELING SALESPERSON problem that, if any of them is in P, then all of them are, and P = NP. These are the NP-*complete* problems.

2.2.3   NP-Completeness and NP-Hardness

Definition 2.8 (NP-Completeness) *A* **polynomial-time transformation** Φ *from a decision problem* $\Pi_1 : I_1 \rightarrow \{0,1\}$ *to a decision problem* $\Pi_2 : I_2 \rightarrow \{0,1\}$ *is a function* $\Phi : I_1 \rightarrow I_2$ *that can be computed by a deterministic polynomial-time algorithm such that, for all* $p \in I_1$ *we have* $\Pi_1(p) = \Pi_2(\Phi(p))$. *If such a transformation* Φ *exists,* Π_1 *is called* **(many-to-one) reducible to** Π_2 *(written* $\Pi_1 \leq_m \Pi_2$). *A problem* $\Pi' \in$ NP *is called* **NP-complete** *if* $\Pi \leq_m \Pi'$ *for all* $\Pi \in$ NP.

A polynomial-time transformation quickly translates an instance of Π_1 into an equivalent instance of Π_2. An NP-complete problem is a problem to which each problem in NP can be reduced. If an NP-complete problem Π' is in P, then P = NP, because then we can solve any problem Π in NP in polynomial time by just translating an instance of Π into an instance of Π' using the polynomial-time transformation, and then solving this instance in polynomial time.

Up to now, several thousand NP-complete problems have been found in such diverse areas as graph theory, logic, algebra, operations research, number theory, and automata theory. Nobody knows a polynomial-time algorithm for any of them, and as the list of problems grows, so does the conviction that P $\neq$ NP. Thus, although nondeterminism is not a realistic model of computation, it can serve to provide evidence that certain algorithmic problems are hard. We will not explore the notion of NP-completeness in detail. For an in-depth discussion of this subject and an extensive list of NP-complete problems, see [133]. D. S. Johnson's regularly appearing NP-completeness column in the *Journal of Algorithms* [209, 210] contains additional information on new results concerning NP-complete problems. For our purposes, we will assume that we cannot solve an NP-complete problem in reasonable time on arbitrary problem instances.

Definition 2.8 immediately suggests a way of proving the NP-completeness of a decision problem Π: Show that $\Pi \in$ NP by exhibiting a polynomial-time nondeterministic algorithm. (In most cases, this task is quite simple.) Then present a polynomial-time transformation from some other NP-complete problem Π' to Π. Since $\leq_m$ is transitive (Exercise 2.5), this procedure proves that

II is NP-complete. We also say that II' has been *reduced to* II in polynomial time. Polynomial-time reductions are the primary method of proving decision problems NP-complete. Of course, not all NP-completeness proofs can be done in this way. For some first problem, the NP-completeness has to be proved by a different method. The method employed here is to prove directly that the problem fulfills Definition 2.8. That is, a reduction is given from every nondeterministic Turing machine that runs in polynomial time. Such reductions are also called *master reductions*.

We will now discuss the consequences that the NP-completeness of the threshold problem has for the related optimization problem. For instance, the NP-completeness of TRAVELING SALESPERSON$_{TH}$ implies that probably the TRAVELING SALESPERSON problem cannot be solved in polynomial time either. If it could be, then we could solve TRAVELING SALESPERSON$_{TH}$ in polynomial time as well: We just compute an optimal tour and check whether its cost exceeds k or not. The same reasoning holds for all optimization problems for which the cost function can be computed in polynomial time in the length of the solution, as is the case for all optimization problems we will consider. We call an optimization problem whose threshold problem is NP-complete NP-*hard*.

2.3 Solving NP-Hard Problems

Most of the optimization problems we will consider are NP-hard. Nevertheless, we have to solve them somehow if we are going to lay out circuits. How can we find good solutions quickly, in spite of the fact that these problems are hard? There are several ways in which we can approach the solution of NP-hard problems; some are preferable to others. We will discuss the most attractive ways first, and then proceed to the less attractive ones.

2.3.1 Restriction

The first thing to check is whether the apparent NP-hardness of the problem is just an artifact of a lax analysis. It may happen that we do really have to solve the problem not on all instances in I, but rather on just a restricted set of instances. It may be that the problem restricted to this subset of instances is solvable in polynomial time. For instance, if we restrict the TRAVELING SALESPERSON problem to upper triangular matrices, then it can be solved in cubic, that is, $O(n^3)$ time [263]. On the other hand, even if a restriction of the instance set is possible, the problem often remains NP-hard on the restricted set. As an example, the TRAVELING SALESPERSON problem remains NP-hard even if the set of instances is restricted to matrices that correspond to distances of points in the plane [263]. Even in such cases, it is a good idea to determine the set of relevant instances as precisely as possible. Although the restricted problem may remain NP-hard, it may become easier in other ways.

2.3.2   Pseudopolynomial Algorithms

The problem may have a *pseudopolynomial* algorithm. This is an algorithm
that runs in polynomial time in the length of the instance *and in the size of the
numbers occurring in the encoding of the instance*. Such an algorithm is efficient
if the numbers involved are not too large; that is, if they are polynomial in the
length of the instance. An NP-complete problem that has a pseudopolynomial
algorithm is called *weakly* NP-*complete*. Several combinatorial problems we
will consider are only weakly NP-complete. Since the numbers involved in
instances of layout problems are usually bounded from above by the total area
of the layout, a weakly NP-complete layout problem has algorithms that run
in polynomial time in the layout area. Such algorithms may well be efficient
enough to be used in practice. On the other hand, no pseudopolynomial-time
algorithm is known for TRAVELING SALESPERSON$_{\text{TH}}$. In fact, this problem is
strongly NP-*complete*: It remains NP-complete even if, in the encoding of the
instances, we code numbers not in binary but rather in *unary*; that is, the
number n is encoded as a string of n 1s. Encoding numbers in unary expands
the length of the instance; thus, even when we use the same function as a
bound on the run time in terms of the length of the instance, more time can
be allocated to solving the instance. The larger the numbers involved, the
more the length of the instance will expand, and the more time we can use for
solving the instance. If a problem is strongly NP-complete, however, then using
only small numbers in the instance does not help. There is no hope of finding
a pseudopolynomial-time algorithm unless P = NP, which, of course, we do
not believe. Thus, strongly NP-complete problems (and also their associated
optimization problems) are unlikely to have polynomial-time algorithms, even
if we restrict the numbers in the instances to be small.

2.3.3   Approximation Algorithms

Sometimes, we can approximate the optimal solution efficiently. The metric by
which we measure the deviation from the optimum may differ. Let us consider
minimization problems. In some cases, we measure the difference between the
cost of the computed solution and the cost of the optimal solution. More often,
we consider the corresponding quotient—that is, the error of algorithm A is
defined by

$$\text{ERROR}(A) := \max_{p \in I} \frac{c(A(p))}{c(\text{opt}(p))}$$

where $A(p)$ is the solution computed by algorithm A, and opt(p) is the optimum
solution for p. Several NP-hard problems have polynomial-time approximation
algorithms with bounded error. For the TRAVELING SALESPERSON problem,
the situation is especially interesting. For the unrestricted problem, it can
be shown that, if there is a polynomial-time algorithm with error $O(1)$, then
P = NP. On the other hand, if we restrict the problem to distance matrices

that fulfill a certain natural triangle inequality stating that the direct path between two points is always shorter than is any detour through other points, then several feasible approximation algorithms exist. One algorithm runs in almost linear time and has an error of 2; another runs in cubic time and has an error of 3/2 [315]. Several layout-optimization problems have approximation algorithms, as well.

2.3.4   Upper and Lower Bounds

Often, we do not succeed in finding a good approximation algorithm. In such cases, we take one further step back in our quest to solve the problem. It is often possible to bound the cost of the solution obtained by the algorithm in terms of some structural parameter related to the problem instance. This parameter could be the cost of an optimal solution of the problem instance, in which case we have an approximation algorithm with some bound on its error. The parameter can also be something weaker. In the case of the TRAVELING SALESPERSON problem, there are analyses of special-case algorithms that involve such parameters as the maximum distance from a specific point to any other point and the cost of an optimal assignment—that is, the value

$$\min\{\sum_{i=1}^n d_{i,\pi(i)}|\pi \text{ is a permutation of } \{1,\dots,n\}\}$$

Both of these values can be computed quickly. Sometimes, such analyses can be complemented with lower bounds that are reasonably tight, at least for some instances. For instance, the cost of an optimal assignment is a lower bound on the cost of an optimal tour (see [263]; Section 3.12).

Many of the algorithms with a more theoretical flavor that we will consider are of this category.

2.3.5   Randomization

We can introduce randomization much as we did in Section 2.1.3. Here, we deal with optimization problems, so in addition to using randomization to decrease the expected run time, we can use it to approximate the optimal cost. Several versions of randomization are possible:

1. The algorithm always computes an optimal solution.

2. The algorithm always computes a solution, and it computes an optimal solution with high probability.

3. If the algorithm computes an optimal solution, then it produces a proof of optimality.

4. The algorithm's run time is small in the worst case.

5. The algorithm's run time in the expected case, averaged over the internal randomization, is small on each problem instance.

Algorithms fulfilling characteristics 1 and 5 are usually called *Las Vegas* algorithms. Algorithms with one of the characteristics 2 or 3 and characteristic 4 are usually called *Monte Carlo* algorithms. In addition, the deviation of the cost of the solution computed by a Monte Carlo algorithm from the optimal cost may be able to be bounded. Las Vegas algorithms differ from Monte Carlo algorithms in that they always produce an optimal solution. Randomization is slowly entering the field of layout. Actually, most of the randomized algorithms used so far in this area are unanalyzable heuristics; that is, they belong to the next category.

2.3.6   Heuristics

The last resort is to proceed heuristically. Promising tricks, perhaps involving randomization, may lead close to the optimal cost, but we have no way of finding out, in absolute terms, just how close we have come and with what probability. Even though we may be quite happy with the solution obtained for some problem instance, we do not know how much better we could have done or how good we will be on the next instance. The heuristic approach is by far the most widespread method in practice today. All of the iterative improvement techniques (deterministic and randomized) fall in this category.

2.4   Summary

We have now defined what we mean by a *hard problem*. We have reviewed some algorithmic techniques for solving hard problems and the promises they bear. In Chapters 4 and 5, we will discuss algorithmic methods for solving hard problems in more detail.

2.5   Exercises

2.1

1. Show that, for every polynomial $p(n)$ of degree d, we have $p(n) = O(n^d)$.

2. Show that $\log n = O(n^\varepsilon)$ for all $\varepsilon > 0$.

3. Show that $\sqrt[n]{n} = O(1)$. Can you give upper and lower bounds for $\lim_{n \to \infty} \sqrt[n]{n}$?

2.2

1. Show that, for fixed α, $0 < \alpha < 1$, and, for any function $f : \mathbb{N} \to \mathbb{N}$,

$$\sum_{i=1}^{f(n)} T(\alpha^i n) = O(T(n)) \tag{2.1}$$

   as long as

$$T(\alpha n) \leq cT(n) \quad \text{for some } c, \ 0 < c < 1. \tag{2.2}$$

2. Give examples of smooth functions $T(n)$ with $T(n) = \Omega(n)$ that fulfill requirement (2.2).

3. Give an example of a polynomially growing function $T(n)$, with $T(n) = \Omega(n)$, that does not fulfill requirement (2.2), and for which equation (2.1) does not hold.

2.3 The notation $f(n) = O(g(n))$ is really a set notation. $O(g(n))$ denotes all functions $f(n)$ such that $f(n) = O(g(n))$. This means that the "=" sign really has the meaning of "$\in$". Nevertheless, it has many properties of the normal "=" as used in set theory.

1. Show that $O(f(n)) + O(g(n)) = O(f(n) + g(n))$.

2. Show that $O(f(n)) \cdot O(g(n)) = O(f(n) \cdot g(n))$.

3. Is $\sum_{i=1}^{n} O(n) = O(n)$?

2.4

1. Show that, for each deterministic choice of the partitioning index i in quicksort, there is an input sequence of length n that needs $\Omega(n^2)$ comparisons.

2. Can you determine the required number of comparisons more exactly?

2.5 Prove that the relation $\leq_m$ defined in Definition 2.8 is transitive.

2.6 The HAMILTONIAN CYCLE problem asks, given an undirected graph G, whether G has a cycle that contains each vertex in G exactly once. HAMILTONIAN CYCLE is NP-complete. Prove that TRAVELING SALESPERSON$_{TH}$ is NP-complete, even if the distances are all 1 or 2. To do so, prove that

HAMILTONIAN CYCLE $\leq_m$ TRAVELING SALESPERSON$_{TH}$

Chapter 3

Graph Algorithms

This chapter summarizes the results from graph theory that will be used in this book. Graphs are one of the fundamental structures treated in discrete mathematics. We begin with an extensive set of classical definitions from graph theory (Section 3.1). Then we review general algorithmic techniques for exploring graphs so as to impose structures on them (Sections 3.2 to 3.7). Subsequently, we discuss important graph problems that are used in many areas of circuit layout. In Sections 3.8 to 3.10, we consider problems of finding paths and more general connection subnetworks in graphs. Such problems occur naturally in wire routing. Path problems are also the basis of many optimization procedures, such as compaction and area minimization in floorplanning. In Sections 3.11 and 3.12, we discuss network flow and matching problems that are of special importance in circuit partitioning and detailed routing, but that also find applications in other aspects of circuit layout. In Section 3.13, we include newer material on processing hierarchically defined graphs. This material gains its significance through being the formal reflection of hierarchical circuit design (see Section 1.2.2.1). Finally, we summarize results on planar graphs (Section 3.14).

3.1 Basic Definitions

We start by giving a synopsis of the basic definitions in the area of graph algorithms. These definitions are included only as a matter of reference and for establishing the notation. Since the reader is assumed to be familiar with them, they will not be explained or made intuitive with examples.

Definition 3.1 (Sets) *In a* **multiset**, *an element can occur several times. The* **multiplicity** *of an element s of a multiset S is the number $\mu(s)$ of times s occurs in S. A multiset in which the multiplicity of all elements is 1 is called a (**proper**) **set**. The* **cardinality** *of a multiset S is defined as*

47

$|S| := \sum_{s \in S} \mu(s)$. *A function* $f : S \to T$ **labels** *the set* S *with elements from* T. $f(s)$ *is the label that* f *assigns to* $s \in S$. *If* T *is a set of numbers, then the labels are also called* **weights**. *In this case, the* **total weight** *of a subset* $S' \subseteq S$ *is defined by* $f(S') := \sum_{s \in S'} f(s)$.

Definition 3.2 (Hypergraphs) *A* **hypergraph** $G = (V, E)$ *consists of a finite set* V *of* **vertices** *and a multiset* $E \subseteq 2^V$ *of* **hyperedges** *or* **nets**. *The elements of hyperedge* $e \in E$ *are called its* **terminals**. *A hyperedge with two elements is sometimes called a* **two-terminal net**; *a hyperedge with more than two terminals is sometimes called a* **multiterminal net**. *We also say that the hyperedge* e *is* **incident** *to the vertex* v *if* $v \in e$. $n := |V|$ *is the* **size** *of* G, *and* $m := |E|$ *is the* **order** *of* G. *The total number of terminals in* G *is denoted by* $p := \sum_{e \in E} |e|$. *The* **degree** *of a vertex* v *is the number of hyperedges incident to* v. *A hypergraph all of whose vertices have the same degree* k *is called a* **regular** *hypergraph of degree* k. *If* E *is a proper set, then* G *is called* **simple**.

A hypergraph is called **connected** *if there is no way of partitioning the set* V *of vertices into two nonempty subsets* V_1, V_2 *such that no hyperedge contains vertices from both* V_1 *and* V_2. *A* **cut** *in a hypergraph is a set of hyperedges whose removal disconnects the graph. The total weight of a cut of weighted hyperedges is also called its* **size**.

A hypergraph $G' = (V', E')$ *such that* $V' \subseteq V$, *and such that* E' *contains only edges all of whose terminals are in* V', *is called a* **subgraph** *of* G. *If* $V' = V$, *then* G' **spans** G. *The hypergraph* **induced** *by a subset* $V' \subseteq V$ *of the vertices is the graph* $G' := (V', E')$, *where* $E' := \{e \in E | w \in V' \text{ for all } w \in e\}$.

Definition 3.3 (Undirected Graphs) *If* $|e| = 2$ *for all* $e \in E$, *then* G *is called an* **(undirected) graph**. *If* $\{v, w\}$ *is an edge, then the vertices* v, w *are called* **adjacent** *or* **neighbors**. *A* **path** p *from vertex* v *to vertex* w *in* G *is a sequence* $v_0, e_1, v_1, e_2, \ldots, v_{k-1}, e_k, v_k$ *such that* $v_i \in V$ *for* $i = 0, \ldots, k$; $e_i \in E$; *and* $e_i = \{v_{i-1}, v_i\}$ *for* $i = 1, \ldots, k$. *Often,* p *is just denoted by its list of vertices* $v_0, \ldots, v_k$. v_0 *and* v_k *are called the* **endpoints** *of* p. k *is the* **length** *of* p. *If* $v_0 = v_k$, *then* p *is called a* **cycle**. *If all vertices in* p *are pairwise distinct—except, perhaps, for the two endpoints—then* p *is called* **simple**. *If there is a path from* v *to* w *in* G, *then* w *is said to be* **reachable** *from* v *in* G. *A* **connected graph** *is a graph such that, for all* $v, w \in V$ *has a path whose endpoints are* v *and* w. *The* **connected components** *of* G *are the maximal connected subgraphs of* G. *A* **forest** *is a graph that contains no cycles. A* **tree** *is a connected forest.*

A **bipartite** *graph is a graph such that* $|V|$ *can be partitioned into two sets of vertices* V_1, V_2, *such that the edges of* G *run only between vertices of* V_1 *and vertices of* V_2.

A graph G *is* **complete** *if all pairs of vertices are adjacent in* G. *A complete subgraph of an undirected graph is called a* **clique**. *A* **coloring** *of an* *undirected graph G is a labeling c(v) of the vertices of G such that, for all edges $\{v, w\} \in E$, $c(v) \neq c(w)$. The complement of an undirected graph $G = (V, E)$ is the graph $\overline{G} = (V, \overline{E})$, such that $\overline{E} = \{\{v, w\} \mid v, w \in V, \{v, w\} \notin E\}$.*

Definition 3.4 (Directed Graphs) *A* **directed graph** $G = (V, E)$ *is a graph such that E is a multiset of* **ordered pairs** *of vertices; that is, each edge is directed. If (v, w) is an edge, then w is called a* **direct successor** *of v, and v is called a* **direct predecessor** *of w. A directed edge $e = (v, w) \in E$ is said to* **leave** *v and to* **enter** *w. The vertex v is denoted with $q(e)$ and is called the* **tail** *of e. The vertex w is denoted with $z(e)$ and is called the* **head** *of e. The* **forward neighborhood** *$N(v)$ of a vertex v is $N(v) := \{w \in V \mid (v, w) \in E\}$. The* **backward neighborhood** *of v is $N^{-1}(v) := \{w \in V \mid (w, v) \in E\}$. The size of the sets $N(v) \setminus \{v\}$ and $N^{-1}(v) \setminus \{v\}$ is called the* **indegree and outdegree** *of v, respectively. The edges (v, w) and (w, v) are called* **antiparallel**. *If $(v, w) \in E$ exactly if $(w, v) \in E$, then G is called* **symmetric**. *A* **directed path** *p from vertex v to vertex w in G is a sequence $v_0, e_1, v_1, e_2, \ldots, v_{k-1}, e_k, v_k$, such that $v_i \in V$ for $i = 0, \ldots, k$, and such that $e_i \in E$, $e_i = (v_{i-1}, v_i)$ for $i = 1, \ldots, k$. Often, p is denoted simply by its list of vertices, $v_0, \ldots, v_k$. v_0 is called the* **starting point** *of p. v_k is called the* **endpoint** *of p. k is the* **length** *of p. If there is a directed path from v to w in G, then w is said to be* **reachable** *from v in G. If $v_0 = v_k$, then p is called a* **circuit**. *A graph in which each pair of vertices lies on a common circuit is called* **strongly connected**.

Definition 3.5 (Directed Acyclic Graphs) *A directed graph that has no circuit is called a* **directed acyclic graph (dag)**. *The vertices of a dag that have indegree (outdegree) 0 are called the* **sources (sinks)** *of the dag. The length of the longest path in a dag G is called the* **depth** *of G and is denoted by* depth(G). *A vertex-weighting function $t : V \to \{1, \ldots, n\}$ of a dag G such that, for all $(v, w) \in E$, we have $t(v) < t(w)$ is called a* **topological ordering** *of G. An edge (v, w) in the dag G is called a* **shortcut** *if we can reach w from v without using the edge (v, w). The graph G' obtained from G by adding (deleting) all shortcuts to G is called the* **transitive closure (transitive reduction)** *of G. A graph G that is equal to its transitive closure is called* **transitively closed**.

Definition 3.6 (Rooted Trees) *If G is a dag and each vertex except for one vertex r has indegree 1, then G is called a* **rooted tree**. *The vertex r is called the* **root** *of G. Let v be a vertex in a rooted tree G. Any vertex that can be reached from v is called a* **descendant** *of v. The direct successors of v are called the* **children** *of v. The number of children of v is called the* **degree** *of v. Any vertex from which v can be reached is called a* **predecessor** *of v. The direct predecessor of v is called the* **parent** *of v. The maximum degree of any vertex in T is called the* **degree** *of T. If T has degree d, then T is also called a* **d-ary tree**. *A vertex in T with degree zero is called a* **leaf**. *A vertex in*

T with degree at least one is called an interior vertex. *A rooted tree all of whose interior vertices have the same degree is called a* complete *tree. The length of the path from the root to a vertex v in T is called the* level *of v. The depth of T as a dag is also called the* height *of T.*

The natural representation for the netlist of a circuit is a hypergraph. The vertices of the hypergraph represent the cells of the circuit. The hyperedges represent the electrical connections. (This is the reason why hyperedges are also called *nets*.) We can associate various weights with the elements of the hypergraph. A vertex labeling may represent something like the inherent size of the vertex, such as the estimated area requirement of the cell the vertex represents. In floorplanning, we may label the vertices with sizing functions. The nets of the hypergraph may be weighted with functions representing inherent wiring costs. For instance, if we represent a bus of several wires—say, 32 of them—with one hyperedge, then this hyperedge may receive a weight of 32 to indicate that wiring is more expensive than is routing a single strand wire.

To represent hierarchically described circuits by hypergraphs, we must make the basic graph definitions hierarchical. We do so in Section 3.13.

3.2   Graph Search

A graph is a rather unstructured object. If we want to traverse the graph to solve some graph problem, we normally do not know where to begin or how to proceed. However, it turns out that, for many graph problems, there is a specific search strategy that makes solving the problem much easier than it would be were we just to traverse the graph in any arbitrary fashion. In this and the following sections, we will introduce some of these strategies, and will point out their relevance to solving graph problems.

Any graph traversal begins at some, usually arbitrary, vertex of the graph, the *start vertex*. The start vertex is marked *visited*. All other vertices of the graph are marked *unvisited*. All edges of the graph are marked *unexplored*. The goal is to visit all vertices of the graph, and, in doing so, to explore all edges of the graph. *Exploring* an edge means walking along it. For directed graphs, there is only one direction in which we can explore an edge (from tail to head), whereas in undirected graphs, an edge can be explored in both directions. Usually, exploring an edge twice does not make much sense, so we will try to avoid exploring edges of undirected graphs in both directions. The general step that is carried out during the graph search is the following.

Search Step: Select some unexplored edge (v, w) such that v has been visited.
        Mark the edge *explored*, and mark w *visited*.

As written here, the graph G is directed; for undirected graphs, an analogous definition holds. Note that, if (v, w) is explored, then both v and w are visited, but the reverse does not have to hold.

So far, there is still substantial freedom as to how we carry out the graph search. For each application of the search step, we have to select a visited vertex v and an unexplored edge leaving v. Let us discuss in more detail how the next explored edge is chosen.

We assume that the graph G is given by an adjacency structure A, which is of the type

array [vertex] of list of vertex A

The list $A[v]$ contains all direct successors of v in some order. If a vertex is a direct successor of v via several multiple edges, then it occurs in $A[v]$ with this multiplicity. In the adjacency structure of an undirected graph, every edge is represented twice, once in the list of each endpoint. Sometimes, it is advantageous to provide cross-links between the two representations. Such an adjacency structure is called a *cross-linked* adjacency structure. Essentially, the search procedure maintains a data structure D, which stores the vertices that have been visited so far. Initially, only the start vertex is in D. At each search step, a vertex v is accessed from D (without deleting it from D), and one of its edges that has not been explored yet, usually the next one in the list $A[v]$, is explored. If this edge is the last unexplored edge from v, then v is deleted from D. If exploring the edge leads to a newly encountered vertex w, then w is put into D.

In addition, instead of just traversing the graph, we will do computations along the way. These computations collect information that is relevant for solving the graph problem at hand. The computations manipulate data that can be thought of as residing in edges or vertices of the graph. In fact, formally these data are labelings of the graph. The computation carried out during each search step involves data local to the neighborhood of the edge explored in this step.

Because of the freedom of choice of the data structure D and of the possibility of attaching essentially unrestricted kinds of computations to the search step, graph search is a powerful paradigm for solving problems on graphs. We will now exhibit several customizations of graph search. First, we will concentrate on D. Each choice of D gives us a special kind of graph search; for instance, depth-first search or breadth-first search. Then, we will choose the attached computations. These choices customize the search to solving specific graph problems, such as the strong-connectivity problem or the shortest-path problem.

The choice of the data structure D has a great influence on the performance of the search. Consider, for instance, a graph that is too large to fit into memory. In this case, D has to be kept on disk. However, not all of the data structure has to be accessible at any one time. Since the computations done in the search step involve only data local to the neighborhood of the edge being explored, only this part of the data structure has to be resident. The size of this part of the data structure varies in different search strategies, and it is a major factor in determining the efficiency of the search strategy on large graphs.

3.3   Depth-First Search

Depth-first search may well be the most popular search strategy. It leads to a large number of efficient, mostly-linear time, graph algorithms. Depth-first search has been used on rooted trees for a long time. Here it is also called *top-down traversal* and was applied, for instance, in top-down parsers and in optimization procedures such as backtracking. Tarjan [440] generalized depth-first search to arbitrary directed, and undirected graphs and thereby boosted the development of efficient graph algorithms.

For depth-first search, the data structure D is a stack. Thus, the depth-first search procedure is a recursive procedure. The stack D maps onto the run-time stack of the recursive execution of the depth-first search procedure and is not explicit in the algorithm.

The algorithm underlying depth-first search is shown in Figure 3.1. It explores edge (v, w) at the time the loop in line 8 is executed for vertex w within the call dfs(v). The first visit of v occurs at the time of the call dfs(v). The values in the array *prenum* indicate in what order the vertices are encountered during the search. The array *postnum* contains the order in which the calls to dfs with the respective vertices as arguments are completed. The procedures initialize, previsit, newvisit, revisit, postvisit, and wrapup contain the accompanying computation that is particular to the specific graph problem. The procedures occurring within the dfs procedure manipulate only data local to the current location of the search—that is, data labeling v or w. If we apply the depth-first search to a tree and customize it by giving it a previsit procedure, a preorder traversal will be carried out executing this procedure. Analogously, we can customize inorder and postorder traversals of trees. Actually, the assignment of the *prenum* and of the *postnum* values can be regarded as a particular preorder or postorder customization, respectively. A lot of the structure imposed on the graph by the search can be exhibited in terms of the *prenum* and *postnum* values. We will now discuss this structure.

Consider a depth-first search on the directed graph depicted in Figure 3.2(a). The result of this exploration is shown in Figure 3.2(b). The edges of the graph fall into the following four categories.

Definition 3.7 (Edge Classification in Directed Graphs) *Let $e = (v, w)$ be an edge in a directed graph. The edge e is called*

- *A* **tree edge***, if prenum[w] > prenum[v] and postnum[w] < postnum[v], and there is no path of length at least 2 from v to w along which the prenum values increase*

```
(1)        precount, postcount : integer;
(2)        prenum, postnum : array [vertex] of integer;

(3)        recursive procedure dfs(v : vertex);
(4)        begin
(5)            prenum[v] := precount;
(6)            precount := precount + 1;
(7)            previsit(v);
(8)            for w ∈ A[v] do
(9)                if prenum[w] = 0 then
(10)                   dfs(w);
(11)                   newvisit(v, w)
(12)               else
(13)                   revisit(v, w) fi od;
(14)           postvisit(v);
(15)           postnum[v] := postcount;
(16)           postcount := postcount + 1;
(17)       end dfs;

(18)       begin
(19)           precount := postcount := 1;
(20)           for v is vertex do
(21)               prenum[v] := postnum[v] := 0 od;
(22)           initialize;
(23)           for v is vertex do
(24)               if prenum[v] = 0 then
(25)                   dfs(v) fi od
(26)       end;
```

Figure 3.1: *Depth-first search.*

- *A* forward edge, *if* $prenum[w] > prenum[v]$ *and* $postnum[w] < postnum[v]$, *and* (v, w) *is not a tree edge*

- *A* back edge, *if* $prenum[w] < prenum[v]$ *and* $postnum[w] > postnum[v]$

- *A* cross-edge, *if* $prenum[w] < prenum[v]$ *and* $postnum[w] < postnum[v]$

The classes of edges defined in Definitions 3.7 are the only ones that can occur. For an undirected graph, we can make similar observations; consider Figure 3.3.

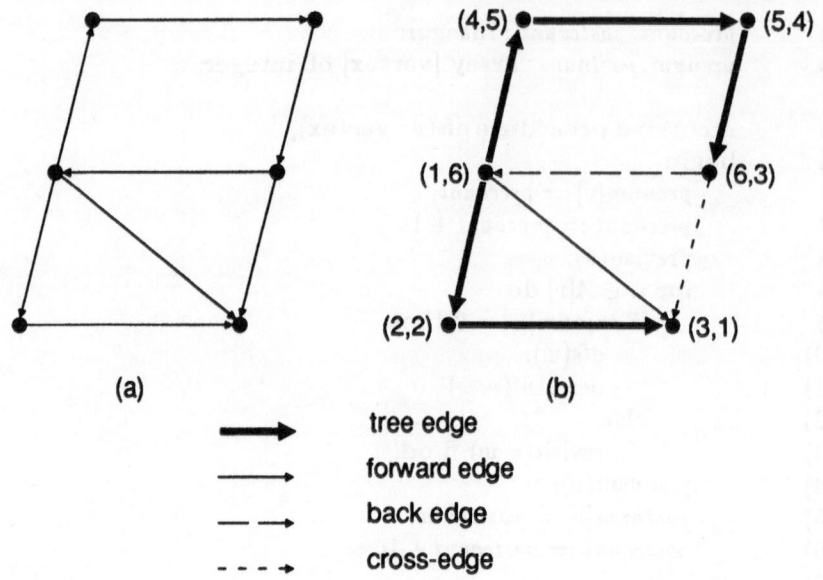

(a)   (b)

$\longrightarrow$ tree edge

$\longrightarrow$ forward edge

$- \longrightarrow$ back edge

$- - \rightarrow$ cross-edge

Figure 3.2: *Depth-first search on a directed graph. (a) The graph before the search; (b) the graph after the search. Vertices v are numbered with (prenum[v], postnum[v]).*

Definition 3.8 (Edge Classification in Undirected Graphs)
An edge $\{v, w\}$ in an undirected graph, such that $prenum[v] < prenum[w]$, is called

- *A **tree edge**, if there is no path of length at least 2 from v to w along which the prenum values increase*

- *A **back edge**, otherwise*

During the search, an edge can be classified at the time that it is explored (in undirected graphs, at the time of its first exploration). An edge (v, w) in a directed graph is

- A tree edge, if at the time of its exploration $prenum[w] = 0$

- A forward edge, if at the time of its exploration $prenum[w] > prenum[v]$

- A back edge, if at the time of its exploration $0 < prenum[w] < prenum[v]$ and $postnum[w] = 0$

- A cross-edge, if at the time of its exploration $0 < prenum[w] < prenum[v]$ and $postnum[w] > 0$

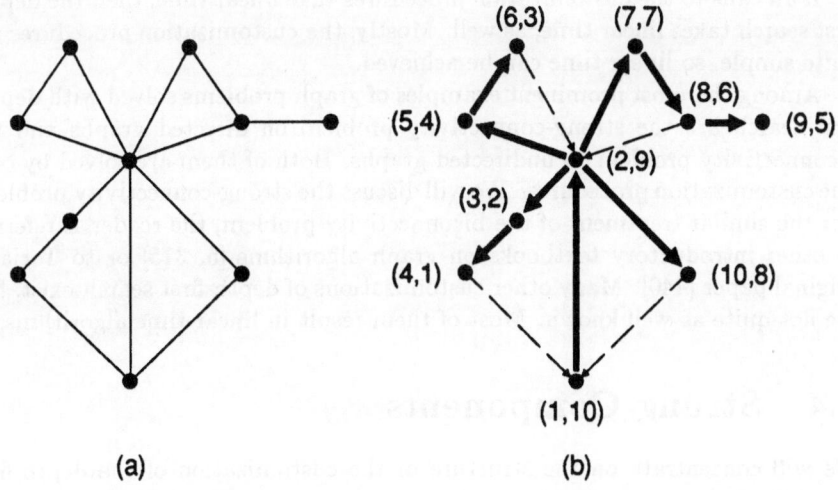

(a)   (b)

Figure 3.3: *Depth-first search on an undirected graph. (a) The graph before the search; (b) the graph after the search. Vertices v are numbered with (prenum[v], postnum[v]).*

An edge $\{v, w\}$ in an undirected graph is

- A tree edge, if at the time of its first exploration $prenum[w] = 0$

- A back edge, if at the time of its first exploration $prenum[w] > 0$

This observation implies that the tree edges indeed form a tree or, more generally, a forest, the so-called DFS *forest*. The DFS forest exhibits the recursive structure of the calls to the dfs procedure that are required for a complete graph traversal. Each DFS tree corresponds to one execution of a call to the dfs procedure on a start vertex. Unless we can reach all vertices in the graph from the first start vertex, more start vertices are necessary. This is the reason for the repeated calls to the dfs procedure in line 25 of Figure 3.1.

Forward and back edges follow the branches of the DFS trees, and cross-edges contribute cross-links within a DFS tree or between two DFS trees. However, such cross-links occur in only one direction; namely, from higher to lower *prenum* values (see Figure 3.2(b)). For undirected graphs, no cross-edges or forward edges exist (see Figure 3.3(b)).

For the purpose of edge classification, we do not need to implement the *postnum* array for undirected graphs. For directed graphs, a Boolean array suffices. In fact, the test $postnum[v] > 0$ just tests whether the call dfs(v) is still active.

If all calls to the customization procedures take linear time, then the depth-first search takes linear time, as well. Mostly, the customization procedures are quite simple, so linear time can be achieved.

Among the most prominent examples of graph problems solved with depth-first search are the strong-connectivity problem on directed graphs and the biconnectivity problem on undirected graphs. Both of them are solved by one-line customization procedures. We will discuss the strong-connectivity problem. For the similar treatment of the biconnectivity problem, the reader is referred to other introductory textbooks on graph algorithms [5, 315] or to Tarjan's original paper [440]. Many other customizations of depth-first search exist, but are not quite as well known. Most of them result in linear-time algorithms.

3.4   Strong Components

We will concentrate on the structure of the customization of the depth-first search to finding strong components.

Definition 3.9 (Strong Connectivity) *Let $G = (V, E)$ be a directed graph. A maximal subset of V whose induced subgraph is strongly connected is a* **strong component** *of G. Equivalently, the strong components are the equivalence classes of the relation*

$$v \sim w :\Longleftrightarrow v \text{ and } w \text{ lie on a common circuit}$$

The graph $\widetilde{G} = (\widetilde{V}, \widetilde{E})$ whose vertex set is the set of strong components of G and whose edge set is

$$\widetilde{E} := \{(C_1, C_2) \mid \quad C_1, C_2 \in \widetilde{G} \text{ and there are vertices} \\ v \in C_1 \text{ and } w \in C_2 \text{ such that } (v, w) \in E\}$$

is called the **supergraph** *of G.*

The supergraph $\widetilde{G}$ of G is a dag: Indeed, the existence of a circuit in $\widetilde{G}$ would contradict the maximality of the strong components. Since G is a dag, its vertices—that is, the strong components—can be topologically ordered. After we have performed a depth-first search on G, the strong components of G are subtrees of the DFS trees. Figure 3.4 illustrates Definition 3.9. The root of such a subtree is called the *root* of the component. The depth-first search is now enhanced with computations that allow it to decide which vertices are roots of strong components. Since the strong components form subtrees of DFS trees, once we know the root, we just have to keep a stack of vertices onto which we push a vertex once it is encountered during the search. We give this stack the name *active*, because it contains all vertices whose processing has begun but has not finished yet. On the completion of a call dfs(v) whose argument is a root of a strong component, we pop the stack up to vertex v. This part of

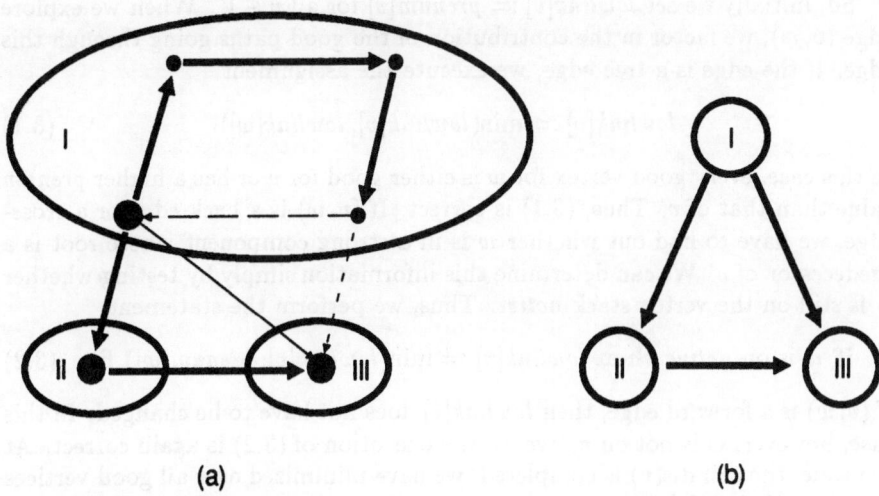

(a)   **(b)**

Figure 3.4: *The strong components of a directed graph. (a) The strong components of the directed graph depicted in Figure 3.2(a); roots of strong components are drawn in bold. (b) The supergraph.*

the stack is the strong component whose root is v. This algorithm outputs the strong components in reverse topological order.

To decide which vertices are roots of strong components, we compute a critical structural value for each vertex—the *lowlink* value. Call a vertex w a *qualifying vertex for a vertex v* if w can be reached from v by following zero or more tree and forward edges—that is, by climbing up a DFS tree—and then moving along at most one back or cross-edge. Call w *good for v* if w is qualifying for v and lies inside a strong component whose root is an ancestor of v in the DFS tree. The corresponding path from v to w is called a *good* path. Now, let *lowlink*$[v]$ be the minimum *prenum* value of all good vertices for v. By definition, $lowlink[v] \leq prenum[v]$, since v is good for v. The following lemma is not difficult to prove.

Lemma 3.1 (Tarjan [440]) v *is a root of a strong component exactly if* $lowlink[v] = prenum[v]$.

It remains to find out how to compute the *lowlink* values. But we can do this iteratively in a rather straightforward way. We do the minimization involved in the computation of the *lowlink* value step by step. At any time during the search, we want the current value of *lowlink*$[v]$ to minimize over all good vertices that can be reached by good paths, leaving v through edges that have already been explored.

So, initially we set $lowlink[v] := prenum[v]$ for all $v \in V$. When we explore edge (v, w), we factor in the contribution of the good paths going through this edge. If the edge is a tree edge, we execute the assignment

$$lowlink[v] := \min(lowlink[v], lowlink[w]) \qquad (3.1)$$

In this case, every good vertex for w is either good for v or has a higher $prenum$ value than that of v. Thus, (3.1) is correct. If (v, w) is a back edge or a cross-edge, we have to find out whether w is in a strong component whose root is a predecessor of v. We can determine this information simply by testing whether w is still on the vertex stack $active$. Thus, we perform the statement

if w is on $active$ **then** $lowlink[v] := \min(lowlink[v], prenum[w])$ **fi**   (3.2)

If (v, w) is a forward edge, then $lowlink[v]$ does not have to be changed. In this case, however, w is not on active, so the execution of (3.2) is again correct. At the time, the call dfs(v) is completed, we have minimized over all good vertices for v, and $lowlink[v]$ has the correct value. Thus, if, just before the completion of dfs(v), $lowlink[v] = prenum[v]$, then we pop the stack $active$ up to v in the postvisit procedure. The complete customization is given in Figure 3.5.

Each of the calls to the procedures previsit, newvisit, and revisit takes constant time. Since these procedures are called once for each explored edge, their total time is $O(m)$. The time for the calls to the procedure postvisit is $O(n)$ plus the time for popping the stack items. Since each vertex is pushed onto the stack only once, the overall time is $O(n)$. The procedure initialize also takes time $O(n)$. Thus, the whole algorithm runs in time $O(m + n)$.

3.5 k-Connectivity

There is a class of notions of connectivity on undirected graphs.

Definition 3.10 (k-Connectivity) *Let $k \geq 1$. An undirected graph G is called k-connected if, for each pair of vertices $v, w \in V$ there are k paths connecting v and w that are vertex-disjoint except for the endpoints.*

An equivalent definition of k-connectivity is the following.

Lemma 3.2 *A graph G is k-connected exactly if no removal of any $k - 1$ vertices together with all of their incident edges disconnects G.*

Obviously, for $k = 1$, the notion of connectivity is as defined in Definition 3.3. A 2-connected graph is also called *biconnected*; a 3-connected graph is also called *triconnected*.

Biconnectivity leads to an interesting decomposition of graphs. We define a relation $\sim$ on the *edges* as follows:

$$e_1 \sim e_2 :\Longleftrightarrow e_1 \text{ and } e_2 \text{ lie on a common simple cycle}$$

Additional global data structures:

(1) z : **vertex**;
(2)   *lowlink* : **array [vertex] of integer**;
(3)   *active* : **stack of vertex**;

Customization procedures:

(4)   **procedure** initialize;
(5)   *active* := emptystack;

(6)   **procedure** previsit(v : **vertex**);
(7)   **begin**
(8)   *lowlink*[v] := *prenum*[v];
(9)   push v onto stack *active*
(10)   **end**;

(11)   **procedure** newvisit(v, w : **vertex**);
(12)   *lowlink*[v] := min(*lowlink*[v], *lowlink*[w]);

(13)   **procedure** revisit(v, w : **vertex**);
(14)   **if** w is on *active* **then**
(15)   *lowlink*[v] := min(*lowlink*[v], *prenum*[w]) **fi**;

(16)   **procedure** postvisit(v : **vertex**);
(17)   **if** *lowlink*[v] = *prenum*[v] **then**
(18)   output("new strong component");
(19)   **repeat** z := pop stack *active*;
(20)   output(z)
(21)   **until** $z = v$ **fi**;

Figure 3.5: *Customization of depth-first search for strong connectivity.*

Definition 3.11 (Biconnected Components) *The equivalence classes of $\sim$ are called the* **biconnected components** *of G.*

A subgraph of G that is formed by all edges in one biconnected component and their endpoints is called a **block**.

A vertex v in G is called a **cutpoint** *if the removal of v and of all its incident edges disconnects G.*

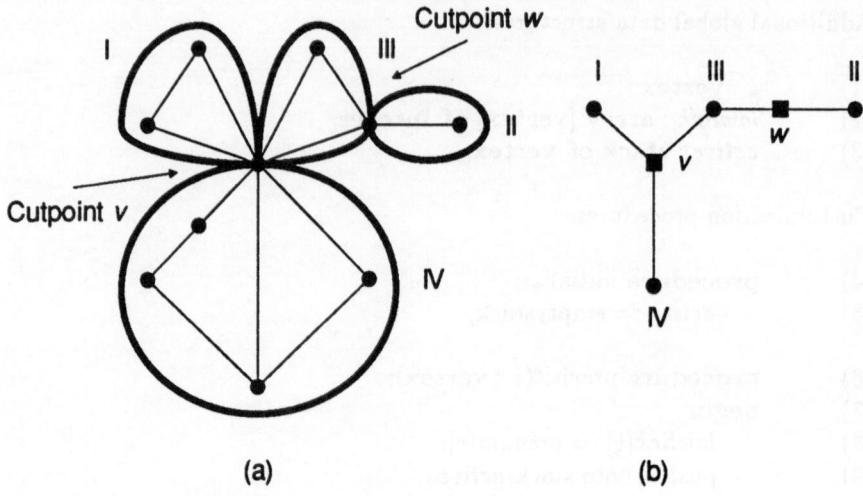

Figure 3.6: *The block-cutpoint tree of the graph depicted in Figure 3.3(a). (a) Cutpoints and biconnected components; (b) the block-cutpoint tree.*

Theorem 3.1

1. *G is biconnected exactly if G has no cutpoint.*

2. *The blocks of G are the maximal biconnected subgraphs of G.*

3. *A vertex in G is in only one block exactly if v is not a cutpoint.*

Theorem 3.1 gives rise to the following definition.

Definition 3.12 (Block-Cutpoint Tree) *The* `block-cutpoint tree` *is a bipartite graph T, with one vertex set being the blocks of G and the other vertex set being the cutpoints of G. There is an edge between the block vertex b and the cutpoint vertex c exactly if cutpoint c is contained in block b.*

For further illustration of Definition 3.12, see Figure 3.6 and Exercise 3.6.

A similar customization of depth-first search as given in Section 3.4 yields a linear-time algorithm that computes the biconnected components of a graph (see Exercise 3.7).

Triconnectivity also gives rise to a decomposition of the graph, which can again be computed in linear time. We will not present details here; see [195, 326].

Biconnectivity and triconnectivity are important concepts in the context of graph planarity (see Section 3.14).

3.6 Topological Search

On dags, a natural kind of search is to visit the vertices in topological order (see Definition 3.5). In this case, the start vertices are the sources of the dag. Topological search can be carried out using depth-first search. The basic observation is that, in a dag, each vertex by itself forms a strong component. The depth-first search algorithm discussed in Section 3.4 finds the strong components in *reverse* topological order. This means that, on a dag, this algorithm encounters the vertices in reverse topological order. Therefore, we can achieve topological ordering of a dag G by reversing all edges in G and then starting a depth-first search on the sinks of G.

3.7 Breadth-First Search

If the data structure D guiding the search is a first=in, first=out (FIFO) queue, then we obtain breadth-first search. The breadth-first search procedure is not recursive. It is detailed in Figure 3.7.

In contrast to depth-first search, we did not augment the breadth-first search algorithm with customization procedures. Such procedures can be attached in appropriate places. The level numbers capture the structure imposed on the graph by a breadth-first search.

Lemma 3.3 *If G is a directed graph, then, for any edge (v, w), we have $level[w] \leq level[v] + 1$.*

If G is an undirected graph, then, for any edge $\{v, w\}$, we have $|level[w]| - |level[v]| \leq 1$.

The proof of Lemma 3.3 uses the fact that on q the vertices appear ordered with respect to increasing *level* numbers, and that at each time only vertices with two different level numbers occur in q.

Definition 3.13 (Edge Classification) *Let G be a directed (undirected) graph, and let $e = (v, w)$ ($e = \{v, w\}$) be an edge of G. If w is encountered edge e is traversed from v to w, we call e a* tree *edge.*

Figure 3.8 shows the result of applying breadth-first search to the graphs in Figures 3.2 and 3.3. Breadth-first search is used in the solution of certain shortest-path problems; see Section 3.8.

3.8 Shortest Paths

3.8.1 Basic Properties

Shortest-path problems are among the most frequently solved graph problems. Many algorithms for wire routing reduce to shortest-path problems. Let us

```
(1)          level : array [vertex] of integer;

(2)          procedure bfs(s : vertex);
(3)          v : vertex;
(4)          q : queue;
(5)          q := s;
(6)          level[s] = 0;
(7)          while q ≠ ∅ do
(8)              v := remove first from q;
(9)              for w ∈ A[v] do
(10)                 if level[w] = −1 then
(11)                     level[w] := level[v] + 1;
(12)                     if w ∉ q then
(13)                         add w to end of q fi fi od od
(14)         end;

(15)         begin
(16)             for v is vertex do
(17)                 level[v] := −1 od;
(18)             for v is vertex do
(19)                 if level[v] = −1 then
(20)                     bfs(v) fi od
(21)         end;
```

Figure 3.7: Breadth-first search algorithm.

first define shortest-path problems.

Definition 3.14 (Shortest Paths) Let $G = (V, E)$ be a directed graph with edge weighting function $\lambda : E \to \mathbf{R}$. The value $\lambda(e)$ is called the **length** of edge e. The **length** of a path $p = v_0, \ldots, v_k$ from $s = v_0$ to $v = v_k$ is defined (somewhat ambiguously) as

$$\lambda(p) := \sum_{i=1}^{k} \lambda(v_{i-1}, v_i)$$

(To avoid confusion, we will call k the **cardinality** of p in the context of shortest-path problems.) The length of the empty path $p = \emptyset$ is $\lambda(\emptyset) := 0$.

The **single-pair shortest-path problem** is, given two vertices $s, t \in V$, to compute the smallest length $d(s, t)$ of any path from s to t. If t cannot be reached from s, then $d(s, t) = \infty$. The value $d(s, t)$ is called the **distance** from s to t. The vertex s is called the **source**; t is called the **target**.

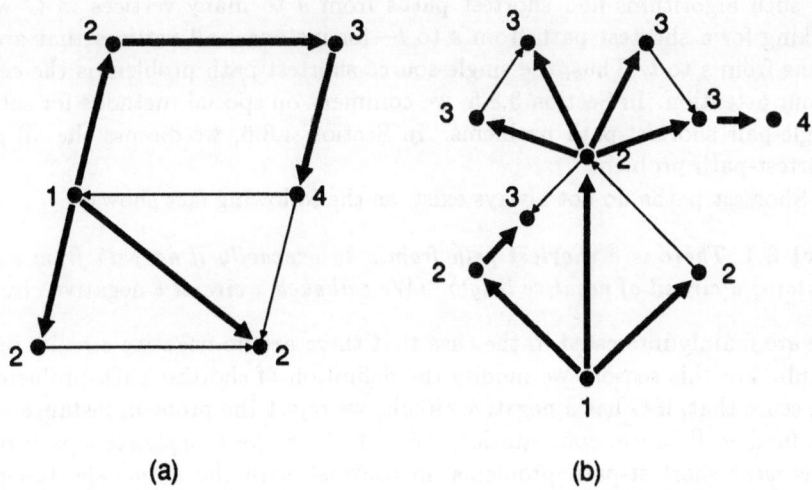

(a)   **(b)**

Figure 3.8: *An example of breadth-first search. Vertices are labeled with their level. (a) Breadth-first search on the directed graph depicted in Figure 3.2(a); (b) breadth-first search on the undirected graph depicted in Figure 3.3(a).*

The `single-source shortest-path problem` *is, given a vertex $s \in V$, to solve for each vertex $v \in V$ the single-pair shortest-path problem with source s and target v. In this case, distances are also written as $d_s(v) := d(s, v)$.*

The `all-pairs shortest-path problem` *is to solve for each vertex $s \in V$ the single-source shortest-path problem with s as a source.*

Note that the single-source shortest-path problem easily generalizes to the problem of finding the shortest distance of each vertex from any of a *set S of source* vertices. We just have to add another vertex to the graph, to make it the source, and to connect it with zero-length edges to all vertices in the set S. By the same token, we can find the shortest path between a selected set of sources S and a selected set of targets T by solving a single-pair shortest path problem with two adjoined vertices s and t, s being the source and t being the target. Zero-length edges run from s to all vertices in S and from all vertices in T to t. This observation will be exploited in Steiner tree heuristics in the context of global routing (Chapter 8).

The single-pair shortest-path problem is a special case of the linear-programming problem (see Section 4.5, Exercise 3.9). Thus, it is solvable in polynomial time (see Section 4.5). There are, however, algorithms for finding shortest paths that are much more efficient than are those for linear-programming.

All such algorithms find shortest paths from s to many vertices in G while looking for a shortest path from s to t—for instance, all vertices that are on paths from s to t. Thus, the single-source shortest-path problem is the center of our attention. In Section 3.8.5, we comment on special methods for solving single-pair shortest-path problems. In Section 3.8.6, we discuss the all-pairs shortest-path problem.

Shortest paths do not always exist, as the following fact shows.

Fact 3.1 *There is a shortest path from s to v exactly if no path from s to v contains a circuit of negative length. (We call such a circuit a* negative *circuit.)*

We are mainly interested in the case that there are no negative circuits in the graph. For this reason, we modify the definition of shortest-path problems in the sense that, if G has a negative circuit, we reject the problem instance without further distance computation. We call the respective shortest-path problems *weak* shortest-path problems, in contrast with the *strong* shortest-path problems defined in Definition 3.14. If the graph has no negative circuits both problem variants are identical. In this section we only discuss weak shortest-path problems and we omit the attribute "weak" if it can be inferred from the context.

Before we discuss algorithms for finding shortest paths, we state a few simple but critical observations. All of our analysis will be based on these observations only.

Lemma 3.4

1. *If there is a shortest path from s to v, then there also is a shortest path from s to v that is simple.*

2. *Let p be a shortest path from s to v and let w be the predecessor of v on p. Then any shortest path from s to w, followed by the edge (w, v), is a shortest path from s to v.*

3. *Let $d : V \to \mathbf{R} \cup \{\infty\}$ be such that, if $d(v) < \infty$, then $d(v)$ is the length of a path from s to v. Then d is the distance function from s exactly if the following holds:*

$$d(s) = 0$$
$$d(w) \leq d(v) + \lambda(v, w) \quad \text{for all } (v, w) \in E \qquad (3.3)$$

The inequality (3.3) is called the *triangle* inequality. Because of Lemma 3.4, parts 1 and 2, we can represent shortest paths from s to all vertices by a tree P rooted at s—the *shortest-path tree*. Each path from s to a vertex in P is a shortest path to this vertex in G. P is not necessarily unique, since shortest paths are not necessarily unique.

(1) d : **array** [**vertex**] **of integer**;
(2) p : **array** [**vertex**] **of vertex**;
(3) e : **edge**;

(4) $d[s] := 0;\ p[s] := s$;
(5)   **for** $v \neq s$ **do** $d[v] := \infty;\ p[v] :=$ nil **od**;
(6)   **while not** finished **do**
(7) $e :=$ some edge (v, w) from E;
(8)   **if** $d[w] > d[v] + \lambda(v, w)$ **then**
(9) $d[w] := d[v] + \lambda(v, w);\ p[w] := v$ **fi od**;

Figure 3.9: *Ford's method for computing shortest paths.*

3.8.2 Ford's Method

Lemma 3.4, part 3 suggests an iterative method proposed by Ford [117] for finding the distance function. We start with a crude approximation to the distance function—namely, $d[s] := 0$ and $d[v] := \infty$ for all other vertices. Then, we test the triangle inequality on different edges (v, w) successively. If it is violated, we reset the value of $d[w]$ such that the triangle inequality holds. An algorithm detailing this method is depicted in Figure 3.9. It is straightforward to see by induction that this method assigns to $d[v]$ only new values that are the lengths of paths from s to v. Indeed, we can show something stronger.

Lemma 3.5 *After testing of the first k edges $e_1, e_2, \ldots, e_k$, for all $v \in V$, $d[v]$ is the length of a shortest path from s to v that is a subsequence of $e_1, \ldots, e_k$. The pointer $p[v]$ points backwards along the last edge of such a path.*

We can stop the tests either after we have ascertained explicitly that all edges satisfy the triangle inequality or if we can be sure that shortest paths from s to all other vertices occur as subsequences of the sequence of tested edges. Then, for all v, $d[v] = d_s(v)$ by Lemma 3.4, part 3, and the p-pointers define a shortest-path tree.

It remains to find an order in which to test the edges such that each shortest path appears as a subsequence, and such that we test as few edges as possible. The simplest solution is to test all m edges and to repeat those tests n times. Here, the order in which the edges are tested in each pass is unimportant. By Lemma 3.4, part 1, the first $n - 1$ passes in this sequence contain shortest paths to all vertices reachable from s, if there are any. The last pass tests whether all triangle inequalities are satisfied. If they are not, the graph contains negative circuits. Since each test takes time $O(1)$, the whole algorithm takes time $O(mn)$. In this way, we obtain a shortest-path algorithm that in the worst case is as fast asymptotically as is any algorithm known for the single-source shortest-path problem.

If the graph is acyclic, we can solve it much more quickly. In this case, there can be no negative circuits. We construct the edge sequence to be used by sorting the vertices topologically and testing all edges out of each vertex in that order. Topological sorting can be done using depth-first search in linear time (see Section 3.6). The overall run time of the algorithm is linear $O(m+n)$ in this case.

This observation is generalizable.

Definition 3.15 (BF-Orderable) *A pair (G, s), where G is a directed graph and s is a distinguished vertex in G, is called* BF-orderable *if the edges of G can be permuted to yield a sequence σ that contains all simple paths in G starting at s. Any such sequence σ is called a* BF-ordering *of (G, s).*

If (G, s) is a BF-orderable graph, then there is a sequence σ of edges such that, no matter what the edge labels are in G, σ is suited for use in Ford's method if s is the source. Thus, given σ, the shortest-path computation can be done in linear time. We first test the edges as arranged in the BF-ordering, and then check the triangle inequality for each edge. The latter check is not necessary if it is certain that no negative cycles exist. As we noted previously, acyclic graphs are BF-orderable. Mehlhorn and Schmidt [321] give necessary and sufficient criteria for (G, s) to be BF-orderable, and present an $O(m^2)$ algorithm for testing for BF-orderability and finding the appropriate sequence σ of edges. Haddad and Schaeffer [160] improve this algorithm to $O(n^2)$. These general results are still somewhat unsatisfying, because finding σ already takes quadratic time, whereas the whole purpose of σ is to do the shortest-path computation in linear time. However, there are interesting classes of graphs for which σ can be found in linear time (see Section 9.3.1). Thus, a linear-time shortest-path algorithm can be obtained for such graphs.

If we want to solve the strong single-source shortest-path problem, we also have to identify the set of all vertices that can be reached from s via some negative circuit (see Exercise 3.12).

3.8.3   The Bellman–Ford Algorithm

So far, we compute off-line the sequence in which edges are tested before solving the shortest-path problem. We can achieve space and time savings on typical graphs if we instead compute the edge sequence on-line. The idea is to maintain a data structure D that contains all vertices that are tails of edges that could possibly violate the triangle inequality. The corresponding algorithm is shown in Figure 3.10. Initially, the only vertex in D is s. Then, we pull vertices v out of D, one by one (line 9), and test all edges out of v. As an edge (v, w) is tested and w receives a new distance value (line 12), we enter w into D (line 13) or

(1) d : **array** [**vertex**] **of integer;**
(2) p : **array** [**vertex**] **of vertex;**
(3) D : **data structure;**
(4) v : **vertex;**

(5) $d[s] := 0$; $p[s] := s$; initialize D;
(6)   insert s into D;
(7)   **for** $v \neq s$ **do** $d[v] := \infty$; $p[v] :=$ nil **od;**
(8)   **while** $D \neq \emptyset$ **do**
(9) $v :=$ extract from D;
(10)   **for** $w \in A[v]$ **do**
(11)   **if** $d[w] > d[v] + \lambda(v, w)$ **then**
(12) $d[w] := d[v] + \lambda(v, w)$; $p[w] := v$;
(13)   **if** $w \notin D$ **then** insert w into D
(14)   **else** relocate w in D **fi fi od od;**

Figure 3.10: *Dynamic version of the shortest-path algorithm.*

move it to its new place in D according to its decreased distance value, if w already was in D (line 14).

It is common nomenclature to refer to the process of entering a vertex into D as *labeling* the vertex, and to that of extracting a vertex from D and processing the edges leaving it as *scanning* the vertex [441].

Instead of using the p-pointers, we can find a shortest path from s to v by tracing from v backward along edges that fulfill the triangle inequality with equality. This reduces the storage but increases the run time for path tracing, because several neighbors of each vertex on the path have to be inspected.

To solve the single-source shortest-path problem without restrictions, Bellman [25] chose a FIFO queue for D. The element at the front of the queue is extracted in line 9 of Figure 3.10. In line 13, elements are inserted at the back of the queue. The relocation of elements in line 14 is a null operation in this case. Then, the shortest-path algorithm becomes a variant of breadth-first search. Here, however, a vertex can enter D several times if there are cycles in the graph. Nevertheless, the basic structure of breadth-first search established in Section 3.7 carries over. Let us call the contents of D at a specific time a *snapshot*. The initial snapshot is called *Snapshot 0*. The period in the Bellman–Ford algorithm that processes a snapshot—that is, that tests the edges in a snapshot—is called a *pass*. The initial pass is called *Pass 0*. The snapshot created after finishing Pass i is called *Snapshot i+1*, and it is processed in *Pass i+1*. The following lemma can be shown by induction.

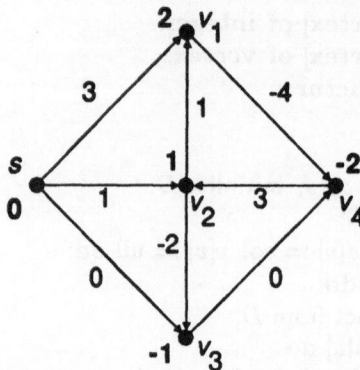

Figure 3.11: *An example of the single-source shortest-path problem.*
Edge lengths are denoted by the numbers in regular type. Distance
values from s are denoted by boldface numbers. The snapshots of D
are as follows (the entries of D are denoted as ordered pairs $(v, d[v])$,
where $d[v]$ is the current distance value of v):

Snapshot Number	Contents of D
0	$(s, 0)$
1	$(v_1, 3)(v_2, 1)(v_3, 0)$
2	$(v_4, -1)(v_1, 2)(v_3, -1)$
3	$(v_4, -2)$
4	$\emptyset$

Lemma 3.6 *If a* FIFO *queue is chosen for D in Figure 3.10, then the following*
holds: After Pass i for all vertices v with a shortest path of cardinality at most
$i + 1$ edges, we have $d[v] = d_s(v)$.

As shown in Figure 3.10, the algorithm does not terminate if there are negative
circuits. To make it stop, we note that, by Lemma 3.4, part 1, if there are no
negative circuits, all distance values are computed correctly and D is empty
after Pass $n - 1$. Each pass has length at most m, since each edge is tested at
most once in a pass. Thus, in the worst case, the algorithm again has run time
$O(mn)$. Figure 3.11 shows an example.

The necessity of repeatedly entering vertices into D arises because of the
existence of circuits in G. If there are few circuits, we would expect the number
of vertex repetitions to decrease. In fact, if there are no circuits—that is, if
G is a dag—then the algorithm carries out a standard breadth-first search.
This means that each vertex is entered into D only once, and the run time is
$O(m + n)$.

This observation leads to the following modification of the shortest-path algorithm. In a first step, we compute the strong components of G. Then, we process them in topological order, starting with the component containing the source s. When processing a component C, we adjoin to C a new source s_C. From s_C, we add edges $e_C = (s_C, w)$ leading to all vertices w in C that are tails of edges $e = (v, w)$ in G such that $v \notin C$. The length of edge e_G is the current value of $d[v]$. Then, we run the Bellman–Ford algorithm with s_C as the source. This computation determines all distance values for vertices in C.

The resulting algorithm reduces to the Bellman–Ford algorithm if G is strongly connected, and reduces to normal breadth-first search if G is acyclic. In the other cases, repeated passes over the graph are made only inside the strong components. Thus, if the strong components are small, then the performance of this algorithm is much better than the performance of the Bellman–Ford algorithm.

The algorithms we have considered so far only prove the *existence* of a negative circuit, if there is one. They do not actually exhibit such a circuit. For diagnostic purposes, exhibiting negative circuits is important. By Lemma 3.5, we encounter a negative circuit by following the p-pointers, starting at the tail of any edge e that does not fulfill the triangle inequality, after the shortest-path computation stops.

3.8.4 Dijkstra's Algorithm

A frequently occurring special case is that in which all edge lengths are positive.

This special case also plays a large role in applications in wire routing and compaction, even though negative edge lengths do occur, as well. If all edge lengths are positive, the appropriate data structure to select for D is a priority queue holding vertices ordered with respect to the (current) d values. The operation in line 9 of Figure 3.10 selects the vertex with the smallest d value. The operation in line 14 relocates w in D according to the new smaller value of $d[w]$. The corresponding algorithm is named after Dijkstra, who was the first person to propose it [90]. The important observation about this version of the shortest-path algorithm is the following.

Lemma 3.7 *The d value of any labeled vertex is no smaller than is the d value of the last scanned vertex.*

Since the d values of vertices can only decrease, Lemma 3.7 implies that every vertex is labeled (and scanned) only once. Thus, each edge is tested only once, and the algorithm takes time $O(m + p)$, where p is the total time for all operations on the priority queue.

Dijkstra's algorithm can be thought of as a process in which a wave is propagated through the graph starting at s. The wavefront—that is, the set of vertices that has been labeled but not scanned—is stored in the priority queue D. Figure 3.12 illustrates this interpretation on a partial grid graph with unit-length edges.

In his original paper, Dijkstra proposed an inefficient implementation of the priority queue that leads to $O(n^2)$ run time of the shortest-path algorithm. Since then, there has been much research on efficient implementations of priority queues, some of these implementations being especially tailored to the shortest-path problem:

- The classical implementation uses a heap [5, 315]. It takes $O(\log n)$ time per operation (enter, extract, or relocate). Thus, the total time of the shortest-path algorithm becomes $O(m \log n)$.

- Johnson [208] suggested basing the heap not on the usual binary tree structure, but rather on k-ary trees for an appropriate value of k, which turns out to be $k = 2 + m/n$. The resulting time per priority queue operation is $O(m \log_{\lceil 2+m/n \rceil} n)$. Note that this time is linear for dense graphs. For sparse graphs, with which we will be working most of the time, this is not much better than the traditional binary heap.

- Fredman and Tarjan [122] were the first researchers to develop a version of the priority queue for which the relocate operations can be done in amortized constant time. (This means that the total time for m relocate operations is $O(m)$.) This data structure is called the *Fibonacci heap*. In addition to being good in theory, it is efficient in practice. Using it, the shortest-path algorithm has a run time of $O(m + n \log n)$.

- If all edges have length 1, then the priority queue in Dijkstra's algorithm becomes a FIFO queue; that is, Dijkstra's algorithm degenerates to a simple breadth-first search of the graph. Thus, the single-source shortest-path problem is solvable in time $O(m + n)$.

- If all edge lengths come from the set $\{0, 1, \ldots, C\}$, then, at any time, D contains only vertices with $C + 1$ adjacent d values. We can exploit this fact to give an implementation of D with which the algorithm runs in time $O(m + nC)$ and in space $O(n + C)$.

  We allocate $C + 1$ doubly linked lists $bucket[i]$, $i = 0, \ldots, C$, each holding the elements of D with a certain d value. Furthermore, there is an integer variable $d_{\min}$ that indicates the size of the smallest d value in D. The operations of the priority queue are implemented as shown in Figure 3.13.

The initialization of D takes time $O(C)$, the operations insert and relocate take time $O(1)$, and the operation extract may take time $O(C)$ in the worst case. Thus, the overall run time of the shortest-path algorithm becomes $O(m + nC)$. This implementation is mentioned in several places in the design-automation literature, especially for the case $C = 1$ [265, 332]. Detailed code for this version of the algorithm is presented in [89].

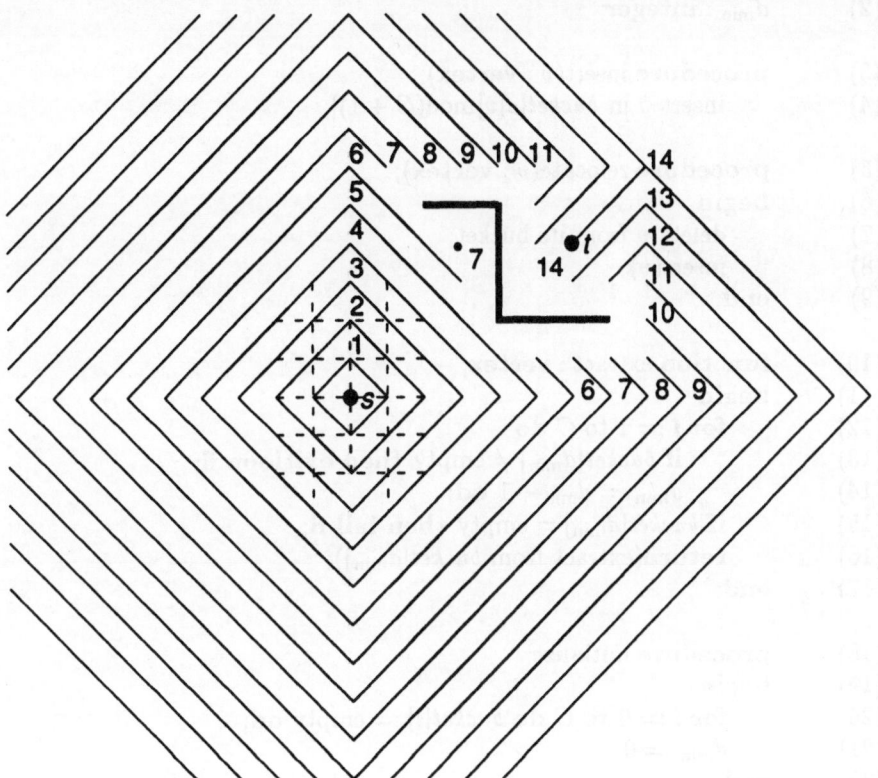

Figure 3.12: *Dijkstra's algorithm as a wave-propagation process. The graph is an implicit symmetric grid graph with unit-length antiparallel edges. The grid is shown locally around s. The obstacle drawn in heavy black lines represents a set of vertices that is deleted from the grid graph. The numbered contours represent wavefronts. Since all edge lengths are 1, the priority queue in Dijkstra's algorithm degenerates to a FIFO queue. Each wavefront is created in a pass of the algorithm, as defined in Section 3.8.3. A wavefront is a set of vertices that have the same distance from s. The wavefronts are labeled with the corresponding distance values. The union of all depicted wavefronts contains the set of all scanned vertices at the time that t is scanned.*

```
(1)        bucket : array [0..C] of doubly linked list of vertex;
(2)        d_min : integer;

(3)        procedure insert(v : vertex);
(4)            insert v in bucket[d[v]mod(C + 1)];

(5)        procedure relocate(v : vertex);
(6)        begin
(7)            delete v from its bucket;
(8)            insert(v)
(9)        end;

(10)       function extract : vertex;
(11)       begin
(12)           for i := 1 to C do
(13)               if bucket[d_min] ≠ empty then exitloop fi;
(14)               d_min := d_min + 1 od;
(15)           if bucket[d_min] = empty then fail fi;
(16)           return(extract from bucket[d_min]);
(17)       end;

(18)       procedure initialize;
(19)       begin
(20)           for i := 0 to C do bucket[i] := empty od;
(21)           d_min := 0
(22)       end;
```

Figure 3.13: *Efficient priority-queue implementation.*

- Ahuja and Orlin improved the priority-queue method by providing $\lceil \log C \rceil + 1$ buckets. Buckets 0 and 1 take all elements with distance $d_{\min}$ and $d_{\min} + 1$, respectively. The following buckets accommodate elements with distances from the interval $[d_{min} + 2^{i-1}, d_{min} + 2^i - 1]$ for $i = 2, \ldots, \lceil \log C \rceil - 2$, respectively, and the last bucket contains all elements with distance at least $d_{min} + 2^{\lceil \log C \rceil - 1}$. The priority-queue operations are modified accordingly. This decreases the time to $O(m + n \log C)$ and the space to $O(m + n + \log C)$ (Exercise 3.14). The algorithm is still quite easy to code and is likely to be quite efficient in practice. We can improve the asymptotic run time of the algorithm by generalizing it to k-ary heaps, and using Fibonacci heaps at a strategic place in the algorithm. The resulting run time for the shortest-path algorithm is $O(m + n\sqrt{\log C})$, but this version of the algorithm is probably too complicated to be competitive in practice. Both modifications are presented in [6].

3.8.5 The Single-Pair Shortest-Path Problem

The single-pair shortest-path problem with source s and sink t gets solved along with the single-source shortest-path problem with source s. Furthermore, in the worst case, if the graph is a path from s to t, solving a single-pair shortest-path problem amounts to solving a strong single-source shortest-path problem. In typical cases, however, the single-pair shortest-path problem can be solved much faster than can the corresponding strong single-source shortest-path problem. In this section, we survey several methods for solving the former problem.

3.8.5.1 Goal-Directed Unidirectional Search

Hart et al. [175] were the first researchers to suggest using lower bounds on distances to the target to direct the search. Let us assume that a lower bound $b(j, t)$ on the distance of each node v to the sink t is available, and assume further that the lower-bound values meet the following *consistency conditions*:

$$\lambda(v, w) + b(w, t) \geq b(v, t) \tag{3.4}$$

In most layout applications of shortest-path algorithms, the graphs have an embedding into Euclidean space, and the edge lengths are the Manhattan or Euclidean distances of their endpoints. In these cases, $b(v, t)$ can be taken to be the metric distance between v and t, and the consistency conditions hold.

If a lower bound that fulfills the consistency conditions is available, then we can relabel the edges such as to guide the search from s more directly toward t. Instead of $\lambda(v, w)$, the edge (v, w) now receives the length

$$\lambda'(v, w) := \lambda(v, w) - b(v, t) + b(w, t)$$

Then, Dijkstra's algorithm is carried out, as usual. (Note that, by the consistency conditions, all modified edge lengths are positive.)

Modifying the edge lengths as such does not change the shortest paths from s to t, because if p is a path from s to t, then

$$\lambda'(p) = \lambda(p) - b(s, t) + b(t, t)$$

Thus, the lengths of such paths change by only a constant additive term. However, the relabeling has two advantages. First, it makes all edge labels positive, such that Dijkstra's algorithm can be employed. Second, the next vertex pulled out of the priority queue D is the vertex v whose distance from s in the reweighted graph is minimum. This distance is

$$d'_s(v) := d_s(v) + b(w, t) - b(s, t)$$

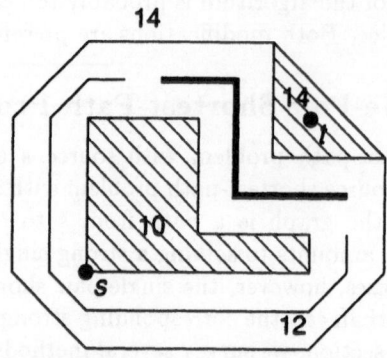

Figure 3.14: *Goal-directed unidirectional search. Wavefronts are sets
of vertices with the same modified distance values. Note that wavefronts can be spread out in area in this case. Wavefronts are labeled
with the modified distance values of their vertices. The union of all
shown wavefronts contains the set of all scanned vertices at the time
that t is scanned.*

where $d_s(v)$ is the distance of v from s in the original graph. Thus, $d_s'(v)$ is
the *estimated distance* of a shortest path from s to t *via* v (minus $b(s,t)$). We
therefore expect that, when we are using the λ' values, t will be scanned much
earlier than with the original edge lengths. It can, in fact, be proved that, in
this *goal-directed* version of unidirectional search, t never appears *later* [264]
(Exercise 3.17). In the worst case, if G is a path, t will not appear earlier; in
typical cases, however, significant speedups are observed. In addition, we can
prove that the average-case behavior of the goal-directed unidirectional search
algorithm is linear for many probability distributions of the input [412]. Such a
result is not known for the standard version of Dijkstra's algorithm. Figure 3.14
shows the region of scanned vertices if unidirectional search with a lower bound
based on the Manhattan distance is used. (This is a typical example from the
application of wire routing in integrated circuits.)

As described so far, goal-directed unidirectional search may assign the same
(modified) distance value $d_s'(v)$ to a large number of vertices v. As an example,
all vertices v in the shaded region in Figure 3.14 have $d_s'(v) = 10$. To orient the
search through such regions toward the target, we can use the value of $b(v,t)$
as a secondary cost measure [384]. Thus, vertices in the priority queue D are
considered ordered according to increasing lexicographic ordering of the values

$$(d_s'(v), b(v,t))$$

The corresponding algorithm still guarantees that we will find a shortest path.

It just eliminates some of the nondeterminism inherent in goal-directed uni-directional search. Note that, in the case of Figure 3.14, however, additional cost measure does not save us scanning all vertices in the wavefronts labeled 10 and 12, because in this case $d(s,t) = 14$. If t were located before the obstacle as seen from s, the goal-directed unidirectional search with a secondary cost measure would find a direct path from s to t without scanning any additional vertices.

3.8.5.2 Bidirectional Search

Another modification is to expand two path segments, one from s and one from t, simultaneously. This is the idea of *bidirectional search methods*. We consider bidirectional search in only the case in which all edge weights are nonnegative. Like unidirectional search methods, bidirectional search methods come in two versions: with lower bounds and without lower bounds.

The idea is to run two (almost) independent unidirectional searches in parallel. One search emanates backward from t; the other moves forward from s. Each search is carried out using Dijkstra's algorithm. (Dijkstra's algorithm can be applied to an arbitrary problem instance without negative cycles, after redefining the edge lengths as discussed in Section 3.8.6.) A vertex that is scanned during the search emanating from s (or from t) is said to have *home s* (or t). Consider the search emanating from s. A vertex v that is extracted from D is processed only if v does not have home t already. Otherwise, the length of a shortest path from s to t via v is known: It is $length[v] := d_s[v] + d_t[v]$, where d_s and d_t are the distance values computed by the searches emanating from s and t, respectively.

Analogous statements can be made about the search emanating from t. Let *minlength* be the minimum over the values $length[v]$ for all vertices $v \in V$. (Initially, $length[v] = \infty$ for all $v \in V$.) At any time during the algorithm, the value *minlength* is an upper bound on the length of a shortest path from the vertex s to the vertex t.

We can obtain lower bound on the length of a shortest path from s to t by just adding the d values of the minimum vertices in the priority queues for both searches. The bidirectional search can be stopped when the upper bound meets the lower bound. This is sure to happen as both searches expand along a shortest path from s to t. It does not have to happen by the time that the first vertex is reached by both searches, in general; if all edges have length 1, however, then it always does happen at this time (Exercise 3.18).

So far, we have not specified how we alternate the control between the two searches. In fact, we can do that in a multitude of ways. We can give one of the searches a higher priority, or run both searches at equal speed. The switch of control can even be decided dynamically. The respective adjustment can be made on the basis of empirical studies and can take into account the structure of the underlying graph. For instance, one good idea is always to expand the search that has labeled fewer vertices so far. The theoretical statements we can make about the superiority of bidirectional search over unidirectionals search are rather weak [264]. For instance, unidirectional search without lower bounds may be more efficient than bidirectional search (even in the goal-directed version described later in this section) in some cases (Exercise 3.19). Empirical studies show that, in practical settings, bidirectional search outperforms unidirectional search for single-source shortest-path problems [359]. Figure 3.15 gives an example of the application of bidirectional search. Luby and Ragde [296] show that, under certain general assumptions on the probability distribution of edges, the bidirectional search algorithm, where both searches progress at the same speed, has an expected run time of $O(\sqrt{n})$. This value is dramatically better than the average-case complexity of any unidirectional search method.

If lower bounds $b(s, v)$ and $b(v, t)$ are available that meet the consistency conditions

$$\lambda(v, w) + b(s, v) \geq b(s, w)$$
$$\lambda(v, w) + b(w, t) \geq b(v, t)$$

then both searches in the bidirectional search can be goal-directed as described in Section 3.8.5.1. The lower bounds are again based on the Manhattan distance.

The speed of the searches are adjusted dynamically. The search that has scanned the fewest vertices is expanded. Luby et al. [264] consider additional ways of incorporating lower bounds into bidirectional search.

3.8.6   The All-Pairs Shortest-Path Problem

We can solve the all-pairs shortest-path problem by solving n single-source shortest-path problems, one with each vertex as the source. This approach yields $O(mn^2)$ run time. We can do better, however.

First, we adjoin a new source s to G with zero-length edges from s to all vertices in G. The first single-source shortest-path problem that we solve has s as its source. Then, we use the resulting distance values to relabel the edges of G such that no negative edge lengths occur. Specifically, we use the new edge-length function $\lambda' : E \to \mathbf{N}$, which is defined as follows:

$$\lambda'(v, w) := \lambda(v, w) + d_s(v) - d_s(w)$$

Because each edge satisfies the triangle inequality (3.3), we have $\lambda'(e) \geq 0$ for all $e \in E$. Furthermore, the length $d(v, w)$ of the shortest path from v to w based on the edge-length function λ can be computed from the distance value $d'(v, w)$ based on λ', as follows:

$$d(v, w) := d'(v, w) - d_s(v) + d_s(w)$$

We compute the values $d'(v, w)$ with n applications of Dijkstra's algorithm. This approach solves the all-pairs shortest-path problem in the time needed for

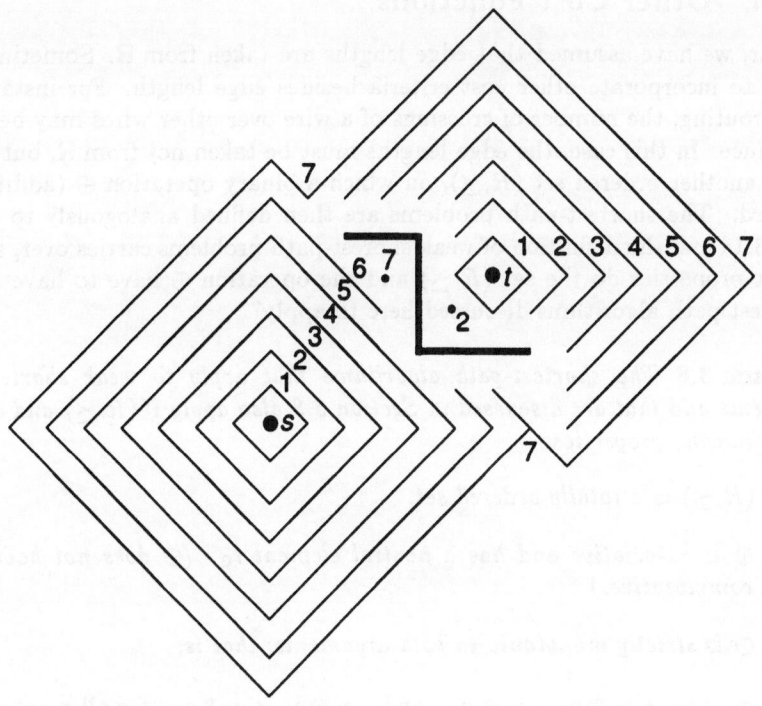

Figure 3.15: *Bidirectional search. The search with the current lowest-numbered wavefront is continued. The wavefronts are labeled with the distance values of their vertices from the source (either s or t). The union of all depicted wavefronts contains the set of all scanned vertices at the time that t is scanned.*

one application of the Bellman–Ford algorithm and n applications of Dijkstra's algorithm. If we use Fibonacci heaps as the data structure for the priority queue, this strategy results in time $O(mn + n^2 \log n)$.

Alternatively, we can solve the all-pairs shortest-path problem with matrix-multiplication methods. This approach leads to $O(n^{2.38})$ run time for the theoretically best methods [74], and to closer to $O(n^3)$ for practical methods. Although this framework is not as appealing for applications involving sparse graphs, it does have the advantage that it is generalizable to other variants of path problems (see the next section).

3.8.7   Other Cost Functions

So far, we have assumed that edge lengths are taken from $\mathbf{R}$. Sometimes, we want to incorporate other cost criteria besides edge length. For instance, in wire routing, the number of crossings of a wire over other wires may be of importance. In this case, the edge lengths must be taken not from $\mathbf{R}$, but rather from another ordered set $(R, \preceq)$, on which a binary operation $\oplus$ (addition) is defined. The shortest-path problems are then defined analogously to Definition 3.14 and the definition of weak shorest-path problems carries over, as well. What properties do the set $(R, \preceq)$ and the operation $\oplus$ have to have for the shortest-path algorithms discussed here to apply?

Lemma 3.8 *The shortest-path algorithms that apply to weak shortest-path problems and that are discussed in Section 3.8 also apply if $(R, \preceq)$ and $\oplus$ have the following properties:*

1. *$(R, \preceq)$ is a totally ordered set.*

2. *$\oplus$ is associative and has a neutral element r_0. ($\oplus$ does not have to be commutative.)*

3. *$\oplus$ is strictly monotonic in both arguments; that is,*

$$r_1 \prec r_2 \Rightarrow r_1 \oplus r_3 \prec r_2 \oplus r_3 \quad \text{and} \quad r_3 \oplus r_1 \prec r_3 \oplus r_2 \quad \text{for all } r_1, r_2, r_3 \in R$$

Proof All of the shortest-path algorithms we considered reside on the properties proved in Lemma 3.4. The proof of this lemma uses only the listed properties of $(R, \preceq)$ and $\oplus$. The neutral element r_0 of $\oplus$ replaces 0. Any element $r \in R$ such that $r \prec r_0$ is *negative*. $\square$

An example for a set $(R, \preceq)$ with addition operation $\oplus$ that Lemma 3.8 is $\mathbf{R}^k$ with the normal addition and the lexicographic ordering, as observed by Rubin [384]. (Note that in this case Fact 3.1 does not hold and the solution of the strong single-source shortest-path problem that is discussed in Exercise 3.12 does not carry over.) Rubin also generalizes goal-directed unidirectional search methods to such cost functions (see Section 3.8.5.1).

We also can come up with simple examples where the premise of Lemma 3.8 does not hold and the shortest-path algorithms discussed here do not work. One example is discussed in Exercise 3.21. Another example is the use of two independent edge weightings $\mu_1, \mu_2 : E \to \mathbf{R}$. We consider the length of a path as the pair $\lambda(p) = (\mu_1(p), \mu_2(p))$ where $\mu_1(p), \mu_2(p)$ are defined as in Definition 3.14. $\lambda(p_1) \preceq \lambda(p_2)$ holds if $\mu_1(p) \preceq \mu_1(p_2)$ and $\mu_2(p_1) \preceq \mu_2(p_2)$. This ordering is not total, so no unique distance values exist. The problem of testing, given $G = (V, E)$, $v, w \in V$, and $k, \ell \in \mathbf{R}$, testing whether there is a path p from v to w such that $\lambda(p) \preceq (k, \ell)$ is a problem that is NP-complete [133].

Thus, a careful study must reveal whether the shortest-path problem with respect to a certain cost function and ordering is feasible. On the way we have to develop new appropriate efficient shortest-path algorithms.

In Chapter 8, we discuss applications of shortest-path algorithms with general cost functions.

For the all-pairs shortest-path problem, there exists a more general framework for varying cost functions, based on the observation that the all-pairs shortest-path problem can also be solved with matrix-multiplication methods. The corresponding generalization is called the *algebraic path problem*. For details see [5, 315].

3.9 Minimum Spanning Trees

The minimum-spanning-tree problem is the easiest of all network problems. It is used in estimating wiring complexity (see Section 7.1.1) and to approximate harder problems that are important in layout, such as the MINIMUM STEINER TREE problem (see Section 3.10). The minimum-spanning-tree problem is defined as follows:

MINIMUM SPANNING TREE

Instance: A connected undirected graph $G = (V, E)$ with edge-weighting function $\lambda : E \rightarrow \mathbf{R}_+$; $\lambda(e)$ is called the *cost* of e

Configurations: All edge-weighted trees

Solutions: All *spanning trees* of G; that is, all subtrees of G that span G

Minimize: $\lambda(T) = \sum_{e \in E_T} \lambda(e)$

We consider only real edge labels here. However, all algorithms and theorems directly apply to the case in which the edge labels come from $\{r \in R \mid 0 \preceq r\}$, where $(R, \preceq, \oplus)$ is an ordered set with addition that fulfills the premise of Lemma 3.8.

3.9.1 A Greedy Method

There is a simple greedy method for constructing minimum spanning trees. This method looks at edges one by one and decides whether to include or exclude them from the minimum spanning tree. Tarjan [441] formulates the method in terms of *edge coloring*.

At the beginning, all edges are uncolored. At each step, we color an edge. We color the edge blue if we decide to include it in the minimum spanning tree, and color it red if we decide to exclude it from the minimum spanning tree. The invariant we want to maintain is the following.

Coloring Invariant: After the ith coloring step, there is a minimum spanning tree T that contains all blue edges and does not contain any red edge.

This invariant is maintained if we apply the following *coloring rules* to carry out coloring steps.

Blue Rule: Applicable If there is a cut (V_1, V_2) through G across which there is no blue edge but at least one uncolored edge.

   Action: Choose the cheapest uncolored edge across the cut, and color it blue.

Red Rule: Applicable If there is a cycle C in G that does not contain a red edge but has at least one uncolored edge.

   Action: Choose the most expensive uncolored edge on C, and color it red.

The following lemma proves the correctness of the coloring method.

Lemma 3.9

   1. *Any application of the blue and red rule preserves the coloring invariant.*

   2. *If there are still uncolored edges, then a coloring rule can be applied.*

The way in which we select applications of the coloring rules determines the particular minimum-spanning-tree algorithm. We will mention two such choices.

3.9.2   Prim's Algorithm

The first method we will examine is Prim's algorithm [366]. Here, we begin at a start vertex v in G and apply first the blue rule to the cut $(\{v\}, V \setminus \{v\})$. In general, after the ith coloring step, there is a tree $T_i = (V_i, E_i)$ with i blue edges. In the $(i + 1)$th coloring step, we apply the blue rule to the cut $(V_i, V \setminus V_i)$. After $n - 1$ applications of the blue rule, we can stop, since we have constructed a blue spanning tree.

   The implementation of Prim's algorithm is shown in Figure 3.16. It uses a priority queue q to find the cheapest uncolored edges across the cuts. At any time, q contains the vertices that are adjacent to the blue tree assembled so far, and the key of a vertex v in q is the cost of the cheapest edge from v to the blue tree. This value is stored in $d[v]$. The corresponding edge is the candidate edge for joining v to the blue tree. Its other endpoint is stored in $T[v]$. At the end, the sets $\{v, T[v]\}$ indicate the edges of the minimum spanning tree, and the $d[v]$ values indicate their respective costs. Note that the coloring is not actually

```
(1)     v, w : vertex;
(2)     T : array [vertex] of vertex;
(3)     d : array [vertex] of integer;
(4)     notintree : array [vertex] of Boolean;
(5)     q : priority queue of vertex;

(6)     for w vertex do
(7)         notintree[w] := true;
(8)         T[w] := nil;
(9)         d[w] := ∞ od;
(10)    notintree[s] := false; d[s] := 0;
(11)    initialize q;
(12)    for w ∈ A[s] do
(13)        d[w] := λ(s, w);
(14)        T[w] := s;
(15)        enter w into q od;
(16)    do n − 1 times
(17)        v := extract minimum from q;
(18)        notintree[v] := false;
(19)        for w ∈ A[v] do
(20)            if notintree[w] and λ(v, w) < d[w] then
(21)                d[w] := λ(v, w);
(22)                T[w] := v;
(23)                if w ∉ q then enter w into q
(24)                else relocate w in q fi fi od od;
```

Figure 3.16: *Prim's algorithm.*

done in the algorithm, but just serves as a paradigm for the development of the algorithm.

The similarity between Prim's algorithm and Dijkstra's algorithm is striking. Both of them test each edge once. The order in which edges are tested is based on a priority queue of vertices. Processing an edge means reevaluating one of its end vertices if some criterion is fulfilled—here, the two algorithms use different but similar criteria. Reevaluating a vertex can only decrease its key in the priority queue. If we implement the priority queue with balanced trees, the run time is $O(m \log n)$. If we use Fibonacci heaps, the run time is $O(m + n \log n)$. We cannot use the priority queue by Ahuja et al. [6] here, since we cannot assume that the d value of a vertex inserted into the queue q is no smaller than the d value of any vertex inside q.

(1) v, w : **vertex**;
(2) L : **sorted list of edge**;
(3) D : **partition of vertex**;
(4)   *tree* : **set of edge**;

(5)   initialize D; *tree* $:= \emptyset$;
(6)   **for** (v, w) **upwards** L **do**
(7)   **if** FIND$(v) \neq$ FIND(w) **then**
(8)   UNION(v, w);
(9)   *tree* $:=$ *tree* $\cup \{\{v, w\}\}$ **fi od**;

Figure 3.17: *Kruskal's algorithm.*

3.9.3   Kruskal's Algorithm

If the edges already happen to come in sorted order according to their cost, then
we can use an algorithm that is faster on sparse graphs. Kruskal [249] presents
an algorithm that colors the edges in order according to increasing cost. In the
course of the computation, a set of blue trees is constructed. To color an edge
e, we inspect whether e joins two blue trees or stays within one blue tree. In
the first case, we color e blue; in the second case, we color e red. The algorithm
is depicted in Figure 3.17. We use a special data structure D for maintaining
the blue trees. D represents the blue forest as a partition of the vertex set such
that each tree is represented by the set of its vertices. Initially, each vertex
forms a singleton set, reflecting the absence of any blue edges. Testing an edge
$e = \{v, w\}$ means finding out whether both endpoints of e belong to the same
set. To this end, we use the operation FIND(v), which returns a name for the
set in which v is contained. If FIND(v) = FIND(w), the edge $\{v, w\}$ is discarded;
otherwise, the two different trees containing v and w are joined with a UNION
operation. The data structure for supporting the UNION and FIND operations
can be implemented efficiently in theory and in practice. The algorithm is short
and has an asymptotic run time of $O(m\alpha(m, n))$ for m FIND operations and at
most $n - 1$ UNION operations ($m \geq n$). Here, $\alpha(m, n)$ is a functional inverse of
Ackermann's function, which means that $\alpha(m, n)$ grows extremely slowly with
m and n [314, 441]. Its value is 3 for all values of m, n that are smaller than the
number of elementary particles in the universe. So, for all practical purposes,
the algorithm runs in linear time, and the constant is very small. If the edges
do not come in sorted order, however, we have to sort them first. This takes
$O(m \log n)$ time, which is worse than the run time of Prim's algorithm.

3.9.4 Faster Minimum-Spanning-Tree Algorithms

There are several minimum-spanning-tree algorithms that are asymptotically faster than Prim's and Kruskal's algorithm. All of these algorithms use data structures that are more complicated than are those of the algorithms that we have discussed. Cheriton and Tarjan [63] present an algorithm that has $O(m \log \log n)$ run time. It is based on a different coloring scheme and uses a more sophisticated implementation of priority queues that, in addition, supports the merging of two priority queues. This algorithm has the promise of being of practical relevance. Galil et al. [125] present the fastest algorithm known so far: Its run time is $O(m \log \beta(m, n))$. Here, $\beta(m, n)$ is defined as follows:

$$\beta(m, n) := \min\{i \mid \log^i(n) \leq \frac{m}{n}\}$$

where $\log^i(n)$ denotes the i-fold application of the base-2 logarithm. $\beta(m, n)$ is a very slowly growing function. Even though it does not grow as slowly as $\alpha(m, n)$ does, its value does not exceed 5 as long as m, n are smaller than the number of elementary particles in the universe. This minimum-spanning-tree algorithm is of only theoretical relevance, however, because it is difficult to implement and runs slowly on small examples.

Whether the MINIMUM SPANNING TREE problem can be solved in linear time in the general case is an open problem. An excellent detailed treatment of minimum-spanning-tree algorithms from which much of the material in this section is taken is given in [441]. However, this reference does not describe more recent developments, such as the minimum-spanning-trees algorithm based on Fibonacci heaps.

3.10 Steiner Trees

The Steiner-tree problem is the first graph problem we consider that is NP-hard. The problem is a direct translation of the routing problem for a single multiterminal net into the context of graphs.

MINIMUM STEINER TREE

Instance: A connected undirected graph $G = (V, E)$ with edge cost function $\lambda : E \rightarrow \mathbf{R}_+$ and a subset $R \subseteq V$ of *required* vertices

Configurations: All edge-weighted trees

Solutions: All *Steiner trees* for R in G; that is, all subtrees of G that connect all vertices in R and all of whose leaves are vertices in R

Minimize: $\lambda(T) = \sum_{e \in E_T} \lambda(e)$

Sometimes, we increase the set of legal configurations in the MINIMUM STEINER TREE problem by dropping the requirement that all leaves of the tree be required vertices, or even that the graph connecting the required vertices be a tree. These modifications do not affect the optimal solutions of the problem, because all edge lengths are nonnegative.

The vertices in a Steiner tree that have degree at least 3 and are not in R are called *Steiner vertices*. The size of R is usually denoted by r. Required vertices will be denoted by s, t, u, whereas other vertices will be denoted by v, w, z.

We consider only real edge labels here. However, all algorithms and theorems directly apply to the case in which the edge labels come from $\{r \in R \mid 0 \preceq r\}$, where $(R, \preceq, \oplus)$ is an ordered set with addition that fulfills the premise of Lemma 3.8.

We have already discussed two special cases of the MINIMUM STEINER TREE problem: $|R| = 2$ and $|R| = n$. The former is the single-pair shortest-path problem (on undirected graphs), and the latter is the MINIMUM SPANNING TREE problem. There are efficient algorithms for both of them. In contrast, the MINIMUM STEINER TREE problem is NP-hard, in general. In fact, the problem remains NP-hard, even if the set of graphs is severely restricted—for instance, to grids, to bipartite graphs, or to graphs with uniform edge costs. On the other hand there are graph families for which the MINIMUM STEINER TREE problem can be solved in polynomial, even linear time. Those are tree-structured graphs, such as series-parallel graphs and others. The solution method is a version of dynamic programming (see Section 4.7.1). For a list of the complexities of special cases of the MINIMUM STEINER TREE problem, see [133, 209, 210].

3.10.1   A Class of Approximation Algorithms

There is a number of approximation algorithms for the MINIMUM STEINER TREE problem that achieve the same error bound $2(1 - 1/\ell)$, where ℓ is the minimum number of leaves in a minimum Steiner tree. All of them aim at finding a good Steiner tree by doing a mixture of minimum-spanning-tree and shortest-paths calculations.

The approximation algorithms we discuss use the following heuristic strategy.

Step1: Compute the complete distance graph G_1 between vertices in R. Each edge $e = \{v, w\}$ in G_1 is weighted with the length $\lambda_1(v, w)$ of the shortest path from v to w in G.

Step 2: Compute a minimum spanning tree G_2 of G_1.

Step 3: Map the graph G_2 back into G by substituting for each edge a corresponding shortest path in G. This yields a G_3.

Step 4: Compute a minimum spanning tree G_4 of G_3.

Step 5: Delete leaves in G_4 that are not vertices in R. Keep doing so until all leaves are vertices in R. This yields graph G_5.

Definition 3.16 (Generalized Minimum Spanning Tree) *A Steiner tree G_5 that can be obtained by the above procedure is called a* **generalized minimum spanning tree.**

We take the generalized minimum spanning tree G_5 as an approximation of the minimum Steiner tree for R.

Theorem 3.2 *Let T_{opt} be a minimum Steiner tree for R, with a minimum number ℓ of leaves. Then*

$$\lambda(G_5) \leq 2(1 - \frac{1}{\ell})\lambda(T_{\text{opt}})$$

Proof Let T_{opt} be a minimum Steiner tree for R, with ℓ leaves. Since T_{opt} has a minimum number of leaves, all leaves of T_{opt} are in R, since all edge costs are positive. Assume that T_{opt} is drawn in the plane. We can traverse T_{opt}—say, in a clockwise manner—going through each edge of T_{opt} once in each direction (see Figure 3.18). This yields a circuit C in T_{opt}. The length of C is twice the cost of T_{opt}. We can partition the circuit C into segments, such that each segment starts and ends at a leaf and does not contain any leaves in the middle. These segments are depicted in Figure 3.18. Now we delete the longest such segment from C. The remainder C' of C still traverses each edge of T_{opt}. The length of C' is at most $2(1 - 1/\ell)\lambda(T_{\text{opt}})$. Now, if a segment in C' runs between the required vertices v and w, then its length is at least $\lambda_1(v, w)$. Thus, if we replace each segment in C' with the edge between its endpoints in G_1, we obtain a graph that spans G_1 and whose cost is at most $2(1 - 1/\ell)\lambda(T_{\text{opt}})$. G_2 is a minimum spanning tree of G_1; thus, the cost of G_2 does not exceed this value. In steps 3 to 5, the cost of the obtained graphs does not increase. $\square$

The bound derived in Theorem 3.2 is tight, since it holds for the case in which we have a complete graph with $V = \{v_0, \ldots, v_{n-1}\}$, $R = \{v_1, \ldots, v_{n-1}\}$, and edge lengths

$$\lambda(v, w) = \begin{cases} 2 & \text{if } v \neq v_0 \neq w \\ 1 & \text{otherwise} \end{cases}$$

Hwang [203] shows that on grid graphs a minimum spanning tree on the complete distance graph has at most one and one-half time the cost of a minimum Steiner tree, thus improving the bound given in Theorem 3.2.

Approximation algorithms that fall into the scheme discussed in this section have been presented in the design-automation literature [9]. The first analysis was by Kou et al. [242]. They applied the traditional algorithms to each step. Step 1 dominates the computation and takes time $O(rm + rn \log n)$ if Fibonacci heaps are used to implement the priority queue in Dijkstra's algorithm.

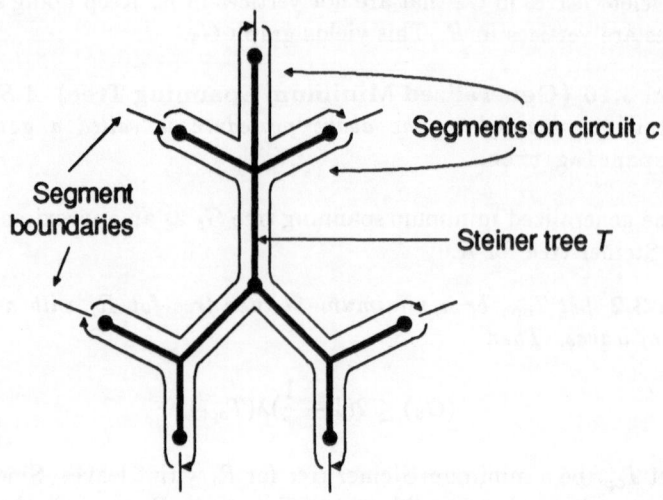

Figure 3.18: *A clockwise circular traversal of a minimum Steiner tree.
Required vertices are indicated by the solid black circles. The minimum Steiner tree T is drawn with thick lines. The circuit C is drawn with normal-width arrows. Segment boundaries are marked on C.*

3.10.2   Using Prim's Algorithm

We can improve the performance of the algorithm detailed in Section 3.10.1 by noting that it is not necessary to compute all edges of G_1. Rather, only $n - 1$ edges of G_1 are eventually used in G_2. There are several versions of the algorithm that are based on this idea.

Several heuristics modify one of the traditional minimum-spanning-tree algorithms. Takahashi and Matsuyama [437] and Akers [9] use Prim's algorithm for this purpose. Their method can be summarized as follows:

Step 1: Choose an arbitrary required vertex. Use Dijkstra's algorithm to propagate a search wave starting at this vertex until it hits another required vertex. Construct the shortest path between the two required vertices.

Step 2: Add all vertices on the path to the set of sources for another unidirectional search. (See Section 3.8.1 for directions on how to do this.) Propagate the search until the next unscanned required vertex is scanned. Construct the respective shortest path.

Step 3: If there is an unscanned terminal, go to step 2. Otherwise, halt.

It is not difficult to prove that the resulting Steiner tree has a cost that does not exceed the cost of a minimum spanning tree of the complete distance graph G_1 defined in Section 3.10.1 (Exercise 3.24). The asymptotic run time of this method is the same as that of the method by Kou et al., but typical run times can be expected to be faster because the searches are terminated when the *first* yet-unscanned required vertex is scanned. Akers suggested the algorithm, and Takahashi and Matsuyama analyzed it. Despite its high complexity when there are many required vertices, the method is very popular in practice; it is a succession of traditional shortest-path computations that are very robust in many applications (see Chapter 8).

3.10.3   Using Kruskal's Algorithm

If we base the Steiner-tree computation on Kruskal's algorithm, we can develop a more efficient method [358, 476]. Here, we construct the Steiner tree G_2 by propagating search wavefronts from each required vertex simultaneously, keeping track of where they intersect, and generating edges of G_2 correspondingly. To find a *minimum* spanning tree, we must not allow the searches to proceed at arbitrary relative speeds; rather, we must ensure that they expand at the same speed (Exercise 3.25). We do so by merging the priority queues of all searches.

We prove that the synchronization of the shortest-path expansions leads to a minimum spanning tree of G_1 by using Kruskal's algorithm. Doing all searches at the same speed ensures that the wavefronts emanating from different required vertices cross in order of increasing distance of the respective vertex pairs. Thus, edges in G_1 are encountered in order of increasing length.

We now describe the algorithm in detail. There are two central data structures. One is a UNION-FIND data structure U whose elements are the required vertices. As in Kruskal's algorithm, this data structure keeps track of the parts of the minimum spanning tree G_2 that have been assembled already. The other data structure is a priority queue D whose entries are triples of the form (s, v, d), where s is a required vertex, v is an arbitrary vertex, and d is the length of a path from s to v. The priority queue D is the result of merging the priority queues of the independent searches from each required vertex. The first entry s in a triple represents the origin of the respective search wavefront. The order of the triples in D is by increasing value of d. This convention synchronizes the different searches. For each vertex v, the vertex $home[v]$ is a closest required vertex to v. The value $length[v]$ denotes the distance between $home[v]$ and v. The algorithm is described in Figure 3.19. (We have to make the technical assumption that each edge (s, t) such that $s, t \in R$ is represented only once in the adjacency structure; that is, one representation is deleted. This assumption is necessary to ensure that, in line 13, no edge is entered into D twice.)

Theorem 3.3 *The multidirectional search algorithm computes a minimum*

(1)   *home* : **array[vertex]** of required vertex;
(2)   *length* : **array[vertex]** of integer;
(3)   *d* : **integer**;
(4)   *s* : **required vertex**;
(5)   *v, w* : **vertex**;

(6)   initialize D; initialize U;
(7)   **for** $s \in R$ **do**
(8)   *home*[s] := s; *length*[s] := 0 **od**;
(9)   **for** $v \in V \setminus R$ **do**
(10)   *home*[s] := **nil**; *length*[s] := ∞ **od**;
(11)   **for** $s \in R$ **do**
(12)   **for** $v \in A[s]$ **do**
(13)   enter $(s, v, \lambda(s, v))$ into D **od od**;

(14)   **while** U has more than one set **do**
(15)   extract minimum element (s, v, d) from D;
(16)   **case** FIND $(home[v])$ **of**
(17)   **nil**:   *home*[v] := s; *length*[v] := d;
(18)   **for** $w \in A[v]$ **and** *home*[w] = **nil do**
(19)   enter $(s, w, d + \lambda(s, w))$ into D **od**;
(20)   FIND (s) :
(21)   **else**:   **if** $v \in R$ **then**
(22)   UNION(FIND(s), FIND(v));
(23)   generate edge $(s, home[v])$ with
                                          length $d + length[v]$ in G_2
(24)   **else**
(25)   insert $(s, home[v], d + length[v])$ into D
                            fi esac od;

Figure 3.19: *The multidirectional search method for computing Steiner trees.*

spanning tree of G_1.

Proof Since the triples (s, v, d) are ordered in D according to increasing distance values, we can establish by induction that d is in fact the length of a path from s to v and *length*[v] is the distance from v to *home*[v].

We still must show that, whenever line 22 is executed, the edge $e = (s, home[v])$ is a shortest edge in G_1 between a pair of required vertices that lie in different sets inside U. Assume indirectly that the shortest edge is $e' = \{t, v\}$, and that it is shorter than e. The edge e' represents a path p of equal length between t and u in G. Since D is ordered according to increasing distance values, lines 23 and 25 should both have been executed with p under consideration. Thus, p should have been found, the corresponding sets should have been merged, and p should not connect required vertices in distinct sets of U. Thus, we have derived a contradiction. □

The complexity of the algorithm is easy to analyze. Lines 1 to 13 take time $O(m \log n)$ if D is implemented using balanced trees. The number of triples generated is at most $2m$, because triples are generated only in lines 19 and 25, and each edge in E can give rise to at most one triple in each of these two statements. Thus, the total run time is $O(m \log n)$.

The algorithm can be enhanced with pointers such that a Steiner tree that is consistent with G_2 can be recovered directly from the data structure (Exercise 3.26).

3.10.4 The Fastest Method

Further improvements on the run time have been obtained by [241, 464]. We will now present the fastest method, which was developed by Mehlhorn [316].

Let the *neighborhood* $N(s)$ of a vertex $s \in R$ be the set of all vertices in G that are closer to v than to any other required vertex. A required vertex s is always put into the neighborhood $N(s)$. If another vertex is equidistant to several required vertices then it is put into a neighborhood of an arbitrary one of these vertices. The required vertex in whose neighborhood v is located is called the *home* of v, and denoted by $h(v)$. The neighborhoods $N(s)$ for $s \in R$ partition the set V. We call two required vertices $s, t \in R$ *adjacent* if there is an edge $\{v, w\} \in E$ such that $v \in N(s)$ and $w \in N(t)$. Let

$$E_1' := \{\{s, t\} \mid s, t \in \mathbf{R} \text{ and } s \text{ and } t \text{ are adjacent}\}$$

We will compute a minimum spanning tree of G_1 that is inside the subgraph $G_1' := (R, E_1')$ of G_1.

To make the computation efficient, we have to introduce one more modification to the algorithm. It will not be possible to compute the shortest distance values $\lambda_1(s, t)$ for all edges in E_1'. Rather, we have to compute an approximation $\lambda_1'(s, t)$ to these distance values, which is defined by

$$\lambda_1'(s, t) := \min\{\lambda_1(s, u) + \lambda(u, v) + \lambda_1(v, t) \mid (u, v) \in E, u \in N(s), v \in N(t)\}$$

As we shall show, we can compute $\lambda_1'(s, t)$ quickly. From now on, we will consider G_1' as being edge weighted with λ_1'. It can happen that $\lambda_1'(s, t) > \lambda_1(s, t)$ (see Figure 3.20). But we can show the following lemma.

Lemma 3.10 (Mehlhorn [316]) *Every minimum spanning tree in G_1' is also a minimum spanning tree in G_1 and has the same cost in both graphs.*

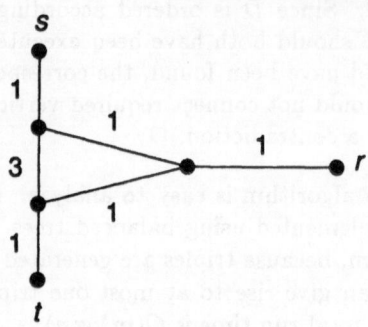

Figure 3.20: *An example in which $\lambda_1'(s,t) > \lambda_1(s,t)$; $R = \{r,s,t\}$.*

Proof First we show the existence of a minimum spanning tree in G_1' that fulfills the lemma. We do this indirectly. Let T_1 be a minimum spanning tree of G_1 such that T_1 has a minimum number δ of edges that are not in E_1'. If there are several such trees, choose one such that $\lambda_1'(T_1)$ is minimum. If there is no spanning tree in G_1 that has the properties stated in the lemma, then one of two cases can occur:

Case 1: $\delta > 0$.

Consider an edge $\{s,t\}$ that is in T_1 but not in E_1'. Let $s = v_0, \ldots, v_k = t$ be a shortest path p from s to t in G. As we go along p, the home of the vertices starts out being s and ends up being t, so it has to change on the way. Let a change happen between vertices v_i and v_{i+1}. By definition of E_1', we have $\{h(v_i), h(v_{i+1})\} \in E_1'$. Furthermore,

$$\lambda_1(h(v_i), h(v_{i+1})) \leq \lambda_1'(h(v_i), h(v_{i+1}))$$

by definition of λ_1

$$\leq \lambda_1(h(v_i), v_i) + \lambda_1(v_i, v_{i+1}) + \lambda_1(v_{i+1}, h(v_{i+1}))$$

by definition of λ_1'

$$\leq \lambda_1(s, v_i) + \lambda_1(v_i, v_{i+1}) + \lambda_1(v_{i+1}, t)$$

since $v_i \in N(h(v_i))$ and $v_{i+1} \in N(h(v_{i+1}))$

$$= \lambda_1(s, t)$$

If the edge $\{s,t\}$ is removed from T_1, the spanning tree is split into two components. Thus, there must be some i such that $h(v_i)$ and $h(v_{i+1})$

are in different components. Let us now add the edge $\{h(v_i), h(v_{i+1})\}$ to T_1. The resulting tree T_1' is again a spanning tree and, by the preceding calculation with respect to λ_1, the length of T_1' does not exceed the length of T_1. Furthermore, T_1' has one more edge in E_1' than does T_1, in contradiction to the definition of T_1.

Case 2: $\lambda_1'(T_1) > \lambda_1(T_1)$.

In this case, the proof is analogous. We consider an edge $\{s, t\}$ in T_1 such that $\lambda_1'(s, t) > \lambda_1(s, t)$. The analysis of case 1 then shows that, if we replace edge $\{s, t\}$ by edge $\{h(v_i), h(v_{i+1})\}$, we end up with a tree that has the same number of edges from E_1' but whose cost with respect to λ_1' is smaller. This result again contradicts the definition of T_1.

Having thus established the existence of at least one tree in G_1' with the properties required by the lemma, we now show that all minimum spanning trees in G_1' have these properties. Let T_1' be an arbitrary minimum spanning tree of G_1', and let T_1 be a minimum spanning tree of G_1 as before. Then,

$$
\begin{aligned}
\lambda_1(T_1') &\le \lambda_1'(T_1') && \text{by definition of } \lambda_1 \\
&\le \lambda_1'(T_1) && \text{since } T_1' \text{ is minimum spanning tree of } G_1' \\
&= \lambda_1(T_1) && \text{by definition of } T_1 \\
&\le \lambda_1(T_1') && \text{since } T_1 \text{ is minimum spanning tree of } G_1
\end{aligned}
$$

Thus, T_1' is a minimum spanning tree of G_1. □

By Lemma 3.10, we can compute a minimum spanning tree of G_1' instead of a minimum spanning tree of G_1. We can do this computation quickly as follows. Adjoin a source vertex s_0 to G, and add edges of length 0 from s_0 to all vertices $s \in R$. Then, solve the single-source shortest-path problem with s_0 as the source, using Dijkstra's algorithm. The resulting distance value of vertex v is $\lambda_1(v, h(v))$. The vertex $h(v)$ is the first vertex on the path from s_0 to v in the resulting shortest-path tree. This computation yields the neighborhoods.

For computing the λ_1' values, we scan all edges $\{u, v\} \in E$ and generate triples $(h(u), h(v), \lambda_1(h(u), u) + \lambda_1(u, v) + \lambda_1(v, h(v)))$. We sort these triples lexicographically with respect to the first two components. Then we scan the sorted list and extract the minimum cost for each edge.

The application of Dijkstra's algorithm takes $O(m + n \log n)$ time; the rest of the computation takes linear time, if we use bucket sort [5]. Thus, the whole algorithm takes $O(m + n \log n)$ time.

3.10.5 Improving Steiner Trees

The Steiner tree obtained by any of the preceding methods can be improved using the following method by suggested by Akers [9]. Delete a long segment between two required vertices from the Steiner tree, splitting the tree into two components. Perform a shortest-path computation whose sources are the vertices of one component, propagating the search until a vertex in the second component is hit. Construct the resulting shortest path between the components. This modification is certain not to increase the cost of the Steiner tree and it may decrease it considerably.

Widmayer [464] has performed a detailed comparison of the results obtained on various graphs using different heuristics yielding the error bound $2(1 - 1/\ell)$. (He did not study the method by Mehlhorn or the iterative improvement technique by Akers.) The conclusion is that no heuristic clearly outperforms any other. Computational experience indicates that, on randomly chosen graphs, all heuristics perform much better than the worst-case error [375]. The problem with all these approximation algorithms is that the worst-case error bound is quite weak. We can improve the error bound if we are willing to spend more time. Plesnik [358] and independently Sullivan [431] have shown that, if in the preceding heuristic G_1 is a complete distance network that contains p vertices that are not required in addition to all required vertices, then there is a choice of the p additional vertices such that reembedding a minimum spanning tree of G_1 into G leads to a Steiner tree with an error of at most $2 - p/(\ell - 2)$. Of course, the number of different G_1 to consider rises exponentially with p, such that the complexity of the algorithms increases quickly.

In Chapter 4, we will report on applications of optimization methods for integer programs to the MINIMUM STEINER TREE problem.

3.11   Network Flow

Network flow is a topic in algorithmic graph theory that is of importance to layout because of its close relationship to circuit partitioning (Chapter 6), to assignment (Section 7.3), and to several detailed-routing problems (Chapter 9).

3.11.1   Maximum Network Flow

Definition 3.17 (Network Flow) *Let $G = (V, E)$ be a directed graph, with an edge-labeling function $c : E \to \mathbf{R}_+$, called the* **capacity** *function. Let $s, t \in V$ be two distinguished vertices in G (the* **source** *and the* **target** *or* **sink***). The tuple (G, c, s, t) is called a* **flow** *graph.*

For a vertex $v \in V$, let $E_{\to v}$ ($E_{v \to}$) be the set of edges entering (leaving) v. A function $f : E \to \mathbf{R}_+$ is called a **flow** *from s to t iff f satisfies the following constraints:*

Capacity Constraints: $0 \leq f(e) \leq c(e)$ *for all $e \in E$.*

Flow-Conservation Constraints: *For $v \in V$, let the* **net flow** *at v be de-*

fined as

$$f(v) := \sum_{e \in E_{\to v}} f(e) - \sum_{e \in E_{v \to}} f(e)$$

Then, for all $v \in V \setminus \{s, t\}$, we require that $f(v) = 0$.

The **value** *of the flow f is defined as $|f| := f(s) = -f(t)$.*

The maximum-network-flow problem is defined as follows.

MAXFLOW

Instance: A flow graph (G, c, s, t)

Configurations: All edge labelings $f : E \to \mathbf{R}_+$

Solutions: All flows f from s to t

Maximize: $|f|$

There is a fundamental relationship between flows and a certain kind of cut in a flow graph.

Definition 3.18 ((s,t)-Cut) *Let (G, c, s, t) be a flow graph. An (s,t)-cut (S, T) is a partition of the vertices of G such that $s \in S$ and $t \in T$. The* **capacity** *of (S, T) is*

$$c(S, T) := \sum_{\substack{e = (v, w) \in E \\ v \in S \\ w \in T}} c(e)$$

The capacity of an (s,t)-cut is a directed version of the cutsize as defined in Definition 3.2. Here the underlying graph is directed, and only the edges leading from S toward T contribute.

Figure 3.21 shows a example of a flow graph. The relationship between flows and (s,t)-cuts is given in the following important theorem.

Theorem 3.4 (Maxflow–Mincut Theorem) *Each flow graph $(G, c, s, t,)$ has a maximum flow f, and the value of f is equal to the minimum capacity of a cut.*

One-half of this theorem is easy to prove.

Lemma 3.11 *If f is a flow and (S, T) is an (s,t)-cut, then $|f| \le c(S, T)$.*

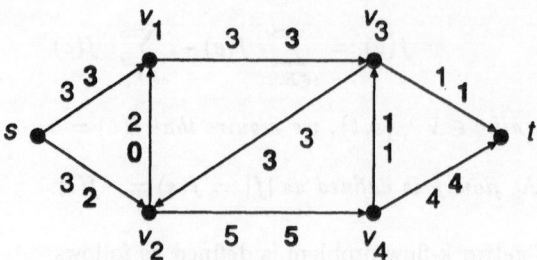

Figure 3.21: *A flow graph. Edges are labeled with their capacities and flow values. Flow values are displayed in boldface type. A minimum cut is $(V \setminus \{t\}, \{t\})$.*

Proof Summing the flow-conservation constraints over all vertices in S yields

$$|f| + \sum_{\substack{e = (w, v) \in E \\ v \in S \\ w \in T}} f(e) = \sum_{\substack{e = (v, w) \in E \\ v \in S \\ w \in T}} f(e)$$

Intuitively, all flow that enters S (through s or from T) has to leave S (toward T). Observing $0 \le f(e) \le c(e)$, we decrease the left-hand side by substituting all flow values with 0, and we increase the right-hand side by substituting all flow values with $c(e)$. This yields

$$|f| \le \sum_{\substack{e = (v, w) \in E \\ v \in S \\ w \in T}} c(e)$$

which proves the lemma. □

To prove the other direction of the maxflow–mincut theorem, we exhibit a relationship between maximum flows and so-called augmenting paths.

Definition 3.19 (Augmenting Path) *Let p be an undirected path from s to t in G. A directed edge on p that points toward t (or s) is called a* forward *(or* backward*) edge. The path p is called* augmenting *if $f(e) > 0$ for every backward edge and $f(e) < c(e)$ for every forward edge on p.*

The significance of an augmenting path p is that we can push flow along p so as to increase the value of f. An amount d of flow pushed along p decreases (or increases) the flow value of the backward (or forward) edges on p by d. Modifying the flow values on the edges of p accordingly yields a new flow whose value is d greater than that of the old one. The maximum amount of flow that can be pushed along p is

$$d_{\max} = \min \left(\min_{e \text{ forward edge}} c(e) - f(e), \min_{e \text{ backward edge}} f(e) \right)$$

After we have pushed a maximum amount of flow through p, the path p is no longer augmenting. It can become augmenting again, however, if pushing flow along other augmenting paths changes the capacities of edges on p.

Figure 3.22 illustrates how we build up a maximum flow of the flow graph in Figure 3.21 by augmenting it along augmenting paths. There is a general characterization of maximum flows based on augmenting paths.

Lemma 3.12 *A flow f is maximum exactly if it does not have an augmenting path. In this case, there is an (s, t)-cut (S, T) such that $|f| = c(S, T)$.*

Proof Clearly, if the flow f has an augmenting path, then it can be increased. Now assume that f does not have an augmenting path. Define S to be the set of all vertices $v \in V$ such that there is an augmenting path from s to v, and include s in S. Let $T := V \setminus S$, and consider the (s, t)-cut (S, T). Then, for all edges $e = (v, w) \in E$ with $v \in S$ and $w \in T$, we have $f(e) = c(e)$; otherwise, w would be in S, as well. Similarly, for all edges $e = (w, v) \in E$ with $v \in S$ and $w \in T$, we have $f(e) = 0$. We can conclude from the flow-conservation constraints that $|f| = c(S, T)$. Thus, it follows from Lemma 3.11 that f is a maximum flow. $\square$

Lemmas 3.11 and 3.12 complete the proof of Theorem 3.4 if the existence of a maximum flow can be ensured. This assertion is proved in the following lemma.

Theorem 3.5 *Each flow graph has a maximum flow.*

Proof The truth of this theorem is easy to see if all capacities are integer. In this case, each augmentation increases the flow by at least 1. However, the flow cannot increase beyond the sum of the capacities of all edges leaving s. Thus, after finitely many augmentations, we find a maximum flow. This procedure also shows that, if all edge capacities are integer, so is the maximum flow value.

This proof generalizes to the case that the capacities are rational, because we can multiply all capacities by the least common multiple M of their denominators to make them integer. Maximum flows stays maximum after this operation. Flow values are multiplied by M.

If irrational edge capacities are involved, the existence of a maximum flow results from Fact 3.2 and from the results presented in Section 4.5. $\square$

Fact 3.2 *The* MAXFLOW *problem is a special case of the linear-programming problem.*

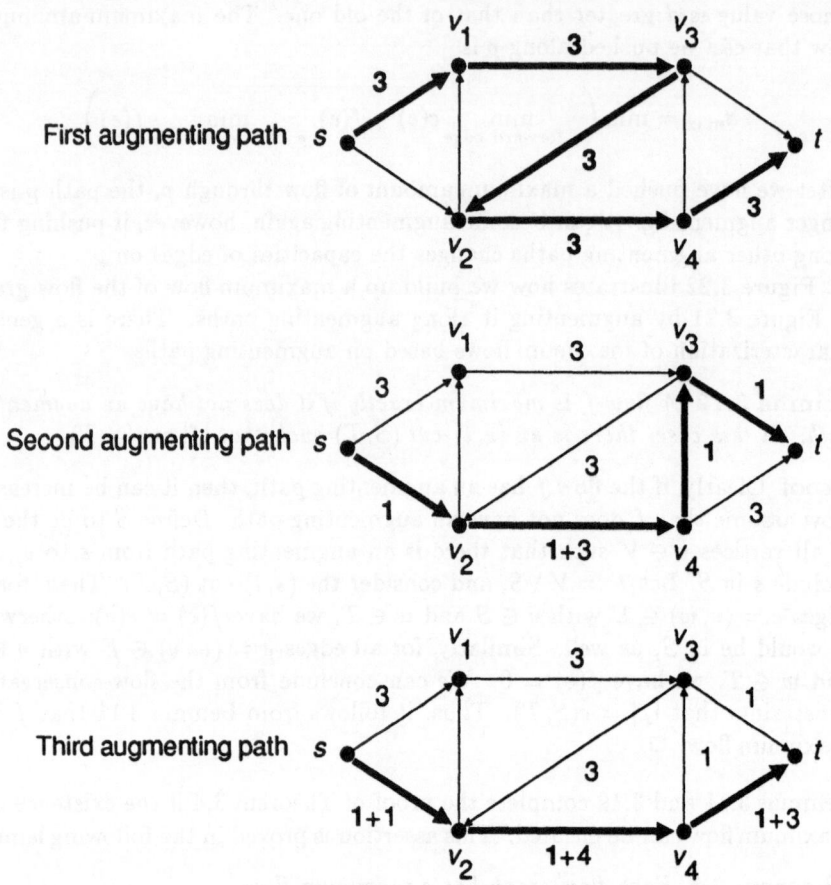

Figure 3.22: *Finding a maximum flow in the flow network depicted in Figure 3.21 by augmentations along paths.*

Proof Exercise 3.28. □

By Fact 3.2, the MAXFLOW problem is in P (see Section 4.5). As in the case of shortest paths, there are much more efficient algorithms for the MAXFLOW problem than there are for general linear programming. A class of efficient methods for computing flows is based on the augmenting-path method. A suitable selection of augmenting paths in addition to the use of appropriate data structures yields algorithms that are efficient in theory and practice. To this class belongs Dinic's algorithm, which works well on small and medium-sized examples and has a run time of $O(n^2m)$ [91], as well as asymptotically faster algorithms. The algorithm by Sleator and Tarjan runs in time $O(mn \log n)$ but is based on a complicated dynamic data structure [424, 425, 441].

A second, more recent, class of efficient methods for computing maximum flows is based in Karzanov's preflow algorithm [217], which runs in time $O(n^3)$. These algorithms do not construct augmenting paths. The most efficient versions are presented in [7, 141, 344]. The algorithm in [141] runs in time $O(mn \log(n^2/m))$.

On special graph classes, more efficient network-flow algorithms exist. Miller and Naor [325] present an algorithm that computes the maximum flow in graphs that become planar if the source and sink, as well as their incident edges, are deleted from the graph. The algorithm runs in time $O(n^{1.5})$.

Tarjan [441] gives an excellent introduction to network-flow methods.

There is an undirected variant of the maximum-flow problem. Here, the flow graph is an undirected graph. Flows are still directed. The edge-capacity constraints now require the flow across an edge to be at most the edge capacity, regardless of the direction in which the flow traverses the edge. This problem is called UNDIRECTED MAXFLOW. We can transform an instance of UNDIRECTED MAXFLOW into an instance of MAXFLOW by replacing each undirected edge $e = \{v, w\}$ in the flow graph with two antiparallel directed edges between v and w, both having the same capacity as e.

3.11.2 Mincost Flow

The MINCOST FLOW problem is of interest to us because it is the linear-programming dual of problems arising in placement and compaction (Chapter 7 and 10). In the MINCOST FLOW problem, we have a specific flow value z that we want to achieve. In addition, each edge e has a cost value $\ell(e) \in \mathbf{R}_+$ associated with it. Pushing d units of flow through e costs $d \cdot \ell(e)$ units. We want to find the flow with flow value z whose cost is minimum.

MINCOST FLOW

> *Instance:* A flow graph (G, c, s, t), a flow value $z \in \mathbf{R}_+$, an edge cost function $\ell :\to \mathbf{R}_+$
>
> *Configurations:* All mappings $f : E \to \mathbf{R}_+$
>
> *Solutions:* All flows F in (G, c, s, t) with $|f| = z$
>
> *Minimize:* $\ell(f) := \sum_{e \in E} f(e) \cdot \ell(e)$

Let an *augmenting cycle* be a closed augmenting path (not necessarily containing s or t). The *cost* of the cycle is based on the edge costs $\ell(e)$ for forward edges e—that is, edges pointing in the direction in which the flow is pushed on the cycle—and on $-\ell(e)$ for backward edges on the cycle. The following theorem characterizes mincost flows.

Theorem 3.6 *A flow f with flow value |f| = z is a mincost flow if and only if f contains no augmenting cycle with negative cost.*

For example, if each edge in the flow network depicted in Figure 3.21 has cost 1, then the flow depicted in Figure 3.21 is not a mincost flow, because the cycle (v_2, v_3, v_4, v_2) is an augmenting cycle along which we can push flow 1 with cost -3. If we push this flow along the cycle, then we obtain a mincost flow with cost 19. We can formulate the MINCOST FLOW problem as a linear program. This program has the form

$$Minimize: \quad c \cdot x$$

$$Subject\ to: \quad A \cdot x = b \tag{3.5}$$

$$0 \le x \le U$$

There is a variable x_e for each edge e. The matrix A is the node–edge incidence matrix $A = ((a_{ve}))$, where $a_{ve} = 1$ if e enters v, $a_{ve} = -1$ if e leaves v, and $a_{ve} = 0$ otherwise. Note that A is totally unimodular (Lemma 4.7). The vector $b = ((b_v))_{v \in V}$ is zero everywhere except at the source s and the sink t. There, we have $b_s = z$ and $b_t = -z$. The constraint system $A \cdot x = b$ ensures the flow conservation. In addition, there are constraints $0 \le x_e \le c(e)$ that maintain the capacity constraints. The objective function that is minimized is defined by the vector $c = (\ell(e_1), \ldots, \ell(e_m))$.

The MINCOST FLOW problem appears to be quite a bit harder than the MAXFLOW problem. It is in P by virtue of being a linear program. There are special versions of the simplex algorithm that are tailored to solving network linear programs efficiently [350]. However, strongly polynomial algorithms— that is, polynomial-time algorithms whose complexity depends only on the size of the flow graph, and not on the size of the numbers labeling the edges—have been found only recently [142, 143, 343, 439]. The complexity of the algorithms is higher than that of algorithms for the MAXFLOW problem. For instance, the complexity of the algorithm presented in [143] is $O(nm^2(\log n)^2)$.

The relationship between the MINCOST FLOW problem and linear programs works the other way, as well. Consider a linear program of the form (3.5), where b can be an arbitrary vector and A is required to be a node–arc incidence matrix (that is, a matrix such that each column has exactly one $+1$ and one -1), and all other entries are 0. The only thing that distinguishes this linear program from the linear-programming formulation of the MINCOST FLOW problem is that the vector b can be arbitrary. In the language of network flow, this generalization amounts to allowing prespecified flows into or out of all vertices. But this more general version of the mincost-flow problem can be reduced to the version defined previously. We just add a new source and a new sink. The source is connected to all vertices with incoming flow. The sink is connected to all vertices with outgoing flow. The capacities of the new edges are the prespecified flow values. The costs of the new edges are zero. Thus, any linear program of the form (3.5) translates into a MINCOST FLOW problem and can be solved with the respective efficient methods. We call a linear program of the form (3.5) with A being a node–arc incidence matrix a *network linear program*.

We can define an UNDIRECTED MINCOST FLOW problem, and can reduce it to the MINCOST FLOW problem, in a manner analogous to that used for the MAXFLOW problem. We replace an undirected edge e with capacity $c(e)$ and cost $\ell(e)$ by a pair of antiparallel directed edges, both with capacity $c(e)$ and length $\ell(e)$.

3.11.3   Multicommodity Flow

In the multicommodity-flow problem, we are pushing through the network not one kind of flow, but rather several kinds. We think of each flow as transporting some *commodity*. Each commodity is supplied at one or more vertices and is demanded at other vertices in the network. The edges have to be able to accommodate all flows simultaneously.

MULTICOMMODITY FLOW

> *Instance:* A directed graph $G = (V, E)$, edge capacities $c(e) \in \mathbf{R}_+$ and edge costs $\ell(e) \in \mathbf{R}_+$ for $e \in E$, demands $d_i(v) \in \mathbf{R}$ for all vertices $v \in V$ and for k commodities $i = 1, \ldots, k$
>
> *Configurations:* All sequences of edge labelings $f_i : E \to \mathbf{R}_+$, $i = 1, \ldots, k$
>
> *Solutions:* All sequences of edge labelings that satisfy the following constraints:
>
> **Capacity Constraints:** For all $e \in E$,
>
> $$\sum_{i=1}^{k} f_i(e) \le c(e)$$
>
> **Flow-Conservation Constraints:** Define the *net flow* of commodity i into vertex v to be
>
> $$f_i(v) := \sum_{e \in E_{\to v}} f_i(e) - \sum_{e \in E_{v \to}} f_i(e)$$
>
> Then, for all $i = 1, \ldots, k$ and $v \in V$, we have $f_i(v) = d_i(v)$.
>
> *Minimize:* $\ell(f) = \sum_{e \in E} \left(\ell(e) \cdot \sum_{i=1}^{k} f_i(e) \right)$

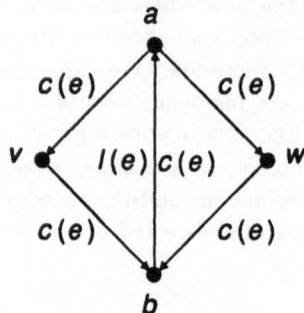

Figure 3.23: *The edge construct replacing an undirected edge e =
{v, w} in the transformation from* UNDIRECTED MULTICOMMODITY
FLOW *to* MULTICOMMODITY FLOW. *The capacities of all edges in the
construct are $c(e)$. The cost of the middle edge is $\ell(e)$. All other edge
costs are 0. The demands on all commodities at the vertices a, b are
0, also.*

The legal configurations are called *multicommodity flows*. The capacity constraints state that no edge can take commodities beyond its capacity. The
demand $d_i(v)$ indicates how much of commodity i is required to flow into vertex v. If $d_i(v) < 0$, then $-d_i(v)$ units of commodity i are supplied at vertex v.
The flow-conservation constraints say that the net flow of a commodity out of
a vertex has to be equal to the amount of the commodity at this vertex. The
function that is minimized formulates a flow cost, as in the MINCOST FLOW
problem. Note that, for a multicommodity flow to exist, $\sum_{v \in V} d_i(v) = 0$ has
to hold for each $i = 1, \ldots, k$.

An undirected variant of the multicommodity-flow problem, the UNDI-
RECTED MULTICOMMODITY FLOW, is of greater importance in layout than is
the directed version. An instance of the UNDIRECTED MULTICOMMODITY FLOW
problem contains an *undirected* graph G. For each edge e, the edge capacity
must not be exceeded by the sum of the flow values for all commodities that go
across the edge, regardless of their direction. We can turn an instance of the
UNDIRECTED MULTICOMMODITY FLOW problem into an instance of the MULTI-
COMMODITY FLOW problem by transforming each undirected edge $e = \{v, w\}$
with capacity $c(e)$ and cost $\ell(e)$ into the construct depicted in Figure 3.23.

It is not surprising that the MULTICOMMODITY FLOW problem can again
be formulated as a linear program (Exercise 3.30). Thus, the problem can
be solved in polynomial time, and algorithms exist that are fast in practice
(see Section 4.5). However, the constraint matrix is not totally unimodular, in
general.

The maxflow–mincut theorem suggests a relationship between multicommodity flows and cuts in G. Let (V', V'') be a cut. Let $c(V', V'')$ denote the capacity of the cut (see Definition 3.18). Define the *density* of the cut as

$$d(V', V'') := \sum_{i=1}^{k} \max\{d_i(V''), 0\}$$

and the *free capacity* of a cut as

$$fc(V', V'') := c(V', V'') - d(V', V'')$$

Lemma 3.13 *For a multicommodity flow to exist, $fc(V', V'') \geq 0$ must hold for all cuts.*

Proof The proof follows directly from the flow-conservation constraints, as in the easy direction of the proof of the maxflow–mincut theorem. $\square$

A cut condition is also a prerequisite for undirected multicommodity flows to exist.

Lemma 3.14 *Lemma 3.13 extends to the* UNDIRECTED MULTICOMMODITY FLOW *problem, if we define $c(V', V'')$ to be the cutsize as defined in Definition 3.2, and $d(V', V'') = \sum_{i=1}^{k} |d_i(V'')|$.*

Proof Exercise 3.29. $\square$

The reverse of Lemmas 3.13 and 3.14, however, need not be true. Figure 3.24 shows an instance of the UNDIRECTED MULTICOMMODITY FLOW problem for which the reverse does not hold.

A few special cases are known in which the nonnegativity of the free capacities of all cuts ensures the existence of a multicommodity flow. They all concern instances of the UNDIRECTED MULTICOMMODITY FLOW problem in which the graph G is planar. Seymour [414] shows this for instances of the UNDIRECTED MULTICOMMODITY FLOW problem in which each commodity has only one vertex with a negative demand (source) and one vertex with a positive demand (sink), and if the source and sink of each commodity are connected with an edge, the resulting graph is planar. We call such instance of the UNDIRECTED MULTICOMMODITY FLOW problem *planar*. Matsumoto et al. [311] use this characterization to provide an efficient $O(n^{5/2} \log n)$ time algorithm for constructing multicommodity flows in this case. The UNDIRECTED MULTICOMMODITY FLOW problem has a close relationship with routing problems involving two-terminal nets. Indeed, we can think of each net as a commodity. One of its endpoints demands one unit of flow; the other one supplies it. The total number of wires running across an edge in the routing graph should not exceed the edge capacity. The cost of routing a unit of flow (wire) across an

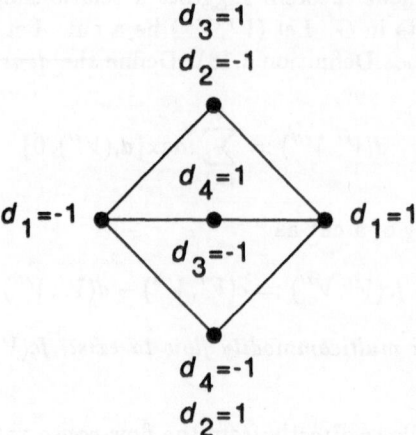

Figure 3.24: *An instance of the* UNDIRECTED MULTICOMMODITY
FLOW *problem for which no flow exists, but for which each cut has
nonnegative free capacity. Nonzero demands label the corresponding
vertices. All edges have capacity 1.*

edge can encode such elements as edge length, resistance of the wiring layer,
and so on.

A difference between wire routing and multicommodity flow is that, in multicommodity flow, the flow values may be fractional, whereas wire routes correspond to integer flow values. Indeed, it does not make sense to speak about
routing a wire 30 percent along one route and 70 percent along another route.
There are instances of the UNDIRECTED MULTICOMMODITY FLOW problem that
have fractional but no integer flows; Figure 3.25 gives an example. In general,
finding integer multicommodity flows is an NP-hard problem, even if only two
commodities are involved or if all edge capacities are 1 [107]. A sufficient
condition for guaranteeing the existence of integer flows in the UNDIRECTED
MULTICOMMODITY FLOW problem is that the problem instance be planar and
$fc(V', V'')$ be even for each cut (V', V'') [342]. Hassin [177] shows that this
result cannot be extended to either nonplanar instances of the UNDIRECTED
MULTICOMMODITY FLOW problem or (directed) planar instances of the MULTI-
COMMODITY FLOW problem.

There is a different and somewhat nonclassical variant of the MULTICOM-
MODITY FLOW problem that plays a role in graph partitioning. We refer the
reader to Section 6.5 for more details.

Figure 3.25: *An instance of the* UNDIRECTED MULTICOMMODITY FLOW *problem that has a rational but no integer solution (notation is the same as in Figure 3.24).*

3.12 Matching

In this section, we describe methods of efficiently pairing up vertices in graphs.

Definition 3.20 (Matching) *Let G be an undirected graph. A **matching** in G is a subset $M \subseteq E$ of edges such that no vertex is incident to more than one edge. The matching is called **perfect** if each vertex is incident to exactly one edge.*

There are two kinds of matching problems. The first kind asks for a matching with a maximum number of edges.

MAXIMUM MATCHING

  Instance: A graph G

  Configurations: All matchings M of G

  Solutions: All configurations

  Maximize: $|M|$

If G is bipartite, the MAXIMUM MATCHING problem reduces to a special case of the MAXFLOW problem. The corresponding instance is constructed as follows. All edges in G are directed from the left to the right side of G. A source s and a target t are added to G. Edges are run from s to all vertices on the left side of G and from all vertices on the right side of G to t. The capacities of all edges are 1. Figure 3.26 illustrates this transformation. It is not difficult to show that each maximum integer flow in the resulting graph corresponds to a maximum matching in G, and vice versa (Exercise 3.31). Thus, network-flow algorithms that compute integer maximum flows, if they latter exist, apply to maximum matching. Since all edge capacities in the instance of the derived MAXFLOW problem are 1, the algorithms can be sped up to run in time $O(m\sqrt{n})$ [441].

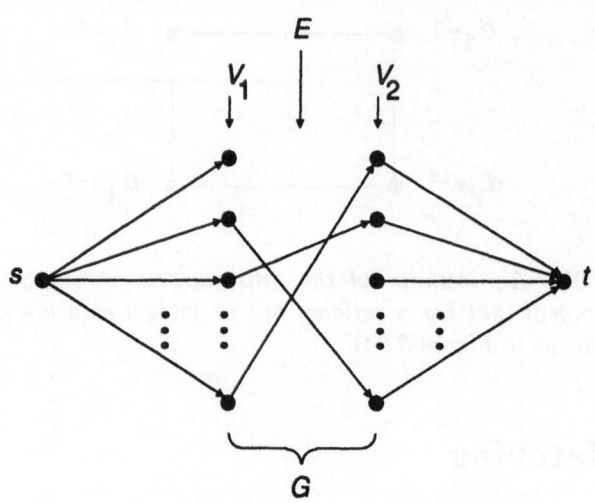

Figure 3.26: *Reducing* MAXIMUM MATCHING *in bipartite graphs to* MAXFLOW. *All edge capacities are 1.*

If G is not bipartite, MAXIMUM MATCHING is not a special case of MAXFLOW, but it is still a linear program. Specialized solution methods take time $O(mn)$. More advanced algorithms get down to $O(m\sqrt{n})$ in this case as well. Tarjan [441, Chapter 9] gives an excellent account of matching algorithms.

The second kind of matching problems assumes the graph to be complete with edge weights and asks for a perfect matching of minimum weight. There are two versions of such problems. One version is based on complete bipartite graphs and is called the LINEAR ASSIGNMENT problem.

LINEAR ASSIGNMENT

Instance: An $n \times n$ matrix $D = ((d_{ij}))$ with value from $\mathbf{R}_+$

Configurations: All permutations $\pi : \{1 \ldots, n\} \to: \{1 \ldots, n\}$

Solutions: All configurations

Minimize: $D(\pi) := \sum_{i=1}^{n} d_{i,\pi(i)}$

D is interpreted as an edge-weighted complete bipartite graph G with n vertices on each side. The weight of the edge between vertex i on the left and vertex j on the right side is given by d_{ij}. The classical solution of the LINEAR ASSIGNMENT problem runs in time $O(n^3)$ [350, Chapter 11]. The fastest algorithm known to date takes time $O(n^2 \log n + mn)$ [122].

In the other version, the base graph is the complete undirected graph.

MINIMUM WEIGHT MATCHING

Instance: A symmetric $n \times n$ matrix $C = ((c_{ij}))$ with values from $\mathbf{R}_+$ indicating the edge weights in K_n

Configurations: All perfect matchings M

Solutions: All configurations

Minimize: $c(M) := \sum_{e \in M} c_e$

It has been known for a long time that this problem also can be solved in time $O(n^3)$ [350, Chapter 11]. Recently, faster algorithms have been developed [123, 124, 127]. The asymptotically fastest algorithm known to date runs in time $O(n^2 \log n + mn)$ [123].

Galil [126] gives an excellent recent survey of matching algorithms. Avis [15] surveys very fast algorithms that find good but not necessarily optimal matchings.

The restriction to nonnegative edge costs in LINEAR ASSIGNMENT and MINIMUM WEIGHT MATCHING is not essential. Also, the corresponding maximization problems are equivalent to the minimization versions (Exercise 3.32). We will use both the maximization and minimization versions of these problems (see Chapters 6 and 7).

3.13 Hierarchical Graph Processing

In Section 1.2.2.1, we discussed the advantages of hierarchical design. Hierarchical design strategies provide expressive power for describing large circuits succinctly. In this capacity, they are the basis for matching today's complex fabrication technologies with suitable design technologies. In addition, hierarchical circuit descriptions should provide the basis for efficient design validation and optimization. This means that the processing of the design should be performed as much as possible on succinct data. Data expansion should be carried out only when absolutely necessary. This section describes how this kind of hierarchical processing can be done.

3.13.1 Hierarchical Graph Definitions

Definition 3.21 (Hierarchical Graph) *A hierarchical (undirected) graph* $\Gamma = (G_1, \ldots, G_k)$ *is a finite sequence of undirected simple graphs* G_i, *called* cells. *For* $1 \leq i \leq k$, *the graph* G_i *has* n_i *vertices and* m_i *edges.* p_i *of the vertices are distinguished and are called* pins; *the other* $n_i - p_i$ *vertices are called* inner *vertices.* r_i *of the inner vertices are distinguished and* *are called* **nonterminals**; *the other* $n_i - r_i$ *vertices are called* **terminals** *or* **proper vertices**. *All neighbors of a nonterminal are terminals.*

Each pin has a unique label, its **name**. *We assume the names to be the numbers between 1 and* p_i. *Each nonterminal has two labels, a* **name** *and a* **type**. *The type of a nonterminal in cell* G_i *is an element of the set* $\{G_1, \ldots, G_{i-1}\}$. *If a nonterminal* v *has type* G_j, *then* v *has degree* p_j *and the neighbors of* v *are bijectively associated with the pins of* G_j *via labels* (v, ℓ) *with* $1 \le \ell \le p_j$.

We assume that Γ *is irredundant, in the sense that each* G_i *appears as the type of some nonterminal. The* **size** *of* Γ *is*

$$n := \sum_{i=1}^{k} n_i$$

The **edge number** *of* Γ *is*

$$m := \sum_{i=1}^{k} m_i$$

Note that, if $\Gamma = (G_1, \ldots, G_k)$ is a hierarchical graph, so is any prefix $\Gamma_i := (G_1, \ldots, G_i)$ of Γ. An example of a hierarchical graph is shown in Figure 3.27.

Definition 3.22 (Expansion) *The* **expansion** $E(\Gamma)$ *of the hierarchical graph* Γ *is obtained as follows:*

$k = 1$: $E(\Gamma) := G_1$. *The pins of* $E(\Gamma)$ *are the pins of* Γ.

$k > 1$: *Repeat the following step for each nonterminal* v *of* G_k —*say, of type* G_j. *Delete* v *and its incident edges. Insert a copy of* $E(\Gamma_j)$ *by identifying pin* ℓ *of* $E(\Gamma_j)$ *with the vertex in* G_k *that is labeled* (v, ℓ). *The size of* $E(\Gamma)$ *is denoted by* N. *The number of edges in* $E(\Gamma)$ *is denoted by* M.

Definition 3.21 defines a certain type of restricted context-free graph grammar. The substitution mechanism glues pins of cells to neighbors of nonterminals representing these cells, as described in Definition 3.22. This type of graph grammar is known as *hyperedge replacement systems* [159] or *cellular graph grammars* [285]. One additional restriction we impose on such a grammar is that, for each nonterminal, there is only one cell that can be substituted. Thus, there are no *alternatives* for substitution. Furthermore, the index of the substituted cell has to be smaller than the index of the cell in which the nonterminal occurs. This *acyclicity* condition implies that the hierarchical definition Γ describes a unique graph—the expansion of Γ. Figure 3.28 gives an example. The parse tree of $E(\Gamma)$ is called the *hierarchy tree*.

Definition 3.23 (Hierarchy Tree) *The* **hierarchy tree** *of* Γ *is a rooted tree* $T(\Gamma)$ *whose vertices are called* **nodes**. *Each node* x *in* $T(\Gamma)$ *represents an instance of a cell* G_i *of* Γ *in* $E(\Gamma)$. G_i *is called the* **type** *of* x. *The root of*

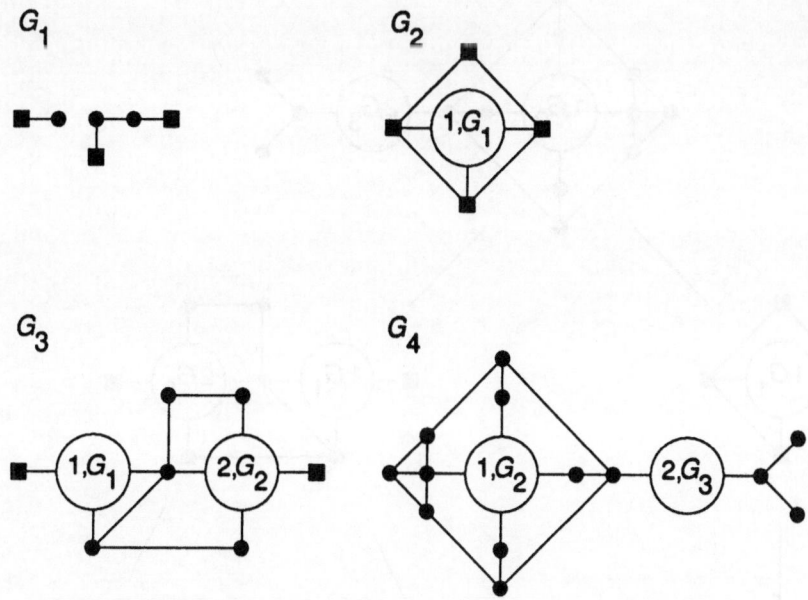

Figure 3.27: A hierarchical graph. Pins are denoted by squares; non-terminals are denoted by large circles, inscribed with their name (a number) and type (a cell). The bijective correspondence between the neighbors of a nonterminal and the pins of the cell that is its type is given via the position of the vertices in the figure.

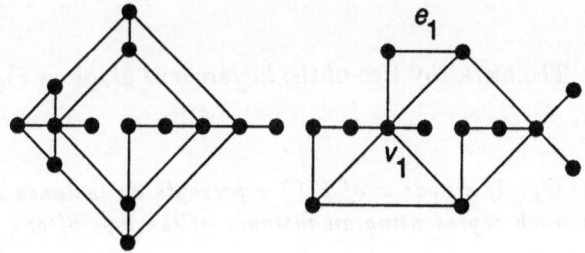

Figure 3.28: The expansion of the hierarchical graph in Figure 3.27.

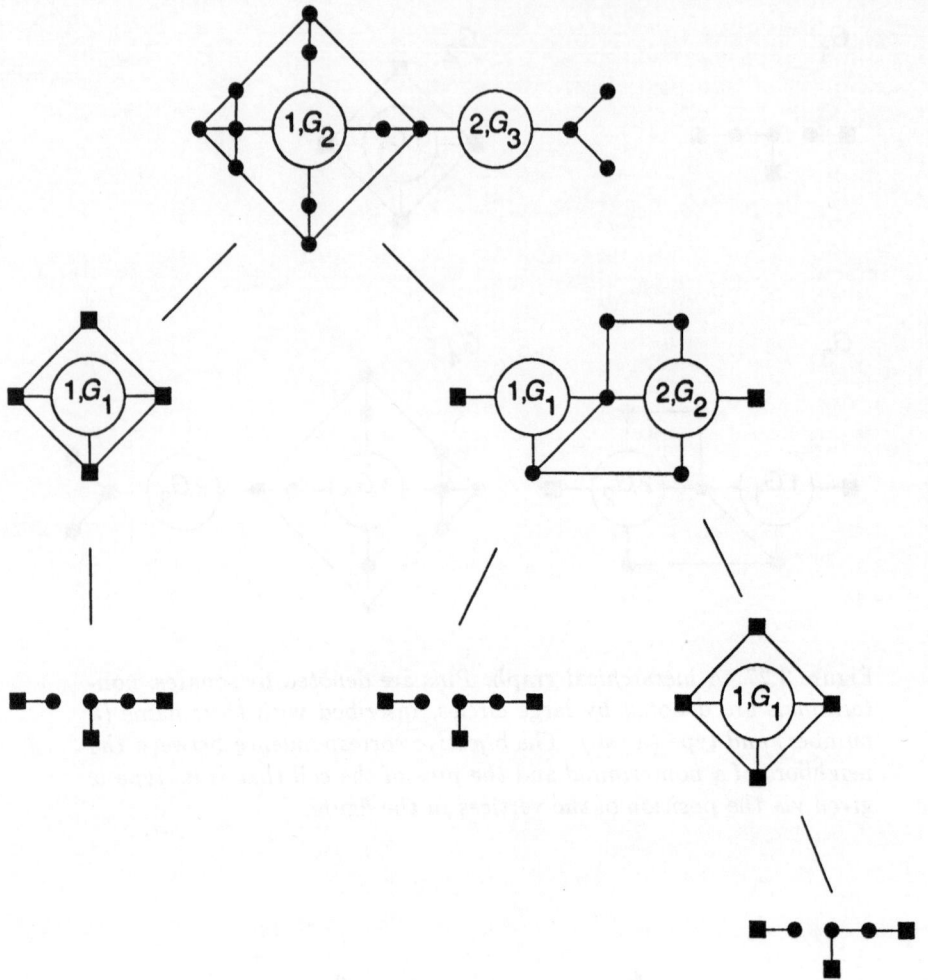

Figure 3.29: *The hierarchy tree of the hierarchical graph in Figure 3.27.*

$T(\Gamma)$ *represents* G_k. *If a node* x *of* $T(\Gamma)$ *represents an instance of* G_i *then* x *has* r_i *children, each representing an instance of the type of one nonterminal of* G_i.

The height of $T(\Gamma)$ is at most k. Figure 3.29 depicts the hierarchy tree of the hierarchical graph shown in Figure 3.27.

Each edge in $E(\Gamma)$ *belongs* to a unique node x in $T(\Gamma)$; namely, to the node that represents the cell instance contributing the edge. Similarly, each vertex in $E(\Gamma)$ belongs to a unique node x in $T(\Gamma)$; namely, to the node representing the cell instance contributing the vertex as an *inner vertex*. For instance, in Figure 3.27, the vertex v_1 belongs to the same node as does the edge e_1, even though v_1 is also contributed to $E(\Gamma)$ by three other cell instances as a pin.

In this way, edges and vertices of $E(\Gamma)$ can be identified by path names without the need of generating $E(\Gamma)$. A path name starts with a sequence of nonterminals denoting a path in $T(\Gamma)$ starting at the root, and ends with an inner vertex or an edge in the last cell of the sequence (or in G_k if the sequence is empty). The nonterminal sequence for both v_1 and e_1 in Figure 3.29 consists of the single item $(2, G_3)$.

3.13.2 The Bottom-Up Method

In general, the size of $E(\Gamma)$, measured in terms of N and M, can be exponential in the size of Γ, measured in terms of n and m (see Exercise 3.33). Thus, the idea is to solve graph problems pertaining to $E(\Gamma)$, but to use only the much smaller number of data in Γ during the solution. If the graph problem is such that the output is not very large, we can, at least in theory, hope for substantial improvement in performance. We will call a graph problem that has a small output, such as a bit or a number, a *decision problem*. An example of a decision problem is the problem: Find out whether $E(\Gamma)$ is connected. There is a linear-time algorithm for deciding whether a graph is connected (see Exercise 3.3). However, this algorithm is linear *in N and M*—which, as we have seen, can be exponential in the input size m, n. If we do not expand the graph, but rather work on the hierarchical definition directly, we can reduce the run time to $O(m + n)$.

The basis of this reduction lies in the concept of replaceability.

Definition 3.24 (Replaceability) *Let G, G' be two cells with p pins each and no nonterminals. (We call such graphs p-graphs.) Let H be a cell with no pins and a single nonterminal of degree p. Let $G : H$ denote the graph resulting from substituting the nonterminal in H with G. G and G' are* **replaceable** *with respect to graph connectivity if, for all H, $G : H$ is connected exactly if $G' : H$ is connected.*

Fact 3.3 *The replaceability relation is an equivalence relation on p-graphs for each p.*

Obviously, there is nothing special about formulating Definition 3.24 in terms of graph connectivity. We could as well have chosen graph problems, such as biconnectivity, strong connectivity (if we extend Definition 3.21 in a natural way to directed graphs), minimum spanning tree (here we require the cost of a minimum spanning tree to be invariant under replacement of cell G by cell G', or vice versa), or other decision problems.

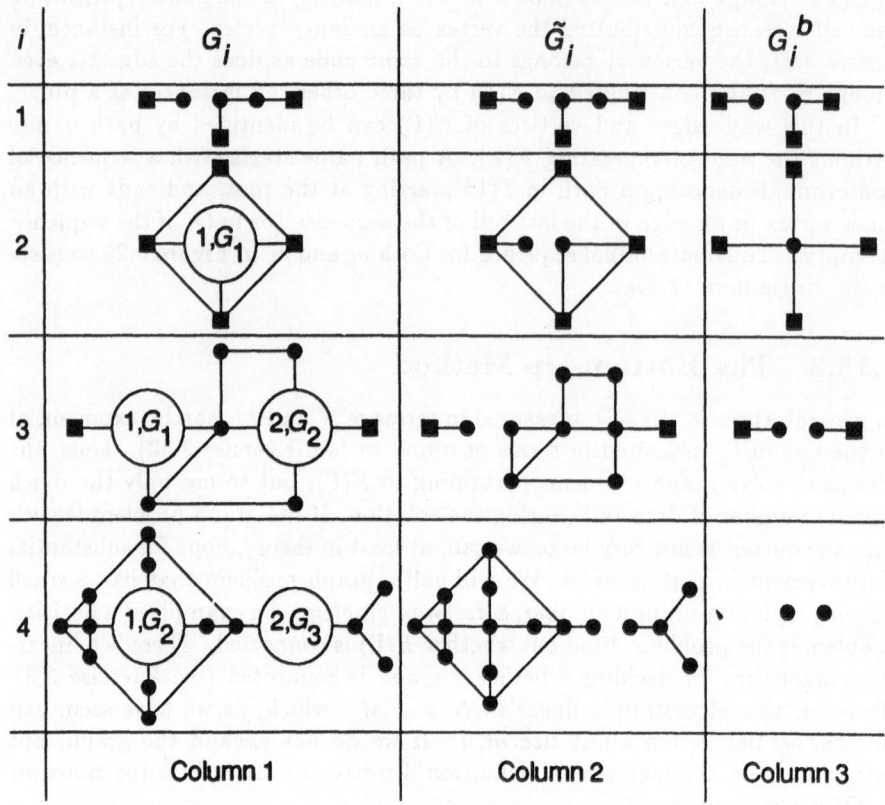

Figure 3.30: BU-*table for analyzing connectivity of the graph in Figure 3.27.*

The aim of the bottom-up method is to find quickly small graphs G_i^b that are replaceable with $E(\Gamma_i)$. Then, we can use G_i^b instead of $E(\Gamma_i)$ in the succeeding analysis. This substitution reduces the number of data on which we must work. The procedure by which we reduce $E(\Gamma_i)$ to the replaceable graph G_i^b is called *burning*; its result G_i^b is called the *burnt graph*.

Let us assume that we have an efficient burner, given a p-graph, that generates small burnt graphs for a certain graph problem. We can use this burner as follows to solve the graph problem on hierarchical graphs. We build a table of graphs with three columns and k rows—the so-called BU-*table* (for bottom-up, see Figure 3.30). Each row is associated with a cell of the hierarchical graph. In particular, row i is associated with cell G_i. Initially, the entry in the first column and in the ith row of the table contains the cell G_i. We now proceed to fill in the table in the same fashion as we write on a page. Let us consider filling row i. At this time, rows 1 to $i - 1$ are filled already. We generate the entry in the second column of row i by substituting for each nonterminal in G_i that has type G_j the burnt graph G_j^b. This graph is found in column 3 of row j. The result of these substitutions is a p_i-graph, which is called the *fooled graph* (because is mimics $E(\Gamma_i)$) and is denoted by $\widetilde{G_i}$. The third column of row i is filled with the burnt graph G_i^b of $\widetilde{G_i}$ (and of $E(\Gamma_i)$). After filling the BU-table, we solve the graph problem on the graph in the last column and row. The result is identical to the solution of the problem on $E(\Gamma)$.

Note that filling the second column of the BU-table is independent of the particular graph problem, whereas the third column is filled using the burning procedure, which depends on the graph problem. Efficient burners lead to efficient hierarchical graph solutions.

Lemma 3.15 *If there is a burner that takes time $O(m + n)$ on a p-graph with n vertices and m edges, and that produces a graph with $O(p)$ vertices and edges, then the graph problem can be solved hierarchically in linear time.*

We call a burner such as that described in Lemma 3.15 *optimal*. Graph connectivity has a very simple optimal burner. To burn a p-graph G, we find its connected components. Then we construct the following graph $G^b = (V^b, E^b)$:

$$V^b := \{v_1, \dots, v_p\} \quad \text{representing the pins of } G$$
$$\cup \{c_1, \dots, c_r\} \quad \text{representing the connected components of G}$$
$$\text{that contain pins}$$
$$\cup \{c_0\} \quad \text{representing the connected components of } G$$
$$\text{that contain no pins}$$

$$E^b := \{\{v_i, c_j\} \mid v_i \text{ is in the component represented by } c_j\}$$

The vertex c_0 is included only if there are connected components in G that do not contain any pins. G^b can certainly be computed in linear time and has $O(p)$ vertices and edges. It is also straightforward to see that G^b and G are replaceable with respect to graph connectivity.

3.13.3 Modifying the Input Hierarchy

The connectivity burner can be used for much more than just deciding graph connectivity on hierarchical graphs. In fact, there is quite a variety of graph problems pertaining to graph connectivity that we may want to solve:

- Construct all connected components.

- Given a vertex v in G, construct the connected component of G containing v.

- Given two vertices v, w in G, decide whether v and w are connected in G. Solve this problem on a sequence of vertex pairs that is not predetermined.

The first two problems are *construction problems*. They require the construction of a potentially large data structure. It may be possible, however, to represent this data structure hierarchically. The third problem is a *query problem*.

The hierarchical definition provided in the input may not be appropriate to solve these problems. An example is given in Figure 3.31. There, the connected components are spread out over the hierarchy tree in quite an unstructured fashion. However, guided by the connectivity burner, the input hierarchy Γ can be modified to a more desirable hierarchy Γ'. To find Γ', we scan top down through the second column of the BU-table. When processing row i, we make a new *component cell* C_j out of each part of $\widetilde{G}_i$ represented by a vertex $c_j \in V_i^b$. The vertex c_j is replaced with a nonterminal v_j of type C_j. The newly created cells C_j for $c_j \in V^b$ are entered into a fourth column attached to the BU-table (see Figure 3.31). The modified hierarchy induces a different hierarchy tree for the same hierarchical graph (see Figure 3.32). The resulting hierarchy Γ' has the following desirable properties. Let us call a component cell *disconnected* if it consists of several isolated nonterminals of degree zero; otherwise, call the cell *connected*.

Lemma 3.16

1. $E(\Gamma') = E(\Gamma)$.

2. Let m', n' be the number of edges and vertices, respectively, in Γ'. Then, $m', n' = O(m + n)$. The depth of $T(\Gamma')$ is at most $k + 1$.

3. A cell in Γ' is connected exactly if its expansion is a connected graph.

4. The hierarchy tree $T(\Gamma')$ has a top part consisting exclusively of nodes whose type is a disconnected cell. The bottom part of $T(\Gamma')$ consists exclusively of nodes whose type is a connected cell. The connected components of $E(\Gamma)$ are the expansions of the topmost nodes in $T(\Gamma')$ whose type is a connected cell.

Proof

1. The proof is obvious.

2. $m', n' = O(m + n)$, since $\widetilde{G}_i$ has size $O(m_i + n_i)$. To see that $T(\Gamma')$ has height $k + 1$, we note that, as we scan a path from a node x in $T(\Gamma')$ to the root of $T(\Gamma')$, we move downward in column 4 of the BU-table, except when we encounter the first disconnected cell. At this time, we scan two cells out of the same entry in column 4. (In Figure 3.31, this entry is in row 4.)

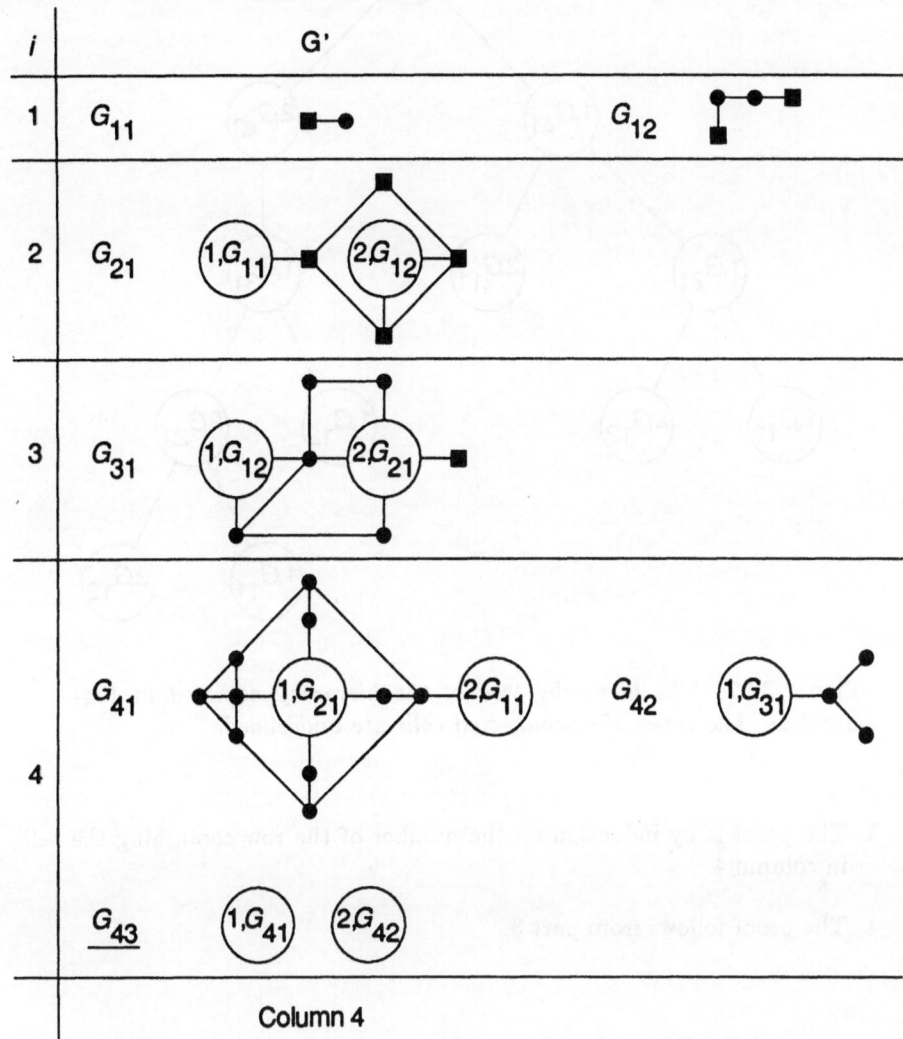

Column 4

Figure 3.31: *The modified hierarchy for the graph in Figure 3.27. The types of disconnected cells are underlined.*

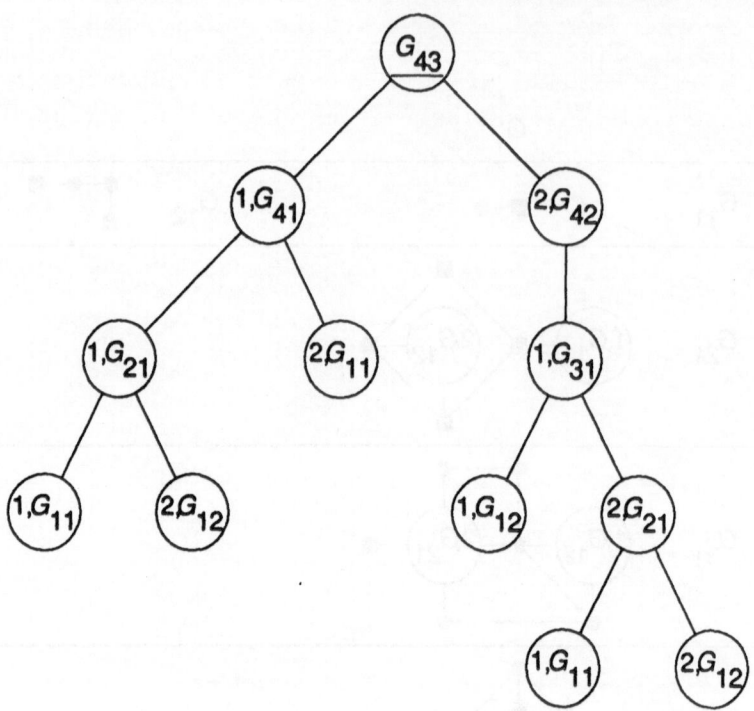

Figure 3.32: *The hierarchy tree of the hierarchy depicted in Figure 3.31. The types of disconnected cells are underlined.*

3. The proof is by induction on the number of the row containing the cell in column 4.

4. The proof follows from part 3.

□

Because of Lemma 3.16, part 4, the path names in $T(\Gamma')$ start with a segment of nonterminals whose type is disconnected, then continue with a segment of nonterminals whose type is connected, and then end with an inner vertex or an edge. We can translate path names in Γ and Γ' into each other by scanning the path name backward and substituting between columns 1 and 4, respectively, of the BU-table.

Using the modified hierarchy we can solve all of the problems pertaining to connectivity that we mentioned previously. Given a vertex v, we can produce a hierarchical description of its connected component by translating the path name of v into the modified hierarchy, finding the leftmost connected cell G'_j in the modified path name, and selecting all component cells that occur in the hierarchy tree rooted at G'_j. We can find these cells by scanning column 4 of the BU-table bottom up, starting at cell G'_j. The whole process takes $O(m+n)$ time. To decide whether two vertices v, w are connected in $E(\Gamma)$, we translate their path names into Γ' and then test whether both modified path names have the same prefix up to the first nonterminal whose type is a connected cell. This operation takes time $O(k)$ per query.

Graph connectivity is one of the simplest problems that we can handle hierarchically. The biconnectivity problem also has an optimal burner and allows for modifying the hierarchy suitably. The strong-connectivity problem on hierarchical directed graphs has an $O(n^3)$ burner that produces graphs with p vertices and $O(p^2)$ edges. The complexity of the hierarchical strong connectivity test based on this burner is $O(\sum_{i=1}^{k} n_i^3)$. Unfortunately, there does not seem to be a good way of modifying hierarchies such that the strong components form subhierarchies. Nevertheless, we can use the hierarchy to gain an improvement over nonhierarchical algorithms for many problems pertaining to strong connectivity. These results on strong connectivity extend to shortest-path problems. This fact is exploited in hierarchical algorithms for layout compaction (see Section 10.4). Details on the hierarchical solution of biconnectivity and strong-connectivity problems can be found in [286]. Efficient burners are discussed for the MINIMUM SPANNING TREE problem in [276], for planarity in [277], for shortest paths in [275], and for the solution of linear systems in [287].

Unfortunately, not all graph problems have efficient burners. Examples of graph problems that become NP-hard when considered on hierarchical graphs can be found in [284].

3.13.4 The Hierarchical Analysis of Graph Families

The theory of hierarchical graph analysis presented in this section has another generalization that is quite important for circuit design. The hierarchical design process allows not only for the succinct description of a single circuit, but also for the unified specification of whole families of related circuits. A circuit family of this sort encompasses all circuits that obey the same law of construction, including circuits of different sizes and circuit variants to provide different options. Such circuit families are described by parameterized cell generators (see Section 2.2).

Figure 1.3 presents a schematic example of a cell generator based on hierarchical circuit design. Since this generator produces hierarchical circuits, we can analyze the output of the generator by submitting it to the methods described in Section 3.13.2 and 3.13.3. However, what we want is to analyze the generator itself, to ensure the correctness and suitability of all the circuits it can generate.

In terms of graphs, a straightforward extension of the definition of a hierarchical graph (Definition 3.21) provides a first step toward describing cell generators. We have only to allow alternatives for types of nonterminals (implementing variants) and to eliminate the requirement that $j < i$ if G_j is the type of a nonterminal in G_i (allowing for cycles in the recursion). When it is generalized in this way, we call the hierarchical definition Γ a *context-free cellular graph grammar*, and the graph family generated by it its *language* $L(\Gamma)$. Figure 3.33 gives an example of a context-free cellular graph grammar that generates an infinite graph language. In the context of design applications, the interesting type of question pertaining to such graph families is whether there is a graph in the family that satisfies a graph property that formalizes some design defect. This is the type of question that has to be answered to ensure correctness (such as absence of short circuits) or performance (such as bounds on delay or area) of all circuits that can be generated. For certain classes of graph properties, the bottom-up method can be extended to answer this question.

Consider again the connectivity property. This graph property is finite in the following sense.

Definition 3.25 (Finite Graph Property) *A decidable graph property* Π *is* p-**finite** *if the replaceability relation with respect to* Π *induces only a finite number of equivalence classes of p-graphs. The classes are called* glue types. *The glue type that contains* G *is denoted by* $[G]_\Pi$, *or by* $[G]$ *for short. The number of glue types for p-graphs is called the* p-**size** *of* Π.

A decidable graph property is called **finite** *if it is p-finite for all p.*

All graph properties that have a burner that produces a graph whose size depends on only p are finite. We call such a burner *bounded size*.

First, we will explain how to test for replaceability of two graphs G and G'. If the burner is well behaved, we may be able just to burn both cells and to test for isomorphy of the result. For the connectivity burner we presented, this method works. If the burner is not well behaved in this sense, we provide ourselves with small representatives of all glue types. One after another, we attach each such representative to G and G', test the graph property on the two resulting graphs, and compare the outcomes. G and G' are replaceable exactly if all comparisons yield identity.

We will now discuss how to extend the bottom-up method to finding out whether there is a graph in $L(\Gamma)$ that satisfies the finite graph property Π for which we have a bounded-size burner. We will again use connectivity as our example property.

Instead of one hierarchy tree, we now have to deal with many trees, one for each graph in $L(\Gamma)$. We will analyze hierarchy trees according to increasing height, starting with trees of height 0. Again, we fill a table of graphs (see Figures 3.34 and 3.35). This *extended* BU-*table* has one row per cell and a number of columns that we will leave unspecified for now. The entry in row i

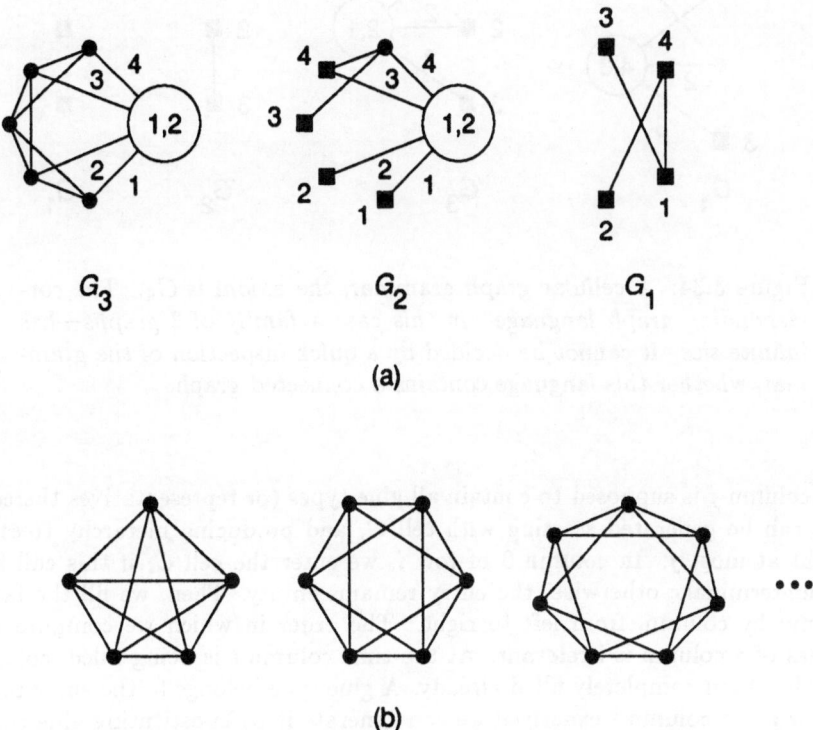

(a)

(b)

Figure 3.33: *A hierarchically generated infinite graph family. (a) The cellular graph grammar; the numbers inside the nonterminals denote the types of the cells which can replace the nonterminal. Numbers labeling edges incident to nonterminals indicate the correspondence between the neighbors of the nonterminal and the pins of a cell replacing this nonterminal. All derivations of graphs start with the cell G_3, the so-called axiom. (b) Graphs in the generated graph family; the family contains all graphs with at least five vertices that form simple cycles augmented with shortcuts across pairs of adjacent edges in the cycle.*

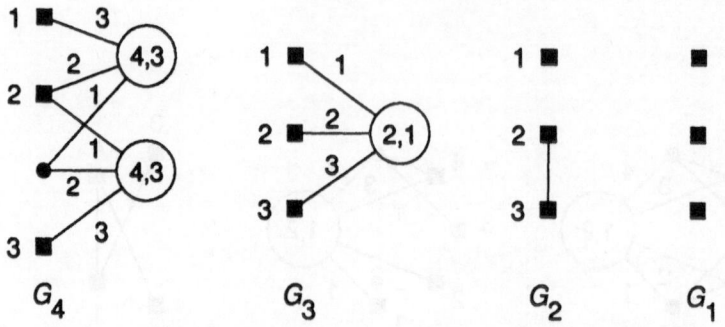

Figure 3.34: *A cellular graph grammar; the axiom is G_4. The corresponding graph language—in this case a family of 3-graphs—has infinite size. It cannot be decided by a quick inspection of the grammar, whether this language contains a connected graph.*

and column j is supposed to contain all glue types (or representatives thereof) that can be generated starting with cell G_i and producing hierarchy trees of height at most j. In column 0 of row i, we enter the cell C_i if this call has no nonterminals; otherwise, the entry remains empty. Then, we fill the table column by column, from left to right. The order in which we compute the entries of a column is irrelevant. At the time column t is being filled, column $t-1$ has been completely filled already. A glue type belongs to the entry table in row i and column t exactly if we can generate it by substituting glue types from the appropriate entries in column $t-1$ for the nonterminals of G_i. Here, the entry in row j and column $t-1$ is appropriate for a nonterminal v of G_i if G_j is in the type set of v. By burning, we can create small representatives of glue types. By testing for replaceability, we can limit ourselves to one representative per glue type.

Does the process of filling the table stop? To answer this question, we note that, as we scan through any row of the table from left to right, the set of glue types in the entries grows. More precisely, the set of glue types in entry (i, j) is a subset of the set of glue types in entry $(i, j + 1)$. Furthermore, as soon as two columns t and $t + 1$ of the table are identical, all columns to the right remain the same. Since the number of glue types is finite for each p, this is sure to happen. As soon as it does happen, we stop the process. We test the graph property on all glue types in the column that we generated last. There is a connected graph in the graph family exactly if one of the tests comes out positive.

As presented so far, this algorithm is very inefficient. The inefficiency has two causes. First, each entry in the extended BU-table can have a large number

j	0	1	2	3	4
$A^j(1)$	1 ■ 2 ■ 3 ■	■ ■ ■	■ ■ ■	■ ■ ■	■ ■ ■
$A^j(2)$	■ ┇ ■	■ ┇ ■	■ ┇ ■	■ ┇ ■	■ ┇ ■
$A^j(3)$		■ ■ ■ ┇ ■ , ■	■ ■ ■ ┇ ■ , ■	■ ■ ■ ┇ ■ , ■	■ ■ ■ ┇ ■ , ■
$A^j(4)$			■●■ ■ , ■ , ■ ■ ■	■●■ ■ , ■ , ■ ■ ■ ,	■●■ ■ , ■ , ■ ■ ■ ,

Figure 3.35: *The extended BU-table for the analysis of the graph property "graph connectivity" on the graph grammar depicted in Figure 3.34. The pins are always arranged such that pin numbers increase from top to bottom, as indicated in the top left table entry. Columns 3 and 4 of the table are identical. Since these columns do not contain connected graphs the grammar generates only disconnected graphs.*

of glue types. Second, the number of columns of the extended BU-table that must be generated may be large. However, there are ways to bound both of these quantities.

Definition 3.26 (Domination Graph) *Let $[G], [G']$ be two glue types of p-graphs. $[G]$ **dominates** $[G']$ with respect to Π if G fulfills Π whenever G' does.*

*The p-**domination graph** for Π is the dag $D_{p,\Pi}$, or D_p for short, whose vertices are the glue types of p-graphs with respect to Π, and such that there is an edge between $[G]$ and $[G']$ exactly if $[G]$ dominates $[G']$.*

Figure 3.36 shows the 3-domination graph for connectivity. It turns out that the domination graph contains substantial information about the performance

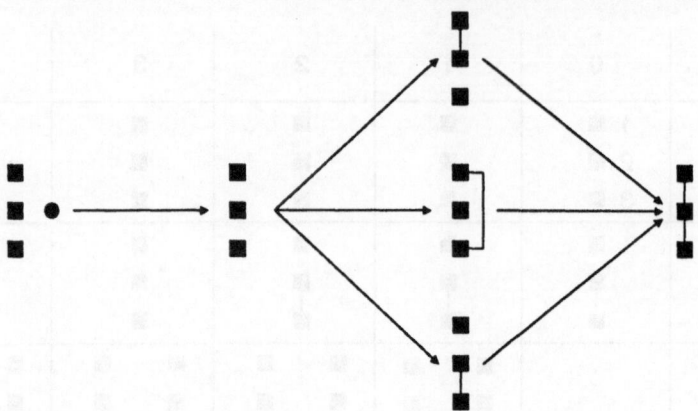

Figure 3.36: *The transitive reduction of the 3-domination graph for connectivity. Each node is represented by a 3-graph that is a representative of the corresponding glue type. The width and depth of the domination graph are both 3.*

parameters for the extended bottom-up method. Let the *width* of D_p be the size of the largest cut through D_p, and let the *depth* of D_p be the length of the longest directed path in D_p.

Lemma 3.17

1. *The number of glue types in each entry in row i is bounded from above by the width of D_{p_i}.*

2. *Let c be an upper bound on the number of glue types in each entry of the extended BU-table. Then the number of columns that have to be generated is bounded from above by $1 + c \cdot \sum_{i=1}^{k} \mathrm{depth}(D_{p_i})$.*

Proof

1. Observe that, in the extended bottom-up method, a glue type $[G]$ can be discarded in an entry in row i of the extended BU-table if there is another glue type $[G']$ in the same entry that dominates $[G]$. The glue types that remain in the entry after we have thus discarded as many glue types as possible form a cut of D_{p_i}.

2. By step 1, we need only to accumulate glue types that are not dominated by other glue types in the same entry. Thus, every time the set of glue types in an entry of row i changes, either a glue type is added or a glue type $[G]$ is exchanged with a glue type that dominates $[G]$. The first case

can happen only c times; the second can happen only $c \cdot \sum_{i=1}^{k} \text{depth}(D_{p_i})$ times.

□

Unfortunately, D_p is usually large. For connectivity, the size of D_p is exponential in p. Thus, either the width of D_p or the depth of D_p is large. For connectivity, the depth of D_p is p, and the width is exponential in p. Practical experience shows, however, that the number of glue types per entry can be contained. For graph connectivity, most examples generate only three or four glue types per entry. Since the depth of D_p is small in this case, the extended bottom-up method is efficient in practice.

An additional method of speeding up the algorithms is based on the realization that, in graphs in $L(\Gamma)$, cells are attached not to arbitrary graphs, but rather only to such graphs that can be generated by the grammar. Thus, the notion of replaceability (Definition 3.24) has to take only such graphs H into account. This optimization decreases the number of possible glue types in theory and practice.

The research on the efficient analysis of graph families is only beginning. Not many graph properties have been investigated as to their suitability for testing on graph families. Furthermore, the graph grammars discussed here are quite restricted. First, they are context-free, which eliminates the possibility of creating balanced hierarchies, such as those depicted in Figure 1.3. To generate the balanced hierarchy of, say, the four-bit adder in Figure 1.4(b), we have to substitute two nonterminals (each representing a two-bit adder) by the same graph. Fortunately, the methods described here can be extended to handle graph grammars that are not context-free [285].

The type of graph substitution that we considered is also quite restrictive. It allows for the description of only *treelike* graphs. Extensions of the methods described here to more powerful graph grammars are discussed in [455].

The hierarchical analysis of graphs and graph families promises to provide important insights into, and practical algorithms for, analyzing and optimizing parameterized circuit families. As of today, this design subtask is an open problem that is considered to be crucial to the effectiveness of the whole design process.

3.14 Planar Graphs

A graph is *planar* if it can be drawn in the plane without edges crossing. Planar graphs have a particular significance in algorithmic graph theory, in general, and in circuit design, in particular. Their general importance stems from the fact that several graph problems that are hard, in general, are easy if restricted to planar graphs. The importance to circuit design is based on two phenomena. First, the quasiplanar characteristic of the fabrication technologies makes planar graphs especially suitable for representing layouts. Second, there is a duality theory of planar graphs that has applications in circuit design. In this chapter, we survey the required results on planar graphs, without going into their proofs in detail.

Definition 3.27 (Crossing Number) *The* **crossing number** *$c(G)$ of an undirected graph $G = (V, E)$ is the minimum number of edge crossings that is required to draw G in the plane.*

Definition 3.28 (Planar Graph) *If the crossing number of G is zero, then G is called* **planar**.

A **drawing** *D of G is a mapping of G into the plane. A vertex v is mapped onto a point $D(v) = (x(v), y(v))$ in the plane and an edge $e = \{v, w\}$ is mapped onto a path $D(e)$ in the plane connecting the points $D(v)$ and $D(w)$. D is* **planar** *if $v \neq w$ implies $D(v) \neq D(w)$ and $e_1 \neq e_2$ implies that $D(e_1)$ and $D(e_2)$ have no common points, except, perhaps, their endpoints. A planar drawing D divides the plane into a finite number of connected regions, called* **facets**. *One of the regions is unbounded and is called the* **outside facet**. *A cycle of edges that bounds a facet is called a* **face**. *The cycle of edges that bounds the outside facet is called the* **outside face**.

The topological properties of a planar drawing are captured with the notion of a planar embedding. We specify a **planar** **embedding** *P of G by giving, for each vertex $v \in V$, the (clockwise) cyclic order of the edges incident to v. A planar drawing D is* **compatible** *with a planar embedding P if D realizes the cyclic orders of edges around vertices that is prescribed by P. Note that, if F is a face in one drawing compatible with P, then it is a face in all drawings compatible with P. F is also called a* **face** *of P.*

If G is a graph and P is a planar embedding of G, then (G, P) is called a **plane graph**.

Figure 3.37 illustrates Definition 3.28. Note that, in a planar embedding, there is no distinguished outside face. For instance, the two planar drawings depicted in Figure 3.37 have different outside faces.

3.14.1 Structural Properties

Kuratowski [251] gives a characterization of planar graphs in terms of forbidden subgraphs.

Definition 3.29 (Homeomorphic Graphs) *Let G and H be graphs. G is called* **homeomorphic** *to H if we can transform G into H by shrinking paths to single edges.*

G is **subgraph-homeomorphic** *to H if there is a subgraph of G that is homeomorphic to H.*

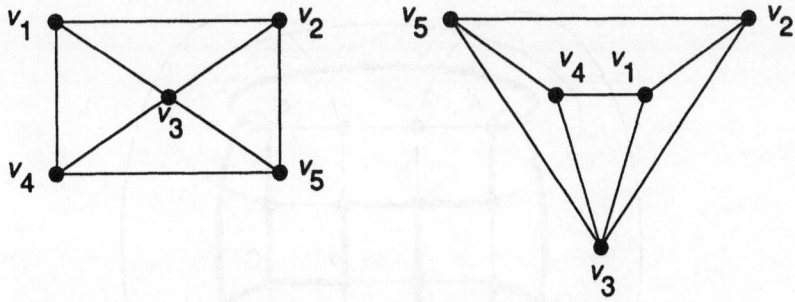

Figure 3.37: *Two planar drawings realizing the same planar embedding but having different outside faces.*

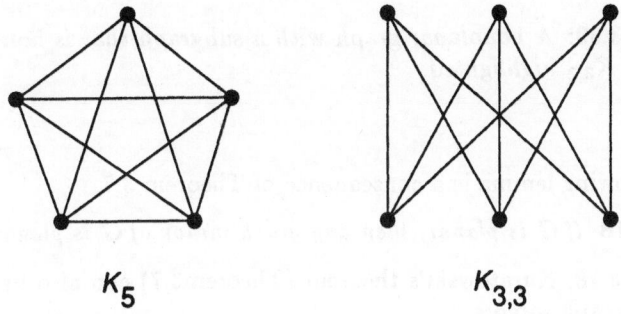

Figure 3.38: *The two smallest nonplanar graphs.*

Theorem 3.7 (Kuratowski [251]) *G is planar if and only if G is not subgraph-homeomorphic to K_5 or $K_{3,3}$.*

Here, K_5 is the complete graph with five vertices, and $K_{3,3}$ is the complete bipartite graph with three vertices on either side (see Figure 3.38). Figure 3.39 shows a nonplanar graph and a subgraph homeomorphic to $K_{3,3}$. The characterization given in Theorem 3.7 is a *forbidden-subgraph* characterization. Similar characterizations hold for other interesting graph classes.

Definition 3.30 (Graph Minor) *A graph H is a* **graph minor** *of a graph G if H can be obtained from G by a sequence of steps of taking a subgraph or shrinking edges to vertices.*

Figure 3.40 shows an example of a sequence of steps yielding a graph minor.

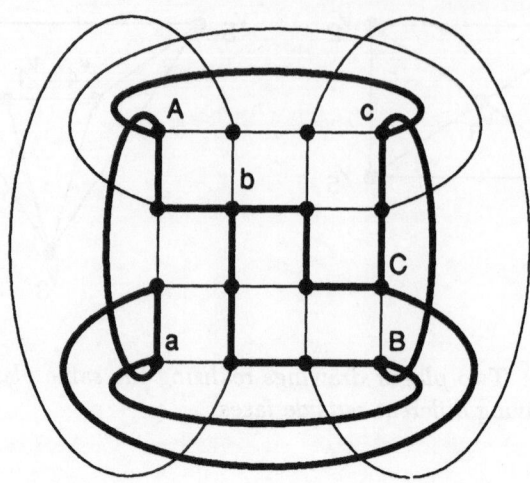

Figure 3.39: *A nonplanar graph with a subgraph that is homeomorphic to $K_{3,3}$ highlighted.*

The following lemma is a consequence of Theorem 3.7.

Lemma 3.18 *If G is planar, then any graph minor of G is planar.*

By Lemma 3.18, Kuratowski's theorem (Theorem 3.7) can also be formulated in terms of graph minors.

Theorem 3.8 *G is planar if and only if G does not contain K_5 or $K_{3,3}$ as a minor.*

An important equation for planar graphs is *Euler's identity.*

Theorem 3.9 (Euler) *If P is a planar embedding of a connected planar graph G with n vertices and m edges, and if P has f faces, then*

$$n - m + f = 2$$

The following lemma is a consequence of Euler's identity.

Lemma 3.19

1. *If G is a planar graph with n vertices and m edges, then $m \leq 3n - 6$. If each face in a planar embedding of G has three edges, then $m = 3n - 6$. If each face in G has four edges, then $m = 2n - 4$.*

2. *If G is a planar graph with $n \geq 4$ vertices, then G has at least four vertices whose degrees are at most 5.*

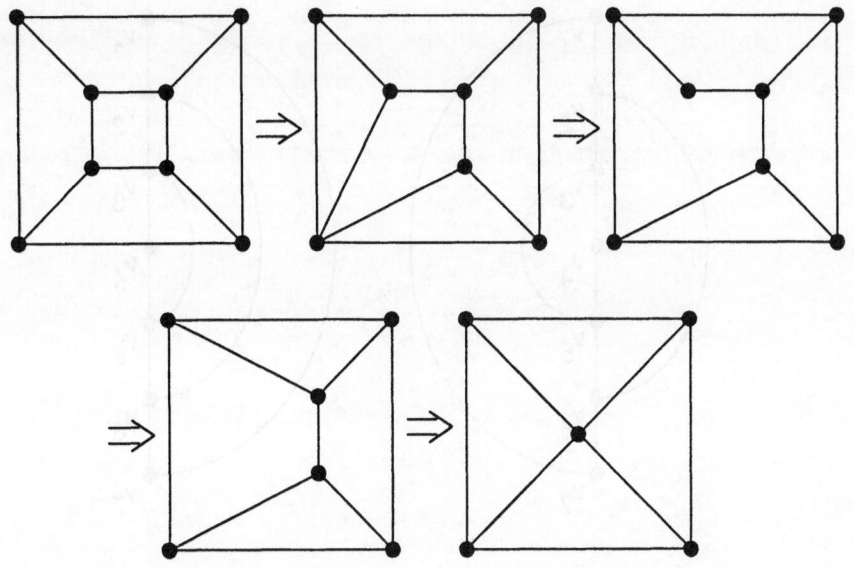

Figure 3.40: *Formation of a graph minor.*

Planar graphs G for which each face in a planar embedding has three edges are called *maximally planar*, because adding another edge destroys the planarity.

Clearly, a disconnected graph is planar exactly if all of its connected components are planar. It is a simple consequence of Definition 3.12 that a connected graph is planar exactly if all of its blocks are planar. A biconnected planar graph can still have many different planar embeddings (see Figure 3.41). The following result was developed by Whitney.

Theorem 3.10 (Whitney [461]) *A triconnected planar graph has a unique planar embedding (up to flipping; that is, reversing the cyclic orders of edges around all vertices).*

Whether a graph is planar can be tested in time $O(n)$. A planar embedding can also be constructed in time $O(n)$ [40, 196]. We will not describe the algorithms here.

3.14.2 Planar Duality

There is a natural notion of a dual for a planar graph. Consider Figure 3.42.

Definition 3.31 (Planar Dual) *Let (G, P) be a plane graph. The (plane) dual graph of $(\widetilde{G}, \widetilde{P})$ of (G, P) has a vertex for each face P. For each edge e*

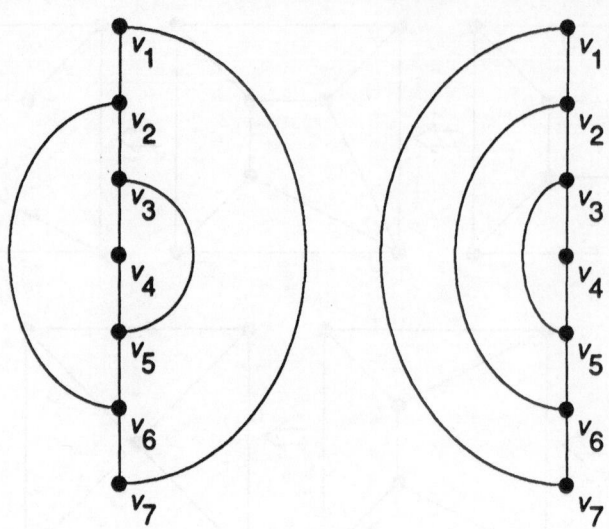

Figure 3.41: *Two different planar embeddings of the same biconnected planar graph.*

in G, there is an edge $\tilde{e}$ in $\tilde{G}$. The edge $\tilde{e}$ connects the vertices that represent the two faces that share edge e in the embedding P. The edge $\tilde{e}$ is called the dual edge of e. In $\tilde{P}$, the edge $\tilde{e}$ crosses the edge e. The mapping $e \mapsto \tilde{e}$ is a bijection between the edges of G and $\tilde{G}$.

If $(\tilde{G}, \tilde{P})$ is the plane dual of (G, P), then $\tilde{G}$ is called a **planar** dual of G.

If G is not biconnected, then some dual edges may be loops. The plane dual of (G, P) is unique. Different planar embeddings, however, may lead to different planar duals of G. The following lemma is obvious by inspection of Figure 3.42.

Lemma 3.20 *The plane dual of the plane dual of (G, P) is identical to (G, P).*

There is an interesting relationship between planar duals [462, 463].

Theorem 3.11 *Let G and $\tilde{G}$ be planar duals. A set $E' \subseteq E$ of edges in G is a cut of G (see Definition 3.4) exactly if $\widetilde{E'} = \{\tilde{e} \mid e \in E'\}$ contains a cycle in $\tilde{G}$.*

Theorem 3.11 is quite obvious from Figure 3.42. As a consequence of Theorem 3.11 and of Lemma 3.20, a set $E' \subseteq E$ is a minimal cut in G exactly if $\widetilde{E'}$ is a cycle in $\tilde{G}$, and vice versa. In fact, the existence of duals that fulfill Theorem 3.11 characterizes planar graphs.

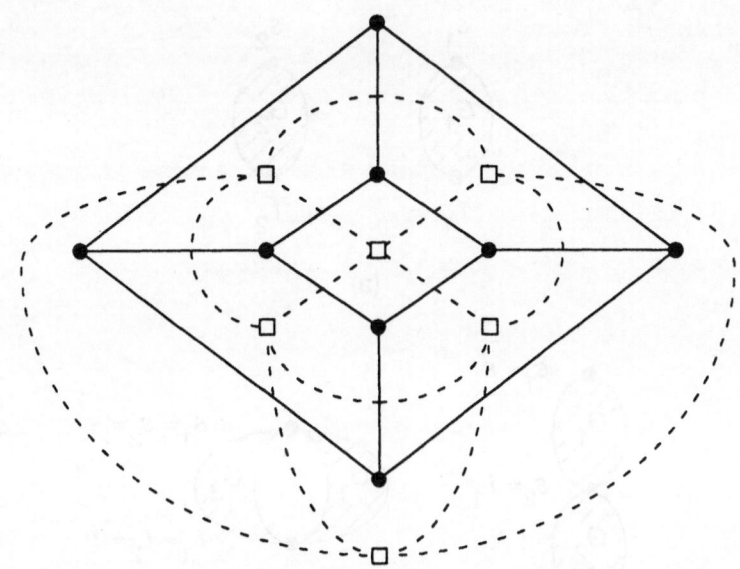

Figure 3.42: *A planar graph and its plane dual. Edges in the original graphs are drawn as solid lines; vertices in the original graph are represented by solid dots. Edges in the plane dual are drawn as dashed lines; vertices in the plane dual are represented by hollow squares.*

Theorem 3.12 *A graph G has a dual $\widetilde{G}$ fulfilling Theorem 3.11 exactly if G is planar.*

For more details on graph planarity, see [106].

3.14.3 Series-Parallel Graphs

Series-parallel graphs are an especially interesting class of planar graphs. They play a major role in circuit design, because they model circuits that compute Boolean formulae, and in layout, because they arise in the representation of floorplans based on slicing trees (Section 7.2.1).

Definition 3.32 (Series-Parallel Graphs) *The set SP of **series-parallel** graphs is the smallest set of graphs $G = (V, E)$ with two distinguished vertices $s, t \in V$ (called the **terminals**) for which the following statements are true.*

Base Graph: *The graph consisting of two vertices and an edge e between them is in SP. The terminals are the endpoints of e.*

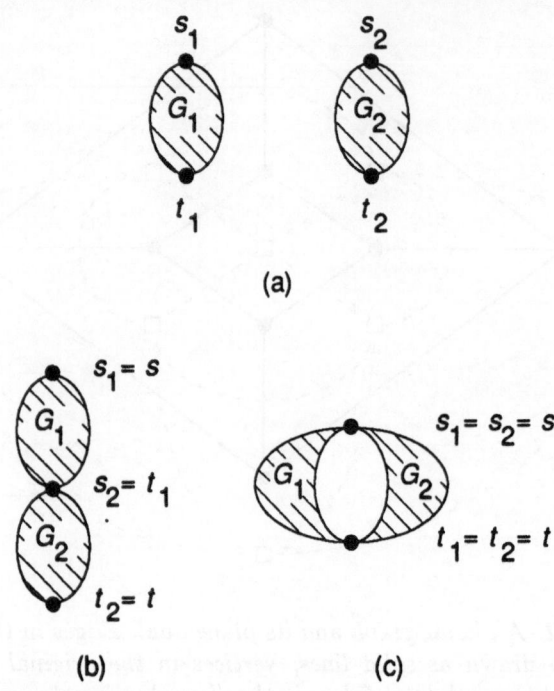

(a)

(b)   (c)

Figure 3.43: *Compositions of series-parallel graphs. (a) Two series-parallel graphs; (b) series composition; (c) parallel composition.*

Series Composition: *If G_1 with terminals s_1, t_1 and G_2 with distinguished vertices s_2, t_2 are in SP, so is the graph G that is obtained by merging t_1 and s_2, and whose terminals are $s = s_1$ and $t = t_2$ (see Figure 3.43(b)).*

Parallel Composition: *If G_1 with terminals s_1, t_1 and G_2 with distinguished vertices s_2, t_2 are in SP, so is the graph G that is obtained by merging s_1 with t_1 to yield the terminal s and by merging s_2 with t_2 to yield the terminal t (see Figure 3.43(c)).*

Series-parallel graphs can be considered hierarchical graphs, and as such they have a hierarchy tree T. Its leaf nodes correspond to edges. Each internal node of T has two children and corresponds to the application of a series or of a parallel composition. Thus, the interior nodes are also called s-nodes and p-nodes, respectively. Figure 3.44 shows a series-parallel graph and its hierarchy tree. Given G and the terminals s and t, there can be several hierarchy trees, since the series composition and parallel composition are both associative. We obtain a unique tree if we merge adjacent interior s- and p-nodes (see

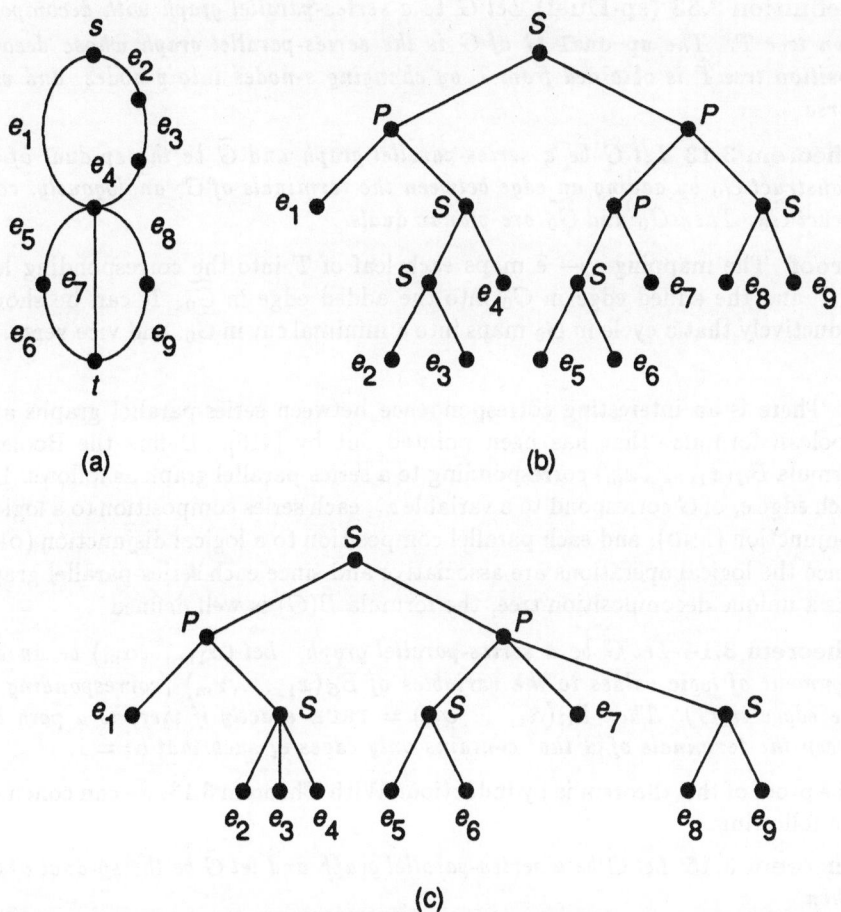

Figure 3.44: *A series-parallel graph and its decomposition tree. (a) A series-parallel graph G; (b) a hierarchy tree of G; (c) the decomposition tree of G.*

Figure 3.44(c)). In this version of the hierarchy tree, s- and p-nodes alternate on paths from the root to a leaf. We call this tree the *decomposition tree* of the series-parallel graph.

Whether a graph is series-parallel can be tested in linear time. Also, the decomposition tree of a series-parallel graph can be found in linear time. [450].

There is a natural concept of duality for series-parallel graphs.

Definition 3.33 (sp-Dual) *Let G be a series-parallel graph with decomposition tree T. The* **sp-dual** *$\widetilde{G}$ of G is the series-parallel graph whose decomposition tree $\widetilde{T}$ is obtained from T by changing s-nodes into p-nodes, and vice versa.*

Theorem 3.13 *Let G be a series-parallel graph and $\widetilde{G}$ be the sp-dual of G. Construct G_0 by adding an edge between the terminals of G; analogously, construct $\widetilde{G}_0$. Then G_0 and $\widetilde{G}_0$ are planar duals.*

Proof The mapping $e \mapsto \widetilde{e}$ maps each leaf of T into the corresponding leaf of $\widetilde{T}$ and the added edge in G_0 into the added edge in $\widetilde{G}_0$. It can be shown inductively that a cycle in G_0 maps into a minimal cut in $\widetilde{G}_0$, and vice versa. □

There is an interesting correspondence between series-parallel graphs and Boolean formulae that has been pointed out by [416]. Define the Boolean formula $B_G(x_1, \ldots, x_m)$ corresponding to a series-parallel graph as follows. Let each edge e_i of G correspond to a variable x_i, each series composition to a logical conjunction (AND), and each parallel composition to a logical disjunction (OR). Since the logical operations are associative and since each series-parallel graph has a unique decomposition tree, the formula $B(G)$ is well defined.

Theorem 3.14 *Let G be a series-parallel graph. Let $(\alpha_1, \ldots, \alpha_m)$ be an assignment of logic values to the variables of $B_G(x_1, \ldots, x_m)$ (corresponding to the edges in G). Then $B_G(\alpha_1, \ldots, \alpha_m) =$ TRUE exactly if there is a path between the terminals of G that contains only edges e_i such that $\alpha_i = 1$.*

The proof of this theorem is by induction. With Theorem 3.13, we can conclude the following.

Theorem 3.15 *Let G be a series-parallel graph and let $\widetilde{G}$ be the sp-dual of G. Then*

$$B_G(x_1, \ldots, x_m) = \overline{B_{\widetilde{G}}(\overline{x_1}, \ldots \overline{x_m})} \tag{3.6}$$

where $\overline{B}$ denotes the negation of B.

Proof Let $(\alpha_1, \ldots, \alpha_m)$ be an arbitrary logic assignment to the variables $(x_1, \ldots, x_m)$. We show one direction of (3.6); the argument is reversible. Let $B_{\widetilde{G}}(\alpha_1, \ldots, \alpha_m) =$ TRUE. By Theorem 3.14, the edges corresponding to the true variables form a path between the terminals of G. With the added edge, this path forms a cycle in G_0. By Theorem 3.13, this cycle translates into a minimal cut in $\widetilde{G}_0$. Deleting the added edge yields a cut in $\widetilde{G}$ that separates the two terminals of $\widetilde{G}$. All variables in $B_{\widetilde{G}}(\overline{x_1}, \ldots \overline{x_m})$ that correspond to edges in the cut have the value FALSE. Thus, there is no path connecting the terminals of $\widetilde{G}$ that consists of only "true" edges. Therefore, $B_{\widetilde{G}}(\overline{x_1}, \ldots \overline{x_m}) =$ FALSE, and thus $\overline{B_{\widetilde{G}}(\overline{x_1}, \ldots \overline{x_m})} =$ TRUE. □

Theorem 3.15 forms the basis for the circuit design in static CMOS logic. Therefore, it is of great importance for cell generators in this technology.

3.15 Exercises

3.1 Let $G = (V, E)$ be an undirected graph. Show that the following statements are equivalent.

1. G is bipartite.

2. There is a coloring of G with two colors.

3. Each cycle in G has even length.

Give a linear time algorithm for deciding whether any of these statements holds.

3.2 Prove Lemma 3.1.

3.3 Let $G = (V, E)$ be an undirected graph. The relation $\sim$ on the vertices that is defined by

$$v \sim w :\Longleftrightarrow v \text{ and } w \text{ lie on a common cycle}$$

is an equivalence relation. Its equivalence classes are called the *connected components* of G. Find the connected components of G in linear time using depth-first search. Can you also use breadth-first search to find the connected components of G?

3.4 Let $G = (V, E)$ be a dag. A vertex pair v, w in G is a *reconvergent fan-in pair* if there are two vertex-disjoint paths from v to w. In this case, w is called a vertex that has *reconvergent fan-in* and v is called a *witness* for w. If there is no path from v to w that contains another vertex u such that (u, w) is a reconvergent fan-in pair, then v is called a *primary* witness for w. Use depth-first search to determine for each vertex whether it has reconvergent fan-in, and, if it does, to find a primary witness. Modify the algorithm such that it finds all primary witnesses. What is the complexity of your algorithms? (This problem has applications in the area of the design for testability.)

3.5 Prove Theorem 3.1.

3.6 Prove that the graph T defined in Definition 3.12 is in fact a tree.

3.7 Customize depth-first search so as to compute the biconnected components of a graph.

Hint: Maintain a stack of edges. Instead of the *lowlink* values, compute values $lowpoint[v]$ such that v is a cutpoint exactly if $lowpoint[v] = prenum[v]$. Pop a biconnected component from the stack if you complete the call dfs(v) where v is a cutpoint of G.

3.8 Prove Lemma 3.3.

3.9 Show that the single-pair shortest-path problem is a special case of the linear-programming problem. Show that the related constraint matrix is totally unimodular (see Section 4.6.4).

3.10 Prove Lemma 3.4.

3.11 Prove Lemma 3.5 by induction on k. Use only the properties proved in Lemma 3.4.

3.12 Discuss an extension of Ford's method that computes distance values $d(s, v) \in \mathbf{R} \cup [-\infty, \infty]$ even if negative circuits exist in the graph.

3.13 Prove Lemma 3.6 by induction on i. Use only facts proved on the basis of Lemma 3.4.

3.14 Detail the implementation of the priority queue in Dijkstra's algorithm that leads to a run time of $O(m + n \log C)$ when all edge lengths are from the set $\{0, 1, \ldots, C\}$.

3.15 Prove Lemma 3.7 by induction on the number of vertices extracted from the priority queue. So that the proof will be extendible to general cost functions (Section 3.8.7), use only Lemma 3.4.

3.16 Develop the details of the algorithm for the single-source shortest-path problem that starts with a strong connectivity analysis (see Section 3.8.3). Embed the shortest-path computation in the computation of the strong components.

3.17 Prove that unidirectional search with lower bounds never processes more vertices than does the standard version of Dijkstra's algorithm.

3.18

1. Give an example where, in bidirectional search without lower bounds, the lower and upper bounds on the length of the shortest path from s to t meet *after* the first vertex has been extracted from the priority queues of both searches.

2. Show that, if all edge lengths are 1, then the upper bound meets the lower bound in bidirectional search when the first vertex has been extracted from the priority queues of both searches.

3.19 Give an example in which unidirectional search without lower bounds outperforms bidirectional search with lower bounds. Use a partial grid graph with lower bounds based on the Manhattan distance.

3.20 Extend the efficient priority-queue implementation of Figure 3.13 to the case in which the d values are k-tuples of positive integers, with the kth integer being from the set $\{1, \ldots, C_k\}$.

3.21 Let $m < 0$, and consider the following definition of $(R, \preceq)$ and $\oplus$:

$$R = \{x \in \mathbf{R} \mid m \le x \le 0\}$$

$\preceq$ is the natural ordering on R

$$a \oplus b = \max\{a + b, m\}$$

Show:

1. $(R, \preceq)$ and $\oplus$ fulfill parts 1 and 2 of the premise of Lemma 3.21.

2. $\oplus$ is monotonic (but not strictly monotonic) in both arguments, that is,

$$r_1 \preceq r_2 \Rightarrow r_1 \oplus r_3 \preceq r_2 \oplus r_3 \text{ and } r_3 \oplus r_1 \preceq r_3 \oplus r_2 \quad \text{for all } r_1, r_2, r_3 \in R.$$

3. $(R, \preceq)$ and $\oplus$ do not fulfill Lemma 3.4, part 1.

3.22 Prove Lemma 3.9.

3.23 Review Section 2.3.3. Develop an efficient approximation algorithm for the TRAVELING SALESPERSON problem with an error of 2. What is the run time of the algorithm?

3.24 Prove that the Steiner tree computed by the method of Takahashi and Matsuyama and of Akers (see Section 3.10.2), has a cost that does not exceed the cost of a minimum spanning tree of the complete distance graph G_1 defined in Section 3.10.4.

3.25 Give an example that shows that emanating search wavefronts from different required vertices at different speeds in the multidirectional search algorithm for Steiner trees may lead to spanning trees of G_1 that are not minimum.

3.26 Augment the multidirectional search algorithm (Figure 3.19) with pointers such that a Steiner tree that is consistent with G_2 can be recovered directly from the data structure.

3.27 Give an example of a flow graph with irrational edge capacities for which there exists an infinite sequence of augmenting paths whose corresponding sequence of flow values converges to a value that is different from the maximum flow value. (This result shows that we have to select special sequences of augmenting paths to construct maximum flows.)

3.28 Prove Fact 3.2. Show that the constraint matrix A in the corresponding linear program is totally unimodular (see Section 4.6.4). This observation proves that, if all edge capacities are integer, so is the maximum flow value.

3.29 Prove Lemma 3.14.

Hint: Use the transformation from the UNDIRECTED MULTICOMMODITY FLOW problem to the directed MULTICOMMODITY FLOW problem depicted in Figure 3.23 to reduce Lemma 3.14 to Lemma 3.13.

3.30 Formulate the MULTICOMMODITY FLOW problem as a linear program. Is the constraint matrix of this program totally unimodular? (See Section 4.6.4.)

3.31 Prove the correctness of the reduction from MAXIMUM MATCHING to MAXIMUM FLOW illustrated in Figure 3.26.

3.32 By adjusting edge costs suitably, prove the following:

1. Each instance of LINEAR ASSIGNMENT and of MINIMUM WEIGHT MATCH-ING with arbitrary edges costs can be transformed into an instance of the same problem with nonnegative edge costs.

2. Each instance of the maximization version of either problem can be transformed into an instance of the minimization version.

3. Each instance of LINEAR ASSIGNMENT can be transformed into an instance of MINIMUM WEIGHT MATCHING.

3.33 Give an example of a hierarchical graph Γ such that $N, M = 2^{\Omega(m+n)}$. Show that, for all hierarchical graphs, $M, N = 2^{O(m+n)}$.

3.34 Develop an optimal burner for bipartiteness.

3.35 Extend Definition 3.21 to directed graphs. Develop a burner for strong connectivity. What is the complexity of the burner, and what is the size of the burnt graphs?

3.36 Construct the domination graph for bipartiteness. What are its width and its depth?

3.37 We call a context-free cellular graph grammar ε-deterministic if each nonterminal v of degree p can have at most two types; further, if v has two types, one of them must be the *empty* p-graph, containing only the p pins and no other vertices or edges.

1. Show that, if Γ is ε-deterministic, then, in the extended bottom-up table for graph connectivity, each table entry has at most one glue type that is not dominated by other glue types in the same entry.

2. Using part 1, show that testing whether the language $L(\Gamma)$ of an ε-deterministic cellular graph grammar Γ has a connected graph can be done in time $O(\sum_{i=1}^{k} p_i^2)$.

3.38 Prove Lemma 3.18: Show that shrinking an edge to a vertex preserves planarity. Use Theorem 3.7.

3.39

1. Prove Theorem 3.9 by induction.

2. Using Euler's identity prove, that, if in a planar embedding of G, all faces have r edges, then
$$m = r(n-2)/(r-2)$$

3. Use part 2 to prove Lemma 3.19

4. Use Lemma 3.19 to prove that K_5 and $K_{3,3}$ are nonplanar.

3.40

1. Prove Theorem 3.11.

2. Prove Theorem 3.12: Show that, if G fulfills the theorem, so does every subgraph of G. Then, show that K_5 and $K_{3,3}$, as well as their homeomorphs, do not fulfill the theorem.

Chapter 4

Operations Research and Statistics

In this chapter, we will look at optimization problems from a different point of view—from one that is, in some sense, more fundamental. We will interpret finding the optimum solution of an optimization problem as a search process on the space of legal configurations. (From now on, we will mostly be concerned with *legal* configurations. Thus, we will omit the word *legal* unless it is not clear from the context what kind of a configuration we mean.) The configuration space can be represented as a set of geometric points or, alternatively, as a directed graph, and we will choose whatever representation is better suited to our purpose. There exists a large number of search heuristics and strategies for optimization problems; we will survey the ones that are most important for layout design.

4.1 Configuration Graphs, Local Search

In this section, we will lay down the fundamental graph-theoretic representation of the configuration space of an optimization problem.

Definition 4.1 (Configuration Graph) *Let* Π *be a minimization problem, let* p *be an instance of* Π *and let* S_p *be the set of legal configurations for* p. *Furthermore, assume that* $|S_p|$ *is finite; that is,* Π *is a* discrete *minimization problem. We consider* S_p *as the set of vertices of a directed graph* $C_p = (S_p, E_p)$, *the* configuration graph. *For a legal configuration* $s \in S_p$, *the set* $N(s) := \{t \in S_p | (s, t) \in E_p\}$ *is called the* neighborhood *of* s. *The configuration* s *is called a* local minimum *exactly if* $c(s) \leq c(t)$ *for all* $t \in N(s)$. *If, in addition, s is an optimal solution for* p, *then* s *is also called a* global minimum.

137

The configuration space of a maximization problem and local and global maxima are defined analogously.

The definition of a local minimum given in Definition 4.1 is an extension of the usual definition for subsets of $\mathbf{R}^n$ to configuration graphs. The notion of an ε-neighborhood is here replaced with the notion of a neighborhood in the configuration graph.

Usually, configuration spaces of instances of optimization problems are very large. As an example, consider the TRAVELING SALESPERSON problem (see Section 2.1). Here, the configuration space consists of all tours, and thus has a size of $2^{\Theta(n \log n)}$. In most cases, the size of the configuration space is at least exponential in the size of the problem instance. (Sometimes, the configuration space of a discrete optimization problem can even be (countably) infinite.) Thus, it is not feasible to search the configuration space exhaustively. Rather, *local* search strategies through the configuration space have to be used.

Definition 4.2 (Search) *A* search *from configuration $s \in S_p$ is a directed path in C_p that starts at s. The endpoint of the path is the solution found by the search. The search is* monotonic *or* greedy *if the costs of the vertices decrease along the search path (for minimization problems). Strategies that solve minimization problems by doing greedy search on C_p are called* iterative improvement strategies.

The goal of each search must be to find a solution that is as close as possible to the optimum. To meet this goal, people frequently employ greedy search strategies. There is, however, an essential problem associated with use of greedy search: This approach can lead to a local optimum that is not a global optimum. The search cannot escape from this local optimum; thus, it is prevented from finding the optimal solution.

There are two ways to deal with this problem. One is to allow for nongreedy search steps; that is, for steps in which the cost function increases. To ensure that we move toward the global optimum eventually, however, we have to take care that such nongreedy moves are not too frequent. There are deterministic as well as probabilistic ways of ensuring this, and we will discuss them in detail in later sections and chapters.

The other way of dealing with the problem of getting trapped in local optima uses the observation that we have many possibilities in choosing the configuration graph for an instance of an optimization problem. If we choose a *dense* graph—that is, a graph that has many edges and large neighborhoods—then we are less likely to hit local optima. In addition, shorter search paths may exist from a given start configuration to a global optimum. If, in the extreme case, C_p is the complete directed graph, then every local optimum is also a global optimum. In fact, a global optimum can be reached from any configuration in a single search step.

On the other hand, the denser the configuration graph is, the more inefficient the search step will be. This relation holds, because, in each search step, we optimize over the neighborhood of the current configuration, and the larger this neighborhood is, the more time we need to find a good configuration to which the search can proceed. Thus, we want to make neighborhoods as small as possible, which means that we do not want to include too many edges in C_p.

For greedy search, the challenge is to select a configuration graph that represents a good tradeoff between these two conflicting goals. On the one hand, it should be possible to optimize easily over a neighborhood. On the other hand, the configuration graph should not contain too many local optima, and should provide short paths to global optima.

If the search is not greedy, we can take a little more liberty in making the configuration graph sparse. In any case, however, whether or not we do greedy search, C_p has to be such that we can reach a global optimum from any configuration. We usually ensure this condition by making C_p strongly connected. In addition, the existence of short paths to global optima increases the efficiency of the search. The most attractive configuration graphs are sparse graphs with the following property.

Definition 4.3 (Convex Configuration Graph) *A configuration graph C_p is* **convex** *if it is strongly connected and each local optimum of C_p is also a global optimum of C_p.*

For convex configuration graphs, a simple greedy search will always find the optimum. In many optimization problems, however, we cannot afford to search greedily through a convex configuration graph, because this graph would be much too dense, thus making the search intractable. In such cases, there are three ways to proceed.

1. Choose a nonconvex configuration graph that is sparse enough to perform a goal-directed search, and perform a (deterministic) greedy search on it. The result is a solution that is not optimal, in general. We can attempt to analyze how far the cost of the obtained solution is from the minimum cost. We can do this analysis if we can compute a bound on the minimum cost.

2. Choose a (in general nonconvex) configuration graph on which to perform nongreedy local search. If the search is deterministic, then the graph as well as the nongreedy search steps have to be chosen carefully so as to get close to the optimal cost.

3. Do the search in a randomized fashion. There is a class of randomized search strategies that ensure finding the optimal solution in the limit, and that also prove quite successful in practice. These search strategies are based on the *probabilistic hill-climbing* paradigm and are discussed in Section 4.3.

Figure 4.1 shows examples of configuration graphs.

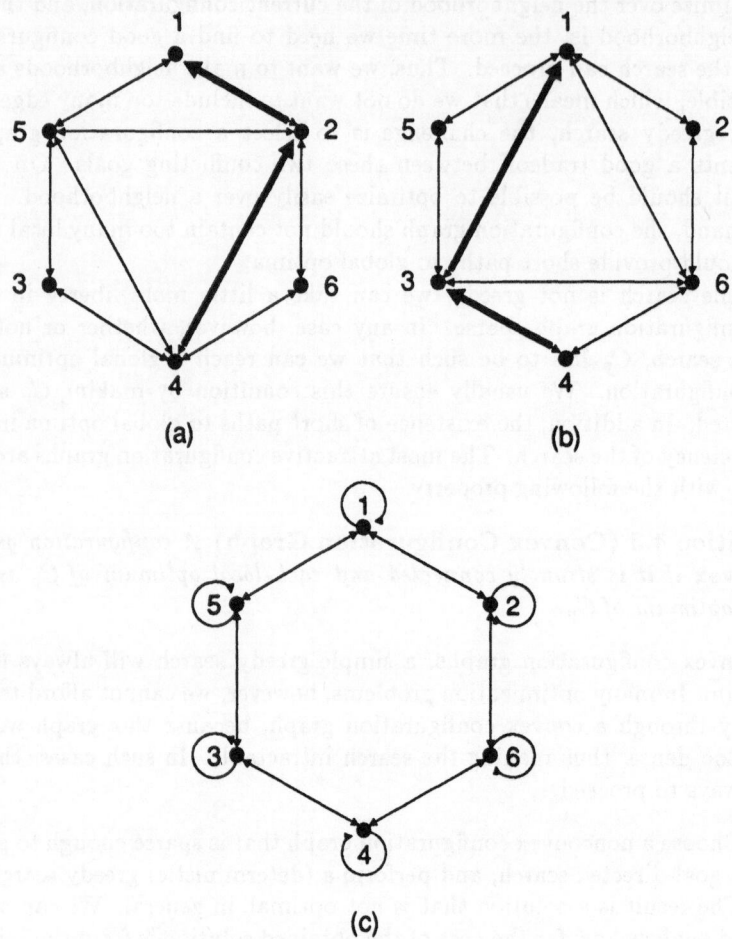

Figure 4.1: *Configuration graphs. (a) A configuration graph on a space of six legal configurations. Configurations are labeled with their costs. We identify the costs with the vertices of the graph. Vertex 1 is the unique global minimum. Vertex 3 is a local minimum. A greedy search path from vertex 4 to vertex 1 is highlighted. (b) A convex configuration graph on the same configuration space. A greedy search path from vertex 4 to vertex 1 is highlighted. (c) A different configuration graph on the same configuration space. Vertices 1 and 3 are local minima. No greedy search path from vertex 4 to vertex 1 exists. Greedy search from vertex 4 ends in the local minimum vertex 3.*

4.2 Markov Chains

In this section, we provide the basis for analyzing randomized local search techniques. In a randomized local search, a particular neighbor of a configuration s is selected as the one to which to move, with a certain probability, in the next search step. Furthermore we assume that the choices of all edges out of s are probabilistically independent. The corresponding mathematical structure is a configuration graph that is enhanced with an edge labeling that assigns to each edge its respective transition probability. Such a configuration graph is called a *Markov chain*.

4.2.1 Time-Homogeneous Markov Chains

Definition 4.4 (Markov Chain) *Let $G = (V, E)$ be a directed graph with $V = \{s_1, \ldots, s_n\}$. Let $c : V \to \mathbf{Z}$ be a vertex-weighting function. We write c_i for $c(s_i)$. Let $p : E \to [0,1]$ be an edge-weighting function (notation: $p_{ij} = p(s_i, s_j)$) such that*

$$\sum_{s_j \in N(s_i)} p_{ij} = 1 \quad \text{for all } s_i \in V$$

We call (G, p) a (finite) time-homogeneous Markov chain. The value p_{ij} is called the transition probability of edge (s_i, s_j). In the context of Markov chains, a configuration is also called a state.

The restriction on the edge labeling p ensures that the transition probabilities of all edges leaving the same vertex add up to 1. In a time-homogeneous Markov chain, the transition probabilities are, in fact, constant; that is, they are the same in each search step. Since the transition probabilities are independent of the search history, the Markov chain has no "memory"; that is, no knowledge of past search steps can be encoded in the transition probabilities for the current search step.

Without loss of generality, we will assume that $p_{ij} > 0$ for all edges $(s_i, s_j) \in E$. If $p_{ij} = 0$ for an edge, we just delete the edge from G. With this provision, we can give the following definition.

Definition 4.5 (Irreducible Markov Chain) *A Markov chain (G, p) is irreducible if G is strongly connected.*

Many properties of a Markov chain can be stated in terms of its *transition matrix*—that is, the matrix $P = ((p_{ij}))$. (By convention, we set $p_{ij} = 0$ if $(s_i, s_j) \notin E$.)

Assume, for instance, that the Markov chain modeling the search is in a state $s_i \in V$ with probability $\pi_i = \pi(s_i)$, $0 \le \pi_i \le 1$, $\sum_{i=1}^n \pi_i = 1$. We call the probability distribution $\pi : \{s_1, \ldots, s_n\} \to [0,1]$ a *state probability distribution*. After a randomized search step guided by the Markov chain (G, p), the state probability distribution is $\pi' = \pi \cdot P$. Indeed, the probability of being in state s_i is

$$\pi_i = \sum_{j=1}^{n} \pi_j \cdot p_{ji}$$

if we take into account that the possible transitions are probabilistically independent.

Our goal is to construct Markov chains that converge to a certain desirable state probability distribution, most favorably to one in which no state that is nonoptimal has a nonnegative probability.

Definition 4.6 (Stationary Distribution) *A state distribution π of a time-homogeneous Markov chain (G, p) is called a* **stationary** *distribution if $\pi \cdot P = \pi$.*

Intuitively, a stationary distribution has the property that it does not change if a randomized local search guided by the Markov chain is performed. Formally, a stationary distribution is a left eigenvector of the transition matrix P for the eigenvalue 1.

We will always be dealing with irreducible Markov chains such that $p_{ii} > 0$ for some i. Let us call such Markov chains *well structured*. The following fundamental theorem on Markov chains is the basis for our analysis of randomized search.

Theorem 4.1 *A well-structured Markov chain (G, p) has a unique stationary distribution $\pi^{(0)}$. Furthermore, if $\Pi^{(0)} = (\pi^{(0)}, \ldots, \pi^{(0)})$ is the matrix whose n columns are all identical to $\pi^{(0)}$, then*

$$\Pi^{(0)} = \lim_{n \to \infty} P^n$$

Here, P^n is the n-fold matrix product of P with itself.

Theorem 4.1 states that, if the Markov chain is well structured, then, if it is started in any state, it converges to the stationary distribution. Indeed, if we start the Markov chain in state s_i, then it will converge to the distribution $\lim_{n \to \infty} e_i \cdot P^n$, where e_i is the ith unit vector. But this limit value is the ith row of Π, which is the stationary distribution (see Exercise 4.1). Figure 4.2 illustrates the relevant definitions and Theorem 4.1.

4.2.2   Time-Reversibility

The following lemma gives a necessary and sufficient condition for a probability distribution π to be the stationary distribution of the Markov chain (G, p).

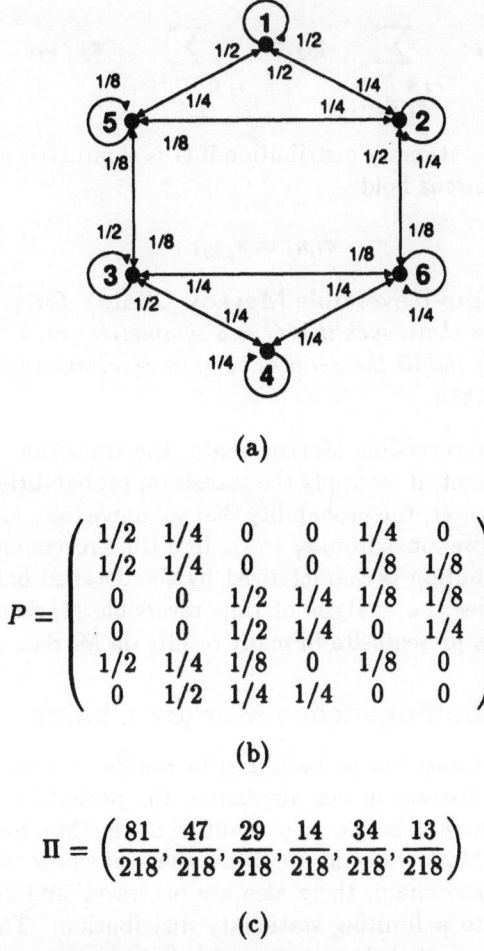

(a)

$$P = \begin{pmatrix} 1/2 & 1/4 & 0 & 0 & 1/4 & 0 \\ 1/2 & 1/4 & 0 & 0 & 1/8 & 1/8 \\ 0 & 0 & 1/2 & 1/4 & 1/8 & 1/8 \\ 0 & 0 & 1/2 & 1/4 & 0 & 1/4 \\ 1/2 & 1/4 & 1/8 & 0 & 1/8 & 0 \\ 0 & 1/2 & 1/4 & 1/4 & 0 & 0 \end{pmatrix}$$

(b)

$$\Pi = \left(\frac{81}{218}, \frac{47}{218}, \frac{29}{218}, \frac{14}{218}, \frac{34}{218}, \frac{13}{218} \right)$$

(c)

Figure 4.2: *An example of a well-formed Markov chain. (a) The edge-labeled configuration graph. Edge labels are located to the left of the arrow pointing to the tail of the edge. (b) The transition matrix. (c) The stationary distribution is the unique solution of the linear system $(P-I)^T \Pi^T = 0$; $\sum_{i=1}^{n} \pi_i = 1$. The probabilities of low-cost states are higher than the probabilities of high-cost states because the transition probabilities of the Markov chain favor transitions to low-cost states. Nongreedy search is possible, however, because some state transitions that increase the cost also have positive probability.*

Lemma 4.1 π *is a stationary distribution for* (G, p) *if and only if, for all* $i = 1, \ldots, n$,

$$\pi_i \cdot \sum_{\substack{s_j \in N(s_i) \\ s_j \neq s_i}} p_{ij} = \sum_{\substack{s_j \in N^{-1}(s_i) \\ s_j \neq s_i}} \pi_j \cdot p_{ji}$$

In particular, π is a stationary distribution if G is symmetric and the following *detailed balance equations* hold:

$$\pi_i p_{ij} = \pi_j p_{ji} \tag{4.1}$$

Definition 4.7 (Time-Reversible Markov Chain) *Let* (G, p) *be a time-homogeneous Markov chain such that* G *is a symmetric graph. If the stationary distribution of* (G, p) *fulfills the detailed balance equations* (4.1), *then* (G, p) *is called* time-reversible.

In general, in a time-reversible Markov chain, the transition probabilities p_{ij} and p_{ji} may be different; if we apply the transition probabilities to the stationary distribution, however, the probability that we move from s_i to s_j is equal to the probability that we move from s_j to s_i. In a time-reversible Markov chain, the stationary distribution is characterized by the detailed balance equations. This greatly simplifies the analysis of time-reversible Markov chains. Thus, time-reversibility is a prerequisite in many results on Markov chains.

4.2.3   Time-Inhomogeneous Markov Chains

In many cases, the transition probabilities in random processes change with time. Since time is discrete in our discussion, the probabilities now take the shape $p(s_i, s_j, t)$, where t is the step number of the Markov process. The corresponding finite Markov chains are called *time-inhomogeneous*. For a time-inhomogeneous Markov chain, there also are necessary and sufficient criteria for its convergence to a limiting stationary distribution. These criteria are somewhat more complicated than they are in the time-homogeneous case. We will not discuss them in detail here. See [413] for an extended discussion of time-inhomogeneous Markov chains.

4.3   Simulated Annealing

Recently, there has been a surge of interest in a special class of randomized local search algorithms called *simulated annealing*. The algorithms are a version of a successful statistical model of thermodynamic processes for growing crystals that has been transformed into computational terms.

A perfectly homogeneous crystal lattice represents a configuration of solid-state material at a global minimum of energy. By experience, investigators have determined that they can get very close to a perfect crystal by applying the process of *annealing*. The solid-state material is heated to a high temperature until it reaches an amorphous liquid state. Then, it is cooled very slowly and according to a specific *schedule* of decreasing the temperature. If the initial temperature is high enough to ensure a sufficiently random state, and if the cooling is slow enough to ensure that thermal equilibrium is reached at each temperature, then the atoms will arrange themselves in a pattern that closely resembles the global energy minimum of the perfect crystal. Thermodynamics teaches that the thermal equilibrium at temperature T is a probability distribution in which a state with energy E is attained with the *Boltzmann* probability

$$\frac{1}{Z(T)} \, e^{-\frac{E}{k_B T}}$$

where $Z(T)$ is a normalization factor and k_B is the *Boltzmann constant*.

In a theoretical model of what occurs inside the material during the annealing process, states are continually perturbed by the introduction of small random changes of the positions of atoms in the matter. If the result is a state of lower energy, then the state perturbation is unconditionally accepted. If not, then the state perturbation may still be accepted with a probability of $e^{-\Delta E/k_B T}$. This probability decreases with the temperature, thus reflecting the higher disorder occurring at high temperatures. Experience has shown, that this method of estimating the results of processes in statistical mechanics is highly accurate [35].

The idea of simulated annealing is to transpose this model into the context of optimization problems. As such, the idea is completely heuristic, and so far has not been backed up with concrete and practically relevant theoretical results. Simulated annealing seems to work well in practice, however, at least if a sufficient amount of computing resource is provided.

In the transformation, the energy is substituted with the cost function that we want to minimize, and states of matter are replaced with the legal configurations of the problem instance. (Sometimes, illegal configurations are allowed but are penalized with high costs.) The temperature is emulated by a control parameter T that is also called the *temperature*. The perturbations that occur if the matter is at high temperature correspond to variations that the legal configurations can undergo in a suitably chosen configuration graph. The general structure of the simulated annealing algorithm is given in Figure 4.3. Such an algorithm is called a *probabilistic hill-climbing* algorithm, because it allows increases in the objective function with a certain probability.

A probabilistic hill-climbing algorithm uses the following auxiliary procedures and parameters:

```
(1)        s, nexts : configuration;
(2)        T : real;
(3)        count : integer;

(4)        begin
(5)            s := random initial configuration;
(6)            T := T₀;
(7)            repeat
(8)                count := 0;
(9)                repeat
(10)                   count := count + 1;
(11)                   nexts := generate(s);
(12)                   if c(nexts) ≤ c(s)
                          or f(c(s), c(nexts), T) > random(0, 1) then
(13)                       s := nexts fi
(14)               until equilibrium(count,s,T);
(15)               T := update(T)
(16)           until frozen(T)
(17)       end;
```

Figure 4.3: *The probabilistic hill-climbing method.*

T_0	Initial temperature
generate(s)	Generates the configuration that the search should proceed to from s
$f(c_i, c_j, T)$	Computes a probability of when to carry out a move to a higher cost configuration
equilibrium($count, s, T$)	Decides whether the equilibrium for T is sufficiently approximated
update(T)	Computes the next value of T
frozen(T)	Decides whether the final value of T has been reached

The procedure generate(s) computes the edges of the configuration graph. The initial temperature T_0 together with the procedures equilibrium($count, s, t$), update(T), and frozen(T) set the parameters that simulate the thermal dynamics. A choice of these parameters is referred to as a *cooling schedule*. The initial configuration is chosen at random, in accordance with the probabilistic character of the method.

We will explain the probabilistic hill-climbing algorithm in terms of Markov chains.

Underlying the search process is a sequence of finite time-homogeneous Markov chains, one for each temperature. The procedure generate(s) selects among the neighbors of s one to which the search should move. This successor state *nexts* is chosen at random according to some probability distribution. Its values are denoted by g_{ij}; that is, g_{ij} is the probability that $s_j := $ generate(s_i). The probabilities g_{ij} define a time-homogeneous Markov chain (G, g) that is independent of T. We call this chain the *generating chain*. We will require that (G, g) be irreducible. We do not actually move to the successor generated in this way in all cases. The function $f(c_i, c_j, T)$ computes a probability, independently of the costs of the respective configurations and the temperature, for whether the move is *accepted*. The expression random$(0, 1)$ generates a random number between 0 and 1. The comparison in line 12 of Figure 4.3 then implements the randomized acceptance of the moves. Thus, at temperature T, the search process behaves like a Markov chain $(G, g'(T))$, where

$$g'_{ij}(T) = \begin{cases} g_{ij}\alpha(i, j, T) & \text{if } i \neq j \\ 1 - \sum_{k \neq i} g'_{ik}(T) & \text{if } i = j \end{cases}$$

Here, $\alpha(i, j, T)$ is the *acceptance criterion*

$$\alpha(i, j, T) = \begin{cases} 1 & \text{if } c_j \leq c_i \\ \min\{f(c_i, c_j, T), 1\} & \text{otherwise} \end{cases} \tag{4.2}$$

The Markov chain $(G, g'(T))$ for temperature T is active until the criterion equilibrium($count, s, T$) is fulfilled. Here, $count$ counts the number of search steps that were executed using the present Markov chain. Then, a new Markov chain for a new temperature is started at the current configuration. The procedure update(T) computes the new temperature. This process is continued until the procedure frozen(T) decides that it should be stopped. Thus, the whole process consists of the consecutive application of a finite number of time-homogeneous Markov chains. Each chain is applied for a finite number of steps.

To customize the probabilistic hill-climbing algorithm to simulate the annealing process from statistical mechanics, we choose the *Metropolis function* to compute the acceptance of a move:

$$f(c_i, c_j, T) := e^{-\frac{c_j - c_i}{cT}} \tag{4.3}$$

Here, c is a control parameter playing the role of the Boltzmann constant. If the Metropolis function is used, the algorithm of Figure 4.3 is called *simulated annealing*. Simulated annealing has been introduced as a tool for optimization by Kirkpatrick et al. [234].

The success of a simulated-annealing algorithm in practice depends on the suitable choice of the cooling schedule. Making this choice turns out to be quite an art, and we will discuss how it is done in Section 4.3.2. There are a certain number of theoretical results on simulated annealing that we will cover in the following section. They do not go as far as we would like, but they show the admissibility of the method in a weak sense that they consider only the convergence behavior, and not the efficiency, of the method.

4.3.1   Convergence of Simulated Annealing

In this section, we summarize analytical results on the appropriateness of using simulated annealing procedures for optimization. We call a probability distribution that assigns zero probability to all states that are not global minima an *optimizing* distribution.

A theoretical justification of the method of simulated annealing that is based on an analysis using Markov chains consists of two steps:

Step 1: Prove that the simulated-annealing procedure, when started in an arbitrary state and given an appropriate cooling schedule, will eventually converge to an optimizing distribution (convergence behavior).

Step 2: Prove that the probability distribution reached after a small number of steps (much smaller than the size of G) does not differ much from an optimizing distribution (efficiency).

The first step shows that simulated annealing is *admissible*; that is, given enough time, simulated annealing will yield a global optimum almost with certainty. Proving this alone does not justify the use of simulated annealing, however. The second step shows that simulated annealing with very high certainty performs better than other search heuristics—for instance, exhaustive search—in that it reaches a global optimum almost with certainty after inspecting only a very small part of the configuration space.

Several versions of a proof of step 1 exist, some of them weaker, some of them stronger. With respect to step 2, the situation is not quite as good. Only few results exist here, and their implications are unclear.

4.3.1.1   Analysis with Time-Homogeneous Markov Chains

Let us first talk about step 1. The most general results here model simulated annealing using time-homogeneous Markov chains. A typical theorem with rather weak preconditions is the following.

Lemma 4.2 *Consider a probabilistic hill-climbing algorithm. Let the function f in the accept expression fulfill the requirements*

$$f(c_i, c_j, T) \cdot f(c_j, c_k, T) = f(c_i, c_k, T)$$

We call such a function f multiplicative. Let (G, g) be a Markov chain for the generate procedure, such that G is a symmetric graph (see Definition 3.4). Let

c_0 be the cost at a global optimum. Then the stationary probability of being at state s_i at temperature T can be bounded as follows:

$$\pi_i^{(T)} \leq \alpha \cdot f(c_0, c_i, T)$$

for all $i = 1, \ldots, n$, where α is a constant depending only on (G, g).

For the proof of Lemma 4.2, see Exercise 4.3. In general, α is not very large. An immediate consequence of Lemma 4.2 is the following theorem.

Theorem 4.2 (Faigle, Schrader [109]) *Consider a probabilistic hill-climbing algorithm with a multiplicative acceptance function f and a generating Markov chain (G, g) such that G is symmetric. Further assume that*

$$\lim_{T \to 0} f(c_i, c_j, T) = 0 \qquad (4.4)$$

whenever $c_i < c_j$. Then,

$$\lim_{T \to 0} \pi^{(T)} = \pi^{(0)}$$

where $\pi^{(0)}$ is an optimizing distribution.

Theorem 4.2 ensures the convergence of the stationary distributions at each temperature under a rather weak set of premises. Many generating mechanisms yield symmetric graphs. The probabilities of antiparallel edges are not restricted in the theorem. Note that the generating Markov chain need not be time-reversible. (A more recent paper by Faigle and Kern [108] even generalizes Theorem 4.2 to the case where the generating chain is weakly reversible; see Definitions 4.8 and 4.9.)

The requirements on f in Theorem 4.2 are met not only by the Metropolis function of simulated annealing, but also by other functions, such as the following function:

$$f(c_i, c_j, T) = \frac{c_i}{c_j} \cdot T^{(c_j - c_i)} \qquad (4.5)$$

Thus, at least as long as we are not concerned with the rate of convergence, models of probabilistic hill-climbing other than simulated annealing are admissible.

Lemma 4.2 even allows us to make statements about how fast states that are not global optima are excluded in the stationary distributions as the temperature is decreased. For instance, if simulated annealing is used, the probability of being in a nonoptimal state decreases exponentially in $1/T$. However, this does not mean much for practical purposes, as long as we do not know how quickly the Markov chains for each temperature converge to their stationary distributions. Also note that Theorem 4.2 does not help us to select cooling schedules.

If we assume time-reversibility of the generating chain, we can compute the optimizing distribution in equation 4.2 exactly.

Theorem 4.3 *Let the generating chain be time-reversible and have station-ary distribution γ. Then, the optimizing distribution of the time-homogeneous Markov chain modeling the probabilistic hill-climbing algorithm at temperature T assigns the following probability to the global minimum s_i:*

$$\pi_i^{(0)} = \frac{\gamma_i}{\sum\limits_{s_j \in V_{\text{opt}}} \gamma_j} \tag{4.6}$$

Here, V_{opt} is the set of global minima. All other states have probability zero.

Proof By verifying the detailed balance equations, we can show that the fol-lowing is an expression for the stationary probability of state s_i at temperature T:

$$\pi_i^{(T)} = \frac{1}{\Lambda} \cdot \gamma_i \cdot \alpha(c_0, c_i, T) \tag{4.7}$$

where

$$\Lambda = \sum_{j=1}^{n} \gamma_j \cdot \alpha(c_0, c_j, T)$$

is a normalizing factor. Note that (4.7) implies, that the Markov chain for each positive temperature is time-reversible. As $T \to 0$, equation (4.7) converges to equation (4.6). $\square$

Theorem 4.3 states that, in the optimizing distribution, each global minimum s_j has a probability that amounts to the conditional probability of s_j in the gen-erating chain (G, g), given that we reside in some global minimum. Restricted cases of Theorem 4.3 have been presented by [164, 301, 329].

Figure 4.4 depicts an example illustrating Theorems 4.2 and 4.3. The sta-tionary distribution Π_T for temperature T is

$$\Pi_T = \left(\frac{180}{N}, \frac{120T}{N}, \frac{80T^2}{N}, \frac{45T^3}{N}, \frac{48T^4}{N}, \frac{40T^5}{N} \right)$$

where

$$N = 40T^5 + 48T^4 + 45T^3 + 80T^2 + 120T + 180$$

Both P and Π_T are functions of T. As Theorem 4.3 predicts, the stationary distribution converges to an optimizing distribution, as $T \to 0$:

$$\lim_{T \to 0} \Pi_T = (1, 0, 0, 0, 0, 0)$$

(see also Exercise 4.4).

Note that, so far, no result distinguishes the Metropolis function from other multiplicative functions that can be used as acceptance functions in probabilis-tic hill-climbing algorithms.

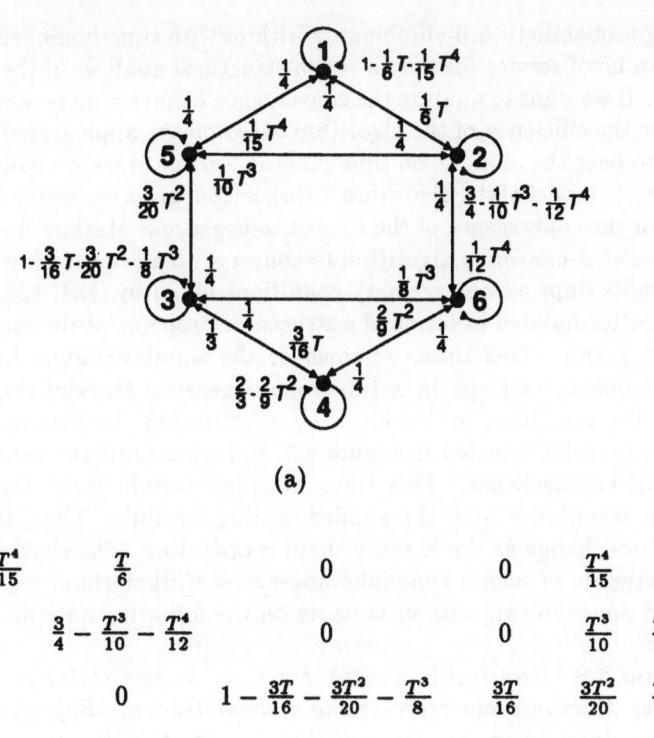

(a)

$$
\begin{pmatrix}
1 - \frac{T}{6} - \frac{T^4}{15} & \frac{T}{6} & 0 & 0 & \frac{T^4}{15} & 0 \\[6pt]
\frac{1}{4} & \frac{3}{4} - \frac{T^3}{10} - \frac{T^4}{12} & 0 & 0 & \frac{T^3}{10} & \frac{T^4}{12} \\[6pt]
0 & 0 & 1 - \frac{3T}{16} - \frac{3T^2}{20} - \frac{T^3}{8} & \frac{3T}{16} & \frac{3T^2}{20} & \frac{T^3}{8} \\[6pt]
0 & 0 & \frac{1}{3} & \frac{2}{3} - \frac{2T^2}{9} & 0 & \frac{2T^2}{9} \\[6pt]
\frac{1}{4} & \frac{1}{4} & \frac{1}{4} & 0 & \frac{1}{4} & 0 \\[6pt]
0 & \frac{1}{4} & \frac{1}{4} & \frac{1}{4} & 0 & \frac{1}{4}
\end{pmatrix}
$$

(b)

Figure 4.4: *Convergence of stationary probability distributions. The configuration graph of Figure 4.2(c) is labeled with probabilities that are composed of uniform generating probabilities over each neighborhood and the acceptance criterion (4.2) based on the function (4.5) for $T \leq 1$. Note that the generating chain is time-reversible. (a) The configuration graph. (b) The transition matrix P. As Theorem 4.3 predicts, the stationary distribution converges to an optimizing distribution, as $T \to 0$.*

4.3.1.2   Analysis with Time-Inhomogeneous Markov Chains

Modeling probabilistic hill-climbing algorithms with time-homogeneous Markov chains can be of service for only a rough structural analysis of the convergence behavior. If we want to analyze the convergence behavior more accurately, or to determine the efficiency of the algorithm based on the applied cooling schedule, we need to base the analysis on time-inhomogeneous Markov chains that represent a precise model of the algorithm. Hajek [164] gives necessary and sufficient criteria for the convergence of the *time-inhomogeneous* Markov chain representing a *simulated-annealing* algorithm to converge to an optimizing distribution. These results improve on necessary conditions found by [134, 135, 329]. These criteria are formulated in terms of a structural property of the underlying configuration graph. This time, we consider the simulated-annealing algorithm to be a sequence of steps in a time-inhomogeneous Markov chain. We still compute the transition probabilities by substituting the Metropolis function into the algorithm depicted in Figure 4.3, and we assume the generating chain to be time-homogeneous. This time, however, we allow the temperature to change in accordance with the applied cooling schedule. Thus, the transition probabilities change as the Markov chain is operating. The characterization of the convergence of such a time-inhomogeneous Markov chain representing the simulated-annealing algorithms is based on the following notion.

Definition 4.8 (Reachable State) *Let s_i, s_j be two states in a time-inhomogeneous Markov chain representing a simulated-annealing algorithm. s_j is* **reachable from** s_i **at height** c *if there is a path in the generator chain of the simulated-annealing algorithm that leads from s_i to s_j and does not go through states whose cost exceeds c.*

Let s_i be a local but not a global minimum. The **depth** *of s_i is the smallest number d_i such that a state whose cost is smaller than c_i can be reached from s_i at height $c_i + d_i$.*

Definition 4.8 defines a quantity that in a certain sense measures the depth of wells in a configuration graph.

Definition 4.9 (Weakly Reversible Markov Chain) *A time-inhomogeneous Markov chain is called* **weakly reversible** *if, for any two states s_i and s_j, s_i is reachable from s_j at height c exactly if s_j is reachable from s_i at height c.*

Theorem 4.4 (Hajek [164]) *Assume that the generating chain is weakly reversible. Furthermore, let T_t be the temperature at time t and assume that the sequence $T_1, T_2, \ldots$ is decreasing and that*

$$\lim_{t \to \infty} T_t = 0$$

Then the time-inhomogeneous Markov chain representing the simulated-annealing algorithm converges to an optimizing distribution exactly if

$$\sum_{t=1}^{\infty} e^{-\frac{D}{T_t}} = \infty \tag{4.8}$$

Here D is the maximum depth of all states that are local but not global minima.

As an example, for the problem depicted in Figure 4.4, we have $D = 2$. Gidas [137] has presented similar necessary and sufficient condition for convergence of the time-inhomogeneous Markov chain, based on different premises.

The proof of this theorem is quite involved and requires arguments about continuous Markov chains [164]. The result, however, is intuitive. If T_t is small, then moves that increase the cost become more and more unlikely, so most of the time during the search is spent in the wells of local minima. The probability of accepting a sequence of cost-increasing moves at temperature T_t that start at the local minimum s_i and escape from its well is e^{-d_i/T_t}. Equation (4.8) gives the cumulative criterion on the size of all of these escape probabilities that ensures the convergence to a global minimum. The condition that the generating chain be weakly reversible is properly weaker than is requiring that of the configuration graph be symmetric, as it was done in the previous theorems.

The convergence criterion (4.8) is satisfied, for instance, if we choose

$$T_t = \frac{D}{\log(t + t_0)} \tag{4.9}$$

with $t_0 > 1$. This is a very slow cooling schedule, however, which is certainly not interesting in practice. Furthermore, the quantity D cannot be effectively estimated or computed for most combinatorial problems. Thus, Theorem 4.4 is a theoretical result that has few implications for the practical choice of cooling schedules. Note also that the result applies only to simulated-annealing algorithms, and not to probabilistic hill-climbing algorithms with other multiplicative acceptance functions.

4.3.2 Cooling Schedules

In the previous section, we saw that the convergence of simulated annealing to an optimizing distribution is ensured if the cooling schedule and the generating chain have certain properties. These properties range from the weak condition on the cooling schedule and the generating chain in Theorem 4.2 and its generalization, to the somewhat stronger criterion in Theorem 4.4.

Convergence results can make only rough structural statements about simulated annealing. If we want to analyze the speed of convergence, we have to take into account the *finite-time behavior* of simulated annealing. That is, we have to account for the fact that only a finite number of temperatures is considered, and the Markov chain corresponding to each temperature is applied for only a finite number of steps. The finite-time behavior of simulated annealing is determined by the cooling schedule and by the generating chain. Therefore, the analysis of the finite-time behavior must depend heavily on both of these components of the algorithm.

As of today, the analysis of the finite-time behavior of simulated annealing has not brought forth results that are strong enough to justify the choice of particular, practically interesting cooling schedules. Therefore, selecting the cooling schedule is still more an art than a science. In this section, we will summarize the general guidelines that have emerged in this art. They are based on experimental evidence and are not strongly supported by analysis. In Section 4.3.3, we will give an overview of what theoretical results are known on the analysis of the finite-time behavior of simulated annealing.

4.3.2.1 Elements of a Cooling Schedule

We will discuss the generating chain first, and then will look at the elements of the cooling schedule.

4.3.2.1.1 Generating Chain We assume that the stationary distribution of the generating chain is reasonably close to uniform. Sometimes, it is artificial to assume that the distribution is *exactly* uniform. Consider, for instance, Figure 4.4. Here, the generating chain is defined on the basis of uniform transition probabilities out of each state. Since the configuration graph is not regular, the stationary distribution $\pi = (3/22, 2/11, 2/11, 3/22, 2/11, 2/11)$ is not uniform. However, having a distribution that is close to uniform allows us to reach all states with about equal probability from a starting state, as long as the temperature is high. It would be desirable to exploit other properties of the generating chain in the analysis of the finite-time behavior of the simulated-annealing algorithm, to distinguish problems on which simulated annealing works well from those on which it does not. This task, however, seems to be quite difficult.

4.3.2.1.2 Initial Temperature The value of the initial temperature should be set high enough that virtually all transitions are accepted, and the generating chain dominates the behavior of the first stages of the algorithm. Let χ be the ratio of generated transitions that should be accepted. Starting from the Metropolis function, we can calculate a reasonable starting temperature using the formula

$$T = \frac{\Delta C}{-\ln \chi}$$

where ΔC is a lower bound on the average cost increase for any cost-increasing transition based on the stationary distribution and on the transition probabilities of the generating chain. Two suggestions have been made for the choice of ΔC. Johnson et al. [211] suggest generating several transitions at random, computing the average cost increase per generated transition during the experiment, and using this value as ΔC. (Transitions that decrease the cost contribute the value 0 to this random experiment.) Huang et al. [202] suggest using $\Delta C = 3\sigma$, where σ is the standard deviation of the cost function with respect to the stationary distribution of the generating chain. This value can be computed using a similar random experiment. Both choices are based on the same intuition of estimating the average cost increase of a cost-increasing transition.

As an example, consider again Figure 4.4. The average cost increase of all transitions is $\Delta C = 10/11$. Using an acceptance ratio of $\chi = 0.9$ yields the starting temperature $T = 8.6$. Increasing the acceptance ratio to $\chi = 0.95$ increases the starting temperature to $T = 17.7$.

4.3.2.1.3 Stopping Criterion The procedure frozen(T) decides when to stop cooling. Here, we can either fix the number of temperatures applied to a constant value of, say, 50, or stop when the average cost over all states reached by the Markov chain is about the same for three or four consecutive temperatures. Another criterion is to require that the acceptance ratio χ be below some prespecified threshold. We can also use the conjunction of the latter two stopping criteria. Huang et al. [202] propose stopping when the difference between the maximum and minimum cost at a temperature is of the same magnitude as the maximum cost change of any accepted transition at that temperature. Stopping criteria based on the variance of the cost distribution are discussed by Otten and van Ginneken [348].

4.3.2.1.4 Chain Length The chain length is determined by the procedure equilibrium($count, s, T$). Here, we want to stop the chain for temperature T if we can assume that the chain is reasonably close to the stationary distribution. In this case, we say that the chain reaches *quasiequilibrium*. Intuitively, a movement toward the stationary distribution is made with each accepted transition. Therefore, we apply some prespecified number h accepted transitions of the Markov chain and stop. As the temperature decreases, accepted transitions become increasingly rare, such that we also limit the number of *generated* transitions by some large constant L. Both h and L depend polynomially on the size of the configuration graph but not on the temperature. Huang et al. [202], and van Ginneken and Otten [139] and Lam and Delosme [257] propose more involved criteria for stopping the chain based on statistics and on information theory, respectively. See also Section 4.3.4.

4.3.2.1.5 Temperature Decrement The rule for decreasing the temperature is embodied in the procedure update(T). Intuitively, the next temperature should be chosen such that the stationary distribution of the Markov chain for the present temperature is not to far from the stationary distribution of the Markov chain for the next temperature. There are two ways of choosing the next temperature, a static and a dynamic one. In the static approach, we choose a rule of the form

$$T_{\text{next}} := \alpha \cdot T$$

where $0 < \alpha < 1$ and α is close to 1. Values between $\alpha = 0.8$ and $\alpha = 0.99$ are popular choices. In the dynamic approach, the choice of the next temperature is determined using the standard deviation of the cost distribution at the current temperature [2, 202].

Different experimentally tuned elements of cooling schedules are presented in [2, 139, 202, 211, 348].

4.3.2.1.6 Starting Solution The philosophy of simulated annealing suggests that we should choose a random starting solution to exploit the statistical properties of the method. In contrast, several researchers have experienced markedly improved behavior by carefully selecting a *good* starting solution [211]. In the absence of detailed theoretical knowledge on simulated annealing, the issue of whether to choose a random or a particular starting solution remains controversial.

4.3.2.2 Experimental Experience

Simulated annealing can produce good results in practice. However, the following qualifying results have been obtained with this method:

1. Simulated annealing works well only if we invest a large amount of computing time. It is essential to approximate accurately the stationary distribution for the Markov chain at each temperature. To do so, we need both small temperature decrements and long runs for the chains at each temperature. Furthermore, if we invest only a small amount of computing time, the results will be both worse with respect to cost and quite different in appearance. Therefore, simulated annealing is more applicable to production runs than it is to prototyping.

2. Any existing knowledge about the specific combinatorial structure of the optimization problem at hand should be worked into the simulated-annealing algorithm—specifically, into the generating chain. Many layout problems have structure based on locality or other phenomena of planar topology and geometry. Layout tools such as the *TimberWolf* set [406, 407, 408, 409] have greatly benefited from incorporation of structural knowledge in the generating chain. In general, simulated annealing should be used only as a last resort, where we do not have enough insight to solve a problem with problem-specific combinatorial understanding.

Simulated annealing has proved to be a powerful tool when it is applied judiciously and is augmented with all available problem-specific expertise. Results on applying simulated annealing to layout problems are presented in [156, 197, 206, 234, 347, 406, 407, 408, 409, 452, 473]. Since simulated annealing requires substantial computing time, parallel versions of the algorithm and their applications in layout design have been of interest [1, 19, 58, 68, 83, 115, 246, 374].

4.3.3 Finite-Time Behavior of Simulated Annealing

In practice, simulated annealing can yield impressive results; it does so, however, at the cost of a large amount of computing time. Determining effective cooling schedules has been an expensive (because experimental) and somewhat mysterious process. Thus, it is not surprising that intensive investigations are underway to prove or disprove the existence of cooling schedules that result in provably efficient algorithms for appropriate optimization problems. Both problem-dependent and problem-independent studies have been made. So far, both encouraging and discouraging results exist, but none of them are consequential enough to justify the use of the method theoretically or disprove its effectiveness.

Let us first discuss of what type of results we might obtain on the finite-time behavior of simulated annealing. We group such results in two classes. In the first class, we strive to find an optimum solution. Thus, we ask the question, how many search steps are necessary (and/or sufficient) for an optimum solution to be found with high probability? In the second class, we give up the demand for an optimum solution and require only that we find a solution that is reasonably close to optimum.

Results on the convergence of simulated annealing can imply results on the finite-time behavior if the rate of convergence is analyzed, as well. Gelfand and Mitter [134] have presented such an analysis. Mitra et al. [329] present similar results. Gelfand and Mitter use the temperature sequence (4.9) and show that, under certain premises, the probability of reaching an optimum solution in t steps is $p = 1 - O(t^{-\gamma/D})$, where D is defined as in Theorem 4.4, and γ is the smallest cost increase of a transition from a global minimum to a state that is not globally optimal. In particular, we can achieve $p \geq 1 - \varepsilon$ for any fixed $\varepsilon > 0$ within $t = O(1/\varepsilon)^{D/\gamma}$ steps. For this number to grow polynomially in the problem size, D/γ has to grow logarithmically in the problem size. However, practical optimization problems tend not to have configuration graphs that are "shallow" enough in this respect. Furthermore, D is quite difficult to compute or estimate. Thus, this result is not likely to lead to proofs of the efficiency of simulated-annealing algorithms on practical optimization problems.

Interestingly, the result obtained by Gelfand and Mitter generalizes to the case in which we require only that we come close to the optimal solution; for practical purposes, however, these results are equally nonconclusive.

The only results on a particular combinatorial problem known so far are by Sasaki and Hajek [393]. These authors consider the MAXIMUM MATCHING problem (see Section 3.12) and prove the following:

1. Any algorithm in a class of natural simulated-annealing algorithms for this problem takes exponential time in the expected case to find a global optimum.

2. There is a derivative of a natural simulated-annealing algorithm in which the temperature is fixed at a constant small value such that the algorithm finds a solution whose cost is within a factor of $1 - \varepsilon$ of optimal in polynomial time in the size of the problem (in the expected case). The exponent of the polynomial depends linearly on ε.

Thus, at least for the MAXIMUM MATCHING problem, aiming at the global optimum is infeasible, whereas coming close to it can be achieved even without cooling. Note also, that the MAXIMUM MATCHING problem can be solved in polynomial time. This shows that simulated annealing can be grossly inferior to other combinatorial algorithms.

Indeed, the theory of simulated annealing is a wide open area.

4.3.4 Probabilistic   Hill-Climbing   Without   Rejected Moves

As the temperature approaches zero, the probability that a move is accepted decreases rapidly. This relation hampers the efficiency of probabilistic hill-climbing algorithms severely.   Greene and Supowit [156] suggest a modification of the probabilistic hill-climbing scheme in which each generated move is also accepted. For this *rejectionless* version to be statistically equivalent to the *original* method, each move in the rejectionless method has to simulate a sequence of stationary moves in the original method that ends in a move in which the state changes. The necessary modification of the probabilistic hill-climbing algorithm is quite straightforward.

Assume that the generating chain is time-reversible, and that the acceptance function is multiplicative, such that the premises of Theorem 4.3 are fulfilled. To calculate the transition probabilities of the rejectionless method, we assume that we are currently in state s_i. The probability that we leave state s_i in the next move at temperature T is

$$\mu_i^{(T)} = \sum_{\substack{s_j \in N(s_i) \\ j \neq i}} g_{ij}\alpha(i, j, T)$$

where $\alpha(i, j, T)$ is the *acceptance criterion* (4.2). The probability that we eventually move from state s_i to state $s_j \neq s_i$ is

$$\sum_{k=0}^{\infty}(1 - \mu_i^{(T)})^k g_{ij}\alpha(i, j, T) = \frac{g_{ij}\alpha(i, j, T)}{\mu_i^{(T)}} =: \tilde{p}_{ij}^{(T)}$$

The rejectionless method makes the move from state s_i to state s_j with probability $\tilde{p}_{ij}^{(T)}$. Note that

$$\sum_{\substack{s_j \in N(s_i) \\ j \neq i}} \tilde{p}_{ij}^{(T)} = 1$$

Thus, the state is changed in each move of the rejectionless method. On the average, each move from state s_i in the rejectionless method simulates

$$\sum_{k=0}^{\infty} k(1 - \mu_i^{(T)})^{k-1} \mu_i^{(T)} = \frac{1}{\mu_i^{(T)}} \qquad (4.10)$$

moves of the original method, only the last of which is a nonstationary move. This intuition is supported by the fact that the stationary distribution $\tilde{\pi}_i^{(T)}$ of the rejectionless method assigns to state s_i the stationary probability

$$\tilde{\pi}_i^{(T)} = \pi_i^{(T)} \mu_i^{(T)} \qquad (4.11)$$

We can prove (4.11) by verifying the detailed balance equations (4.1) using Theorem 4.3. Since $\mu_i^{(T)} \to 0$ as $T \to 0$, $\tilde{\pi}_i^{(T)}$ need not converge to an optimizing distribution as $T \to 0$. If, however, we weigh $\tilde{\pi}_i^{(T)}$ with the average time of residence in state s_i in the original method (4.10), we obtain $\pi_i^{(T)}$. In this sense, the rejectionless method and the original method are statistically equivalent.

If, in the applied cooling schedule for the original algorithm, the length of the Markov chain at temperature T is determined by the number of accepted moves, the rejectionless method has to run the Markov chain for just this number of moves. If the cooling schedule uses the number of generated moves, some calculation of the number of moves of the original method simulated by the rejectionless method has to be performed. To apply the transition probability $\tilde{p}_{ij}^{(T)}$, we must compute the acceptance probabilities for transitions to all states in the neighborhood of s_i. Greene and Supowit [156] discuss efficient ways of doing this computation. The task does, nevertheless, involve considerable overhead. Thus, the application of the rejectionless method makes sense only if the temperature is sufficiently small that accepted moves are rare. In some cases, the change in the neighborhood after a nonstationary move can be captured efficiently. In such cases, the explicit evaluation of the transition probabilities over the whole neighborhood can be avoided in all but the first move, thus making the rejectionless method more efficient. Greene and Supowit [156] discuss an example from the area of circuit partitioning (see also Section 6.3).

4.4   Geometric Configuration Spaces and Polytopes

In Section 4.1, we introduced a graph-theoretic way of representing *discrete* (that is, finite or at most countably infinite) configuration spaces. Often, the more convenient representation is geometric. Here, configurations are points in m-dimensional Euclidean space. In this case, the set of configurations can, in principle, be uncountably infinite. We will not be interested in such *continuous* optimization problems per se. It turns out, however, that continuous optimization methods based on geometric concepts are quite helpful in solving some discrete optimization problems. We will discuss this subject in more detail in this and the following sections.

A particular geometric structure that is of interest in optimization is the *convex polytope*. A wide body of knowledge from polyhedral combinatorics is available for dealing with polytopes. A special class of discrete optimization problems whose configuration spaces have polyhedral structure comprises problems in which the configurations are subsets of a common universe.

Definition 4.10 (Set-Construction Problem) *Let Π be a discrete optimization problem such that, for each instance p, S_p is a subset of the power set 2^{U_p} of some finite universe U_p. In this case, we call Π a **set-construction problem**.*

All optimization problems at which we have looked so far are strict set-construction problems. In the MINIMUM STEINER TREE problem and the MINIMUM SPANNING TREE problem, U_p is the set of edges of the graph. We can also modify the definition of the TRAVELING SALESPERSON problem syntactically so as to make each tour also a set of edges, thus getting a set-construction problem with the same universe.

It is convenient to view the configurations $s \subseteq U_p = \{u_1, \ldots, u_n\}$ of an instance of a set-construction problem as characteristic vectors $s = s_1 \ldots s_n$ such that

$$s_i = \begin{cases} 1 & \text{if } u_i \in s \\ 0 & \text{otherwise} \end{cases}$$

Each of the characteristic vectors is a string of n 0s and 1s and can thus be regarded as a corner of the n-dimensional unit cube. As such, a set of points S_p spans a convex polytope. To discuss this phenomenon in detail, we survey the fundamental definitions underlying polyhedral combinatorics.

4.4.1   Convex Sets and Functions

Definition 4.11 (Convex Sets and Polytopes) *A set P in n-dimensional Euclidean space $\mathbf{R}^n$ is called **convex** if which each pair of points $s, t \in P$ all convex combinations $\lambda s + (1 - \lambda)t$, $0 < \lambda < 1$, are also in P. A point v in a* *convex set P such that there is no pair of vertices $s, t \in P$ with $v = \lambda s + (1-\lambda)t$, $0 < \lambda < 1$ is called an* **extreme point** *of P.*

The **convex hull** $CH(P)$ of a set P of points is the smallest convex set that contains P. If $P' = CH(P)$, then P **spans** P'.

A set that is spanned by a finite point set is called a **convex polytope**. The extreme points of a convex polytope are its **corners**.

Definition 4.12 (Affine Subspaces) *An* **affine subspace** *H is a linear subspace S translated by a vector b, $H = S + b$. Equivalently,*

$$H = \{s \in \mathbf{R}^n \mid A \cdot s = b\}$$

for some $m \times n$ matrix A, $m > 0$, and some n-vector b.

The **dimension** *of H is the dimension of S, or equivalently, $n - \mathrm{rank}(A)$. The* **dimension** *of a convex subset $P \in \mathbf{R}^n$ is the dimension of the smallest affine subspace containing P.*

Intuitively, a set is convex if it does not have any indentations; that is, if, given two points in P, the total straight-line segment between those points is also in P. The extreme points of P are points that do not lie in the interior of any straight-line segment that is contained in P. Thus, all extreme points lie (in a topological sense) on the boundary of P, but the boundary of P may contain points that are not extreme points. Figure 4.5 shows examples. We give the following basic geometric statements without proof.

Lemma 4.3

1. *A point v is in $CH(P)$ exactly if there is a finite number of points $s_1, \ldots, s_k$ in P such that $v = \sum_{i=1}^{k} \lambda_i s_i$ with $\sum_{i=1}^{k} \lambda_i = 1$ and $0 \leq \lambda_i \leq 1$ for all $i = 1, \ldots, k$. In fact, k can be restricted to $k \leq n + 1$.*

2. *Let X be the set of extreme points of the convex set P. Then, for each $x \in X$, we have $x \notin CH(X \setminus \{x\})$.*

3. *Each convex polytope is spanned by its extreme points.*

Definition 4.13 (Half-Space) *A* **half-space** *S is the set of points $s \in \mathbf{R}^n$ that satisfy an inequality of the type $a \cdot s \leq b$. Here, a is an n-vector, the dot indicates the inner product, and b is a real number.*

The set H of points s in the half-space that fulfill $a \cdot s = b$ is the **hyperplane** **supporting the half-space**. *Note that a hyperplane is an affine subspace of dimension $n - 1$.*

A hyperplane $\{x \mid a \cdot x = b\}$ *separates two points s, t if one point lies on either side of the hyperplane; that is, if $a \cdot s < b$ and $a \cdot t > b$.*

There is a fundamental connection between convex polytopes and half-spaces.

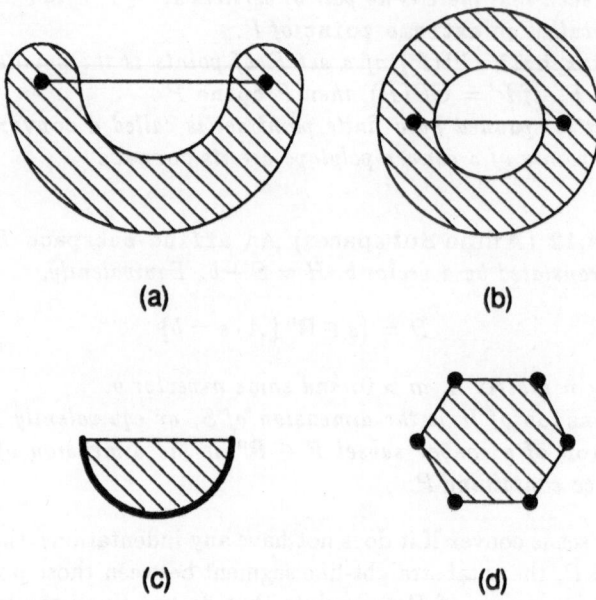

Figure 4.5: *Example of convex and nonconvex subsets of* $\mathbf{R}^2$. *(a and b) Nonconvex sets. (c) A convex set. (d) A convex polytope. In (a) and (b), a straight-line segment is depicted whose endpoints lie in the set, but that is not totally contained in the set. In (c) and (d), the set of extreme points is highlighted.*

Lemma 4.4 *A set P is a convex polytope exactly if P is bounded and P is the intersection of a finite set of half-spaces.*

Lemma 4.4 suggests the following definition.

Definition 4.14 (Convex Polyhedron) *A* convex polyhedron *is the intersection of finitely many half-spaces.*

A convex polyhedron need not be bounded. The first quadrant Q_1 in the plane is an example of a convex polyhedron. The definition of Q_1 as an intersection of half-spaces is

$$Q_1 = \{(x, y) \mid x \geq 0\} \cap \{(x, y) \mid y \geq 0\}$$

The half-spaces whose intersection is a convex polyhedron P can be visualized as follows.

Definition 4.15 (Supporting Hyperplane) *Let S be a half-space that contains the convex polyhedron P and whose supporting hyperplane H has a nonempty intersection with P. S and H are said to* support *P.*

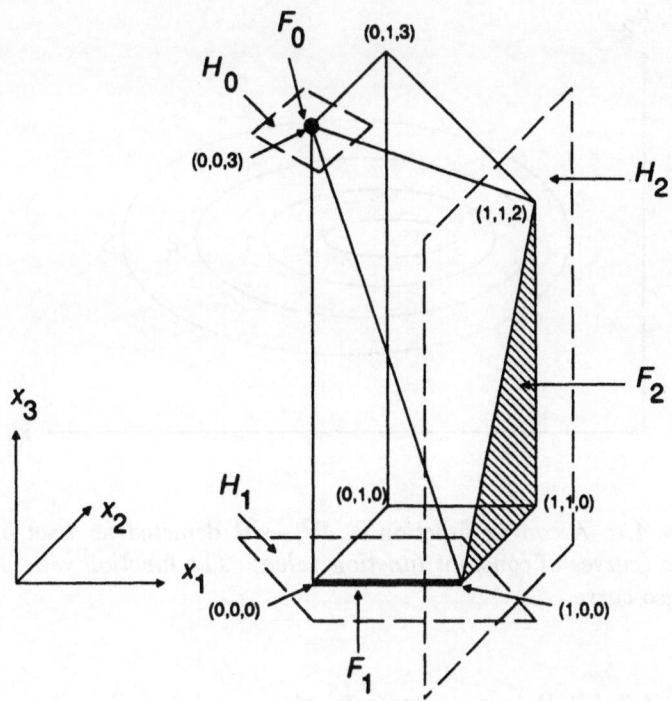

Figure 4.6: *Faces of a polytope in* $\mathbf{R}^3$. *For* $i = 0, 1, 2$ *the face* F_i *is defined by the supporting hyperplane* H_i. *Face* F_i *has dimension* i.

Definition 4.16 (Facets) *Let* H *be a hyperplane supporting the n-dimensional polyhedron* P. *The set* $H \cap P$ *is called a* **face** *of* P. *A face that has dimension* $n - 1$ *is called a* **facet**. *The* **corners** *of* P *are the faces of dimension 0. A face of dimension 1 is called an* **edge** *of the polyhedron (see Figure 4.6). Two corners of the polyhedron that lie on the same edge are called* **adjacent** *or* **neighbors**.

Figure 4.6 illustrates Definition 4.16. For convex polytopes, the definitions of a corner given in Definition 4.11 and 4.16 coincide.

The following theorem is quite intuitive.

Theorem 4.5 *The faces of a polyhedron* P *are again polyhedra. The faces of faces of* P *are also faces of* P.

There is an important statement on convex polyhedra that generalizes to arbitrary closed convex subsets of $\mathbf{R}^n$. (For an illustration, see Figure 4.9.)

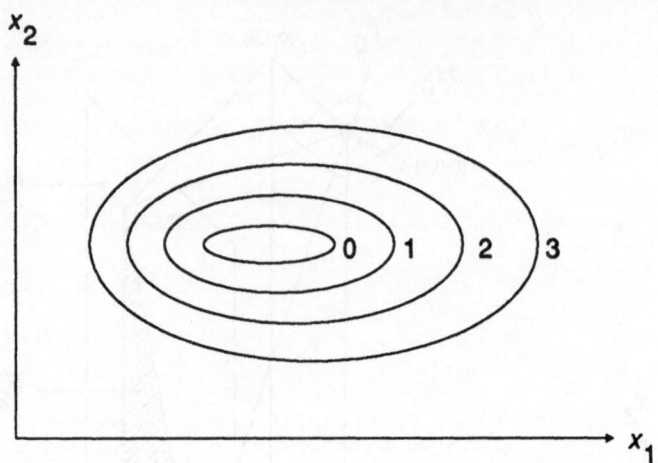

Figure 4.7: *A convex function $c : \mathbf{R}^2 \to \mathbf{R}$ depicted as a set of iso-curves (curves of constant function value). The function value labels each iso-curve.*

Theorem 4.6 *Let P be a convex polyhedron, and $s \notin P$. Then, there is a hyperplane H that separates s from (all points in) P.*

We now establish the connection between set-construction problems and polytopes.

Lemma 4.5 *Let Π be a set-construction problem. The configurations of an instance p of Π are the corners of an m-dimensional polytope.*

Proof By Lemma 4.3, part 3, we have to show that no configuration of p is the convex combination of the other configurations. This statement follows directly from the fact that all configurations are binary vectors. $\square$

There is a natural way in which a polytope induces a configuration graph: We just define the (bidirectional) edges to be the edges of the polytope. The resulting graph is called the *skeleton*. If, in addition, we add a cost function $c(s)$ on the corners of the polytope, we have represented a set-construction problem as a discrete optimization problem on the corners of a polytope. Often, the cost function $c(s)$ can be written as $c(s_1, \ldots, s_n)$, where $s_1, \ldots, s_n$ are the components of the characteristic vector representing the configuration s. But then the cost function can be naturally extended to a function $c : \mathbf{R}^n \to \mathbf{R}$. The properties of the *skeleton* depend on the properties of the cost function c.

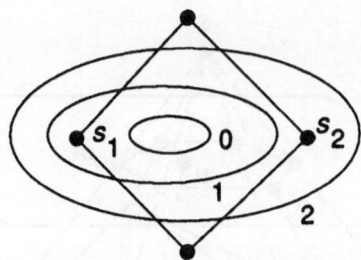

Figure 4.8: *A convex polytope and a convex cost function (represented by a set of iso-curves) such that the skeleton of the polytope is not convex with respect to the given cost function. The vertices s_1 and s_2 are two local minima, but only s_1 is a global minimum.*

Definition 4.17 (Convex Function) *A function $c : P \to \mathbf{R}$ on a convex set P is* **convex** *if, for all $s, t \in S$, we have $c(\lambda s + (1 - \lambda)t) \leq \lambda c(s) + (1 - \lambda)c(t)$. A function $c : P \to \mathbf{R}$ on a convex set P is* **concave** *if $-c$ is convex.*

Theorem 4.7 *If P is a convex set, $c : P \to \mathbf{R}$ is a convex function, and $t \in \mathbf{R}$, then $\{s \in \mathbf{R}^n | c(s) \leq t\}$ is a convex set.*

By Theorem 4.7, a convex function can be illustrated as in Figure 4.7. The following lemma shows that greedy search is applicable to convex functions that are defined on convex sets.

Theorem 4.8 *Let $c : P \to \mathbf{R}$ be a convex function defined on the convex set P. Then, each local minimum (as defined in $\mathbf{R}^n$) of c on P is also a global minimum.*

We will discuss optimization problems with a convex cost function in Section 4.8. Unfortunately, Theorem 4.8 does not extend to the case where P is only the set of *corners* of a convex polytope, and the neighborhood is defined via the skeleton of the polytope. Figure 4.8 shows a counterexample; that is, a skeleton that is not convex with respect to a convex cost function. If the cost function is *linear*, however—that is, if, for a corner $s = (s_1, \ldots, s_n) \in \mathbf{R}^n$, $c(s) = \sum_{i=1}^{n} c_i s_i$ for appropriate values $c_i \in \mathbf{R}$—then the convexity of the skeleton can be proved.

Theorem 4.9 *Let P be a polytope. Assign costs to the corners of P with respect to a linear cost function c. Then the skeleton of P is a convex configuration graph.*

Proof Let s, t be two different corners of P, and let $c(s) > c(t)$. We have to show that there is a neighbor s_i of s with $c(s_i) < c(s)$. Assume, indirectly, that

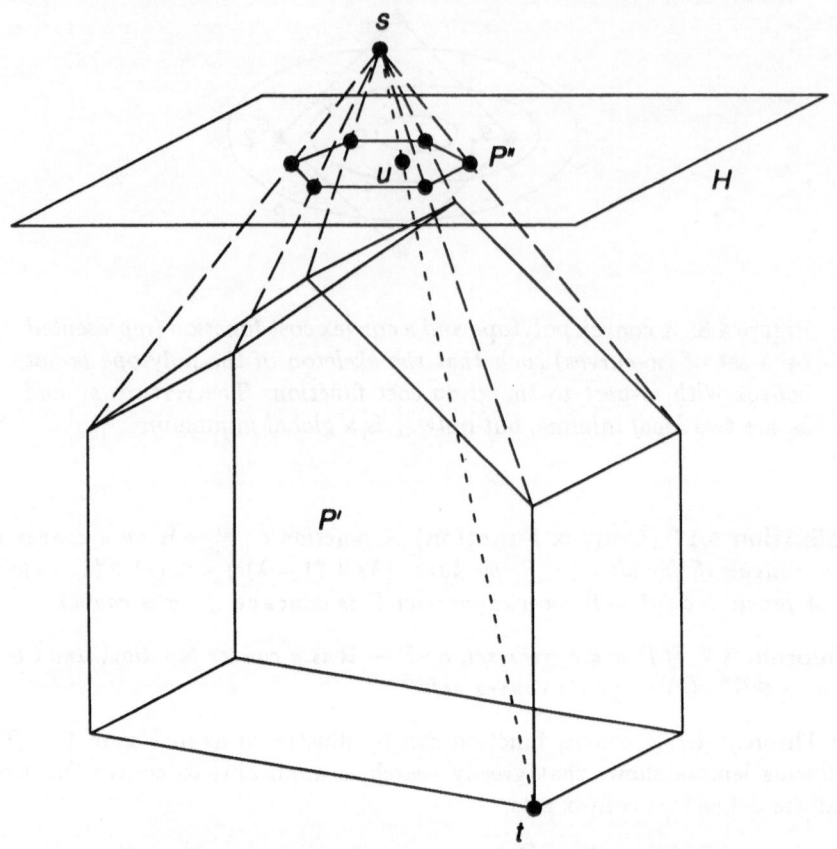

Figure 4.9: *A hyperplane H separating a point $s \notin P'$ from a convex polytope P'. The convex hull of $P' \cup \{s\}$ is the convex polytope P. The intersection of H with P is another convex polytope P''. Note that the corners of P'' are exactly the intersection points of H with the edges of the skeleton of P that are incident to s.*

this is not the case. Then, t is not a neighbor of s in P. Let P' be the polytope spanned by all corners of P except s (see Figure 4.9). By Theorem 4.5, there is a hyperplane H that separates s from P'. Let $P'' = P' \cap H$. P'' is again a convex polytope. The corners of P'' are exactly the intersection points of H with the edges of the skeleton of P that are incident to s (Exercise 4.11). Now consider the intersection point u of H with the straight-line segment from s to t. Clearly, $u \in P''$. Since the cost function c is linear, we have $c(u) < c(s)$. At the same time, however, by the preceding characterization of the corners of

P'', for all these corners, s_i, $c(s_i) > c(s)$, which is a contradiction. $\square$

By Theorem 4.9, we can use greedy search methods on the skeleton to find the optimum corner of P with respect to a linear cost function. Greedy optimization on polytopes is a well-developed field, and we will devote to it a special section (Section 4.5).

Definition 4.18 (Linear Set-Construction Problem) *We call a set-construction problem* linear *if its cost function is of the form* $c(s) = \sum_{i=1}^{n} c_i s_i$, *where* $s = (s_1, \ldots, s_n)$ *is the binary vector representing the configuration and* $c_i \in \mathbf{R}$.

All set-construction problems that we have mentioned so far, and most of the ones that we will consider in this book, are linear set-construction problems. Thus, the results we have surveyed and will continue to survey on polytopes apply directly to them. Figure 4.10 illustrates the concepts linking set-construction problems with polytopes.

The book by Grünbaum [157] is an excellent reference on convex sets and polytopes.

4.5 Linear Programming

In the preceding section, we considered a kind of discrete optimization problem, namely, set-construction problems. We represented the configurations of such a problem with the corners of a convex polytope. We noted that the cost function can often be extended naturally to a function on $\mathbf{R}^n$; in fact, most often this function is linear. Thus, the corresponding optimization problem can be considered not only as a discrete optimization problem (on the corners of the polytope), but also as a continuous optimization problem (on all points inside the polytope). This problem is called a *linear programming* problem or, for short, a LINEAR PROGRAM.

LINEAR PROGRAM

Instance: A real $m \times n$ *constraint matrix* A, a real m-vector b, and a real n-vector c

Configurations: All real n-vectors

Solutions: All real n-vectors x that satisfy the *linear constraints* $A \cdot x \leq b$ and $x \geq 0$

Minimize: $c^T \cdot x$

The discrete and continuous optimization problem on a convex polytope are equivalent.

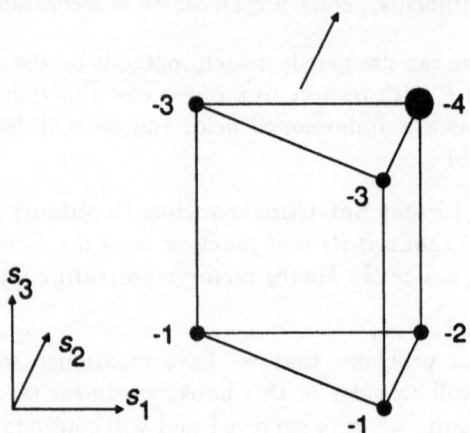

Figure 4.10: *The convex polytope of the following set-construction problem. The universe is $U = \{u_1, u_2, u_3\}$. The configuration space is $S = \{\{u_1\}, \{u_2\}, \{u_1, u_2\}, \{u_1, u_3\}, \{u_2, u_3\}, U\}$. The linear cost function is $c(s) = -s_1 - s_2 - 2s_3$. In the figure, the cost function is illustrated by a vector that points in the direction of steepest descent. Corners of the polytope are labeled with their cost. The minimum-cost corner of the polytope is highlighted.*

Lemma 4.6 *Let P be a polytope and $c(x) = c \cdot x$ be a linear cost function. Then, the minimum cost on P with respect to c is attained in a corner of P.*

Proof Assume that P has corners s_i, $i = 1, \ldots, \ell$, and that the s_i are ordered according to increasing cost $c_i = c(s_i)$. An arbitrary point $x \in P$ can be written as a convex combination $x = \sum_{i=1}^{\ell} x_i s_i$, with $\sum_{i=1}^{\ell} x_i = 1$. The cost of x is $c(x) = \sum_{i=1}^{\ell} c_i x_i$. Obviously, the cost is minimized by maximizing x_1. Thus, s_1 is a point of minimum cost in P. $\square$

Because of Lemma 4.6, we can use linear-programming techniques to solve discrete optimization problems such as set-construction problems. This is why we are interested in linear programming.

In this section, we will summarize results on linear programming that serve to exhibit the connections of this subject to the optimization problems considered in this book. For more detailed accounts of linear programming and the simplex algorithm, we refer to textbooks such as [350].

To adjust the notation, we remark that in the context of linear programming what we call a legal configuration is called a *feasible solution*. An illegal configuration is called an *infeasible solution*. The cost function—that is, the vector c—is also called the *objective function*. A is called the *constraint matrix*; b is the *constraint vector*.

Theorem 4.10

1. *The set of feasible solutions of a linear program is either empty (in which case the program is* infeasible*) or a convex polyhedron in $\mathbf{R}^n$. In the latter case, if there is no feasible solution with minimum cost, then the polyhedron is unbounded, and the linear program is also called* unbounded. *If there is a feasible solution with minimum cost, then the polyhedron may or may not be bounded. In the latter case, the linear program can be modified such that the set of optimum solutions does not change, but the polyhedron is bounded. We will be interested in only this case. The set of feasible solutions may be degenerate; that is, it may be contained in a subspace of $\mathbf{R}^n$ with dimension less than n.*

2. *If the linear program is feasible, the inequalities of the system $A \cdot x \leq b$ define half-spaces that support the set of feasible solutions. Thus, they define faces of the polyhedron. However, a face can correspond to several inequalities in this way. An inequality that corresponds to a facet is called* facet-inducing. *Deleting the inequalities that are not facet-inducing does not change the polytope of the linear program, but deleting all inequalities that induce a given facet does.*

3. *The number of corners of the polyhedron of feasible solutions can be as large as $\binom{n}{m}$.*

4. *The most popular algorithm for solving linear programs is the simplex algorithm. The simplex algorithm performs a greedy search on the skeleton of the polytope of a linear program. (Thus, by part 3, its running time may be exponential.)*

5. *A linear program can also be defined in any of the following ways:*

   a. *We may substitute "$\leq$" with "$\geq$" or "$=$" in some of the inequalities in $A \cdot x \leq b$.*

   b. *Some of the variables in x may be unrestricted or restricted to be nonpositive.*

   c. *The objective function $c^T \cdot x$ may be maximized.*

   Any of these forms of linear programs can be transformed into any other form without asymptotic increase in the size of the linear program.

6. *Each linear program L has a dual linear program $\overline{L}$ that is obtained as depicted in Figure 4.11. In this context, L is also called the* primal. *Figure 4.11 depicts how the transformation is made.*

Primal		Dual		
minimize $c^T \cdot x$		maximize $b^T \cdot \pi$		
$a_i \cdot x$	$=$ b_i	π_i		unrestricted
$a_i \cdot x$	$\geq$ b_i	π_i	$\geq$	0
x_j	$\geq$ 0	$A_j^T \cdot \pi$	$\leq$	c_j
x_j	unrestricted	$A_j^T \cdot \pi$	$=$	c_j

Figure 4.11: *A pair of dual linear programs L and $\overline{L}$. We obtain the dual $\overline{L}$ from the primal L by substituting the elements of the primal (left column) with the respective elements of the dual (right column). The variable vectors of L and $\overline{L}$ are x and π, respectively. The ith row of the constraint matrix A of L is denoted with a_i. The jth column of A is denoted with A_j.*

7. *The dual of the dual is the primal.*

8. *If L is infeasible (bounded, unbounded), then $\overline{L}$ is unbounded (bounded, infeasible).*

9. *Let L and $\overline{L}$ be a primal-dual pair of the form depicted in Figure 4.11. Let x be an n-vector and π be an m-vector.*

   a. *If x is a feasible solution of L and π is a feasible solution of $\overline{L}$, then $c^T \cdot x \geq b^T \cdot \pi$; that is, the cost of x is greater than the cost of π.*

   b. *$c^T \cdot x = b^T \cdot \pi$ exactly if x and π are optimal solutions of L and $\overline{L}$, respectively.*

   c. *(Complementary slackness conditions) If x is a feasible solution of L and π is a feasible solution of $\overline{L}$, then x and π are optimal solutions of their respective linear programs exactly if the following holds:*

$$\pi^T \cdot (A \cdot x - b) = 0$$
$$(c - A^T \cdot \pi)^T \cdot x = 0$$

   Thus, if a constraint in the primal is binding, the corresponding variable in the dual is unrestricted, and vice versa.

10. *Let the matrix A in the linear program with constraint system $A \cdot x = b$ and $x \geq 0$ have full rank. Then, each corner of the polytope of the linear program is given by $B^{-1} \cdot b$, where B is a nonsingular submatrix of A.*

We have seen in Lemma 4.6 that linear programs are, in effect, discrete optimization problems. Theorem 4.10, part 3, however, shows that the configuration space can be very large. Thus, exhaustive search on the corners of the polytope does not qualify as an appropriate solution technique.

There is a problem with evaluating the efficiency of an algorithm for linear programming. This problem occurs because, so far, we have assumed the numbers to be real; however, we were not explicit about how to devise an encoding for them. In general, since we can encode the numbers in only a finite string, we have to deal with rational numbers of some finite precision. With this observation, we can now ask the question about the efficiency of linear programming precisely. We assume that the coefficients of the linear program are all integer. (If they are rational, then we multiply with the least common denominator of all of them. This is just a scaling step that can be inverted after we solve the linear program. The problems of numerical stability that enter the discussion if we have to deal with linear programs that have coefficients of widely differing sizes represented in floating-point notation are not discussed here.) The question is then whether we can find an optimal solution of the linear program whose numbers are rationals that are represented by their denominator and numerator in polynomial time in the length of the encoding of the linear program. In principle, this may be possible, since the optimal solution of a linear program involving only integers is a vector of rational numbers. However, it could be that, to encode these numbers, we have to use such a large precision that the encoding of the solution has a length that is exponential in the length of the encoding of the linear program. Even if this were not the case, it might take exponential time to find the optimal solution.

The simplex algorithm is not a good candidate for a polynomial-time algorithm. Even though it runs in almost linear time on most practical examples, we have seen that the polytope may have an exponential number of corners in the length of the linear program. There is still some nondeterminism in the simplex algorithm. Resolving it means choosing the exact greedy search path through the skeleton. To date, however, nobody has found a way of choosing the path that ensures a polynomial run time of the simplex algorithm. (Nobody has proved that there are polytopes on which this is not possible, either.) So, whether linear programming can be done in polynomial time was for a long time an open question.

In 1979, Khachian [232] proved that linear programming can in fact be done in polynomial time. Khachian's method is called the *ellipsoid method*. It has a strong geometric flavor. We will not describe the method in detail here (see [350]). We just mention that the method can be shown to run in polynomial time because it uses a polynomial number of steps and it uses only polynomial precision (even though it performs real arithmetic in this precision). In practice, this method cannot compete with the simplex algorithm. It can be extended, however, to yield polynomial-time algorithms for the optimization of more general convex functions on more general convex sets [295].

More recently, Karmarkar [215] has developed a polynomial-time algorithm for linear programming that is also claimed to be competitive with, if not better than, the simplex algorithm on most practical examples. Karmarkar starts with a point in the interior of the polytope and iteratively finds new interior points that in polynomial time converge in an optimal solution. For this reason, his method is called an *interior-point method*. Interior-point methods seem to be able to combine polynomial complexity in the worst case with competitive behavior in typical cases. One distinct advantage of the simplex algorithm that interior-point methods do not have is robustness. Due to the fact that the simplex algorithm is nothing but a greedy search on the skeleton, and that its search steps are based on fundamental concepts of linear algebra, the simplex algorithm behaves nicely if the linear program is changed only slightly. Then, we do not have to run the algorithm again from the start, but rather can do incremental computation to adjust to the new linear program. Also, there are versions of the simplex algorithm that tie together the primal and the dual. Furthermore, at any time in the simplex algorithm, we have complete information about the corner of the polytope in which we are currently located. This is very helpful if we want to modify the search by means of other heuristics we might develop.

4.6   Integer Programming

Often, we have to add to the constraints of a linear program the restriction that all variables be integer (*integrality constraint*). Then we get an integer program.

INTEGER PROGRAM

> *Instance:* A real $m \times n$ matrix A, a real m-vector b, and a real n-vector c
>
> *Configurations:* All integer n-vectors
>
> *Solutions:* All integer n-vectors x that satisfy $A \cdot x \leq b$ and $x \geq 0$
>
> *Minimize:* $c^T \cdot x$

There are several other varieties of integer programs:

- In a 0, 1 *integer program*, all integer variables are restricted to be binary.

- In a *mixed integer program*, only some of the variables are restricted to be integer.

- In a *mixed* 0, 1 *integer program* some of the variables are binary, whereas the others are real.

The linear program that we get if we omit the integrality constraint is called the *linear relaxation* of the integer program.

Integer and linear programs are highly correlated. Therefore, it may at first be surprising that, whereas linear programming can be done in polynomial time, integer programming is an NP-hard problem [133]. The intuitive reason is that linear programming is an inherently continuous problem. The faces and, in particular, the corners of the polytope of a linear program come about by calculations involving only linear combinations of components of the linear program. Such calculations can be done quickly, as Khachian proved.

For an integer program, the polytope of the linear relaxation provides only a hull in which all integer vectors have to be contained that are solutions to the integer program. The corners of this polytope, however, are not integer vectors in general. Thus, we cannot find the optimal solution just by optimizing over corners of the polytope. Rather, the part of the integer lattice that is contained in the polytope has to be searched. This set is determined by additional phenomena, such as rounding. Therefore, its structure is considerably more complicated, and searching through it becomes a hard problem.

Even though integer programming is NP-hard, the close relation to linear programming helps us in many ways to find a good solution to integer programs. We will now discuss several such methods.

4.6.1 Polytopes of Integer Programs

Let us consider integer programs with a finite configuration space—that is, integer programs that are *bounded*. The configuration space of a bounded integer program spans a convex polytope, which is called the *polytope of the integer program*. In general, the polytope P_I of the integer program I is not the same as the polytope P_L of its linear relaxation L , but $P_I \subseteq P_L$ always holds.

By Theorem 4.10, part 2, each facet in the polytope P_L is induced by a constraint of the linear program L. Thus, there can be only as many facets in P_L as there are constraints in L. In contrast, the polytope P_I can have a very large number of facets that may be exponential in the number of constraints in I. This is the polyhedral version of expressing the fact that polytopes of integer programs are much more complicated than are polytopes of linear programs.

As an example, the TRAVELING SALESPERSON problem can be expressed as an integer program as follows. There are n^2 variables x_{ij}, $i, j = 1, ..., n$, one for each edge (v_i, v_j). The cost vector is the vector of all edge lengths. The variables are constrained to be $0, 1$. The variable x_{ij} is 1 exactly if the edge (v_i, v_j) is included in the tour. Thus, we have a $0, 1$ integer program. However, we can also make the 0, 1 constraint implicit, as follows:

$$\sum_{j=1}^{n} x_{ij} = 1 \quad \text{for all } i = 1, \ldots, n$$

$$\sum_{i=1}^{n} x_{ij} = 1 \quad \text{for all } j = 1, \ldots, n \tag{4.12}$$

$$x_{ij} \text{ integer} \quad \text{for all } i, j = 1, \ldots, n$$

The constraints (4.12) ensure that the variables are all 0, 1, and that the indegrees and outdegrees of all vertices are 1 in each tour. We still have to ensure by additional constraints that the tour is a single circuit, not a collection of disjoint circuits. We do that by adding only $(n-1)^2$ more constraints, as follows [323]. The constraints

$$u_i - u_j + n x_{ij} \leq n - 1 \quad \text{for all } i, j = 2, \ldots, n \tag{4.13}$$

where u_i and u_j are additional integer variables exclude all tours that do not contain vertex v_1. Thus, after incorporation of these constraints, all tours must consist of a single circuit. Furthermore, the constraints (4.13) still allow for all such tours: If v_i is the kth vertex visited after v_1, then setting $u_i = k$ satisfies all constraints (see also Exercise 4.15). In fact, the polytope of the TRAVELING SALESPERSON problem has an exponential number of facets and, intuitively, that is what makes the problem NP-hard [263].

In general, it is not a good idea to solve an integer program I by solving its linear relaxation L and then rounding the solution to the nearest integer. Figure 4.12 shows that this may not even yield a feasible solution of I. In fact, the minimum cost of L may be far below the minimum cost of I, as Figure 4.12 also illustrates. In Section 8.6, we will discuss special cases of integer programs, where rounding is effective. In general, however, this is not the case.

One way of solving an integer program I is to start with the polytope P_L of its linear relaxation L and to add new constraints to L that cut off parts from P_L but not from P_I. In the best case, those constraints are facet-inducing constraints of the polytope P_I of the integer program. To do this, we solve L. If the optimal solution s_0 is integer, we are finished. Otherwise, we generate a (we hope facet-inducing) constraint of I whose corresponding hyperplane separates s_0 from P_I. We add this constraint to L and solve again. In this way, we proceed until the optimal solution of the augmentation of L is integer. Algorithms that follow this strategy are called *cutting-plane algorithms* because they induce hyperplanes to cut off excess from the polytope P_L. Gomory [148, 149] introduced a systematic way of solving any linear program in this way, but his method is not efficient enough in general to be used in practice.

Recently, special integer programs have been studied in detail with the objective of quickly finding facet-inducing constraints that separate a given point from a polytope. The most detailed knowledge of this kind has been acquired about the polytope of various variants of the TRAVELING SALESPERSON problem [263, Chapters 8,9]. Here, several classes of constraints have been identified

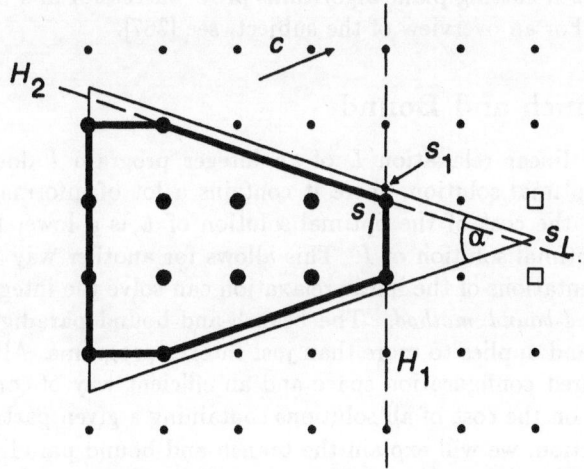

Figure 4.12: *An integer program I and its linear relaxation. P_I is outlined with thick black lines, P_L is outlined with normal-width lines. The cost function c is indicated by a vector pointing in the direction of steepest descent. The integer points closest to the optimal solution s_L of L are denoted by hollow squares. Both of these points are infeasible solutions of I and, in fact, of L. The optimal solution of I is s_I. As the angle α is decreased, s_L and s_I separate more. Adding the facet-inducing inequality corresponding to the hyperplane H_1 converts P_L into a polytope whose optimal solution is s_1. Adding a second inequality corresponding to H_2 leads to a polytope that has s_I as the optimal solution.*

that result from simple combinatorial properties of tours, and the following has been exhibited about them:

- Each such constraint is facet-inducing.

- Given a point, it can be decided whether x violates any such constraint, and, if it does, then a constraint can be found quickly that supports this fact.

These constraints are used as a basis for the cutting-plane algorithm. If we do not find an optimal solution quickly, we resort to branch-and-bound methods to do the final phase of the solution (see Section 4.6.2).

For the TRAVELING SALESPERSON problem, the progress that has been made with this method is tremendous. However, it remains to enlighten the structure of the polytopes of other important integer programs and to show that these customized cutting-plane algorithms prove successful in a broad range of applications. For an overview of the subject, see [367].

4.6.2   Branch and Bound

Although the linear relaxation L of an integer program I does not always lead to the optimal solution, there it contains a lot of information about I. In particular, the cost of the optimal solution of L is a lower bound on the cost of the optimal solution of I. This allows for another way that carefully chosen augmentations of the linear relaxation can solve the integer program— the *branch-and-bound method*. The branch-and-bound paradigm is actually very general and applies to more than just integer programs. All it requires is a tree-structured configuration space and an efficient way of computing tight lower bounds on the cost of all solutions containing a given partial solution.

In this section, we will explain the branch-and-bound paradigm in general terms. We will give several examples, among them integer programming.

Definition 4.19 (Branch and Bound) *The configuration space of a minimization problem is* **tree-structured** *if we can define a rooted tree T with whose vertices z (which we call* **nodes***) we can associate subsets S_z of configurations such that the following is true:*

1. *If r is the root of the tree, then S_r contains an optimal solution. (Mostly, we ensure that this is true by choosing S_r to be the total configuration space.)*

2. *If z is an interior node in T and $z_1, \ldots, z_k$ are the children of z, then the sets $S_{z_1}, \ldots, S_{z_k}$ cover the set S_z. (That is, $\cup_{i=1}^{k} S_{z_i} = S_z$. The method is most efficient if the $S_{z_1}, \ldots, S_{z_k}$ partition S_z.)*

3. *If z is a node in T, then there is an efficiently computable lower bound value L_z such that $L_z \leq c(s)$ for all configurations $s \in S_z$. The tighter the lower bound L_z is, the more efficiently the branch-and-bound method will work. $L_z = \infty$ implies that $S_z = \emptyset$.*

4. *If z is a node in T, then there is an efficiently computable upper-bound value U_z such that either $U_z = \infty$ (trivial upper bound) or $U_z = c(s_z)$ for some $s_z \in S$. (s_z is not necessarily a configuration in S_z. s_z is called the* **sample solution** *for z.) Furthermore, if z is a leaf, then $U_z \leq L_z$ has to be the case.*

4.6.2.1   A Tree-Structured Formulation of Integer Programming

Let I be an integer program and let L be its linear relaxation. We assume that L is bounded. The tree T can be constructed as follows.

With each node z in T, we associate an integer program $I^{(z)}$. The set S_z is the set of feasible solutions of $I^{(z)}$. We set $I^{(r)} = I$.

We now discuss the other elements of the tree T; that is, the bounds and the tree edges. Let $L^{(z)}$ be the linear relaxation of $I^{(z)}$. We distinguish several cases:

Case 1: $L^{(z)}$ is unsolvable. In this case, $U_z = L_z = \infty$, and z is a leaf.

Case 2: $L^{(z)}$ is solvable. Let s_z be the optimal solution of $L^{(z)}$ obtained by the employed optimization procedure. We set $L_z = c(s_z)$.

  Case 2.1: s_z is integer. In this case, s_z is the sample solution for z, and $U_z = c(s_z)$. The node z is again a leaf.

  Case 2.2: Otherwise, there is a variable $x_i = \alpha_i$ in s_z such that α_i is not integer. In this case, $U_z = \infty$. Furthermore, z has two children $z_{(<)}$ and $z_{(>)}$. We obtain the integer program $I^{(z_{(<)})}$ from $I^{(z)}$ by adding the constraint $x_i \leq \lfloor \alpha_i \rfloor$. We obtain the integer program $I^{(z_{(>)})}$ from $I^{(z)}$ by adding the constraint $x_i \geq \lfloor \alpha_i \rfloor + 1$. If there are several noninteger variables, any of them can be chosen.

Since L is bounded, the height of the tree T is finite. Indeed, by splitting the polytope for $L^{(z)}$ into the polytopes of $L^{(z_{(<)})}$ and $L^{(z_{(>)})}$, we exclude a finite part (namely, a hyperplane strip of width 1), from the smallest integer hypercube containing the polytope of L. The number of leaves of T is bounded by the size of this hypercube, which can be shown to be at most exponential in the size of the integer program [350].

In principle, any number can be chosen in place of α_i for expanding a tree node. This branch-and-bound procedure is especially effective if all variables are binary (0, 1 integer programs). In this case we fix the value of one binary variable x_z at each interior node z of the branch-and-bound tree. The two children of z correspond to the choices $x_z = 0$ and $x_z = 1$.

4.6.2.2 A Tree-Structured Formulation of the Minimum-Steiner-Tree Problem

Consider the MINIMUM STEINER TREE problem in its original formulation. As the set of legal configurations, we choose all subtrees that connect all required vertices but whose leaves may be nonrequired vertices. For each node z in T, we maintain pairs of edge sets $(\text{IN}_z, \text{OUT}_z)$. S_z is the set of all Steiner trees that contain all edges in IN_z and no edge in OUT_z. $(\text{IN}_r, \text{OUT}_r) := (\emptyset, \emptyset)$. An interior node of T has two children. In one of them, an arbitrary edge out of $E \setminus (\text{IN}_z \cup \text{OUT}_z)$ is added to IN_z; in the other, the same edge is added to OUT_z. A leaf is a node z at which one of the following holds:

Case 1: The edges in $E \setminus \text{OUT}_z$ form a tree T_z that connects all required vertices.

Case 2: IN_z contains a cycle of edges.

Case 3: The removal of all edges in OUT_z from G separates two required vertices.

The lower bound L_z is computed as follows. If case 1 applies, L_z is the cost of T_z after successive deletion of leaves of T_z that are neither required vertices nor incident to an edge in IN_z, until this is no longer possible. In cases 2 and 3, $L_z = \infty$. Otherwise, $L_z = c(IN_z)$.

The upper bound U_z is computed as follows. In cases 2 and 3, $U_z = \infty$. Otherwise, U_z is the cost of the sample solution s_z for z that is constructed by the following procedure. Find a minimum spanning tree T_z of G that contains all edges in IN_z and no edge in OUT_z (see Exercise 4.16). Successively delete leaves of T_z that are not required vertices, until this is no longer possible. The resulting Steiner tree is s_z. Note that s_z need not belong to S_z. If case 1 applies, the lower bound L_z always exceeds the upper bound U_z. (Because this is known in advance, L_z need not be computed by the branch-and-bound algorithm if case 1 applies; see Section 4.6.2.3.)

The bounds and the sample solution can be computed in almost linear time. In fact, we can make the algorithm even faster if we use incremental methods (see Exercise 4.17).

Alternatively, we can use one of the approximation algorithms for the MINIMUM STEINER TREE problem, modified suitably to generate only Steiner trees that include edges in IN_z and that exclude edges in OUT_z (see Section 3.10 and Exercise 4.18). Tighter lower bounds will be discussed in detail in Section 4.6.3.

The tree T described here is not the only way to tree structure the MINIMUM STEINER TREE problem. We can make the generation of tree nodes z more complicated, but decrease the overall size of T. For instance, instead of considering single edges for inclusion in IN_z and OUT_z one at a time, we sometimes can include several edges along a path p at once. This is permissible if p runs exclusively across unrequired vertices v such that the only edges incident to v that are not in OUT_z are on p. Modifying the tree T in such a way eliminates from T many nodes z with $S_z = \emptyset$.

We can also define a completely different tree T that is based on the inclusion or exclusion of vertices, rather than edges into the Steiner tree [23]. We see that tree structuring a configuration space is a paradigm that entails much room for heuristic tuning.

4.6.2.3   Details of the Branch-and-Bound Algorithm

We now describe how the branch-and-bound method optimizes over a tree-structured configuration space.

The branch-and-bound method does a top-down expansion of the tree T. In its course, it aims at expanding the part of the tree that shows the greatest promise of leading to an optimal solution. To find this location, we compute at each node z a sample solution s_z (if it can be found), and determine the bounds U_z and L_z. The best of all sample solutions computed so far is remembered. Once L_z is at least as large as the cost of this solution, the expansion of the node z can be stopped, since S_z contains no optimal solution. Such a node is called *fathomed*. By stopping the tree expansion at fathomed nodes, we exclude large parts of the tree T from the exploration, if the lower bounds L_z are tight. The details of the method are as follows.

Nodes that have been generated during the tree expansion and that are not fathomed, but none of whose children have been generated yet, are called *active*. The algorithm maintains a data structure D in which it stores the active nodes. The active and fathomed nodes together form a cut in the tree that marks the frontier between explored parts of the tree (above the cut) and unexplored parts of the tree (below the cut) (see Figure 4.13). The tree expansion happens exclusively at active nodes. The untraversed parts of the tree either are not yet explored, or are subtrees rooted at fathomed nodes. The algorithm is shown in Figure 4.14.

The loop in line 7 of Figure 4.14 implements the general tree-expansion step. At any time, a node extracted from D in line 7 is an interior node z with $U_z > L_z$. This means that the optimal configuration in S_z is not known at the time of the extraction. Thus, z is *expanded*: Its children are generated one by one. The test in line 11 finds out whether the sample solution of a child z' is a new best solution. If it is, the nodes that become fathomed by virtue of the newly found solution are deleted from D (line 14). The test in line 15 checks whether z' is fathomed. If it is not, then z' is inserted into D. The minimum of L_z over all active nodes z is a lower bound on the cost of an optimal solution.

The most popular choices for D are a stack and a priority queue ordered according to increasing lower-bound values. The stack implements a depth-first traversal of the tree. It is used whenever there is not much space to store generated configurations. In this case, the branch-and-bound algorithm is usually modified such that the children of a node are generated not at the time the node is expanded, but rather at the time that they would be retrieved from the stack. This *incremental* expansion scheme conserves even more space, because now only as many configurations have to be kept in D as the length of the path from the root of the branch-and-bound tree to the node currently expanded. This parameter is usually small; for example, it is linear in the size of the problem instance. The incremental expansion scheme can be refined further. First, we may introduce heuristics that generate more promising children—that is, those with smaller lower bounds. Second, let z_i be the ith child of z that is generated. If $U_{z_i} = L_z$, then we need not generate any more children of z, and z is fathomed. The branch-and-bound algorithm can be modified to test for this condition. The version of the branch-and-bound method in which D is a stack and we use incremental expansion can be implemented as a recursive algorithm (see Exercise 4.19).

If more space is available for storing configurations, a priority queue is used

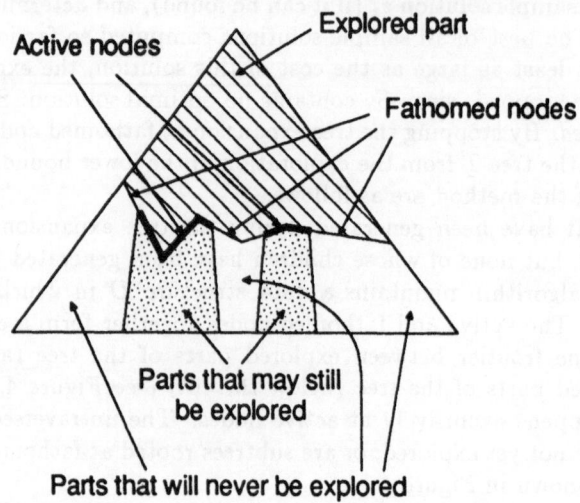

Figure 4.13: *The tree T at some point during the branch-and-bound algorithm. The traversed parts are hatched. The set of active nodes is drawn in thick black lines. The tree expansion occurs exclusively at these nodes. The active nodes together with the fathomed nodes form a cut of the tree separating the explored upper part (cross-hatched) from the unexplored lower part. Parts of the tree that lie below fathomed nodes will never be explored. Parts of the tree that lie below active nodes will still be (partly) explored.*

for D. If the lower bounds are tight enough to give a reasonable estimate of the promise that a tree node z shows of S_z containing an optimal solution, then this choice for D is in some sense the optimal one.

Figure 4.16 applies this branch-and-bound algorithm to the MINIMUM STEINER TREE problem as described in Section 4.6.2.3. The relevant problem instance and the notation used to depict the branch-and-bound tree are described in Figure 4.15. Children of a node are generated from left to right in Figure 4.16.

Branch and bound is a method for enumerating the complete configuration space. Parts of the configuration space are not explored only if it has been proved that they cannot contain optimal solutions. Thus, given enough time, branch and bound always finds an optimal solution. Of course, this time may be large—for example, exponential—if the problem we have to solve is NP-hard. If there is not enough time to complete the exploration, then branch and bound can be stopped early. Unless the exploration is interrupted so early that

(1)	u, ℓ : **integer**;
(2)	*CurrentBest* : **configuration**;
(3)	D : **data structure**;
(4)	$u := \infty$;
(5)	*CurrentBest* := **nil**;
(6)	**insert** (r, L_r) into D;
(7)	**repeat extract** next (z, L_z) from D;
	comment z is not fathomed yet;
(8)	$\ell := \infty$;
	comment expand z;
(9)	**for** z' child of z **do**
(10)	compute $L_{z'}$ and $U_{z'}$;
(11)	**if** $U_{z'} < u$ **then**
(12)	*CurrentBest* := $s_{z'}$;
(13)	$u := U_{z'}$;
	comment delete fathomed vertices;
(14)	delete all $(z'', L_{z''})$ with $L_{z''} \geq u$ from D **fi**;
(15)	**if** $L_{z'} < u$ **then**
	comment z' is not fathomed yet, and thus active;
(16)	**insert** $(z', L_{z'})$ into D;
(17)	$\ell := \min(\ell, L_{z'})$ **fi od**;
(18)	**until** $D = \emptyset$;

Figure 4.14: *The branch-and-bound algorithm.*

not a single sample solution has been found, the branch-and-bound algorithm
is able to display the configuration stored in *CurrentBest* as a solution. Thus,
if stopped early, the branch-and-bound algorithm is a heuristic. In general, no
asymptotic worst-case upper bounds can be derived for this heuristic. But each
application of the branch-and-bound algorithm to a specific problem instance
comes up with a numerical lower bound on the cost of an optimal solution of
that instance. This is the minimum of the lower bound of all active nodes. If
the branch-and-bound algorithm is allowed to proceed far enough, this bound
can be quite close to the upper bound u provided by the cost of *CurrentBest*
such that, on an instance by instance basis, branch and bound is able to present
a provably good solution.

Branch and bound is a very powerful paradigm that entails a lot of room
for heuristic tuning. Degrees of freedom are the choice of the tree T, and of
the upper and lower bounds and the data structure D. This robustness, and
the fact that the result of a branch-and-bound algorithm is, if not an optimal

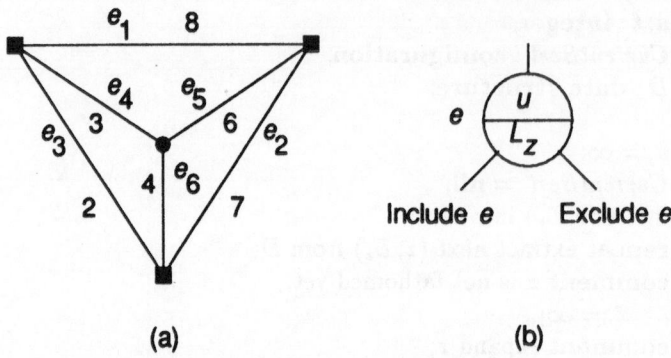

(a)   (b)

Figure 4.15: *The branch-and-bound algorithm on an instance of the*
MINIMUM STEINER TREE *problem. The problem instance* (G, R) *is
depicted in (a). The minimum spanning tree has a cost of 11; the
minimum Steiner tree has a cost of 9. Nodes z of the branch-and-
bound tree are represented as shown in (b). u is the upper bound at
the time of expansion of the node. L_z is the lower bound of node z.
e is the edge that is included (left branch) or excluded (right branch)
to expand z. The branch-and-bound tree for this instance is shown in
Figure 4.16.*

solution, a good solution paired with a lower bound, are what make branch
and bound so attractive.

4.6.3 Lagrangian Relaxation

The main problem in the customization of the branch-and-bound algorithm is
the fast computation of good lower bounds. Simple bounds, such as the ones
presented in Sections 4.6.2.1 and 4.6.2.2, come to mind rather quickly. Such
bounds, however, tend not to be good enough in practice.

We can obtain better results if we go into a little more trouble computing a
lower bound for a vertex z in the branch-and-bound tree. This usually involves
solving a relaxed version of the optimization problem.

Definition 4.20 (Relaxation) *Let* $\Pi : I \to 2^S$ *be a hard—for instance, NP-
hard—minimization problem with cost function c. A* **relaxation** *of* Π *is a
minimization problem* $\Pi' : I' \to 2^S$ *with cost function* c' *such that there is an
easily computable function* $f : I \to I'$ *such that, for all* $p \in I$, *the following
holds:*

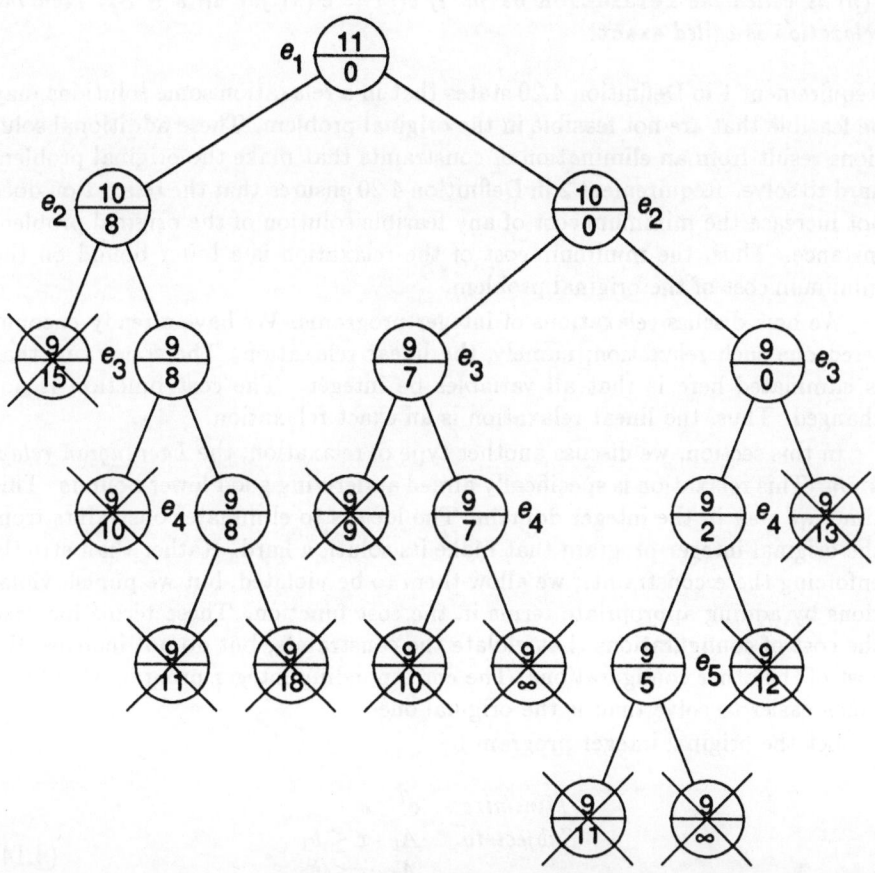

Figure 4.16: *The branch-and-bound tree at the time of completion of the algorithm run on the instance (G, R) of the* MINIMUM STEINER TREE *problem depicted in Figure 4.15(a). Children of z are generated left to right. Edges are picked in the order $e_1, \ldots, e_6$. Bounds are computed as discussed in Section 4.6.2.2.*

1. $S_p \subseteq S'_{f(p)}$.

2. For all $s \in S_p$, we have $c(s) \geq c'(s)$.

$f(p)$ *is called the* **relaxation of** *p. If $c(s) = c'(s)$ for all $s \in S_p$, then the relaxation is called* **exact**.

Requirement 1 in Definition 4.20 states that in a relaxation some solutions may be feasible that are not feasible in the original problem. These additional solutions result from an elimination of constraints that make the original problem hard to solve. Requirement 2 in Definition 4.20 ensures that the relaxation does not increase the minimum cost of any feasible solution of the original problem instance. Thus, the minimum cost of the relaxation is a lower bound on the minimum cost of the original problem.

We now discuss relaxations of integer programs. We have already encountered one such relaxation; namely, the linear relaxation. The constraint that is eliminated here is that all variables be integer. The cost function is not changed. Thus, the linear relaxation is an exact relaxation.

In this section, we discuss another type of relaxation, the *Lagrangian relaxation*. This relaxation is specifically aimed at deriving good lower bounds. This time, we stay in the integer domain. The idea is to eliminate constraints from the original integer program that make its solution hard. Rather than strictly enforcing these constraints, we allow them to be violated, but we punish violations by adding appropriate terms in the cost function. These terms increase the cost of configurations that violate the constraints, but do not increase the cost of the other configurations. The corresponding integer program should be much easier to solve than is the original one.

Let the original integer program be

$$\begin{aligned}
Minimize: \quad & c^T \cdot x \\
Subject\ to: \quad & A_1 \cdot x \leq b_1 \\
& A_2 \cdot x \leq b_2 \\
& x \text{ integer}
\end{aligned} \tag{4.14}$$

The constraints $A_1 \cdot x \leq b_1$ are assumed to be the ones that make the integer program hard to solve. These constraints are eliminated in the relaxation. Thus, solutions are allowed that violate these constraints. Such violations, however, are punished in the cost function. Specifically, we add a term $s^T \cdot (A_1 \cdot x - b_1)$ to the cost function. The penalty increases linearly with the amount of violation. s is a vector of *nonnegative* numbers that determine the penalty for violating the constraints $A_1 \cdot x \leq b_1$. The components of s are called the *Lagrangian multipliers*; the relaxed problem is called the *Lagrangian problem*; and its cost function is called the *Lagrangian*. In summary, the Lagrangian problem that results from eliminating the constraints $A_1 \cdot x \leq b_1$ from (4.14) looks like this:

$$\begin{aligned}
Minimize: \quad & c^T \cdot x + s^T \cdot (A_1 \cdot x - b_1) \\
Subject\ to: \quad & A_2 \cdot x \leq b_2 \\
& x \ \text{integer} \\
& s \geq 0
\end{aligned} \qquad (4.15)$$

Let us first check that (4.15) is indeed a relaxation of (4.14). Requirement 1 of Definition 4.20 holds, since (4.15) results from (4.14) by the elimination of constraints. Requirement 2 holds because $s \geq 0$, and therefore the added term in the cost function is nonpositive for any feasible solution of (4.14). In general, the relaxation is not exact.

If a relaxed constraint of (4.14) is an equality constraint $a_i \cdot x = b_i$, the corresponding Lagrangian multiplier s_i need not be nonnegative. If all relaxed constraints are equality constraints, (4.15) is an exact relaxation of (4.14).

The Lagrangian multipliers are constants as far as the optimization is concerned. Of course, the quality of the lower bound presented by the cost of the optimal solution of (4.15) depends on this choice. We can either take an educated guess here, or resort to systematic methods (see Section 4.6.3.2).

4.6.3.1 An Example: Steiner Trees

We exemplify the method of Lagrangian relaxation in the following integer-programming formulation of the MINIMUM STEINER TREE problem, taken from [23]. We start by formulating the MINIMUM STEINER TREE problem as a $0, 1$ integer program. As we will see, we do not even have to complete the formulation.

Note that a Steiner tree on a graph $G = (V, E)$ with edge-weighting function $\lambda : E \to \mathbf{N}$ and required vertex set $R = \{v_1, \ldots, v_r\}$ can be characterized as a subgraph T of G that is connected, that spans R, and such that, between each two vertices in R, there is exactly one path in T. Let $R_1 := \{2, \ldots, r\}$. We provide binary variables y_e for all edges $e \in E$. The binary values of the y_e denote in the canonical way the presence or absence of the edge e in the Steiner tree. Furthermore, we provide binary variables $x_{e,k}$ for all $e \in E$ and $k \in R_1$. The value of $x_{e,k}$ is supposed to be 1 exactly if the edge e is on the unique path from v_1 to v_k in the Steiner tree represented by the values y_e.

The corresponding cost vector is the vector of the edge lengths in G,

$$c(y_{e_1}, \ldots, y_{e_m}, x_{e_1,2}, \ldots, x_{e_m,r}) = \sum_{e \in E} \lambda(e) \cdot y_e \qquad (4.16)$$

Let us now define the set of constraints of the integer program that ensure that the optimal solution represents a minimum Steiner tree.

$$y_e \geq x_{e,k} \quad \text{for all } e \in E, \ k \in R_1 \qquad (4.17)$$

$$P_\ell \le \sum_{e \in E} y_e \le P_u \tag{4.18}$$

the edges e with $x_{e,k} = 1$ connect v_1 with v_k for all $k \in R_1$ $\qquad$ (4.19)

$$x_{e,k} \in \{0,1\} \quad \text{for all } e \in E,\ k \in R_1 \tag{4.20}$$

$$y_e \in \{0,1\} \quad \text{for all } e \in E \tag{4.21}$$

The constraint set (4.17) ensures that the paths indicated by the $x_{e,k}$ are realized by edges e for which $y_e = 1$. The constraint set (4.18) actually is redundant; that is, eliminating it still results in a correct integer program for the MINIMUM STEINER TREE problem. If we choose $P_\ell = r - 1$ and $P = n - 1$ in (4.18), however, we strengthen the lower bounds that we will compute. (Remember that n is the number of vertices of G.) The constraint (4.19) is not made formal here, since in a moment we will use special solution methods to enforce it. Let us denote by I_{ST} the integer program minimizing (4.16) subject to constraints (4.17) to (4.21).

The feasible solutions of I_{ST} correspond exactly to the subgraphs of G that connect all required vertices. Since all edge weights are positive, an optimal solution of I_{ST} necessarily represents a minimum Steiner tree.

What makes the integer program I_{ST} hard is the dependency between the $x_{e,k}$ and the y_e exhibited in the constraint set (4.17). If we eliminate these constraints, then the choices of the $x_{e,k}$ and the y_e are independently governed by the constraint sets (4.19) and (4.20) and by (4.18) and (4.21), respectively. The resulting Lagrangian problem I_{LR} has the constraints (4.18), (4.19), (4.20), and (4.21), and has the cost function

$$c(y_{e_1}, \ldots, y_{e_m}, x_{e_1,2}, \ldots, x_{e_m,r}) =$$

$$\sum_{e \in E} \left(\lambda(e) - \sum_{k \in R_1} s_{e,k} \right) \cdot y_e + \sum_{\substack{e \in E \\ k \in R_1}} s_{e,k} \cdot x_{e,k} \tag{4.22}$$

The $s_{e,k}$ are the Lagrangian multipliers. They have to be nonnegative.

We now describe how I_{LR} can be solved efficiently. Here, we come back to the observation that the $x_{e,k}$ and the y_e are decoupled in I_{LR}. In fact, the part of I_{LR} concerning the $x_{e,k}$ asks for just a few shortest-path computations. Specifically, for $k \in R_1$, the value of $x_{e,k}$ in an optimal solution of I_{LR} is 1 exactly if e is on an arbitrarily selected shortest path from v_1 to v_k inside the graph G, modified such that each edge e' has length $s_{e',k}$. Thus, we can compute the $x_{e,k}$ quickly using k applications of Dijkstra's algorithm (see Section 3.8.4). The minimization of the part of the Lagrangian involving the variables y_e amounts to sorting the y_e according to increasing the size of their coefficients $(\lambda(e) - \sum_{k \in R_1} s_{e,k})$, and then scanning the list, setting the first P_ℓ of them to 1 and continuing to set at most $P_u - P_\ell$ more to 1, as long as their Lagrangian coefficients are negative.

So far, we have considered only the Lagrangian problem of the unadulterated MINIMUM STEINER TREE problem. The branch-and-bound procedure discussed in Section 4.6.2.2 requires solving problems at an interior tree node z that are modified MINIMUM STEINER TREE problems. Here, some edges are required to stay inside, and some outside, the Steiner tree. We can adapt the preceding relaxation method to solving these problems by setting

$$y_e = \begin{cases} 1 & \text{if } e \in \text{IN}_z \\ 0 & \text{if } e \in \text{OUT}_z \end{cases}$$

$$s_{e,k} = \begin{cases} 0 & \text{if } e \in \text{IN}_z, \ k \in R_1 \\ \infty & \text{if } e \in \text{OUT}_z, \ k \in R_1 \end{cases}$$

Figure 4.17 shows the branch-and-bound tree of the example in Figure 4.15(a), if the preceding Lagrangian relaxation is used with all $s_{e,k} = 0$ to compute the lower bounds for all nodes z in T for which cases 1 to 3 in Section 4.6.2.2 do not apply (See Exercise 4.20 for a graph-theoretic formulation of this lower bound.) Note that only one-half of the tree depicted in Figure 4.16 is generated, because the lower bounds computed using Lagrangian relaxation are tighter. This particular choice of the values of $s_{e,k}$ is certainly not an optimal one (see also Section 4.6.3.2). There are several other methods for finding Lagrangian problems for some integer-programming formulation of the MINIMUM STEINER TREE problem (see [469]). Lagrangian relaxation has also been used with great success on the TRAVELING SALESPERSON problem [182, 183].

4.6.3.2 The Lagrangian Dual and its Optimization

Of course, the quality of the lower bound computed with Lagrangian relaxation depends on the choice of the values for the $s_{e,k}$. Since we want the lower bound to be large, we get a maximization problem. We want to choose an assignment to the Lagrangian multipliers $s_{e,k}$ such as to maximize the optimal cost of the Lagrangian problem (4.15). Let us discuss this maximization problem in general.

Let F be the set of feasible solutions of the Lagrangian problem. F is independent of the choice of s. Bringing the minimum cost of the Lagrangian problem as close as possible to the minimum cost of the original integer program amounts to maximizing the following function:

$$L(s) = -s^T \cdot b_1 + \min_{x \in F}(c^T + s^T \cdot A_1) \cdot x \qquad (4.23)$$

subject to the constraint $s \geq 0$.

This maximization problem is called the *Lagrangian dual* of the original integer program. In many cases, for instance, in the Lagrangian relaxation for the MINIMUM STEINER TREE problem above F is finite. Then, the Lagrangian dual

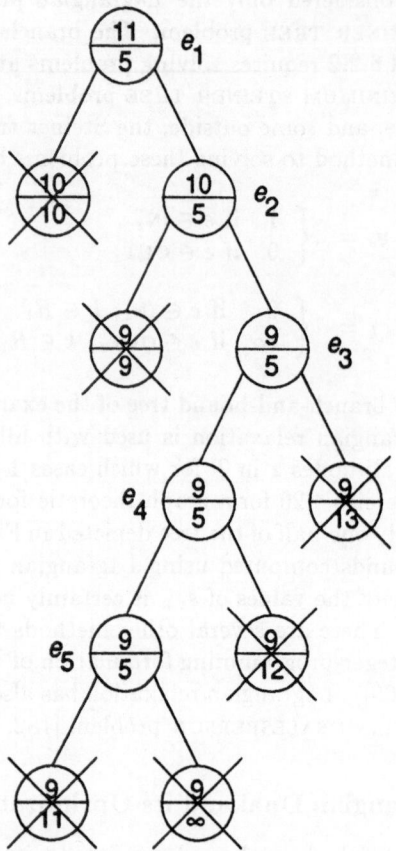

Figure 4.17: *The branch-and-bound tree of the example in Figure 4.15(a), where the lower bounds are computed with Lagrangian relaxation, $s_{e,k} = 0$ for all $e \in E, k \in R_1$. Only about one-half of the nodes of the tree in Figure 4.16 are generated.*

is actually a linear program, because the minimization part can be transformed into a set of linear constraints:

$$
\begin{aligned}
&Maximize: &&w \\
&Subject\ to: &&w + s^T \cdot (b_1 - A_1 \cdot x) \le c^T \cdot x, \quad x \in F \qquad (4.24)\\
& &&s \ge 0
\end{aligned}
$$

If F is infinite, then $L(s)$ is the minimum of an infinite set of linear functions. As such, $L(s)$ is concave and piecewise linear, but in general is nondifferentiable.

There are two different ways of computing $L(s)$, one pertaining to each of the characterizations (4.23) and (4.24) of the Lagrangian dual.

4.6.3.2.1 Subgradient Optimization The maximization procedure for general F is called *subgradient optimization*, and is based on an adaptation of straightforward steepest-ascent methods for differentiable functions to the nondifferentiable function (4.23). In general, much as in linear programming, the maximum is attained at a point of nondifferentiability. There is no gradient at such a point, but we can define a *subgradient* at point $s^{(0)}$ to be a vector t such that for all $s \in \mathbf{R}$

$$L(s) \le L(s^{(0)}) + (s - s^{(0)})^T \cdot t$$

The set of all subgradients at $s^{(0)}$ is called the *subdifferential* at $s^{(0)}$.

The significance of a subgradient t is that the half-space $\{s|(s-s^{(0)})^T \cdot t \ge 0\}$ contains all solutions of the Lagrangian dual whose cost values are larger than the cost at $s^{(0)}$. Thus, any subgradient points to a direction of ascent of $L(s)$ at $s^{(0)}$ (see Figure 4.18). (If $L(s)$ is differentiable at s_0, then the unique subgradient at $s^{(0)}$ is the gradient of $L(s)$ at $s^{(0)}$.)

The idea of the subgradient methods is to choose a subgradient at the current solution s_0, and to walk along it a sufficiently small step such that $L(s)$ still increases. We can generate a subgradient by computing

$$t = A_1 \cdot x - b_1$$

at any point $x \in F$ that achieves the minimum in (4.23) for $s = s^{(0)}$. The size $\theta > 0$ of the step taken in the direction of this subgradient is chosen heuristically. A common choice for θ is

$$\theta = \beta \frac{\overline{L} - L(s^{(0)})}{||t||^2} \tag{4.25}$$

where $\varepsilon < \beta \le 2$ for some fixed $\varepsilon > 0$, $||t||$ is the Euclidean norm of t, and $\overline{L}$ is an estimate for the maximum value of $L(s)$. Usually, we start with $s^{(0)} = 0$. The new assignment to the Lagrangian multipliers is then

$$s^{(1)} = \max\{0, s^{(0)} + \theta t\}$$

This procedure is iterated until the changes in s are sufficiently small. Actually, it can be shown that, with the choice (4.25) of θ, this method converges to $\overline{L}$ as long as this value is a feasible value of $L(s)$ [183]. Convergence occurs, even though the sequence of cost values obtained need not be monotonically increasing. Of course, we want to choose the maximum value of $L(s)$ for $\overline{L}$, but since this is the value we want to compute, we can give only heuristic estimates here. Thus, applying the subgradient method entails a residual of heuristic choice.

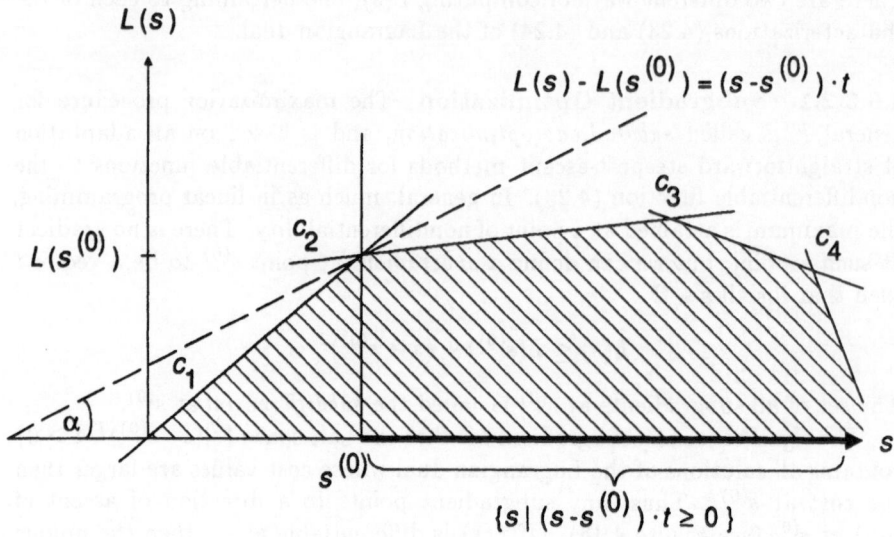

Figure 4.18: *Subgradients. The function $L(s)$ is the minimum of four linear functions c_1 to c_4, each contributed to (4.23) by one element $x \in F$. A subgradient t at a nondifferentiable point $s^{(0)}$ is $t = \tan \alpha$, where α is the angle of the dashed line with the positive x axis. The half-space $\{s|(s - s_i^{(0)}) \cdot t \geq 0\}$ is highlighted. It contains all values of s, such that $L(s) - L(s^{(0)}) \geq 0$. The subdifferential is obtained by "rotating" the dashed line between the lines for c_1 and c_2.*

Computational experience suggests that one can also use upper bounds on the cost values for $\overline{L}$. This is convenient since such upper bounds arise naturally in the branch-and-bound process. For instance, the current upper bound in the branch-and-bound tree can be used for $\overline{L}$ when computing lower bounds for a node in the branch-and-bound tree with Lagrangian relaxation. The robustness of the method with respect to the choice of $\overline{L}$ is achieved by letting the step size β decrease to 0 with increasing k. Held et al. [184] report on computational experiments here, and suggest to start with $\beta = 2$ for a number of iterations that compares with the problem size, and then halving α at each step until some sufficiently small threshold.

4.6.3.2.2   Solving the Lagrangian Dual with Linear Programming   If F is finite, then the Lagrangian dual is a linear program (4.24). In general, this program is very large, since it entails constraints for each feasible solution of the Lagrangian problem (4.15). For instance, for the relaxation of the MINIMUM

STEINER TREE problem minimizing (4.22) subject to (4.18) to (4.21), there is a constraint for each subgraph of G connecting all required vertices. This number can be exponential in the problem size.

The dual of (4.24) in the linear-programming sense is the following linear program:

$$Minimize : \sum_{x \in F} (c^T \cdot x)^T \cdot \pi_x$$

$$Subject\ to : \sum_{x \in F} (A_1 \cdot x) \cdot \pi_x \leq b_1$$

$$\sum_{x \in F} \pi_x = 1 \tag{4.26}$$

$$\pi_x \geq 0 \quad \text{for all } x \in F$$

Note that, if we impose the additional constraint on (4.26) that all π_x be integer, then (4.26) becomes the same as the original integer program (4.14). The formulation of (4.26) clearly displays the way in which we have converted the original integer program into a linear program by imposing additional convexity constraints.

The linear program (4.26) can be solved with techniques specifically designed to solve large linear programs (see [350, Chapter 4]).

In practice, the linear-programming techniques have not performed as well on optimizing the Lagrangian dual as have the subgradient methods [417].

4.6.4 Simple Integer Programs

Not all integer programs are hard. In fact, we have already seen examples of simple integer programs. In Section 4.6.3.1, we relaxed an integer program of the MINIMUM STEINER TREE problem to a simple integer program that could be solved in almost linear time. In this section, we will state some properties of integer programs that ensure that these programs are easily solvable. In fact, we will characterize a class of integer programs that have the property that their linear relaxation has an optimal solution that is integer. Thus, we can solve such integer programs simply by solving their linear relaxation. The integer programs in which we are interested have special constraint matrices.

Definition 4.21 (Unimodular Matrix) *A square integer matrix A is called* **unimodular** *if $|\det(A)| = 1$.*

A matrix A is called **totally unimodular** *if every square nonsingular submatrix of A is unimodular.*

Theorem 4.11 *Let A be an integer $m \times n$ matrix. If A is totally unimodular, then, for all integer m-vectors b, all corners of the polytope defined by $Ax = b, x \geq 0$ are integer.*

Proof Without loss of generality we can assume that A has full rank. By Theorem 4.10, part 10, each corner of the polytope is of the form $B^{-1} \cdot b$, where B is a nonsingular square submatrix of A. By Cramer's rule,

$$B^{-1} \cdot b = \frac{B^{adj} \cdot b}{\det(B)}$$

where B^{adj} is the adjoint of B. Since A is totally unimodular, all these vectors are integer. □

The opposite direction of Theorem 4.11 does not hold (see Exercise 4.21, [453]). A corresponding result can be shown for the constraint set $A \cdot x \leq b$ and $x \geq 0$ (see Exercise 4.22). In this case, the opposite direction does hold [453].

When is a matrix totally unimodular? The following theorem gives a set of matrices that often comes up in practice and that is totally unimodular.

Lemma 4.7 *Let A be a matrix with coefficients from the set $\{-1, 0, +1\}$. Assume that the rows of A can be partitioned into two sets R_1 and R_2 such that the following holds:*

1. *If a column has two entries of the same sign, their rows are in different sets.*

2. *If a column has two entries of different signs, their rows are in the same set.*

Then, A is totally unimodular.

Among the matrices that fulfill Lemma 4.7 are various types of adjacency matrices of graphs such as

- Node–edge incidence matrices of directed graphs: These are matrices with rows corresponding to vertices, columns corresponding to edges, and $a_{ij} = +1 \ (-1)$ if node v_i is the tail (head) of edge e_j. Here, we set $R_1 = \emptyset$. Such matrices come up in the integer-program formulation of the single-pair shortest-path problem and in the maximum-network flow and mincost-flow problems (see Exercise 4.24).

- Node–edge incidence matrices of undirected bipartite graphs: These matrices are defined analogously. Let V_ℓ and V_r be the two classes of nodes. Then, we set $R_1 = V_\ell$ and $R_2 = V_r$. Such matrices are the basis of the integer-programming formulation of the LINEAR ASSIGNMENT problem.

In general, the total unimodularity of A is not necessary to ensure that the solution of a linear program is integer. In fact, we need not assert that *all* corners of the polytope be integer for *all* right sides b. For more results on linear programs with integer optimum solutions, see [367].

4.7 Dynamic Programming

Dynamic programming is a bottom-up technique that can be applied to solving optimization problems that are decomposable. Since dynamic programming is quite well known, we take the liberty of introducing the subject in a slightly more formalized fashion than is usual, and illustrating it with a somewhat advanced example. Furthermore, we will outline areas of application of this technique.

4.7.1 Decomposing Optimization Problems

First, we define precisely what premises an optimization problem must meet for dynamic programming to apply. Essentially, we require that each problem instance can be broken up into several smaller problem instances such that the solution of the subproblems quickly leads to the solution of the original problem instance.

Definition 4.22 (Decomposable Problem) *A minimization problem Π is* **decomposable** *if each instance p of the problem can be decomposed (perhaps in multiple ways) into "smaller" problem instances $p_1, \ldots, p_k$ such that we can obtain a solution s_p (not necessarily an optimal one) of p by combining optimal solutions $s_{p_1}, \ldots, s_{p_k}$ of the $p_1, \ldots, p_k$. There must be a monotonically increasing function $f : \mathsf{N}^k \to \mathsf{N}$ such that*

$$c(s_p) \leq f(c(s_{p_1}), \ldots, c(s_{p_k})) \tag{4.27}$$

Furthermore, there must be at least one way of breaking up p into $p_1, \ldots, p_k$ such that an optimal solution s of p results from combining arbitrary optimal solutions of the subproblems $p_1, \ldots, p_k$ and

$$c(s) = f(c(s_{p_1}), \ldots, c(s_{p_k})) \tag{4.28}$$

In addition, we must be able to compute each such s quickly by combining the s_{p_i}. (4.27) and (4.28) are also called the **principle of optimality**.

In a similar way, we can also define sets of optimization problems $\Pi_1, \ldots, \Pi_k$ to be decomposable. Here, we require that each instance of each problem can be decomposed into smaller instances of any of the problems.

We can formalize the notion of "smallness" by giving each subproblem a positive **decomposition size (d-size)** *that decreases in each decomposition step. Problem instances whose d-size is below some threshold—say, 0—are required to be directly solvable.*

4.7.2 An Example: Minimum Steiner Trees

We exemplify Definition 4.22 with the MINIMUM STEINER TREE problem [98]. Intuitively, the decomposability of this optimization problem hinges on the following fact.

Lemma 4.8 *Assume that a minimum Steiner tree T has an interior vertex v that is required. By deleting v, we cut T into several connected components. Let R_1 be the set of required vertices in one of the components, and let $R_2 = R \setminus R_1$. Then, T is a union of two minimum Steiner trees, one over $R_1 \cup \{v\}$, and the other over R_2.*

Of course, it can also be the case that all required vertices are leaves of all minimum Steiner trees. In this case, the decomposition in Lemma 4.8 does not help us to solve the problem. We can use the decomposition as the basis for the definition of a second optimization problem that is only a slight derivative of the MINIMUM STEINER TREE problem. We call this derivative the R-STEINER TREE problem. An instance (G, R, v) of the R-STEINER TREE problem is a graph G with required vertex set R and a distinguished vertex $v \in V \setminus R$. The aim is to find a minimum Steiner tree T for the required vertex set $R \cup \{v\}$ such that v is an interior vertex of T.

We will write instances of the MINIMUM STEINER TREE problem as triples (G, R, v) with $v \in V \setminus R$. Given such an instance of the MINIMUM STEINER TREE problem, we are looking for a minimum Steiner tree in G with required vertex set $R \cup \{v\}$. We can put the normal instance of the MINIMUM STEINER TREE problem into this form by selecting a vertex $v \in R$ and forming the instance $(G, R \setminus \{v\}, v)$. We perform this redefinition of the problem instances of the MINIMUM STEINER TREE problem so that we will be able to define decomposability.

We now show that the MINIMUM STEINER TREE problem and the R-STEINER TREE problem together are decomposable. The d-size of an instance (G, R, v) of the R-STEINER TREE problem is defined to be $2|R|$. The d-size of an instance (G, R, v) of the MINIMUM STEINER TREE problem is defined to be $2|R|+1$. Note that instances of d-size at most 3 either can be solved directly by shortest-path computations (MINIMUM STEINER TREE problem) or are trivial (R-STEINER TREE problem).

We now outline the decomposition. An instance (G, R, v) of the R-STEINER TREE problem with $|R| \geq 2$—that is, with d-size at least 4—can be decomposed into two instances (G, R_1, v) and (G, R_2, v) of the MINIMUM STEINER TREE problem. Here, R_1 and R_2 are two nonempty sets that partition R. We obtain the solution by forming the union of optimal solutions of the two subproblems, and by deleting edges to break all cycles, if there are any.

We can decompose an instance (G, R, v) of the MINIMUM STEINER TREE problem with $|R| \geq 2$—that is, with d-size at least 5—by choosing an arbitrary vertex $w \in V$. Then, first a shortest path p is computed between v and w. Finally, we make a case distinction:

Case 1: $w \in R$. In this case, we solve the instance $(G, R \setminus \{w\}, w)$ of the MINIMUM STEINER TREE problem.

Case 2: $w \notin R$. In this case, we solve the instance (G, R, w) of the R-STEINER TREE problem.

The solution of (G, R, v) is the tree that we obtain by merging the optimal solution of the R-STEINER TREE or MINIMUM STEINER TREE subproblem with the shortest path between v and w, and by deleting edges to break cycles.

Let us now check that the decomposition satisfies the criteria in Definition 4.22. First, it is easy to check that the d-sizes of nontrivial subproblems created by the decomposition are strictly less than the d-size of the original problem instance.

We now show that we can always compose an optimal solution for a problem instance from optimal solutions for the subproblems. For the R-STEINER TREE problem, this follows from Lemma 4.8. For the MINIMUM STEINER TREE problem, consider the instance (G, R, v). There can be two cases:

Case 1: There is a minimum Steiner tree T for (G, R, v) that has v as an interior vertex. In this case, choosing $w = v$ yields a decomposition into the instance (G, R, v) for the R-STEINER TREE problem (whose optimal solution is also an optimal solution of the original problem instance) and the shortest path between v and v (whose cost is 0).

Case 2: Otherwise, v is a leaf in all minimum Steiner trees. Let T be such a tree. Choose w to be the first vertex that we reach when we walk along the tree path from v such that either of the following

Case 2.1: $w \in R$.

Case 2.2: w has degree at least 3.

This leads to a decomposition into a shortest-path problem and an instance of the MINIMUM STEINER TREE problem in case 2.1 an instance of the R-STEINER TREE problem in case 2.2. Using Lemma 4.8, it can again be shown that there are optimal solutions to these subproblems whose cycle-free merging yields T.

Finally, we check (4.27) and (4.28). As function f, we choose the sum of the minimum costs for the subproblems. Obviously, (4.27) holds. For the R-STEINER TREE problem and for the MINIMUM STEINER TREE problem in case 2.2, (4.27) may hold strictly, since positive length edges may be deleted to break cycles in the graph obtained by merging the solutions of the subproblems. If the solution obtained is optimal, however, then (4.28) holds, because an optimal solution for the problem instance can be obtained by *cycle-free* merging of optimal solutions for the subproblems.

4.7.3 The Dynamic-Programming Algorithm

Of course, there are many ways we can choose to decompose a problem instance. In the example discussed in Section 4.7.2, there is an exponential number of ways of partitioning R. Decomposability states only that *one* of these ways will work. We still have to find out which one it is.

The first idea of dynamic programming is to do a recursive decomposition of the original problem instance until all subproblems are trivial. Here, we have to try all choices of decomposing a problem instance, because we have no clue as to which is the right one. This can potentially result in a very large search tree and a very inefficient algorithm.

The second idea of dynamic programming, and the one that makes the method efficient, starts with the observation that many subproblems that are generated during the search process are identical. Thus, it is not a good idea to solve problems as they are generated by a recursive search that considers all possible ways of decomposing the problem. Rather, we should process the subproblems from smaller to larger—that is, according to increasing d-size. In this way, a subproblem has been solved before its solution is demanded as part of the solution of larger subproblems. If we store the optimal costs for all subproblems in a table, then, rather than repeatedly solving a subproblem, we just solve it once and perform table lookups to fetch the optimal cost multiple times. In fact, the table will not contain an explicit description of the optimal solution for each instance. Rather, only the optimal costs are given, and pointers are provided to the solutions of subproblems whose combination yields the optimal solution of the problem instance. The details of the method are as follows.

All trivial subproblems (with d-size below the threshold) can be solved directly. Their costs are stored in the table. Now assume that the table contains the optimal costs of all subproblems up to d-size k. Then, we scan through all problem instances of d-size $k + 1$. For each instance, we consider all ways of decomposing it into subproblems. The optimal costs of all subproblems are contained in the table, so we can easily combine them to evaluate the costs of different decomposition alternatives for the instance. The best one is the optimal cost of the instance, by (4.27) and (4.28). We can store the corresponding optimal solution implicitly by providing pointers to the subproblem instances whose combination forms the optimal solution. We can then recover The optimal solution by doing a final pass over the table that assembles the optimal solution from its components.

The complexity of this method is determined by the number of different subproblems that are contained in a given problem instance. Of course, if the problem we want to solve is NP-hard, we still expect this number to be large. With simpler optimization problems, however, this number can be small enough to yield fast algorithms where, on first inspection, no such algorithms are evident. In general, the fact that multiply occurring subproblems are solved only once implies that dynamic-programming algorithms are much faster than are schemes based on less sophisticated exhaustive search methods.

4.7.4 Application to Minimum Steiner Trees

The decomposition of the MINIMUM STEINER TREE problem together with the R-STEINER TREE problem immediately leads to a dynamic programming algorithm. This algorithm fills two tables, *Steiner* and *RSteiner*, that contain the optimal costs of the subproblem instances of the MINIMUM STEINER TREE problem and of the R-STEINER TREE problem, respectively. Let us discuss the index set of these tables.

The input to the algorithm is an instance (G, R, v) of the MINIMUM STEINER TREE problem. Let $|R| = r$. (Note that there are $r + 1$ required vertices.) A simple inductive argument shows that, if we decompose this instance recursively, we generate only those instances (G, R', v') of the MINIMUM STEINER TREE problem or R-STEINER TREE problem for which $R' \subseteq R$. Thus, the tables *Steiner* and *RSteiner* are indexed by a pair (R', v'), where $R' \subseteq R$ and $v' \in V \setminus R'$.

The algorithm is detailed in Figure 4.19. It uses a function $d(v, w)$ to compute the length of shortest paths between two vertices v and w in G. The arrays *BestS* and *BestRS* contain pointers that implicitly store the optimal solution. The optimal cost of the problem instance (G, R, v) is stored in $Steiner[R, v]$ after the algorithm terminates.

We can recover the optimal solution by recursively navigating along the pointers provided by the arrays *BestS* and *BestRS*, and combining the partial solutions appropriately. The respective recursive function $MakeSteiner[R, v]$ is shown in Figure 4.20. It uses the following auxiliary functions:

ShortestPath(v, w) Computes the shortest path between
                         two vertices in G

MergeTrees$(T_1, T_2, \ldots)$ Merges the trees given as arguments
                         and deletes edges to break all cycles

The call MakeSteiner(R, v) produces the optimal solution.

Let us now discuss the complexity of this algorithm. Providing all information on the shortest paths takes time $O(mn + n^2 \log n)$ (see Section 3.8.6). The part of the algorithm that computes the solutions of instances of the R-STEINER TREE problem (lines 8 through 16) does a constant amount of work for each selection of $R' \subseteq R$, $R'' \subseteq R'$, and $v' \notin R'$. We choose R' and R'' by determining, for each vertex w in R, whether $w \in R''$, $w \in R' \setminus R''$, or $w \in R \setminus R'$. There are 3^r such choices. Thus, in total, the run time for this part is $O(n \cdot 3^r)$.

The part of the algorithm that computes the solutions of instances of the MINIMUM STEINER TREE problem (lines 19 through 28) does a constant amount of work for each selection of $R' \subseteq R$, $v' \notin R'$, and $w \in V$. Thus, the run time of this part is $O(n^2 \cdot 2^r)$.

Note that the total run time of the algorithm is polynomial in n, but is exponential in r. Thus, the algorithm runs fastest if the set R is small. We

```
(1)      New, Current : integer;
(2)      Steiner, RSteiner : array[subset of R, vertex] of integer;
(3)      BestRS : array[subset of R, vertex] of set of vertex;
(4)      BestS : array[subset of R, vertex] of vertex;

(5)      for v vertex and v ∈ R, w vertex do
(6)          Steiner[{v}, w] := d(v, w) od;

         comment Process subsets of R in order of increasing size;
(7)      for i := 2 to r do
         comment Compute cost of minimum R-Steiner tree;
(8)          for R' ⊆ R and |R'| = i do
(9)              for v' vertex and v' ∉ R' do
(10)                 Current := ∞;
(11)                 for R'' ≠ ∅ and R'' ⊂ R' do
(12)                     New := Steiner[R'', v'] + Steiner[(R' \ R''), v'];
(13)                     if New < Current then
(14)                         Current := New;
(15)                         BestRS[R', v'] := R'' fi od;
(16)                 RSteiner[R', v'] := Current od od;

         comment Compute minimum Steiner trees;
(17)         for R' ⊆ R and |R'| = i do
(18)             for v' vertex and v' ∉ R' do
(19)                 Current := ∞;
(20)                 for w vertex do
(21)                     if w ∈ R' then
(22)                         New := d(v', w) + Steiner[R' \ {w}, w]
(23)                     else
(24)                         New := d(v', w) + RSteiner[R', w] fi;
(25)                     if New < Current then
(26)                         Current := New;
(27)                         BestS[R', v'] := w fi od;
(28)                 Steiner[R', v'] := Current od od od;
```

Figure 4.19: *The dynamic-programming algorithm for the* MINIMUM
STEINER TREE *problem.*

(1)   **recursive function** MakeSteiner
                    $(R' : \textbf{subset of } R, v' : \textbf{vertex}) : \textbf{tree};$
(2) $w : \textbf{vertex};$

(3)   **if** $|R'| < 2$ **then**
(4)   solve the corresponding shortest-path problem;
(5)   **else**
(6) $w := BestS[R', v'];$
(7)   **if** $w \in R'$ **then**
(8)   MergeTrees(MakeSteiner($R' \setminus \{w\}, w$),
                            ShortestPath(v', w))
(9)   **else**
(10)   MergeTrees(MakeSteiner($BestRS[R', w], w$),
                            MakeSteiner($R' \setminus BestRS[R', w], w$),
                            ShortestPath(v', w)) **fi fi**;

Figure 4.20: *The construction of the minimum Steiner tree.*

note that there also are algorithms for computing minimum Steiner trees that run in time polynomially in r but exponentially in $n - r$. Such algorithms are especially efficient if R is close to n. Not surprisingly, they reduce the MINIMUM STEINER TREE problem to a set of instances of the MINIMUM SPANNING TREE problem (see Exercise 4.25).

4.7.5 Dynamic Programming on Hierarchical Graphs

Dynamic programming takes on a particular worth in the case that the graphs that form problem instances are hierarchical graphs, as defined in Definition 3.21. Many optimization problems on hierarchical graphs are decomposable in a quite natural manner: The problem on the given hierarchical graph decomposes into subproblems, each pertaining to one nonterminal in the root cell G_k. If the cells of the hierarchical graph are small, we can solve many optimization problems in linear time in the size of the (expanded) graph by using dynamic programming based on this decomposition.

Throughout this section, we assume that a small constant s is an upper bound on the number of terminal vertices in each cell. The number of nonterminals is unrestricted. We call a graph that can be expressed hierarchically in this way a *partial k-tree*, where $k = s - 1$. (For alternative equivalent definitions of partial k-trees see, for instance, [381].)

For the optimization problem Π to be solvable with the dynamic-programming method outlined in Section 4.7, we must be able to devise a family

$\Pi_1, \ldots, \Pi_\ell$ of optimization problems such that $\Pi = \Pi_1$, and such that together these optimization problems are decomposable in the sense of Definition 4.22. Here, the smaller problem instances are instances pertaining to nonterminals of the root of the hierarchy tree. If the combination of the solutions of subproblems is computable in linear time in the number of nonterminals inside a cell, then dynamic programming finds the optimum solution in linear time in the size of the graph. We call each problem Π_j a *derivative* of the problem Π. The number ℓ of derivatives generally depends on s; in fact, it grows dramatically with s. For a fixed s, however, the number ℓ is constant and does not grow with the size of the problem instance.

Optimization problems with this property have been discussed, for instance, in [13, 29]. It turns out that a large number of optimization problems that are hard on general graphs are decomposable in this sense and therefore can be solved quickly on hierarchical graphs. In fact, [13] proves that a problem that can be stated in a certain restricted version of second-order logic is always decomposable in this sense. The definition of the derivatives of the problem can be deduced from the logic formula. This approach is not always practical, but very often we can find a direct route from the problem definition to a fast linear-time optimization algorithm.

As an example, consider the following optimization problem on graphs.

CHINESE POSTMAN

Instance: An undirected graph $G = (V, E)$ with edge-weighting function $\lambda : E \to \mathbf{N}$.

Configurations: All paths p in G

Solutions: All cycles p in G that contain each edge at least once

Minimize: $\sum_{e \in p} \lambda(e)$

The CHINESE POSTMAN problem comes up in the area minimization of CMOS functional cells in certain fabrication technologies. The CHINESE POSTMAN problem is not NP-hard; it can be solved in time $O(n^3)$ [66]. On partial k-trees that are given in terms of their hierarchical descriptions, however, the problem can be solved in linear time for all fixed k. Let us discuss how this can be done.

The following reformulation of the CHINESE POSTMAN problem will help us. A cycle that traverses each edge of the graph G exactly once is called an *Euler cycle*. A graph is called *Eulerian* if it has an Euler cycle. A graph G is Eulerian exactly if all vertices in G have even degree. We can find an Euler cycle in linear time by starting at any vertex and proceeding along edges that have not been traversed yet in an arbitrary fashion.

Solving the CHINESE POSTMAN problem is tantamount to finding a minimum-sized set D of edges in G whose duplication causes G to have no odd-degree vertices. (By *duplication* of an edge e, we mean the introduction of another edge that has the same endpoints as e.)

Before we discuss the set of problem derivatives that we need, we note that we can change any hierarchical graph definition Γ of a partial k-tree to a hierarchical graph definition Γ' such that

1. $E(\Gamma') = E(\Gamma)$.

2. Each cell G_i' in Γ' has at most two nonterminals (that is, the decomposition is *binary*).

3. If G_i' contains a nonterminal, then G_i' does not contain any edges (that is, the decomposition is *leaf-oriented*).

4. The size of Γ' is linear in the size of Γ.

A hierarchical definition with property 3 is called *leaf-oriented*, because only the leaves of the hierarchy tree contributes edges to $E(\Gamma)$. The procedure by which we transform Γ into Γ' with these properties essentially converts the hierarchy tree of Γ that has arbitrary degree vertices to a leaf-oriented binary tree (Exercise 4.27) piece by piece in a straightforward fashion.

Consider a cell C corresponding to an interior node in a leaf-oriented binary hierarchy tree for a partial k-tree. G_i has two nonterminals n_1 and n_2, at most $s = k + 1$ pins, and no edges. Let S be the set of pins of C. An optimal solution D for C partitions D into subsets D_1 and D_2, pertaining to n_1 and n_2, respectively. These subsets are solutions not of the original CHINESE POSTMAN problem, but rather of problem derivatives, in which an (even) number of pins can have odd degree. Thus, they are solutions of problem derivatives from the following set.

For each $S' \subseteq S$, $|S'|$ even, let $\Pi_{S'}$ be the problem of minimizing the size of an edge set D whose duplication makes exactly the vertices in S' have odd degree. Clearly, $\Pi_\emptyset$ is the original CHINESE POSTMAN problem. Also, for fixed s, there is a bounded number of derivatives; namely, 2^{s-1} of them.

We can obtain optimal solutions for C of all problem derivatives $\Pi_{S'}$ by simply analyzing all combinations of optimal solutions for n_1 and n_2 of all problem derivatives. These are at most 4^{s-1} such combinations. This number is constant for fixed s. Thus, a linear-time dynamic-programming algorithm results.

We can organize the process of combining optimal solutions of the nonterminals to solutions of the cell G_i by using a *multiplication table*. This is a matrix $M = ((m_{ij}))$ whose rows and columns correspond to problem derivatives pertaining to the two nonterminals n_1 and n_2, respectively, of C. The entry m_{ij} depicts the problem derivative that we solve by combining the optimal solution s_1 of problem derivative i for n_1 with the optimal solution s_2 of problem derivative j for n_2. The entry m_{ij} also contains a formula computing the value $f(s)$ (see Definition 4.22) of the solution s for C, dependent on the corresponding values $f(s_1)$ and $f(s_2)$. Figure 4.22 shows an example of a multiplication table for the derivatives of the CHINESE POSTMAN problem for partial k-trees.

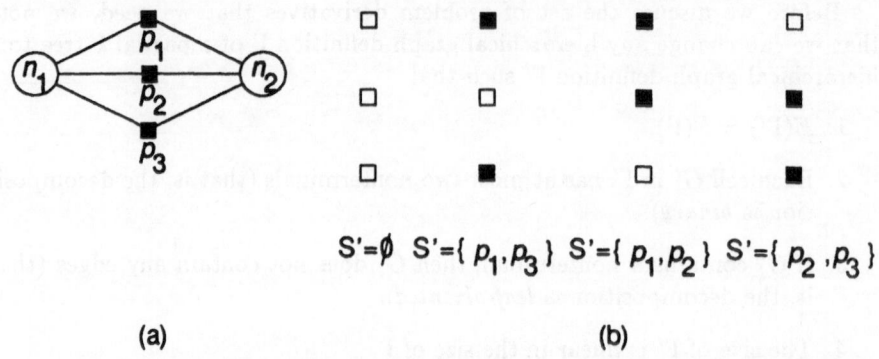

$$S'=\emptyset \quad S'=\{\,p_1,p_3\,\} \quad S'=\{\,p_1,p_2\,\} \quad S'=\{\,p_2,p_3\,\}$$

(a)   (b)

Figure 4.21: *The derivatives of the* CHINESE POSTMAN *problem. (a)*
A cell. The cell has three pins, p_1, p_2, p_3, *two nonterminals,* n_1 *and*
n_2, *and no interior edges. (b) The problem derivations. Odd-degree*
pins are drawn in black; even-degree pins are drawn in white.

The underlying cell with three pins and the corresponding problem derivatives
are depicted in Figure 4.21. In this example, $f(s) = f(s_1) + f(s_2)$ for all table
entries. Therefore, in Figure 4.22, the table entries do contain not this formula,
but rather example sizes.

When processing C, the dynamic programming algorithm fills the multiplication table, using the information on n_1 and n_2 that it has already computed.
Then, it selects for each problem derivative the minimum-cost solution entered
into the table.

Obviously, this kind of decomposition applies to many other optimization
problems, among them the TRAVELING SALESPERSON problem and the MINI-
MUM STEINER TREE problem (see Exercises 4.28 and 4.29).

Often, the hierarchical definition of a partial k-tree is not known. There is,
however, an $O(n^{k+2})$-time parsing algorithm that finds a hierarchical definition
that is leaf-oriented and has a binary hierarchy tree for every partial k-tree
[13].   Robertson and Seymour [379] prove the existence of an $O(n^2)$-time
algorithm for this purpose. However, the algorithm cannot be constructed and
its run time has an impractically large constant factor. For many practically
interesting subclasses of partial k-trees, however, simple parsing algorithms
exist.

4.7.5.1   Relation to the Bottom-Up Method

The bottom-up method discussed in Section 3.13 is closely related to dynamic
programming on hierarchical graphs. There, however, the accent is slightly
different.

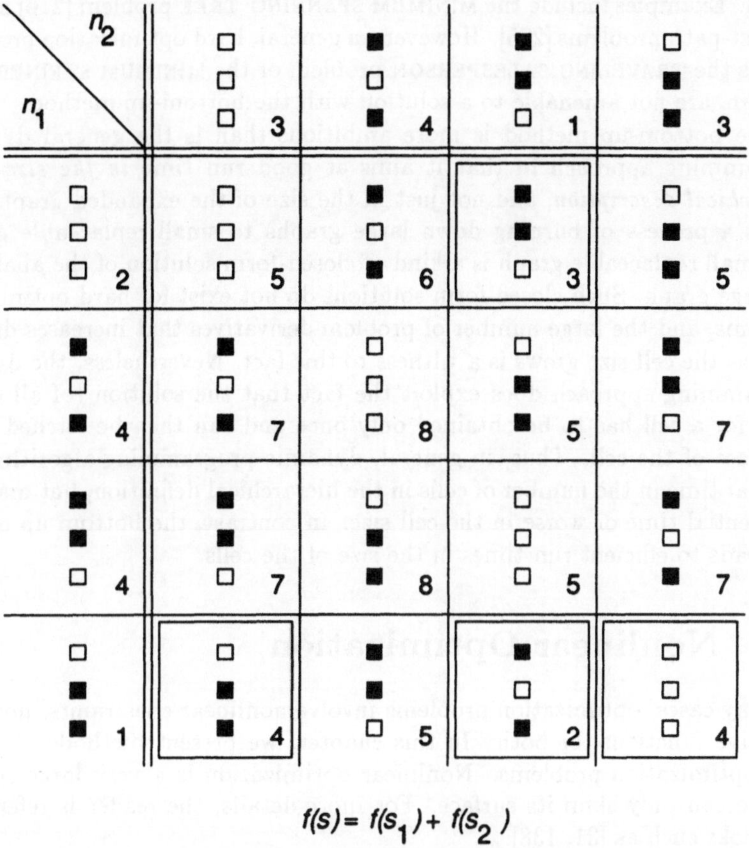

$$f(s) = f(s_1) + f(s_2)$$

Figure 4.22: *The multiplication table for computing the* CHINESE POSTMAN *problem on the cell depicted in Figure 4.21(a). Rows correspond to solutions of* n_1. *Columns correspond to solutions of* n_2. *For all table entries, the size of the solutions for* C *is the sum of the sizes of the optimal solutions* s_1 *for* n_1 *and* s_2 *for* n_2 *that combine to form s. The table contains example sizes of optimal solutions. The optimal solutions for cell* C *of the four problem derivatives are highlighted.*

First, with the bottom-up method, we did not discuss the solution of optimization problems. Rather, we solved decision, construction, and query problems. Some optimization problems can be solved with the bottom-up method, as well. Examples include the MINIMUM SPANNING TREE problem [276] and the shortest-path problems [275]. However, in general, hard optimization problems, such as the TRAVELING SALESPERSON problem or the MINIMUM STEINER TREE problem, are not amenable to a solution with the bottom-up method.

The bottom-up method is more ambitious than is the general dynamic-programming approach in that it aims at good run time *in the size of the hierarchical description*, and not just in the size of the expanded graph. This entails a process of burning down large graphs to small replaceable graphs. The small replaceable graph is a kind of closed-form solution of the analysis of the large graph. Such closed-form solutions do not exist for hard optimization problems, and the large number of problem derivatives that increases dramatically as the cell size grows is a witness to this fact. Nevertheless, the dynamic programming approach does exploit the fact that the solution (of all derivatives) for a cell has to be obtained only once and can then be fetched for all instances of the cell. Thus, in general, dynamic-programming algorithms run in linear time in the number of cells in the hierarchical definition, but may need exponential time or worse in the cell size. In contrast, the bottom-up method also leads to efficient run times in the size of the cells.

4.8   Nonlinear Optimization

In many cases, optimization problems involve nonlinear constraints, nonlinear objective functions, or both. In this chapter, we present methods of solving such optimization problems. Nonlinear optimization is a very large subject, and we can only skim its surface. For more details, the reader is referred to textbooks such as [31, 138].

4.8.1   The Gradient-Projection Method

As an example of greedy local search methods for minimizing nonlinear functions, we discuss the gradient method. The gradient method starts with an arbitrary point in the domain of a nonlinear function and makes improvements in the direction of steepest descent.

Definition 4.23 (Gradient) *Let $f : \mathbf{R}^n \to \mathbf{R}$ be a function that is continuously differentiable at point x_0. The **gradient** of f at point x_0 is defined as*

$$\nabla f(x_0) := \left(\frac{\partial f}{\partial x_1}(x_0), \ldots, \frac{\partial f}{\partial x_n}(x_0) \right)^T$$

The gradient is a vector that is normal to the iso-surface (surface of constant function value) defined by f at point x_0 and points in the direction of the steepest ascent of f at x_0. Thus, the direction $-\nabla f(x_0)$ is the preferred direction of movement from x_0 if we want to find a minimum of f. This observation is the basis of a class of greedy local search methods. They differ in how large a step they take in this direction.

One possible choice is to find the minimum of the function on the ray emanating from x_0 in the direction of $-\nabla f(x_0)$. Finding this point involves a minimization of a one-dimensional nonlinear function, so it may be expensive. Weaker criteria for the choice of the step length are that the function must decrease by some threshold amount if we take the step. We will not detail the possible choices and their relative worth here, but instead refer the reader to the related literature [31, 138].

If f has to be minimized subject to nonlinear constraints, we have to modify the gradient method. We will illustrate this modification, the *gradient-projection method*, using a single nonlinear equality constraint of the type $g(x) = 0$, where g is a function that is continuously differentiable in x_0. This constraint can be envisioned as a smooth surface on which the feasible points are located. We start with a point x_0 on this surface. Walking in the direction of $-\nabla f(x_0)$, we would probably leave the constraint surface, so we project this vector onto a hyperplane that lies tangential to the constraint surface at x_0 (see Figure 4.23). This is the vector

$$\overline{\sigma} = -\nabla f(x_0) + \frac{\nabla f(x_0)^T \nabla g(x_0)}{\nabla g(x_0)^T \nabla g(x_0)} \nabla g(x_0)$$

Indeed, we obtain the projection of $-\nabla f(x_0)$ by subtracting the part that points in the direction of $\nabla g(x_0)$. It is easy to check that $\overline{\sigma}^T \nabla g(x_0) = 0$.

Walking in the direction of $\overline{\sigma}$ still will get us away from the constraint surface, but only tangentially. So, after we decide on a step length t, we have to project the resulting point $x_0 + t\overline{\sigma}$ again on the constraint surface. A natural direction in which we walk for this purpose is $\nabla g(x_0)$. We can obtain the distance d we have to walk in this direction by solving a system of nonlinear equations. Heuristically, we can choose an appropriate step length t and the corresponding value of d, again in an iterative process.

If several equality constraints are involved, then the dimension of the constraint surface decreases. The structure of the gradient-projection method stays the same, but the projection of the gradient—that is, the computation of σ—now involves solving a system of linear equations.

If inequality constraints occur, the set of feasible points is not a surface in $\mathbf{R}^n$, but has nonzero volume. Its boundary is defined by surfaces of *active constraints*. These are inequality constraints that are fulfilled with equality. The gradient-projection method can be modified to keep track of which constraints are active in a given point. For more details, see [138].

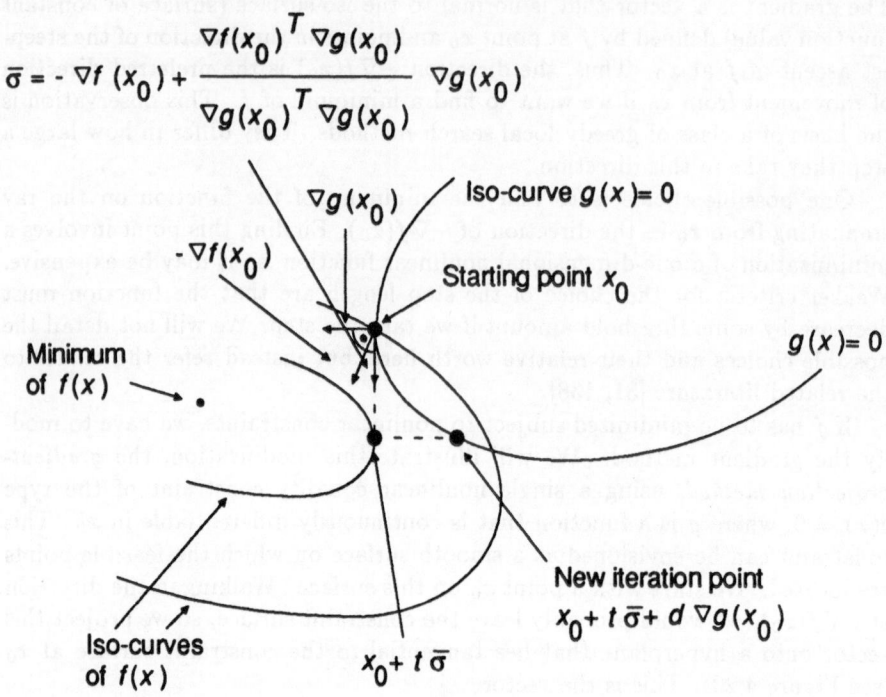

Figure 4.23: *The gradient-projection method illustrated at an example function $f : \mathbf{R}^2 \to \mathbf{R}$ with constraint $g(x) = 0$, $g : \mathbf{R}^2 \to \mathbf{R}$, t is chosen such that $f(x_0 + t\overline{\sigma})$ is minimized. The values for t and d are computed iteratively.*

The gradient-projection method is an inherently greedy method. As such, it involves the danger of being trapped in local minima. However, it can be shown that, with appropriate choices of the step lengths, a local minimum can always be found.

4.8.2   Penalty-Function Methods

Penalty-function methods carry the idea of Lagrangian relaxation into the context of nonlinear optimization. They convert a nonlinear optimization problem with equality constraints into an unconstrained nonlinear optimization problem not by *forbidding* infeasible solutions, but by *punishing* them with extra terms in the cost function. Assume that we want to minimize the cost function $f(x)$ subject to nonlinear constraints $g(x) = 0$, where $g : \mathbf{R}^n \to \mathbf{R}$. Instead of ensuring the constraints, a penalty function method performs unconstrained optimization on the modified cost function

$$P(x, \rho) = f(x) + \frac{\rho}{2} g(x)^T g(x) \qquad (4.29)$$

We will not go into the details on how to ensure the necessary properties of continuous differentiability and so on, but will just note that, under the proper circumstances, there is an unconstrained minimum $x^*(\rho)$ for each value of ρ such that

$$\lim_{\rho \to \infty} x^*(\rho) = x_{\min}$$

where $x_{\min}$ is a constrained minimum of $f(x)$. In practice, we start with a small value for ρ and then increase ρ geometrically, at each time performing an unconstrained minimization of $P(x, \rho)$. It is not advisable to start with a large value for ρ and just to perform one or a few minimizations, because of the possibility of numerical instabilities [138].

Instead of using the quadratic penalty function defined in (4.29), we could use penalty functions based on absolute values or logarithms.

For more details on penalty-function methods, see [138].

4.8.3 Quadratic Optimization

Quadratic objective functions and constraints play a major role in nonlinear layout-optimization problems. The reason is that quadratic objective functions have a natural matrix representation. This allows us to use concepts from linear algebra for their solution. Furthermore, with quadratic objective functions, most of the subproblems that have to be solved during an iterative minimization process have closed-form solutions.

Definition 4.24 (Quadratic Function) *A function* $f : \mathbb{R}^n \to \mathbb{R}$ *is quadratic if it can be written as*

$$f(x_1, \ldots, x_n) := \sum_{i,j=1}^{n} g_{ij} x_i x_j + \sum_{I=1}^{n} c_i x_i + a \qquad (4.30)$$

where

$$g_{ij} = g_{ji} \quad \text{for all } i, j = 1, \ldots, n. \qquad (4.31)$$

The terms involving the g_{ij} *are called the* **quadratic terms**. *The terms involving the* b_i *are the* **linear** *terms.*

In matrix form, a quadratic function can be written as

$$f(x) = x^T G x + c^T x + a$$

The matrix G *is called the* **Hessian matrix** *of* f. *It has property (4.31), that is, it is* **symmetric**. *If* $a = 0$ *and* $c = 0$, *then* f *is called a* **quadratic form**. *In this case,* f *and* G *are often identified.*

If $x^T G x \geq 0$ *for all* x, *then* G *is called* **positive semidefinite**.

Quadratic functions are just polynomials of degree 2 in n variables (Exercise 4.30). Usually, the Hessian matrix is defined as the matrix of second partial derivatives of a function in two variables. In the case of quadratic forms, this would be $2G$, but we drop the factor of 2 for simplicity. The quadratic-programming problem asks us to minimize a quadratic function subject to a system of linear inequalities. The positive semidefiniteness of G is of particular importance in this context, because it implies the convexity of the cost function (Theorem 4.13).

QUADRATIC PROGRAM

Instance: A real $m \times n$ matrix A, a real m-vector b, a real $n \times n$ matrix G, and a real n-vector c

Configurations: All real n-vectors

Solutions: All real n-vectors x that satisfy $A \cdot x \geq b$

Minimize: $f(x) = x^T G x + c^T x$

The concept of duality also applies to quadratic programs. The dual of the quadratic program defined here maximizes the objective function

$$F(y, w) = b^T y - w^T G w \qquad (4.32)$$

subject to the constraints

$$-A^T y \;=\; Gw + c \qquad (4.33)$$
$$y \;\geq\; 0 \qquad (4.34)$$

Note that (4.32), (4.33), and (4.34) are a generalization of rows 1, 5, and 3 in Figure 4.11. General conditions for the existence of solutions of quadratic programs can be found in [138].

Sometimes, even the constraints are quadratic. We call such a problem a QUADRATICALLY CONSTRAINED QUADRATIC PROGRAM . QUADRATICALLY CONSTRAINED QUADRATIC PROGRAMS can be solved with the gradient-projection method. The method works especially well here, because the constraint surfaces and the objective function are convex (see Section 4.8.5). A special case of a QUADRATICALLY CONSTRAINED QUADRATIC PROGRAM comes up particularly frequently. This is the case in which there is only one constraint; namely, the *spherical* quadratic constraint $x^T x = 1$. We call a quadratic program of this form a SPHERICAL QUADRATIC PROGRAM.

Often, integer constraints are involved, as well. A particularly important case is the following.

QUADRATIC ASSIGNMENT

Instance: A distance matrix $D = ((d_{ij}))_{i,j=1,\ldots,n}$, and a cost matrix $C = ((c_{ij}))_{i,j=1,\ldots,n}$

Configurations: All permutations $\pi : \{1, \ldots, n\} \to \{1, \ldots, n\}$

Solutions: All configurations

Minimize: $C(\pi) = \sum_{i,j=1}^{n} d_{ij} c_{\pi(i),\pi(j)}$

Intuitively, the distance matrix D describes the mutual distances between pairs of slots among n *slots* in some geometry. D is arbitrary; thus, the geometry is not restricted in any way. The cost matrix C describes the pairwise connection costs between n *modules*. The entries in C are assumed to denote connection cost per unit distance. The QUADRATIC ASSIGNMENT problem asks us to minimize the total connection cost by choosing an appropriate mapping of modules to slots.

A formulation of the QUADRATIC ASSIGNMENT problem as a QUADRATIC PROGRAM with integer constraints asks us to minimize the objective function

$$\sum_{i,j,k,\ell=1}^{n} d_{ik} c_{j\ell} x_{ij} x_{k\ell}$$

subject to the constraints

$$\sum_{i=1}^{n} x_{ij} = 1 \quad j = 1, \ldots, n$$

$$\sum_{j=1}^{n} x_{ij} = 1 \quad i = 1, \ldots, n$$

$$x_{ij} \in \{0, 1\} \quad i, j = 1, \ldots, n$$

Actually, in this case we can even linearize the quadratic terms in the objective function, such that the QUADRATIC ASSIGNMENT problem becomes a special integer program. There is a rather straightforward way of doing this that yields n^4 additional variables and $O(n^4)$ constraints (Exercise 4.31). Kaufman and Broeckx [218] present a formulation with n^2 new variables and $O(n^2)$ constraints.

The QUADRATIC ASSIGNMENT problem is NP-hard [133]. Overviews of this optimization problem are given in [54, 116, 171, 172]. Hanan and Kurtzberg [171] concentrate on practical issues. Hanan et al. [172] report on the comparative experiments. Burkard [54] also considers theoretical results. Heuristic methods for solving versions of the QUADRATIC ASSIGNMENT problem with regular distance matrices are discussed in Section 6.6.

4.8.4   Quadratic Forms and Eigenvalues

If f is a positive semidefinite quadratic form, then there is a relationship between the solution of a SPHERICAL QUADRATIC PROGRAM minimizing f and eigenvalues of G.

Definition 4.25 (Eigenvalue) *Let G be an $n \times n$ matrix. A value λ such that there is a vector $x \neq 0$ with $G \cdot x = \lambda \cdot x$ is called an **eigenvalue** of G. The vector x is an **eigenvector** of G with eigenvalue λ.*

*The **characteristic polynomial** for A is the polynomial*

$$P_G(\lambda) = \det(\lambda \cdot I - G)$$

The following lemma summarizes basic facts on eigenvalues and eigenvectors [12].

Theorem 4.12

1. *The eigenvalues are exactly the roots of the polynomial $P_G(\lambda)$. This polynomial has degree n.*

2. *The eigenvectors for each eigenvalue λ form a vector space. Its dimension is the multiplicity of λ as a root of $P_G(\lambda)$.*

3. *If G is symmetric, then it has an* orthonormal basis *of eigenvectors $e_1, \ldots, e_n$—that is, a basis that fulfills*

$$e_i^T e_j = \begin{cases} 0 & i \neq j \\ 1 & i = j \end{cases}$$

Let us assume that G is symmetric. Let $e_1, \ldots, e_n$ be an orthonormal basis of eigenvectors ordered such that, for the corresponding eigenvalues, $\lambda_1 \leq \cdots \leq \lambda_n$ holds. If we use $e_1, \ldots, e_n$ as a basis—that is, if we express the vector x as

$$x = \sum_{i=1}^{n} x_i e_i$$

then the cost of x is

$$
\begin{aligned}
f(x) =\ & x^T G x \\
=\ & \left(\sum_{i=1}^{n} x_i e_i^T \right) G \left(\sum_{j=1}^{n} x_j e_j \right) \\
=\ & \sum_{i,j=1}^{n} x_i x_j \cdot e_i^T G e_j \\
=\ & \sum_{i,j=1}^{n} x_i x_j \cdot \lambda_j e_i^T e_j \qquad \text{since the } e_i \text{ are eigenvectors} \\
=\ & \sum_{i=1}^{n} x_i^2 \lambda_i \qquad\qquad\quad \text{since } e_1, \ldots, e_n \text{ is an orthonormal basis}
\end{aligned}
$$

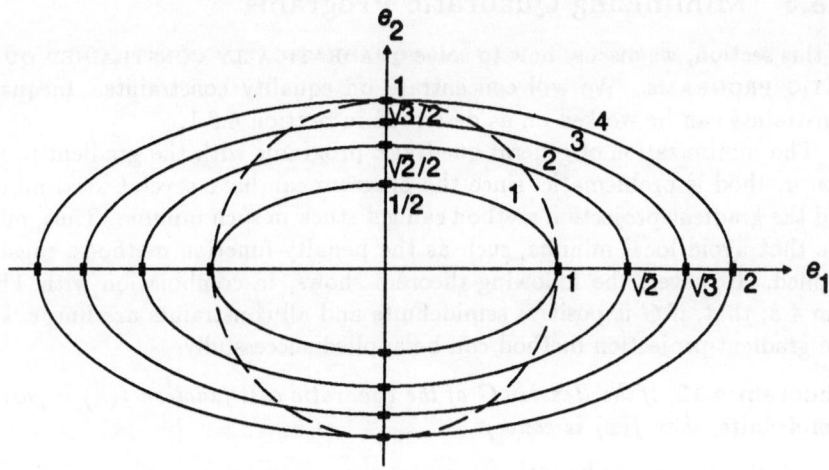

Figure 4.24: *Iso-curves of a positive semidefinite quadratic form* $x^T G x$, *with*

$$G = \begin{pmatrix} 1 & 0 \\ 0 & 4 \end{pmatrix}$$

with eigenvalues $\lambda_1 = 1$ *and* $\lambda_2 = 4$. *The coordinate axes* e_1 *and* e_2 *are defined in terms of an orthonormal eigenvector basis. The unit circle* $\{x^T x = 1\}$ *is outlined by the dashed line.*

Thus, the mixed contributions to the objective functions disappear, and each component x_i makes an independent contribution to the cost that is weighted with the value λ_i. This independence between the contributions of different basis vectors to the objective function is the prime advantage of using $e_1, \ldots, e_n$ as a basis.

Theorem 4.12, part 3 affords us the following geometric intuition about *positive semidefinite* quadratic forms. If G is a symmetric positive semidefinite matrix, then iso-curves of the quadratic form $x^T G x$ are ellipsoids that are centered around the origin (see Figure 4.24). If an orthonormal basis of eigenvectors of G is chosen, then the axes (of symmetry) of the ellipsoids are aligned with the coordinate axes. For each ellipsoid, the ratio of the lengths of two axes is the inverse square root of the ratio of the eigenvalues of the respective eigenvectors.

For more details on eigenvalues, see [147].

4.8.5   Minimizing Quadratic Programs

In this section, we discuss how to solve QUADRATICALLY CONSTRAINED QUAD-RATIC PROGRAMS. We will concentrate on equality constraints. Inequality constraints can be worked on as described in Section 4.8.1.

The minimization of general quadratic programs with the gradient-projection method is problematic, since the program can have several local minima and the gradient-projection method can get stuck in such minima. Thus, methods that avoid local minima, such as the penalty-function methods, must be applied. However, the following theorem shows, in combination with Theorem 4.8, that, if G is positive semidefinite and all constraints are linear, then the gradient-projection method can be applied successfully.

Theorem 4.13 *If the Hessian G of the quadratic cost function $f(x)$ is positive semidefinite, then $f(x)$ is convex.*

Proof See Exercise 4.32. $\square$

Based on the intuition from Figure 4.24, we can prove the following theorem, which shows that, for SPHERICAL QUADRATIC PROGRAMS with a cost function that is a positive semidefinite quadratic form, we can find the minimum cost using eigenvalues. This theorem includes the case that the quadratic form is not positive semidefinite.

Theorem 4.14 *Let $x^T G x$ be a quadratic form where G is a symmetric matrix with eigenvalues $\lambda_1 \leq \lambda_2 \leq \cdots \leq \lambda_n$. Then the following holds:*

1. *The minimum of $\frac{x^T G x}{x^T x}$ is λ_1.*

2. *If the additional constraints $x^T c_i = 0$ are imposed, where c_i, $i = 1, \ldots, k$ are arbitrary vectors, then the minimum of $\frac{x^T G x}{x^T x}$ is no larger than λ_{k+1}.*

Proof Part 1 follows from the following observation. Note that

$$x^T x = \sum_{i=1}^{n} x_i^2$$

since $e_1, \ldots, e_n$ is orthonormal. Thus, $\frac{x^T G x}{x^T x}$ is the weighted average of the eigenvalues, where the weights are chosen to be $x_i^2 / x^T x$. Clearly, the vector x such that $x_1 = 1$ and $x_2 = \cdots = x_n = 0$ is an optimal solution in this case. This is the normalized eigenvector of the smallest eigenvalue.

For the proof of part 2, see Exercise 4.33. $\square$

By Theorem 4.14, part 1, we can minimize a SPHERICAL QUADRATIC PROGRAM directly by finding a normalized eigenvector of λ_1. Efficient methods from linear algebra can be employed here [147].

If the quadratic cost function involves a positive semidefinite matrix G and an additional linear term, then the following theorem applies.

Theorem 4.15 (Blanks [37])

1. *A* SPHERICAL QUADRATIC PROGRAM *with a cost function that involves a positive semidefinite matrix G with distinct eigenvalues and an additional linear term has exactly two local minima.*

2. *If some eigenvalues are identical, then there can be infinitely many local minima, but each local minimum has one of at most two function values.*

Proof Exercise 4.34. □

Thus, the gradient-projection method can be used with reservations or if supplemented with an experimental or theoretical analysis about the location and relative cost of the two minima.

In some cases, quadratic programming provides suitable relaxations for the QUADRATIC ASSIGNMENT problem. Section 6.6 presents examples from the area of graph partitioning.

4.9   Exercises

4.1 Theorem 4.1 is a special case of the *ergodic theorem* on Markov chains. To state the theorem, we need a few more definitions. Let $P^k = ((p_{ij}^{(k)}))$. Note that $p_{ij}^{(k)}$ denotes the probability of moving from s_i to s_j in exactly k search steps. We define the *period* of a state s_i to be the greatest common divisor of all values k such that there exists a path of length k from s_i to s_i with nonzero transition probability. A state of k with period 1 is called *aperiodic*. An irreducible Markov chain is called *ergodic* if it has a unique stationary distribution $\pi^{(0)}$, and for the matrix $\Pi^{(0)}$ consisting of n copies of columns $\pi^{(0)}$ we have

$$\Pi^{(0)} = \lim_{n \to \infty} P^n$$

We can now state the ergodic theorem.

Ergodic Theorem: (G, p) is ergodic exactly if it is irreducible and if all states are aperiodic.

For a proof of the ergodic theorem see, for instance, [110, Chapter XV].

Prove that, if $p_{ii} > 0$ for some $i = 1, \ldots, n$ and (G, p) is irreducible, then all states of (G, p) are aperiodic. Thus, Theorem 4.1 is a special case of the ergodic theorem.

4.2 Prove that the ergodic Markov chain in Figure 4.2(a) has the stationary probability distribution depicted in Figure 4.2(c). Analyze how quickly a search starting at the state with cost 4 converges to this probability distribution. (Use a formula manipulator, such as MACSYMA or MAPLE.)

4.3 Prove Lemma 4.2.

Hint: Let $p(\sigma)$ be the product of all transition probabilities along a path σ in the generating chain G. Show that there is a particular bijection $\sigma \leftrightarrow \widetilde{\sigma}$ on all paths (simple and nonsimple) in G such that $p(\sigma) \leq \alpha \cdot p(\widetilde{\sigma})$ for all paths σ for an appropriate constant $\alpha > 0$. The bijection is related to the concept of path reversal. If σ is a simple path, choose $\widetilde{\sigma}$ to be the reversal of σ. Extend this mapping appropriately to nonsimple paths. Use the bijection to prove the lemma.

4.4 Consider Figure 4.4. Replace the acceptance function (4.5) with the function $f(c_i, c_j, T) = \frac{c_i}{c_j}T$. Note that $\lim_{T \to 0} f(c_i, c_j, T) = 0$ if $c_i < c_j$. But $f(c_i, c_j, T)$ is not multiplicative. Compute the stationary probability distribution for the Markov chain in Figure 4.4(a) that is based on this acceptance function. (Use a formula manipulator, such as MACSYMA or MAPLE.) Show that, as $T \to 0$, the stationary probability of vertex 3 converges to a nonnegative value. This is an indication that the multiplicativity of $f(c_i, c_j, T)$ may be necessary in some cases for the acceptance function to be admissible.

4.5 Prove Lemma 4.3.

4.6 Prove Lemma 4.4.

4.7 Prove Theorem 4.5.

4.8 Prove Theorem 4.6.

4.9 Prove Theorem 4.7.

4.10 Prove Theorem 4.8.

4.11 Show that, in the proof of Theorem 4.9, the corners of P'' are exactly the intersection points of H with the edges of the skeleton of P that are incident to s.

4.12 Prove Theorem 4.10.7.

4.13 Prove Theorem 4.10.8.

4.14 Prove Theorem 4.10.9.

4.15 The TRAVELING SALESPERSON problem can be formulated as an integer program with only $O(n)$ constraints. To do this reformulation, modify an instance of the TRAVELING SALESPERSON problem such that the optimality of a tour is not changed and minimizing the length of the tour is tantamount to minimizing the longest edge in the tour. Then, formulate the corresponding problem, the BOTTLENECK TRAVELING SALESPERSON problem, as an integer program. Does the asymptotic size of a problem instance change during this transformation from the TRAVELING SALESPERSON problem to the integer program?

4.16 Modify the minimum-spanning-tree algorithms by Prim and Kruskal such that the only spanning trees generated are those that include all edges in a set IN and exclude all edges in a set OUT.

4.17 Use incremental methods to compute U_z and L_z quickly, based on the bounds for the parent of z in the branch-and-bound tree.

4.18 Modify the approximation algorithm for the MINIMUM STEINER TREE problem by Kou et al. [242] such that the only Steiner trees generated are those that include all edges in a set IN and exclude all edges in a set OUT.

4.19

1. Formulate the recursive version of the branch-and-bound algorithm for integer programming that results from implementing D as a stack and using the incremental scheme for generating configurations.

2. Discuss what characteristics a problem must have so that the nodes of T can be generated efficiently using the incremental scheme. Give examples for such problems.

4.20 To what does the Lagrangian lower bound for the MINIMUM STEINER TREE problem discussed in Section 4.6.3.1 reduce if we choose the following values for $s_{e,k} = 0$, $e \in E$ and $k = 2, \ldots, r$?

1. $s_{e,k} = 0$ for all $e \in E$ and $k = 2, \ldots, r$

2. $s_{e,k} = \lambda(e)/(r - 1)$ for all $e \in E$ and $k = 2, \ldots, r$

3. $s_{e,k} = 0$ if $e \in \text{IN}_z$, $s_{e,k} = \infty$ if $e \in \text{OUT}_z$; $s_{e,k} = \lambda(e)/(r - 1)$ otherwise

4.21 Give an example of an integer matrix A that is not totally unimodular, but such that, for all integer vectors b, all corners of the polytope $Ax = b, x \geq 0$ are integer.

4.22 Show that, if A is an integer $m \times n$ matrix and A is totally unimodular, then, for all integer m-vectors b, all corners of the polytope of $Ax \leq b, x \geq 0$ are integer.

4.23 Prove Lemma 4.7 by induction on the size of the unimodular submatrices.

4.24 Derive the integer program for the single-pair shortest-path problem.

4.25

1. Prove the following lemma.

Lemma *Let $G = (V, E)$ be a complete graph with edge-weight function λ that satisfies the triangle inequality*

$$\lambda(\{u, w\}) \le \lambda(\{u, v\}) + \lambda(\{v, w\}) \text{ for all } u, v, w \in V$$

Then, for each set $R \subseteq V$ of required vertices, there is an minimum Steiner tree with at most $2r - 2$ vertices.

2. Consider the following algorithm for the MINIMUM STEINER TREE problem. Let (G, R) be the problem instance, $|R| = r$.

Step 1: Compute the shortest distance network D between all vertices in G. (This is a complete undirected graph in which the edge $\{v, w\}$ has the length of the shortest path between v and w in G.)

Step 2: For each possible set $R' \subseteq V \setminus R$ of $r - 2$ required vertices, find the minimum spanning tree in the graph induced in D by $R \cup R'$. Let T' be the smallest such spanning tree for any R'.

Step 3: Map T' back into G, transforming edges in T' into their corresponding paths in G and breaking cycles. This yields a minimum Steiner tree T.

Prove the correctness of this algorithm. What is the asymptotic run time of the algorithm?

4.26 Modify the dynamic-programming algorithm for the MINIMUM SPANNING TREE problem such that it runs in polynomial time in the number of required vertices on one face of a planar graph and in exponential time in the number of required vertices on all other faces of the graph.

Hint: Solve the special case that all required vertices are on the same face first (see [28]).

4.27 Let Γ be a hierarchical graph. Transform Γ into a hierarchical graph Γ' such that

1. $E(\Gamma) = E(\Gamma')$.

2. In Γ', each cell that contains nonterminals does not contain any edges.

3. The hierarchy tree of Γ' is binary.

How large is Γ' in general? How large is Γ' for partial k-trees?

4.28 Develop a decomposition of the TRAVELING SALESPERSON problem for dynamic programming on partial k-trees. How many derivatives do you need?

4.29 Develop a decomposition of the MINIMUM STEINER TREE problem for dynamic programming on partial 2-trees. How many derivatives do you need?

4.30 Prove that the set of quadratic functions (Definition 4.24) is exactly the set of polynomials of degree at most 2 in n variables. To do so, show that each such polynomial can be written as in (4.30), where $g_{ij} = g_{ji}$ for all $i, j = 1, \ldots, n$.

4.31 Formulate the QUADRATIC ASSIGNMENT problem as an integer program by using n^4 new variables and $O(n^4)$ constraints.

4.32 Prove Theorem 4.13.

4.33 Prove Theorem 4.14.2 by induction on k.

Hint: Show that the case in which the minimum is largest occurs when $c_i = e_i$, where e_i is the eigenvector of λ_i in an orthonormal eigenvector basis $(1 \le i \le k)$ of G. In this case, the minimum of $\frac{x^T G x}{x^T x}$ subject to the constraints is exactly λ_{k+1}.

4.34 Prove Theorem 4.15. To this end, extend the intuition given in Figure 4.24 from quadratic forms to quadratic functions.

Part II

Combinatorial Layout Problems

Chapter 5

The Layout Problem

5.1 Requirements

As we already mentioned in Chapter 1, the layout problem is a constrained optimization problem. There is a multitude of cost functions that is of interest in layout. The most important ones are the following.

- *Layout area:* This cost measure has a large effect on chip delay, but it is even more important because of its dramatic influence on chip yield. The *yield* of a chip is the percentage of fabricated chips that are working. The yield depends on the number of defects that are introduced during chip fabrication. Defects are introduced by statistical phenomena such as dust particles or defects in deposited materials. In many technologies, even a single defect causes the chip to be completely nonfunctional. Statistical models for defects of fabrication imply that the yield Y depends critically on the chip area A. Often, the dependence is a dramatic, as $Y = e^{-cA}$ for a suitable constant c [140].

- *Maximum wire length:* The wire length is of importance because it influences the delay of signals sent along the wire. There are several models for wire delay. Depending on what phenomena they incorporate (resistance, capacitance, and so on), they range from a logarithmic delay in the wire length to a delay that is quadratic in the wire length [140]. The assumption that the wire delay does not depend on the wire length is valid only if either the scale of integration is quite small (not more than a few ten thousand devices per chip), or we formulate the delay in terms of the number of clock cycles in synchronous circuits. In circuit design, the wire delay cannot be neglected, in general.

  In many formulations of the layout problem that involve wire length, the length of the longest wire is minimized. Often, this is not suitable in

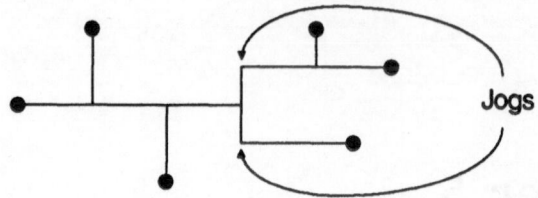

Figure 5.1: *Jogs in a wire with several terminals.*

practical situations. Rather, wire-length minimization is important only
for the small part of the circuit that is critical for the circuit delay.

- *Crossing number:* The number of wire crossings in a circuit layout may
  affect the electrical robustness of the circuit. This cost measure is not
  of major importance for practical circuit layout. However, it serves as
  a combinatorial parameter elucidating the structure of the layout (see
  Section 5.7).

- *Via count:* A *via* or *contact* is a change of wiring layers in a wire. Vias usually introduce electrical instabilities that effectivity decreases the yield.
  Even if this is not the case, vias usually have an adverse effect on the
  delay along the wire. The via count should therefore be minimized. This
  is especially important in printed-circuit-board layout.

- *Maximum number of jogs per wire:* A *jog* is a bend in a wire. Jogs and
  vias are related if there are preferred directions of wiring in wiring layers,
  but in general they need not be correlated. A jog has adverse effects on
  the delay and other electrical properties of a wire. Often, these effects are
  of minor importance, so this cost measure does not have a high priority in
  layout design. However, the number of jogs is considered in some channel
  routers (see Chapter 9). Note that Steiner trees for hyperedges have to
  bend at vertices of degree 3. So the only significant jogs are those where
  the Steiner tree (see Figure 5.1) changes direction at a vertex of degree
  2.

In different applications, these cost measures are minimized with different priorities. Area always tends to be the most important one. Therefore, in many
combinatorial algorithms for circuit layout, only area is considered. However,
the other cost measures do play a role, as well, and combinatorial solutions to
the layout problem should incorporate them more than they usually do today.

The minimization of the cost measures has to be done under certain constraints imposed by the design or fabrication technology. These constraints

are supposed to ensure that the circuit is fabricated correctly. They include minimal sizes for features such as transistors and wires (to ensure proper fabrication) and minimum separations between different features (to eliminate the danger of short circuits). There can be a multitude of other constraints. Examples are the preferred direction of wiring in certain wiring layers (horizontal or vertical); the exclusion of certain geometric shapes—for instance, of 45° angles (mostly dictated by the design technology); and the requirement that two power-supply wires not cross.

Thus, there is a large number of variants of the layout problem; for a particular application, even selecting the right formulation is difficult. Consequently, we cannot expect a single layout algorithm or method to be satisfactory for all applications.

The combinatorial difficulty of most of the formulations of the layout problem has as a consequence that many layout algorithms can deal with only a restricted class of layout problems. The fact that many layout algorithms consider area as the only cost measure is only one facet of this problem. In general, layout methods are highly customized to certain technological boundary conditions, and changes in the technology require a fundamental revision of the corresponding layout systems, if they do not render the systems completely useless. One of the major research goals in circuit layout is to improve the *robustness* of layout algorithms and technologies. The layout algorithm should

- Be adaptable to a wide range of technological boundary conditions

- Allow us to incorporate several cost functions of varying degrees of importance

We usually have to pay a price for robustness. A robust algorithm has to deal with more freedom in the problem formulation. Usually, this can be achieved only at the cost of increased run time or, even worse, of reduced quality of the constructed layout. With the fast pace of the technological progress in microelectronics, the robustness of a layout method is well worth a reasonable sacrifice in the performance of the algorithms or in the quality of the output. Today, however, not many layout algorithms are robust.

In this chapter, we elucidate the combinatorial structure that is common to all variants of the layout problem. To this end, Section 5.2 contains a combinatorial definition of the layout problem that abstracts from requirements particular to a given fabrication technology. The remaining sections of this chapter develop and discuss an algorithm that produces provably good layouts. This algorithm is not practical, but it exhibits a tight correlation between the various cost measures of an optimal layout and a certain structural parameter of the circuit.

5.2   Definition of the Layout Problem

In this section, we give a definition of the layout problem that formalizes the major characteristics common to all integrated-circuit fabrication technologies. These are as follows:

- *Finiteness of resolution:* Each fabrication process for integrated circuits has a finite minimum feature size. We take this fact into account by formulating the layout problem as a discrete optimization problem.

- *Quasiplanarity:* In integrated circuits, wiring can be done in only a re-stricted number of wiring layers. This number is larger than 1, allowing for the realization of nonplanar circuits, but it is small compared to the circuit size (between 2 and 4 on chips, and up to about 30 on printed circuit boards). Thus, the layout problem retains the flavor of a planar embedding problem.

We now give the formal definition of the layout problem.

LAYOUT

Instance: A hypergraph $G = (V, E)$, $|V| = n$, $|E| = m$, each vertex has degree at most 4

Configurations: All embeddings of G into an infinite grid graph (also called a *mesh*), such that the vertices of G are placed on nodes of the grid, and the nets of G are realized as Steiner trees in the grid, and such that a net $e \in E$ is not incident to a vertex $v \in V$ unless the grid node corresponding to v lies on the Steiner tree corresponding to e

Solutions: All embeddings in which no two Steiner trees share the same grid edge

Minimize: The size of the smallest rectangular grid subgraph containing the embedding

The cost measure minimized in the LAYOUT problem is called *area*. Figure 5.2 gives an example of a legal embedding for a small hypergraph.

The LAYOUT problem was first defined in [442]. It abstracts from such issues as details of the admissible geometries, number of wiring layers, and restrictions on wiring on different layers. Thus, it only roughly captures the requirements on layout in practical situations. This makes this problem formulation impractical for optimizing real-world layouts, in which many of the features not modeled here have to be taken into account. The LAYOUT problem is, however, well suited for an investigation of the basic combinatorial properties of circuit layout.

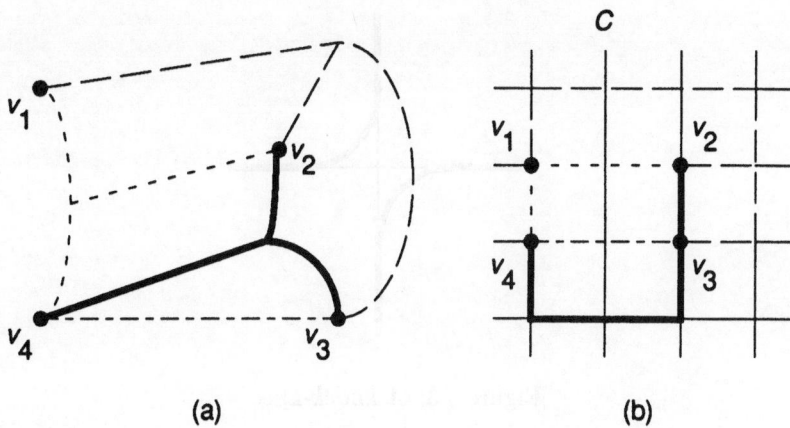

(a)   (b)

Figure 5.2: *A legal embedding of a hypergraph. (a) A hypergraph G with four vertices and three hyperedges. (b) A legal embedding of G with area 16. By eliminating the column labeled C, we can reduce the area to 12. No rows can be eliminated without introducing overlap of different nets.*

Given the technology-independent nature of the LAYOUT problem, it is appropriate to analyze the problem asymptotically with a rough estimate of the constants involved. Before we investigate the solution of the LAYOUT problem, we will mention which of the restrictions inherent in its definition are only technical and do not significantly change the combinatorial structure of the problem, as long as we are concerned with only asymptotic layout costs (see also Exercises 5.1 and 5.2).

- *Grid graph:* The fact that the host graph of the embedding is a grid restricts the model to *Manhattan style* layouts. These are layouts in which only features aligned with the Cartesian coordinate axes are allowed. We can model 45° angles or other finite-resolution geometries by enlarging the grid by a constant factor in each dimension. This modification changes the layout area by only a constant factor.

- *Degree 4 vertices:* This restriction has been made to allow for at least one grid embedding. The restriction is not significant, given the fact that the LAYOUT problem does not exhibit any hierarchical structure, and thus vertices of G represent small circuit elements, such as transistors. On the other hand, there are straightforward extensions of the model that allow for larger degrees. Here, a vertex of degree d is embedded not into a single grid node, but rather into a square or rectangle of grid nodes whose

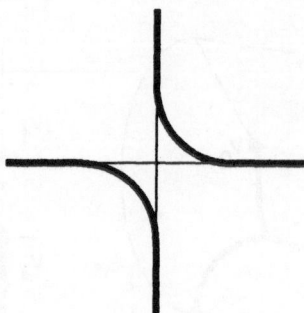

Figure 5.3: *A knock-knee.*

periphery can accommodate pins of d wires. We achieve the results on this extension of the LAYOUT problem by generalizing the solution method presented in this chapter [418].

- *Edge-disjointness:* This requirement formalizes the quasiplanarity of the layout. The number of available layers or technological restrictions on routing in a layer affect only the constant factor of the layout area, if the number of wiring layers is bounded by a constant. For instance, if two wiring layers are available, one for vertical wiring and one for horizontal wiring, then *knock-knees* (see Figure 5.3) cannot be realized. We can, however, remove knock-knees from the layout model by introducing new gridlines between each pair of adjacent horizontal and vertical gridlines, modifying the layout and assigning layers appropriately.   As another example, we can model the existence of k wiring layers by expanding the grid by a factor of k in each direction, and assigning one of the k interleaved original-sized subgrids inside the expanded grid to each layer. This again does not change the asymptotic layout area.

Each of these adaptations to technological constraints changes the layout area by a large amount; in general by at least a factor of 4. Thus, these adaptations are completely out of the question for practical purposes. In this chapter, however, we are concerned with only the asymptotic growth in the layout area (and other cost measures). For this purpose, all changes of the grid size by a constant factor are admissible.

Other cost measures that are interesting in the context of the LAYOUT problem are the crossing number C—that is, the number of crossings occurring on grid vertices (excluding knock-knees!); the maximum number J of jogs per wire; and the maximum wire length L, where each edge of the grid is assumed to have unit length. All of these cost measures should be minimized. This optimization is discussed in more detail in Section 5.7. Since the layout model does not assign wiring layers to wires or wire segments, the via count cannot be represented.

Kramer and van Leeuwen [245] proved that the LAYOUT problem is NP-hard, even if G is restricted to be a connected undirected graph. (Formann and Wagner [119] pointed out a mistake in the proof by Kramer and van Leeuwen and corrected it.) Thus, we must investigate suboptimal methods for solving the LAYOUT problem.

Our goal for this chapter is to present an approximation algorithm for the LAYOUT problem. This algorithm has been presented by [34]. It has a provably small, but still unbounded, error. As presented, the algorithm handles only graphs, but extensions to hypergraphs are possible. The algorithm is a refined version of a popular placement heuristic. This heuristic is discussed in the next section.

5.3 The Mincut Heuristic

A widely used heuristic for circuit placement is based on the hypothesis that what governs the area needs of a circuit is a certain measure of congestion. The somewhat surprising implication of the theory presented in this chapter is that, up to a small error, this is *the only parameter* that governs the area needs of a circuit. In this section, we describe the heuristic intuitively. In Section 5.4, we precisely define the notion of congestion that we need to get tight worst-case bounds. In Section 5.5, we present the algorithms and proofs. Improvements on the layout algorithm are discussed in the remaining sections of this chapter.

Consider the following recursive method for laying out hypergraphs. We cut the hypergraph G into two pieces that have approximately the same number of vertices. We can do this by severing certain hyperedges, such that two disconnected subgraphs G_1 and G_2 result. The severed hyperedges form the *cut S* (see Definition 3.2). We can obtain a layout of G by laying out G_1 and G_2 recursively, and then routing the hyperedges in S, expanding the grid where necessary (see Figure 5.4). Here, care has to be taken that the two recursively computed sublayouts have approximately the same length in the dimension in which they abut. This recursive partitioning scheme can be continued until a subgraph becomes trivial to lay out—for instance, until it is a single vertex.

In Figure 5.4, the cutline is oriented vertically in the layout. Of course, we can also orient cutlines horizontally. As we descend recursively, it makes sense to alternate the directions of the cutlines such that the layouts of the subgraphs stay roughly square. Thus, we get a picture like that in Figure 5.5.

The size of the layout critically depends on the amount of grid expansion that we have to do in each recursive call to the layout procedure. Here, we must expect to require a lot more space for the grid expansion for calls at a small recursion depth than we require for calls at a large recursion depth. This

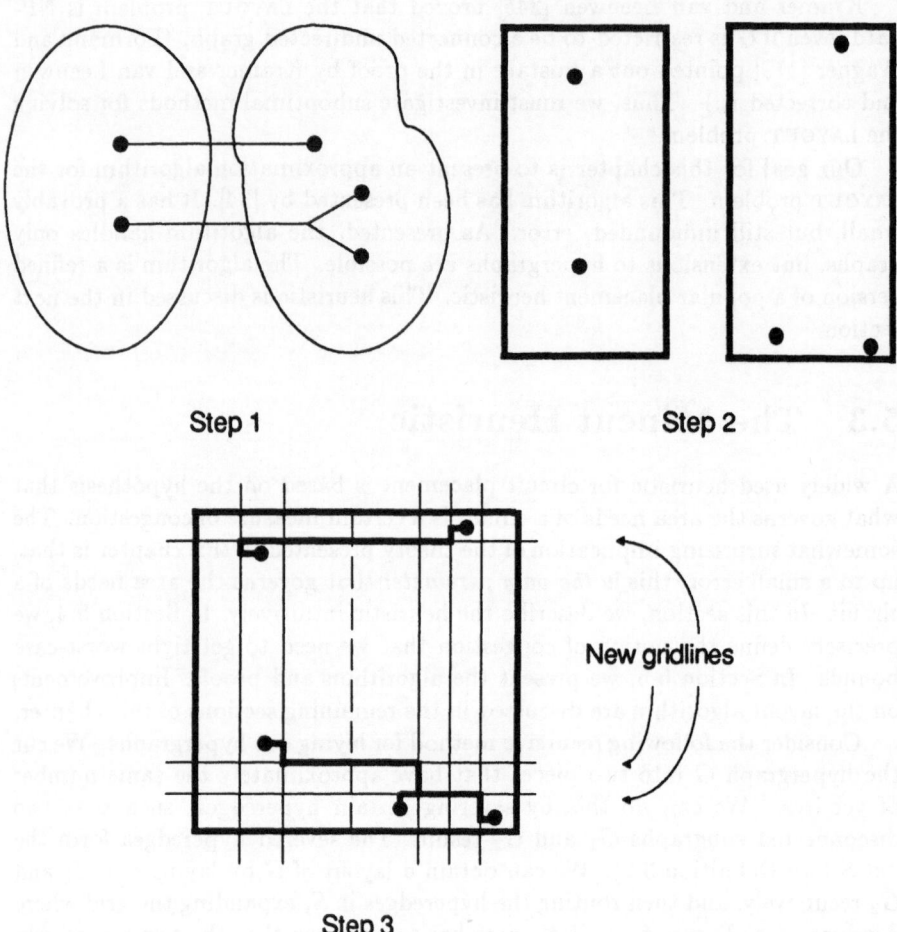

Figure 5.4: *The mincut heuristic. Step 1: Split the hypergraph into two roughly equal-sized parts such that few wires cross the cut. Step 2: Lay out the parts recursively. Step 3: Route the wires across the cut by grid expansion.*

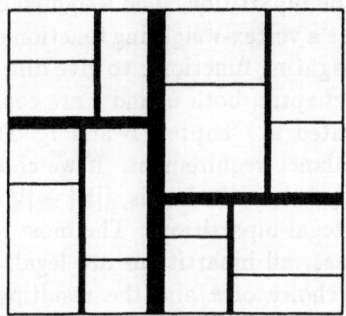

Figure 5.5: *A mincut partition with several levels of recursion. The thickness of the cutline indicates the recursion depth at which the cutline was generated. The thickness decreases with increasing recursion depth.*

is because the hyperedges combining large subgraphs must use longer wires than do the hyperedges combining small subgraphs, in general. Thus, we aim at making all subgraphs small quickly. This results in the requirement that the cut be *balanced*; that is, that both subgraphs in a cut be of about the same size. Balancing the cut together with a carefully choosing the directions of the cutline also ensure that layouts of subgraphs have a roughly square shape. Such layouts consume less area than do skinny layouts.

In addition to balancing the cut, we want to cut as few hyperedges as possible, in order to reduce the amount of wiring necessary for combining the subgraphs. Formally, the problem we want to solve at each recursive call is the following:

BALANCED BIPARTITION

Instance: A hypergraph $G = (V, E)$, with vertex-weighting function
$w : V \rightarrow \mathbb{N}$ and edge-weighting function $c : E :\rightarrow \mathbb{N}$; and a number k

Configurations: All partitions $V' \subset V$, $V' \neq \emptyset$ of vertices

Solutions: All *balanced* partitions $V' \subset V$ such that $w(V'), w(V \setminus V') \leq k$

Minimize: $c(S)$, where $S = \{e \in E|$ there are $v' \in V', v \in V \setminus V'$ such that $v, v' \in e\}$

The BALANCED BIPARTITION problem asks us to cut a hypergraph into two about equal-sized pieces with a minimum-cost cut. The sets V' and $V \setminus V'$ are also called the *sides* of the bipartition. The weighted cost of a cut is also called the *cutsize*. We introduce a vertex-weighting function w to give different sizes to vertices, and an edge-weighting function c to give different costs to hyperedges. For the purpose of this chapter, both w and c are constant 1; other definitions of w and c are investigated in Chapters 6 and 7. The number k determines the stringency of the balance requirement. If we choose $k = \lceil w(V)/2 \rceil$, then the bipartition is *strictly balanced*; that is, $||V| - |V \setminus V'|| \leq 1$. Any smaller value of k allows for no legal bipartitions. The most relaxed balance condition is $k = w(V)$. In this case, all bipartitions are legal. Obviously, there is also a tradeoff between the choice of k and the resulting cutsize. If we make k small, the variety of legal bipartitions decreases, and thus the minimum cutsize increases. On the other hand, relaxing the balance condition potentially leads to larger subgraphs, which are harder to cut in subsequent recursive calls. The analysis of Section 5.4 will resolve how to make this tradeoff so as to achieve tight worse-case bounds on the deviation from the optimum area.

The mincut heuristic has been used in block placement for circuit layout for a long time (see Section 7.2.1). The heuristic was introduced by Breuer [49], and it is still one of the most promising approaches to circuit placement. This is so despite the fact that the BALANCED BIPARTITION problem is NP-hard. Fortunately, quite successful heuristics exist for this problem (see Section 6.2). In practice, the choice of k is usually made heuristically, and k is chosen close to $w(V)/2$—that is, close to one-half of the total vertex weight, so as to achieve balanced partitions.

5.4   Graph Bifurcation

Although minimizing the cutsize works well in practice, it is not amenable to analysis, because the cutsizes at each level are difficult to evaluate. Therefore, to get an analyzable layout algorithm, we will correlate the cutsizes on different levels of the recursion. The basic idea is to make an allowance for the cutsizes at each level that ensures that, on each level, wiring all cuts takes about the same amount of area. We will now make this idea precise and discuss why it works.

To investigate this issue further, we study the recursion tree of the bipartitioning process. In the rest of this chapter, we consider only graphs.

Definition 5.1 (Cut Tree) *A* **cut tree** *T for a graph G is the recursion tree of a bipartitioning process on G. The root of T represents G. Each interior node v of T represents a subgraph G_v of G with at least two vertices. The two children v_1, v_2 of v represent the sides of a bipartition of G_v into two nonempty graphs G_{v_1} and G_{v_2}. The node v corresponds to the bipartition itself. Leaves in T represent single vertices in G.*

Figure 5.6 displays a cut tree. Obviously, any recursive bipartition induces a

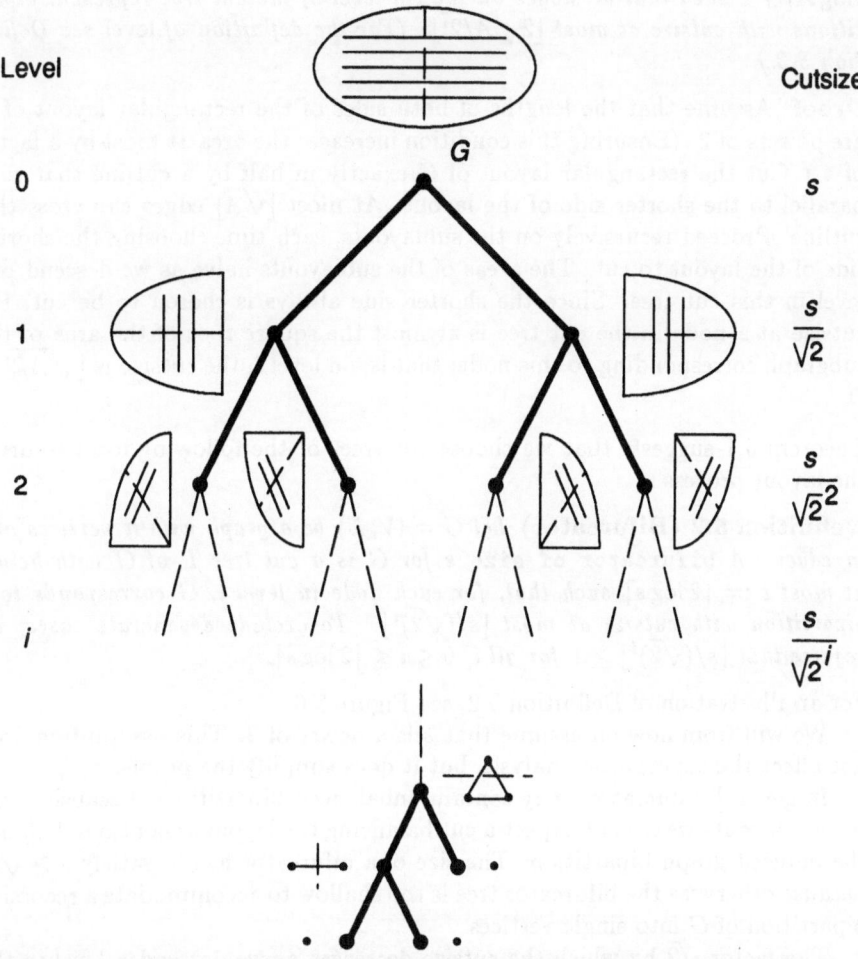

Figure 5.6: *A cut tree with appropriate cutsize for a bifurcator of size s.*

cut tree.

Rather than devising a tradeoff between choices of cutsizes and balance criteria for nodes in the cut tree explicitly, we declare an allowance for the cutsize at each node in the tree.

We will find a suitable choice for this allowance by reversing the direction of our investigation. Instead of turning a cut tree into a layout, let us turn a layout into a cut tree.

Theorem 5.1 *If G can be laid out in area A, then G has a cut tree of depth*

$\lceil \log A \rceil + 1$ *such that all nodes on the ith level of the cut tree represent bipartitions with cutsize at most* $\lfloor 2\sqrt{A/2^i} \rfloor$. *(For the definition of level see Definition 3.3.)*

Proof Assume that the lengths of both sides of the rectangular layout of G are powers of 2. (Ensuring this condition increases the area at most by a factor of 4.) Cut the rectangular layout of G exactly in half by a cutline that runs parallel to the shorter side of the layout. At most $\lfloor \sqrt{A} \rfloor$ edges can cross this cutline. Proceed recursively on the sublayouts, each time choosing the shorter side of the layout to cut. The areas of the sublayouts halve as we descend one level in this cut tree. Since the shorter side always is chosen to be cut, the cutsize at a node in the cut tree is at most the square root of the area of the subgraph corresponding to this node; that is, on level i, the cutsize is $\lfloor \sqrt{A/2^i} \rfloor$. □

Theorem 5.1 suggests that we choose cut trees of the following form to drive the layout process.

Definition 5.2 (Bifurcator) *Let* $G = (V, E)$ *be a graph with n vertices and m edges. A* **bifurcator of size** s *for* G *is a cut tree* T *of* G *with height at most* $t := \lceil 2 \log s \rceil$ *such that, for each node in level i, G corresponds to a bipartition with cutsize at most* $\lfloor s/(\sqrt{2})^i \rfloor$. *To exclude degenerate cases, we require that* $\lfloor s/(\sqrt{2})^i \rfloor \geq 1$ *for all i*, $0 \leq i \leq \lceil 2 \log s \rceil$.

For an illustration of Definition 5.2, see Figure 5.6.

We will from now on assume that s is a power of 2. This assumption does not affect the asymptotic analysis, but it does simplify the proofs.

In general, bifurcators may contain unbalanced bipartitions, because in arbitrary layouts we cannot expect a cut balancing the layout area also to balance the induced graph bipartition. The size of a bifurcator has to satisfy $s \geq \sqrt{n}$, because otherwise the bifurcator tree is too shallow to accommodate a recursive bipartition of G into single vertices.

The factor $\sqrt{2}$ by which the cutsize decreases as we descend a level in the bifurcator tree is a reflection of the planar flavor of the layout problem. If we were to lay out graphs in k-dimensional rectangular grids, then the cutsize would have to be smaller than the $(k-1)$-dimensional volume of the smallest face of the layout. This volume is at most $V^{(k-1)/k}$, where V is the size of the grid. Thus, as we descended a level in the bifurcator, the cutsize would have to decrease by a factor of $2^{(k-1)/k}$.

5.5   An Approximation Algorithm Based on Bifurcators

If we use bifurcators to drive the layout process, then Theorem 5.1 automatically produces a lower bound on the layout area: A graph whose minimum bifurcator has size s needs layout area $s^2/4$. (Remember that s is assumed to be a power of 2.) How close can we get to achieving this area with a layout algorithm that uses the bifurcator to drive a recursive bipartitioning and embedding of the graph?

The leaves of the bifurcator tree are embedded into single grid vertices. We obtain the layouts for internal nodes on level i of the bifurcator tree by abutting the respective two sublayouts in one dimension, and then expanding the grid with $2\lfloor s/(\sqrt{2})^i\rfloor$ properly placed horizontal and vertical lines to accommodate the wires for wiring the cut (see step 3 in Figure 5.4, and remember that G is a graph now). Thus, for each cut wire that has to be routed, two new horizontal and two new vertical gridlines suffice. The direction in which cells are abutted alternates between horizontal and vertical. We get the following recurrences for a subgraph corresponding to a node at level i in the bifurcator tree. $L(i)$ denotes the length of the side of the layout along which the abutment of the subgraphs takes place, and $W(i)$ denotes the length of the other side.

$$
\begin{aligned}
L(t) &= 1 \\
W(t) &= 1 \\
L(i) &\leq 2W(i+1) + 2\left\lfloor \frac{s}{(\sqrt{2})^i} \right\rfloor, \quad 0 \leq i < t \\[2mm]
W(i) &\leq L(i+1) + 2\left\lfloor \frac{s}{(\sqrt{2})^i} \right\rfloor, \quad 0 \leq i < t
\end{aligned}
\tag{5.1}
$$

The coupling of the recurrences reflects the fact that cut directions alternate. The second additive term accounts for the $2s'$ horizontal and $2s'$ vertical gridlines that have to be created in order to wire the cut of size s' that corresponds to a node at level i, $s' = \lfloor s/(\sqrt{2})^i \rfloor$. We can decouple the recurrences (5.1) by unfolding them once. This yields

$$
\begin{aligned}
L(t) &= 1 \\
L(i) &\leq 2L(i+2) + 2(2+\sqrt{2})\frac{s}{(\sqrt{2})^{i+1}}, \quad 0 \leq i < t \\[2mm]
W(t) &= 1 \\
W(i) &\leq 2W(i+2) + (2+\sqrt{2})\frac{s}{(\sqrt{2})^{i+1}}, \quad 0 \leq i < t
\end{aligned}
\tag{5.2}
$$

$L(0)$ and $W(0)$ sum to

$$
\begin{aligned}
L(0) &\leq 2(2+\sqrt{2})\sum_{i=0}^{t/2} 2^i \cdot \frac{s}{(\sqrt{2})^{2i+1}} \\[2mm]
W(0) &\leq (2+\sqrt{2})\sum_{i=0}^{t/2} 2^i \cdot \frac{s}{(\sqrt{2})^{2i+1}}
\end{aligned}
\tag{5.3}
$$

Since $t = 2 \log s$, the solution to both recurrences is $L(0), W(0) = O(s \log s)$, and thus $A = L(0) \cdot W(0) = O(s^2 \cdot (\log s)^2)$. This upper bound is only a factor of $(\log s)^2$ away from the lower bound proved in Lemma 5.1. The mincut algorithm based on bifurcators is thus an approximation algorithm for the LAYOUT problem.

Note also that it follows from (5.1) that $1 \leq L(i)/W(i) \leq 2$ for $0 \leq i < t$. Thus, the layouts produced do not deviate too much from a square shape. Also, it can be shown that, at each level i, the area allocated for wiring all cuts in level i is the same; namely, $O(s^2 \log s)$ (Exercise 5.3).

The algorithm does, however, have a few drawbacks:

1. For all sensible values of s, the size of the error grows as $(\log n)^2$. In particular, if the bifurcator has size $O(n)$, then the area of the layout is $O(n^2(\log n)^2)$. However, it is simple to lay out any graph in area $O(n^2)$: We just put the vertices on the main diagonal of a $2n \times 2n$ grid, and route the wires using the grid as a general switchbox. We allow two rows and columns of the grid around each vertex to route the incident edges to this vertex (see also Exercise 5.6). Since communication networks such as appear frequently on chips tend to have large separators, it is worthwhile to investigate whether the error of the approximation algorithm can be improved so as to *decrease* as the size of the bifurcator increases.

2. Other cost functions besides the area, such as maximum wire length or crossing number, are not minimized very well. A trivial upper bound on the crossing number in our layout model is the layout area. This is because of the quasiplanarity of the model: Only a single crossing can appear on a grid vertex. We would hope to be able to ensure that many fewer crossings exist in a layout, but we cannot do so with the present approximation algorithm. (Note, however, that the number of jogs per wire is at most 5, which is satisfactory.)

3. The maximum wire length in the layout produced by the approximation algorithm can be limited to $O(\sqrt{A}) = O(s \log s)$, since wires never meander around the grid. Again, we would hope to show that we can get by without running some wires across all of the layout.

In the next two sections, we discuss how these drawbacks can be eliminated.

5.6 Balancing Bifurcators

In this section, we remove the first drawback. To this end we introduce two additional ideas: Balancing the bipartitions and stopping the recursion when graphs get small enough to be laid out directly.

Definition 5.3 (Strictly Balanced Cut Tree) *A cut tree is* strictly balanced *exactly if the sizes of the sides of all bipartitions differ by at most* 1. *A
bifurcator is strictly balanced if its cut tree is strictly balanced.*

For simplicity, let us again assume that the size n of G is a power of 2. If we
have a strictly balanced bifurcator, then we do not need to do the recursive
cutting up to level $t = 2 \log s$. We can stop at the smallest level t' such that
the allowed cutsize at level t' exceeds the size of the subgraphs at level t'. At
this level, we can embed the whole subgraph directly (see Exercise 5.6).

Since the cutsize decreases by a factor of $\sqrt{2}$, whereas the size of the subgraphs decreases by a factor of 2, as we descend in the cut tree, we have

$$\frac{s}{(\sqrt{2})^{t'}} \geq \frac{n}{2^{t'}}$$

Hence,

$$t' = 2 \log \frac{n}{s}$$

(By the assumptions on n and s, t' is even.) At level t', there are s^2/n vertices
per subgraph. The initial values for the recurrences (5.2) thus change to

$$\begin{aligned}
L(t') &= 2\,s^2/n \\
W(t') &= 2\,s^2/n
\end{aligned}$$

If we modify the expressions (5.3) accordingly, we obtain

$$L(0) \;\leq\; 2\,(2 + \sqrt{2})\left(\sum_{i=0}^{t'/2-1} 2^i \cdot \frac{s}{(\sqrt{2})^{2i+1}}\right) + 2^{t'/2} \cdot 2\,\frac{s^2}{n} \qquad (5.4)$$

$$W(0) \;\leq\; (2 + \sqrt{2})\left(\sum_{i=0}^{t'/2-1} 2^i \cdot \frac{s}{(\sqrt{2})^{2i+1}}\right) + 2^{t'/2} \cdot 2\,\frac{s^2}{n} \qquad (5.5)$$

The second terms on the right-hand sides of (5.4) and (5.5) account for the
area consumed by all subgraphs on level t'. The first terms add up the space
needed for grid expansion. The asymptotic growth of (5.4) is $L(0) = W(0) =
O(s \log(n/s))$. Thus, the error in the layout area is $O((\log n/s)^2)$. Since $s \geq
\sqrt{n}$, we always gain an improvement over that obtained in Section 5.5. If
$s = O(n)$, then $A = O(n^2)$; thus, drawback 1 has been removed.

We still have to show that we can balance bifurcators without increasing
their size too much. Indeed, the work by Bhatt and Leighton [34] contains a
result to this effect, but the proof is quite involved. Here, we present a result
that is based on the following somewhat stronger concept of a bifurcator. This
concept is also useful in the following discussion (see Sections 5.7 and 5.8).

Definition 5.4 (Strong Bifurcator) *A* **strong bifurcator** *of size s is a cut tree such that, for each node x at level i, the number of wires that run between G_x and $G \setminus G_x$ is at most $s/(\sqrt{2})^i$.*

In a strong bifurcator, we count all wires leaving a subgraph, not just the ones entering its brother in the cut tree. Therefore, the following fact is obvious.

Fact 5.1 *A strong bifurcator of size s is a bifurcator of size $s/\sqrt{2}$.*

Because of Fact 5.1, the layout algorithms discussed so far achieve the same asymptotic upper bounds in terms of strong bifurcators as in terms of bifurcators. Furthermore, Theorem 5.1 carries over to strong bifurcators (see Exercise 5.7). In fact, Bhatt and Leighton show the following.

Theorem 5.2 (Bhatt, Leighton [34]) *If G has a bifurcator of size s, then G has a strictly balanced strong bifurcator of size $O(s)$.*

The constant implied by [34] is about 20. The proof of Theorem 5.2 is quite involved. Conceptually—even though it is not done in this way in [34]—we could break up the proof into the following two steps.

Lemma 5.1 *If G has a bifurcator of size s, then G has a strong bifurcator of size $O(s)$.*

Lemma 5.2 *If G has a strong bifurcator of size s, then G has a strictly balanced strong bifurcator of size $O(s)$.*

Here, we will explicate a proof of Lemma 5.2. This establishes the asymptotic area bound in terms of strong bifurcators. We will start by explaining the proof method intuitively.

Let G be a graph with a bifurcator of size s, and let T be the related cut tree. T has height at most $2\log s$, but T is not necessarily a complete binary tree. The first step is to make T a complete binary tree of height $2\log s$. We do that by expanding leaves v in T that are on levels i with $i < t$ into subtrees. One (arbitrary) leaf of such an expanded subtree represents the vertex of G represented by v; all other leaves do not represent any vertex in G. The newly created interior vertices represent trivial bipartitions at least one of whose sides is the empty graph. In this way, we obtain a complete binary tree $\widetilde{T}$ with height $2\log s$.

Only some of the leaves of $\widetilde{T}$ represent vertices in G. Those leaves are colored black; the other leaves are colored white. Thus, n leaves are black, and $s^2 - n$ leaves are white. Also, we think of the leaves of $\widetilde{T}$ as being ordered in a natural sense from left to right (see Figure 5.7).

We now generate a balanced cut tree T' top down, starting with the root and proceeding level by level. When we generate a node x in T', we also determine the subgraph G_x that x represents. Rather than partitioning G, we

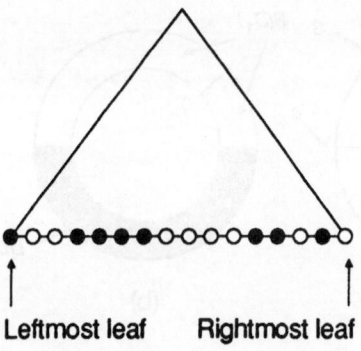

Figure 5.7: *The tree $\widetilde{T}$.*

partition the leaves of $\widetilde{T}$. Thus, each node x in T' also represents a subset L_x of the leaves of $\widetilde{T}$. G_x is induced by the black leaves in L_x. Furthermore, we take care that both children of each node in T have the same number of black leaves. This ensures that bipartitions in T' are strictly balanced.

We limit the sizes of all cuts by making sure that, for each x in T', the set L_x consists of at most two sets of *consecutive* leaves with respect to the ordering of leaves in $\widetilde{T}$. The following technical lemma proves that this can be done.

Lemma 5.3 (Leighton et al. [268]) *Let L and S be two strings of black and white pebbles, both with an even number of black as well as white pebbles. By making at most two cuts, we can obtain at most four strings that can be arranged in two sets, each of which contains at most two strings and has exactly one-half of the black and one-half of the white pebbles.*

Proof Let b be the sum of the number of black pebbles in both strings. Assume without loss of generality that L is the longer string. Concatenate the strings L and S into a circular binary string C (see Figure 5.8). Let D be a double cut of C—that is, a cut in two places of C. Let $L(D)$ and $R(D)$ be the two sides of bipartition induced by the cut. Let $f(D)$ be the number of black pebbles in $R(D)$. We find a proper double cut D by an application of a discrete version of the mean value theorem. We construct a finite sequence $D_1, \ldots, D_k$ of double cuts of C such that each D_i fulfills the following:

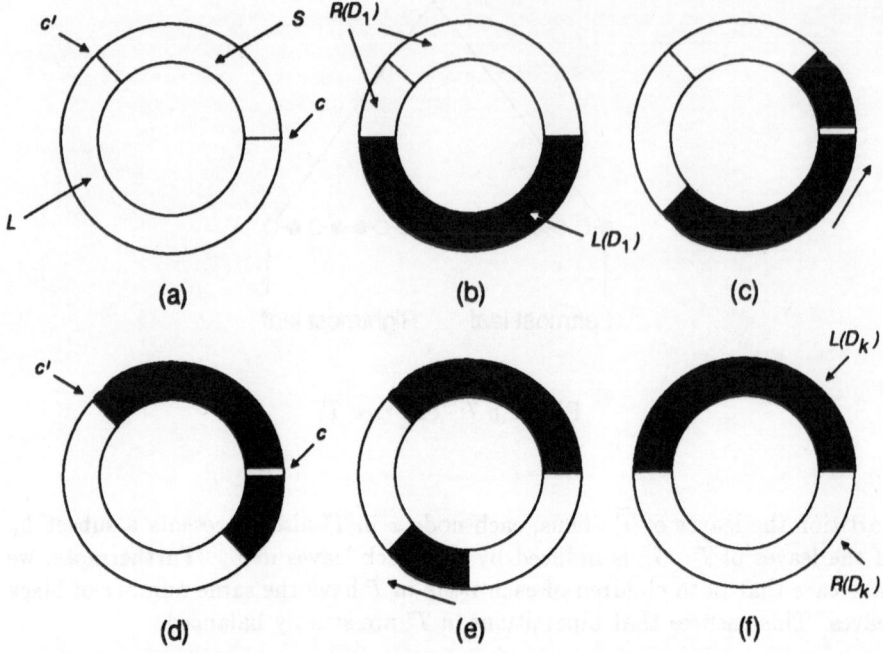

Figure 5.8: *An illustration of the proof of Lemma 5.3.*

1. $|R(D_i)| = |L(D_i)|$.

2. Both $R(D_i)$ and $L(D_i)$ consist of two segments of consecutive pebbles from L and/or S.

3. $|f(D_i) - f(D_{i-1})| \leq 1$ for all $i = 2, \ldots, k$.

4. $R(D_1) = L(D_k)$.

By property 2, each double cut in the sequence fulfills all requirements of the lemma except the balance of the black and white pebbles. By property 1, a double cut D_i fulfills the balance requirement for the black and white pebbles if $f(D_i) = b/2$. If $f(D_1) = b/2$, then we can choose D_1 as the proper double cut. Assume without loss of generality that $f(D_1) > b/2$. By property 4, we have $f(D_k) = b/2 - f(D_1) < b/2$. Because of property 3, there must be a value i, $1 < i < k$ for which $f(D_i) = b/2$. D_i is the cut we want.

The sequence $D_1, \ldots, D_k$ of cuts is illustrated in Figure 5.8. In this figure, $L(D_i)$ is depicted in black, and $R(D_i)$ is depicted in white. We start by cutting D into two half circles, where one of the cuts is located at the clockwise first

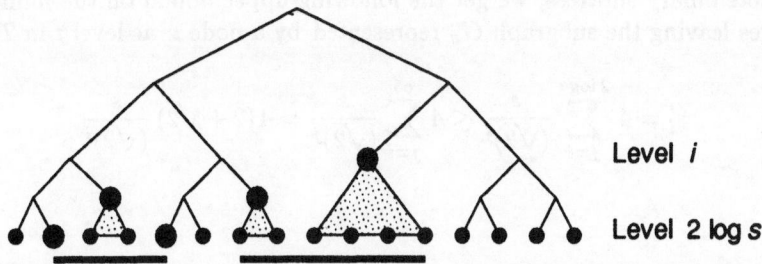

Figure 5.9: *The edges leaving G_x all extend from subgraphs of G associated with the highlighted roots of complete binary subtrees of $\widetilde{T}$. The two segments that make up L_x are depicted by bold lines below the tree $\widetilde{T}$.*

end of L. We denote this location with c. The other end of L we denote with c'. Then, we generate double cuts by moving both cutlines counterclockwise across one pebble at a time. When one cutline hits c' (Figure 5.8(d)), we split $L(D_i)$ into two pieces. One is S and the other is a segment of L that is moved clockwise across one pebble at a time until it hits c'. This results in double cut D_k. Properties 1 to 4 are easily checked by inspecting Figure 5.8. $\square$

An inductive application of Lemma 5.3 proves that this procedure for constructing T' yields a strictly balanced cut tree. We now have to bound the cutsizes occurring in T'. We prove the following strengthened version of Lemma 5.2.

Theorem 5.3 *Let G be a graph with a strong bifurcator of size s. The preceding construction yields a strictly balanced strong bifurcator of size at most*

$$4\left(2 + \sqrt{2}\right)s$$

Proof We have seen that, for each node x in T', L_x consists of two segments of consecutive leaves in $\widetilde{T}$. Consider a node x in level i of T'. We have $|L_x| = s^2/2^i$. The leaves in L_x occur in two segments inside $\widetilde{T}$, each of length at most $s^2/2^i$ (see Figure 5.9). Each such segment is composed of leaves that can be partitioned into disjoint sets of leaves making up complete binary subtrees of $\widetilde{T}$. Because of the size of L_x the roots of all such subtrees are nodes at levels j with $i \leq j \leq 2\log s$. Also, because the leaves in each segment are adjacent, for each segment there are at most two such subtrees of each height. The edges leaving the subgraph G_x of G all leave one of the subgraphs of G associated with the roots of such complete binary subtrees. If the root of the relevant complete binary subtree is at level j then number of edges leaving the corresponding subgraph of G is at most $s/(\sqrt{2})^j$. If we add up these terms for all relevant

complete binary subtrees, we get the following upper bound on the number s'_i
of edges leaving the subgraph G_x represented by a node x at level i in T':

$$s'_i = 4 \sum_{j=i}^{2\log s} \frac{s}{(\sqrt{2})^j} \leq 4 \sum_{j=i}^{\infty} \frac{s}{(\sqrt{2})^j} = 4\,(2+\sqrt{2})\,\frac{s}{(\sqrt{2})^i}$$

□

5.7   Improving Secondary Cost Functions

In Section 5.6, we decreased the error in layout area to $O((\log n/s)^2)$. This is
the best bound for the layout of general graphs known today. The number of
jogs can be kept at 5 per wire (see Exercise 5.6). However, the crossing number
and maximum wire length are not improved. In this section we improve the
bounds on the crossing number C and the maximum wire length L. We achieve
these improvements by departing from the layout method used in Section 5.6.

The problem with the layout method in Section 5.6 is that cut wires are
routed on top of the layouts of the subgraphs. Since the exact structure of
these layouts is not taken into account, we can potentially cross many wires
when routing a cut wire. This has as a consequence that the upper bound on
the crossing number is the same as the upper bound on the area. Similarly, it
may happen that a wire must be routed across the whole layout. This leads to
the trivial upper bound on maximum wire length, which is $L = O(s\log(n/s))$.

5.7.1   Lower Bounds

Let us discuss lower bounds for the crossing number C and the maximum wire
length L.

Theorem 5.4 *Let G be a graph whose minimum bifurcator has size s. Then,
the crossing number C of G fulfills $C + n = \Omega(s^2)$.*

Proof A layout Λ of G with C crossings in a natural manner induces a planar
graph G' whose vertices are the vertices of G and the crossings in Λ. It is a
well-known result, shown by Lipton and Tarjan [291], that a planar graph with
n vertices and bounded vertex degree has a bifurcator of size $O(\sqrt{n})$. Thus, G'
has a bifurcator of size $O(\sqrt{C+n})$ (see also Section 6.4.3). □

Theorem 5.5 *Let G be a graph whose minimum bifurcator has size s. Then,
the maximum wire length L of any layout of G fulfills $L = \Omega(s^2/n)$.*

Proof The sum of all wire lengths in a layout is at least as large as $2C + n/2$, because each crossing can be charged with two units of wire length contributing to the crossing, and each vertex can be charged with at least a half-unit of wire attaching to it. By Theorem 5.4, the total wire length is therefore $\Omega(s^2)$. However, there can only be $2n$ wires in the layout, since node degrees are bounded by 4. This observation yields the theorem. $\square$

The error that the layout algorithm from Section 5.6 achieves for $C + n$ is the same as the error for the layout area. The error for L is $O((n/s) \log(n/s))$, so it is not very good. In the following section, a factor of $\log(n/s)$ is eliminated from the error for $C + n$, and a factor of $\log \log(n/s)$ is eliminated from the error for L. These improvements are not dramatic by any means. Still, the reduction for $C + n$ implies that graphs can be laid out such that the number of crossings is small compared to the area. Furthermore, the layout method is interesting in its own right.

The main idea of the method is to set aside special regions in the layout in order to route buses that consist of the wires in a cut. This approach has two advantages. First, the resulting layouts are very regular, which leads to the possibility of decreasing wire length in bus channels. Second, the number of crossings is reduced, since the buses are routed in specifically assigned regions and not in an uncontrolled fashion.

We define the regions for routing the cut wires by using a two-phase layout process. In the first phase, G is embedded into a special non–grid graph, the *tree of meshes*. This graph separates the routing of wires in a cut from the layouts of the two subgraphs connected by the cut. The second phase embeds the tree of meshes into the square grid in a very regular fashion. The following two sections explain these two phases in more detail.

5.7.2 Embedding Graphs in the Tree of Meshes

Not surprisingly, the intermediate graph that we use in the two-phase layout process is reminiscent of the cut tree for the graph G. It is defined as follows.

Definition 5.5 (Tree of Meshes) *Let r be a power of 2. We obtain the* **tree of meshes** H_r *of size r by providing an $r \times r$ grid whose left and right boundary vertices are connected with edges to the top vertices of two $r/2 \times r$ grids. The left and right boundary vertices of these grids are in turn connected with edges to the top vertices of four $r/2 \times r/2$ grids. In this way, we proceed until all grids are 1×1. Thus, there are $2 \log r + 1$ levels in the tree of meshes H_r. The tree that we obtain by expanding only p levels (p even) is denoted by $H_{r,p}$. The grids in the bottom level of $H_{r,p}$ have size $r/(\sqrt{2})^p$.*

Figure 5.10 shows the tree of meshes of size 4.

Analogously to the discussion in Section 5.6, the suggestion is to embed a graph G with a balanced bifurcator T of size s in the tree of meshes $H_{cs, 2 \log(n/s)}$,

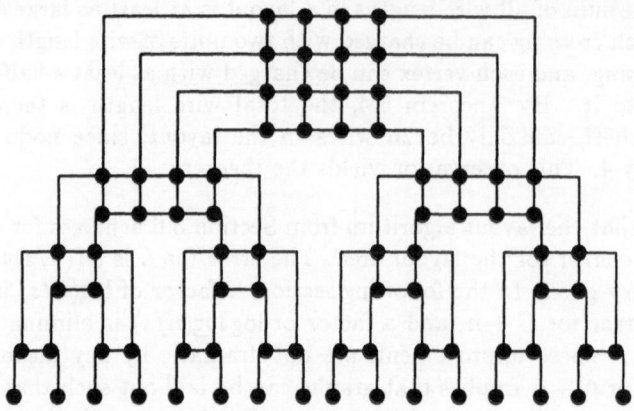

Figure 5.10: *The tree of meshes of size 4.*

where c is an appropriate constant. We embed the subgraphs represented by
the nodes at level $t' = 2\log(n/s)$ in the grids at the leaves of $H_{cs,2\log(n/s)}$.
Then, we use the grids at interior nodes of $H_{cs,2\log(n/s)}$ to route cuts.

There is one problem with this approach. The suggested embedding embeds
all vertices of G into the grids at the leaves of the tree of meshes. This means
that the routing grids corresponding to interior nodes of the tree of meshes have
to accommodate not only wires of the cut represented by the corresponding
node x in T, but also wires of cuts represented by nodes *above* x. Thus, in
this context, the concept of a strong bifurcator must be used. In fact, using
Theorem 5.2, we can prove the following.

Theorem 5.6 *If G has a strictly balanced strong bifurcator of size s, then there
is a constant c such that G can be embedded in $H_{cs,2\log(n/s)}$, where $c \leq 4$.*

Proof The grids at the leaves of $H_{cs,2\log(n/s)}$ are cs^2/n on a side. If $c \geq 2$,
they accommodate the subgraphs of G at level $2\log(n/s)$, which have s^2/n
vertices each. Since there are at most twice as many wires as vertices in the
subgraphs, all wires that have to leave the subgraph can do so through the top
edge of the grid.

In the ith level of $H_{cs,2\log(n/s)}$, the grids have side length at least
$cs/(\sqrt{2})^{i+1}$. Thus, if $c \geq \sqrt{2}$, all edges entering from the two children as
well as from the parent can do so through their respective sides of the grid.

We still have to show that the grids corresponding to interior nodes in
$H_{cs,2\log(n/s)}$ are large enough to route all wires running through them. Here,
we do not take too much care to keep the constants involved small, but rather
aim at simplifying the argument. Three types of wires occur in an interior grid.
One type connects the parent and the left child, another the parent and the

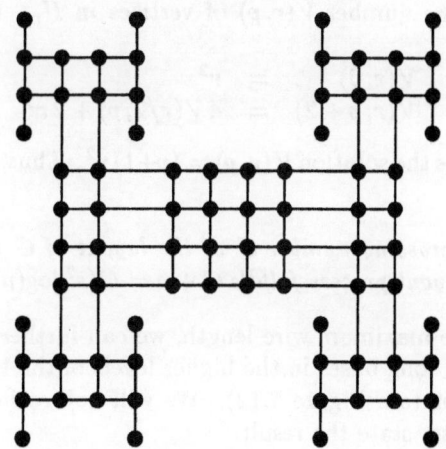

Figure 5.11: *An H-tree layout of the tree of meshes.*

right child, and a third the two children. The first two types can be routed inside the mesh in switchbox style, if we double the size of the grid in each dimension in order to supply additional gridlines for jogs. The third type of wire can be routed by itself in such a grid, as well. Thus, all wires can be routed if the grid is expanded by a factor of 4 in each dimension. □

5.7.3 Efficient Embeddings of the Tree of Meshes

The tree of meshes is especially well suited for a planar embedding. Figure 5.11 shows such an embedding for H_4, the *H-tree* layout. It is apparent from the figure that, if r is a power of 2 and p is even, the side length $L(r, p)$ of a layout for $H_{r,p}$ obeys the following recurrence:

$$
\begin{aligned}
L(r, 0) &= r \\
L(r, p+2) &= 2\,L(r/2, p) + r
\end{aligned}
\tag{5.6}
$$

The solution of recurrence (5.6) is $L(r, p) = (1+p/2)\,r$. This yields the following theorem.

Theorem 5.7 *The tree of meshes $H_{cs,2\log(n/s)}$ can be laid out in area $O(s^2(\log n/s)^2)$.*

This area bound is the same as that in Section 5.6.

Since the layout in Figure 5.11 is planar, in any layout of G constructed with this two-phase procedure, $C + n$ cannot exceed the number of vertices in $H_{cs,2\log(n/s)}$. The number $V(r,p)$ of vertices in $H_{r,p}$ follows a recurrence similar to (5.6):

$$
\begin{aligned}
V(r,0) &= r^2 \\
V(r,p+2) &= 4V(r/2,p) + 2r^2
\end{aligned}
\tag{5.7}
$$

Recurrence (5.7) has the solution $V(r,p) = (p+1)r^2$. Thus, we get the following lemma.

Lemma 5.4 *The crossing number C of the layout of G obtained by the preceding two-phase layout process fulfills $C + n = O(s^2 \log(n/s))$.*

With respect to the maximum wire length, we can further improve the H-tree layout by pulling in long buses in the higher levels of the tree of meshes where the long wires occur (see Figure 5.12). We will not go into the detail of the analysis, but will just state the result.

Theorem 5.8 (Bhatt, Leighton [34]) *Using the layout of the tree of meshes indicated in Figure 5.12, we can achieve simultaneously*

$$
A = O(s^2(\log n/s)^2)
$$

$$
C + n = O(s^2 \log(n/s))
$$

$$
L = O\left(\frac{s\log(n/s)}{\log\log(n/s)}\right)
$$

The only cost measure that is not optimized by the two-phase layout process is the number of jogs. It increases from 5 to $\Omega(\log n/s)$ per wire. This may amount to a disadvantage with respect to electrical soundness in some technologies. With long wires, however, the delay is a problem, anyway. We can solve this problem by providing wires with driver chains (see Exercise 5.8). If this is done, the wire is split up into sections, each of which has only $O(1)$ jogs.

5.8   Discussion

The layout algorithms developed in this chapter are not immediately applicable to practical layout problems for the following reasons:

1. As it stands, the discussion is limited to graphs. However, both the model and the upper and lower bounds that are based on strong bifurcators extend directly to hypergraphs. An extension of the bounds based on bifurcators to hypergraphs is also possible (see Exercise 5.9).

2. Since all results are asymptotic with constants that are at times quite large, the layouts produced are of far lower quality than that needed in practice.

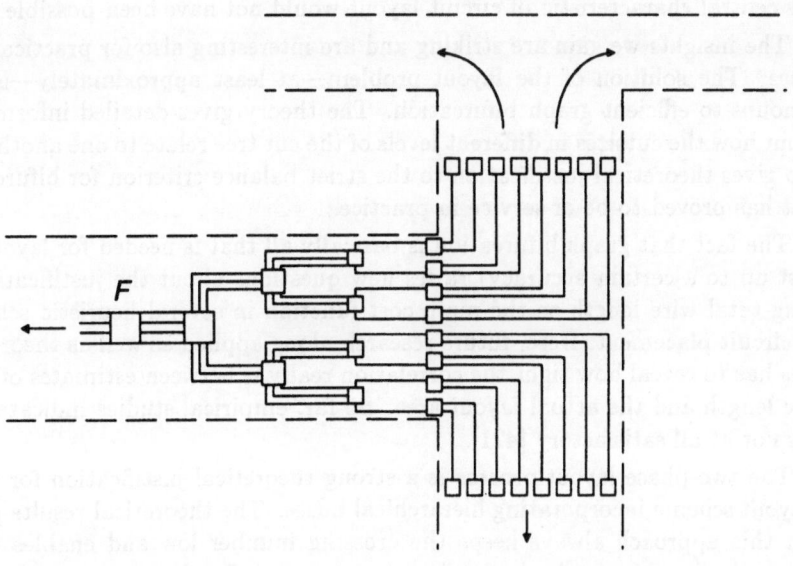

Figure 5.12: *Modified H-tree layout to shorten long wires.*

3. Graph bifurcation is hard in itself. Until recently, no approximation algorithm with an appreciably small error (even asymptotically) was known. Now Leighton and Rao [269] have presented a polynomial-time approximation algorithm for finding minimum bifurcators that has an error of $O((\log n)^{5/2})$. This algorithm exploits the connection between graph bifurcation and multicommodity flow problems (see Sections 3.11.3 and 6.5). By applying these approximation methods appropriately in the preceding layout framework, Leighton and Rao can in polynomial time achieve an error of $O((\log n)^6)$ for layout area and of $O((\log n)^4)$ for $C+n$. These errors are still much too large to be useful in practice. There are also heuristics on graph bifurcation that work well in practice (see Section 6.2). It is doubtful, however, that the layout algorithms would be competitive in practice, even if a very good bifurcator could be found.

Still, a theoretical investigation of this sort reveals important insights into the combinatorial structure of the layout problem. In fact, the reason that we can gain such insights at all is that we abstract enough from the details of a particular technology. By using asymptotic analysis, we can abstract the layout problem far enough that we do capture not the characteristics of a specific technology, but only the structure common to all finite-resolution quasiplanar technologies. If we had insisted on a more exact evaluation, concentrating on this central characteristic of circuit layout would not have been possible.

The insights we gain are striking and are interesting also for practical purposes: The solution of the layout problem—at least approximately—is tantamount to efficient graph bifurcation. The theory gives detailed information about how the cutsizes in different levels of the cut tree relate to one another. It also gives theoretical justification to the strict balance criterion for bifurcators that has proved to be of service in practice.

The fact that graph bifurcation is basically all that is needed for layout (at least up to a certain accuracy) raises new questions about the justification of using total wire length as the main cost function in several heuristic schemes for circuit placement. Here, future research along applied as well as theoretical lines has to reveal how tight the correlation really is between estimates of total wire length and the actual layout area. So far, empirical studies indicate that it is not at all satisfactory [421].

The two-phase layout process is a strong theoretical justification for using a layout scheme incorporating hierarchical buses. The theoretical results prove that this approach always keeps the crossing number low and enables us to reduce the maximum wire length by some amount. Furthermore, no wires are routed across sublayouts and no sublayouts are expanded while they are being connected to other parts of the graph. Thus, the electrical properties of a layout, such as delay, are much more amenable to control and prediction. This is bound to increase the robustness of the layout with respect to circuit failure. In addition, the regularity of the layout enables us to put amplifying drivers on long buses at geometrically increasing intervals, in order to reduce the delay of the circuit (see Exercise 5.8).

The regularity of the two-phase layout process is interesting in another respect. Its suggests that we use the tree of meshes as a universal processor network in which specific processor-network topologies can be embedded by customizing—for instance, through microprogramming. This idea has been studied extensively by Leiserson [154, 155, 271].

The theory presented in this chapter can be modified to handle three-dimensional circuit layout. Bisecting a cube with volume V takes a cutting plane with area $V^{2/3}$. Thus, the factor of $1/\sqrt{2}$ in the definition of a bifurcator (Definition 5.2) has to be replaced with $1/2^{2/3}$. The layout scheme can be lifted to three dimensions. The results are modified accordingly. For details, see [270].

In summary, the theory developed in this chapter gives us a good structural basis for investigating the layout problem. As we enter the following chapters, which detail methods used today for optimized layout design, we will use the insights gained here as a theoretical counterpoint that motivates us to ask continually the question whether the success of a layout method can be backed up by a tight correlation between the method and the combinatorial structure of the problem.

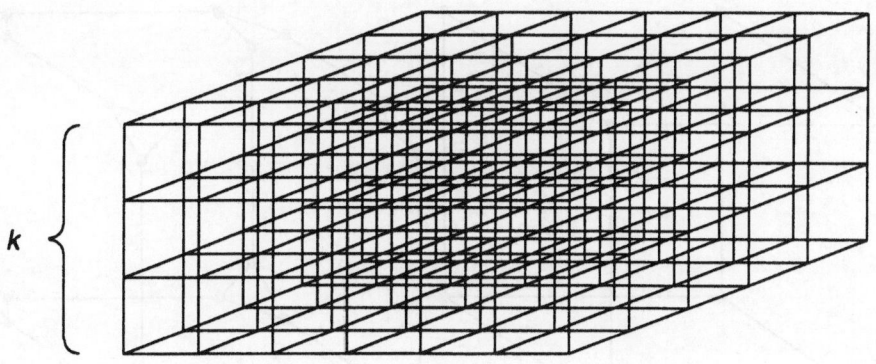

Figure 5.13: *A k-layer grid graph.*

5.9 Exercises

5.1 Let *k*-LAYOUT be a variant of the LAYOUT problem, in which we are looking for an embedding of a hypergraph G into a *k-layer grid graph*, that we form by stacking k grids on top of each other and providing vertical connections between corresponding grid nodes on adjacent layers (see Figure 5.13). Again, a reasonable embedding rule requires the *vertex*-disjointness of different Steiner trees. Also, we assume that all vertices of G are embedded into the bottom layer. (This models the fact that switching devices are fabricated in the bottom layer of multilayer circuits.) Furthermore, the maximum vertex degree of G is again 4 (although 5 could be achieved here).

1. Prove that the minimum area and maximum wire length of a graph are asymptotically the same in the LAYOUT and *k*-LAYOUT problems.

2. Define a reasonable version of the crossing number for the *k*-LAYOUT problem and exhibit its relation to the crossing number in the LAYOUT problem.

3. Some technologies contain specific rules for vias. One possibility is that each via cuts through *all* layers. Model this restriction with an appropriate embedding rule in the *k*-LAYOUT problem. Some technologies prefer vertical and horizontal wires on alternating layers. Model such restrictions in the *k*-LAYOUT problem. Argue that this and other restrictive versions of the *k*-LAYOUT problem do not invalidate the proof in part 1.

5.2 Model layout technologies that allow 45° angles, using a version of the LAYOUT problem that does not have asymptotically different area and maximum wire length.

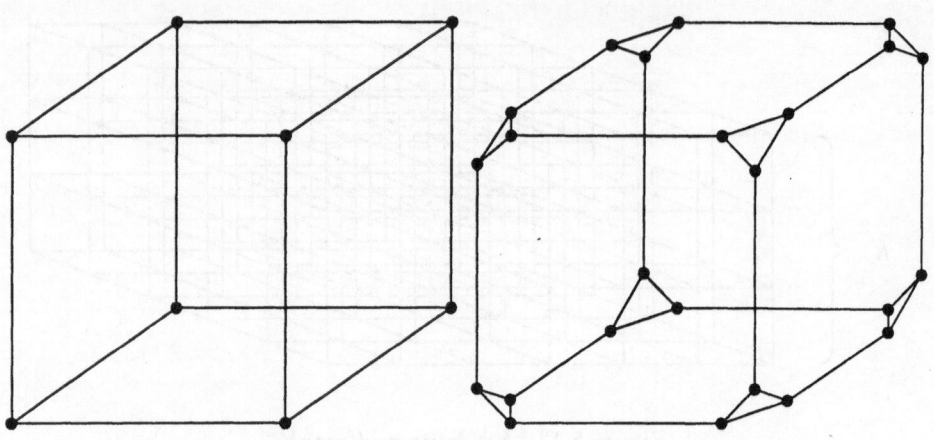

Figure 5.14: *A hypercube and a cube-connected cycles network, CCC_d.*

5.3 Using the recurrences (5.1), show that, for each level i in the cut tree, the area allocated for wiring cuts in level i is the same—namely, $O(s^2 \log s)$.

5.4 Determine a strictly balanced strong bifurcator for the complete binary tree of height r. Analyse its size. What level in the cut tree determines the bifurcator size?

5.5 The d-*dimensional hypercube* K_d is a graph with 2^d vertices numbered from 0 to $2^d - 1$, and edges joining vertices the binary representations of whose numbers differ by a single bit. The hypercube is an extremely useful network topology. Unfortunately, K_d has unbounded degree. Thus, the graph does not fit our layout model, and the high degree also stands in the way of the applicability of the hypercube as a processor network. We can eliminate this problem by expanding each vertex of K_d into a cycle of d vertices. Thus, we obtain the *cube-connected cycles network* CCC_d. The CCC_d has $n = d \cdot 2^d$ vertices labeled with pairs (p, q), where $0 \leq p \leq 2^{d-1}$ and $O \leq q \leq d - 1$. Edges exist between vertices (p_1, q) and (p_2, q) if the binary representations of p_1 and p_2 differ in the q-order bit. Furthermore, edges exist between the vertices labeled (p, q) and $(p, (q + 1) \bmod d)$. The CCC_d network is a regular graph of degree 3. Figure 5.14 illustrates the hypercube and the CCC_d network.

Find a strictly balanced strong bifurcator S_d of CCC_d that has size $O(2^d) = O(n/\log n)$.

1. What is the constant of the size of S_d as a strong bifurcator?

2. What is the constant of the size of S_d as a bifurcator?

3. Which level in the cut tree determines the size of the bifurcator?

Comment: One way of proving a lower bound on the size of bifurcators of CCC_d is by considering functions that are computed on the CCC_d processor network. For instance, the discrete Fourier transform of a $(d \cdot 2^d)$-vector can be computed on a CCC_d in time $T = O(d)$ [365]. Thompson [442] proves that $AT^2 = \Omega(d^2 \cdot 2^{2d})$ for any Boolean circuit that is laid out using the model described in this chapter. This result implies that the area of any layout of CCC_d is $\Omega(2^{2d}) = \Omega(n^2/(\log n)^2)$. By Theorem 5.1, this lower bound on the area implies that any bifurcator of CCC_d has to have size $\Omega(2^d) = \Omega(n/(\log n))$. This lower bound matches the upper bound given in part 2 above asymptotically.

5.6 Prove that any graph with maximum vertex degree 4 and n vertices can be laid out inside a $2n \times 2n$ grid with at most five jogs per wire.

Hint: Lay out the vertices along the diagonal. Prove and use the fact that a layout can be found in which at most two edges leaving vertex v leave the neighborhood of v horizontally, and at most two leave it vertically.

5.7

1. Modify the proof of Theorem 5.1 so as to show that each graph G that has a layout with area A also has a strong bifurcator of size $O(\sqrt{A})$. What is the constant factor?

2. Extend Theorems 5.4 and 5.5 to strong bifurcators. In addition to Theorem 5.3, these results imply that the layout algorithm presented in this chapter and its analysis extend to the case in which strong bifurcators are used instead of bifurcators, even without using Theorem 5.2.

5.8 A well-known delay model for an integrated circuit is the *capacitive* model. Here, we abstract from resistive and inductive phenomena. A driver stage in the capacitive model consists of a transistor with gate aspect ratio D, a wire with length ℓ, and a capacitive load of size L that is connected to the driver by the wire. The delay of the driver stage is proportional to $(\ell + A)/D$.

We can extend our layout model to accommodate this kind of driver stage. We think of G as being directed. Each edge represents a driver stage. The head of the edge is the load, the edge is the wire, and the tail is the transistor. To model large capacitive loads, we allow vertices of G to be spread over several nodes and edges of the grid. We will always embed vertices in *straight* paths in the grid. The capacitance A of the vertex (as a load) and also its driving capability D (as a transistor) is the length of the path in which the vertex is embedded. To ensure compatibility with technological restrictions, we do not allow a vertex that is stretched in this way to be crossed over by another stretched vertex or by a wire. The *delay* of the circuit is the maximum of the delay across any of its driver stages.

Modify the two-phase layout process without increasing the asymptotic bounds on the layout area and crossing number, such that a layout results that has a delay that is bounded by a constant.

Hint: Use the two-phase layout method. Break up wires that run through several meshes of the tree of meshes into portions, one for each mesh. Make each portion into a driver stage. Expand the meshes by a constant factor. Use the free space thus created in each mesh to accommodate appropriately sized vertices to keep the delay through each driver stage bounded by a constant.

5.9 * Extend the two-phase layout process presented in Section 5.7 to hypergraphs while maintaining the asymptotic bound on layout area in terms of bifurcators.

1. Extend Theorem 5.2. Consult [34].

2. Extend Theorem 5.6, increasing the constant c appropriately.

3. What is the problem with extending the layout method of Section 5.5 to hypergraphs with hyperedges of unbounded size?

5.10 * Try to decrease the constants involved in the asymptotic analysis. You can approach this problem both by sharpening the analysis and by modifying the algorithms.

Chapter 6

Circuit Partitioning

Circuit partitioning is the task of dividing a circuit into smaller parts. Circuit partitioning is an important aspect of layout for several reasons. Partitioning can be used directly to divide a circuit into portions that are implemented on separate components, such as printed circuit boards or chips. Here the objective is to partition the circuit into parts such that the sizes of the components are within prescribed ranges and the complexity of connections between the components is minimized. A natural way of formalizing the notion of wiring complexity is to attribute to each net in the circuit some connection cost, and to sum the connection costs of all nets connecting different components. The connection cost of a net may express such parameters as the buswidth of the net—that is, the number of bits that have to be sent across the net in parallel. Or the connection cost may serve to assign higher priorities to nets that should not run across component boundaries.

A second use of circuit partitioning that is even more important is to divide up a circuit hierarchically into parts within divide-and-conquer algorithms for placement, floorplanning, and other layout problems. Here cost measures to be minimized during partitioning may vary, but mainly they are similar to the connection cost measure described previously. We shall see in Chapters 7 and 8 that divide-and-conquer strategies based on circuit partitioning provide a robust framework for solving many layout problems and integrating several phases of the layout process. Therefore, such strategies hold much promise of advancing the design technology in circuit layout in the future.

Circuit partitioning is formalized as an operation on graphs and hypergraphs. This operation is of interest in layout not only as applied to circuits; it is useful to be able to partition other kinds of graphs as well. Specifically, partitioning *routing graphs* (see Chapters 8 and 9) serves as a basis for divide-and-conquer algorithms for wire routing. In this application, the routing graph represents the topology of the routing area on a chip or printed circuit board, on which several wires have to be routed. Rather than routing one wire after the other through the complete routing region, divide-and-conquer algorithms for routing successively simplify the routing region by breaking it up into pieces. Once a piece becomes simple enough, all wires are routed through it simultaneously. Finally, the solutions of the simple subregions are combined to form the solution of the original routing problem.

Thus, the circuit-partitioning problem, formalized as a problem on graphs and hypergraphs, is a basic problem occurring in many places of the layout process. In this chapter, we shall define variants of the circuit-partitioning problem that are of interest in layout and survey methods for their solution. There are quite a few such variants. We shall give a taxonomy of the corresponding definitions in Section 6.1, which also summarizes known complexity results.

Sections 6.2 to 6.5 discuss partitioning into two components. This *bipartitioning* problem plays a central role among partitioning problems occurring in divide-and-conquer algorithms. Bipartitioning can also be used as a basis for heuristics for partitioning into many components—that is, for *multiway partitioning*. In Section 6.2, we survey iterative improvement heuristics for graph and hypergraph bipartitioning that to date are the basis of most practical partitioning procedures. Section 6.3 discusses the neighborhoods used in simulated annealing algorithms for circuit bipartitioning. In Section 6.4, we discuss results on partitioning special classes of graphs, such as trees and planar graphs. Section 6.5 presents a natural relationship between circuit bipartitioning and network flow that is the basis of more recent developments in partitioning.

Sections 6.6 considers both bipartitioning and multiway partitioning to an equal degree. In that section, we discuss the partitioning problem as a QUADRATIC ASSIGNMENT problem. Section 6.7 discusses a version of multiway partitioning in which a circuit is divided into a large number of small components.

6.1   Definitions and Complexities

All variants of the partitioning problem that we discuss in this book aim at dividing a graph or hypergraph into several disconnected components. The graph or hypergraph represents either a circuit, where the vertices are circuit elements and the (hyper-) edges are wires or a routing region where the vertices represent *switch boxes* for wire routing and the edges represent *routing channels*. In the second application, the graph is usually planar. This observation explains the significance of Section 6.4.

We can obtain the partitioning of the graph by either removing (hyper-) edges or removing vertices (and their incident edges). The latter operation is considered only in the context of graphs, and is called *graph separation* (see Section 6.1.3). The cost of the partition is a weighted sum of the number of the elements removed from the graph.

Partitioning problems may have to obey different constraints. Such constraints may require that the parts of the circuits have prescribed or similar sizes, and that the number of disconnected parts be in a prespecified range. Here, dividing a circuit into *two* pieces (*bipartitioning*) is done especially often. This special case is discussed in Sections 6.1.2 and 6.1.3.

We now define the variants of the partitioning problem that are of significance in circuit layout in detail.

6.1.1 Multiway Partitions

In this section, we discuss the problem of partitioning a hypergraph into a prespecified number m of disconnected parts by removing hyperedges.

MULTIWAY PARTITION

Instance: A hypergraph $G = (V, E)$ with n vertices, a vertex weight function $w : V \to \mathbb{N}$, an edge cost function $c : E \to \mathbb{N}$, a number $r \in \mathbb{N}$ of *parts*, maximum part sizes $B(i) \in \mathbb{N}$, minimum part sizes $b(i) \in \mathbb{N}$, $i = 1, \ldots, r$

Configurations: All *r-way partitions* $\Pi = (V_1, \ldots, V_r)$ of V, with $V_i \subset V$, $V_i \neq \emptyset$, and $V_i \cap V_j = \emptyset$ for $i, j \in \{1, \ldots, r\}$ and $\bigcup_{i=1}^{r} V_i = V$

Solutions: All partitions $\Pi = (V_1, \ldots, V_r)$ such that $b(i) \leq w(V_i) \leq B(i)$ for all $i = 1, \ldots, r$

Minimize: $c(\Pi) := \frac{1}{2} \sum_{i=1}^{r} \sum_{e \in E_{\text{ext},i}} c(e)$,
where $E_{\text{ext},i} = \{e \in E \mid e \cap V_i \neq \emptyset, e \setminus V_i \neq \emptyset\}$ is the set of *external edges* of part V_i

The partition has to obey the constraint that the size of part V_i be in the range between $b(i)$ and $B(i)$. Minimum part sizes are imposed to achieve a *balanced* partition, where each part has about the same size. This requirement is advantageous for divide-and-conquer algorithms based on circuit partitioning (see Chapter 5, 7). The size of a part is the sum of its vertex weights. This provision enables us to account for circuits having elements of different sizes, such as they occur in placement and floorplanning.

The hyperedges that connect different parts in Π are said to *contribute to the cut*. The cost $c(\Pi)$ of the partition is also called the *cutsize*. The cutsize is a weighted sum of the external connections of a part, summed over all parts. The weights $c(e)$ allow us to incorporate different costs for cutting different edges. This flexibility enables us to account for technological constraints that may make cutting some edges preferable to cutting other edges. For instance, cutting edges representing wide buses or cutting edges representing wires whose delay is critical may not be advisable. The factor of $1/2$ is solely for the purpose of giving external *edges* a contribution of $c(e)$ to the cost.

This definition of the cutsize is not the only one possible. Donath [96] suggests having a hyperedge e connecting k parts contribute an amount of $(k-1)c(e)$. Many of the methods we describe in this chapter, especially the important heuristics of Section 6.2, are independent of such minor differences in the definition of the cutsize.

We now discuss the complexity of the MULTIWAY PARTITION problem.

Not surprisingly, the MULTIWAY PARTITION problem is NP-hard, even if it is severely restricted. For instance, the threshold version of the restriction of the MULTIWAY PARTITION problem in which G is a graph, $w \equiv 1$, $c \equiv 1$, $b(i) = 1$, and $B(i) = n$ for all $i = 1, \ldots, r$; that is, the bounds on the part sizes are trivial, is strongly NP-complete. If r is fixed, then there exists an algorithm for this restriction that runs in time $O(n^{r^2})$ [146].

If, in addition, we select distinguished vertices $v_i \in V$ such that v_i has to belong to V_i, then the problem becomes even harder. Now it is NP-hard even for $r = 3$, and G having vertex degree at most 4, in addition to the restrictions discussed previously. If G is planar, then this restriction of the problem can be solved in polynomial time. However, if r is arbitrary, then the problem is again NP-hard, even with the additional restriction that G be planar [76].

Applications of the MULTIWAY PARTITION problem in layout belong to two classes. One class of applications decomposes the circuit into very small parts first, lays them out, and then combines these cells in small numbers to create larger assemblies, until the total circuit has been laid out. Such methods, as well as the corresponding partitioning algorithms, are called *clustering methods*. The typical instance of the MULTIWAY PARTITION in clustering methods has unit vertex weights, a very small uniform maximum part size $B(i) = B$ (usually $B \leq 10$), a uniform minimum part size $b(i) = B - 1$, and $r = \lceil n/B \rceil$. Unfortunately, the algorithm by [146] is not efficient here, since r is large. In fact, this version of the MULTIWAY PARTITION problem is NP-hard for any fixed $B \geq 3$ (see Exercise 6.1). Heuristic clustering methods are discussed in Section 6.7.

The complementary class of applications consist of top-down partitioning schemes for layout. We have already seen an example in Chapter 5. Such partitioning schemes also occur in heuristic schemes for block placement (see Chapter 7). Here r is small (usually $r \leq 10$), since r defines the degree of the cut tree. Furthermore, part sizes are supposed to be about equal—namely, approximately W/r. In other words, the partitions are required to be *balanced*. This requirement ensures that the divide-and-conquer scheme of top-down layout design is well balanced and efficient. The balance requirement is also the reason why the algorithm by [146] does not apply here. The case of $r = 2$ (bipartitioning) plays a special role. For one thing, in top-down partitioning schemes, we often use only bipartitions. Furthermore, solutions for other small values of r can be reduced to the case of $r = 2$, at least heuristically.

6.1.2 Bipartitions

The special case of the MULTIWAY PARTITION problem in which $r = 2$ is called the BIPARTITION problem. The set of hyperedges that contribute to the cut is called the *cut* in this case. The two parts are also called *sides* of the cut. The cutsize sums $c(e)$ over all hyperedges e in the cut. In the bipartition problem, the bounds on the part sizes are usually chosen uniformly, $B(1) = B(2) = \lfloor \alpha w(V) \rfloor$ for some α, $1/2 \leq \alpha < 1$ and $b(1) = b(2) = 1$. The number α is called the *balance factor*. We call a partition whose parts have sizes within these bounds α-*balanced*. The additional restrictions—n even, $w \equiv 1$, and $\alpha = 1/2$— define the BISECTION problem. In the BISECTION problem, the hypergraph has to be cut into two parts whose sizes are identical. Section 5.6 presented an application in which this kind of exact bisection is important. If the vertex weights are not 1, then exact bisections cannot be achieved. In this case, we speak of an *almost exact bisection* if the sizes of both parts of the bipartition differ by at most the maximum vertex weight $w_{\max}$; that is, if the size of each part is at most $\lfloor (w(V) + w_{\max})/2 \rfloor$.

We now discuss the complexity of the BIPARTITION problem. The BIPARTITION problem is also NP-hard. In fact the BISECTION$_{\text{TH}}$ problem restricted to graphs is strongly NP-complete, even if the graphs are regular, and are of some fixed degree $d \geq 3$ [52]. For trees—and, in fact, for partial k-trees that are given by their hierarchical descriptions—a quadratic-time algorithm can be obtained by a direct application of the dynamic-programming methods discussed in Section 4.7.5 [53, 145, 302]. The constant factor grows exponentially with k. A linear-time algorithm that computes good but not optimal bipartitions of trees is described in Section 6.4.1. The complexity of the BIPARTITION problem restricted to *planar* graphs is open (see Section 6.4).

If the part size is trivial, $B = w(V)$, then the problem can be solved with $n - 1$ applications of a maximum network-flow algorithm (see Section 3.11). Apparently, what makes the BIPARTITION problem hard on general graphs is the requirement that partitions be balanced. Here, it does not matter too much what balance factor we choose, as the following theorem shows.

Theorem 6.1 *Let Q be a class of hypergraphs that is closed with respect to forming subgraphs. Assume that we have an algorithm A_{part} that finds an α-balanced bipartition with cutsize $c(n)$ for a hypergraph $G \in Q$ with n vertices and unit vertex weights $w \equiv 1$ in time $T(n)$. Then, we can find an almost exact bisection of any hypergraph $G \in Q$ with arbitrary vertex weights that has cutsize*

$$c'(n) \leq \sum_{i=0}^{\lceil -\log n / \log \alpha \rceil} c(\alpha^i n) \tag{6.1}$$

in time $O(T(n) \log n)$.

Proof We construct the bipartition by iterating algorithm A_{part}. Specifically, we compute a sequence of partitions of the vertices into three sets (LS_i, SS_i, U_i)

and cuts C_i, $i = 0, \ldots, \lceil - \log n / \log \alpha \rceil$. The sets LS_i (large side) and SS_i (small side) contain vertices for which the side of the partition has been decided. Here, LS_i will always have the larger total vertex weight. U_i is the set of vertices whose side has not been determined yet. The initial partition is $LS_0 = SS_0 = \emptyset$ and $U_0 = V$, and the initial cut is $C_0 = \emptyset$. We maintain the following invariant:

1. All hyperedges running between any of the sets LS_i, SS_i, and U_i are in C_i.

2. $w(SS_i) \leq w(LS_i) \leq w(SS_i \cup U_i)$.

3. $|U_i| \leq \alpha |U_{i-1}|$, $(i > 0)$.

By (3), after $\lceil - \log n / \log \alpha \rceil$ steps, $|U_i| \leq 1$. At this time, $(w(LS_i), w(SS_i) \cup U_i)$ is an almost exact bisection by (2). By (1), the cutsize is at most $|C_i|$.

We maintain the invariant as follows. In step i, apply A_{part} to the subgraph of G induced by U_{i-1} *with unit vertex weights*. This computation returns a partition (L, R) of U_{i-1}. with cut C. Assume without loss of generality that $w(L) \leq w(R)$, where w is the vertex weight function of G. We define LS_i to be the set of $SS_{i-1} \cup L$ and LS_{i-1} with larger total vertex weight, and SS_i to be the other set. $C_i := C_{i-1} \cup C$ and $U_i := R$. It is straightforward to check that this process maintains the invariant. The cutsize—that is, the total cost of C—amounts to (6.1).

This algorithm finds an exact bisection if there is one. The run time of the algorithm is

$$O \left(\sum_{i=1}^{\lceil - \log n / \log \alpha \rceil} T(\alpha^{i-1} n) \right) = O(T(n) \log n)$$

If $T(\alpha n) \leq c T(n)$ with $c < 1$, then the run time is $O(T(n))$. This is the case for smooth functions $T(n) = \Omega(n)$ (see Exercise 2.2). $\square$

Here are a few examples of functions $c(n)$ and the corresponding sizes of bisections:

$c(n)$			$c'(n)$		
$c(n)$	$=$	$O(1)$	$c'(n)$	$=$	$O(\log n)$
$c(n)$	$=$	$O(\log n)$	$c'(n)$	$=$	$O((\log n)^2)$
$c(n)$	$\leq$	$c_0 \, n^\varepsilon$	$c'(n)$	$\leq$	$\frac{c_0}{1 - \alpha^\varepsilon} n^\varepsilon$

6.1.3 Separators

In this section, we discuss partitioning graphs by removing vertices.

SEPARATION

Instance: A hypergraph $G = (V, E)$, a vertex weight function $w :$
$V \rightarrow \mathbf{N}$, a maximum part size $B \in \mathbf{N}$

> *Configurations:* All tripartitions (L, R, C) of V with $L, R, C \subset V$, L, R, and C are mutually disjoint, and such that no edge runs between L and R

> *Solutions:* All partitions (L, R, C) such that $w(L), w(R) \leq B$

> *Minimize:* $|C|$

The SEPARATION problem asks us to remove as few vertices as possible such that the graph is disconnected into two parts (sides), each of which has a total vertex weight not exceeding B. A side may also be empty. The set C of vertices to be removed (and sometimes also the triple (L, R, C)) is called a *separator*, or, more specifically, an α-*separator*, where $\alpha = B/w(V)$. Note that we do not use edge costs in the SEPARATION problem.

The SEPARATION problem can be of help in solving the BIPARTITION problem. Assume that the vertex degree of the hypergraph is bounded by d and the maximum cost of a hyperedge is $c_{\max}$. Let (L, R, C) be an α-separator of G with size $|C| = s$. Assume without loss of generality that $w(L) \leq w(R)$. Then $\Pi := (L \cup C, R)$ is an α'-balanced bipartition of G, where

$$\alpha' = \max\{\alpha, \frac{1}{2}\left(1 + \frac{w(C)}{w(V)}\right)\}$$

If the vertex weights in G are approximately uniform—that is, if the maximum vertex weight is at most M times the average vertex weight for some constant $M > 0$—

$$\max_{v \in V} w(v) \leq M \cdot \frac{w(V)}{n}$$

then α' can be bounded by

$$\alpha' \leq \max\{\alpha, \frac{1}{2}\left(1 + \frac{s}{n}M\right)\}$$

The cutsize of Π is at most $c_{\max} d \, s$. If the constants d and $c_{\max}$ are large, then the cutsize of Π is not very good, and if α is close to $1/2$ and s and M are large, then α' differs from α substantially. Otherwise, however, graph separation may actually provide good bipartitions. Many of the more theoretical results on partitioning that are concerned with only asymptotic behavior solve the BIPARTITION problem via the SEPARATION problem.

One advantage of separators is that there are small separators for graphs that do not have bipartitions with small cutsize. An example is the star graph (see Figure 6.1). It has an $1/2$-separator of size 1—namely, the central vertex—but each bisection has size $\lceil (n-1)/2 \rceil$.

The SEPARATION problem is NP-hard (Exercise 6.2). It is not known whether the SEPARATION problem restricted to *planar* graphs is NP-hard. Lipton and Tarjan [291] present a linear-time algorithm for separating planar

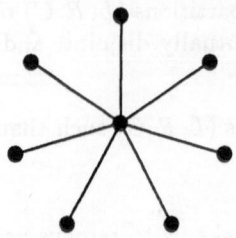

Figure 6.1: *A star graph.*

graphs. The size of the separator is $O(\sqrt{n})$. The obtained separator may be highly suboptimal in some cases, but there are graphs for which it is optimal up to a constant [353]. Rao [373] presents a polynomial-time approximation algorithm for separating planar graphs. The error of the algorithms is $O(\log w(V))$ (see Section 6.4). Leighton and Rao recently extended these results to the BI-PARTITION and BISECTION problems without the restriction that the graph be planar. For a discussion of these results, see Section 6.5. The complexities of these algorithms are still too great for practical applicability.

6.1.4   Free Partitions

There is a variant of the MULTIWAY PARTITION problem in which the number of parts is not specified. In this case, the bounds on part sizes cannot vary from part to part. Usually, only a uniform *upper* bound $B \in \mathbb{N}$ for all part sizes is imposed. We call this variant the FREE PARTITION problem.

The FREE PARTITION problem is also NP-hard. In fact, the decision problem FREE PARTITION$_{TH}$ is strongly NP-complete, even if $B \geq 3$ is fixed and $w \equiv 1$, $c \equiv 1$ [204]. The problem is only weakly NP-complete if G is restricted to be a tree [30]. In this case, a pseudopolynomial-time algorithm exists that find the optimum partition in time $O(nB^2)$ [300]. If G is a tree and all edge weights are identical [161] or all vertex weights are the same [133], then the FREE PARTITION problem is in P. This variant of the partition problem is not discussed in detail in this book.

6.1.5   Modeling Hypergraphs with Graphs

All interesting variants of the partitioning problem that involve hypergraphs are NP-hard, and no approximation algorithms exist. This problem can be overcome in two ways. One way is to transform a hypergraph instance of a partition problem into a graph instance. The most natural way of doing this transformation is to represent each hyperedge by a clique of its terminals.

However, this approach changes the structure of the problem instance quite a bit. For instance, the cost contribution of a unit cost hyperedge across a cut of a bipartition is 1. In contrast, the cost contribution of a clique that is evenly split across a cut rises quadratically in the size of the clique. This quadratic growth does not adequately reflect the costs arising in practice, for instance, when we wire the hyperedge with a Steiner tree [404].

A straightforward way out of this dilemma is to reduce the edge costs on the clique edges. (If this reduction yields fractional edge costs, we can scale the problem appropriately to represent the numbers involved with integers.) Let us consider only bipartitions. Assume that the unit cost hyperedge e with k terminals is represented by a clique over the k terminals of e. Denote the contribution of the clique to the cutsize of bipartition Π with $g(\Pi)$. We want to choose costs $c(e')$ for the clique edges e' such that $g(\Pi)$ is close to 1 for all bipartitions Π. Choosing $c(e') = 1/(k-1)$ for each clique edge e' yields cost 1 for all bipartitions of the clique in which a single vertex is on one side. But if Π bisects the clique, then $g(\Pi) \approx k/4$. On the other hand, choosing $c(e') := 4/k^2$ yields cost 1 for bisections, but $g(\Pi) \approx 4/k$ if only one vertex is on one side of Π.

We can prove that, no matter how we choose the costs $c(e')$ of the clique edges, there always is a bipartition Π such that

$$\max\left(\frac{1}{g(\Pi)}, g(\Pi)\right) = \Omega(\sqrt{k})$$

that is, the deviation from the desired cost 1 is large (see Exercise 6.3). Therefore, in general there is no good way of choosing the cost of clique edges to represent the hyperedge.

The clique has been chosen as a substitute for the hyperedge because of its symmetry. Our conjecture is that every other choice, such as substituting the hyperedge with a tree, or even just choosing different costs for different clique edges, presupposes an asymmetric topology, which is also bad for certain partitions (see Exercise 6.4). There just is no good way of mapping a hyperedge into a graph.

A completely different way of dealing with hyperedges in partition problems is to use robust heuristics that also work on hypergraphs. This approach is quite successful in practice. We shall discuss it in detail in the next section.

The BIPARTITION problem takes a central place in the area of hypergraph partitioning. It has been studied most extensively, and multiway partitioning has been heuristically reduced to bipartitioning. In Section 6.2, we present several bipartitioning heuristics, and we comment on how they can be used for doing multiway partitioning.

6.2   Iterative Improvement Heuristics for Bi-partitioning Hypergraphs

The combinatorial structure of partitioning problems is just beginning to be elucidated. To date, iterative improvement techniques that make local changes to an initial partition are still the most successful partitioning algorithms in practice. Even though we cannot say much in theory about the performance of these heuristics, they perform quite satisfactorily in practice.

The advantage of these heuristics is that they are quite robust. In fact, they can deal with hypergraphs (for which no feasible algorithm or approximation algorithm is known), as well as arbitrary vertex weights, edge costs, and balance criteria.

The heuristics are especially frequently used in divide-and-conquer algorithms for placement and floorplanning that are variants of the mincut strategy described in Section 5.3. Such algorithms are detailed in Chapter 7.

6.2.1   The Kernighan–Lin Heuristic

Let us start by considering graph bisection. A natural local search method for solving this problem is to start with an initial bisection and to exchange pairs of vertices across the cut of the bisection if doing so improves the cutsize. The corresponding configuration graph for a problem instance G is denoted by $C_{\mathrm{PI}}(G)$; the subscript PI stands for *pairwise interchange*. Consider the example given in Figure 6.2. This graph has 18 vertices and 34 edges. There are $\binom{18}{9}/2 = 24310$ bisections. The configuration graph $C_{\mathrm{PI}}(G)$ has that many nodes and $\binom{18}{9} \cdot 9^2/4 = 984555$ edges.

If, to explore this configuration graph, we choose a greedy search strategy that chooses the next vertex to move to as the neighbor of the current vertex that affords the largest decrease in the cutsize, we are in danger of being trapped in a local minimum. In fact, the bisection

$$C_0 = (\{1, 2, 3, 4, 14, 15, 16, 17, 18\}, \{5, 6, 7, 8, 9, 10, 11, 12, 13\})$$

illustrated in Figure 6.2 is an example of such a local minimum with the large cutsize of 10.

To reduce the danger of being trapped in local minima, Kernighan and Lin modify the search procedure. Assume that the current bisection is C_0. We determine the vertex pair whose exchange results in the largest decrease of the cutsize *or* in the smallest increase, if no decrease is possible. In Figure 6.2, such a pair is, for instance, the vertex pair $\{4, 10\}$, and the corresponding exchange increases the cutsize by 2. This exchange is made only tentatively, however. We now *lock* the exchanged vertices. This locking prohibits them from taking part in any further tentative exchanges. Then we look for a second vertex pair whose exchange improves the cut cost, and do the same for it. In our example,

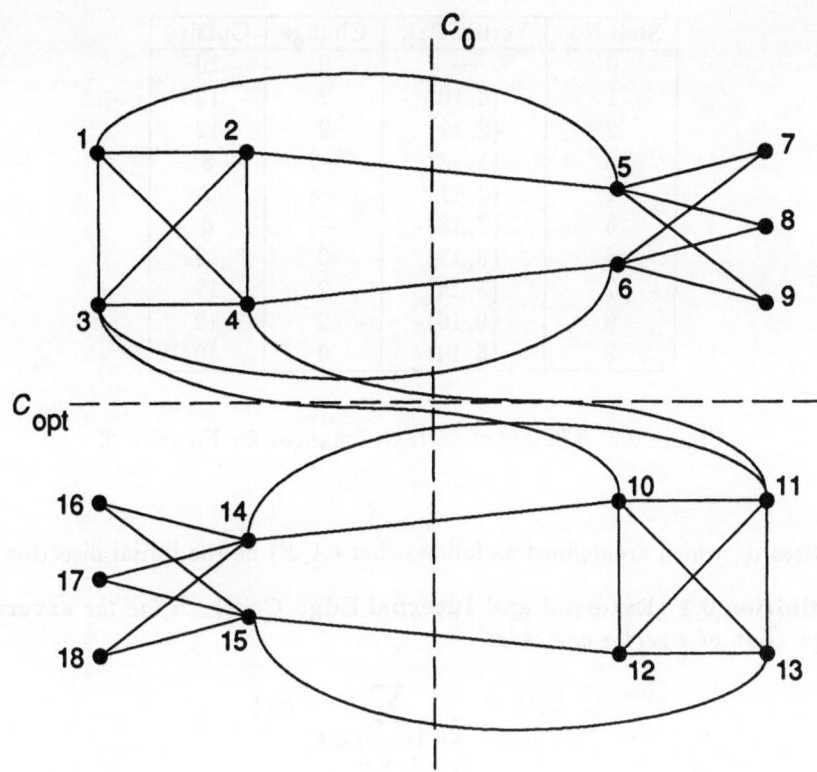

Figure 6.2: *A graph to be bisected.*

a suitable vertex pair is the pair $\{2, 13\}$. We continue in this way, keeping a log of all tentative exchanges and the resulting cutsizes. We finish when all vertices are locked. At this time, we have exchanged both sides of the bisection and are back to the original cutsize. Figure 6.3 shows a possible choice of a vertex pair for the example in Figure 6.2. The entry $(s, 3)$ in the table displays the change in the cutsize if the first s exchanges in the table are performed. The fourth column of the table contains the respective new cutsize.

The move we make from the original bisection now exchanges those pairs of vertices that result in the smallest cutsize in the log. In the example, we exchange the two vertex sets $\{1, 2, 3, 4\}$ and $\{10, 11, 12, 13\}$, and arrive at a global optimum.

We now explain the algorithm for the Kernighan–Lin heuristic in detail, referring to Figure 6.4. Vertices are numbered from 1 to n. The algorithm computes the improvement *Dcost* of the cutsize using the gain values $D(i)$ of

Step No.	Vertex Pair	Change	Cutsize
0	—	0	10
1	$\{4, 10\}$	2	12
2	$\{2, 12\}$	2	12
3	$\{1, 13\}$	-2	8
4	$\{3, 11\}$	-8	2
5	$\{7, 18\}$	-4	6
6	$\{8, 17\}$	0	10
7	$\{5, 15\}$	2	12
8	$\{9, 16\}$	2	12
9	$\{6, 14\}$	0	10

Figure 6.3: *The log of vertex exchanges for Figure 6.2.*

vertices v_i, which are defined as follows. Let (A, B) be the initial bisection.

Definition 6.1 (External and Internal Edge Cost) *Define the* **external edge cost** *of a vertex* $v_i \in A$ *as*

$$E(i) := \sum_{\substack{e = \{v_i, v_j\} \in E \\ v_j \in B}} c(e)$$

and the **internal edge cost** *as*

$$I(i) := \sum_{\substack{e = \{v_i, v_j\} \in E \\ v_j \in A}} c(e)$$

Analogous definitions are made if $v_i \in B$.

Definition 6.2 (Gain) *The* **gain** *of* v_i *is defined as*

$$D(i) := E(i) - I(i)$$

$D(i)$ is the amount by which the cutsize decreases if v_i changes sides in the partition. Then obviously an exchange of the vertex pair $\{v_i, v_j\}$ is associated with a decrease in the cutsize of $D(i) + D(j) - 2c(i, j)$, where $c(i, j)$ is the cost of the edge $\{v_i, v_j\}$ if that edge exists, and $c(i, j) = 0$ otherwise. Thus, the higher the D values are, the smaller cutsize is. The tentative exchange step that results in the currently best cutsize is kept in *BestChange*; its cost is in *BestCost*. The Boolean array *Locked* contains the information on which vertices are locked. The log information on the vertex pairs selected for tentative moves

```
(1)      Pair : array [1 : n/2] of pair of [1 : n];
(2)      Cost : array [0 : n/2] of integer;
(3)      Locked : array [1 : n] of Boolean;
(4)      D : array [1 : n] of integer;
(5)      c : array [1 : n, 1 : n] of integer;
(6)      BestChange : [1 : n/2];
(7)      BestCost : integer;
(8)      imin, jmin : [1 : n];

(9)      compute the c and D values;
(10)     for i from 1 to n do
(11)         Locked[i] := false od;

(12)     BestCost := Cost[0] := cutsize(A, B);
(13)     BestChange := 0;
(14)     for s from 1 to n/2 do
(15)         Cost[s] := ∞;
(16)         for i, j from 1 to n such that vᵢ ∈ A and Locked[i] = false
                                  and vⱼ ∈ B and Locked[j] = false do
(17)             if 2c[i, j] − D[i] − D[j] < Cost[s] then
(18)                 Pair[s] := (i, j);
(19)                 Cost[s] := 2c[i, j] − D[i] − D[j] fi od;
(20)         (imin, jmin) := Pair[s];
(21)         Locked[imin] := Locked[jmin] := true;
(22)         for i from 1 to n such that Locked[i] = false do
(23)             if vᵢ ∈ A then
(24)                 D[i] := D[i] − c[i, jmin] + c[i, imin]
(25)             else
(26)                 D[i] := D[i] − c[i, imin] + c[i, jmin] fi od;
(27)         Cost[s] := Cost[s − 1] + Cost[s];
(28)         if Cost[s] < BestCost then
(29)             BestChange := s;
(30)             BestCost := Cost[s] fi od;

(31)     for s from 1 to BestChange do
(32)         exchange Pair[s] od;
```

Figure 6.4: A pass of the Kernighan–Lin heuristic.

and the corresponding cutsizes (columns 2 and 4 of Figure 6.3) are kept in the arrays *Pair* and *Cost*. Lines 9 to 11 initialize the computation. Lines 12 to 30 contain the computation of the log. Specifically, lines 14 to 21 select the vertex pair that results in the maximum decrease or minimum increase of the cutsize, lines 21 to 26 update the D values according to the tentative exchange, and lines 27 to 30 fill the log tables. Finally, lines 31 and 32 recover the best exchange from the log.

There are two ways in which we can view the Kernighan–Lin heuristic as a local search procedure. One is as a nongreedy search on the configuration graph $C_{PI}(G)$. In fact, some exchanges in the exchange sequence that increase the cutsize are allowed, even if in the end only an exchange sequence that decreases the cutsize is performed. Alternatively, we can consider the Kernighan–Lin heuristic as a greedy procedure on a denser configuration graph $C_{CI}(G)$, where the subscript CI stands for *Complex Interchange*. In $C_{CI}(G)$, all bisections encountered during the construction of the log are considered neighbors of the initial bisection. This modification increases the outdegree of the configuration graph from $n^2/8$ to $O(n^3)$.

Let us analyze the complexity of the Kernighan–Lin heuristic. The minimization process in lines 16 to 19 dominates the run time of the whole algorithm. If all possible pairs are considered in the minimization process, the overall complexity is $O(n^3)$. In practice, we can speed up the algorithm, but a run time of $O(n^2 \log n)$ still results in typical cases (see Exercise 6.6).

Figure 6.4 shows only one pass of the Kernighan–Lin heuristic. Since the heuristic is still greedy, this pass may end in a local optimum that is not a global optimum. A repetition of the pass starting with the new bisection can lead to further improvement. After a few passes, usually about five, no more improvement is possible, and the Kernighan–Lin heuristic terminates. The bisection computed is not a global optimum, in general, and we do not know in a specific case how close it is to being one. In bad cases, it can be quite far away from a global optimum (see Exercise 6.7). The only thing that can be said about the obtained bisection is that it is a local optimum for the last pass and the best bisection encountered. In their original paper [229], Kernighan and Lin give further heuristic improvements of this procedure (such as finding a good starting partition) that are aimed at finding better bisections.

The Kernighan–Lin heuristic is quite a robust algorithm. We can accommodate additional constraints, such as required sides for certain vertices. This feature is important in layout because we may want some blocks of the circuit to be in prespecified parts. If partitioning is used for placement, then vertices representing preplaced pads may be required to be on the side on which the pad is located. In the Kernighan–Lin heuristic, we can take care of this requirement simply by assigning the corresponding side to the vertex and locking it.

Besides the fact that its outcome is not analyzable, the main disadvantages of the Kernighan–Lin heuristic are the following:

1. The heuristic handles only unit vertex weights. This restriction is not suitable for applications in circuit layout, since here vertex weights represent block sizes, which usually differ between different blocks. Kernighan and Lin suggest that we represent a vertex v with integer vertex weight $w(v)$ by a clique (or a similar kind of vertex cluster) with $w(v)$ vertices and edges with a high cost. This modification increases the size of the graph substantially.

2. The heuristic handles only exact bisections. We can eliminate this restriction by adding dummy vertices that are isolated from all other vertices. Depending on the number of dummy vertices, more or less unbalanced bipartitions can be obtained. However, the heuristic does not prefer more balanced bipartitions to less balanced ones with the same cutsize.

3. The heuristic cannot handle hypergraphs. However, in applications in circuit design, we have to be able to deal with multiterminal nets that translate into hyperedges. In Section 6.1.5, we saw that, in general, it is not possible to transform a hypergraph instance of a partition problem into a graph instance that approximates it well. Thus, we are in need of heuristics that apply to hypergraphs directly.

4. The complexity of a pass is still too high. We would like it to be linear in the size of the hypergraph.

5. The heuristic still incorporates some degree of nondeterminism, which makes bad choices possible. For instance, in Figure 6.2, we could have chosen the pair $\{4, 5\}$ for the first exchange. This choice would have prevented us from finding the global optimum. It would be desirable to resolve the nondeterminism inherent in the heuristic so as to make further optimizations and to exclude bad choices.

The first four disadvantages are eliminated in a modification of the Kernighan–Lin heuristic that is discussed in the next section.

6.2.2 The Fiduccia–Mattheyses Heuristic

Fiduccia and Mattheyses [114] introduce the following new elements to the Kernighan–Lin heuristic:

1. Only a single vertex is moved across the cut in a single move. This modification allows for handling unbalanced partitions and nonuniform vertex weights. A minimum balance of the bipartition is maintained throughout the process.

2. The concept of the D value is extended to hypergraphs. This extension in conjunction with (1) allows for handling hypergraphs. It has already been observed by Schweikert and Kernighan [404].

3. A specially tailored algorithm for selecting vertices to be moved across the cut saves run time.

As before, the moves in a pass are tentative and are followed by locking the moved vertex. They may increase the cutsize. However, just as in the Kernighan–Lin heuristic, at the end of a pass, when no more moves are possible, the sequence of moves is realized that achieves the best cutsize, and it is realized only if it decreases the cutsize. Otherwise, the pass is ended.

Let us now detail the concepts of the algorithm. The extension of the D value to hypergraphs is quite straightforward. We have only to redefine the internal and external hyperedge costs as follows. Let, without loss of generality, $v_i \in A$.

Definition 6.3 (External and Internal Hyperedge Cost) *The external hyperedge cost of vertex* v_i *is defined as*

$$E(i) := \sum_{e \in E_{\text{ext},i}} c(e)$$

where

$$E_{\text{ext},i} := \{e \in E \mid \{v_i\} = e \cap A\}$$

Analogously, the internal hyperedge cost of vertex v_i *is defined as*

$$I(i) := \sum_{e \in E_{\text{int},i}} c(e)$$

where

$$E_{\text{int},i} := \{e \in E \mid v_i \in e \text{ and } e \cap B = \emptyset\}$$

With these provisions, the gain $D(i)$ *is defined as in Definition 6.2.*

Intuitively, $E_{\text{ext},i}$ is the set of hyperedges that is removed from the cut if v_i changes sides, and $E_{\text{int},i}$ is the set of hyperedges that is added to the cut if v_i changes sides. With these observations, it is obvious that, when we move v_i from A to B, the cutsize changes by an amount of $-D(v_i)$. The hyperedges in $E_{\text{ext},i} \cup E_{\text{int},i}$ are also called *critical hyperedges for* v_i.

The configuration graph $C_{\text{VM}}(G)$ that underlies the Fiduccia–Mattheyses heuristic has a directed edge for each allowable move of a vertex across the cut—that is, for each such move that does not violate the minimum balance criterion. The subscript VM stands for *vertex move*. The size of the neighborhood of a bipartition in $C_{\text{VM}}(G)$ is $O(n)$. Thus, $C_{\text{VM}}(G)$ is a much sparser graph than is $C_{\text{PI}}(G)$. The Fiduccia–Mattheyses heuristic is a nongreedy local search method on $C_{\text{VM}}(G)$ in the same sense in which the Kernighan–Lin heuristic is a nongreedy local search method on $C_{\text{PI}}(G)$.

Figure 6.5 gives a rough outline of the Fiduccia–Mattheyses heuristic. The maintenance of the log tables (line 5) and the determination of the final bipartition (line 6) can be handled as in the Kernighan–Lin heuristic We shall now discuss in detail how lines 1, 3 and 4 are implemented to run in linear time.

(1)   initialize D−values;
(2)   **while** moves of vertices are possible **do**
(3)   select an unlocked vertex v;
(4)   update the data structures to reflect the tentative move of v;
(5)   extend the log of tentative moves od;
(6)   compute the prefix of the sequence of tentative moves
                      that achieves the smallest cutsize;

Figure 6.5: *A rough outline of the Fiduccia–Mattheyses heuristic.*

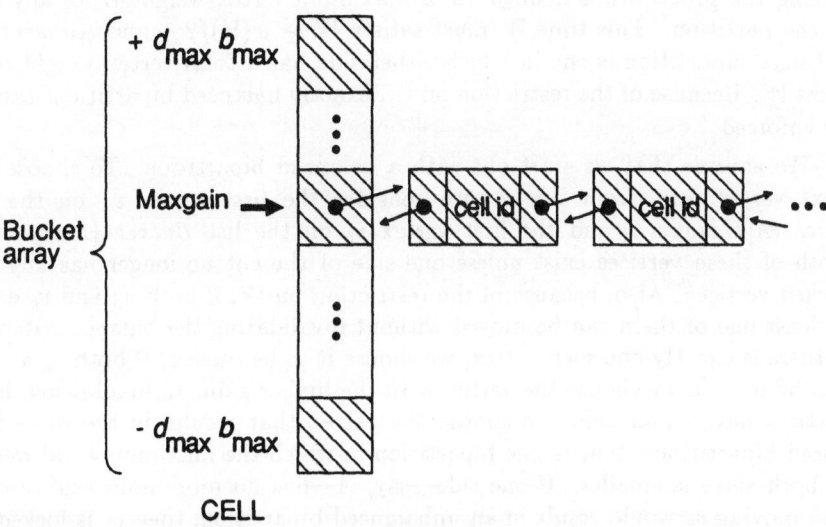

Figure 6.6: *The data structure for choosing vertices in the Fiduccia–Mattheyses heuristic.*

6.2.2.1   Choosing A Vertex To Be Moved

To choose the next vertex to be moved, we use the data structure depicted in Figure 6.6. The data structure consists of two one-dimensional pointer arrays *BucketA* and *BucketB* indexed by the set $[-d_{max} \cdot b_{max}, d_{max} \cdot b_{max}]$. Here, d_{max} is the maximum vertex degree in the hypergraph, and b_{max} is the maximum cost of a hyperedge. Each pointer in the array represents a possible D value for some vertex. (It is straightforward to see that the D values have to lie inside this interval.) The pointers in the array *BucketA* point to linear lists of items representing unlocked vertices inside A with the corresponding gain. In addition, there is an integer *MaxgainA* that is set to the maximum value i such that $BucketA[i] \neq$ **nil**—that is, to the maximum gain of any vertex on side A. Analogous statements hold for *BucketB*.

We also maintain lists *LockedA* and *LockedB* containing all items representing locked vertices. Initially, *LockedA* contains all (unlocked) vertices on side A; *LockedB* contains those for side B. In line 1 of Figure 6.5, these two lists are emptied and their elements are used to build up the bucket data structures. As vertices become locked, their items are put on the corresponding list of locked vertices and are saved for the next pass of the heuristic.

As defined in the BIPARTITION problem, the balance criterion we maintain during the procedure is defined via a maximum vertex weight W of any side of the partition. This time W must satisfy $W \geq w(V)/2 + \max_{v \in V} w(v)$. A *balanced* bipartition is one in which either side has a total vertex weight of at most W. Because of the restriction on W, exactly balanced bipartitions cannot be enforced.

We assume that we start out with a balanced bipartition. To choose the next vertex, say, from side A, we consider the first vertex v_a on the list $BucketA[MaxgainA]$ and the first vertex v_b on the list $BucketA[MaxgainB]$. Both of these vertices exist unless one side of the cut no longer has any unlocked vertices. Also, because of the restriction on W, if both v_a and v_b exist, at least one of them can be moved without invalidating the balance criterion. If there is exactly one such vertex, we choose it to be moved. If both v_a and v_b can be moved, we choose the vertex with the higher gain. If, in addition, both vertices have equal gain, we choose the vertex that results in the more balanced bipartition—that is, the bipartition in which the maximum total weight of both sides is smaller. If one side—say, A—has no more unlocked vertices and moving v_b would result in an unbalanced bipartition, then v_b is locked on side B. The pass ends if there are no more unlocked vertices.

This procedure amounts to a heuristic search through the graph $C_{\text{VM}}(G)$. $C_{\text{VM}}(G)$ is sparser than is $C_{\text{PI}}(G)$, so it is not surprising that the number of passes that are necessary before the method stabilizes is slightly greater than in the Kernighan–Lin heuristic. But experience shows that the number of passes still is bounded by a small constant—say, about 7.

We are left with showing that the necessary initialization and update procedures take linear time. The size of the encoding of G is $O(p)$, where p is the number oft terminals in G (see Definition 3.2). In fact, we shall show that the run time of the algorithm is $O(p \cdot b_{\max})$. This quantity is linear only if $b_{\max} = O(1)$. In fact, this criterion is met in circuit design. In contrast to the vertex weights, which can be quite large, hyperedge costs encode parameters such as word length and are limited to small constant values. Thus, the Fiduccia–Mattheyses heuristic runs in linear time.

```
(1)        for i such that vᵢ ∈ LockedA do
(2)            D[i] := 0;
(3)            for e hyperedge such that vᵢ ∈ e do
(4)                if Aside[e] = 1 then D[i] := D[i] + c(e) fi;
(5)                if Bside[e] = 0 then D[i] := D[i] − c(e) fi od od;
```

Figure 6.7: *Initial computation of the D values.*

6.2.2.2 Initializing the Data Structure

The computation and manipulation of the D values presupposes two more arrays:

$$Aside, Bside : \textbf{array [hyperedge] of } [1:p]$$

$Aside[e]$ contains the number of terminals of hyperedge e on side A. $Bside[e]$ is defined analogously. Obviously, $e \in E_{\text{ext},i}$ exactly if $Aside[e] = 1$, and $e \in E_{\text{int},i}$ exactly if $Bside[e] = 0$.

Moreover, we provide two vertex lists $UnlockedA[e]$ and $UnlockedB[e]$ for each hyperedge e containing unlocked terminals of e on side A and side B, respectively. These data structures can be computed in time $O(p)$ by a simple pass over the adjacency structure encoding G.

Figure 6.7 shows how to do the initial computation of the D values of vertices on side A. Side B is handled analogously. Remember that $LockedA$ initially contains all vertices on side A. The time for this computation, once $Aside$ and $Bside$ are computed, is $O(m + n)$. Given the D values, the bucket data structures can be generated in a straightforward manner in time $O(n)$. Note that the D values are computed using a scan of the hyperedges, and not of the vertices.

Given the D values, the bucket data structures can be initialized in time $O(n + d_{\max} b_{\max})$. The second term accounts for the initialization of the arrays $BucketA$ and $BucketB$, and the first term accounts for generating the linear lists of vertex items linked to the term bucket arrays.

6.2.2.3 Updating the Data Structure

Assume now, without loss of generality, that a vertex v_i on side A has been chosen. The vertex is then *locked*—that is, its item is removed from its bucket list and is attached to the list $LockedB$, where it awaits the next pass.

Then, adjustments of D values become necessary, since during the move of v_i hyperedges become or cease to be critical, or their contribution changes from positive to negative or vice versa. Let us call a hyperedge for which this happens *changing*. For a hyperedge e to be changing in the move of v_i, it is

(1)	**for** e **hyperedge such that** $v_i \in e$ **do**
(2)	remove i from *UnlockedA*$[e]$;
(3)	*Aside*$[e]$:= *Aside*$[e]$ − 1;
(4)	**if** *Aside*$[e]$ = 0 **then**
(5)	**for** $j \in$ *UnlockedB*$[e]$ **do**
(6)	$D[j]$:= $D[j]$ − $c(e)$ **od**;
(7)	**else if** *Aside*$[e]$ = 1 **then**
(8)	**for** $j \in$ *UnlockedA*$[e]$ **do**
(9)	$D[j]$:= $D[j]$ + $c(e)$ **od fi**;
(10)	*Bside*$[e]$:= *Bside*$[e]$ + 1;
(11)	**if** *Bside*$[e]$ = 1 **then**
(12)	**for** $j \in$ *UnlockedA*$[e]$ **do**
(13)	$D[j]$:= $D[j]$ + $c(e)$ **od**;
(14)	**else if** *Bside*$[e]$ = 2 **then**
(15)	**for** $j \in$ *UnlockedB*$[e]$ **do**
(16)	$D[j]$:= $D[j]$ − $c(e)$ **od fi od**;

Figure 6.8: *Updating the D values.*

necessary that $v_i \in e$, otherwise the values *Aside*$[e]$ and *Bside*$[e]$ do not change. However, not all such hyperedges are changing. A convenient way of viewing the changes is to think of the move of v_i as a two-phase procedure. First, we delete v_i from side A, then we add it to side B. Figure 6.8 details the respective algorithm for changing the D values. In line 3, v_i is effectively deleted from A. Through this action, some hyperedges become critical, but none of them cease to be critical. Lines 4 to 6 process the hyperedges that make a new negative contribution to the D values. Lines 7 to 9 process the edges whose new contribution to the D values is positive. In line 10, v_i is effectively added to side B, which causes some hyperedges to cease to be critical. Lines 11 to 12 make the respective positive changes, and lines 14 to 16 make the respective negative changes, to the D values.

Lemma 6.1 *The total run time for all invocations of the procedure updating D values in one pass of the Fiduccia–Mattheyses heuristic is $O(p)$.*

Proof The total run time for updating D values is a sum of terms, one for each invocation of the algorithm in Figure 6.8, associated with a vertex move. One execution of lines 7 to 9 and of lines 14 to 16, respectively takes $O(1)$ time, since the loops are executed for at most one and two vertices, respectively. In total, these lines are executed once for each terminal of a hyperedge—that is, p times. Lines 4 to 6 and 11 to 13 take longer, since they may involve changing the D values of many terminals of the hyperedge. However, these lines are executed only *once* and *twice* per hyperedge, respectively. To see this, we observe that, after the number of terminals of a hyperedge e on some side is decreased to 0 by a vertex v_i being moved, this can never happen again on either side, since vertices are locked after they are moved. Similarly, the number of terminals of a hyperedge e can increase from 0 to 1 only once on each side. Thus, the total run time for lines 4 to 6 and 11 to 13 is again $O(p)$. □

To conclude that one pass of the Fiduccia–Mattheyses heuristic runs in linear time, we are left with showing that updating the bucket data structure is achievable in this time.

Lemma 6.2 *All updates of the bucket data structure that take place in one pass of the Fiduccia–Mattheyses heuristic take time* $O(p \cdot b_{max})$.

Proof The update of an item representing a vertex v_j after $D[j]$ changes obviously takes time $O(1)$. Since the number of changes of the D values is $O(p)$ by Lemma 6.1, the total number of all these updates is $O(p)$. We still have to bound the run time for updating the value of *MaxgainA* (and analogously for *MaxgainB*). Two cases can arise.

Case 1: *MaxgainA* increases. This happens if some D value is increased to a higher value than the current value of *MaxgainA*. We reset *MaxgainA* to this new value.

Case 2: *MaxgainA* decreases. This happens if the D value of the last vertex in *BucketA[MaxgainA]* is decreased or if this vertex is locked. We decrement the value of *MaxgainA* until *BucketA[MaxgainA]* becomes nonempty.

In case 1, the value of *MaxgainA* can increase by at most b_{max}. Also, case 1 can happen only once per D value update—that is, $O(p)$ times in total. The total number of decrements in all occurrences of case 2 cannot exceed the initial size of the array *BucketA* plus the sum of all increments, which amounts to $O(d_{max} b_{max} + p \cdot b_{max})$. Since each occurrence of case 1 and each decrement take time $O(1)$, the lemma follows. □

We can extend the Kernighan–Lin heuristic directly to hypergraphs by using the generalized notion of gain that underlies the Fiduccia–Mattheyses heuristic. Dunlop and Kernighan [100] report on experiments comparing this version of the Kernighan–Lin heuristic with the Fiduccia–Mattheyses heuristic. As is expected, the Fiduccia–Mattheyses heuristic performs much faster than does the Kernighan–Lin heuristic, and its results, although close in quality, are not always quite as good.

6.2.3 Higher-Order Gains

One disadvantage that the Fiduccia–Mattheyses heuristic shares with the Kernighan–Lin heuristic is that there is a large amount of unresolved nonde-

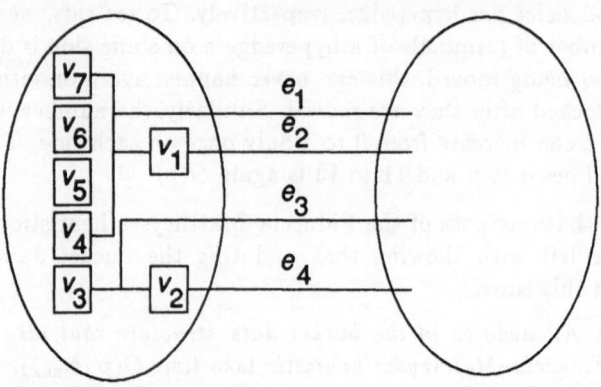

Figure 6.9: *Vertices that the Fiduccia–Mattheyses heuristic cannot distinguish.*

terminism. The heuristics choose arbitrarily between vertices that have equal gain and equal weight. If we introduce more *look-ahead* into the heuristic we can distinguish between such vertices with respect to the gains they make possible in later moves. Krishnamurthy [247] extends the Fiduccia–Mattheyses heuristic in this direction.

Specifically, Krishnamurthy extends the gain value to a sequence of numbers $(D_1(i), \ldots, D_k(i))$. Here, $D_1(i)$ is the original gain, and the following numbers $D_2(i), \ldots, D_k(i)$ define secondary and lower-order criteria for distinguishing vertices with the same D_1 value. When substituted in the Fiduccia–Mattheyses heuristic, the extended gain values $(D_1(i), \ldots, D_k(i))$ are compared according to the lexicographic ordering. This provision places the highest priority on the value of $D_1(i)$, and successively lower priorities on the following components of the extended gain value.

What could be a second-order criterion for distinguishing vertices with the same D_1 value? We want to favor vertices that are likely to reduce cutsizes in future moves. Consider Figure 6.9. Here, v_1 and v_2 have the same gain $D_1(1) = D_1(2) = 0$. However, moving v_2 increases the gain of v_3 from 0 to 1, whereas moving v_1 does not change any gains. Therefore v_2 should be preferred to v_1 as the next vertex to be moved. Thus, $D_2(2)$ should be larger than $D_2(1)$.

A generalization of this concept to higher degrees of look ahead is based on the following definition of gain.

Definition 6.4 (Binding Number) *The* **binding number** $\beta_A(e)$ *of a hyperedge e on side A is defined to be the number of unlocked terminals of e on side A, unless there is a locked terminal of e on side A, in which case $\beta_A(e) = \infty$.*

The binding number indicates the degree of difficulty of eliminating all terminals of e from side A.

Definition 6.5 (kth-Order Gain) *The kth-order gain of a vertex v_i on side A is defined as*

$$D_k(i) := \sum_{e \in E_{pos,k}(i)} c(e) - \sum_{e \in E_{neg,k}(i)} c(e)$$

where

$$E_{pos,k}(i) := \{e \in E \mid v_i \in e, \ \beta_A(e) = k, \ \beta_B(e) > 0\}$$
$$E_{neg,k}(i) := \{e \in E \mid v_i \in e, \ \beta_A(e) > 0, \ \beta_B(e) = k - 1\}$$

An inspection of Definition 6.5 shows that $D_1(i)$ is the gain value defined in Definition 6.2. The value $D_2(i)$ is the sum of the following contributions. Each hyperedge that *after the move of v_i* has a single terminal v_j on side A, and thus contributes positively to the gain of v_j, also contributes positively to $D_2(i)$. On the other hand, a hyperedge that before but not after the move of v_i contributes positively to the gain of a (single) vertex on side B contributes negatively to $D_2(i)$. The kth-order gain extends this analysis to k moves into the future.

For $1 \leq j \leq k$, we choose the jth-order gain as the jth item in the extended gain sequence. If we modify the gain concept in this way in the Fiduccia–Mattheyses heuristic, we achieve the desired removal of nondeterminism. The larger value we choose for k, the more vertices this heuristic can distinguish. For $k = 1$, we get the gains originally used in the Fiduccia–Mattheyses heuristic. Already choosing $k = 2$, we can distinguish the vertices v_1 and v_2 in Figure 6.9, since $D_2(1) = 0$ and $D_2(2) = 1$.

The data structures in the Fiduccia–Mattheyses heuristic can be extended so as to maintain and update extended gains in time $O(p)$, as long as k is a constant. We leave the details to the reader (Exercise 6.8).

Using kth-order gains makes sense only if G has a hyperedge with at least k terminals. As we have seen in Section 6.2.1, there also is undesirable nondeterminism if G is a graph. It is useful to investigate similar heuristic techniques for resolving this nondeterminism in an appropriate manner (Exercise 6.9).

6.2.4* Improving Bipartitions

Experiments run by several authors [52, 144, 211] indicate that iterative improvement heuristics yield very good bipartitions if the minimum vertex degree or the density of the graph is large. This result suggests the conjecture that, for graphs with large minimum vertex degree or for dense graphs, the configuration graph of the partitioning problem has few local optima that are not global optima. This conjecture has not been proved, so far. But we can use the experimental observation to tune bipartitioning heuristics.

The following heuristic for graph bisection has been suggested in a similar form by Goldberg and Burstein [144], Bui et al. [52], and Jones [212].

Step 1: Find a matching M in G.

Step 2: Contract the edges in M in order to increase the vertex degree and density of G. This step yields a graph G'.

Step 2: Run any iterative improvement heuristic to find an (almost) optimal bisection (V_1', V_2') of G'.

Step 3: Uncontract the edges and thereby extend (V_1', V_2') to a bisection (V_1, V_2) of G.

Step 4: Run the iterative improvement heuristic again, starting with the bisection (V_1, V_2).

Reports on experiments with such a heuristic are encouraging [52, 212]. Goldberg and Burstein suggest replacing Step 4 with the following sequence of steps:

Step 4': Run the heuristic on the subgraphs $G|_{V_1}$ and $G|_{V_2}$ yielding almost optimal bisections (V_{11}, V_{12}) and (V_{21}, V_{22}).

Step 5: Let M_{ij} be matchings in V_{ij} for $i, j = 1, 2$. Let $\tilde{M} = M_{11} \cup M_{12} \cup M_{21} \cup M_{22}$. Execute steps 2 and 3 on G, using $\tilde{M}$ instead of M.

This version of the heuristic has the property that the iterative improvement heuristic is executed only on subgraphs of G with one-half the number of vertices. This property is an advantage if the iterative improvement heuristic is time or space consuming, as in simulated annealing.

6.2.5* Extension to Multiway Partitioning

There are heuristic ways of reducing the MULTIWAY PARTITION problem to the BIPARTITION problem. Kernighan and Lin suggest using their heuristic to achieve balanced multiway partitions as follows. Start with an arbitrary partition of the graph into r equal-sized sets. Then, apply the heuristic to pairs of subsets so as to improve the partition. Any pair of subsets in which at least one set has been changed since the last time the heuristic was applied to the pair qualifies for an application of the heuristic. For small r, this process can be run until stabilization occurs. Even if we stop early, the largest improvements in the cost of the partition are made in the first passes.

The same idea applies to the Fiduccia–Mattheyses heuristic and its variants.

6.2.6 Summary

The heuristics for bipartitioning presented in Section 6.2 are still the main methods for partitioning graphs and hypergraphs in practice. They are robust and quickly generate satisfactory solutions for many applications. The heuristics are mainly used for bipartitioning, but can also be applied to partitioning hypergraphs into a larger number of parts. The heuristics use iterative improvement and are not based on a global understanding of the combinatorial nature of the partitioning problem. Krishnamurthy and Mellema [248] state that the evaluation of mincut heuristics such as those presented in this section has to be done experimentally, rather than analytically. They present benchmark examples with known optimal cutsize.

Barnes et al. [22] present a heuristic for determining exchanges of sets of vertices that are more complicated than two-way, in order to improve multiway partitions. Their method is based on the formulation of the MULTIWAY PARTITION problem as a QUADRATIC ASSIGNMENT problem (see also Sections 6.6.1 and 6.6.3.1).

6.3 Simulated Annealing

Johnson et al. [211] suggest the following customization of the simulated-annealing algorithm to the BIPARTITION problem. States are all bipartitions of G. Edges of the configuration graph are undirected, and they connect bipartitions that can be obtained from each other by moving a single vertex across the cut. The cost of a bipartition (V_1, V_2) is defined as

$$c(V_1, V_2) = \sum_{\substack{e = \{v_1, v_2\} \in E \\ v_1 \in V_1, \ v_2 \in V_2}} c(e) + \beta(w(V_1) - w(V_2))^2 \qquad (6.2)$$

where $\beta > 0$ is the *imbalance factor*. The size of a neighborhood in a graph with n vertices is n.

This customization does not model the BIPARTITION problem exactly, since it does not incorporate restrictions on the part sizes. Rather, it can be understood as a relaxation of the BIPARTITION problem that penalizes for the imbalance of a partition in the cost function, similarly to the idea on which Lagrangian relaxation and the penalty methods in nonlinear optimization are based (see Sections 4.6.3 and 4.8.2). To solve the BIPARTITION problem for graphs with unit vertex weight, we can legalize a partition arrived at with simulated annealing using the following greedy heuristic: Iteratively move vertices from the larger side to the smaller side that imply the least increase of (6.2).

Johnson et al. report on detailed experiments using this customization together with appropriate cooling schedules to bipartition graphs with unit vertex weights. The results show that simulated annealing usually needs *much* more running time than does the Kernighan–Lin heuristic. On random graphs, however, the partitions obtained by simulated annealing are likely to have a smaller cutsize than can be expected to be obtained by the Kernighan–Lin heuristic, even if equal (large) amounts of run time are invested. The authors of this study also identify classes of graphs for which the Kernighan–Lin heuristic consistently finds partitions with smaller cutsize. Thus, the worth of simulated annealing has to be studied carefully in each application.

Greene and Supowit [156] identify the preceding customization of simulated annealing to graph bipartitioning as one example of an application in which the rejectionless method for simulated annealing can be implemented efficiently (see Section 4.3.4).

6.4 Partitioning Planar Graphs

Planar graphs are a quite restricted graph family, but they are the class of graphs that we know best how to partition. Clearly, circuits are not likely to be planar. However, as we mentioned in the introduction, graph partitioning also has application in divide-and-conquer algorithms for wire routing. In such applications, the graph to be partitioned represents the routing region and is planar.

In this section, we shall focus on subfamilies of planar graphs that have simple partitioning algorithms, but shall also point to references detailing more powerful partitioning algorithms for planar graphs.

6.4.1 Separating Trees

In this section, we discuss how to solve the SEPARATION problem on trees. The algorithm is simple and yields extremely good separators.

Theorem 6.2 *Let T be an undirected tree with n vertices and arbitrary vertex weights. Then there is a vertex v in T such that $\{v\}$ is a 2/3-separator of T. The vertex v can be found in linear time.*

Proof Pick a root r in T arbitrarily, thus converting T into a rooted tree. In a preorder traversal of T, label each vertex v with the number $\ell[v]$ denoting the total weight of all vertices in the subtree rooted at v. Now execute the algorithm depicted in Figure 6.10. The algorithm walks down a path in the tree starting at the root. In line 9, the algorithm arrives at a separating vertex v. The sides L, R of the partition are computed in lines 10 to 16. To prove the correctness of this procedure, we make a case distinction.

Case 1: $w(v) \geq w(V)/3$. Then, $w(L), w(R) \leq 2w(V)/3$ because, in fact, $w(L) + w(R) \leq 2w(V)/3$, and the partition is legal.

(1) v, w : **vertex**;
(2)   *weight* : **integer**;
(3) L, R, C : **set of vertex**;

(4) $v := r$;
(5)   **while** v is not a leaf **and** $\ell[v] > 2w(V)/3$ **do**
(6) $w :=$ child of v maximizing $\ell[w]$;
(7)   **if** $\ell[w] \leq 2w(V)/3$ **then** exit;
(8) $v := w$ **od**;
(9) $C := \{v\}$;

(10) $L := \emptyset$;
(11)   *weight* $:= 0$;
(12)   **for** w child of v in order of decreasing $\ell[w]$ **do**
(13)   *weight* $:=$ *weight* $+ \ell[w]$;
(14)   **exit if** *weight* $\geq w(V)/3$;
(15) $L := L \cup$ subtree rooted at w **od**;
(16) $R := V \setminus L \setminus \{v\}$;

Figure 6.10: *Separating trees.*

Case 2: $w(v) \leq w(V)/3$. Then, v is not a leaf, because otherwise the algorithm would already have stopped moving down the tree at some ancestor of v. Therefore, $\ell[w] \leq 2w(V)/3$ for each child w of v, by line 7.

    Case 2.1: $\ell[w] \geq w(V)/3$ for some child w of v. Then, the algorithm sets L to the subtree rooted at the child w of v maximizing $\ell[w]$, and the resulting partition is legal.

    Case 2.2: $\ell[w] < w(V)/3$ for all children w of v. Then, the algorithm collects subtree rooted at children of v into L, until $w(L) \geq w(V)/3$. At the end, we again get a legal partition, since at all times $w(L) \leq 2w(V)/3$.

□

The algorithm in Theorem 6.2 can be modified to yield 2/3-balanced bipartitions of trees with vertex degree at most 3 that consist of a single edge. (The cost of the edge cannot be minimized.) Together with Theorem 6.1, this observation yields a linear-time algorithm for solving the BISECTION problem on trees and finding almost exact bisections of trees with unit edge cost. The corresponding cutsizes are $O(\log n)$ (see Exercise 6.10). (Note that the BIPAR-

TITION and BISECTION problem for trees can be solved exactly in quadratic time; see Section 6.1.2.)

6.4.2 Partitioning Partial Grid Graphs

In this section, we discuss a linear-time algorithm for partitioning subgraphs of a square grid. The algorithm and its proof contain the main elements of a more general result by Lipton and Tarjan [291] (see Section 6.4.3). We discuss the restricted version of the algorithm here, because the proof is much simpler. An application of the partitioning algorithm presented in this section to a problem in the area of detailed wire routing is discussed in Exercise 9.17.

The input to the partitioning algorithm is an embedding of the partial grid graph G into the planar square grid. That is, vertices are embedded in points of the two-dimensional integer lattice, and edges are pairs of neighboring vertices in the lattice (see Figure 6.11). All edge weights are 1, but vertices v may have different weights $w(v)$.

Theorem 6.3 *Let G be a vertex-weighted partial grid graph with n vertices such that for no vertex $v \in V, w(v) > w(V)/3$. A 2/3-balanced partition of a vertex-weighted partial grid graph G with n vertices that has cutsize at most $2\sqrt{n} + 3$ can be computed in time $O(n)$.*

Proof We shall first explain the intuition behind the partitioning algorithm. We refer to Figure 6.11 for the purpose of illustration.

The easiest way to cut G would be to sever all edges running vertically between two adjacent rows about in the middle of the graph. Let us call *layer* L_i all edges running vertically between vertices whose y coordinates are i and $i+1$. A layer that does not have any edges is called *empty*. The removal of layer L_i partitions the vertex set V into a bottom part B_i and a top part T_i. Let m be minimum i such that $w(B_i) \geq w(V)/2$. Eligible layers for the horizontal cut lie close to layer L_m. Unfortunately, all such layers may have too many edges (as is the case in Figure 6.11). In fact, we may not be able to partition the graph with a single horizontal or vertical cut.

Therefore, we make a second attempt. We try to find *two small layers* L_t and L_b, one above and one below layer L_m. Clearly, such layers can be found: The empty layers at the top and bottom side of the grid graph are suitable here. However, the part of the graph that is between the two layers L_t and L_b—that is, the vertex set $T_b \cap B_t$—may still have a large weight. If we can make sure that layer L_t and L_b are close enough to each other, then the vertex set $T_b \cap B_t$ can be partitioned into two subsets of roughly equal vertex weight with a *short vertical* cut. The two horizontal cuts in layers L_t and L_b and the vertical cut together make up the final partition of the graph.

There is a tradeoff between the size limit for the layers L_t and L_b and their distance. Both quantities enter the cutsize; thus, both have to be chosen to be

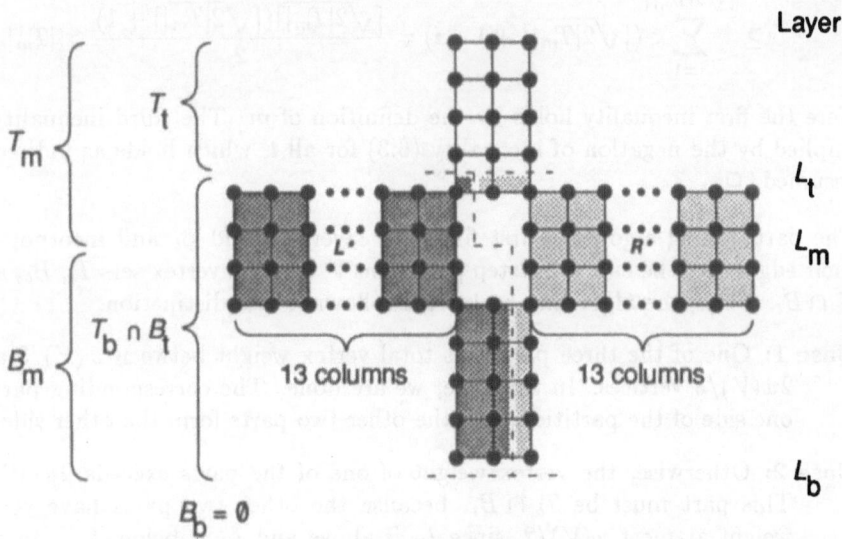

Figure 6.11: *A partial grid graph and its bipartition by the algorithm discussed in the proof of Theorem 6.3. The graph has 130 vertices. The bottommost eligible layers L_t and L_b are chosen. Case 2 of the proof applies; thus, a vertical cut becomes necessary. The parts L^* and R^* are shaded. The cutsize is $9 \leq 2\sqrt{130} + 2 = 25.8$. This value is not the minimum cutsize possible; we can find a cut with size 4.*

small. Working through the arithmetic of this problem and optimizing constant factors yields the following lemma.

Lemma 6.3 *For each partial grid graph G, there are two layers L_t and L_b, $t > m > b$, such that*

$$|L_t| + (t - m) < \sqrt{2|T_m|} + 1 \qquad (6.3)$$
$$|L_b| + (m - b) < \sqrt{2|B_m|} + 1 \qquad (6.4)$$

Proof We prove only inequality (6.3); inequality 6.4) can be proved analogously.

Assume inequality (6.3) did not hold. Then, we would get the following contradiction:

$$|T_m| \geq \sum_{i=1}^{\infty} |L_{m+i}| \geq \sum_{i=1}^{\lceil \sqrt{2|T_m|} \rceil} |L_{m+i}|$$

$$\geq \sum_{i=1}^{\lceil\sqrt{2|T_m|}\rceil} (\lceil\sqrt{2|T_m|}\rceil + 1 - i) = \frac{\lceil\sqrt{2|T_m|}\rceil(\lceil\sqrt{2|T_m|}\rceil + 1)}{2} > |T_m|$$

Here the first inequality holds by the definition of m. The third inequality is implied by the negation of inequality (6.3) for all t, which holds as indirectly assumed. $\square$

The partitioning algorithm first finds the layers L_t and L_b and incorporates their edges into the cut. This step partitions V into the vertex sets T_t, B_b, and $T_b \cap B_t$. Then, the algorithm makes the following case distinction.

Case 1: One of the three parts has total vertex weight between $w(V)/3$ and $2w(V)/3$ vertices. In this case, we are done. The corresponding part is one side of the partition, and the other two parts form the other side.

Case 2: Otherwise, the vertex weight of one of the parts exceeds $2w(V)/3$. This part must be $T_b \cap B_t$, because the other two parts have vertex weight at most $w(V)/2$, since L_t is above and L_b is below L_m. In this case, we partition $T_b \cap B_t$ with a vertical cut between L_t and L_b. If we allow one horizontal edge in the middle of the cut, we can bipartition $T_b \cap B_t$ into two parts L^* and R^* whose total vertex weights differ by at most $w(V)/3$. Without loss of generality, let $w(L^*) \leq w(R^*)$. We conclude that $w(V)/3 \leq w(R^*) \leq 2w(V)/3$. Thus, R^* can be taken as one side of the partition, and the rest of V can be taken as the other side.

The cut partitioning $T_b \cap B_t$ into L^* and R^* has a size of at most

$$(t - m) + (m - b) + 1$$

The total size of the cut is

$$|L_t| + |L_b| + (t - m) + (m - b) + 1 < \sqrt{2}(\sqrt{|T_m|} + \sqrt{|B_m|}) + 3$$

Since $|T_m| + |B_m| = n$ and the function $\sqrt{x} + \sqrt{n - x}$ in the variable x is maximized at $x = n/2$, the cutsize is bounded by $\lceil 2\sqrt{n}\rceil + 2$. The linear-time implementation of the algorithm is left to the reader. $\square$

The cutsize of the partition found by the algorithm presented in the proof of Theorem 6.3 is asymptotically optimal (see part 3 of Exercise 6.5).

6.4.3* Separating General Planar Graphs

The algorithm presented in Section 6.4.2 is a simplification of a more general algorithm by Lipton and Tarjan [291]. Their algorithm finds a *2/3-separator* of size at most $2\sqrt{2}\sqrt{n}$ in a vertex-weighted *arbitrary* planar graph with n vertices, in which w vertex weight exceeds one-third of the total vertex weight. The run time of the algorithm is $O(n)$. The algorithm uses the same basic cut strategy. The role of the grid embedding is played by a breadth-first traversal of the graph (see Section 3.7). The layers in the breadth-first traversal replace the layers of grid edges. Layers L_t and L_b can be found as described in Section 6.4.2. Making the vertical cut here turns into the task of cutting a planar graph that has a shallow breadth-first search tree. This task is more complicated, and a powerful theorem in topology, the Jordan curve theorem [165], must be employed to prove the correctness of the cutting process. Details can be found in [291].

The algorithm by Lipton and Tarjan does not necessarily find balanced *bipartitions* with small cutsize. In fact, such bipartitions may not exist if the vertex degrees are large. (A star graph is an example; see Figure 6.1.)

Since Lipton and Tarjan's paper, several improvements on the separator algorithm have been presented. Djidjev [92] improved the constant factor in the separator size from $2\sqrt{2}$ to $\sqrt{6}$. Miller [324] found small separators that have a particular shape; namely, they are simple cycles.

The NP-hardness of the problem of finding a 2/3-separator for a planar graph has been strongly conjectured, but has not yet been established. Rao [373] presents a polynomial-time approximation algorithm for separating planar graphs. The error of the algorithm is $O(\log w(V))$.

6.5 Graph Partition and Network Flow

In this section, we discuss the application of network-flow techniques graph bipartitioning. This approach is suggested by the maxflow–mincut theorem (Theorem 3.4), which relates cuts to flows.

6.5.1* Maxflow Techniques

Unfortunately, the maxflow–mincut theorem applies in only a restricted setting. Specifically, it involves bipartitions without restrictions on the part size and with two distinguished vertices, one on each side. By computing maximum network flows for all pairs of vertices, we can remove the requirement of the distinguished vertices. However, most interesting cases of circuit partitioning involve either multiway partitions or balanced bipartitions; the maxflow–mincut theorem is not directly applicable to either.

Nevertheless, maxflow techniques can be used to find good *bisections*. Bui et al. [52] suggest one approach: Do the following for each pair of distinct vertices v_i and v_j in the graph. Collapse the neighborhoods of some specific size around v_i and v_j to single vertices v_i' and v_j'. (The neighborhood of size s of a vertex v_i consists of all vertices that can be reached from v_i along paths of length at most s.) From the resulting graph, construct an UNDIRECTED MAXFLOW problem with v_i' as the source and v_j' as the sink. The capacities of the edges incident to v'_i and v'_j are infinite; all other edge capacities are identical to the edge costs. Find a maximum flow of this problem, which also yields a minimum cut. Output the smallest minimum cut over all vertex pairs that is a bisection, or a suitable collection of edges that is in many minimum cuts.

This rough outline of the algorithm has to be enhanced with several details, such as the choice of the neighborhood size and the exact construction of the bisection. We shall not go into detail here; we shall just mention that it can be proved that, for certain subclasses of graphs that are distributed uniformly in some natural sense, a suitably tuned algorithm using this approach finds an optimal bisection with high probability. The probability distribution has to be chosen carefully, because on uniformly distributed random graphs all bisections have a cutsize of about $n/2$ with high probability [39]. For this reason, random analyses of graph partitioning consider uniform distributions over classes of graphs whose minimum bisections have some preset *small* cutsize.

Even with these provisions, the relevance of average-case analyses of partitioning algorithms to practical applications is questionable. The reason is that, in practice, it is difficult to ascertain the probability distribution of graphs and, even if we can do so, the distribution is unlikely to be uniform (see also Section 2.1.2). In addition, the graph classes for which the analysis holds are restricted to be *regular*; that is, all vertices in the graph have to have the same degree. For details, refer to [52].

Experiments run by Bui et al. indicate that using network flow for graph partitioning can be competitive, especially if the degree of the graph is small. For larger-degree graphs, even quite simple-minded greedy heuristics appear to yield better—and in fact, very good—results (see Section 6.2.4).

Kahng [214] presents a heuristic algorithm based on similar ideas that finds with high probability the optimum bisection of a (not necessarily regular) hypergraph $G = (V, E)$ with unit weight vertices and edges, according to a natural uniform probability distribution. Kahng exploits the robustness of the random analysis and does not stick to the network-flow idea closely. Therefore, he can devise a simple $O(m^2)$-time algorithm.

6.5.2 Multicommodity-Flow Techniques

In this section, we introduce a variant of the MULTICOMMODITY FLOW problem that has a close relationship with balanced graph bipartitioning. This relationship has first been observed by Matula and Shahrokhi [312].

In our instances of the balanced graph bipartitioning problem, we require all vertex weights to be unity. Let G be the undirected graph and c its edge-cost function. In the corresponding instance of the multicommodity-flow problem, we assign a special commodity to each ordered pair (v, w) of vertices. We require that f units of this commodity flow from v to w. Here, the edge capacities have to be obeyed, and z is some number independent of (v, w).

What we want is to maximize the amount z of flow of each commodity that we can push through the network.

UNIFORM MULTICOMMODITY FLOW

Instance: An undirected graph $G = (V, E)$ with edge capacity function $c : E \to \mathbf{R}_+$

Configurations: All mappings $f = (f_{v,w})_{v,w \in V}$ such that $f_{v,w} : E \to \mathbf{R}_+$

Solutions: All f that are multicommodity flows for the instance $G = (V, E)$ of the UNDIRECTED MULTICOMMODITY FLOW problem that consists of G with capacities c, edge cost $\ell \equiv 0$, demands $d_{v,w}(v) = -z/2$ and $d_{w,v} = z/2$ for all $v, w \in V$; $z \in \mathbf{N}$ is an arbitrary constant

Maximize: z

In the UNIFORM MULTICOMMODITY FLOW problem, we are again maximizing flow values. What does this problem have to do with graph partitioning? Lemma 3.14 suggests a relationship. It says that a necessary condition for z to be the value of a multicommodity flow is

$$0 \le fc(V', V'') = c(V', V'') - d(V', V'') \qquad (6.5)$$

for all cuts (V', V''). The density of the cut is easily determined:

$$d(V', V'') = z \cdot |V'| \cdot |V''|$$

Thus, (6.5) translates to

$$z \le \min_{(V',V'') \text{ cut}} \frac{\displaystyle\sum_{\substack{e = \{v, w\} \in E \\ v \in V', \, w \in V''}} c(e)}{|V'| \cdot |V''|} \qquad (6.6)$$

The expression that is minimized over cuts in (6.6) is also called the *sparsity* of a cut and is denoted with $sp(V', V'')$. A cut with minimum sparsity is called a *sparsest* cut. The notion of a sparsest cut relaxes the balance criterion in the sense of Section 4.6.3. Cuts of any balance are allowed, but the more balanced a cut is, the higher its density is, and the smaller the weight factor in the sparsity of the cut. Hopefully, this fact makes it unlikely that a sparse cut is unbalanced. If G is a complete graph, then, as a cut becomes more balanced, the density increases by the same amount as the capacity of the cut. Thus, balanced cuts are not favored. But if G is sparse, such as in layout applications, then balanced cuts are favored over unbalanced cuts. Thus, a sparsest cut is likely to be an attractive cut for balanced bipartitioning.

Since for multicommodity flows only one direction of the maxflow–mincut theorem holds, the maximum value of a multicommodity flow is only a lower bound on the minimum sparsity over all cuts. Leighton and Rao [269] show that it is not far off.

Theorem 6.4 (Leighton, Rao [269]) *Let f be a maximum flow for an instance of the* UNDIRECTED MULTICOMMODITY FLOW *problem. Let the flow value of f be z. Then, there exists a cut (V', V'') with sparsity s such that*

$$s = O(z \log n)$$

and such a cut can be found in polynomial time.

The run time of the corresponding solution algorithm finding a cut with sparsity s is dominated by the solution of the involved instance of the UNDIRECTED MULTICOMMODITY FLOW problem.

The following lemma relates sparse cuts to cuts with small cutsize.

Lemma 6.4 *Let (V_1', V_1'') be a sparsest cut, and let the balance of (V_1', V_1'') be α_1—that is, $\alpha_1 n = \max(|V_1'|, |V_1''|)$. Let $\alpha_2 \geq \alpha_1$. Then, the cutsize of (V_1', V_1'') is within a factor of $\alpha_1(1 - \alpha_1)/\alpha_2(1 - \alpha_2)$ of the minimum cutsize over all α_2-balanced cuts.*

Proof Note that $\alpha_1 \geq 1/2$. Let (V_2', V_2'') be a cut with balance $\alpha \leq \alpha_2$.

$$
\begin{aligned}
c(V_1', V_1'') &= sp(V_1', V_1'') \cdot \alpha_1(1 - \alpha_1)n^2 \\
&\leq sp(V_2', V_2'') \cdot \alpha_1(1 - \alpha_1)n^2 \\
&= c(V_2', V_2'')\alpha_1(1 - \alpha_1)/\alpha(1 - \alpha) \\
&\leq c(V_2', V_2'')\alpha_1(1 - \alpha_1)/\alpha_2(1 - \alpha_2)
\end{aligned}
$$

The last inequality holds because the function $f(x) = x(1 - x)$ is concave on $[0, 1]$ and has a maximum in $x = 1/2$. $\square$

Lemma 6.4 can be generalized to yield that a cut that is α_1-balanced and whose sparsity is within a factor of β of optimum has a cutsize that is within a factor of $\beta\alpha_1(1-\alpha_1)/\alpha_2(1-\alpha_2)$ of optimum among all α_2-balanced cuts. If the sparse cut (V', V'') obtained by the algorithm mentioned in Theorem 6.4 satisfies the balance criterion of the instance of the BIPARTITION problem, then Lemma 6.4 bounds the distance of the cutsize from the optimum. If it does not, then we can iterate the procedure of finding sparse cuts similar to Theorem 6.1 to find a cut that satisfies the balance criterion and whose cutsize is at most a factor of $O(\beta \log n)$ away from the optimum [373].

The investigation of the connections between partitioning and multicommodity flow is still in its infancy. The future will show how amenable to practical application are the results obtained in this area.

6.6 Partitioning as a Nonlinear Optimization Problem

There are several ways in which partitioning problems can be regarded as non-linear optimization problems. In this section, we review such formulations of partitioning problems, and we discuss their relaxations to quadratic optimization problems. We consider partitioning of only a simple graph in this section. To apply the methods presented here to hypergraph partitioning, we have to resort to the reduction discussed in Section 6.1.5.

In Section 6.6.1, we formulate a version of the MULTIWAY PARTITION problem as a QUADRATIC ASSIGNMENT problem. In Section 6.6.2, we present lower bounds on the cutsize of optimal partitions that are based on quadratic relaxations of this formulation. In Section 6.6.3, we discuss heuristics complementing these lower bounds.

6.6.1 Partitioning by Quadratic Assignment

We consider the MULTIWAY PARTITION problem on graphs with unit vertex weights $w \equiv 1$ and fixed part sizes $b(i) = B(i)$ for $i = 1, \ldots, r$ and $\sum_{i=1}^{r} B(i) = n$. Without loss of generality, we assume that $B(1) \geq B(2) \geq \cdots \geq B(r)$. This problem can be formulated as a QUADRATIC ASSIGNMENT problem (see Section 4.8.3) as follows. The cost matrix $C = ((c_{ij}))_{i,j=1,\ldots,n}$ is the weighted adjacency matrix of the graph G based on the edge-cost function c; that is,

$$c_{ij} = \begin{cases} c(\{i, j\}) & \text{if } \{i, j\} \in E \\ 0 & \text{otherwise} \end{cases}$$

The distance matrix $D = ((d_{ij}))_{i,j=1,\ldots,n}$ is $D = 1 - \Delta$, where 1 is the $n \times n$ matrix containing all 1s, and Δ is the following block matrix:

		$B(1)$				$B(2)$				$B(r)$	
$B(1)$	1	1	1	1	0	0	0	$\cdots$	0	0
	1	1	1	1	0	0	0	$\cdots$	0	0
	1	1	1	1	0	0	0	$\cdots$	0	0
	1	1	1	1	0	0	0	$\cdots$	0	0
$B(2)$	0	0	0	0	1	1	1	$\cdots$	0	0
	0	0	0	0	1	1	1	$\cdots$	0	0
	0	0	0	0	1	1	1	$\cdots$	0	0
	$\vdots$	$\vdots$	$\vdots$	$\vdots$	$\vdots$	$\vdots$	$\vdots$		$\vdots$	$\vdots$
$B(r)$	0	0	0	0	0	0	0	$\cdots$	1	1
	0	0	0	0	0	0	0	$\cdots$	1	1

The general methods for solving the QUADRATIC ASSIGNMENT problem that are referred to in Section 4.8.3 apply to this formulation. However, the QUADRATIC

ASSIGNMENT problem is quite a difficult combinatorial problem, so this formulation per se is not of much help in practice. Yet it serves as a starting point for relaxations that yield lower bounds on the minimum cutsize and suggest heuristic algorithms for solving partitioning problems.

6.6.2 Lower Bounds

In this section, we develop lower bounds on the minimum cutsize of graph partitions. The bounds are based on relaxations of the problem formulation given in Section 6.6.1. They are stated in terms of eigenvalues. For this purpose and throughout the remainder of Section 6.6, we shall denote the *ith largest* eigenvalue of a matrix A with $\lambda_i(A)$.

6.6.2.1 Bounds Based on the Hoffman–Wielandt Inequality

In this section we discuss lower bounds that are based on the following variant of a fundamental theorem by Hoffman and Wielandt [194].

Theorem 6.5 *Let A and B be two real symmetric $n \times n$ matrices. Then,*

$$\text{Trace}(AB^T) \le \sum_{i=1}^{n} \lambda_i(A) \cdot \lambda_i(B)$$

We use several matrices to state the lower bounds. The weighted adjacency matrix C of the graph G and the block matrix Δ are defined as in Section 6.6.1. We allow for permuting the columns and rows of C, effectively renumbering the vertices of the graph. This operation facilitates the choice of the permutation π needed in the solution to the QUADRATIC ASSIGNMENT problem, and it is formulated via a *permutation matrix* $P = ((p_{ij}))_{i,j=1,...,n}$—that is, by an $n \times n$ matrix with 1s in the entries $p_{\pi(i),i}$ and 0s everywhere else. The matrix $\Delta' = P\Delta P^T$ is the result of applying the permutation π^{-1} to Δ; that is, $\Delta' = ((\delta'_{ij}))_{i,j=1...,n}$ with $\delta'_{ij} = \delta_{\pi^{-1}(i),\pi^{-1}(j)}$. In our context, P will generate the partition Π_P of G that puts exactly the vertices

$$v_{\pi\left(1+\sum_{j=1}^{i-1} B(j)\right)}, \ldots, v_{\pi\left(\sum_{j=1}^{i} B(j)\right)}$$

into part V_i. Thus, Π_P results from permuting the vertices of the graph using the permutation π and then assigning vertices to the parts $V_1, \ldots, V_r$ in ascending order. Of course, each partition of G with the required part sizes can be generated in this way. The free choice of the partition is therefore equivalent to the free choice of P.

E_{ext,Π_P} and E_{int,Π_P} denote the sets of edges that run between different parts and within a part in Π_P, respectively. Finally, we shall use a diagonal matrix U such that $\text{Trace}(U) = 2c(E)$. We denote the diagonal entries of U with $u_1, \ldots, u_n$.

We begin by relating the cutsize of Π_P to the matrices C and Δ'.

Lemma 6.5

$$c(\Pi_P) = c(E) - \frac{1}{2}\text{Trace}(C\Delta'^T) \tag{6.7}$$

Proof

$$
\begin{aligned}
c(\Pi_P) &= c(E_{\text{ext},\Pi_P}) = c(E) - c(E_{\text{int},\Pi_P}) \\
&= c(E) - \frac{1}{2}\sum_{i,j=1}^{n} c_{ij}\delta'_{ij} = c(E) - \text{Trace}(C\Delta'^T)
\end{aligned}
$$

$\square$

By using the matrix U, we can bring the term $c(E)$ in Lemma 6.5 into the matrix whose trace is taken:

$$c(\Pi_P) = -\frac{1}{2}\text{Trace}(M\Delta'^T) \tag{6.8}$$

where $M = C - U$.

If we now apply Theorem 6.5 to (6.7) and (6.8), respectively, we get two different lower bounds. To do so, let us note that the eigenvalues of Δ'^T are the same as the eigenvalues of Δ; namely,

$$\lambda_i(\Delta) = \begin{cases} B(i) & \text{if } 1 \le i \le r \\ 0 & \text{if } r+1 \le i \le n \end{cases}$$

Applying Theorem 6.5 to (6.7), we get the following bound on the optimum cutsize c_{opt}.

Theorem 6.6 (Barnes [21])

$$c_{\text{opt}} \ge c(E) - \frac{1}{2}\sum_{i=1}^{r} B(i)\lambda_i(C) \tag{6.9}$$

Applying Theorem 6.5 to (6.8), we get the following bound.

Theorem 6.7 (Donath, Hoffman [97])

$$c_{\text{opt}} \ge -\frac{1}{2}\sum_{i=1}^{r} B(i)\lambda_i(C - U) \tag{6.10}$$

Theorem 6.6 is weaker than Theorem 6.7 is, because the latter allows for a suitable choice of the matrix U. Donath and Hoffman observe that the right-hand side of (6.10) is a concave function of the variables $u_1, \ldots, u_n$ (see Exercise 6.12). This fact allows us to employ methods of nonlinear optimization to maximize the right-hand side of (6.10).

6.6.2.2   The Bound by Boppana

Boppana [41] extends the approach of Theorem 6.7 to derive tighter lower bounds on the cutsize of graph *bisections*. A bisection can be represented by a vector $(x_1, \ldots, x_n)$ such that

$$x_i = \left\{ \begin{array}{ll} 1 & \text{if } v_i \in V_1 \\ -1 & \text{if } v_i \in V_2 \end{array} \right.$$

If we assume that n is even, then the following lemma holds.

Lemma 6.6 *If x represents a bisection Π of G, then for arbitrary $u \in \mathbf{R}^n$*

$$f(G, u, x) = \frac{1}{4} \left(\sum_{i,j=1}^{n} c_{ij}(1 - x_i x_j) + \sum_{i=1}^{n} u_i(x_i^2 - 1) \right) \tag{6.11}$$

is the cutsize of the bisection Π.

Proof The first term of the right-hand side of (6.11) is exactly $c(\Pi)$; the second term is zero for all $u \in \mathbf{R}^n$. $\square$

The idea of deriving the lower bound is now to relax the requirement that x represent a bisection to the following two sets of constraints:

$$\sum_{i=1}^{n} x_i = 0 \tag{6.12}$$

$$\sum_{i=1}^{n} x_i^2 = n \tag{6.13}$$

and to find the minimum of $f(G, u, x)$ over all such x for fixed u—that is, the value

$$g(G, u) = \min\{f(G, u, x) \mid \sum_{i=1}^{n} x_i = 0, \ \sum_{i=1}^{n} x_i^2 = n\} \tag{6.14}$$

The function $g(G, u)$ is concave in u, since it is a minimum of linear functions in u—namely $f(G, u, x)$ for fixed x. Thus, efficient methods of nonlinear optimization can be employed to maximize $g(G, u)$. We can do so iteratively using the methods described in Section 4.8.1 in order to approximate the maximum. The shape of $g(G, u)$ also allows us to apply special methods based on the ellipsoid method that is also used in linear programming [295, Chapter 2] to determine the maximum to arbitrary precision in polynomial time.

The advantage of using the relaxed sets of constraints (6.12) and (6.13) is that we can compute $g(G, u)$ exactly using the classical method of Lagrangian multipliers for the minimization of differentiable functions with constraints.

To this end, we note that $f(G, u, x)$ can be written in terms of the weighted adjacency matrix C of G and the diagonal matrix U with diagonal u as follows:

$$f(G, u, x) = \frac{1}{4} \left(x^T(U - C)x + \sum_{i,j=1}^{n} c_{ij} - \sum_{i=1}^{n} u_i \right)$$

The latter two terms on the right-hand side of this equation are constant in x; the first term is a quadratic form in x. This term has to be minimized with respect to the constraints (6.12) and (6.13). The constraints (6.12) and (6.13) suggest using Theorem 4.14 to get the lower bounds. Unfortunately, we cannot use this theorem, because the matrices that are involved are not positive semi-definite. However, a direct application of the method of Lagrangian multipliers yields the following result: Compute the restriction of $x^T(U-C)x$ on the linear subspace $S = \{x \in \mathbf{R}^n \mid \sum_{i=1}^{n} x_i = 0\}$. Call the resulting $(n-1) \times (n-1)$ matrix D. Let B be the corresponding restriction of the (identity) matrix representing the spherical constraint. Let $A := B^{-1}D$. Then

$$g(G, u) = \frac{1}{4} \left(n\lambda_{n-1}(A) + \sum_{i=1}^{n} c_{ij} - \sum_{i=1}^{n} u_i \right)$$

Note that $\lambda_{n-1}(A)$ is the smallest eigenvalue of A. In summary, we obtain the following theorem.

Theorem 6.8 *With the preceding notation, define*

$$h(G) = \max_{u \in \mathbf{R}^n} g(G, u)$$

where

$$g(G, u) = \frac{1}{4} \left(n\lambda_{n-1}(A) + \sum_{i=1}^{n} c_{ij} - \sum_{i=1}^{n} u_i \right)$$

Then $h(G)$ is a lower bound on the cutsize of an optimal bisection of G.

Compared to the lower bounds discussed in Section 6.6.2.1, $h(G)$ is quite difficult to compute. Its computation involves an iteration to maximize $g(G, u)$ with respect to u. For each iteration step, we have to compute the matrix A and its smallest eigenvalue. But $h(G)$ turns out to be a tight bound in many cases (see Section 6.6.3.2).

This lower bound relies heavily on the fact that only bipartitions are considered. An interpretation of the vector x as a multiway partition is not obvious. We can generalize the bound to part sizes $B(1) = k$, $B(2) = n - k$ by replacing the constraint (6.12) with

$$\left| \sum_{i=1}^{n} x_i \right| = n - 2k \tag{6.15}$$

and to part sizes $B(1) \leq k$, $B(2) \leq k$ by replacing (6.12) with

$$\left| \sum_{i=1}^{n} x_i \right| \leq n - 2k \qquad (6.16)$$

The constraints (6.15) and (6.16) are more difficult to handle, however, since Theorem 4.14 can no longer be applied.

6.6.3   Heuristic Algorithms

The lower bounds derived in Section 6.6.2 suggest several heuristic algorithms for computing partitions.

6.6.3.1   The Algorithm by Barnes

Barnes [21] observes that an equivalent statement to (6.7) is that the difference between the matrices C and $\Delta' = P\Delta P^T$ in the well-known *Frobenius norm* $\|A\|_F = \sqrt{\sum_{i=1}^{n} |a_i|^2}$ is minimized if Π_P is an optimal partition.

Lemma 6.7

$$\|C - \Delta'\|_F^2 = 4c(\Pi_P) - \varepsilon$$

where ε is a value that depends only on the graph G and not on the matrix P.

Proof

$$
\begin{aligned}
\|C - \Delta'\|_F^2 &= \|C\|_F^2 - 2 \sum_{i,j=1}^{n} c_{ij} \delta'_{ij} + \|\Delta'\|_F^2 \\
&= \|C\|_F^2 - 2 \operatorname{Trace}(C \Delta'^T) + \|\Delta\|_F^2 \\
&= \|C\|_F^2 + 4c(\Pi_P) - 4c(E) + \|\Delta\|_F^2
\end{aligned}
$$

The last identity holds because of Lemma 6.5. $\square$

Thus, we can find good partitions by choosing P such that $\|C - \Delta'\|_F$ is small. We can give an equivalent formulation of Theorem 6.5 involving the Frobenius norm. This formulation is commonly known as the *Hoffman–Wielandt Inequality.*

Lemma 6.8 *If A and B are real symmetric $n \times n$ matrices, then*

$$\|A - B\|_F^2 \geq \sum_{i=1}^{n} (\lambda_i(A) - \lambda_i(B))^2$$

Proof

$$\begin{aligned} \|A - B\|_F^2 &= \|A\|_F^2 - 2\operatorname{Trace}(AB^T) + \|B\|_F^2 \\ &= \sum_{i=1}^n \lambda_i(A)^2 - 2\operatorname{Trace}(AB^T) + \sum_{i=1}^n \lambda_i(B)^2 \\ &\geq \sum_{i=1}^n (\lambda_i(A) - \lambda_i(B))^2 \end{aligned}$$

Here, we can derive the second identity by transforming A and B to diagonal matrices using orthonormal eigenvector bases for both matrices. The Frobenius norm is invariant under orthogonal transformations such as these [147]. The last inequality is an application of Theorem 6.5. $\square$

Applied to partitioning, Lemma 6.8 yields

$$\|C - \Delta'\|_F^2 \geq \sum_{i=1}^r (\lambda_i(C) - B(i))^2 + \sum_{i=r+1}^n \lambda_i(C)^2 \qquad (6.17)$$

To find a good approximation of C by a permutation of Δ, we consider orthonormal eigenvector bases of both matrices. Let $a_1, \ldots, a_n$ be such a basis for C, and let $A = (a_1, \ldots, a_n)$ be the matrix composed of the columns $a_1, \ldots, a_n$. There is not much we know about A except that it is an orthogonal matrix. On the other hand, we can give orthonormal eigenvectors for the positive eigenvalues of Δ' explicitly, by choosing

$$d'_j = \begin{pmatrix} d'_{1j} \\ \vdots \\ d'_{nj} \end{pmatrix} = \pm \frac{1}{\sqrt{B(j)}} \begin{pmatrix} x_{1j} \\ \vdots \\ x_{nj} \end{pmatrix} \qquad j = 1, \ldots, r \qquad (6.18)$$

To get orthonormal eigenvectors for the matrix $\Delta' = P\Delta P^T$, where $P = ((p_{ij}))_{i,j=1,\ldots,n}$ is the permutation matrix such that there are 1s in the entries $p_{\pi(i),i}$, $i = 1, \ldots, n$ and 0s in all other entries, we have to set to 1 the $x_{\pi^{-1}(\sum_{i=1}^{j-1} B(i)+1)}, \ldots, x_{\pi^{-1}(\sum_{i=1}^{j} B(i))}$ and to set to 0 all other x_{ij}. The signs of the d_i can be chosen freely.

Let $D' = (d'_1, \ldots, d'_r)$ be the $n \times r$ matrix whose columns are $d'_1, \ldots, d'_r$. Furthermore, let Λ be the diagonal $n \times n$ matrix with the diagonal $\lambda_1(C), \ldots, \lambda_n(C)$, and let B be the diagonal $r \times r$ matrix with the diagonal $B(1), \ldots, B(r)$. Then,

$$\|C - \Delta'\|_F^2 = \|A\Lambda A^T - D'BD'^T\|_F^2 = \|\Lambda - (A^T D')B(A^T D')^T\|_F^2$$

The last identity holds because of the invariance of the Frobenius norm under orthogonal transformations.

Since we are allowed to choose P freely, the constraints on the x_i take the following shape:

$$x_{ij} \in \{0,1\} \qquad i,j = 1,\ldots,n$$

$$\sum_{i=1}^{n} x_{ij} = B(j) \qquad j = 1,\ldots,r \tag{6.19}$$

$$\sum_{j=1}^{r} x_{ij} = 1 \qquad i = 1,\ldots,n$$

If we could find a matrix D' satisfying (6.18) and (6.19) such that

$$A^T D' = \begin{pmatrix} 1 & 0 & \cdots & 0 \\ 0 & 1 & \cdots & 0 \\ \vdots & \vdots & & \vdots \\ 0 & 0 & \cdots & 1 \\ 0 & 0 & \cdots & 0 \\ \vdots & \vdots & & \vdots \\ 0 & 0 & \cdots & 0 \\ 0 & 0 & \cdots & 0 \end{pmatrix} =: J$$

then (6.17) would be fulfilled with identity using the corresponding matrix $\Delta' = D'BD'^T$. In this case, we would have found an optimum partition that can be reconstructed from D'. In general, such a D' cannot be found because of the constraints (6.18) and (6.19), so we approximate J by $A^T D'$ in the Frobenius norm. To do so, let us first fix the signs of the d'_i to be all positive. We show later how to incorporate different signs.

Approximating J by $A^T D'$ amounts to minimizing

$$\|A^T D' - J\|_F^2 = \|D' - AJ\|_F^2 = \sum_{j=1}^{r} \|d'_j - a_j\|_F^2$$

$$= \sum_{j=1}^{r} \left(\sum_{i=1}^{n} d'^2_{ij} - 2\sum_{i=1}^{n} d'_{ij}a_{ij} + \sum_{i=1}^{n} a^2_{ij} \right) \tag{6.20}$$

The part of (6.20) that depends on P is

$$-2\sum_{i=1}^{n} d'_{ij}a_{ij} \tag{6.21}$$

This function has to be minimized with respect to (6.19). (Remember that we assume all signs in (6.18) to be positive, for now.) This problem is a linear program. In fact, relaxing the first constraint in (6.19) to $x_{ij} \geq 0$ yields a linear program that is a generalization of the LINEAR ASSIGNMENT problem (see Section 3.12), which it reduces to if $B(j) = 1$ for all j. The constraint matrix of this linear program is totally unimodular, and because of the third constraint in (6.19) all integer solutions are 0,1. Thus, there are optimal solutions of this program that are 0,1. This program is called the *transportation problem*. It can be solved with general linear-programming techniques, or with special algorithms using network flow techniques [336].

We still have to discuss how to choose an appropriate assignment of signs in (6.18). Each such assignment yields a different instance of the transportation problem, and we are looking for the instance with the smallest minimum cost. By substituting (6.18) into (6.21), we see that the objective function to be minimized with respect to (6.19) is

$$-2\sum_{i=1}^{n}\sum_{j=1}^{r}\pm\frac{a_{ij}}{\sqrt{B(j)}}x_{ij} \qquad (6.22)$$

where the signs are adjusted according to the chosen assignment. A heuristic choice of the signs aims at minimizing (6.22), including a reasonable estimate of the x_{ij} (which we determine *afterward* by solving the transportation problem). Barnes suggests that we maximize the projection of the vector a_j on the nonnegative orthant for each j, $1 \le j \le r$ independently. This procedure amounts to choosing the sign of a_j to be positive exactly if the sum of the squares of the positive components of a_j exceeds the sum of the squares of the negative components of a_j. Other heuristics are conceivable here, as well (see Exercise 6.13).

The resulting partition is not necessarily even a local optimum. It can be used as a starting point for iterative improvement procedures, such as those discussed in Section 6.2. Barnes et al. [22] present an improvement method that is also based on the formulation of the MULTIWAY PARTITION problem as a QUADRATIC ASSIGNMENT problem. The problem of improving the partition is again reduced to a transportation problem.

6.6.3.2 The Algorithm by Boppana

The lower bound by Boppana (see Section 6.6.2.2) suggests a simple heuristic algorithm for computing bisections: Find the vector u^* that maximizes $g(G, u)$ (see (6.14)). Let

$$g(G, u^*) = \frac{1}{4}\left(n\lambda_{n-1}(A^*) + \sum_{i,j=1}^{n}c_{ij} - \sum_{i=1}^{n}u_i^*\right)$$

Find an eigenvector a^* for the eigenvalue $\lambda_{n-1}(A^*)$ of A^*. Map a^* back into $\mathbf{R}^n$ to yield a vector $\tilde{a} \in \mathbf{R}^n$. Now assign to the largest $n/2$ components of $\tilde{a}$ the value 1, and to the remaining components the value -1. This rounding procedure yields a vector x^* that represents the bisection computed by the algorithm. We justify intuitively the choice of x^* by observing that, in x^*, the value 1 is assigned to those components of the relaxed solution $\tilde{a}$ that are farthest on the right side—that is, the "1 side"—of the real line.

This algorithm in many cases computes an optimum bisection together with a proof of its optimality, if we use methods that compute $h(G)$ to a sufficient precision.

Lemma 6.9 *If $h(G)$ is a tight lower bound—that is, if $h(G)$ is the minimum cutsize of any bisection in G—and if, in addition, the multiplicity of $\lambda_{n-1}(A)$ is 1, then x^* represents an optimum bisection.*

Proof Scaling $\tilde{a}$ to the norm $\sqrt{n}$ yields an exact bisection. Also in this case, this scaling is equivalent to the rounding of $\tilde{a}$ done by the algorithm. $\square$

If the multiplicity of $\lambda_{n-1}(A)$ exceeds 1, then it is more difficult to find an eigenvector of $\lambda_{n-1}(A)$ that represents a bisection. However, such an eigenvector can still be found in polynomial time, if G has a unique bisection [42] (see also Exercise 6.14).

Boppana [41] shows that, if the probability distribution of the graphs G that are input to the algorithm is uniform in a natural sense, then almost certainly $h(G)$ is tight and G has a unique minimum bisection. Therefore, if we compute $h(G)$ precisely using the ellipsoid method [295, Chapter 2], we have a polynomial-time algorithm that with very high probability computes an optimum bisection.

About the relevance of this average-case analysis, we note the same reservations as those we made in Section 6.5. Nevertheless, the algorithm by Boppana may perform quite well on practical examples.

6.7   Clustering Methods

In this section, we discuss methods for solving the special case of the MULTIWAY PARTITION problem in which G is a graph, $n = rB$, and $B(i) = b(i) = B$ for some small value of B. Furthermore, instead of requiring the cost of a partition to be minimum, we require the sum of the costs of all edges that run inside a part to be maximum. In other words, instead of minimizing $c(\Pi)$, we maximize $c(E) - c(\Pi)$. Let us call this problem the CLUSTERING problem.

The optimal solutions in the CLUSTERING problem and in the MULTIWAY PARTITION problem are the same. But if we approximate optimal costs, then the change from a minimization to a maximization problem is significant. Intuitively, in the CLUSTERING problem, we focus our attention on the insides of parts, whereas in the MULTIWAY PARTITION problem we concentrate on connections between parts. A solution that is close to the optimum in one problem may not be close to the optimum in the other problem. Specifically, if $c(\Pi) \gg c(E)/2$, then a solution that approximates the CLUSTERING problem well also approximates the MULTIWAY PARTITION problem well, but not the other way around. If $c(\Pi) \ll c(E)/2$, this situation is reversed. Fortunately, if B is small, we can expect most of the edges to run between parts; that is, $c(\Pi) \gg c(E)/2$. Thus, good approximations for the CLUSTERING problem lead to good approximations for the MULTIWAY PARTITION problem.

We now discuss an approximation algorithm for the CLUSTERING problem. We can assume, without loss of generality, that the graph in the problem instance is the complete graph K_n. If this is not the case, we add zero-cost edges. This does not change the cost of any partition.

For $B = 2$, this problem is also called the MAXIMUM WEIGHT MATCHING problem (see Section 3.11). If $B > 2$, then we can heuristically reduce the CLUSTERING problem to the MAXIMUM WEIGHT MATCHING problem. We handle the technically simpler case that B is even. For the case that B is odd, see Exercise 6.15. The basic idea is very simple: We first find a maximum weight matching M_{opt} of G. Then we choose $B/2$ edges at a time out of M_{opt} and form a cluster out of the vertices incident to these edges. Note that we do not specify any particular way how the portions of $B/2$ edges are chosen. Nevertheless, we can show that this process has bounded error if the edge weights satisfy the triangle inequality

$$c(\{u, v\}) + c(\{v, w\}) \geq c(\{u, w\}) \quad \text{for all } u, v, w \in V$$

This observation follows from the fact that a maximum weight matching contains a large part of the total edge weight (for the notation, see Section 3.12).

Lemma 6.10 *If M_{opt} is a maximum weight matching for G, then $c(E) \leq (n-1)c(M_{\text{opt}})$.*

Proof The edges of the complete directed graph on n vertices can be partitioned into $n - 1$ perfect matchings. $\square$

Let Π be the partition found by the preceding heuristic procedure. Denote the cost of Π with respect to the CLUSTERING problem with $c(\Pi)$.

Theorem 6.9 *Let c_{opt} be the cost of an optimal solution of the CLUSTERING problem.*

1. $c_{\text{opt}} \leq (B - 1)c(M_{\text{opt}}) \leq (B - 1)c(\Pi)$.

2. *If the edge weights in G satisfy the triangle inequality, then*

$$c_{\text{opt}} \leq \frac{2(B - 1)}{B}c(\Pi)$$

Thus, in the general case, the error of the heuristic algorithm is bounded by $B-1$; in the case that the triangle inequality holds, it is bounded by $2(B-1)/B$.

Proof For a partition $(V_1, \ldots, V_r)$ denote with $c(V_i)$ the total cost of all edges inside part V_i.

1. The right-hand inequality of part 1 is trivial. To prove the left-hand inequality, we let $V_{\text{opt},i}$ be the ith part in the optimal partition. Let $M_{\text{opt},i}$ be a maximum matching of this part. By Lemma 6.10,

$$c(V_{\text{opt},i}) \leq (B - 1)c(M_{\text{opt},i}) \qquad (6.23)$$

The matchings $M_{\text{opt},i}$ combine to a matching of the whole graph whose total cost is at most $c(M)$. Thus, the inequalities (6.23) add up to yield the theorem.

2. We show that for each part V_i in the partition Π computed by the heuristic algorithm

$$c(V_i) \geq \frac{B}{2} \cdot c(M_{\text{opt}}|_{V_i}) \qquad (6.24)$$

Then the theorem follows by summing the inequalities (6.24) for $j = 1, \ldots, r$ and applying the left-hand inequality of part 1.

To see (6.24), we number the vertices in part V_i from 1 to B, such that the edges in $M_{\text{opt}}|_{V_i}$ are the edges $\{1, 2\}, \{3, 4\}, \ldots, \{B - 1, B\}$. Focus on edge $\{2j - 1, 2j\}$. Summing up $B - 2$ applications of the triangle inequality, one for each $\ell \neq j, j + 1$, yields

$$(B - 2)c(\{2j - 1, 2j\}) \leq \sum_{\ell \neq 2j-1, 2j} c(\{2j - 1, \ell\}) + c(\{\ell, 2j\}) \qquad (6.25)$$

Summing (6.25) over all edges in $M_{\text{opt}}|_{V_i}$ yields

$$(B - 2)c(M_{\text{opt}}|_{V_i}) \leq \sum_{j=1}^{B/2} \sum_{\ell \neq 2j-1, 2j} c(\{2j - 1, \ell\}) + c(\{\ell, 2j\}) = 2c(\overline{M_{\text{opt}}|_{V_i}})$$

where $\overline{M_{\text{opt}}|_{V_i}}$ is the set of edges inside part V_i that is not in $M_{\text{opt}}|_{V_i}$. Since $M_{\text{opt}}|_{V_i}$ and $\overline{M_{\text{opt}}|_{V_i}}$ comprise all edges inside part V_i, (6.24) follows.

$\square$

Exercise 6.16 shows that there are cases in which this error is achieved. The results in this section are taken from [113].

6.8 Summary

In this chapter, we discussed the problem of partitioning graphs and hypergraphs into several pieces. The bipartitioning case (dividing a graph into two pieces of roughly equal size) is especially interesting. Another special case we discussed is partitioning a graph into many small pieces of roughly equal size (clustering).

We alluded to how partitioning can aid in the process of layout design. A theoretical foundation of the relationship between partitioning and layout has been given in Chapter 5. More details on how partitioning is used in layout practice will be discussed in the succeeding chapters.

The main emphasis of this chapter was on presenting methods for solving partitioning problems. We have seen that, in general, partitioning problems are NP-hard, even in strongly restricted versions. In layout practice today, iterative improvement methods such as those presented in Section 6.2 are still the most popular, in spite of the fact that they lead to partitions whose quality cannot be analyzed with respect to the optimum.

There are special cases in which finding optimal, or at least provably good, partitions is possible. The corresponding graph classes (for instance trees, grids, or planar graphs; see Section 6.4) are not relevant for circuit partitioning, since they present restrictions that are too strong. However, in some divide-and-conquer layout procedures, planar routing graphs need to be partitioned. In such applications, these provably good and efficient algorithms can be of service.

For general graphs and hypergraphs, the development of provably good algorithms is still in its infancy. Approaches via network-flow techniques are promising and have provided the first known polynomial-time approximation algorithm for bipartitioning (Section 6.5.2). There are several polynomial-time bisection algorithms that can be proved to give optimum bisections of graphs and hypergraphs with high probability, given a certain uniform random graph model. The relevance of the underlying random graph model is questionable, since graphs occurring in layout practice tend to have certain symmetries that the random graph model deems unlikely. Furthermore, the fact that optimal bisections can be obtained with high probability using a variety of techniques suggests that these results exhibit as much a property of the random graph model as of the applied algorithms. In fact, it is a reasonable question to ask how hard one must try to get an algorithm that performs much worse than optimal with high probability.

For the analysis of the quality of graph partitions, the development of lower bounds is important. In conjunction with suitable algorithms, such lower bounds can provide algorithmic proofs of optimality for the constructed graph partitions. We discussed several lower bounds based on the formulation of the graph-partitioning problem as a QUADRATIC ASSIGNMENT problem (Section 6.6).

6.9   Exercises

6.1 Show that the restriction of MULTIWAY PARTITION$_{TH}$ to $B(i) = b(i) = B$, $n = rB$ is strongly NP-complete for all fixed $B \geq 3$. For this purpose, give a polynomial-time transformation from FREE PARTITION$_{TH}$ (Section 6.1.4) to MULTIWAY PARTITION.

6.2 Prove that SEPARATION$_{TH}$ is NP-complete by giving a polynomial-time transformation from the NP-complete problem BISECTION$_{TH}$ restricted to regular graphs of degree 3 to SEPARATION$_{TH}$. What restrictions of the problem SEPARATION$_{TH}$ does your reduction prove NP-complete ?

1. Bounded degree k? What is k?

2. Restriction on vertex weights? Which one?

3. Restriction on α? Which one?

(Using the methods of [52], you can prove that SEPARATION$_{TH}$ is NP-complete for $\alpha = 1/2$, for regular graphs of fixed degree $d \geq 3$, and for unit vertex weights.)

6.3 Let e be a hyperedge with k terminals that is to be represented by a clique. Let $BIP(e)$ be the set of all bipartitions of e.

1. Prove that, for each choice of a uniform cost assignment c to the clique edges,

$$\Delta(c) = \max_{\Pi \in BIP(e)} \left(\frac{1}{g(\Pi)}, g(\Pi) \right) = \Omega(\sqrt{k})$$

   Hint: Consider only bisections and the bipartition that puts a single terminal of e on one side.

2. Show that for each nonuniform cost assignment c to the clique edges that represent a hyperedge, we have $\Delta(c) = \Omega(\sqrt{k})$.

   Hint: Consider average edge costs.

6.4 * Prove the *conjecture* that adding "virtual" vertices v' with $w(v') = 0$ to the clique discussed in Exercise 6.3 does not increase the quality of the representation. Let C be a clique of the terminals of e with an in general nonuniform edge-cost assignment c. If Π is a partition of the terminals of e, let an *extension* of Π be a partition of the vertices of C that has the same number of parts as Π and reduces to Π if restricted to the terminals of e. $E(\Pi)$ denotes the set of extensions of Π. Let

$$g'(\Pi) := \min_{\Pi' \in E(\Pi)} g(\Pi')$$

and

$$\Delta'(c) = \max_{\Pi \in BIP(e)} \left(\frac{1}{g'(\Pi)}, g'(\Pi) \right)$$

Show that always $\Delta'(c) = \Omega(\sqrt{k})$ holds.

6.5 For some regular graphs with unit vertex weights and edge costs we can determine the minimum cutsize of a bisection or α-balanced bipartition exactly. The method is as follows. Let $G = (V, E)$ be the graph that has to be bipartitioned. Assume G has n vertices. Let K_n^r be the complete graph on n vertices in which each edge is replicated r times. The cutsize of a minimum bisection of K_n^r is exactly $rn^2/4$. The minimum cutsize of a bipartition with maximum part size B is exactly $r(n - B)B$.

We embed K_n^r into G. Specifically, different vertices of K_n^r are embedded in different vertices of G, and edges in K_n^r are embedded in paths connecting the images of their endpoints in G. We choose an embedding in which each edge of G accommodates at most k edges of K_n^r for some (hopefully small) k. Now we can argue about cutsizes in G from our knowledge of the exact minimum cutsizes in K_n^r. For instance, any bisection of G must have at least $rn^2/4k$ edges. Otherwise, the embedding of K_n^r in G would give us a bipartition with less than $rn^2/4$ edges.

Use this method to prove that the natural way of bisecting the following graphs is the best one.

1. The n-dimensional hypercube (see Exercise 5.5). To do this exercise, find a suitable embedding of K_n^1 into the hypercube.

2. The cube-connected cycles (see Exercise 5.5). Extend the result of part 1 of this exercise.

3. The $n \times n$ grid. Embed the $K_{n^2}^2$.

4. The k-dimensional grid with side length n. Embed $K_{n^k}^k$.

6.6 Speed up the minimization process in lines 16 to 19 of the Kernighan–Lin heuristic (Figure 6.4) by sorting the D values of the vertices in A and B, respectively, in decreasing order and searching for minimizing pairs by scanning the sorted lists. What is the total complexity of all selections in the worst case and in the typical cases, respectively? (Note that $c(e) \geq 0$ for all edges $e \in E$.)

6.7 This exercise shows that the Kernighan–Lin heuristic can yield bad results. Consider the *ladder graph* G with $2n$ vertices depicted in Figure 6.12.

1. Show that the highly suboptimal bisection $V_1 = \{1, \ldots, n\}, V_2 = \{n + 1, \ldots, 2n\}$ is a local minimum in the configuration graph $C_{CI}(G)$.

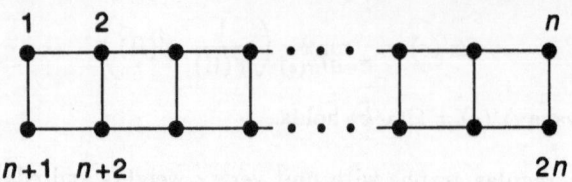

Figure 6.12: *A ladder graph.*

2. Run computer experiments and give structural arguments to support the thesis that the Kernighan–Lin heuristic performs poorly on G, given a random starting partition.

6.8 Modify the data structures used in the Fiduccia–Mattheyses heuristic so as to maintain kth-order gains for some fixed k in time $O(p)$. How does the constant factor grow with k?

6.9 Suggest your own heuristics for resolving nondeterminism in iterative improvement algorithms for graph partition.

6.10 Modify Theorem 6.2 so as to show that we can cut trees with vertex degree at most 3 into two parts of size at most $2w(V)/3$ by deleting one edge.

6.11 The algorithm by Lipton and Tarjan has a natural geometric variant. The corresponding theorem is the following.

   Theorem: *Let M be a compact region in the Euclidean plane with $\mu(M) := \int_{x \in M} \mu(x)d\mu = 1$ according to some measure function μ, and with area A. Prove that we can cut M into two pieces M_1 and M_2 such that $\mu(M_1), \mu(M_2) \leq 2/3$ by using straight cuts whose total length is at most $2\sqrt{A}$.*

1. Prove this theorem.

   Hint: Use the geometric outline of the algorithm by Lipton and Tarjan as a guide to your proof.

2. The theorem can be generalized to d dimensions ($d \geq 2$). Here, the cut consists of regions of hyperplanes, and the length of the cut changes to the total volume in $(d-1)$-dimensional space of the hyperplane regions that compose the cut. This total volume can be bounded by

$$(2d - 3/2) \cdot 2^{1/d} \cdot V^{(d-1)/d}$$

where V is the d-dimensional volume of set M.

Prove this generalization.

6.12 Prove that the right-hand side of (6.10) is a concave function of the variables $u_1, \ldots, u_n$. Use the inequality

$$\sum_{i=1}^{\ell} \lambda_i(A+B) \leq \sum_{i=1}^{\ell} \lambda_i(A) + \sum_{i=1}^{\ell} \lambda_i(B)$$

which holds for arbitrary symmetric real-valued $n \times n$ matrices A and B, and for $\ell \leq n$.

6.13 Suggest your own heuristics for assigning signs to the d'_i in Barnes' algorithm.

6.14 * This exercise discusses different variants of Lemma 6.9.

1. Use methods from linear algebra to show that, as long as the multiplicity of $\lambda_{n-1}(A^*)$ is $O(\log n)$, if $h(G)$ is the cutsize of a minimum bisection, then an eigenvector x of $\lambda_{n-1}(A^*)$ representing a minimum bisection can be found in polynomial time.

2. Show that, if $h(G)$ is the cutsize of a minimum bisection and G has a unique minimum bisection x^*, then x^* can be found in polynomial time, given $\lambda_{n-1}(A^*)$.

6.15 Modify the reduction of the CLUSTERING problem to the MAXIMUM WEIGHT MATCHING problem to handle the case that B is odd. Modify the analysis and show that in this case the error is $2B/(B+1)$.

6.16 Give an example of a graph in which the error stated in Theorem 6.9 is achieved.

Chapter 7

Placement, Assignment, and Floorplanning

In real-life applications, the provably good layout algorithm presented in Chapter 5 does not perform well enough to be of service. Thus, in practice, the layout problem is solved in a sequence of phases that attack subproblems with exact algorithms or heuristically. These phases are combined with a suitable usage of the circuit hierarchy. The first layout phase usually solves some variant of the placement problem.

Depending on the design style, placement has different goals. In general-cell and standard-cell placement, we want to position the components of the circuit such that the layout area is minimized. The area measure used here comprises the area taken up by the circuit components as well as the area needed for wiring the circuit components together (*wiring area* or *routing area*). Since hierarchy is employed in practical circuit layout, the circuit components to be placed are not necessarily single devices such as transistors, but may represent large chunks of circuitry, such as complete adders or control subcircuits. As in Chapter 1, a circuit component to be placed is also called a *block*; its layout is called a *cell*.

In gate-array layout, the master provides a fixed amount of routing area, and placement tries to find locations for the circuit elements that ensure routability within this area. Here, minimizing the area actually needed for routing is not as important as is ensuring routability by minimizing the congestion that may appear in spots inside the routing channels.

Thus, the placement problem has the dual flavor of a two-dimensional packing problem and a connection-cost optimization problem. The packing problem is concerned with fitting a number of cells of different sizes and shapes tightly into a rectangle. The connection-cost optimization aims at minimizing the amount of wiring necessary. The main difficulty in posing a suitable precise

303

formulation of the placement problem is that the area claimed by the subsequent routing stage cannot be estimated easily. Thus, it may happen that the estimates of wiring area that enter the cost function for the placement differ significantly from the wiring area actually claimed by the following routing phase. This problem places a particular significance on the choice of the cost function minimized in placement. Several different cost functions exist for this purpose. They can be roughly classified into cost functions that measure *total wire length* and cost functions that measure *congestion*.

In this chapter, we shall discuss the connection cost and the packing aspect of the placement problem. Section 7.1 considers homogeneous placement, such as occurs in gate-array layout. This variant of the placement problem involves only connection cost; packing is not an issue. Section 7.2 discusses general-cell placement and floorplanning. Here, packing problems have to be solved, as well. In both sections, we first define the basic optimization problems, and then survey several existing methods of solution. The packing optimization usually follows the optimization of connection cost. Therefore, we discuss the topics in this order. As of today, there is no systematic way of dealing simultaneously with connection cost and packing problems in placement.

In addition to positioning of the circuit components, there may be other tasks involved in placement. In full-custom layout, if the layout is done hierarchically in a top-down fashion, then at the time of placement the exact layout of the blocks is not known. Indeed, the interior of a cell belongs to the next lower level of the hierarchy, which in top-down layout will be processed later on. This lack of information complicates the packing aspect of the placement problem, because the only information on which we can base the placement of the blocks is statistical or experimental data on what the shapes of the cells are estimated to be in different layout alternatives. Cells whose shape is not known at the time of placement are represented by *variable cells* in the layout.

Placement involving variable cells is usually called *floorplanning*. Floorplanning is more difficult than placement is, because the different layout alternatives for the blocks involve new degrees of freedom and increase the configuration space over which the minimization has to take place. Floorplanning is discussed in Section 7.2, as is general-cell placement.

In semicustom layout, placement involves more than just determining locations for the circuit components. For instance, in standard-cell placement, the netlist has to be mapped onto a set of (instances of) standard cells, which then are placed onto the silicon. This mapping already involves optimization problems: If the primitive components of the netlist do not coincide with standard cells, a *gate assignment* of parts of the netlist to standard cells has to be found such that all components of the netlist are covered and the resulting set of standard cells allows for a placement with minimum area. If gate arrays are used, a similar assignment has to take place. Here, the mapping of netlist components onto gates entails the placement of the component on the gate-array master. Similarly, predefined cells often have several pins that carry the same signal. The *pin-assignment* problem is to connect to the pin that optimizes wiring cost. Assignment problems of this nature are discussed in Section 7.3.

Placement of components on printed circuit boards can be done with a mixture of techniques for general-cell placement and assignment.

Finally, in Section 7.4, we discuss simulated annealing approaches to placement.

7.1 Gate Arrays: Regular Placement

In this section, we ignore the packing aspect and concentrate on the connection-cost aspect of the placement problem. To this end, we introduce a number of abstract placement problems for identically sized cells. These problems are a somewhat stylized abstraction of gate-array placement, and they do not apply to all design and fabrication technologies. In particular, they do not model general-cell placement well. But they capture the essential combinatorial features of the part of circuit placement concerned with *connection cost*.

7.1.1 Definitions and Complexity

All placement problems we define in this section are versions of a simplified model of a regular gate array. Here, the master consists of an $r \times s$ array of identical *islands* that are arranged in a grid (see Figure 7.1). (In practice, the islands consist of structures that can be connected to form various types and assortments of gates on an island.) We assume that the blocks of the circuit map one-to-one onto arbitrary islands on the master. (This assumption means that, in practice, each island can be custom-tailored such that it implements any block.) Thus, packing the blocks is not an issue. The positions of the islands on the master are also called *slots*. We want to place the blocks such that the selected cost function is minimized.

UNCONSTRAINED RECTANGULAR ARRANGEMENT

Instance: A hypergraph $G = (V, E)$ with edge costs $\ell(e) \in \mathbf{R}_+$ for $e \in E$, $|V| = n$, constants $r, s \in \mathbf{N}$, $r \cdot s \geq n$ indicating the size of the master

Configurations: All *placements*—that is, injective mappings $p : V \to [1, r] \times [1, s]$ of circuit components into slots on the master. The slot that a component is placed into is called its *site*

Solutions: All configurations

Minimize: The cost function $c(p)$ in question

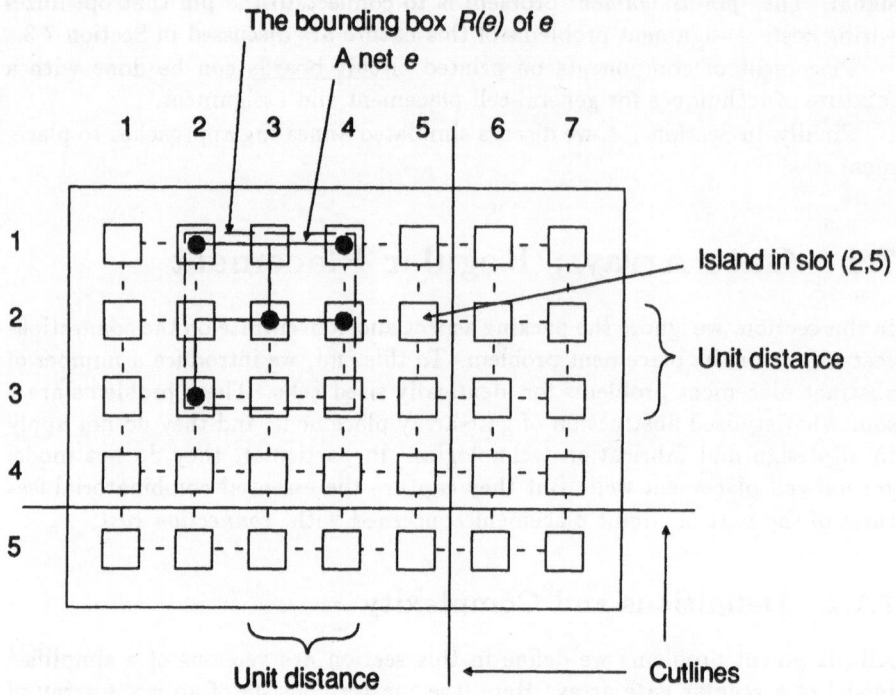

Figure 7.1: *An r × s gate-array master; r = 5, s = 7. The grid graph representing the master is drawn with dashed lines. The figure depicts a net e with five terminals, a routing for e (solid lines), and the bounding box R(e) of e. The half-perimeter of R(e) (4) is only a lower bound on the length of the depicted routing for e (6) and also on the minimum wire length of e (5). The figure also depicts a horizontal and a vertical cutline through the master.*

In the UNCONSTRAINED RECTANGULAR ARRANGEMENT problem, the hypergraph G represents the circuit, the vertices of G represent the blocks, and the edges of G represent the nets. We allow for different wiring costs $\ell(e)$ for the nets e. All blocks are assumed to have the same size, so we do not need a vertex-weighting function.

The UNCONSTRAINED RECTANGULAR ARRANGEMENT problem is a *grossly simplified* version of gate-array placement. In practice, a block may have to be spread over several adjacent islands, or several blocks can share an island. No such gate-assignment problems are addressed. (We discuss assignment problems in Section 7.3.) Also, no constraints on the legality of placements, such as

requiring routability, are incorporated. This optimization problem will serve
the purpose of exhibiting relationships between cost functions that minimize
congestion and cost functions that minimize wire length.

7.1.1.1 Cost Functions Based on Wire Length

For the purpose of defining cost functions that minimize wire length, we normalize the distances between adjacent islands on the master to be 1. The cost
functions are based on an estimate of the wire length of a net, given the placement of its terminals. For the sake of simplicity we locate all terminals to an
island in the center of the island.

One estimate of wire length that is quite popular is the *half-perimeter* of
the smallest rectangle $R(e)$ containing the sites of all terminals of the net e
(the *bounding box* of the net) [403]. Equivalently, we can define this estimate
to be the maximum L_1-norm of any vector of absolute differences between the
coordinates of the sites for any pair of terminals of the net. Formally, this is
the value

$$c_1(e, p) := \max_{v, w \in e} |p(v)_x - p(w)_x| + |p(v)_y - p(w)_y| \qquad (7.1)$$

Here, $p(v) = (p(v)_x, p(v)_y)$ is the site for terminal v—that is, a point with
integer coordinates in the plane. This estimate of wire length corresponds to a
wiring model in which vertices of the graph G are mapped onto the nodes of a
grid graph representing the master (see Figure 7.1). The islands on the master
are represented by the grid nodes. Nets are wired with Steiner trees. This
wiring model is only a gross approximation of the real situation. First, the grid
graph does not accurately indicate the wiring topology on the master. Second,
the wiring model does not interrelate different wires. In fact, we consider
each net independently of the other nets. Therefore, the model is not strong
enough to evaluate the routability of placements. In this context, the model
suffices, because we do not actually want to route wires, but rather want just
to estimate lengths of individual wires. However, the fact that routability is
not incorporated in the estimate is a central weakness of the whole approach
of evaluating placements by wire-length estimation. We shall comment on this
problem in more detail in Section 7.1.1.3.

The cost measure $c_1(e, p)$ is clearly a lower bound on the wire length for the
net e inside the grid graph representing the master. It may not be the exact wire
length, however. It is smaller if the wire has to leave $R(e)$, because the wiring
regions inside $R(e)$ are overly crowded. Even if the wire stays inside $R(e)$,
the half-perimeter may be too small, because the wire may have to circumvent
obstructions inside $R(e)$. However, if no obstructions exist and a minimum
Steiner tree in the grid graph representing the master can be realized for the net,
then the half-perimeter measure gives the exact wire length for two-point and
three-point nets [170]. Wiring nets with many terminals usually necessitates

more wire than is indicated by the half-perimeter, even if no obstructions exist (see Figure 7.1).

A wire-length estimate that deals better with large nets is the cost of a *rectilinear minimum spanning tree* of the net—that is, the minimum spanning tree in the complete graph whose vertices are the terminals of the net and whose edges are weighted with the L_1-distances between their endpoints on the master. (This concept was introduced in Section 3.10.) This estimate provides neither a lower nor an upper bound on the actual wire length, however. (If we could be sure that there are no obstructions on the master that force the net to take detours, then the net could be wired with a minimum Steiner tree, and the cost of the minimum spanning tree would be an upper bound for general nets.) Two-thirds of this estimate is a lower bound on the wire length for the net, however (see Section 3.10.1).

Sometimes, each net is modeled with a *star graph* (see Figure 6.1). This approach is tantamount to placing not G, but rather the graph $\tilde{G}$ obtained from G as follows: Provide a vertex $\tilde{v}$ in $\tilde{G}$ for each vertex v in G, and a vertex $\tilde{e}$ for each hyperedge e in G. Connect $\tilde{v}$ with $\tilde{e}$ in $\tilde{G}$ exactly if $v \in e$. Now place the resulting graph $\tilde{G}$. Kleinhans et al. [235] use this metric in conjunction with a quadratic wire-length estimate (see also Section 7.1.5.4).

If we eliminate the vertices $\tilde{v}$ from $\tilde{G}$ and connect vertices $\tilde{e_1}$ and $\tilde{e_2}$ if the corresponding hyperedges e_1 and e_2 in G are incident to a common vertex, we get a graph G'. Placing G' instead of G is referred to as placing according to the *nets-as-points* metric. This procedure places nets instead of blocks (see Section 7.1.5.4).

We shall use the half-perimeter measure from now on because it is easy to compute and provides a good lower bound for small nets. Thus, the cost function representing the total wire length is

$$c_1(p) := \sum_{e \in E} \ell(e) c_1(e, p)$$

The edge lengths are weighted with the edge-cost factors $\ell(e)$ in $c_1(p)$. In gate-array layout, we mostly use $\ell \equiv 1$, but some technologies may ask for different weights on different nets. For instance, different wire widths may lead to different weights on the nets.

The placement problem that uses $c_1(p)$ as the cost function is called the OPTIMAL RECTANGULAR ARRANGEMENT problem. The restriction $r = 1$, $s = n$ is called the OPTIMAL LINEAR ARRANGEMENT problem.

Cost functions estimating wire length are very common in placement. In gate-array placement, the rationale is that small wiring area should make the placement more routable. In full-custom layout, total wire length is assumed to relate directly to wiring area. There are two problems with this assumption:

1. Not all wires have uniform width. In other words, there may be wires whose width varies along the course of the wire. This problem is not

significant, however. Wires with nonuniform width typically occur in only the routing of power supply nets. This subproblem of the layout problem has other restrictions that require that it be solved separately after placement. On gate-array masters, power and ground nets are even built in from the beginning. Thus, it is not a great mistake to disregard the wide power and ground nets during placement.

2. Minimizing total wire length may neither minimize area nor maximize routability. Wires could still be distributed quite unevenly over the chip surface, such that some regions are quite congested while others are empty of wires. This phenomenon may decrease the wirability in gate-array layout or cause waste of layout area in full-custom layout. Thus, in addition to keeping the total wire length small, we want to generate an even distribution of wires [351].

Some approaches to the placement problem place a priority on eliminating long wires. We perform this weighting by having the wire-length estimate enter the cost function in some prespecified power. Specifically, the cost function in this case is

$$c_k(p) := \sum_{e \in E} \ell(e) c_k(e, p)$$

where we obtain $c_k(e, p)$ by replacing the L_1-norm in (7.1) with the kth power of the L_k-norm,

$$c_k(e, p) := \max_{v, w \in e} |p(v)_x - p(w)_x|^k + |p(v)_y - p(w)_y|^k$$

Of course, this cost function is nonlinear, and thus the solution of the placement problem becomes more difficult. Most often, the choice is $k = 2$, because for this case the best optimization methods are known. The corresponding version of the UNCONSTRAINED RECTANGULAR ARRANGEMENT problem is called the QUADRATIC RECTANGULAR ARRANGEMENT. (The one-dimensional version is called the QUADRATIC LINEAR ARRANGEMENT problem.) We discuss solution methods for this problem in Section 7.1.4.2. It is doubtful whether the QUADRATIC RECTANGULAR ARRANGEMENT problem does a better job at distributing wires evenly than does the OPTIMAL RECTANGULAR ARRANGEMENT problem. In fact, there is some need for hard evidence for the suitability of choosing $k = 2$ in the cost function.

Note that, if G is a graph, the OPTIMAL RECTANGULAR ARRANGEMENT and QUADRATIC RECTANGULAR ARRANGEMENT problem are both special cases of the QUADRATIC ASSIGNMENT problem (see Section 4.8) that use special distance matrices. In the first case, the distance matrix contains the half-perimeter or L_1-distances of slots on the master. In the second case, the distance matrix is based on the squares of the Euclidean distances. In both cases, the connection cost consists of two *independent* terms, one for each dimension.

7.1.1.2 Cut-Based Cost Functions

A more direct way of evening out wire distributions is to choose a cut-based cost function. Cut-based cost functions measure *congestion*, that is, the number of nets crossing certain cutlines in the master. There are $r - 1$ straight horizontal cutlines and $s - 1$ straight vertical cutlines through an $r \times s$ master (see Figure 7.1). Denote the set of all these cutlines with H. As in Chapter 6, we define the cutsize $c(h)$ across a cutline $h \in H$ to be the sum of the weights of all nets that have terminals on both sides of the cutline. It is not difficult to show that the sum of the cutsizes over all $r - 1$ horizontal and $s - 1$ vertical cutlines is exactly $c_1(p)$ (see Exercise 7.1). Thus, there is a close correlation between cutsizes and wire lengths.

We achieve a better wire distribution if we minimize not the *sum* of the cutsizes, but rather the *maximum* cutsize. The corresponding cost function for a quadratic master is

$$c_{\mathrm{MC}}(p) := \max_{h \in H} c(h)$$

(For a nonquadratic rectangular master it is sensible to divide $c(h)$ by the length of the corresponding cutline h.) This cost function is called the *density*. Minimizing the density aims at distributing wires across the chip evenly. With the cost function $c_{\mathrm{MC}}(p)$, the UNCONSTRAINED RECTANGULAR ARRANGEMENT problem is called the MINCUT RECTANGULAR ARRANGEMENT problem. The case $r = 1$, $s = n$ is called the MINCUT LINEAR ARRANGEMENT problem. The layout problem it models particularly well is the placement of a single row of equally sized standard cells. Here, $c_{\mathrm{MC}}(p)$ indicates the density of the single accompanying horizontal wiring channel (see also Section 9.6.1). In this case, the minimum value of the cost function over all legal placements is called the *cutwidth*.

As we have seen, the summation of cutsizes yields the total wire length. However, this is true only if wire lengths enter the summation in the cost function linearly, as it happens in $c_1(p)$. Wire-length estimates based on the L_k-norm bear no direct relationship to cutsize measures (see also Exercise 7.2).

Both the OPTIMAL LINEAR ARRANGEMENT problem and the MINCUT LIN-EAR ARRANGEMENT problem are NP-hard even when restricted to simple graphs [133]. The MINCUT LINEAR ARRANGEMENT problem is NP-hard even if restricted to planar graphs with maximum vertex degree 3 and $\ell \equiv 1$ or to trees [331]. Polynomial-time algorithms have been found for special cases, but they are not of practical significance, either because the special cases are too restricted or because the exponent or constant factor in the run time of the algorithm is too high. We summarize the results but do not discuss them in detail.

The OPTIMAL LINEAR ARRANGEMENT problem can be solved on trees with $\ell \equiv 1$ in time $O(n^{1.58})$ [67]. However, an analogous result is not known if $c_k(p)$ for some $k > 1$ is used as the cost function.

The MINCUT LINEAR ARRANGEMENT problem is solvable in time $O(n \log n)$ when restricted to trees with $\ell \equiv 1$ [479]. For complete m-ary trees, a simpler algorithm that runs in linear time has been given in [274]. The MINCUT LINEAR ARRANGEMENT problem restricted to hypergraphs with cutwidth k can be solved in time $O(n^{p(k)})$ with dynamic-programming techniques [327]. Here, $p(k)$ is a quadratic polynomial in k. Furthermore, the far-reaching results on efficient graph algorithms by Robertson and Seymour imply the existence of an $O(n^2)$ algorithm for this restriction of the MINCUT LINEAR ARRANGEMENT problem [111]. However, the constant is astronomical, and the result is by no means practical. To the contrary, the result raises questions as to whether asymptotics can be used to define the feasibility of an algorithm (see also [112] and Section 2.2.1).

7.1.1.3 Remarks on Estimates of Connection Cost

All the cost functions we have discussed so far are only imprecisely related to the cost measure in which we are really interested. In particular, they do not take wiring into account directly. Let us again consider gate-array placement. A regular master provides wiring channels (between or on top of the gate cells) that have some fixed uniform width. We are looking for a placement that is routable in the routing area provided. A more accurate way of modeling this situation is to give each edge of the grid representing the master the same capacity s. This value is related to the channel width on the master. Then we are looking for a placement that is routable such that no grid edge accommodates more than s wires. Actually this wiring model is still not quite accurate since it does not take into account the assignment of wiring layers (see Chapter 9). But the model is much closer to what we want since it does consider wiring capacity explicitly. On the other hand the resulting placement problem is substantially more difficult to solve, because we can evaluate a placement only by testing it for routability, and this is a difficult chore (see Chapter 8).

Patel [351] takes a step into the direction of using a cost function that considers routability explicitly. He evaluates a placement by routing Steiner trees for all nets independently and using the number of Steiner trees going across each grid edge as a contribution to the cost function. Of course, this cost function is quite difficult to compute. Patel bases a heuristic placement algorithm on this cost function. A similar suggestion has been made by Goto and Matsuda [153]. They develop a cost function that is easier to compute by basing the congestion across grid edges on a heuristic version of the probability that a net uses a grid edge. Currently, no cost function is known that gives a reliable estimate of wiring area or wirability and can be evaluated and optimized efficiently enough to serve in placement. In fact, an experimental study by Shiple et al. [421] indicates that wire-length estimates are quite inaccurate as far as the final layout area is concerned. Shiple et al. consider transistor placement in a style that is akin to sea-of-gates layout. They consider both the wire-length estimates that we discussed as well as wire-length estimates that are specifically tailored to this layout style. In all cases, the authors find substantial deviations of the estimates from the final area requirements. In fact, the best wire-length estimate was able to determine with only a 90 percent confidence whether the final layout area of a placement would be within 30 percent of the best known value. This unreliability of wire-length estimates is the major stumbling block in the way of significant progress on this crucial subproblem of the layout problem.

Given the fact that wiring-directed placement is not at our disposal today, the results of Chapter 5 show that, at least to a rough approximation, congestion measures have a strong correlation with layout area. Indeed, they involve both the minimization of wiring area and an even distribution of wires. The relationship between the total wire length and the density of a placement is not yet sufficiently well understood. There are examples of graphs where minimizing one of these cost functions makes the other one large, and vice versa (see Exercise 7.2), but it is not clear how much of this phenomenon occurs in practical applications.

Ultimately, the best way to estimate routability accurately during the placement phase seems to be to integrate the placement and the routing phase. Approaches to this goal are discussed in Section 8.8.

In the following sections, we shall use the problems defined here to present fundamental optimization strategies for circuit placement. Later on, we shall introduce other versions of the placement problem that are more closely tailored to full-custom layout.

7.1.2 Linear Arrangement

There is a large body of literature on linear arrangement problems. These problems are interesting as a simplified case of rectangular arrangement problems and because of their affinity to standard-cell placement. We shall concentrate on two-dimensional placement because of its greater relevance in circuit layout. For this reason, we shall not discuss linear arrangement problems in detail, but shall just provide pointers into the literature. References to NP-completeness proofs and polynomial-time algorithms for special cases have already been given in Section 7.1.1.1.

Adolphson and Hu [4] relate linear arrangement problems to network flow. They show that, if s and t are two arbitrary vertices in G, there always is an optimal arrangement of the vertices in G with respect to the OPTIMAL LINEAR ARRANGEMENT problem such that a minimum (s, t)-cut is realized in the arrangement. Specifically, there is a number j such that the j leftmost vertices in the arrangement form one side of the cut. Adolphson and Hu also give a structural condition that is sufficient to generate optimal solutions for both the OPTIMAL LINEAR ARRANGEMENT problem and the MINCUT LINEAR

ARRANGEMENT problem using network-flow techniques. Lawler [262] relates linear placement problems to job-sequencing problems. Here, the direction of the arrangement is considered as the time axis, and the vertices in G are tasks. The edges represent interdependencies between the tasks. Cheng [62] presents optimal solution algorithms and heuristics based on Adolphson and Hu's and on Lawler's work.

Chowdury [65] applies nonlinear optimization methods to the QUADRATIC LINEAR ARRANGEMENT problem. The same problem is heuristically solved by the *linear probe technique* developed by Frankle and Karp (see Section 7.1.6).

7.1.3 Mincut-Based Placement of Gate Arrays

The MINCUT LINEAR ARRANGEMENT problem is NP-hard, but the version of the recursive mincut heuristic (see Section 5.3) that bisects the subgraphs at each stage provides an approximation algorithm that gets to within a factor of $\log n$ of the optimum if the minimum cut is found at each stage (see Exercise 7.3). For the MINCUT RECTANGULAR ARRANGEMENT problem, the mincut heuristic is no longer necessarily an approximation algorithm. The cutsizes have to be balanced, as discussed in Section 5.4, to make the mincut heuristic an approximation algorithm. In practice, placement is often done by recursive graph bipartitioning, but bifurcators are not used. Rather, the cut is minimized at each stage. All versions of the mincut heuristic generate the cut tree (Definition 5.1) in a top-down fashion. Each node v in the cut tree represents a subgraph G' of the hypergraph G to be placed, and a rectangular subblock R of the master in which G' is to be placed. R is also called a *room*. The root of the cut tree represents the total hypergraph G and the complete master. A node v is processed by selecting a horizontal or vertical cutline that subdivides R into two subrooms R_1 and R_2. Then, the subgraph represented by v is bipartitioned into two pieces whose sizes have the same ratio as the sizes of R_1 and R_2. This process creates two children of v in the cut tree. In this way, we proceed until each subgraph consists of a single vertex.

Two popular traversals for the generation of the cut tree are depth-first and breadth-first. Among all possible choices of cutlines for bipartitioning the master, the bisection lines play a major role. In fact, a frequently used strategy is to bisect the subblocks in alternating dimensions (see Figure 7.2). Sometimes, however, cutlines are chosen that *slice* off a constant small portion of the master and therefore result in unbalanced bipartitions. Figure 7.3 illustrates a strategy in which we cut the master by choosing first the vertical cutlines in left-to-right and then bisecting each slab recursively.

Suaris and Kedem [430] propose that we make two orthogonal cuts simultaneously in each step. This procedure *quadrisects* the circuit. Suaris and Kedem apply this technique to standard-cell layout. In standard-cell layout, as in gate-array layout, the packing problem is secondary. Thus, the shape of subfloorplans does not have to incorporate cell shapes. Suaris and Kedem

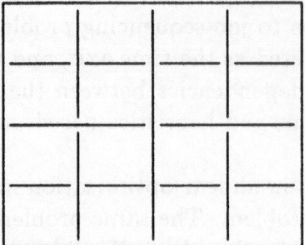

Figure 7.2: *A frequently used mincut strategy. The cutlines depict the temporal succession in which the cuts are made.*

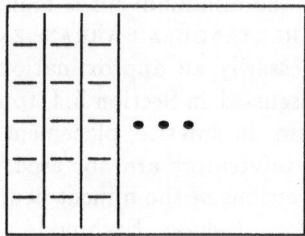

Figure 7.3: *Mincut slicing. The cutlines depict the temporal succession in which the cuts are made.*

report on favorable comparisons with classical bisection techniques (see also Section 7.2.1).

7.1.4   Wire-Length Minimization

In this section, we present methods for solving the wire-length minimization versions of rectangular arrangement problems. Section 7.1.4.1 is concerned with the OPTIMAL RECTANGULAR ARRANGEMENT problem, and Section 7.1.4.2 is concerned with the QUADRATIC RECTANGULAR ARRANGEMENT problem.

We already know from Section 7.1.1.1 that both problems are NP-hard. Thus, we define *relaxations* of the problems that can be solved efficiently. The solutions of these relaxation heuristics will take us, we hope, to good solutions of the problems themselves.

In both cases, the idea of the relaxation is to do away with the concept of a *slot*. Rather than requiring that vertices (blocks) be put on grid nodes, we allow vertices to be put anywhere in the plane. Indeed, we can put vertices very close to each other and, in the extreme case, on top of each other. To prevent the trivial placement with connection cost zero, where all vertices are placed on top of one another, we shall impose constraints that require some vertices to be spaced apart. Still, the fact that vertices can be put close to or on top of each other effectively does away with the *no-overlap* constraint that was implicit in the rectangular arrangement problems and is critical for all circuit-placement problems. This constraint makes the placement problem hard. The subsequent heuristic procedures have to resolve conflicts introduced by overlapping blocks.

The fact that both distance measures are composed of independent terms for the two dimensions suggests that the solution of the relaxation of the rectangular arrangement problems be done independently for both dimensions.

We shall also discuss generalizations of the rectangular placement problems defined in Section 7.1. Specifically, we shall allow preplaced components, and where appropriate we shall mention extensions to general-cell placement and floorplanning.

Note that, with wire-length minimization methods, there are similar problems of representing hypergraphs by graphs as with partitioning problems (Section 6.1.5). The natural way of representing a hyperedge is again as a clique with certain edge costs. No matter how $k \geq 1$ is chosen, and no matter what the costs of the clique edges are, there are always placements in which the cost contribution from the clique deviates substantially from the cost contribution of the hyperedge (see Exercise 7.5). Thus, there is a definite preference for methods of wire-length minimization that can deal with hypergraphs directly.

We now discuss relaxed placement problems minimizing linear and quadratic wire-length estimates, respectively.

7.1.4.1 Linear Wire Length

The one-dimensional version of the relaxation of the OPTIMAL RECTANGULAR ARRANGEMENT problem is the following:

RELAXED PLACEMENT

> *Instance:* A hypergraph $G = (V, E)$ with $|V| = n$, edge weights $\ell(e) \in \mathbf{R}_+$ for $e \in E$, a few *fixed* vertices $v_1, \ldots, v_k$ with positions $x_1, \ldots, x_k$
>
> *Configurations:* All placements $p : V \to \mathbf{R}$
>
> *Solutions:* All placements p such that $p(v_i) = x_i$ for $1 \leq i \leq k$
>
> *Minimize:* $c_1(p) = \displaystyle\sum_{e \in E} \ell(e) \max_{v,w \in e} (p(v) - p(w))$

The vertices $v_1, \ldots, v_k$ represent *pads* of the circuit whose placement coordinates are fixed. Having these vertices be in different locations forces the placement to be nontrivial.

The RELAXED PLACEMENT problem is readily translated into a linear program. We reserve two variables $z_{l,e}$ and $z_{r,e}$ for each net e. These variables represent the location of the leftmost and rightmost terminal of e. Furthermore, we have variables z_v for each vertex representing the location of this vertex. The constraints are the following:

$$z_{l,e} \leq z_v \leq z_{r,e} \quad \text{for all } e \in E,\ v \in e$$
$$z_{v_i} = z_i \quad\quad \text{for } 1 \leq i \leq k$$

The first set of constraints ensures the proper meaning of $z_{l,e}$ and $z_{r,e}$. The second set of constraints ensures that the preplaced vertices are in their correct positions (*preplacement constraints*). The objective function of the linear program is

$$\sum_{e \in E} \ell(e)(z_{r,e} - z_{l,e})$$

This linear program is the linear-programming dual of a network linear program (see Exercise 7.6). Thus, mincost-flow techniques can be used to solve it (see Section 3.11.2). In addition, the layout community has contributed several heuristic approaches to solving this problem [394, 451, 458, 459]. These methods do not always find optimal solutions and have no bounded error, but they may work faster on typical problem instances.

Interest in the RELAXED PLACEMENT problem has been spawned by the fact that a variant of this problem also occurs in wire-length minimization for one-dimensional compaction (see Chapter 10).

The RELAXED PLACEMENT problem has straightforward extensions to the placement of blocks of different size, if overlap constraints are not considered (see Exercise 7.7).

7.1.4.2 Quadratic Wire Length

In the literature, there are several motivations for using quadratic wire-length estimates in placement. We mentioned already the desire to eliminate long wires by punishing their use with a disproportionate factor. This motivation calls for a wire estimate $c_k(p)$ with $k > 1$. The value $k = 2$ suggests itself, because it is the only such value for which the contributions of the dimensions are independent.

Another motivation for using the quadratic wire-length measure $c_2(p)$ comes from the physical analogy, in which each wire is considered as a spring with a certain spring constant [368]. For this reason, this type of placement is also called *force-directed* placement. Since the energy of a spring is the square of that spring's length, multiplied by the spring constant, $c_2(p)$ is exactly the energy of the spring system corresponding to placement p. With this analogy, finding an optimum placement is akin to letting a spring system find its state of lowest energy. This analogy is quite intuitive and appealing, but equally is mysterious, because it is not supported by any analytical evidence.

The main reason why quadratic wire-length estimation is so popular is that the method leads to quadratic cost functions that can be minimized easily (see Section 4.8.3). If we are careful, the function remains convex and therefore has a single global minimum. The function is continuously differentiable, and eigenvector methods can be applied for its minimization. Thus, the final motivation for using quadratic wire length is driven by methodical, not modeling, arguments.

A straightforward way of formulating the RELAXED PLACEMENT problem corresponding to the quadratic wire-length estimate is simply to substitute the quadratic distance measure for the linear distance measure in the definition of the RELAXED PLACEMENT problem. This substitution yields the QUADRATIC RELAXED PLACEMENT problem, which is a quadratic program with linear constraints. This problem can be solved iteratively—that is, with an especially efficient version of the gradient-projection method that can handle several inequality constraints (see Section 4.8.1). Note that this method does not need to project points onto the constraint surface, because this surface is linear. If we are willing to turn the quadratic program into a general nonlinear optimization problem we can incorporate no-overlap constraints and other types of floorplanning constraints and cost functions (see Section 7.2.5).

The QUADRATIC RELAXED PLACEMENT problem has two prime advantages. The first is that it considers hypergraphs directly; the second is that it can take into account preplacements of components. Nevertheless, the author knows of no instance where this problem has been applied in circuit placement.

Rather, the published approaches to placement that use quadratic objective functions start from a graph instance of the placement problem and do not incorporate preplaced components. They solve a quadratic placement problem with quadratic constraints, but the quadratic constraints serve only to exclude the trivial placement that places all components on top of one another. Many authors impose a spherical constraint [120, 121, 166, 345]. This trick allows us to use eigenvector methods to solve the respective quadratic programs (see Section 4.8.5).

This approach starts with the adjacency matrix $A = ((a_{ij}))$ of the edge-weighted graph, where a_{ij} is the weight of the edge between vertices i and j. Since the graph is undirected, A is symmetric. The cost function to be minimized is

$$c_2(p) = \frac{1}{2} \sum_{i,j=1}^{n} a_{ij} \left(p(i) - p(j) \right)^2$$

Expanding the terms yields

$$c_2(p) = - \sum_{i,j=1}^{n} a_{ij} p(i) p(j) + \frac{1}{2} \left(\sum_{i=1}^{n} p(i)^2 \sum_{j=1}^{n} a_{ij} + \sum_{j=1}^{n} p(j)^2 \sum_{i=1}^{n} a_{ij} \right)$$

Since A is symmetric, the ith row sum $a_{i.} = \sum_{j=1}^{n} a_{ij}$ is the same as the ith column sum $a_{.i} = \sum_{j=1}^{n} a_{ji}$; thus,

$$c_2(p) = -p^T A p + \sum_{i=1}^{n} p(i)^2 a_{i.} = p^T C p$$

where $C = -A + D$ and D is a diagonal matrix of the row sums of A.

Obviously, $c_2(p) \geq 0$ for all placements p; thus, C is positive semidefinite. We now impose the spherical constraint $p^T p = 1$. This constraint does not change the set of optimal placements in the original (nonrelaxed) placement problem, because

1. All *legal* placements—that is, all placements that place vertices precisely onto slots—have the same norm.

2. The norm of p can be factored out of the cost function:

$$c_2(p) = ||p||^2 c_2(p/||p||)$$

However, the introduction of the constraint enables us to use eigenvector methods to solve the relaxed placement problem. In fact, by part 1 of Theorem 4.14, the quadratic program is minimized if p is the eigenvector e_1 in an orthonormal basis $e_1, \ldots, e_n$ of eigenvectors e_i corresponding to eigenvalues λ_i such that $\lambda_1 \leq \cdots \leq \lambda_n$. Here, we have $\lambda_1 = 0$, and the corresponding eigenvector is the vector $e_1 = (1/\sqrt{n}, \ldots, 1/\sqrt{n})$; that is, all vertices are placed on top of one another. This is an uninteresting solution. To exclude any part of this solution in the placement p, we require that $p^T e_1 = 0$, which is equivalent to $\sum_{i=1}^{n} p_i = 0$. This constraint does not exclude good placements, since each placement p can be made to meet the constraint by a shift that centers p around the origin and does not change. Hall [166] shows that, as long as the graph G is connected, the matrix C has rank $n - 1$. This means that the eigenvalue 0 has multiplicity 1, and e_2 is an eigenvector for a different eigenvalue. By part 2 of Theorem 4.14, $p = e_2$ is a minimum placement such that $p^T e_1 = 0$. This placement solves the one-dimensional case.

In the two-dimensional case we have to solve two quadratic programs, one for each dimension. Since the constraints do not formulate preplacements but are just intended to exclude the trivial solution, we have to take care that we do not get correlations between the coordinates in each dimension. If we simply took the same spherical constraint for the x and the y dimensions, then both quadratic programs would be identical and therefore would have the same solution. Thus, the vertices of the graph would all be placed on the main diagonal. This is clearly not a good way of using the chip surface. Thus, the constraint set for the quadratic program corresponding to the y dimension contains the spherical constraint and, in addition, the constraint

$$p_x^T p_y = 0 \qquad (7.2)$$

which ensures that p_x and p_y are not collinear. The respective minimum for the y dimension occurs in an eigenvector for the third-smallest eigenvalue, and the two eigenvectors together minimize the two-dimensional problem (part 2 of Theorem 4.14).

Blanks [36, 37] extends the discussion of two-dimensional placement with a quadratic cost function and spherical constraints to the case where preplacements exist. Then, the variables of the preplaced components turn into constants with the respective coordinate values. This modification decreases the number of variables and introduces a linear term in the cost function. Blanks suggests that we use gradient-projection methods to solve the corresponding quadratic programs. This approach is possible because Blanks can show that the local minima of a spherical quadratic program involving a cost function with a positive semidefinite Hessian take on at most two different eigenvalues (see Theorem 4.15). Experimental studies indicate that these two values are close together. As more components become preplaced, it is not clear why the spherical constraint or the constraint (7.2) is still needed.

7.1.4.3 Incorporating Timing Optimization in Placement

Timing estimates along a wire in a circuit are closely correlated with geometrical properties of the circuit. Therefore, timing constraints translate into geometrical constraints that we can enforce by adding them to the constraints already present in the relaxed placement problem. We can perform an optimization with respect to timing issues by modifying the cost function suitably, to include timing characteristics.

We can obtain timing constraints for a circuit by performing a static-delay analysis on the circuit, translating the resulting constraints into geometrical constraints on the wires and adding those to the optimization problem. Mostly these constraints will restrict the length of the wires, and are therefore linear. Hauge et al. [178] discuss methods to compute the additional constraints.

Jackson and Kuh [205] modify the cost function of the RELAXED PLACEMENT problem to incorporate delay optimization. The modified problem remains a linear program.

7.1.5   Elimination of Overlap

The solutions of the relaxed problems discussed in Sections 7.1.4.1 and 7.1.4.2
may place vertices (blocks) very close together. Thus, in a subsequent phase,
the vertices have to be rearranged so as to be equispaced on the master. We
discuss three methods of doing this rearrangement.

7.1.5.1   Linear Assignment

Rearranging the location of a vertex v in a placement computed by the optimization of a relaxed placement problem influences—generally, increases—the
cost function. The amount of this increase depends on the edges connecting
to v. As we mentioned in Section 7.1.1.1, finding an optimal placement on the
master amounts to a QUADRATIC ASSIGNMENT problem, if G is a graph. Of
course, this problem is NP-hard.

In this section, we present a relaxation that disregards the interdependence
between vertices in the placement through edges. Rather, we assume that
each vertex, when moved, changes the cost function by an amount that is
independent of the location of any other vertex. The placement p computed
during the optimization of the relaxed placement problem serves as a basis for
the rearrangement. We now try to move each vertex to a slot that is as close as
possible to the location of the vertex in p. Of course, several vertices may be in
conflict, and we have to resolve such conflicts. The appropriate optimization
problem for modeling this situation is the LINEAR ASSIGNMENT problem (which
is solvable in time $O(n^3)$; Section 3.12).

We obtain the corresponding cost matrix by placing a suitably scaled copy
of the master on top of the placement p. One way of doing this placement is
to assume that the master extends between the extreme placement coordinates
in each dimension (see Figure 7.4). Then, distances from each vertex location
in the placement to each slot on the master are computed in an appropriate
distance metric. In the corresponding matrix D, the entry d_{ij} indicates the
distance between slot i and the location of vertex j in the placement. Solving
the LINEAR ASSIGNMENT problem on this matrix computes an assignment of
vertices to slots that minimizes the sum of the distances between the vertices
and their assigned slots.

The linear assignment approach is quite flexible, since we can base the
construction of D on an arbitrary metric. For instance, if some slots are forbidden for some components, the corresponding matrix entries can be set to
infinity. However, this approach does not easily generalize to the case where
blocks have different sizes. The problem here is that large blocks have to be
spread over several islands. These islands should be adjacent on the master. In
fact, they should form a rectangular subblock. There is no appropriate way to
custom-tailor the LINEAR ASSIGNMENT problem to ensure this constraint (see
also Section 7.2).

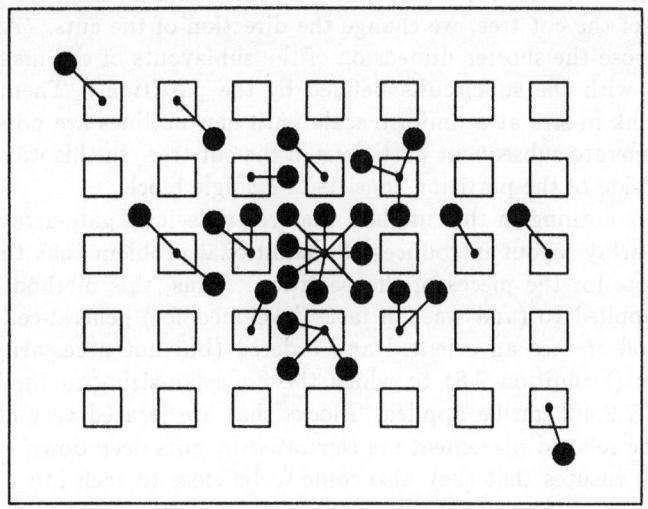

Figure 7.4: *A copy of a gate-array master with a relaxed placement. The dots represent the block locations in the relaxed placement. From each dot, a straight line is drawn to its nearest slot. Different lines that lead to the same slot represent conflicts.*

Akers [10] discusses other methods for using LINEAR ASSIGNMENT to construct initial placements or improve given placements.

7.1.5.2 Mincut

Otten [345] presents a heuristic for eliminating overlap in relaxed placements that is based in the mincut procedure. We describe the method conceptually. We begin with the relaxed placement and assume that blocks are centered at the positions specified by the relaxed placement and have their actual size. Overlap has to be expected in this placement. We now introduce cutlines along the shorter dimension of the layout—say, along the vertical dimension. We scan an imaginary vertical cutline across the layout from left to right. Whenever we find a spot where the line does not slice through a block, we stop and introduce a cut. Since we are doing gate-array layout, we have to ensure that the sizes of the parts of the partition are multiples of the height of the master. We can do this by shifting a few blocks across the cutline. Here, the candidate blocks are those that lie closest to the cutline. This procedure defines the first partition. We continue scanning the cutline across the layout and creating new cuts until we reach the right side of the layout. The corresponding partition defines the first level of a cut tree. Note that multiway cuts can arise here. To generate the next level of the cut tree, we change the direction of the cuts. (Alternatively, we can choose the shorter dimension of the sublayouts of the master that are associated with the subcircuits defined by the partition.) Then, we let the blocks shrink in size at a uniform scale until new cutlines are possible. These cutlines generate subsequent partitions in the cut tree. In this way we proceed until each side of the partition consists of a single block.

There is nothing in this method that restricts it to gate-array layout. In fact, gate-array layout introduces the additional problem that there are size requirements for the pieces of the partition. Thus, this method can be successfully applied to (and was, in fact, developed for) general-cell placement. The method creates an oriented and ordered (but not necessarily balanced) slicing tree (Definition 7.8) to which the floorplan-sizing methods discussed in Section 7.2.4.2 can be applied. Blocks that are located very close to each other in the relaxed placement are separated by cuts deep down in the slicing tree, which ensures that they also come to lie close to each oth r in the final placement.

Other authors [448, 470] use a different method for determining the cuts. Here, a binary cut tree is generated top-down and balancing the sides of a partition with respect to area is taken into account explicitly. Orientations of cutlines alternate in adjacent levels of the tree, and the location of the cutline is chosen to be the point at which one-half of the block area resides on either side of the cut. The side of a block is determined by the position of the vertex representing the block in the relaxed placement with respect to the cutline. This procedure does not ensure that blocks that are close in the relaxed placement also come to lie close to each other in the final placement.

7.1.5.3 Quadratic Optimization

Blanks [36, 37] suggests an optimization technique that is especially tailored to a solution method of the QUADRATIC RECTANGULAR ARRANGEMENT problem that starts out by solving the QUADRATIC RELAXED PLACEMENT problem (see Section 7.1.4.2). We discuss only the one-dimensional version of this technique. The method rests on the following assumptions that Blanks derives from measurements (see Figure 7.5). Consider a placement of n blocks as an n-dimensional vector p.

1. The Euclidean distance d_0 of the optimal placement p_0 with respect to the QUADRATIC RELAXED PLACEMENT problem to the closest *legal* placement p_1 is much larger than are the Euclidean distances between adjacent legal placements.

2. Locally around p_1, the set of legal placements is located near a surface with a low curvature, that is, they are approximately contained in a hyperplane at distance d_0 from p_0.

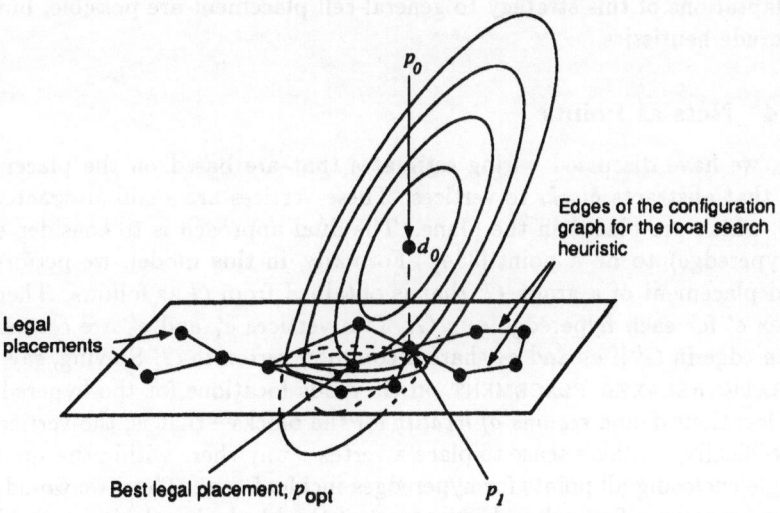

Figure 7.5: *Structure of the quadratic placement problem. p_0 is the optimum placement of the* QUADRATIC RELAXED PLACEMENT *problem. The ellipsoids indicated in the figure represent iso-surfaces of the cost deviation $p^T C p - p_0^T C p_0$ from the optimum cost of the* QUADRATIC RELAXED PLACEMENT *problem. (Exercise 7.11 proves that the surfaces are indeed ellipsoids.) p_1 is the legal placement that is closest to p_0. Locally around p_1, the legal placements are located approximately in a surface with low curvature. The axes of the ellipsoids may be tilted with respect to this surface. Thus, p_1 may not be an optimum legal placement. In the figure, the optimum legal placement is p_{opt}. We can move from p_1 to p_{opt} using local search. The figure indicates the underlying configuration graph.*

As Figure 7.5 suggests, p_1 is not necessarily the optimal placement with respect to the QUADRATIC RECTANGULAR ARRANGEMENT problem.

Blanks suggests that we first find an approximation to p_1 heuristically, and then use greedy local search to approximate p_{opt}. The search is based on a basic step in which a pair of modules is interchanged between slots to reduce the connection cost of the placement. Furthermore, Blanks derives approximate values of the cost of p_{opt} based on particular probabilistic assumptions. These assumptions imply that, on the average, $p_{\text{opt}}^T C p_{\text{opt}} - p_0^T C p_0$ approaches zero as the number n of blocks increases—that is, that with growing problem size, the eigenvector solution presents a sharp lower bound.

Adaptations of this strategy to general-cell placement are possible, but involve crude heuristics.

7.1.5.4   Nets as Points

So far, we have discussed wiring estimates that are based on the placement model that abstracts *blocks* to vertices. These vertices are again abstracted to points, which are placed in the plane. The dual approach is to consider each *net* (hyperedge) to be a point [354]. Formally, in this model, we perform a relaxed placement of a graph G' that is obtained from G as follows: There is a vertex e' for each hyperedge e in G. Two vertices e'_1 and e'_2 are connected with an edge in G' if e_1 and e_2 share a common vertex in G. Solving, say, the QUADRATIC RELAXED PLACEMENT on G' yields locations for the hyperedges. These locations define *regions of locality* for the blocks—that is, the vertices of G. Specifically, it makes sense to place a vertex v anywhere within the smallest rectangle enclosing all points for hyperedges incident to v. Thus, we would, for instance, achieve a first relaxed placement of the blocks by placing each block in the center of its region, or alternatively, as the center of gravity between the points representing its incident hyperedges. The linear assignment approach (see Section 7.1.5.1) can be used to assign vertices to slots.

Zhang et al. [483] discuss other methods of placing the blocks based on the nets-as-points placement. They also report on experiments that show the superiority of this approach to methods using the model that represents blocks as points. Still, the methods of evaluation are highly heuristic, and in the light of the results by Shiple et al. [421] this metric has to be considered with as much caution as are the more traditional wire-length estimates.

7.1.6   The Linear Probe Technique

The linear probe technique was invented by Frankle and Karp [120, 121]. It is a heuristic method of solving the QUADRATIC RECTANGULAR ARRANGEMENT problem restricted to graphs and some generalizations of this problem. The technique does not resort to a relaxed problem formulation, but exploits the relationship between quadratic programs and eigenvectors. We first discuss the technique in the restricted case of the QUADRATIC LINEAR ARRANGEMENT problem.

The linear probe technique starts out by turning this minimization problem into a maximization problem. In the transformed problem, we are looking for a legal point of *maximum* norm—that is, a point that is farthest away from the origin. Then, linear probing is introduced as an efficient heuristic for finding such a point.

The problem transformation is simple. With the notation of Section 7.1.4.2, let $p = \sum_{i=1}^{n} p_i e_i$ be a placement, represented by its coordinates with respect to an orthonormal eigenvector basis for $C = -A + D$. Remember that we imposed the constraints $p^T p = 1$ (unit norm) and $\sum_{i=1}^n p_i = 0$ (placement centered around the origin). Both constraints did not truly restrict the problem.

Instead of minimizing

$$c_2(p) = \sum_{i=1}^n p_i^2 \lambda_i$$

we maximize the value of

$$H - c_2(p) = \sum_{i=1}^n p_i^2 (H - \lambda_i) = \sum_{i=1}^n (p_i \sqrt{H - \lambda_i})^2 = \|p^T E\|^2 \qquad (7.3)$$

where

$$E = \left(\sqrt{H - \lambda_1} e_1, \ldots, \sqrt{H - \lambda_n} e_n \right)$$

and $H \geq \lambda_n$ is chosen arbitrarily. By (7.3), $H - c_2(p)$ is the square of the Euclidean norm of the vector $p^T E$, and we are seeking to maximize this norm with a legal placement. Remember that, in the original problem domain, a legal placement p is just a permutation π of the numbers $1, \ldots, n$. The meaning of $\pi(i) = j$ is that block i be put in slot j. However, the original problem domain has been transformed into the orthonormal eigenvector basis, then scaled to unit norm, and centered around the origin. We can formalize this situation by writing $p(i) = \widetilde{\pi(i)}$. Here the operator $\sim$ performs the basis transformation, normalization, and centering of the slot positions.

A *linear probe* finds a legal placement p for which $p^T E$ is farthest away from the origin in a prespecified direction, defined by a directional vector d (see Figure 7.6). This point is the legal point that maximizes $p^T E d$. Such a placement can be found as follows. We compute Ed. This procedure takes time $O(nk)$, where k is the number of nonzero elements of d. The maximal placement is obviously the one that orders the vertices according to increasing value of the corresponding component of Ed. That is, if component i of Ed has the smallest value, then vertex i is put in the slot j for which $\tilde{j}$ is smallest, and so on. This ordering process can be done in time $O(n \log n)$. (A similar optimization technique was used in the proof of Theorem 4.14.)

Thus, a linear probe can be done efficiently. How do we choose the directions of the probes? Frankle and Karp suggest that we iterate probes as shown in Figure 7.7, and they elaborate on versions of this idea.

In addition to computing a placement, every linear probe yields a proof that in the specified direction d there is no farther placement. This observation can be used as the basis for the computation of lower bounds on the minimum cost of a placement. For instance, covering the whole space evenly with probe directions leads to a lower bound whose quality depends on the number of probe directions. As an example, if we execute probes in the directions of all coordinate axes, we can determine the smallest n-dimensional Cartesian brick containing all legal placements. A *Cartesian brick* is just a hypercube that

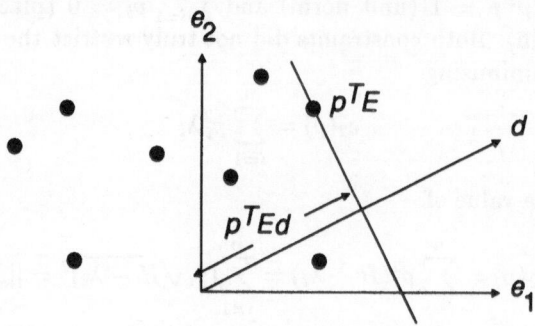

Figure 7.6: *A linear probe; the coordinate axes correspond to the orthonormal eigenvectors of C. The dots indicate the legal placements. The direction of the probe is the vector d.*

can be translated and scaled differently in each dimension. The maximum cost of any point in this brick is an upper bound on the maximum cost of any legal placement in the transformed problem In fact, it is not difficult to show that this upper bound is never more than a factor of $\sqrt{n}$ away from the true maximum cost (see Exercise 7.8). This upper bound turns into a lower bound on the minimum cost of any legal placement in the original problem.

By increasing the number of probe directions, we can further decrease the distance of the upper bound from the optimum for the transformed problem. Choosing directions randomly yields good placements with high probability (for details, see [120]).

As we noted in Section 7.1.4.2, preplaced components add a linear term to the objective function. After the problem transformation, this linear term translates into a constant vector z; that is, instead of maximizing $||p^T E||^2$, we are now trying to maximize $||p^T E + z^T||^2$ (see Exercise 7.10). The corresponding linear probe in direction d now asks for maximizing $(p^T E + z^T)d$. Since z is a constant vector, the addition of $z^T d$ does not affect the linear probe.

An extension of the linear probe technique to the QUADRATIC RECTANGULAR ARRANGEMENT problem is possible. Here, we have to maximize the sum of two terms $||p_x^T E||^2 + ||p_y^T E||^2$, one for each dimension. One possibility is to work on each dimension alternately. When we are considering the x dimension, we divide the set of blocks into classes, one for each row of the master in the current placement. Then, we probe each class separately with the same direction vector. We handle the y dimension analogously.

This approach has the disadvantage of separating the dimensions. The following more powerful approach handles both dimensions jointly. We specify a *two-dimensional probe* by giving two direction vectors d_x and d_y. The probe

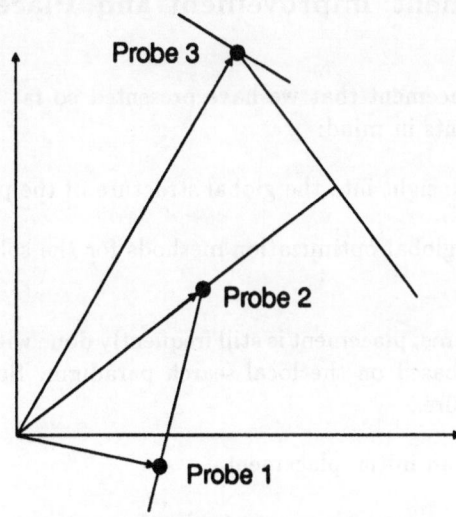

Figure 7.7: *Iteration of linear probes.*

asks that we maximize $p_x^T E d_x + p_y^T E d_y$ over all legal placements p. This problem is an instance of the LINEAR ASSIGNMENT problem. To see why it is, we recall that $n = r \cdot s$, where $r \times s$ are the dimensions of the master, and let $j \mapsto (r_j, s_j)$ be any bijective mapping from slot numbers to slot coordinates. The matrix $D = ((d_{ij}))_{i,j=1,\ldots,n}$ for the instance of the LINEAR ASSIGNMENT problem has the entries

$$d_{ij} = \tilde{r}_j \cdot (E d_x)_i + \tilde{s}_j \cdot (E d_y)_i$$

denoting the cost contribution of putting block i into slot j. (Remember that the operator $\sim$ performs the basis transformation, normalization, and centering of the slot positions.) The problem instance asks us to maximize the cost function.

Note that the positions of the slots on the master do not need to be in a grid pattern for the linear probe technique to work. Arbitrary positions can be assigned to the slots. The numbers r_j and s_j receive the corresponding slot coordinates.

There is, however, no apparent way in which the linear probe technique can address the packing problems that arise from blocks of different sizes (see Section 7.2).

7.1.7*   Placement Improvement and Placement Heuristics

The methods of placement that we have presented so far have been selected with two requirements in mind:

1. They exhibit insight into the global structure of the placement problem.

2. They employ global optimization methods for the solution of the placement problem.

In current CAD systems, placement is still frequently done with more elementary heuristics that are based on the local search paradigm. Such heuristics have the following structure:

Step 1: Construct an initial placement.

Step 2: Improve the given placement by local search, mostly using a greedy search strategy.

Actually, elements of some of the methods we have described also have this flavor. Blanks' method of solving the QUADRATIC RECTANGULAR ARRANGEMENT problem is an example (see Section 7.1.5.3). In fact, any placement method that we have discussed may be used for step 1 of this two-phase procedure, and it is often necessary to employ local search techniques to improve the placement generated in this way.

There is a whole set of local search strategies available for improving placements. None of them generates placements with analyzable quality. We shall not discuss them in detail here; for extensive discussions of this subject, see [153, 172, 362].

Simulated annealing—the randomized version of local search techniques—is discussed in Section 7.4.

7.2   General Cells: Floorplanning

Packing problems enter the discussion if cells have different shape and size, such as in general-cell placement and floorplanning. (Recall that general-cell placement is the special case of floorplanning in which all cells have a fixed shape.) The data structure manipulated in case is the *floorplan*. A floorplan exists in two variants, *sized* and *unsized*. The unsized floorplan only determines the relative positions of cells and captures the topology of a placement. The sized floorplan, in addition, determines the actual sizes of the floorplan features and the area of the placement. The topology optimization is mainly concerned with connection cost, whereas the sizing considers packing problems.

Definition 7.1 (Floorplan) *An* **unsized floorplan** *for a circuit G is a labeled directed planar graph F representing the subdivision of a rectangle called the* **base rectangle** *into subrectangles called* **rooms***. Each room accommodates a unique block of G; that is, there is a bijection between vertices in G and rooms in F. From now on, we identify the rooms in F with the vertices in G.*

The vertices of F are the corners at which different rooms meet and the corners of the base rectangle. The edges of F are the straight-line segments in the floorplan that run between corners. These segments are also called **walls***. The edges of the floorplan representing horizontal (vertical) walls are directed from left to right (top to bottom). Each edge e is labeled with an* **orientation label** *dir(e) ∈ {hor, ver} indicating whether it represents a horizontal or a vertical wall.*

A floorplan in which no vertex has degree 4 is called **normal***.*

A **sizeable** *floorplan is a floorplan in which there is a sizing function* $s_v : \mathbf{R}_+ \to \mathbf{R}_+$ *attached to each room v (see Definition 7.2).*

A function $\ell : E \to \mathbf{R}_+ \cup \{0\}$*, such that $\ell(e)$ reflects the length of e in a geometric sizing of the rectangular subdivision is called a* **sizing** *of F. The pair (F, ℓ) is called a* **sized floorplan***. The height (width) of room r in sizing ℓ is denoted with $h_\ell(v)$ $(w_\ell(v))$. Similarly, w_ℓ and h_ℓ denote the width and height of the whole floorplan.*

Figure 7.8(a) shows a sized floorplan that is not normal. The notion of normality of a floorplan helps us to simplify concepts and algorithms. In practice, it does not present a restriction, because after sizing the floorplan depicted in Figure 7.8(a) can also be represented by either of the normal floorplans depicted in Figure 7.8(b) or (c). Therefore, we shall from now on assume that floorplans are normal. We shall also omit directions and orientation labels from edges of a floorplan in the figures, because they can be inferred from the figure.

The unsized floorplan specifies the topology of a general two-dimensional arrangement of cells. The sized floorplan determines the exact geometry. A common strategy for general-cell placement is to determine, in a first phase, the floorplan topology based on connection-cost criteria. This step assigns a room to each block. In a second phase, the geometry is specified; that is, *floorplan sizing* is performed with the goal of minimizing the packing area. Sizing of a floorplan has to take into account the shapes of the cells. These shapes are attached to the corresponding rooms in the sizeable floorplan. The goal of sizing is to minimize area while maintaining the floorplan topology generated in the first step.

In floorplanning, each block can have multiple layout alternatives. A block with a single or multiple layout alternatives is described by a *sizing function*.

Definition 7.2 (Sizing Function) *A* **sizing function** $s : [w_0, \infty) \to \mathbf{R}_+$ *is a monotonically decreasing piecewise continuous function. If all continuous segments of s are constant and all function values and points of discontinuity of s are integer, then s is called a* **sizing step function***. The way in which s*

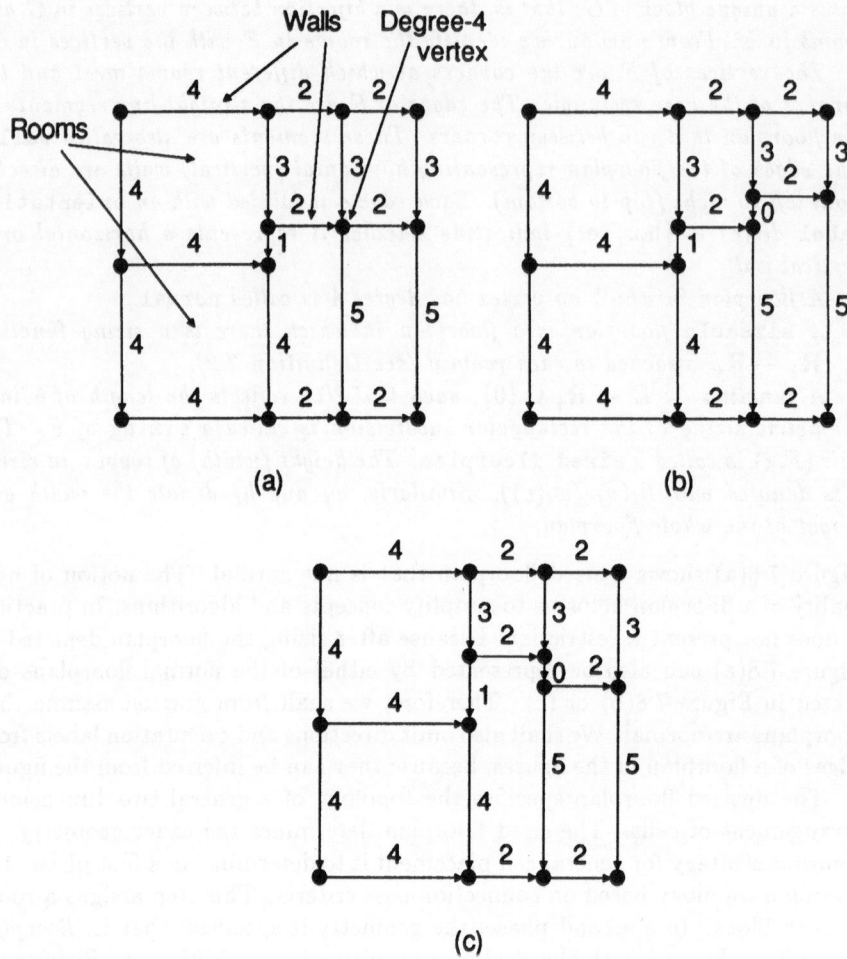

Figure 7.8: *A sized floorplan that is not normal; sizing labels are attached to the walls. Orientation labels are omitted. (a) The sized floorplan; (b) and (c) two sized normal floorplans that are equivalent to that in (a).*

describes the sizes of layout alternatives of a block is that the minimum height of a layout with width w is $s(w)$. The minimum width of a layout is w_0.

Figure 1.15 shows several sizing functions. The sizing of a floorplan topology entails picking a layout alternative for each block.

Definition 7.3 (Notions in Floorplanning) *Let F be a sizeable floorplan. A choice function $a : V \rightarrow \mathbf{R}_+$ for F is a function that selects a layout alternative for each room.*

A sizing ℓ of F is legal for a choice function a, if for each room v in F, ℓ satisfies $w_\ell(v) \geq a(v)$ and $h_\ell(v) \geq s_v(a(v))$; that is, the room v accommodates the layout alternative with the proportions $a(v) \times s_v(a(v))$. Here, s_v is the sizing function of room v.

A floorplan size measure is a function $\phi : \mathbf{R}_+ \times \mathbf{R}_+ \rightarrow \mathbf{R}_+$ that is monotonic in both arguments. The arguments w and h in $\phi(w, h)$ denote the width and the height of a sized floorplan.

Sizing a floorplan amounts to finding a choice function a and a legal sizing ℓ for a such that $\phi(w_\ell, h_\ell)$ is minimum. Popular floorplan size measures are the area $\phi(w, h) = w \cdot h$ and the *half-perimeter* $\phi(w, h) = w + h$.

For most floorplanning methods, the unsized floorplan is too restrictive a structure to describe the floorplan topology. The first phase that minimizes connection cost determines less information than is contained in an unsized floorplan. For instance, mincut methods do not determine all adjacencies between rooms, but determine only adjacencies between rooms and cutlines (see Section 7.2.1.1). The floorplan-sizing operation therefore should not be restricted to a single unsized floorplan, but rather should optimize the floorplan size over a class containing all floorplans that are indistinguishable by the first phase of the floorplan generation. We can achieve this goal by using a *similarity relation* $\sim$ on sizeable floorplans that groups indistinguishable floorplans together.

Definition 7.4 (Floorplan Similarity) *An equivalence relation $\sim$ on sizeable floorplans is called a floorplan similarity relation. If $F \sim F'$ for two floorplans F, F', then F and F' are similar. The equivalence classes of $\sim$ are called floorplan topologies.*

We are now ready to formulate the general floorplan-sizing problem formally.

FLOORPLAN SIZING

Instance: A floorplan topology $[F]$, a floorplan size measure ϕ

Configurations: All triples (F', ℓ, a), where $F' \in [F]$, ℓ is a sizing of F' and a is a choice function for F'

Solutions: All (F', ℓ, a) such that ℓ is legal for a

Minimize: $\phi(h_\ell, w_\ell)$

In general, each floorplanning algorithm is based on a certain floorplan similarity relation. Furthermore, each floorplan topology $[F]$ receives a certain connection cost; that is, all floorplans in $[F]$ have the same connection cost. In a first phase, the floorplanning algorithm selects the floorplan topology with the minimum connection cost (or with a cost close to the minimum). In the second phase, the FLOORPLAN SIZING problem is solved on the resulting floorplan topology.

Different floorplanning algorithms are based on different floorplan similarity relations. We shall discuss several such relations in detail. The larger the floorplan topologies are, the less information the first phase of floorplan generation computes, and the more powerful (and complex) the sizing operation is.

There are two ways of arriving at a floorplanning method. One is to choose an algorithm for generating floorplan topologies. Thereby we define a specific floorplan similarity relation, and this relation, in turn, defines which variant of the FLOORPLAN SIZING problem we have to solve. This *operational approach* is the path taken by mincut floorplanning (Section 7.2.1). The other possibility is to start with a natural floorplan similarity relation and to build a floorplanning method around that relation. This *declarative* approach involves developing suitable algorithms for *finding and sizing* floorplan topologies. This approach has been taken in rectangular dualization (Section 7.2.4).

7.2.1   Floorplanning Based on Mincut

The mincut heuristic can be applied successfully to general-cell placement and floorplanning. The first phase of the mincut method recursively bipartitions the circuit G. This bipartitioning process mimics the construction of a sized floorplan. We require a vertex weight for each block in G. This weight roughly estimates the area taken up by the block. In general-cell placement, this weight can be taken to be the area of the corresponding cell. In floorplanning, where no definite cell shapes are available, we can choose measures such as transistor count or the average area over all layout alternatives, perhaps biased toward layouts that are almost square. The initial sized floorplan represents an empty base rectangle whose area is the total vertex weight of G.

The recursive cutting process generates a binary cut tree in a top-down fashion. Most often, the cut tree is generated in a breadth-first fashion. However, the tree-generation procedure also allows for different traversal schemes, such as depth-first. Each node in the tree represents a subgraph of the circuit and a rectangular room in the layout area. When node v in the tree is processed, a bipartition of the corresponding subgraph of G is generated that is balanced with respect to the vertex weights and that minimizes the cutsize. This bipartition can be computed with the Fiduccia–Mattheyses heuristic, for instance (see Section 6.2.2). Then a horizontal or vertical cutline is chosen that subdivides the room corresponding to v according to the size ratio of the two halves of the partition. Each subroom is assigned to its respective subgraph. This process generates two children of v. In this way, we continue generating vertices in the cut tree and subdividing the floorplan until each subgraph consists of a single vertex.

In the following two sections, we describe two versions of mincut-based floorplanning. In the *oriented-mincut method*, we use heuristics to orient the cutlines already in the first phase of the floorplanning algorithm—that is, the phase that generates the floorplan topology. As a consequence, the orientations of the cutlines are the same for all floorplans in a floorplan topology, and the phase in which the FLOORPLAN SIZING problem is solved cannot change the orientation of cutlines in order to decrease the floorplan-size measure (Section 7.2.1.1). The more powerful *unoriented-mincut method* leaves the orientations of the cutlines undetermined in the first phase. The orientations of the cutlines can be left undetermined because they do not influence the connection cost of a floorplan topology. The result is a floorplan similarity relation with properly larger floorplan topologies. Therefore, the FLOORPLAN SIZING problem solved in the second phase entails a more powerful optimization than is used in the oriented-mincut method, but it is also more difficult to solve (Section 7.2.1.3).

7.2.1.1 Floorplanning Based on Oriented Mincut

In addition to generating the cut tree, the oriented-mincut method orients and orders the tree. *Orienting* the tree amounts to choosing a suitable orientation for each cutline. La Potin and Director [255] apply oriented mincut to floorplanning with general sizing step functions and suggest that cut directions be alternated in adjacent levels of the tree. The goal of this heuristic is to keep rooms about square. Another heuristic that achieves this goal is always to cut along the shorter side of a room. Lauther [258] applies oriented mincut to general-cell placement. Since here the shapes of the cells are known, he can orient cutlines corresponding to nodes low in the cut tree according to a different heuristic that aims at fitting the cells nicely into rooms.

Ordering the cut tree means that, at each cut, a choice is made regarding which subcircuit to put on which side of the cutline. In the oriented-mincut

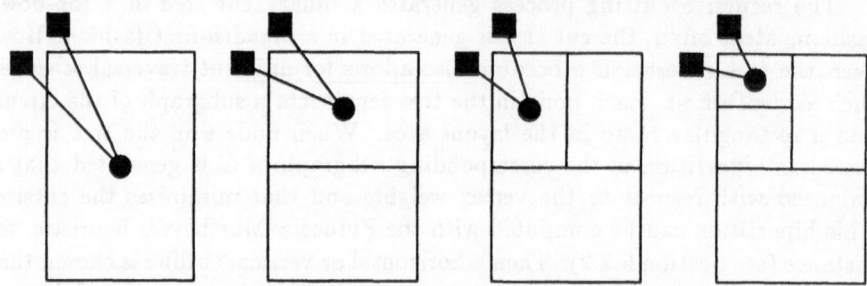

Figure 7.9: *Heuristic ordering of the cut tree; the figure depicts a vertex representing a circuit block that is connected to two preplaced pads. These connections guide the ordering process of the cut tree heuristically, such that the block is eventually assigned to a room close to the pads. Four consecutive stages during the generation of the cut tree are displayed.*

method, this choice is again done heuristically during the generation of the tree. The choice can be made on the basis of the number of connections from the block to be placed to blocks whose position is known already. Such preferences with respect to which side of the cut to choose for a block can easily be incorporated into the Fiduccia–Mattheyses heuristic. They can even take the form of strict requirements, such as needed for incorporating preplaced pads (see Figure 7.9).

Ordered mincut is a popular method in general-cell placement and floorplanning. Lauther [258] was the first person to use this method.

7.2.1.1.1 Similarity Relation for Oriented Mincut As described, the output of the first phase of the oriented-mincut method is a cut tree *and* a sized floorplan. However, the floorplan contains some irrelevant information. For instance, the sizing of the floorplan is computed on the basis of rough estimates for the sizes of the blocks in the circuit. The subsequent sizing operation will change these estimates, and therefore the present sizing of the floorplan does not convey much information. In fact, it may be illegal. Even the unsized floorplan is not unique. It also depends on the estimates of the block sizes. For instance, the two unsized floorplans F_1 and F_2 in Figure 7.10 can arise from an oriented mincut on the same circuit with different estimates for the block sizes. The relative sizes of the blocks determine whether the wall w_1 comes to lie higher than the wall w_2, or vice versa. Since the sizes of rooms will change in the second phase of the floorplanning algorithm, floorplan F_1 can be transformed into floorplan F_2, and vice versa. Therefore, both floorplans should belong

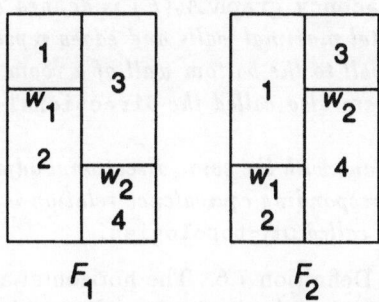

Figure 7.10: *Two floorplans that are indistinguishable by oriented mincut. Whether the wall w_1 is located higher or lower than the wall w_2 depends on the size of the respective rooms. Thus, during sizing, floorplan F_1 can be turned into floorplan F_2 without the cut tree being changed. Therefore, F_1 and F_2 belong to the same floorplan topology.*

to the same floorplan topology. In fact, a floorplan topology corresponding to the oriented-mincut method should contain exactly all floorplans for which the *oriented adjacencies* between rooms and cutlines are identical. In other words, two floorplans F_1 and F_2 should be in the same floorplan topology if, for each room r and for each side s ($s \in \{\text{top}, \text{right}, \text{bottom}, \text{left}\}$), the same cutline forms side s of room r in F_1 and F_2. (The two floorplans F_1 and F_2 in Figure 7.10 have this property.)

In general, cutlines are composed of many walls. The vertical cutline in both floorplans in Figure 7.10 is composed of three walls.

Definition 7.5 (Maximal Wall) *A **maximal wall** is a nonextendable directed path of edges in F that have the same orientation label.*

If F is normal, then each cutline generated by the mincut method is a maximal wall. The theory presented here can be extended to floorplans that are not normal (see Exercise 7.14), but we shall restrict our discussion to normal floorplans.

The following definition captures the oriented adjacencies between rooms and cutlines in a floorplan.

Definition 7.6 (Directional Adjacency Graphs) *The **horizontal adjacency graph** $A_h(F)$ is a planar dag that has a vertex for each vertical maximal wall and a directed edge for each room. The edge for room v runs from the maximal wall forming the left side of v to the maximal wall forming the right side of v. If F is a sizeable floorplan, then the edge for room v is labeled with the sizing function s_v.*

The **vertical adjacency graph** $A_v(F)$ *is defined analogously. Here, vertices represent horizontal maximal walls and edges represent rooms. Edges are directed from the top wall to the bottom wall of a room.*

$A_h(F)$ *and* $A_v(F)$ *are also called the* **directional adjacency graphs** *of F.*

Two sizeable floorplans with the same directional adjacency graphs are called OM-**similar***. The corresponding equivalence relation is denoted with* $\sim_{OM}$*. Its equivalence classes are called* OM-**topologies***.*

Figure 7.11 illustrates Definition 7.6. The horizontal adjacency graph $A_h(F)$ models the horizontal adjacencies in the floorplan, and analogously for $A_v(F)$. Let $s_\ell, s_r, s_t,$ and s_b be the vertices representing the left, right, top, and bottom side of the base rectangle. Then, s_ℓ is the unique source and s_r is the unique sink of $A_h(F)$; s_t is the unique source and s_b is the unique sink of $A_v(F)$. Also both $A_h(F)$ and $A_v(F)$ are planar. We can construct planar embeddings by placing the vertices on the respective maximal walls and routing the edges through the rooms. If F is normal, a structural relationship exists between $A_h(F)$ and $A_v(F)$.

Lemma 7.1 *Let* $A_h^*(F)$ *and* $A_v^*(F)$ *be the two graphs resulting from* $A_h(F)$ *and* $A_v(F)$ *if an outside edge is added between* s_ℓ *and* s_r *in* $A_h(F)$ *and between* s_t *and* s_b *in* $A_v(F)$*, and, if all edges are considered undirected. If* F *is normal, then* $A_h^*(F)$ *and* $A_v^*(F)$ *are planar duals of each other.*

We can think of the outside edge as representing the *outside room*—that is, the part of the plane that is outside the base rectangle.

Proof Consider the planar embeddings of $A_h(F)$ and $A_v(F)$ just described, and enhanced with the added outside edges. Because F is normal, no two maximal walls cross in F. Thus, the edges in $A_h(F)$ representing rooms that are adjacent to the same maximal horizontal wall—say, s—form a face. This face corresponds to the vertex in $A_v(F)$ that represents s. The edge bijection between the two dual graphs maps the edges for room r in both graphs onto each other, and maps the two outside edges onto each other (see Figure 7.11(b)). $\Box$

An inspection of Figure 7.8(a) reveals that Lemma 7.1 does not hold if F is not normal.

7.2.1.1.2 Slicing Floorplans So far, we have discussed OM-similarity of arbitrary floorplans. Arbitrary floorplans can be generated with normal cut methods only if we use multiway partitions that result in high-degree cut trees. As described, the mincut method is based on bipartitioning. This fact properly restricts the set of floorplans that can be generated. In fact, the floorplan depicted in Figure 7.11(a) cannot be generated with mincut bipartitioning. An inductive argument shows that all floorplans that can be generated with mincut bipartitioning are *slicing floorplans* in the following sense (see Exercise 7.13).

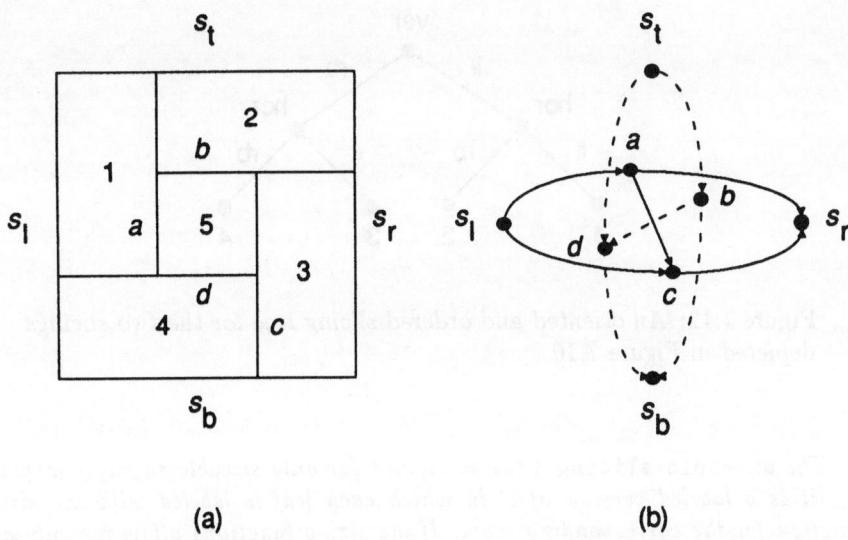

(a)   **(b)**

Figure 7.11: *A wheel floorplan and its directional adjacency graphs. (a) The floorplan. Rooms are labeled with numbers, cutlines (maximal walls) are labeled with letters. (b) The directional adjacency graphs. The edges of $A_h(F)$ are solid lines; the edges of $A_v(F)$ are dashed lines.*

Definition 7.7 (Slicing Floorplan) *A* **slicing floorplan** *is a normal floorplan F such that $A_h(F)$ and $A_v(F)$ are dual series-parallel graphs.*

The cut tree of a mincut algorithm that generates the slicing floorplan is a binary hierarchy tree for the two series-parallel graphs $A_h(F)$ and $A_v(F)$ (see Section 3.14.3). In this case, an OM-topology can be alternatively described by a labeled version of the cut tree, instead by a pair of directional adjacency graphs.

Definition 7.8 (Slicing Tree) *The* **slicing tree** *T of a slicing floorplan F is a binary tree in which each leaf represents a room and each interior node represents a maximal wall.*

    The **oriented slicing tree** *is a labeled version of T in which each interior node receives an* **orientation label** *from the set {hor, ver}, depending on whether the associated cutline runs horizontal or vertical.*

    The **ordered slicing tree** *is a labeled version of T in which each edge receives an* **ordering label** *from the set {lt, rb}, depending on whether the room to which it points lies to the left/on top or to the right/below of the cutline represented by the node from which the edge diverges.*

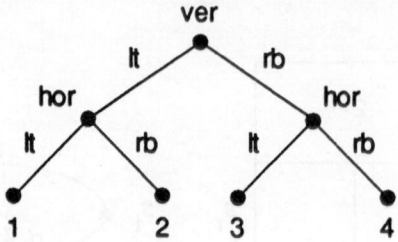

Figure 7.12: *An oriented and ordered slicing tree for the two slicings depicted in Figure 7.10*

The **sizeable slicing tree** *is defined for only sizeable slicing floorplans F. It is a labeled version of T in which each leaf is labeled with the sizing function for the corresponding room. If the sizing functions allow for only one layout alternative per room, then the slicing tree is called* **sized**.

Figure 7.12 illustrates the notion of a slicing tree.

An OM-topology of slicing floorplans is described by an oriented, ordered, and sizeable slicing tree T. From T, we can construct the directional adjacency graphs $A_h(F)$ and $A_v(F)$. The orientation labels govern what type of composition should be used at each interior node of T. In $A_h(F)$, the label hor translates to the parallel composition; the label ver translates to the vertical composition. In $A_v(F)$, the relationship is reversed. The ordering labels determine what graphs to put on what side of an operation.

7.2.1.2   Floorplan Sizing

In this section, we discuss how to size an OM-topology. We first discuss the case that all cells are fixed; that is, only one layout alternative exists per block. Then, we consider the case that cells can have variable shapes. Note that we discuss the sizing of general floorplans—not just slicing floorplans—unless otherwise specified.

7.2.1.2.1   Sizing of Fixed Cells
In this case, each cell has a specified height and width. Let us denote the height of the cell for room v with $h(v)$ and its width with $w(v)$. We assume that cells may not be rotated during the placement. This means that the sizing function for each room is constant (see Figure 1.15(a)). We find the optimal configuration (F', ℓ, a) for the FLOORPLAN SIZING problem as follows.

Choosing a is easy. The choice function that assigns to each room v the minimum point in the domain of s_v suffices.

To determine F' and ℓ, we label the edge for room v in $A_h(F)$ and $A_v(F)$ with $-w(v)$ and $-h(v)$. In the following discussion, we shall concentrate on $A_h(F)$. Analogous facts hold for $A_v(F)$.

We solve the single-source shortest-path problem on $A_h(F)$ (weighted as before) with source s_ℓ. The resulting distance function is $d(s)$, where s is a vertical maximal wall in F.

Lemma 7.2 *If all segments of the vertical maximal wall s in F receive the x coordinate $-d(s)$, then a sized floorplan (F', ℓ) results such that*

1. *$F' \sim_{\text{OM}} F$.*

2. *ℓ is legal with respect to a.*

3. *The area of (F', ℓ) is minimum, subject to part 1.*

Proof Requirement (1) holds trivially, since we do not change $A_h(F)$ and $A_v(F)$. By the triangle inequality for shortest-path distance functions (see Lemma 3.4, part 3), each cell fits into its room. Thus, requirement (2) holds. The cells on a shortest path from s_ℓ to s_r fit into their rooms tightly. Thus, the width of the base rectangle cannot be decreased without having some cells not fit horizontally into their rooms. Therefore, requirement (3) holds. □

Analogously, we can determine y coordinates for the horizontal segments that minimize the height of the base rectangle while fitting all cells vertically into their rooms. Since $A_h(F)$ and $A_v(F)$ are dags, the shortest-path problems are solvable in linear time (see Section 3.8.2).

This method is called the *critical-path method*.

7.2.1.2.2 Sizing of Variable Cells

In this section, we assume cells to be variable. For ease of presentation, we consider only sizing *step* functions. At the end of this section, we comment on how to extend the sizing method to other classes of sizing functions.

Let us first discuss what the length of the input is in this case. An instance of the FLOORPLAN SIZING problem consists of two directional adjacency graphs and sizing functions for the rooms represented by the edges in the graphs. The graphs can be defined in space $O(n)$ where n is the number of rooms—that is, blocks in the circuit. We can define a sizing step function by listing the coordinates of its steps, say, sorted according to increasing x coordinate. Let the total number of steps in all sizing functions be q $(q \geq n)$. Then, in the unit cost measure, the length of the input is $O(n + q)$.

One way of solving this variant of the FLOORPLAN SIZING problem would be to employ the method of Section 7.2.1.2.1 to each choice function. However, if the sizing function for room r has q_r steps, then there are $\prod_{v \in V} q_v$ different choice functions. This number is too large for an efficient algorithm. For instance, if each sizing function has two steps, such as is the case if the blocks are nonsquare and rotatable, then the number of choice functions is 2^n. Indeed, Stockmeyer [429] has shown that the FLOORPLAN SIZING problem based on OM-similarity and restricted to sizing functions of rotatable fixed cells is NP-hard.

There is a special case in which this problem can be solved efficiently, however; namely, if we restrict floorplans to be slicing floorplans. In this case, an OM-topology is specified by a sizeable oriented and ordered slicing tree T. The algorithm works in two passes.

In the first pass, T is processed bottom-up. Sizing functions are computed for all interior nodes of the tree. These nodes represent rectangular subfloorplans. Let z be an interior node of T, and let z_1 and z_2 be the two children of z. The sizing function s_z for z is computed from the sizing functions s_{z_1} and s_{z_2} for the two children of z. Two cases must be distinguished.

Case 1: z represents a horizontal cutline. In this case, we can obtain s_z by summing s_{z_1} and s_{z_2} on the intersection of their domains (Figure 7.13).

Case 2: z represents a vertical cutline. In this case, the inverse of s_z is the sum of the inverses of s_{z_1} and s_{z_2} on the intersection of their domains. (Flip Figure 7.13 around the line $y = x$.)

This process is a parameterized version of the critical-path method introduced in Section 7.2.1.2.1 applied to slicing floorplans.

Lemma 7.3 *The first pass computes a sizing function s_z for the root z of T such that the following facts hold:*

1. *Let a be an arbitrary choice function for the rooms, and let (F', ℓ, a) be a configuration that minimizes the floorplan-size measure using a. Let the width and height of (F', ℓ) be w and h. Then, $h \geq s_z(w)$.*

2. *Let w be a value in the domain of s_z. Then, there is a choice function a for the rooms and a corresponding sized floorplan (F', ℓ) such that ℓ is a legal sizing for a, the floorplan's width is w, and its height is h.*

Part 1 of Lemma 7.3 states that the sizing function s_z is *correct*; that is, it does not include illegal sized floorplans. Part 2 of Lemma 7.3 states that s_z is *exact*; that is, it does not exclude legal sized floorplans.

Proof The proof is by induction on the height of T. The base case follows from Definition 7.3.

Now assume that the root z has two children z_1 and z_2. Further assume, without loss of generality, that z represents a horizontal cutline. A sized floorplan (F, ℓ) for z with width at most w and height at most h can be composed from two sized floorplans (F_1, ℓ_1) for z_1 and (F_2, ℓ_2) for z_2 whose heights add up to at most h and whose maximum width is at most w. Since sizing functions are monotonically decreasing, we achieve the smallest height for width w if the width of the floorplans for z_1 and z_2 is w. The inductive hypothesis states that

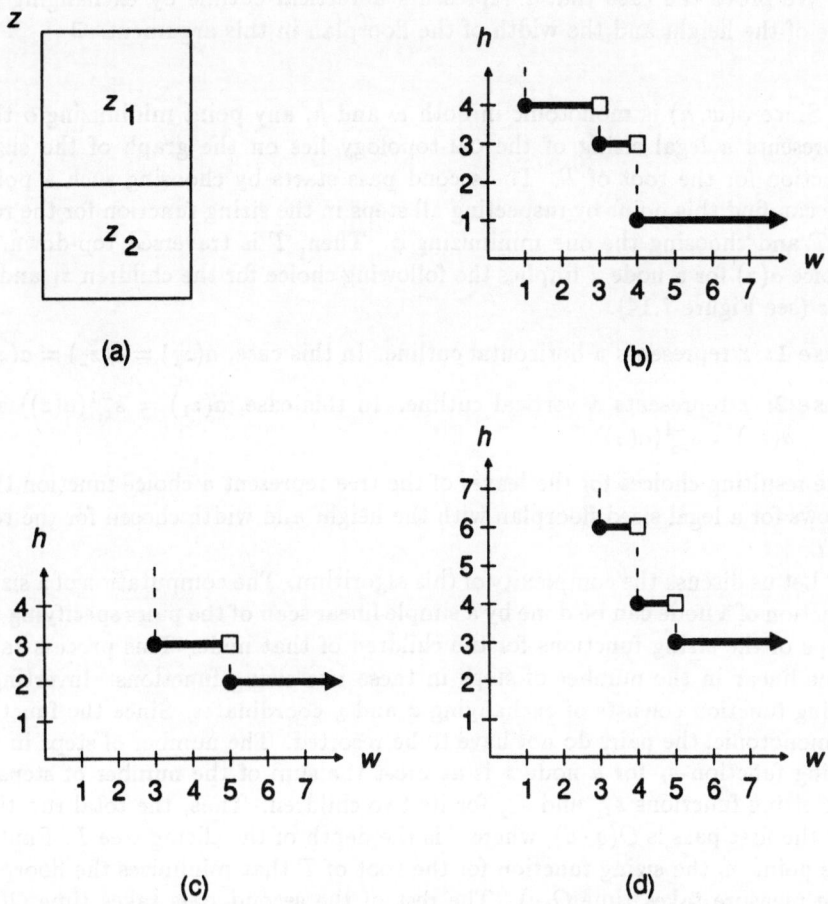

Figure 7.13: *Computing the sizing function of a node representing a horizontal cutline. (a) Floorplan diagram; (b) sizing function for z_1; (c) sizing function for z_2; (d) sizing function for z.*

the height of F_1 and F_2 can be made as small as $s_{z_1}(w)$ and $s_{z_2}(w)$, respectively, but not smaller. Thus, the height of F can be decreased to $s_{z_1}(w) + s_{z_2}(w)$, but not smaller.

We prove the case that z represents a vertical cutline by exchanging the role of the height and the width of the floorplan in this argument. $\square$

Since $\phi(w, h)$ is monotonic in both w and h, any point minimizing ϕ that represents a legal sizing of the OM-topology lies on the graph of the sizing function for the root of T. The second pass starts by choosing such a point. We can find this point by inspecting all steps in the sizing function for the root of T and choosing the one minimizing ϕ. Then, T is traversed top-down. A choice $a(z)$ for a node z implies the following choice for the children z_1 and z_2 of z (see Figure 7.14).

Case 1: z represents a horizontal cutline. In this case, $a(z_1) = a(z_2) = a(z)$.

Case 2: z represents a vertical cutline. In this case, $a(z_1) = s_{z_1}^{-1}(a(z))$ and $a(z_2) = s_{z_2}^{-1}(a(z))$.

The resulting choices for the leaves of the tree represent a choice function that allows for a legal sized floorplan with the height and width chosen for the root of T.

Let us discuss the complexity of this algorithm. The computation of a sizing function of a node can be done by a simple linear scan of the pairs specifying the steps of the sizing functions for the children of that node. This process takes time linear in the number of steps in these two sizing functions. Inverting a sizing function consists of exchanging x and y coordinates. Since the function is monotonic, the pairs do not have to be resorted. The number of steps in the sizing function s_z for a node z is at most the sum of the number of steps in the sizing functions s_{z_1} and s_{z_2} for its two children. Thus, the total run time for the first pass is $O(q \cdot d)$, where d is the depth of the slicing tree T. Finding the point on the sizing function for the root of T that minimizes the floorplan size measure takes time $O(q)$. The rest of the second pass takes time $O(n)$. Thus, the total run time is $O(q \cdot d)$; that is, it is $O(q \cdot n)$ in the worst case, and $O(q \cdot \log n)$ if the oriented slicing tree is node-balanced, such as occurs if the mincut method is used on a circuit with blocks that have roughly equal area estimates.

The algorithm does not depend on the fact that the sizing functions are step functions. The only requirement is that adding and inverting sizing functions should be simple. This requirement also holds, say, for piecewise linear functions. We could represent such functions by listing their points of discontinuity (*breakpoints*) and the slopes of the line segments starting in these points. Adding two functions consists of scanning the list of breakpoints ordered according to their x coordinates, computing the respective y coordinates, and adding slopes. We can invert a function by exchanging x and y coordinates of breakpoints and inverting the slopes. These operations take linear time in the number of breakpoints if we can multiply and divide in one step. Note, however, that we can get rational numbers with large denominators this way, even

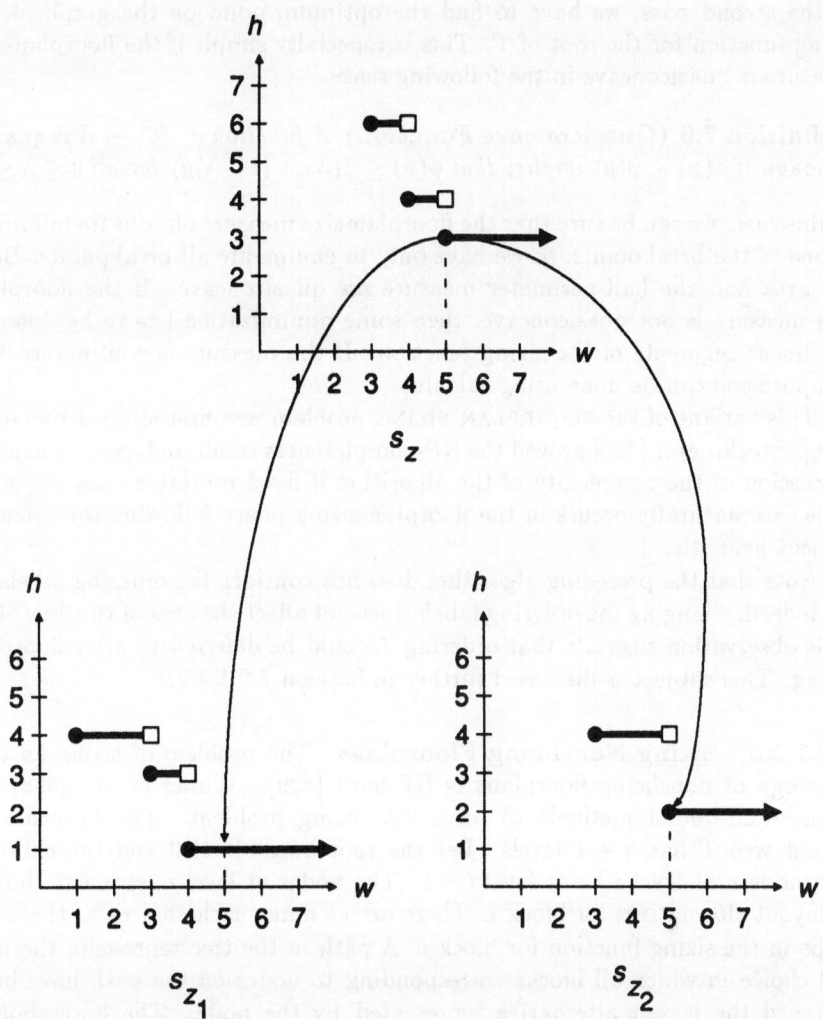

Figure 7.14: *Propagating a choice of minimum area for a horizontal cutline. For a horizontal cutline, we propagate the x coordinates: $a(z) = a(z_1) = a(z_2) = 5$. You can see what happens if the cutline is vertical by rotating by 90 degrees the floorplan in Figure 7.13(a) and all sizing functions. In this case, we must propagate the y coordinates in this figure: $a(z) = 3, a(z_1) = 1, a(z_2) = 2$.*

if we start out with integers. Thus, arithmetic precision may become an issue. In the second pass, we have to find the optimum point on the graph of the sizing function for the root of T. This is especially simple if the floorplan-size measure is quasiconcave in the following sense.

Definition 7.9 (Quasiconcave Function) *A function $\phi : \mathbf{R}^n \to \mathbf{R}$ is* quasi-concave *if $\phi(x) \leq \phi(y)$ implies that $\phi(x) \leq \phi(\lambda x + (1 - \lambda)y)$ for all $0 \leq \lambda \leq 1$.*

In this case, we can be sure that the floorplan-size measure obtains its minimum in one of the breakpoints, so we have only to enumerate all breakpoints. Both the area and the half-perimeter measure are quasiconcave. If the floorplan-size measure is not quasiconcave, then some minimization has to be done on the linear segments of the sizing function. If the measure is continuous, this computation can be done using calculus.

This variant of the FLOORPLAN SIZING problem was first studied by Otten [346]. Stockmeyer [429] proved the NP-completeness result and gave an explicit discussion of the complexity of the algorithm if fixed rotatable cells are used. This case naturally occurs in the floorplan-sizing phase following the oriented mincut heuristic.

Note that the preceding algorithm does not consider the ordering labels on T. Indeed, changing the ordering labels does not affect the area of the floorplan. This observation suggests that ordering T could be deferred to *after* floorplan sizing. This subject is discussed further in Section 7.2.1.3.2.

7.2.1.2.3 Sizing Nonslicing Floorplans

The problem of sizing an OM-topology of nonslicing floorplans is NP-hard [429]. Wimer et al. [468] use branch-and-bound methods to solve this sizing problem. The branch-and-bound tree T has $n + 1$ levels. Let the root be at level 0 and the children of a node v at level i be at level $i + 1$. The nodes at level i represent choices of layout alternatives for block i. There are as many such choices as there are steps in the sizing function for block i. A path in the tree represents the partial choice in which all blocks corresponding to nodes on the path have been assigned the layout alternative represented by the node. The lower bound for a partial choice function is the area of the smallest floorplan that accommodates the selected layout alternatives for the blocks on the path and some ficticious minimal layout alternative for all other blocks. This ficticious layout alternative has the minimum height and width possible for the block. Computing the lower bound involves shortest-path computations in the two respective directional adjacency graphs (Section 7.2.1.2.1). These graphs are called *lower-bound adjacency graphs*. Similarly, we can compute upper bounds for the nodes of T (see Exercise 7.15).

How should we assign the blocks to levels? A natural choice is to process the largest blocks first. Here, largeness could be represented by some area estimate drawn from the sizing function.

If general sizing step functions are used, the branch-and-bound tree has such a high degree that we have to use a stack as the data structure D for space reasons, and the children of interior nodes in the branch-and-bound tree have to be generated incrementally (see Section 4.6.2.3). Therefore, the ordering in which the children of a node in T representing the layout alternatives for a block are generated is of importance. The following order is a suitable choice.

Definition 7.10 (Increasingly Interlaced Order) *A sequence* $(w_1, h_1), \ldots,$ (w_n, h_n) *is said to be in* **increasingly interlaced order** *if* $w_i \leq w_j$ *for all odd i and $j > i$, and $h_i \leq h_j$ for all even i and $j > i$.*

The sequence $(1, 64), (64, 1), (2, 31), (31, 2), (3, 20), (20, 3), (4, 13), (13, 4), (7, 7)$ is in increasingly interlaced order. In a sequence in increasingly interlaced order, the highly nonsquare layout alternatives occur before the more squarelike layout alternatives. If the layout alternatives of the blocks are in increasingly interlaced order at every node in T, then we can prove an efficient criterion for pruning the branch-and-bound tree (see Exercise 7.16).

Often, a nonslicing floorplan deviates only slightly from a slicing floorplan. We can find the parts of a nonslicing floorplan that have the form of slicing floorplans by decomposing the adjacency graphs into their triconnected components (see Section 3.5). Then, the more efficient algorithm of Section 7.2.1.2.2 can be used on the slicing parts, and the more inefficient branch-and-bound method needs to be used on only the nonslicing parts of the floorplan.

Wong and Sakhamuri [474] present a heuristic scheme for sizing nonslicing floorplans. However, this algorithm only approximates the true sizing function.

In general, the floorplan-sizing problem is a QUADRATICALLY CONSTRAINED QUADRATIC PROGRAM (see Section 4.8). Rosenberg [383] exploits the special structure of this problem to apply efficient nonlinear optimization techniques to it.

7.2.1.3   Floorplanning Based on Unoriented Mincut

The oriented-mincut method orients and orders a sized slicing tree heuristically. The method may yield quite suboptimal results, because there is an intricate interplay between cell shape and cell size that the common heuristics for assigning orientations to cutlines do not address. However, the formulation of the FLOORPLAN SIZING problem and the results in Section 7.2.1.2.2 suggest that we modify the definition of OM-similarity to reflect a variant of the mincut heuristic, in which the orientation of the cutlines is done not while the cut tree is generated, but rather only in the second phase of the floorplanning algorithm—that is, when floorplan sizing takes place. The reason why we can do wait until the second phase is that the orientation of cutlines does not affect the cutsizes of any interior nodes of the slicing tree. And these cutsizes are our measure of connection cost in mincut methods. The resulting more powerful approach to mincut floorplanning is discussed in this section.

The basic idea is not to do any orientation of the cutlines during the first phase of the floorplanning algorithm, which generates the slicing tree. In the corresponding notion of floorplan similarity, the structure defining a floorplan topology is a slicing tree without orientation or ordering labels.

Definition 7.11 (UM-Similarity) *Two sizeable slicing floorplans are called* UM-similar *if they have the same sizeable but unoriented and unordered slicing tree. The corresponding similarity relation is denoted with* $\sim_{UM}$. *The respective floorplan topologies are called* UM-topologies.

The term UM-*topology* is short for *unoriented mincut* topology. A UM-topology is a union of several OM-topologies; thus, floorplan sizing based on UM-similarity is more powerful than is floorplan sizing based on OM-similarity. Given a circuit G, we can generate an unoriented slicing tree T defining a UM-topology for G by doing the recursive mincut process described in Section 7.2.1.1 but disregarding the orientation of the cutlines.

7.2.1.3.1 Sizing Based on Unoriented Mincut

We shall now extend the algorithm in Section 7.2.1.2.2 to size UM-topologies. The corresponding sizing procedure computes orientation labels for the slicing tree and a choice function for the leaves of the slicing tree (that is, it *sizes* the slicing tree). However, it does not yet compute ordering labels. Ordering labels cannot be determined at this stage, because all slicing floorplans that have the same oriented sized slicing tree have the same height and width independent of their ordering labels. Thus, an optimization of $\phi(w, h)$ cannot distinguish between different ordering labels. Ordering labels are assigned in Section 7.2.1.3.2.

We keep the main idea of composing exact sizing functions for the interior node z of T from exact sizing functions for the children z_1 and z_2 of z. This time, however, we do not know the orientation of the cutline represented by the interior node; thus, we have to consider both possibilities. Denote the sizing function resulting if the cutline is horizontal with $s_{z,h}$, and define $s_{z,v}$ analogously. The functions $s_{z,h}$ and $s_{z,v}$ can be computed as described in Section 7.2.1.2.2. The sizing function s_z is then defined on the union of the domains of $s_{z,h}$ and $s_{z,v}$ as

$$s_z(w) := \min(s_{z,h}(w), s_{z,v}(w))$$

An argument analogous to the proof of Lemma 7.3 shows that s_z is a correct and exact sizing function for the subfloorplan corresponding to the node z. In the first pass, we compute $s_z(w)$ for each node z of the slicing tree in a bottom-up fashion.

In the second pass, we again determine choices top-down. This time, choosing a point w on the abscissa of the sizing function s_z entails making a choice of an orientation of the cutline represented by z—namely, the orientation that achieves the minimum of $s_{z,h}(w)$ and $s_{z,v}(w)$. Which orientation this is can be stored during the computation of s_z from $s_{z,h}$ and $s_{z,v}$, using appropriate pointers.

The complexity of this algorithm is higher than that in the oriented case, because this time the number of steps of s_z can be as large as *twice* the sum of the number of steps of s_{z_1} and s_{z_2}. The additional factor of 2 arises from the minimization over the two orientations, and it can actually occur (Figure 7.15). Thus, in general, the run time is $O(q \cdot 2^d)$. This quantity is exponential in $n + q$ if the slicing tree has a large height. In fact, Lengauer and Müller [280] have shown that the threshold version of this variant of the FLOORPLAN SIZING problem is weakly NP-complete if sizing step functions are used. However, as we noted in Section 7.2.1.2.1 we can often assume the slicing tree to be node-balanced—that is, to have a height of $O(\log n)$. In this case, the run time still grows polynomial in $q \cdot n$. The degree of the polynomial depends on the constant factor in the height of the slicing tree. This fact makes it especially desirable to generate shallow slicing trees. We can achieve this goal by making blocks approximately the same size and by balancing the cuts as much as possible.

Even if the slicing tree is not balanced, there is a way of making this algorithm efficient in practice. This observation is based on the fact that the problem is only *weakly* NP-complete. If sizing functions are step functions, then the maximum number q_{max} of steps that a sizing function for an interior node of T can have is related to the size of the largest number L occurring in the description of a sizing function. In fact, $q_{max} \leq n \cdot L$. Thus, the algorithm runs in time $O(n^2 L)$. If L is large, this algorithm is quite inefficient. In this case, we can eliminate steps from the step function to speed up the sizing operation. But this method can only approximate the minimum-area floorplan, since the sizing functions are no longer exact. Instead of uniformly deleting, say, every other step, we can develop heuristics that eliminate steps from the step function selectively in places where we do not need high resolution (see Exercise 7.17).

The sizing of UM-topologies also works with sizing functions that are not step functions, but the run-time analysis has to be modified. For instance, in the case that sizing functions are piecewise linear, taking the minimum of two sizing functions can result in *quadrupling* the number of breakpoints (Figure 7.16).

Floorplan sizing based on UM-topologies was introduced in Zimmermann [484] and analyzed in [280].

7.2.1.3.2 Ordering Mincut Floorplans

The approach described in Section 7.2.1.3.1 does not yet determine a unique minimum-area floorplan of a UM-topology. Specifically, no ordering labels are computed. (We have seen that these labels do not affect the floorplan-size measure.) Thus, in effect, the UM-topology is reduced to the following small set of floorplans (see Figure 7.17).

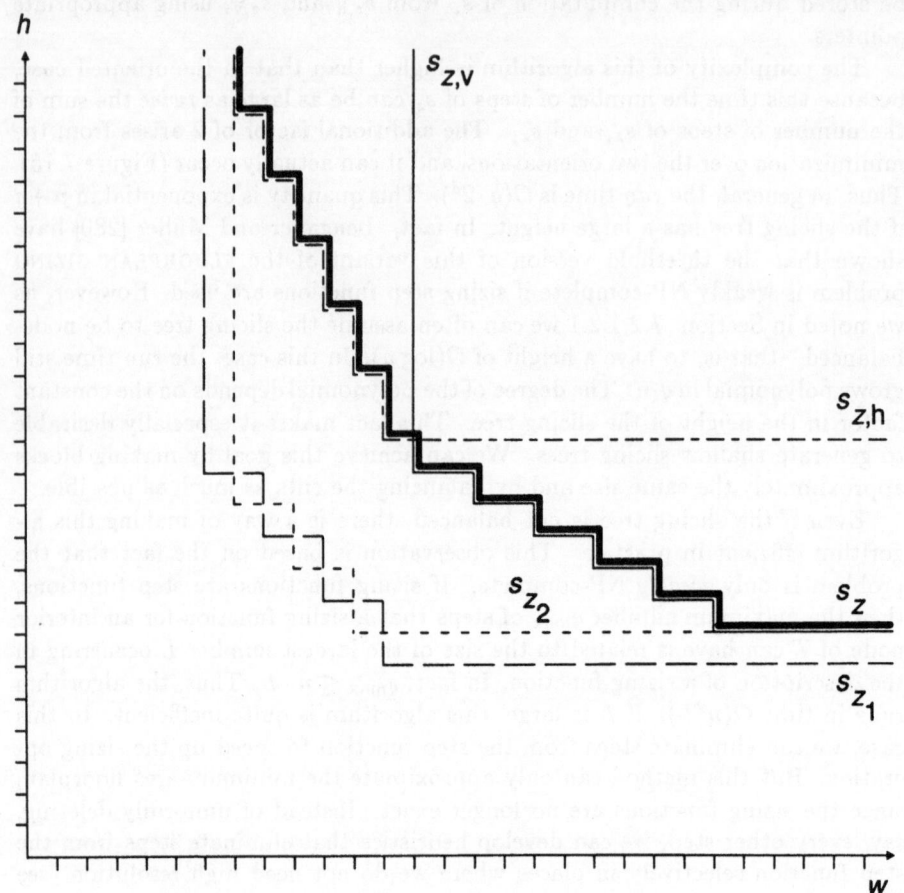

Sizing function	Number of steps
s_{z_1}	4
s_{z_2}	3
$s_{z,h}$	6
$s_{z,v}$	6
s_z	12

Figure 7.15: *Doubling the number of steps in a minimization. The step functions are drawn in a somewhat imprecise fashion, including vertical segments in the function graphs. This representation has been chosen to make the step function more easily distinguishable.*

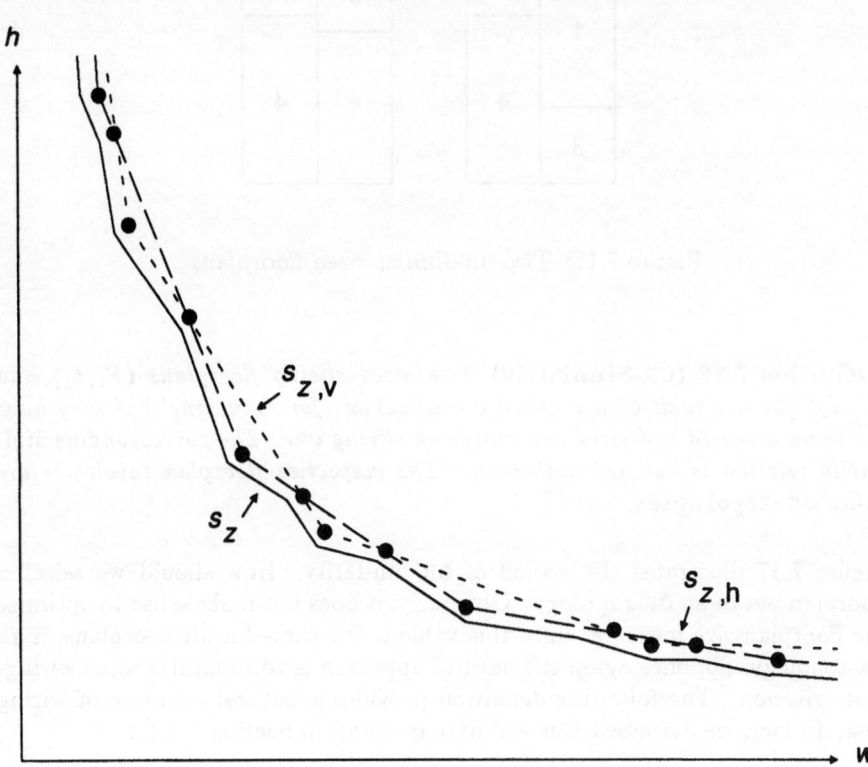

Sizing function	Number of breakpoints
$s_{z,h}$	4
$s_{z,v}$	3
s_z	13

Figure 7.16: *The minimum of two piecewise linear functions during the minimization. The number of breakpoints can quadruple. In general, if $s_{z,h}$ and $s_{z,v}$ have q and $q-1$ breakpoints; then s_z can have as many as $4q-3$ breakpoints.*

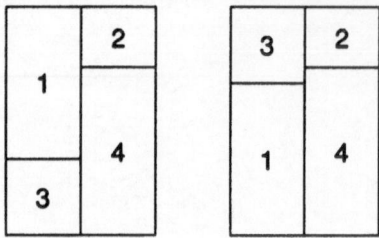

Figure 7.17: *Two* OR-*similar sized floorplans.*

Definition 7.12 (OR-Similarity) *Two sized slicing floorplans* (F_1, ℓ_1) *and* (F_2, ℓ_2) *for a circuit* G *are called* OR-**similar** *(for "ordering") if they have the same oriented and sized but unordered slicing tree. The corresponding similarity relation is denoted with* $\sim_{OR}$. *The respective floorplan topologies are called* OR-**topologies**.

Figure 7.17 illustrates the notion of OR-similarity. How should we select a floorplan out of an OR-topology? Obviously, it does not make sense to optimize the floorplan-size measure, since this value is the same for all floorplans in an OR-topology. So, once again the natural approach is to minimize some wiring-cost criterion. The following definition provides a natural estimate of wiring cost. In fact, we discussed this estimate in detail in Section 7.1.1.1.

Definition 7.13 (Wiring Cost) *Let* F *be a sized floorplan for a circuit* G. *The* **wiring cost** $c_e(F, G)$ *of a net* e *of* G *in floorplan* F *is the half-perimeter of the smallest rectangle containing the rooms for all blocks that are terminals of* e.

The **wiring cost** *of* (F, G) *is* $c(F, G) := \sum_{e \in E} c_e(F, G)$.

The structure of the problem of minimizing wiring cost after sizing is similar to the structure of the FLOORPLAN SIZING problem. We just use a different cost function and a new similarity relation—namely, OR-similarity. Specifically, we define the SLICING ORDERING problem as the variant of the FLOORPLAN SIZING problem, in which only slicing floorplans are considered, $\sim_{OR}$ is used as the similarity relation, and the cost measure minimized is $c(F, G)$.

Lengauer and Müller [281] prove that the SLICING ORDERING problem is NP-hard, even if the slicing tree is required to be a complete binary tree—that is, if it must be completely balanced. However, there are powerful heuristics that integrate the solution of this problem with hierarchical global routing (see Section 8.5).

7.2.2* Floorplanning Based on Multiway Mincut and Clustering

Multiway mincut and clustering methods both generate cut trees of degree higher than 2. Multiway mincut does the generation top down. Clustering does the generation bottom up. A node in the cut tree now represents not a bipartition but a k-way partition, with $k \geq 2$ (see Chapter 6). Figure 7.18 shows a cut tree and a corresponding floorplan that can be generated with multiway mincut or clustering methods.

7.2.2.1* Floorplanning by Clustering

The partitioning methods used in clustering are discussed in Section 6.7. We generate a cut tree by clustering the blocks of the circuit into small groups, which then are considered as supernodes and are clustered again. Each time we perform clustering, the internal edge cost is maximized; that is, the cutsize is effectively minimized. Clustering does have the disadvantage compared to multiway cut that preferences or requirements on locations of certain blocks cannot be incorporated elegantly. The reason is that clustering, working bottom-up, fixes affinities between blocks without any global knowledge of the connections in the circuit. On the other hand, with clustering we have a more direct control over block adjacencies in the floorplan. Specifically, clustering methods can group together blocks not only on the basis of interconnection cost, but also on the basis of how their shapes fit together. Thus, the bottom-up development of the cut tree and the computation of the sizing functions can influence each other. For this to happen, clustering must be based on the sizing functions in addition to connection cost; that is, the methods described in Section 6.7 have to be modified to take into account the sizing functions as a parameter influencing the cost measure.

The cut tree defines a natural *multiway similarity* or MW-*similarity* relation $\sim_{MW}$ on sizeable floorplans. We describe $\sim_{MW}$ informally. Two sizeable floorplans are multiway similar if we can build up both of them by processing the *same* cut tree bottom-up and composing clusters of already created subfloorplans in patterns out of a prescribed set of *admissible floorplan patterns*. Here, the set of allowed floorplan patterns for a cluster is the second ingredient, besides the cut tree, that defines an MW-topology.

In Figure 7.18, floorplan patterns label the nodes of the cut tree. The set of admissible floorplan patterns is the set of floorplan patterns from which these labels can be taken.

If the cut tree is generated bottom-up by clustering methods, we can integrate aspects of floorplan sizing into the generation of the cut tree. In fact, the generation of the cut tree, and the selection of floorplan patterns for each interior node of the tree, can be regarded as a variant of the FLOORPLAN SIZING problem. The floorplan-size measure in this case also incorporates connection

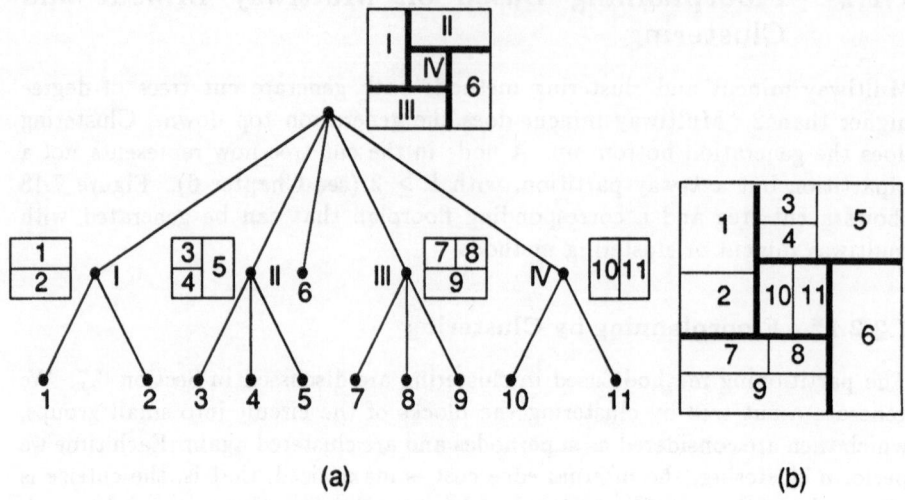

Figure 7.18: *Floorplanning by multiway partitioning. (a) A multiway cut tree; (b) the corresponding floorplan.*

cost aspects. This provision is necessary, because the choice of a suitable floorplan pattern for a cluster must also be made on the basis of connection cost.

The degree of the cut tree must be kept low for several reasons. First, the sizes of MW-topologies increase dramatically with the cluster size, and thus the sizing problem becomes increasingly complex. Second, as the cluster size rises, connection-cost aspects become increasingly dominant, and the floorplan sizing problem becomes as complex as the general floorplanning problem. Usually, the maximum cluster size is limited to about 5.

7.2.2.2* Floorplanning by Multiway Mincut

We now describe the natural extension of the sizing method for unoriented mincut presented in Section 7.2.1.3.1 to multiway mincut. We can again understand it as a procedure that labels the cut tree with admissible floorplan patterns together with assignments of blocks (represented by sizing functions) to rooms in these floorplan patterns. Whereas in Section 7.2.1.3.1 a minimum of *only two* sizing functions had to be computed per node, now *each possible floorplan pattern* and room assignment contributes a sizing function that has to be incorporated in the minimization.

Incorporating rotational and reflectional symmetry reduces the number of possibilities to be checked. With this observation for two blocks, just one possible floorplan pattern remains. For three blocks, the number of floorplan patterns rises to six. For four blocks, the number of floorplan patterns rises dramatically to 72. As the number of rooms increases the number of admissible floorplan patterns rises sharply (see Exercise 7.18).

We can reduce further the time for processing admissible floorplan patterns by noting that each floorplan pattern with up to four rooms is a slicing floorplan. Therefore, the respective nodes in the cut tree can be expanded into binary subtrees. Using this observation the methods discussed in Section 7.2.1.3 can be extended to handle floorplans with cut trees with degree up to 4.

As the number of rooms increases to and beyond five, we also get nonslicing floorplans, such as the floorplan depicted in Figure 7.11. Besides the fact that the number of admissible floorplan patterns increases sharply with the number of rooms, it is difficult to compute sizing functions for nonslicing floorplan patterns based on sizing functions for the block. The sizing method discussed in Section 7.2.1.2.3 is a candidate, but if the cluster size is small, there may be specially tailored, more efficient sizing methods. (see Section 7.2.1.2.3).

As the node degree in the cut tree increases, it is useful to decompose floorplan patterns into parts that are slicing floorplans and parts that are not. Performing this decomposition amounts to expanding the node in the cut tree into a subtree; therefore this technique enhances the efficiency of the algorithm (see, however, Exercise 7.19).

Lengauer and Müller [282] base floorplanning on these ideas.

7.2.2.3* Combining the Methods

The preceding discussion shows that clustering is a preferable approach for generating the bottom part of the cut tree. With clustering, block adjacencies can be controlled well. On the other hand, mincut partitioning is the suitable approach for generating the top part of the cut tree, since mincut can incorporate preplaced components and has a global view of the circuit. Thus, it is desirable to combine both strategies. Efforts in this direction are presented in [282]. Dai and Kuh [80] present a heuristic floorplanning method that generates a multiway cut tree bottom up by clustering and perform the selection of the floorplan patterns top down. A mixed top-down and bottom-up strategy is applied to floorplan orientation in [105]. There the cut tree is generated bottom up. Neither of the latter two references uses sizing functions.

7.2.3* Combination of Mincut and Wire-Length Minimization in Floorplanning

Kleinhans et al. [235, 236] present a placement method that combines wire-length minimization with mincut techniques in a more natural way than does the two-pass procedure described in Section 7.1.5.2. They alternate cuts with wire-length–minimization steps.

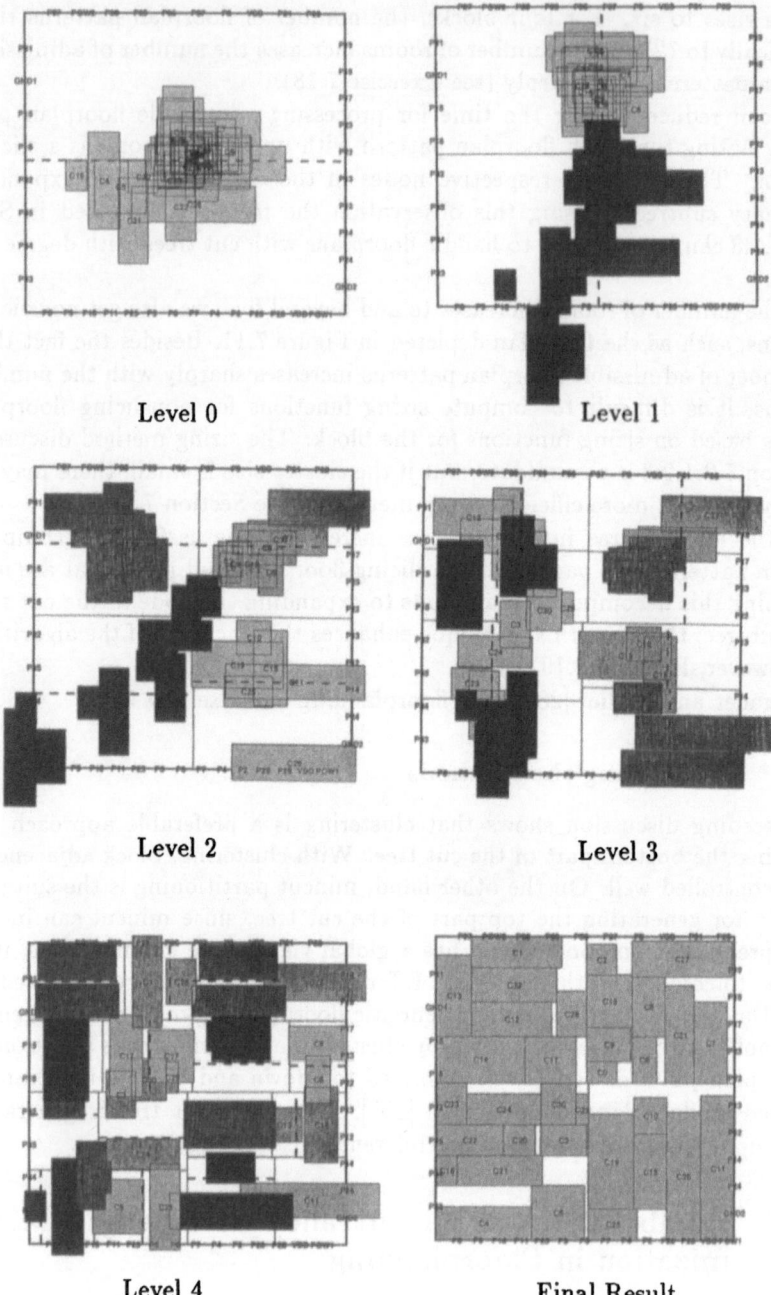

Level 0   Level 1

Level 2   Level 3

Level 4   Final Result

Figure 7.19: *A sequence of placements obtained with the method by Kleinhans et al. [235, 236].*

First, a relaxed placement is computed by a solution of the QUADRATIC RE-LAXED PLACEMENT problem. The resulting placement induces a cut through the floorplan that bisects the placement. This cut can be found as described in Section 7.1.5.2. The cut induces a second relaxed placement problem. This problem attempts to minimize the same cost function as the first one, but it imposes the additional constraints that the "center of gravity" of the vertex sets on either side of the cut reside in the center of the corresponding room. These constraints are all linear. The resulting placement is the basis of two additional cuts, one in each room. In this way we proceed, alternating wire-length minimizations with cuts until the complete floorplan is determined. Figure 7.19 shows a sequence of placements obtained by this method. The corresponding wire-length optimization problems minimize a quadratic positive semidefinite cost function with respect to a set of linear constraints. By Theorems 4.8 and 4.13, there is a unique global minimum for such an optimization problem.

This method has the advantage over pure cutting methods that blocks in different subfloorplans can still influence each other's location. Therefore, the lack of control over adjacencies in the floorplan has been eliminated. The resulting slicing floorplan can be subjected to the floorplan-sizing methods described in Section 7.2.1.1.

7.2.4* Floorplanning Based on Rectangular Dualization

Rectangular dualization is a method of floorplanning that is based on a concept of adjacency between rooms different from OM-similarity.

Definition 7.14 (Rectangular Dual) *Let F be a floorplan. The **rectangular adjacency graph** $A_r(F)$ for F is a plane graph (Definition 3.28) that has a vertex for each room and edges connecting two rooms that are adjacent to each other across a wall. The planar embedding is obtained by each vertex being put into its corresponding room and edges being drawn across their corresponding walls. F is called a **rectangular dual** of $A_r(F)$. Two edges in $A_r(F)$ and F that cross are called **dual edges**.*

*The equivalence relation $\sim_{RD}$ that relates two floorplans if they have the same rectangular adjacency graph is called RD-**similarity**; its floorplan topologies are called RD-**topologies**.*

Figure 7.20 shows two floorplans with the same rectangular adjacency graph. One advantage of RD-similarity over OM-similarity is that RD-similarity does not separate the dimensions of the floorplans. Thus, turning horizontal into vertical adjacencies and vice versa is possible. The definition of UM-similarity in Section 7.2.1.3 accomplishes the same goal, at least to some extent. On the other hand, UM-similar floorplans can have quite different appearances. The fact that RD-similarity is based on walls instead of maximal walls ensures that adjacencies between rectangles are kept invariant. This is not the case with OM-similarity or UM-similarity.

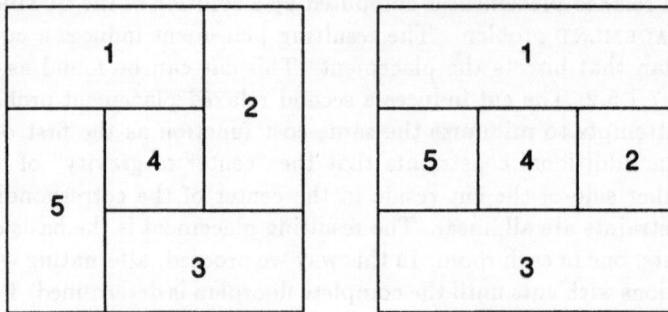

Figure 7.20: *Two floorplans with the same rectangular adjacency graph.*

Rectangular dualization first generates a rectangular adjacency graph that has small connection cost, and then sizes the corresponding RD-topology. The connection-cost measures suitable in this context are arithmetic combinations (sum, maximum, and so on) of the numbers $\#(s)$ for each wall s, where $\#(s)$ denotes the number of nets that run between the blocks that s separates. The method is not as well developed as is mincut-based floorplanning, and many steps are governed by crude heuristics.

7.2.4.1*   Planarization of the Circuit

Finding a suitable RD-topology is especially not well understood. This process involves turning the hypergraph representing the circuit into a plane graph such that some appropriate connection-cost measure is optimized. One way of doing this transformation is to turn the hypergraph into a graph, use some placement method minimizing wire length (Section 7.1.4), and then to planarize the resulting embedding by making vertices out of all crossings [16]. These new vertices represent new *routing blocks* that contain only wires. This heuristic has several problems. One is that the placement procedure minimizes total wire length, but it is not likely that this cost measure is the one we want to minimize with respect to the following steps. Furthermore, by modifying the edges of the embedding, we can achieve a large variety of distributions of edge crossings, and it is not clear what is the best distribution in this context. Lokanathan and Kinnen [294] present a heuristic method for planarizing the circuit that aims at minimizing the number of crossings. This problem is NP-hard, so no fast solution algorithm exists. More fundamentally, it is not clear that the number of crossings is the only parameter that governs wiring complexity. In short, there is no outstanding candidate for a cost function that we want to optimize in the first phase of floorplanning.

Another problem is that the plane graph generated by the first phase cannot be arbitrary, but it must have a rectangular dual. Kozminski and Kinnen [243] give a characterization of such graphs as well as a quadratic recognition algorithm. Bhasker and Sahni [32, 33] present a linear-time algorithm for deciding whether a planar graph has a rectangular dual, and if it does, for constructing one.

7.2.4.2* Floorplan Sizing

The methods for optimizing the RD-topology enumerate all rectangular duals of the plane graph corresponding to the RD-topology. For each dual, they do a sizing operation, such as the critical-path method described in Section 7.2.1.2.1 [185]. Kozminski and Kinnen [244] describe a branch-and-bound algorithm that is more efficient on typical examples. This algorithm generates walls of the rectangular duals incrementally. If all cells are fixed, then we can again obtain lower bounds on the sizes of such partially specified rectangular duals by using the methods of Section 7.2.1.2.1. If some cells are variable, the sizing method discussed in Section 7.2.1.2.3 can be used.

7.2.5* Nonlinear Methods for General-Cell Placement

There are also methods for general-cell placement that are based on wiring-area estimates. We have already alluded to such methods in Section 7.1.5. There, methods were discussed that combine relaxed placement to minimize wire length with mincut methods to exclude overlap. We remarked that these methods naturally extend to general-cell placement.

Another set of methods for general-cell placement that is based on wire-length minimization formulates the placement problem as a general nonlinear optimization problem. Convexity of the cost function cannot be assumed, in general. Thus, general nonlinear optimization techniques, such as the penalty-function methods (Section 4.8.2), must be employed.

The idea of the formulation of the placement problem for general cells as a nonlinear optimization problem is to extend the QUADRATIC RELAXED PLACEMENT problem by incorporating other cost functions and constraints. Cost functions include wire length (such as used in the QUADRATIC RELAXED PLACEMENT problem), chip area (a nonconvex quadratic cost function), and possibly others. Constraints include size constraints and no-overlap constraints for the blocks (with different sizes), and possibly area and aspect-ratio constraints. Rotation of blocks can also be allowed. As an example, we detail the formulation of size and no-overlap constraints for a pair of blocks. Each block v is represented by four variables $x_{\ell,v}$, $x_{r,v}$, $x_{t,v}$ and $x_{b,v}$, representing the coordinates of its four sides. A block of width w and height h gives rise to the size constraints

$$x_{r,v} - x_{\ell,v} \geq w \quad x_{t,v} - x_{b,v} \geq h$$

The no-overlap constraint between two blocks v_1 and v_2 can be formulated as follows:

$$x_{\ell,v_1} - x_{r,v_2} \leq 0 \quad \text{or} \quad x_{\ell,v_2} - x_{r,v_1} \leq 0$$

and

$$x_{b,v_1} - x_{t,v_2} \leq 0 \quad \text{or} \quad x_{b,v_2} - x_{t,v_1} \leq 0$$

For each dimension, one alternative excludes the other. Thus, the constraints can also be written as

$$(x_{\ell,v_1} - x_{r,v_2})(x_{\ell,v_2} - x_{r,v_1}) \leq 0 \text{ or } (x_{b,v_1} - x_{t,v_2})(x_{b,v_2} - x_{t,v_1}) \leq 0$$

These two constraints can be combined into

$$\left(\max\{0, (x_{\ell,v_1} - x_{r,v_2})(x_{\ell,v_2} - x_{r,v_1})\} \cdot \max\{0, (x_{b,v_1} - x_{t,v_2})(x_{b,v_2} - x_{t,v_1})\}\right)^2 = 0$$

Note that the left-hand side of this constraint is continuously differentiable everywhere.

The formulation of general-cell placement as a nonlinear optimization problem is clearly a very general one. It allows us to incorporate all sorts of constraints and cost functions. On the other hand, the related solution methods are also quite general and do not exploit any special structure of the placement problem.

It is not obvious how we would handle general sizing functions in this approach. The sizing function $s(w) = A/w$ can be incorporated with a quadratic area constraint. However, incorporating multiple layout alternatives leads to high-degree polynomials in the constraints. This fact makes the optimization very difficult.

Formulations of general-cell placement as a nonlinear optimization problem that incorporate cell shapes and sizes have been suggested in [82, 187, 415].

7.2.6*   Estimation of Wiring Area in Floorplan Sizing

All floorplan-sizing methods that we have described so far take into account only the shape of blocks. The actual chip area has a large contribution of wiring that is not accounted for, and the wiring area does not need to be minimized by the choice function that minimizes the floorplan-size measure. Thus, wiring-area estimates have to be incorporated into the sizing process. Conceptually, we achieve this goal by adding a wiring-cost contribution on top of the sizing function for a block or node in the slicing tree. The question is how we determine exactly how much to add. Since the exact block shapes are not known during the bottom-up computation of the sizing function, we have to resort to methods of estimation. The problem is complicated further by the fact that, in soft cells, the pin locations are not known. We shall not discuss this subject in detail, but refer the reader to an overview on this subject [173].

Wiring-area estimation for the special case of sizing UM-topologies is discussed in [484]. In this case, the sizing process proceeds bottom-up in the slicing tree T; thus, data about the wiring between the blocks representing children of a node v in T can be gathered as the sizing function of v is computed. However, global wiring that uses the vicinity of a block to connect far-away blocks still has to be estimated statistically.

More accurate wiring area estimates are achieved if the global-routing phase is integrated with the floorplanning phase (see Section 8.8.1).

7.2.7* Other Issues in Floorplanning

In the preceding sections, we gave a somewhat restricted picture of the floorplanning problem. The rationale was to impose a mathematical structure on the problem around which we could classify the existing floorplanning techniques. In this section, we shall make up for this restriction, and shall comment on floorplanning techniques that extend the approaches discussed so far.

7.2.7.1* Rectagonal Blocks

There are several geometric restrictions that are problematic low in the hierarchy of the floorplan as, for instance, described by a slicing tree, but not higher up. One is the fact that the slicing tree be binary. Low in the hierarchy, the shapes of blocks may lead us to wheel floorplans (Figure 7.11), which are not slicing floorplans. But higher up, as subfloorplans have more and more layout alternatives, we can expect the subfloorplans to be roughly square, and the restriction to slicing floorplans is probably not significant. A similar observation applies to the restriction that blocks be rectangular. In the low levels of the cut tree, this restriction probably excludes some efficient layout alternatives, but higher up it makes less and less of a difference. On the other hand, none of the algorithms we described extend easily to rectagonal blocks. Preas and Chow [361] present a graph representation of the floorplan topology that allows for rectagonal blocks that arise from biting off one or more corners of a rectangular block. They define the respective directional adjacency graphs of the maximal walls in the floorplan, and also the rectangular adjacency graph. However, much of the structure that was present with rectangular blocks is now missing. For instance, the directional adjacency graphs are not duals of each other. Therefore, rather than basing optimization algorithms on this graph representation, it is just used as a data structure, which can be modified by local search heuristics or even by interactive designer intervention. Dai et al. [81] extend this idea to developing a data structure that conveniently describes both the topology *and* the geometry of the floorplan. Edge-labeled adjacency graphs can be used for this purpose, but a more explicit tilelike representation of the empty space between blocks is advantageous for the efficiency of dynamic updating operations.

7.2.7.2*   Information Flow in Floorplanning

The approach to floorplanning that we have described in Sections 7.2.1 to 7.2.6 is based on a two-step procedure. In the first step, the floorplan topology is generated. In the second step, floorplan sizing is carried out.

These two steps do not necessarily have to be done separately. As we mentioned in Section 7.2.2, when clustering methods are used, the generation of the floorplan topology can be done simultaneously with floorplan sizing. This method allows us to use sizing *and* connection-cost aspects in both floorplanning steps.

Finally, the following routing phase may partly invalidate the floorplan, because the routing area provided on the basis of wiring-area estimates may not be sufficient, or deviations of the actual wiring area from the estimates may change the floorplan topology. Thus, floorplanning may have to be reiterated after the routing has been completed. The source of this problem is, of course, that the cost functions optimized in floorplanning do not represent the actual routing costs sufficiently exactly.

There are several models of how to structure the information flow of floorplanning and the following global-routing phase. Breuer and Kumar [50] give a quite general model for this problem. Other approaches are presented in [80, 255, 372, 402, 434, 435].

The dynamic data structures for representing floorplan topology and geometry can be extended to incorporate routing information [81]. This extension allows for a large set of heuristics for integrated floorplanning and global routing based on local modifications.

We shall discuss the integration of floorplanning and global routing further in Section 8.8.

7.2.7.3*   Other Constraints and Cost Functions

We have viewed floorplanning as primarily a layout task. In fact, floorplanning can also be considered as a first rough planning stage in the whole *design* process. Some of the floorplanning methods we have discussed support this view. In general, sizing functions can represent layout alternatives for *different designs* of a block. For instance, if a block represents an adder, the sizing function can include layout alternatives that incorporate fast adders such as the carry-lookahead adder, and slower adders such as Manchester carry-chain adders [313]. Faster adders will usually take up more area than slower adders will. If each point on the graph of the sizing function is augmented with the necessary information about the relevant design alternative, then in the sizing procedure outlined in Sections 7.2.1.2.2, 7.2.1.2.3, and 7.2.1.3, the choice of a point on the graph of the sizing function for the whole chip, implies choices of design alternatives for all blocks. In this scenario, these choices are not made simply in terms of area, but rather are influenced by performance parameters as well. For this purpose, we need models for composing performance parameters, such as delays, of subcircuits to yield appropriate parameters for the composed circuit. Such models are provided by rigorous methods of hierarchical circuit design. Still, basing floorplanning on slicing trees is more restrictive than is desirable in some applications.

Nonlinear floorplanning (Section 7.2.5) allows for incorporation of quite general cost functions and constraints [360]. However, the properties of smoothness that many nonlinear optimization methods require restrict the flexibility of the floorplan topology and the accuracy with which it can be modeled.

As the flexibility of the floorplan geometry and topology as well as the degrees of freedom involved in the choice of design alternatives, cost functions, and constraints increase, the floorplanning problem becomes so complex that using artificial-intelligence techniques to solve it seems justified [3].

7.3 Assignment Problems

Assignment problems mostly occur in semicustom design styles such as gate-array or standard-cell design. The origin of assignment problems is that there are several equivalent *slots* on the chip that can receive a specific circuit element. The assignment problem is to assign to each circuit element a slot such that some cost function is minimized. Even with only this general description, we note the close relationship between assignment problems in the placement domain and the LINEAR ASSIGNMENT problem as well as the QUADRATIC ASSIGNMENT problem. The permutations that are asked for in these combinatorial optimization problems represent the assignment of circuit elements to slots in our application. Depending on whether the cost function only contains additive terms dependent on single circuit elements or also links pairs of circuit elements, the LINEAR ASSIGNMENT problem or the QUADRATIC ASSIGNMENT problem is the correct formalization.

There are two varieties of assignment problems in circuit layout: *gate assignment* and *pin assignment*. We shall discuss both version of the assignment problem in turn.

7.3.1 Gate Assignment

In gate assignment, the slots are library cells or portions thereof. The circuit elements are (small) blocks. We call these blocks *gates*. The gates implement simple functions, such as the logical AND, or a flip-flop. The gate-assignment problem asks us to assign the gates to suitable cells or cell portions. It occurs in all semicustom design styles.

7.3.1.1   Gate-Array Design

In gate-array design, the cells are prefabricated and are arranged on the master. The function to be performed by the gate can be implemented by certain cells or by cell portions on the master. The locations of these structures on the master are the set of legal locations for the gate. We have to select, for each gate, one location out of the set of legal locations for the gate. This selection has to be done so as to minimize a certain cost function.

Let us simplify the problem with the assumption that each cell on the master can accommodate exactly one gate. In this case, the gate-assignment problem is a generalization of the UNCONSTRAINED RECTANGULAR ARRANGE-MENT problem. In fact, the UNCONSTRAINED RECTANGULAR ARRANGEMENT problem is exactly the gate-assignment problem, if all slots are legal for all gates. In general, each gate can be assigned to only a certain subsets of the slots. This set augments the description of the problem instance.

GATE ASSIGNMENT

    Instance: A hypergraph $G = (V, E)$ with edge costs $\ell(e) \in \mathbf{R}_+$ for $e \in E$, $|V| = n$, constants $r, s \in \mathbf{N}$, $r \cdot s \geq n$ indicating the size of the master; for each $v \in V$, a set $S(v) \subseteq [1, r] \times [1, s]$ of *legal slots*

    Configurations: All *placements*—that is, injective mappings $p : V \rightarrow [1, r] \times [1, s]$—of circuit components into slots on the master. The slot into which a component is placed is called that component's *site*.

    Solutions: All placements such that, for all $v \in V$, we have $p(v) \in S(v)$

    Minimize: The cost function $c(p)$ in question

The close relationship between the GATE ASSIGNMENT problem and the UNCON-STRAINED RECTANGULAR ARRANGEMENT problem suggests that we modify techniques for solving the latter problem to solve the former problem. Indeed, this modification is possible.

Placement techniques that first solve the relaxed placement problem can be adapted as follows. The relaxed problem is solved as in the placement phase. During the elimination of overlap—that is, in the final phase that maps gates to slots—we account for the legal slots. It is reasonable to do elimination of overlap by linear assignment (Section 7.1.5.1); here, the cost of putting a gate in an illegal slot is set to infinity. Elimination of overlap by mincut methods (Section 7.1.5.2) is not adaptable to the GATE ASSIGNMENT problem. If the set of legal slots for a vertex is restricted to a certain local area on the master, linear constraints that are added to the relaxed placement problem can force the gate to lie within this area after the relaxed placement.

The linear probe technique (Section 7.1.6) can also be adapted to the GATE ASSIGNMENT problem. In this case, the LINEAR ASSIGNMENT problem that has to be solved for each probe attributes *negatively infinite* cost to putting a vertex in an illegal slot.

More ad hoc methods for gate assignment carry over from the applications in printed-circuit-board and standard-cell design (see Section 7.3.1.2). Note that straightforward heuristic methods for improving placements, such as component exchange, can easily be adapted to accept only those modifications that result in legal placements.

If islands on the master can accommodate several gates, the gate assignment problem becomes more complicated. We can distinguish several cases.

The simplest case is that each island effectively contains slots for several gates. In this case, the same approaches as are used for the GATE ASSIGNMENT problem work. We just have to modify the number and locations of the slots.

In many applications, however, there are several ways to break up an island into gates. For instance, an island could contain either a four-input NAND-gate or two two-input NAND-gates or, say, two inverters and one two-input NAND-gate. There could also be several ways of arranging the same set of gates on an island. In this case, the definition of the gate-assignment problem must be extended. The instance of the problem now has to describe all possibilities of arranging slots on an island, and the solution has to select one such possibility for each island, in addition to mapping gates onto slots.

7.3.1.2 Layout of Printed Circuit Boards and Design of Standard Cells

In layout of printed circuit boards and design of standard cells, the gate-assignment problem takes a slightly different form. Here, the library components (chips and standard cells, respectively) are not yet placed at the time of assignment. Thus, the gate-assignment problem merely asks for the *selection* of suitable components out of the library, such that the subsequent placement is efficient. The figure of merit used here is usually an estimate involving parts use—for instance, the number of the used chips on the boards or the total area of all standard cells used—and connection cost. Sometimes, a gate can be realized by components with different power consumption or different layout characteristics, such as aspect ratio. In this case, the cost function to be optimized during gate assignment should take these aspects into account, as well. As a result, the formulation of the suitable gate-assignment problem for a given application can be quite application-specific.

Only highly heuristic methods are reported for solving variants of the gate-assignment problem that are not direct extensions of placement problems. Most heuristics are based on greedy local search [297, 339, 427].

If the circuit contains complex blocks that do not trivially map to single library components or to small sets of library components, then a decomposition of the functions performed by the blocks to subcircuits made up of library components has to be performed. This process is called *technology mapping*. We do not discuss it in detail here, because it entails aspects of logic synthesis that are not covered in this book, see [27, 87, 231, 292].

7.3.2*   Pin Assignment

In pin assignment, the slots are pin locations on predesigned (library) cells— or are existing chips in printed-circuit-board layout—and the circuit elements are net terminals. Such cells often provide several logically equivalent pins. For instance, all input pins to an n-input AND-gate are functionally equivalent. Furthermore, several pins may occur multiply in the cell. Such pins are called *equipotential* pins. Equipotential pins form disjoint *equipotential classes*. Functionally equivalent equipotential classes are then, in turn, grouped to form *functional classes*. Thus, the set of pins has a two-level structure imposed on it by two equivalence relations. Figure 7.21 shows an example.

Pin assignment takes place after placement, and sometimes even after global routing. The objective is to assign pins to the net terminals so as to minimize a given measure of wiring cost or routability. Here, terminals of different nets may be permuted among functionally equivalent equipotential classes, and each net terminal must be assigned a pin in the respective equipotential class. Different nets must be assigned pins in different equipotential classes, however. In addition, we allow the same terminal to be realized through *several* pins in an equipotential class. For instance, the same terminal may be realized at pins B and B' in Figure 7.21. This net does not need to connect between the pins B and B' outside of the AND-cell, since the connection has already been made inside. In standard-cell layout, this use of equipotential pins is called a *feedthrough*. Since no wiring around the cell is needed, the wire length decreases with the use of feedthroughs. The cost function should reflect this phenomenon.

Formally, then, the pin-assignment problem looks as follows.

PIN ASSIGNMENT

Instance: 1. A hypergraph $G = (V, E)$; let T be the multiset of terminals of all nets; $T = \bigcup E$

2. A set P of *pins* $p_i = (x_i, y_i)$, described by their location in the plane

3. A set of *equipotential classes* determined by an equivalence relation $\sim_\chi$ on P derived, for instance, from the cell library. $[p_i]_{\sim_\chi}$ is the class of pins that are *equipotential* with p_i. (This notation extends naturally to sets of equipotential pins.) Let $P/\sim_\chi$ be the set of equivalence classes of P with respect to $\sim_\chi$.

4. An equivalence relation $\sim_f$ on $P/\sim_\chi$, whose classes are the

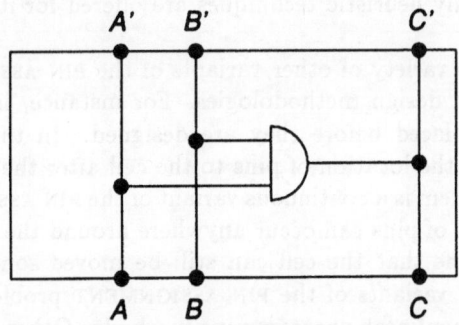

Figure 7.21: *A standard cell with several logically equivalent pins. The cell implements an* AND-*gate. Equipotential classes are* $\{A, A'\}$, $\{B, B'\}$, *and* $\{C, C'\}$. *The classes* $\{A, A'\}$ *and* $\{B, B'\}$ *are functionally equivalent.*

sets of *functionally equivalent* pins. The equivalence classes with respect to $\sim_f$ are called *functional classes*. Each terminal $v \in T$ corresponds to a functional class $[v]_{\sim_f}$ of pins. This class is the class of pins that can realize the terminal v.

Configurations: All *pin assignments*—that is, all mappings $a : T \to 2^P$. The pin assignment a assigns to terminal v the pins in the set $a(v)$.

Solutions: All pin assignments such that

1. For all $v \in T$ and for all $p_1, p_2 \in a(v)$ we have $p_1 \sim_\chi p_2$, that is, all pins in $a(v)$ belong to the same equipotential class.

2. For all $e_1, e_2 \in E$, if $e_1 \neq e_2$ and $v_1 \in e_1$ and $v_2 \in e_2$, then $[a(v_1)]_{\sim_\chi} \neq [a(v_2)]_{\sim_\chi}$; that is, different nets map onto different equipotential classes of pins.

3. For all $e \in E$ and $v \in e$, we have $[a(v)]_{\sim_\chi} \in [v]_{\sim_f}$; that is, the pin assignment assigns each terminal to a class of equipotential pins in the functional class of the terminal.

Minimize: The cost function in question

Most authors discuss the PIN ASSIGNMENT problem with the restriction that no equipotential pins exist; that is, $\sim_\chi$ is the identity relation. The functional pin classes typically group together subsets of pins that belong to the same cell (in general-cell layout) or chip (in printed-circuit-board layout). The PIN ASSIGNMENT problem is believed to be NP-hard, even if $\sim_\chi$ is the identity relation. Only highly heuristic techniques are offered for its solution [45, 240, 449].

There is a large variety of other variants of the PIN ASSIGNMENT problem, dictated by specific design methodologies. For instance, in top-down design, variable cells are placed before they are designed. In this design style, we can still determine the location of pins to the cell after the cell is placed. The corresponding problem is a continuous variant of the PIN ASSIGNMENT problem, where the locations of pins can occur anywhere around the boundary of a cell. We can even assume that the cell can still be moved somewhat. Yao et al. [480] consider such variants of the PIN ASSIGNMENT problem and solve them heuristically with nonlinear-programming methods. Other variants of the PIN ASSIGNMENT problem are discussed in [398, 466]; see also Section 8.7.

The scope of such variants is limited to a rather narrow application domain. Therefore, formulating the appropriate variant of the PIN ASSIGNMENT problem is an application-specific task.

Much work needs to be done in the area of pin assignment. The problem complexities are not determined, a suitable taxonomy of pin-assignment problems needs to be created, and advanced optimization methods have to be applied to pin-assignment problems.

7.4   Simulated Annealing

Several authors have applied simulated annealing to placement and floorplanning problems. Wong et al. [473] give an excellent overview over the use of simulated annealing in circuit layout. Sechen [407] summarizes the Timber-Wolf toolset. Newer contributions are presented in [307, 382, 406]. We shall not attempt a comprehensive discussion. Rather, we shall concentrate on one problem that exhibits an interesting relationship between algebraic expressions and simulated annealing: the optimization of slicing floorplans. The contents of this section are adapted from [473].

7.4.1   Arithmetic Expressions for Slicing Floorplans

We have seen in Section 7.2.1.1 that slicing floorplans can be effectively sized, even if the cells have variable shape. Furthermore, a slicing floorplan is uniquely determined by its oriented and ordered slicing tree (Definition 7.8). Thus, the floorplan optimization problem amounts to selecting a suitable oriented and ordered slicing tree. Slicing trees, on the other hand, can be efficiently encoded in *Polish notation*. This encoding represents the slicing tree as an arithmetic expression over two operators. Each internal tree node represents a binary operator acting on rectangular subfloorplans. The operators are from the set {hor, ver} representing a horizontal and a vertical cut. The left (right) operand of hor is the subfloorplan to above (below) the horizontal cutline. The left (right) operand of ver is the subfloorplan to the left (right) of the vertical cutline. The distinction between the two operators accounts for the orientation of the slicing tree; the distinction between left and right operands accounts for the ordering.

Before we encode the slicing tree, we make another transformation. As described so far, the slicing tree is just a binary hierarchy tree for the directional adjacency graphs of the slicing floorplan (see Section 3.14.3). As such, several nodes with orientation labels hor and ver, respectively, can be adjacent in the tree. Furthermore, for a given slicing floorplan, there may be many slicing trees. To restrict the size of the configuration graph underlying the simulated annealing algorithm, we require that, on a path in the slicing tree composed solely of edges with the ordering label rb, hor and ver nodes alternate. It is not difficult to see that, for an unsized slicing floorplan, we can always find a *unique* normalized slicing tree (see Exercise 7.20).

We can obtain the Polish notation for a slicing tree by traversing the tree in postorder. At each node, we explore the child with the label *lt* first. Figure 7.22 gives an example.

Polish expressions of slicing trees with k rooms are strings of length $2k - 1$, in which each of the numbers $1, \ldots, k$ (operands) occurs exactly once and there are exactly $k - 1$ operators from the set $\{hor, ver\}$. Furthermore, each prefix of the string has more operands than operators. If the slicing tree is normalized, then in the Polish notation adjacent operators are different. We call the set of corresponding strings $S_{\mathrm{norm},k}$, and its elements *normalized k-slicing expressions*.

The following lemma is not difficult to prove.

Lemma 7.4 *Oriented and ordered normalized slicing trees with k rooms correspond one-to-one to normalized k-slicing expressions.*

Proof Exercise 7.21 □

7.4.2 The Configuration Graph

The set of vertices of the configuration graph that is underlying the simulated-annealing algorithm is $S_{\mathrm{norm},k}$. Clearly, the size of $S_{\mathrm{norm},k}$ is exponential in k (see also Exercise 7.22).

We now describe the edges in the configuration graph. Call a substring of a normalized k-slicing expression that consists of only operators and has maximal length a *chain* (see Figure 7.22). For strings in S_{norm}, there are only two kinds of chains—namely, hor, ver, hor, ver, ... and ver, hor, ver, hor .... We define three kinds of (undirected) edges in the configuration graph:

M1 (Operand Swap): Connects two normalized k-slicing expressions that we can obtain from each other by swapping two adjacent operands, that

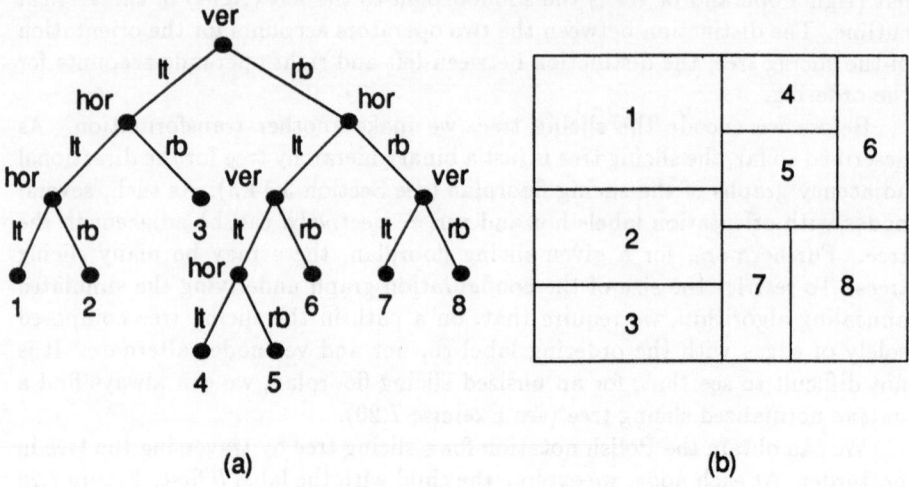

(a)   (b)

1 2 hor 3 hor 4 5 hor 6 ver 7 8 ver hor ver

(c)

Figure 7.22: *Normalized slicing expressions. (a) A normalized oriented and ordered slicing tree; (b) the corresponding floorplan; (c) the Polish notation—the longest chain is* ver hor ver *at the end of the string.*

is, two operands that are adjacent after deleting all operator symbols from the expression.

M2 (Chain Invert): Connects two normalized k-slicing expressions that we can obtain from each other by complementing a chain (interchanging hor and ver).

M3 (Operator/Operand Swap): Connects two normalized k-slicing expressions that we can obtain from each other by swapping an operand with an adjacent operator from {hor, ver}, if this results in a legal normalized k-slicing expression.

Figure 7.23 shows a path in the configuration graph. We now discuss a few structural parameters of the configuration graph.

Lemma 7.5 *All vertices in the configuration graph have degree less than* $4k-6$.

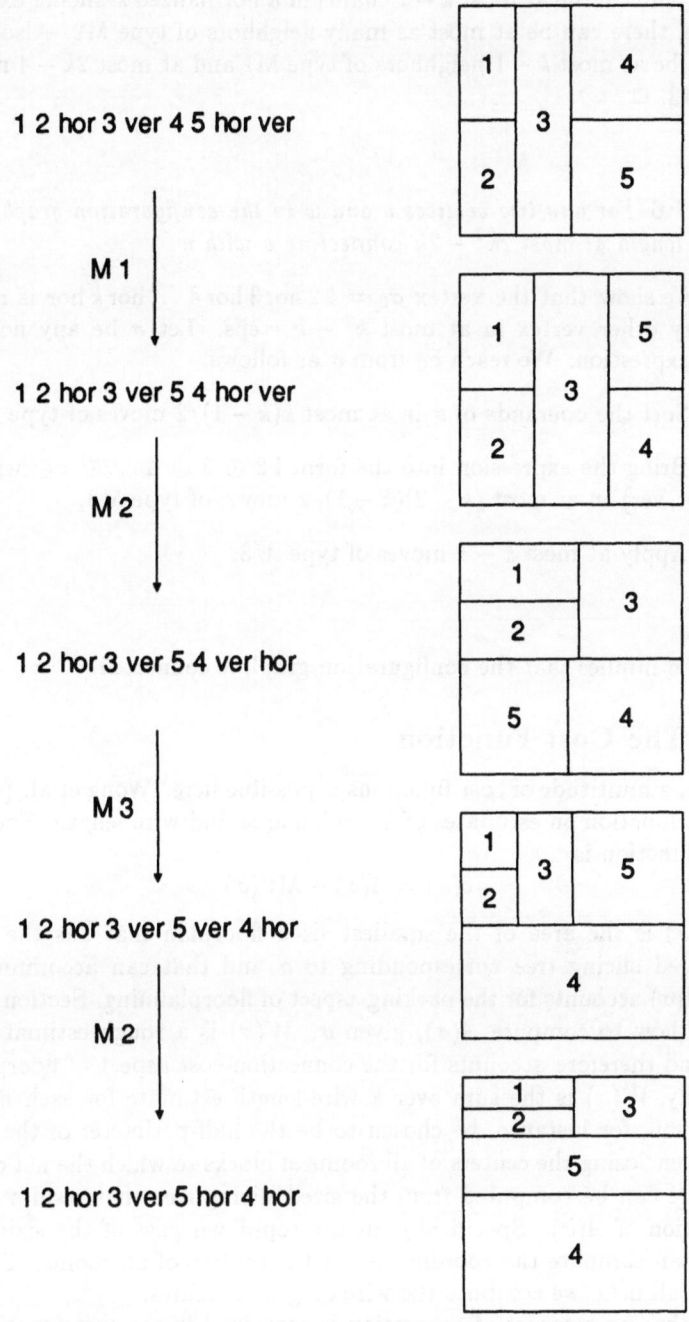

Figure 7.23: A sequence of moves.

Proof There can be at most $k - 1$ chains in a normalized k-slicing expression. Therefore, there can be at most as many neighbors of type M2. Also, in total there can be at most $k - 1$ neighbors of type M1 and at most $2k - 4$ neighbors of type M3. □

Lemma 7.6 *For any two vertices v and w in the configuration graph, there is a path of length at most $2k^2 - 2k$ connecting v with w.*

Proof We show that the vertex $\sigma_0 = 1\,2\,\mathrm{hor}\,3\,\mathrm{hor}\,4\ldots\mathrm{hor}\,k\,\mathrm{hor}$ is reachable from every other vertex in at most $k^2 - k$ steps. Let σ be any normalized k-slicing expression. We reach σ_0 from σ as follows:

Step 1: Sort the operands of σ in at most $k(k-1)/2$ moves of type M1.

Step 2: Bring the expression into the form $1\,2 \odot 3 \odot 4 \ldots \odot k \odot$ where $\odot \in \{\mathrm{hor}, \mathrm{ver}\}$ in at most $(k-2)(k-1)/2$ moves of type M3.

Step 3: Apply at most $k - 1$ moves of type $M3$.

□

Lemma 7.6 implies that the configuration graph is connected.

7.4.3   The Cost Function

Of course, a multitude of cost functions is possible here. Wong et al. [473] base their cost function on estimates of floorplan area and wire length. Specifically, the cost function is

$$c(\sigma) = A(\sigma) + \lambda W(\sigma)$$

Here, $A(\sigma)$ is the area of the smallest sized floorplan that has the oriented and ordered slicing tree corresponding to σ and that can accommodate all blocks. $A(\sigma)$ accounts for the packing aspect of floorplanning. Section 7.2.1.2.2 describes how to compute $A(\sigma)$, given σ. $W(\sigma)$ is a rough estimate of wire length, and therefore accounts for the connection-cost aspect of floorplanning. Specifically, $W(\sigma)$ is the sum over a wire-length estimate for each net. This estimate can, for instance, be chosen to be the half-perimeter of the smallest rectangle enclosing the centers of all rooms of blocks to which the net connects. This figure can be computed from the sized slicing tree that results from the computation of $A(\sigma)$. Specifically, in the top-down pass of the sizing of the tree, we can compute the coordinates for the centers of all rooms. Then, in a pass over all nets, we compute the wire-length estimates.

Note that no measure of congestion is contained in the cost function.

The parameter λ can be set to adjust the relative priority between the packing and connection-cost aspects in the cost function.

Given σ, the computation of the cost $c(\sigma)$ takes time $O(T_s + p)$, where T_s is the time for normal floorplan sizing and p is the number of terminals in the circuit. If we modify slicing trees by moving along edges in the configuration graph, we want to compute the cost function *incrementally*. That is, we hope that the small modification made in each move will also result in a quicker update of the cost. It turns out that it does. The reason is that each move changes only a small portion of the slicing tree. In particular, each of the moves changes the information in the slicing tree along only one or two paths from the root to a leaf. In the operand swaps (M1), those two paths lead to the adjacent leaves that are swapped. In the chain inverts (M2), only the path in the slicing tree that corresponds to the chain is affected. In the operator–operand swaps (M3), the two paths lead to the operator and operand involved.

The first pass of the incremental computation of the cost recomputes only the sizing functions pertaining to the (few) updated nodes bottom-up in the tree. The second pass has to process the whole slicing tree top-down, since all rooms may have changed their location. The third pass again scans the list of terminals. The incremental computation of the cost reduces only the first term T_s in the estimate of the run time. If sizing functions are complex, this term can be expected to be dominant. Otherwise, we can make additional modifications to the algorithm in order to render the computation of the wirelength estimates more incremental (see Exercise 7.24). The time contribution of this procedure is especially large if the tree is shallow. However, normalized slicing trees tend not to be shallow if most operators are of the same kind.

7.4.4 Randomization

The assignment of the probabilities of the generating chain deserves some experimentation.

We learned in Section 4.3.2 that the generating chain should assign to each edge out of a vertex roughly the same probability, because then all vertices in the configuration graph are accessible with about the same probability at high temperatures. We can do this assignment by first choosing a type of move (M1, M2, or M3) randomly, and then randomly selecting the operands and operators involved. For vertices in the configuration graph for which there are greatly differing numbers of moves of the different types, it is useful to bias the random selection of the type according to the number of moves of that type. For instance, if there is only one chain, then there is only one move of type M2. So, type M2 should be selected with only a small probability.

One the other hand, a move of type M2 on a long chain may change the floorplan substantially, and thus it can be understood as a "long" step in the configuration graph. Intuitively, we may want long steps to occur less often than do short steps—that is, steps that do not change the floorplan so much. Then, we want to select the move types with equal probability. Wong et al. take this approach. We should note carefully, however, that such arguments are largely intuitive and are not backed by analysis.

Section 4.3.2 gives detailed information on how to select an appropriate cooling schedule.

7.4.5　Experimental Results

Wong et al. [473] report on experiments with this configuration graph. They find that incorporating wire length into the cost function does not cause the area to increase very much, but it does reduce the wire length greatly. Still, even if $\lambda = 0$, the wire length of the floorplan computed with the simulated-annealing algorithm is substantially smaller than the wire length of a floorplan based on a random slicing tree. This observation is intuitive, because we would expect area-efficient floorplans to be efficient in wire length, as well.

7.4.6　Extensions

We provided a configuration graph for simulated-annealing algorithms for floorplanning by identifying floorplans with certain algebraic expressions and defining moves that modified the tree representation of these algebraic expressions locally. This paradigm is quite general and applies to many other optimization problems.

An immediate extension is the one to nonslicing floorplans and to floorplans with nonrectangular rooms. We can perform this extension by identifying new operators that compose subfloorplans. The algebraic expressions constructed with these operators describe general floorplans; in fact, they impose a tree structure on these general floorplans. A configuration graph can now be defined, whose edges represent local modifications within the algebraic expressions.

Wong et al. [473] present a detailed account of one such definition. They also report on applications of simulated annealing to gate-array, standard cell, and general-cell placement.

7.5　Summary

In this chapter, we discussed techniques for placing circuit blocks on the surface of a chip or printed circuit board. This problem is attacked in the first phase of most layout systems. The problem is not well understood, in several respects.

First, all versions of the placement problem are NP-hard combinatorial problems. Solution methods come from a wide variety of areas in discrete and continuous optimization. In practice, however, the solution of placement problems is still quite heuristic. Sometimes, even the formal definition of the placement problem is governed more by the restrictions of the available solution methods (for instance, convexity) than by the characteristics of the application domain. In short, there is much work to be done to tailor methods of global optimization to placement problems.

Aside from these methodological difficulties, it is not even clear what the appropriate definition of the placement problem should be in different applications. The main difficulty here is that the cost function to be optimized has to take into account subtle wirability issues. Most cost functions used today do not meet this requirement. Specifically, we can distinguish classical cost functions that measure a certain estimate of *wire length* and cost functions that measure some estimate of *congestion*. It turns out that neither type of cost function is fully appropriate, since the *distribution of wires* has to be taken into account. There are attempts to integrate such aspects into the cost function [153, 351]. However, the cost function then becomes more complex, and the use of global optimization methods becomes more difficult or impossible. The more promising approach is to do away with estimates of wirability or wiring area altogether, and to let the global-routing phase itself provide the wiring estimates needed during placement. This method amounts to an integration of the global-routing phase into the (classically preceding) placement phase (see Section 8.8).

Placement by partitioning is especially suitable in this context, because it imposes on the circuit a natural hierarchy that can be used as a structural scaffolding for the integration of global routing into the placement phase. An additional advantage of placement by partitioning is that the packing aspect of placement—that is, the selection of appropriate layout alternatives for the blocks and their area-efficient arrangement—can be incorporated naturally. Furthermore, the results of Chapter 5 supply a theoretical justification for the experimental observation that partitioning is a key element of placement.

Placement is the first phase of layout, and in some sense it is the least understood. As we discuss subsequent phases of layout in later chapters, we shall find that, although the corresponding optimization problems are still difficult, model issues become less and less controversial. This shift is only natural: As we move on through the phases of layout, cost estimates we have to perform point less and less far into the future of layout phases still to be executed, and therefore our results tend to be more accurate.

7.6 Exercises

7.1 Prove that the sum of the cutsizes over all $r - 1$ horizontal and $s - 1$ vertical cutlines across an $r \times s$ gate-array master is the same as the total wire length $c_1(p)$ according to the half-perimeter estimate.

7.2 This exercise is concerned with finding graph instances of placement problems that exhibit the difference between the different cost functions. Consider placing graphs into an $r \times s$ master with $r = 1$ and $s = n$. Find a graph

G_1 all of whose placements minimizing c_1 do not minimize c_2 and c_{MC}. Find corresponding graphs G_2 for c_2 and G_{MC} for c_{MC}.

7.3 Show that the mincut heuristic provides an approximation algorithm for the MINCUT LINEAR ARRANGEMENT problem with error $\log n$ if the exact minimum cutsize is found at each stage.

7.4 Prove that, if the mincut method is based on a balanced bifurcator of minimum size, then it is also an approximation algorithm for the MINCUT RECTANGULAR ARRANGEMENT problem that comes to within an error of size $O(\log \max(r, s))$ of the optimum cost.

7.5 This exercise discusses the problems of representing hypergraphs with graphs in placement problems minimizing wire length. Assume that the hyperedge e with r terminals is represented by a clique with uniform edge cost c, and that $c_k(p)$ is minimized.

1. Show that, no matter how c is chosen, there always are placements p_1 and p_2 such that $c_k(p_1)/c_k(p_2) = \Omega(r)$.

2. Show that choosing nonuniform edge costs does not improve the representation.

7.6 Show that the RELAXED PLACEMENT problem is the linear-program dual of a network linear program.

7.7 Generalize the RELAXED PLACEMENT problem to incorporate nonrotatable cells of different size. Again, overlap is not excluded. Show that the resulting linear program is still the linear-programming dual of a network linear program.

7.8 Prove that, by executing linear probes in the directions of all coordinate axes, we can compute an upper bound for the transformed placement problem that is at most a factor of $\sqrt{n}$ away from the maximum cost of a legal placement.

7.9 Assume that the two-dimensional configuration space for an instance of the QUADRATIC LINEAR ARRANGEMENT problem with two blocks is covered with k linear probes at angular distances of $2\pi/k$. Let p be the farthest placement found in all probes—that is, the placement with the largest value of $||p^T E||^2$.

1. What is an upper bound on the size of $||p_{\text{opt}}^T E||^2$, where p_{opt} is the optimum legal placement?

2. How large does k have to be such that $||p_{\text{opt}}^T E||^2/||p^T E||^2$ is at most the constant $\delta > 0$?

7.10 Show that adding a linear term to the quadratic cost function in the QUADRATIC LINEAR ARRANGEMENT problem amounts to adding a constant vector z to the vector $p^T E$ whose Euclidean norm is to be maximized in the linear probe technique.

7.11 Prove that the surfaces containing placements p with equal values of $p^T C p - p_0^T C p_0$ are ellipsoids. Here, p_0 is the optimal placement with respect to the QUADRATIC RELAXED PLACEMENT problem. What are the directions of the axes of the ellipsoid?

7.12 When is a pair of dual directed planar graphs the pair of directional adjacency graphs for a floorplan?

7.13 Show that a normal floorplan is a slicing floorplan exactly if it can be generated by the mincut method based on bipartitioning.

7.14 Extend the OM-similarity to floorplans that are not normal. What replaces the notion of *maximal wall*? Does Lemma 7.1 still hold?

7.15 Develop a good upper bound for the area resulting from choice functions represented by a node in the branch-and-bound tree for the FLOORPLAN SIZING problem of OM-topologies.

7.16 The following observation is the basis for an efficient pruning method of the branch-and-bound tree T in the algorithm for solving the FLOORPLAN SIZING problem on OM-topologies: Let (w_k, h_k) be a layout alternative for room i in some sized floorplan F with horizontal adjacency graph $A_h(F)$. If the edge for room i is not on the shortest path in $A_h(F)$, then choosing any layout alternative for room i with width smaller than w_k cannot decrease the floorplan area (Figure 7.24).

1. Turn this observation into a pruning criterion by proving the following lemma.

   Lemma: Let the layout alternatives for a block in T be ordered in increasingly interlaced order. Consider a node v in T corresponding to a layout alternative (w_k, h_k) of block i, and assume that the children $v_1, \ldots, v_\ell$ of v have been explored completely. Let $z_1, \ldots, z_h$ be the descendants of $v_1, \ldots, v_\ell$ at which backtracking was initiated. Let $A_h(z_j)$ and $A_v(z_j)$, $1 \leq j \leq h$, be the lower-bound adjacency graphs for node z_j. Then the following holds:

   a. If k is odd and the edge representing block i is not on a shortest path in $A_v(z_j)$ for all j, then v is fathomed.

   b. If k is even and the edge representing block i is not on a shortest path in $A_h(z_j)$ for all j, then v is fathomed.

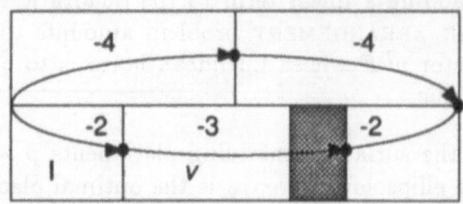

Figure 7.24: *A floorplan and its horizontal adjacency graph. Exchanging the layout alternative for block v with another alternative whose width is less than 3 does not decrease the width of the whole floorplan.*

2. Discuss how to extend the branch-and-bound algorithm so as to implement this pruning criterion.

7.17 Develop heuristics for the sizing of UM-topologies that selectively eliminate steps of sizing functions based on "resolution criteria".

7.18 This exercise asks you to count the number of possible numbered floorplan patterns that can occur at an interior node of a cut tree with high degree (see Section 7.2.2.2). A *numbered k-room floorplan pattern* is a normal unsized floorplan with k rooms that are numbered from 1 to k in some manner.

We can identify floorplan patterns that can be transformed into each other by reflection and rotation, because the orientation of the pattern inside its neighborhood in the floorplan can be determined by an optimization while the corresponding node in the cut tree is processed (see Section 7.2.1.3).

1. Determine the number $A(k)$ of all nonisomorphic numbered floorplan patterns (up to rotational and reflectional symmetry) for $k = 3, 4, 5$ rooms.

2. Develop a recursion formula for $A(k)$, and derive a closed form of the solution.

7.19 Present an example of a floorplan with k rooms, k arbitrarily large, that cannot be decomposed into floorplans with a smaller number of rooms.

This exercise shows that, as the number of rooms increases, expanding nodes in the cut tree so as to decrease the node degree does not cover all possible floorplan patterns.

7.20 Show that any unsized slicing floorplan has a unique normalized oriented and ordered slicing tree.

Hint: Use the relationship between slicing floorplans and series-parallel graphs.

7.21 Prove Lemma 7.4.

7.22 Determine the size of the set S_{norm} as accurately as you can.

7.23 Derive an upper bound on the degree of the configuration graph for the simulated annealing algorithm for optimizing slicing floorplans that is tighter than the one given in Lemma 7.5.

7.24 Discuss possibilities of speeding up the incremental computation of wire-length estimates in the simulated-annealing algorithm for optimizing slicing floorplans. You may also consider methods that redefine the wire-length estimate.

Chapter 8

Global Routing and Area Routing

The routing phase follows the placement phase. It determines the course of the wires that connect the cells laid out during the placement. The structure of the routing phase depends greatly on the design and fabrication technology. There are two approaches to routing—*two-phase* and *area routing*. In area routing, the routing process is carried out in one phase that determines the exact course of the wires. In two-phase routing the routing phase is subdivided into the *global* (or *loose*) *routing phase*—which determines how wires maneuver around and through cells—and the *detailed* (or *local* or *homotopic*) routing phase—which determines the exact course of the wires. We now describe both approaches to routing intuitively.

Two-phase routing is usually done in chip layout. Here, the placement phase computes a sized floorplan that determines the approximate locations of the cells. As a byproduct of the floorplanning phase, estimates on the sizes of routing subregions are computed (Section 7.2.6). The routing phase determines the exact course of the wires around or through the cells. As a byproduct, it computes the exact sizes and shapes of the routing subregions. In this setting, the routing phase is divided into a global and a detailed-routing phase. During global routing, the way in which each wire maneuvers through or around the cells is determined. We make this determination by finding paths in an appropriate *routing graph* whose edges represent subregions of the total *routing region*. The edges have labels called *capacities* that indicate how many wires the corresponding routing region can accommodate. The capacity of an edge is only an estimate of the actual number of wires that can be placed in the routing region. This number depends on such factors as the *detailed-routing model*, which specifies what geometric features are allowed for the wires, and on the exact location of the terminals of the wires. The evaluation of routing alternatives for each wire is made on the basis of the size estimates computed in the floorplanning phase. The exact layout of the wire is not computed, however. The following detailed-routing phase determines the exact courses of the wires, and the sizes and shapes of the routing subregions. As a byproduct, the exact locations of the cells are also computed in this phase.

There are several reasons why the routing task is subdivided into the two phases of global routing and detailed routing. One is that each phase is more tractable than is the whole routing problem in one phase. In fact, many versions of the detailed-routing problem are in P (see Chapter 9). Even though most interesting versions of the global-routing problem are NP-hard, there are close ties between this problem and special cases of integer programming. We can exploit these connections to develop several efficient algorithms and heuristics for global routing.

Another reason is that, depending on the design and fabrication technology, there are several different detailed-routing models. The detailed-routing model influences the size of the routing space needed, but it is assumed to have only a secondary influence on the global routing. Thus, the global routing is a somewhat more technology-independent form of layout description than is the detailed routing.

For the two-phase approach to work, the output of the placement problem must allow a suitable definition of a routing graph. This requirement entails that the routing region on the chip can be broken down into subregions that can be assigned to edges in the routing graph. Whether the routing region is *decomposable* in this sense depends on the fabrication and design technology. On the fabrication-technology side, if there are few wiring layers available, then the topological freedom for wires is restricted. The resulting planar flavor of the routing problem tends to increase the decomposability of the routing region. On the design-technology side, floorplanning methods incorporating wiring-area estimation lead to decomposable routing regions. Both features are found most commonly in chip layout. This is the reason why the two-phase approach to routing is used primarily in chip layout.

In printed-circuit-board layout, the locations of components on the board are usually determined exactly during the placement phase. Thus, the routing region is known in detail after the placement phase. Furthermore, a larger number of wiring layers is available. This feature imposes an intricate nonplanar topology on the routing region. As a result, the routing region is usually not easily decomposable. Consequently, the two-phase approach to routing is not used; rather, the routing is done directly on a graph that represents the routing region in complete detail. We call this version of routing *area routing*. Many authors do not distinguish area routing from the detailed-routing phase in the two-phase approach, because area routing also determines exact wire routes. In area routing, however, no global routes are given as part of the input. This fact causes area routing to resemble global routing more than it does detailed routing, from a mathematical point of view. For this reason, we discuss global routing and area routing in one chapter, and devote another chapter (Chapter 9) to detailed routing.

8.1 Definitions and Complexity

In general, routing problems are sets of interdependent Steiner tree problems in appropriate routing graphs. In this section, we discuss various formalizations of routing problems.

8.1.1 Routing in Floorplans

The following graph-oriented version of the routing problem is useful for global routing in floorplans.

CONSTRAINED GLOBAL ROUTING

Instance: A *routing graph* $G = (V, E)$, labeled with edge *capacities* $c : E \to \mathbf{R}_+$ and edge *lengths* $\ell : E \to \mathbf{R}_+$; a multiset $N \subseteq 2^V$ of *nets*, where each net is a subset of the vertices of G, the multiplicity of a net $n \in N$ is denoted with k_n, and the instances are denoted with $(n, 1), \ldots, (n, k_n)$. (We shall use the notations $n \in N$ and $(n, i) \in N$ interchangeably.) The nets (n, i) are labeled with cost factors $w(n, i) \in \mathbf{R}_+$; a set of *admissible routes*—that is, Steiner trees $T_n^1, \ldots, T_n^{I_n}$—for each net $n \in N$.

Configurations: All *routings*—that is, families $R = (T_{n,i})_{(n,i) \in N'}$ of routes. Here, $N' \subseteq N$ is a subset of N that contains all nets that are routed, and $T_{n,i}$ is an admissible route for net n.

Solutions: All routings R such that $U(R, e) \leq c(e)$ for all $e \in E$. Here, the *traffic* $U(R, e)$ on edge e in routing R is the weighted sum of all routes that contain edge e. Formally,

$$U(R, e) = \sum_{\substack{(n, i) \in N' \\ e \in T_{n,i}}} w(n, i)$$

Minimize: $c_1(R) := \left(\sum_{(n,i) \in N \setminus N'} w(n, i), W(R) \right)$, where $W(R)$ is the

total weighted wire length of routing R,

$$W(R) = \sum_{e \in E} \ell(e) \cdot U(R, e)$$

A routing R is a set of Steiner trees—the *routes*—one for each net in N'. The route for a net has to be taken from a specified set of *admissible* routes. (This restriction allows us to prohibit routes that are not compatible with the design or fabrication technology or that are otherwise undesirable.) The nets in $N \setminus N'$ are not routed. A legal routing is a routing in which, for each edge in the routing graph, the traffic on the edge does not exceed that edge's capacity. This constraint is called the *capacity constraint* for the edge. The *traffic* $U(R, e)$ on an edge e in routing R is computed by a weighted sum over all nets that go along the edge. Here, each net (n, i) contributes a cost factor $w(n, i)$ that indicates how many *tracks* are taken up by routing the net. Incorporating weighted nets allows us to treat wide buses as single net. The cost factor can be set to the width of the bus. Thus, we can ensure that all strands of the bus have the same route. This property of a routing may be desirable or even necessary for reasons of electrical soundness of the layout. The goals are to route as many nets as possible within the capacity constraints on the edges, and to achieve a minimum total wire length. Routing as many nets as possible takes priority over minimizing the wire length. Thus, the cost function is a pair of numbers, the first of which is concerned with the completeness of the routing and the second of which is concerned with the total wire length. (Sometimes, the second term is the *maximum* length of each route.) The pairs are considered to be ordered lexicographically.

The routing graph is computed from the floorplan, and it differs in different design styles. In general-cell layout, we have to distinguish the cases of mostly fixed and mostly variable cells, respectively. If mostly variable cells are used, then *channel-free* routing through the cells is possible, and a suitable routing graph is the rectangular adjacency graph (Definition 7.14). Figure 8.1 illustrates this graph. The vertices of the graph are thought of as being located in the centers of the respective rooms. The edge lengths can be based on the Euclidean norm or on the Manhattan distance. The routing specifies how wires maneuver through rooms and across walls. The capacity of an edge determines how many wires can be run across the corresponding dual wall. Therefore, to a first approximation, the capacity of an edge can be taken to be the length of the corresponding dual wall, taken with respect to a unit of measure that accounts for desired minimum spacing between wires crossing the wall. (In fact, estimates of the routing area available inside the rooms whose centers are connected by edge e should enter the definition of the capacity of e, as well.)If a capacity constraint is violated, a wall has to accommodate more wires than it can with its given length. Thus, such a violation implies a floorplan modification that increases the length of the wall. This modification may change the floorplan topology. Viewed in this light, the capacity constraints that a legal routing has to obey enforce the compliance with the topology determined by the preceding floorplanning phase.

Note that, in the preceding scenario, a global routing also determines on which side of a cell we should put a pin—namely, on the side represented by the

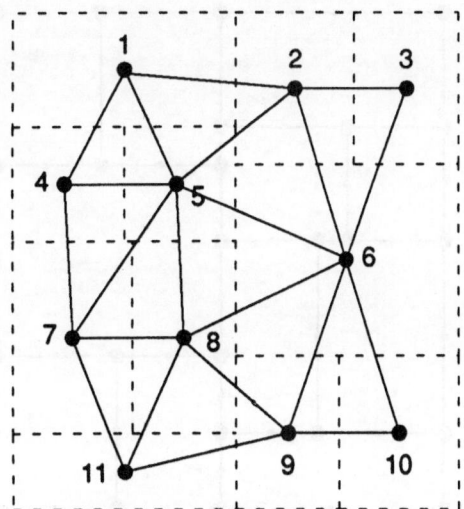

Figure 8.1: *The routing graph in a channel-free design style is the rectangular adjacency graph of the floorplan (see Definition 7.14). Vertices are thought of lying in the center of their respective room. Edge lengths are based on the Manhattan or Euclidean distance between the room centers. Edge capacities are based on the lengths of the dual walls. Note that the routing graph is planar.*

wall through which the respective net leaves the room for the cell. However, the global routing phase does not determine the exact location of the pin or wiring layers for routing segments.

If only fixed cells are used, then the pin locations are known beforehand. Also, wiring through a cell is often undesirable or impossible, such that wiring has to be done around the cells in channels that follow the course of the walls in the floorplan. The appropriate routing graph for modeling this situation is a close derivative of the floorplan itself. This graph is sometimes called the *channel intersection graph*. Vertices are added to the floorplan to account for the approximate positions of terminals. In Figure 8.2, one vertex has been added in the middle of each wall to represent pins on the respective sides of the cell. Here, the capacity constraints result directly from the wiring-area estimation for the respective routing channels that is computed during floorplanning (see Section 7.2.6).

If some of the cells are fixed and others are variable, then an appropriate routing graph is the union of the channel-intersection graph (Figure 8.2) and the rectangular adjacency graph of the floorplan (Figure 8.1). The pins whose

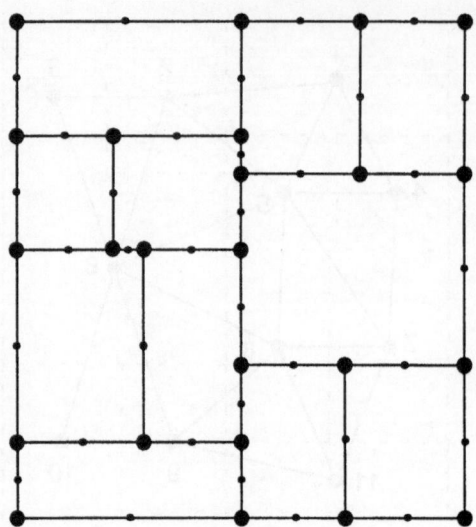

Figure 8.2: *The routing graph in a design style with channels is a close derivative of the floorplan. Some vertices (drawn large) represent points at which walls intersect. Other vertices (drawn small) bisect walls. These vertices are added to represent approximate pin locations (see also Section 8.7). If the layout of the cells is known, then pin locations can be approximated more accurately by correspondingly positioned vertices. The edge lengths can be derived directly from the figure. Edge capacities reflect the wiring-area estimates computed during floorplanning.*

locations are known are assigned to vertices in the channel-intersection graph (see Figure 8.3). The pins, whose locations have yet to be determined, are assigned to the center of the room for their cell. The degree to which routing through a cell is possible can be controlled via the choice of appropriate edge capacities for the edges in the rectangular adjacency graph. Another way of influencing the tradeoff between routing through cells and routing in channels is via the relative sizes of edge lengths. For instance, if routing in channels is to be enforced, the lengths of the edges in the rectangular adjacency graph can be set much larger than the other edge lengths; if it is to be discouraged, then the sizes can be assigned the other way around. We shall not go into detail on how to assign edge lengths and capacities in this framework. This problem is closely linked to that of wiring-area estimation in floorplans (Section 7.2.6). Statistical arguments have to be used here.

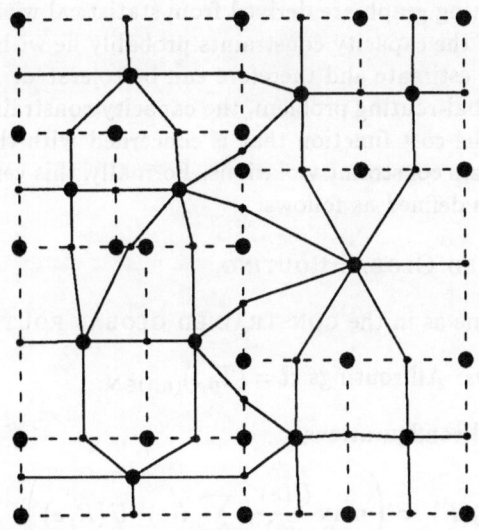

Figure 8.3: *Routing graph combining channel-free routing with routing in channels. The part of the graph originating from the channel-intersection graph is dashed.*

Note that routing in a channel can be viewed as a special case of routing through a cell, in which we decide to route the wire along the cell boundary. However, the explicit introduction of channels into the routing graph has the advantage of modeling known pin positions more precisely and of representing the estimated channel widths more accurately.

We incorporate the total wire length into the cost function for the CON-STRAINED GLOBAL-ROUTING PROBLEM because we hope that minimizing the total wire length will also reduce the area necessary for routing. The same comments that we made in Section 7.1.1.1 about the correlation between wire length and wirability also apply to this cost function.

The solution of the CONSTRAINED GLOBAL ROUTING problem may not route all nets. In this case, the designer has to step in manually, to fix up the unrouted nets. To do this routing, she or he may have to increase some edge capacities judiciously such that a revision of the floorplan topology is avoided or is kept at a minimum. Note, however, that the edge capacities are only an estimate of the expected wiring area. This is a point in favor of having the manual wiring of the unrouted nets follow the *detailed-routing phase*, which determines the exact wiring area.

If manual intervention is not an option, we can get rid of the capacity constraints altogether. The basic observation here is that, because the edge capacities in the routing graph are derived from statistical wiring-area estimates, small violations of the capacity constraints probably lie within the uncertainty of the overall area estimate and therefore can be tolerated. In the correspondingly modified global-routing problem, the capacity constraints are eliminated, and the term in the cost function that is concerned with the completeness of the routing penalizes constraint violations. Formally, this version of the global-routing problem is defined as follows.

UNCONSTRAINED GLOBAL ROUTING

    Instance: Same as in the CONSTRAINED GLOBAL ROUTING problem

    Configurations: All routings $R = (T_{n,i})_{(n,i) \in N}$

    Solutions: All configurations

$$\textit{Minimize: } c_\infty(R) := \left(\max_{e \in E} \frac{U(e)}{c(e)}, \ \sum_{e \in E} \ell(e) \cdot U(R,e) \right)$$

In the UNCONSTRAINED GLOBAL ROUTING problem, all nets are always routed. The expression $U(e)/c(e)$ is called the *load* of edge e. Thus, the first term in the cost function $c_\infty(R)$ measures the maximum load over all edges—that is, we attempt to meet the capacity constraints with top priority. A routing that obeys all capacity constraints yields a maximum load value of at most 1. Note that, in the UNCONSTRAINED GLOBAL ROUTING problem, not only meeting the capacity constraints, but also *minimizing the maximum load of any edge*, takes priority over minimizing the total wire length. This choice is especially well suited in the context of routing on gate arrays, because on fixed masters meeting all capacity constraints is all that we need to ensure routability. But also in general-cell placement the even wire distribution that is obtained by minimizing the maximum load is more likely to reduce the wiring area than is a small total wire length.

    The size of an instance of a global-routing problem is the sum of the size of the routing graph and the number of nets. The set of admissible routes is usually very large, but it does not affect the size of the global-routing problem, because it is usually defined implicitly by predicates such as *all routes are admissible* or *all k-bend routes are admissible.*

8.1.2   Area Routing

In area routing, the routing graph models the routing region in complete detail. This means that the routing graph must be dependent on characteristics of the fabrication technology. The approach to dealing with fabrication-specific details is to solve a standardized version of the area-routing problem first, and then to run postprocessors on the output. This postprocessing phase smoothes out bends in wires, makes local changes to the routing in the vicinity of vias (contact holes on a printed circuit board), and satisfies other local constraints imposed by the fabrication technology.

The standardized routing problem uses k-layer grid graphs, or k-*grids*, for short, as routing graphs (Figure 5.13). Each edge in the k-grid represents a track that can accommodate exactly one wire; thus, the capacities of all edges are 1. If obstacles obstruct the routing region, the obstructed edges are deleted from the graph and a subgraph of the k-grid—a *partial k-grid*—results. In this application, we set the cost factors for all nets to 1. Furthermore, it is required that routes for different nets be *vertex-disjoint*. We call the corresponding version of the CONSTRAINED GLOBAL ROUTING problem (with $c \equiv 1$, $w \equiv 1$, and vertex-disjointness of the routing required) the AREA ROUTING problem. A routing that solves the AREA ROUTING problem specifies both the detailed course of the wires and a layer assignment.

The routing model offered by the AREA ROUTING problem is quite robust. The cost of layer changes can be encoded with the lengths of the vertical *via edges* in the k-grid. We can specify preferred wiring directions on a layer by suitably enlarging the lengths of the edge running in the orthogonal direction. Additional rules can incorporate the sizes of vias and other parameters of the technology.

The distinction between area routing and some of the detailed-routing problems discussed in Chapter 9 is fuzzy. For instance, channel-routing problems can be viewed as special area-routing problems. As defined here, the area-routing problem models each layer explicitly. In some detailed-routing models, wiring layers are not modeled explicitly, and layer assignment has to follow after the routing phase. We describe channel-routing problems in Chapter 9 because their solution and some of the routing models are derived from other versions of the detailed-routing problem, and because they appear in the context of the two-phase approach to routing. In this chapter, we restrict ourselves to area-routing problems that deal with general partial k-grids and whose solution methods are related to global-routing methods.

8.1.3 Complexity of Global Routing Problems

All versions of the global-routing problem are NP-hard. In fact, Kramer and van Leeuwen [245] show that the threshold versions of all problems are strongly NP-complete, even if the set of nets is a proper set, all nets are two-terminal nets, all routes are admissible, $c \equiv 1$, $\ell \equiv 0$ (that is, wire-length minimization is not an issue), $w \equiv 1$, and the routing graph is a 1-grid. Karp et al. [216] prove the strong NP-completeness of the version of the problem in which only one-bend routes (of two-terminal nets) are admissible. On the other hand, restricting N to a single multiterminal net and fixing $c \equiv 1$, $\ell \equiv 1$, and $w \equiv 1$ reduces all versions of the routing problem to the NP-hard MINIMUM STEINER TREE problem. If this net is required to be a two-terminal net, then the routing problem reduces to a shortest-path problem and can be solved efficiently (see Section 3.8). If the net touches every vertex, then the routing problem reduces to the MINIMUM SPANNING TREE problem, which is also easy to solve (Section 3.9). These observations are the basis of a whole set of popular routing methods (see Section 8.4). Among the extensions of these special cases of the CONSTRAINED GLOBAL ROUTING problem that are still in P are the restrictions to a single net with at most r or at least $n - r$ terminals, where r is fixed. (These are Steiner tree results; see Section 4.7 and Exercise 4.25.)

Three special cases of global-routing problems come up particularly often in practice. We mention them here because there are special solution methods pertaining to these restrictions. The first one is the case $w \equiv 1$. This case occurs if we do route not buses, but rather just single strand wires. The second case is where $\ell \equiv 0$. In this case, we are concerned not with total wire length, but rather just with routing the maximum number of wires in the CONSTRAINED GLOBAL ROUTING problem and minimizing the maximum load in the UNCONSTRAINED GLOBAL ROUTING problem, respectively. The third special case is that where the routing graph is a subgraph of a (1-)grid or even a complete grid graph. Note that, by the results of Kramer and van Leeuwen, all these special cases are NP-hard, even if only two-terminal nets are involved.

8.2   Global Routing and Integer Programming

In this section, we formulate all versions of the global-routing problem as integer programs.

We first describe the elements common to all versions of the problem. There is a variable $x_{n,i,j}$ for each net (n, i) and each j, $1 \le j \le I_n$. This variable is supposed to have the following meaning:

$$x_{n,i,j} = \begin{cases} 1 & \text{if net } (n, i) \text{ uses the route } T_n^j \\ 0 & \text{otherwise} \end{cases}$$

A routing R is described by an assignment to the variables $x_{n,i,j}$. Then, the traffic on edge e takes the form of the following expression:

$$U(x, e) := \sum_{\{(n,j) \mid e \in T_n^j\}} \sum_{i=1}^{k_n} w(n, i) \, x_{n,i,j}$$

The total wire length is

$$W(x) = \sum_{e \in E} \ell(e) \, U(x, e)$$

8.2.1 The Integer Program for the Constrained Global Routing Problem

The constraints of the integer program for the CONSTRAINED GLOBAL ROUTING problem are

$$x_{n,i,j} \in \{0,1\} \quad \text{for all nets } (n,i) \text{ and all } j, \ 1 \leq j \leq I_n$$

$$\sum_{j=1}^{I_n} x_{n,i,j} \leq 1 \quad \text{for all nets } (n,i) \tag{8.1}$$

$$U(e) \leq c(e) \quad \text{for all edges } e \in E$$

The first two sets of constraints ensure that at most one admissible route is chosen for each net. The third constraint set implements the capacity constraints on all edges. The cost function to be minimized is

$$c_{1,\ell}(x) = \lambda \sum_{(n,i) \in N} w(n,i) \left(1 - \sum_{j=1}^{I_n} x_{n,i,j}\right) + W(x)$$

$c_{1,\ell}(x)$ is the linearized version of the function $c_1(x)$. We achieve the linearization by multiplying the first term in $c_1(x)$ with a suitably large constant λ and adding both terms. Any constant λ that exceeds the maximum value of the second term in any feasible solution is appropriate. A possible value of λ is $\lambda = \sum_{(n,i) \in N} w(n,i) \cdot \sum_{e \in E} \ell(e)$.

Note that this integer program is a 0, 1-integer program.

8.2.2 The Integer Program for the Unconstrained Global Routing Problem

For the UNCONSTRAINED GLOBAL ROUTING PROBLEM, here we eliminate the capacity constraints from the constraint set. We ensure that all nets are routed by modifying the second set of constraints in (8.1) to

$$\sum_{j=1}^{I_n} x_{n,i,j} = 1 \quad \text{for all nets } (n,i)$$

We add a new variable x_L that is supposed to denote the maximum load of any edge. This meaning is ensured by the following constraints on x_L:

$$\frac{U(e)}{c(e)} \leq x_L \quad \text{for all } e \in E$$

The cost function to be minimized is

$$c_{\infty,\ell}(x) = \lambda x_L + W(x)$$

The factor λ is chosen as in Section 8.2.1 for the CONSTRAINED GLOBAL ROUT-ING problem.

8.2.3   Special Cases

If $\ell \equiv 0$, the cost functions given in the previous sections simplify to

$$c_{1,\ell}(x) = \sum_{(n,i) \in N} w(n,i) \left(1 - \sum_{j=1}^{I_n} x_{n,i,j} \right)$$

and

$$c_{\infty,\ell}(x) = x_L$$

If $w \equiv 1$, we can reduce the number of variables in the integer programs. Since copies of a net that are routed the same way are indistinguishable now, we just need to know how many of them there are. Thus, we reserve an integer variable $x_{n,j}$ for each net $n \in N$ and each admissible route T_n^j for n, $1 \le j \le I_n$. The value of $x_{n,j}$ is the number of copies of net n that are routed using route T_n^j. The constraints are modified accordingly. For the CONSTRAINED GLOBAL ROUTING problem, they become

$$x_{n,j} \text{ integer} \quad \text{for all } n \in N, \ 1 \le j \le I_n$$

$$\sum_{j=1}^{I_n} x_{n,j} \le k_n \quad \text{for all } n \in N$$

$$U(e) \le c(e) \quad \text{for all edges } e \in E$$

The cost function is then

$$c_{1,\ell}(x) = \sum_{n \in N} \left(k_n - \sum_{j=1}^{I_n} x_{n,j} \right)$$

The corresponding integer program is no longer a $0, 1$-integer program.

The modifications of the integer program for the UNCONSTRAINED GLOBAL ROUTING problem are analogous.

This merging of variables reduces the size of the integer programs substantially. In particular, that size is now *independent* of the number of nets. The size still depends on the number of routes and on the size of the routing graph.

8.2.4 The Integer Program for the Area-Routing Problem

We obtain the integer program for the AREA ROUTING problem by adding *vertex-disjointness* constraints to the integer program for the CONSTRAINED GLOBAL ROUTING problem with $w \equiv 1$ that was described in Section 8.2.3. These constraints are defined as follows:

$$\sum_{\{(n,j) \mid v \in T_n^j\}} x_{n,j} \leq 1 \quad \text{for all } v \in V$$

Note that the planarity constraints imply that the integer program for the AREA ROUTING problem is again a $0, 1$-integer program.

8.2.5 Solution of the Integer Programs

At first glance, the formulation of routing problems as integer programs seems to offer us little help in solving them. The integer programs are huge; often, their size is exponential in the size of the corresponding instance of the global-routing problem. The reason is that the number of admissible routes does not enter the size of the problem instance, but it affects the number of variables of the integer program. If we restrict ourselves to few admissible routes, such as one-bend routes for two-terminal nets, then the corresponding integer programs at least have a size that is a small-degree polynomial in the size of the problem instance. Even then, integer programming is a very difficult problem, and we cannot expect to gain much ground by applying general integer-programming techniques to the integer programs derived from global-routing problems. This observation applies particularly to the AREA ROUTING problem, since large routing graphs occur in this problem.

Small routing problems can be solved efficiently, however. This observation is the basis of all hierarchical routing algorithms (see Section 8.5). Figure 8.4 depicts a small floorplan F and its associated routing graph G for channel-free routing in general-cell layout. If $w \equiv 1$, the corresponding integer program has less than 20 constraints and about 60 variables, independent of the number of nets (Exercise 8.1). Integer programs of this size can be solved efficiently with standard methods, such as branch-and-bound based on the linear relaxation (see Section 4.6.2.1).

Furthermore, the special structure of the integer program can be exploited to find specially tailored fast solution algorithms. These algorithms are based on the observation that, on small routing graphs, we can partition the set of admissible routes into classes that can be routed separately in a specific sequence. In many cases, for instance, all two-terminal nets whose terminals are in adjacent rooms can be routed before all other nets (see Exercise 8.3). Such observations lead to a decomposition of the integer program for the routing

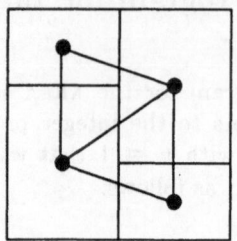

Figure 8.4: *A floorplan F with a small routing graph G.*

problem into smaller integer programs that can be solved in succession. Details can be found in [181].

8.2.6   Extensions of the Problem Formulation

An advantage of the integer program formulation is that constraints that are imposed in addition to the capacity constraints can be incorporated painlessly. Such constraints may arise from geometrical restrictions due to the origin of the routing graph (see Exercise 8.2, [299]) or due to technological rules for placing vias [266]. The additional constraints increase the size of the integer programs somewhat, but not forbiddingly so.

8.3   Overview of Global Routing Methods

Solution methods for routing problems can be grouped into three classes. The first class, and the one that has received the most attention in practice, is the class of *sequential routing algorithms*. Here, the Steiner tree problems are decoupled. The nets are ordered in some way, and a MINIMUM STEINER TREE problem is solved for finding the route of each net. The Steiner tree for the net has to be routed, given the already-existing routes. Thus, the edge capacities change after each routing of a net to accommodate the space taken up by the net just routed. The ordering of the nets, as well as the assignment of edge capacities, is done heuristically.

Since the MINIMUM STEINER TREE problem is NP-hard, heuristics have to be used for solving it (Section 3.10). In area routing, the routing graphs are so large that the Steiner tree heuristics have to be tuned further to reduce their resource requirements. Space is a primary bottleneck here. Sequential area-routing algorithms are discussed in Section 8.4.

The problem with this sequential approach to global routing is that it is inherently unfair to the nets routed late in the game. These nets have to

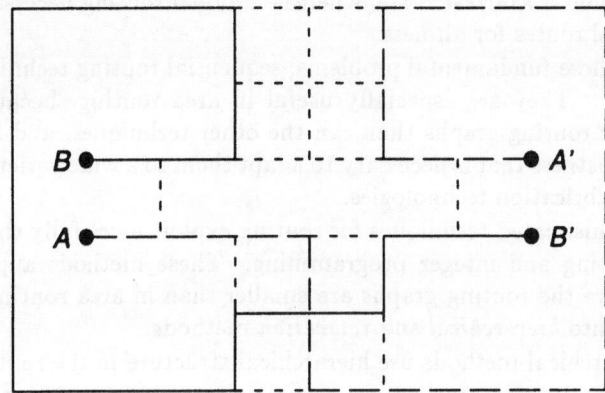

Figure 8.5: *An instance of the* CONSTRAINED GLOBAL ROUTING *problem that cannot be solved sequentially. The edges have lengths as drawn in the figure. All edge capacities are 1. There are two two-terminal nets* $n_1 = \{A, A'\}$ *and* $n_2 = \{B, B'\}$. *All routes are admissible. Observe that choosing a shortest route for either net precludes the choice of any route of the other net within the capacity constraints. However, there is a legal routing in which both nets have nonoptimal routes. This routing is highlighted in the figure. Note that the routing graph is a partial grid.*

maneuver through a largely filled-up routing graph. Thus, the routing depends highly on the order in which the nets are processed. There are $|N|!$ different orderings, and it is a difficult task to find an ordering that promises a good routing.

The approach used to tackle this problem in practice is the *rip-up and reroute* method. It deletes nets that have already been routed if they pose obstacles to nets that are still to be routed. The deleted nets have to be rerouted later on [38]. Other heuristics can be used to decide which nets to rip up at what time [85, 86]. A variation—the *rip-up and shove-aside* method—locally modifies the courses of nets routed earlier to make room for the net to be routed currently [85]. However, rip-up and reroute strategies do not provide a remedy in all cases.

Figure 8.5 illustrates an example for a small instance of the CONSTRAINED GLOBAL ROUTING problem that cannot be solved with a sequential routing algorithm. For the UNCONSTRAINED GLOBAL ROUTING problem, Figure 8.5 shows that there are instances for which no *optimal* solution can be found by a sequential routing algorithm. Note that, for Figure 8.5, rip-up and reroute strategies do not help either. The problem lies in the fact that for each net one attempts to find a *shortest route*, whereas a global routing necessarily consists of suboptimal routes for all nets.

Despite these fundamental problems, sequential routing techniques are still very popular. They are especially useful in area routing, because they can handle larger routing graphs than can the other techniques, and because they have the robustness that is necessary to adapt them to a wide variety of different design and fabrication technologies.

More sophisticated techniques for routing exploit more fully the connection between routing and integer programming. These methods apply to global routing, where the routing graphs are smaller than in area routing. They can be grouped into *hierarchical* and *relaxation* methods.

The hierarchical methods use hierarchical structure in the routing graph to decompose the integer program into small pieces that can be solved quickly. This hierarchical structure is imposed on the routing graph by the preceding floorplanning phase, or in a first phase of the global-routing procedure. The solutions of the partial integer programs are then tied together with approaches that are sometimes sequential and sometimes parallel. This approach does not lead to algorithms with provable quality, but does lead to a quite powerful set of heuristics. We discuss hierarchical global-routing algorithms in Section 8.5.

Relaxation methods actually attack the total integer program. They use some relaxation of it as the basis for the computation of a routing. A prominent example is provided by the randomized techniques. Here, the relaxation used is the linear relaxation. Its solution provides a probabilistic bias for a randomized experiment that aims at finding a good routing. Randomized routing methods can be proved to perform well with high probability, and, in fact, they can be made deterministic. Relaxation methods are discussed in Section 8.6.

8.4   Sequential Routing Algorithms

The seminal work that started research on sequential routing algorithms is that of Lee and Moore [265, 332]. Both had area routing on printed circuit boards in mind. Lee considers partial 1-grids. Moore considers arbitrary graphs. In both cases all edges have length 1 and the algorithm uses a wire routing method that is a version of Dijkstra's algorithm that incorporates the linear-time implementation of the priority queue of Figure 3.13 and an efficient coding scheme for the information associated with each vertex. Although Moore considered arbitrary graphs most of the further developments of the algorithm have been obtained with partial k-grids or even partial 1-grids in mind. In this context, the algorithm is called *Lee's algorithm*.

Since Lee's and Moore's initial papers, research has progressed in several directions. The algorithm has been made more efficient on practical area-routing problems. This streamlining is done using clever coding tricks that exploit the fact that all edges have length 1 and that the underlying graph is a partial k-grid. Furthermore, the goal-directed search techniques described in Section 3.8.5 have been tuned for use in Lee's algorithm. The corresponding efficient versions of Lee's algorithm are called *maze-running algorithms*; we discuss them in detail in Section 8.4.1. To reduce the space requirement incurred by maze-running algorithms, some algorithms do not represent all parts of the routing region explicitly. Rather, only the obstacles are represented. The paths between the obstacles are computed with geometrical path-searching methods. The corresponding path algorithms are called *line-searching algorithms*; they are discussed in Section 8.4.2. Furthermore, Lee's algorithm has been extended to cope with Steiner tree problems for multiterminal nets (see Section 8.4.3.1). Finally, the robustness of Dijkstra's algorithm has been exploited to allow for optimizing cost functions different from path length; this topic is the subject of Section 8.4.4.

8.4.1　Maze-Running Algorithms

Ideas for making Lee's algorithm more efficient while maintaining the condition that shortest paths are found can be grouped into two classes. One class reduces the space requirement of Dijkstra's algorithm by minimizing the amount of information that has to be stored for each vertex (see Section 8.4.1.1). The other class tries to decrease the number of vertices scanned by Dijkstra's algorithm (see Section 8.4.1.2).

8.4.1.1　Coding Techniques

In area routing, very large routing graphs can occur. Space requirements become a critical bottleneck in such applications. In Dijkstra's algorithm, every scanned vertex has a defined d value and p-pointer. The p-pointer can be eliminated if the d value is used for path tracing. However, in area routing, the number of scanned vertices can become so large that we cannot afford to reserve a whole word for each scanned vertex to store that vertex's d value. There are a number of packing techniques that reduce the space requirement for the d values. All of them are based on the observation that, since we are solving only single-*pair* shortest-path problems, the d values for vertices that *are not in the priority queue D* are not of interest per se, but are of interest only insofar as they enable us to trace back the shortest path. Thus, the d values for such vertices take over the role of the p-pointers, and therefore they can be represented with a small number of bits. (The d values for vertices *inside* the priority queue are still stored in their original form.) We now present some of the most popular packing techniques. For ease of presentation, we consider partial 1-grids. We shall remark on the extensions to k-grids, if they are not obvious.

Coding Technique I In a partial 1-grid, $d[v]$ can be taken from the set $\{up, down, left, right\}$; that is, it can be coded in two bits. This suggestions was made by Lee in his original paper [265].

Coding Technique II If the routing graph is a partial k-grid and all edge lengths are 1, then a very efficient packing strategy can be devised [8]. This strategy is based on the following observation.

Lemma 8.1 *Let G be a partial k-grid with unit length edges, and let s, v be vertices in G. Then, for each neighbor w of v in G, either $d(s, w) = d(s, v) - 1$ or $d(s, w) = d(s, v) + 1$.*

Proof It is sufficient to show that the parities of $d(s, v)$ and $d(s, w)$ differ. Then the lemma follows since, for all neighbors w of v, we have $d(s, v) - 1 \leq d(s, w) \leq d(s, v) + 1$.

We now show inductively on the length of a shortest path from s to v that the parity of $d(s, v)$ is the parity of the Manhattan distance $||s - v||_1$ between s and v. Clearly, this fact holds for $v = s$. If $d(s, v) > 0$, let w be the predecessor of v on a shortest path from s to v. By the inductive hypothesis, $d(s, w) \bmod 2 = (d(s, v) - 1) \bmod 2 = ||s - w||_1 \bmod 2$. Also, $||s - w||_1 \bmod 2 \neq ||s - v||_1 \bmod 2$. Thus, $d(s, v) \bmod 2 = ||s - v||_1 \bmod 2$. This proves the lemma. $\Box$

Each neighbor w of v such that $d(s, w) = d(s, v) - 1$ is a predecessor of v on a shortest path from s to v. If we use a binary label to distinguish these neighbors from the other neighbors of v, then $d[v]$ has only to take on a binary value. Specifically, we set $d[v] = \lfloor (d(s, v) \bmod 4)/2 \rfloor$. As $d(s, v)$ increases from 0, the labels go through the sequence $\sigma = 0, 0, 1, 1, 0, 0, 1, 1, \ldots$. Because of Lemma 8.1, with this labeling we have

$$d(s, w) = d(s, v) - 1 \Leftrightarrow \begin{array}{l} d(s, v) \text{ even and } d[v] \neq d[w] \\ \text{or } d(s, v) \text{ odd and } d[v] = d[w] \end{array}$$

Thus, we can trace the path by observing the distance of t from s and following along a sequence of labels that is the reverse of an appropriate prefix of σ.

Although it compacts the storage space, this labeling strategy has several drawbacks. First, it makes the process of path recovery more complicated, because several neighbors of each vertex may have to be probed. Second, if several shortest paths are computed on the same routing graph, then all d values have to be erased before the next path is searched. The reason is that the succinct labeling entails a lack of redundancy that can lead to erroneous results in later shortest-path computations. We must decide for each application whether the savings in space are worth the additional run time.

Coding Technique III If the routing graph has unit length edges but is not a partial k-grid, then we can only assume that, for each neighbor w of v, $d(s,v) - 1 \leq d(s,w) \leq d(s,v) + 1$. In this case, we need to distinguish between three different distance values in the neighborhood of a vertex, and thus we can use the labeling $p[v] = d(s,v) \bmod 3$. The corresponding labeling sequence as the distance increases is $0, 1, 2, 0, 1, 2, \ldots$. A predecessor of v on a shortest path is any neighbor w such that $p[w] = (p[v] - 1) \bmod 3$ [332]. In this case, d values are ternary. This may still be better than the quaternary scheme of Coding Technique I. In general, other information has to be attached to a vertex, such as whether that vertex is already occupied with a wire. This additional value can also be accommodated in two bits with the ternary packing scheme presented here, but not with the quaternary packing scheme of Coding Technique I.

8.4.1.2 Goal-Directed Search Techniques

Several authors discuss goal-directed search techniques in the context of Lee's algorithm [161, 193, 384]. In general, the methods described in Section 3.8.5 can be employed to reduce the number of scanned vertices. However, if the routing graph is a partial k-grid and all edges have unit length or the edge lengths are integer and are bounded by a small constant C, we can improve the efficiency of the implementation.

Rubin [384] incorporates the lower bound $b(v,t)$ on the distance from v to the target t as a secondary cost measure, as described in Section 3.8.5.1. Rubin also proposes a heuristic way of keeping the number of wire bends low. In Section 8.4.4, we shall describe a way of minimizing the number of bends exactly with second priority—that is, among all shortest paths.

Hadlock [163] provides an efficient implementation of the priority queue in the case of k-grids with unit edge lengths. He observes that, if all edge lengths in the original routing graph are 1, then, after the modification of the edge lengths for goal-directed unidirectional search, all edge lengths are 0 (if the edge leads toward the target) or 2 (if the edge leads away from the target). After dividing edge lengths by 2—this does not change shortest paths—we can use the efficient priority-queue implementation of Figure 3.13.

Hoel applies this efficient priority-queue implementation to the case that edge lengths are from the set $\{1, \ldots, C\}$.

Other heuristic suggestions for speeding up the Lee's algorithm are presented in [9, 136, 341].

8.4.1.3 Finding Suboptimal Paths

All algorithms in Section 8.4.1 are variations on the breadth-first search scheme used by Dijkstra's algorithm. As a consequence, they find a shortest path

between s and t, if one exists. Although goal-directed search techniques can
reduce the number of vertices that are scanned, we may still end up looking at
many vertices. Incorporating depth-first search techniques into the algorithm
may lead to a further reduction of the number of scanned vertices. In general,
however, depth-first search will not find the shortest path. This disadvantage
may not be critical, since we have seen that the total wire length does not
always have a close correlation with routability. In fact, choosing the shortest
path can preclude routability (see Figure 8.5). On the other hand, it is of more
importance to ensure that a path is found whenever one exists. Giving up *this*
property of the algorithm is a great sacrifice and should be considered only if
the corresponding efficiency gain is tremendous. Let us first discuss versions of
path algorithms that always find a path if there is one, but not necessarily the
shortest path.

    If we do not need to find a *shortest* path, then we can improve the goal-
directed unidirectional search strategy whose secondary cost measure is the
lower bound value $b(v, t)$ (see Section 3.8.7) by making it the *only* cost measure.
The idea is to let the search be guided not by increasing distance from s, but
rather by the decreasing lower bound $b(v, t)$. Thus, we always proceed at the
currently most-promising vertex. To this end, we order the vertices in D *only*
according to increasing value of the lower bound $b(v, t)$. The path algorithm
has to be adapted to this change to make sure that a path is always found
if one exists, and that as few vertices as possible are visited. The modified
algorithm uses a Boolean array *seen* to remember which vertices have been
visited. The pointer array p stores the trace pointers through which the path
can be recovered. The pointer array r is used to scan the adjacency lists of each
vertex during the search. The algorithm is shown in Figure 8.6. The general
step in lines 8 to 17 begins with *inspecting* the vertex v with the smallest lower
bound from the priority queue D (line 9). It is important to note that, at this
point, the vertex v is not yet removed from D. The next neighbor $r[v]$ on the
adjacency list $A[v]$ of v is inspected. If $r[v]$ has not been seen before, then
its trace pointer is set to v (line 15) and it is entered into D (line 16). If all
neighbors of v have been scanned, then v is removed from D. Note that not
all edges out of v need to be inspected at once. If one edge leads us closer to
the target, its other endpoint moves to the front of D and is considered right
away.

    Unless the target is encountered earlier, this algorithm eventually visits all
vertices that are reachable from v. Thus, it finds a path from s to t if there is
one. Figure 8.7 gives an example in which a path that does not have minimum
length is found. The figure demonstrates that the number of scanned vertices
can be reduced below what the shortest-path algorithms achieve (compare with
Figure 3.14). If the routing graph is a partial k-grid, there is a sense of *direction*
of a edge (horizontal, vertical, and so on). Here, a useful heuristic to adopt is
to continue moving into the same direction. This means that the first neighbor
of v scanned in lines 10 ff. lies opposite to the neighbor of v from which v

```
(1)       seen : array [vertex] of Boolean;
(2)       p, r : array [vertex] of vertex;
(3)       D : priority queue;
(4)       v : vertex;

(5)       seen[s] := true;  p[s] := s;  r[s] := A[s];  D := empty;
(6)       enter s into D;
(7)       for v ≠ s do seen[s] := false; p[v] := nil; r[v] := A[v] od;
(8)       while D ≠ ∅ do
(9)            v := extract minimum from D;
(10)           if r[v] = t then succeed fi;
(11)           if r[v] = nil then
(12)               remove v from D
(13)           else if seen[r[v]] = false then
(14)               seen[r[v]] = true;
(15)               p[r[v]] = v;
(16)               enter r[v] into D;
(17)               r[v] := next(r[v]) fi od;
(18)      fail;
```

Figure 8.6: *Fast path algorithm.*

was reached. This heuristic gives the algorithm the flavor of searching along straight lines (see also Section 8.4.4). Also, note that the pointers $p[v]$ and $r[v]$ can be packed into a few bits using the tricks discussed in Section 8.4.1.1. In addition, the pointer $r[v]$ has to be maintained only as long as v is in D.

If all edges in the routing graph have unit length and if the lower bounds fulfill the consistency conditions (3.4) (see Section 3.8.5.1), then we can implement the priority queue D as a pair of a list M that holds all minimum vertices in D and a stack S that holds all other vertices in D ordered by increasing value, smallest values on top. To see why this is true, we note that, because of the consistency conditions for each vertex $v \in V$ and for each neighbor w of v, we have $b(v,t) - 1 \leq b(v,t) \leq b(v,t) + 1$. Entering w into D is done as follows:

Case 1: $b(v,t) = b(v,t) - 1$. Push all elements of M onto S. Make w the only element in M.

Case 2: $b(v,t) = b(v,t)$. Insert w into M.

Case 3: $b(v,t) = b(v,t) + 1$. Push w onto S.

The minimum vertex in D can be chosen to be any vertex in M. Soukup [426]

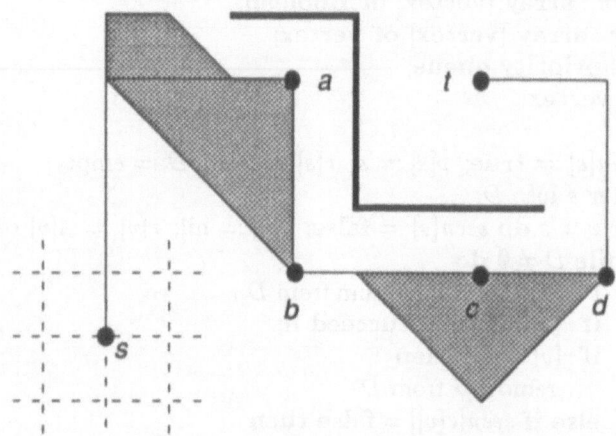

Figure 8.7: *Unidirectional search for a path without minimum length.*
We assume that the adjacency lists contain the neighbors of a vertex in
clockwise order, starting with the top neighbor. The search proceeds
along a path to vertex a, then backs up along wavefronts until vertex
b is found. Then, the search proceeds to vertex c. After a second
back-up phase, the search proceeds from vertex d to vertex t along a
direct path. The shaded area covers all vertices encountered (that is,
seen) during the search. The path found has a length of 20. This is 6
units above the minimum length. Compare with Figure 3.14.

incorporates this idea into a maze-running algorithm for partial k-grids. On
partial k-grids, case 2 never happens.

This implementation of the priority queue, which allows each operation to
be performed in time $O(1)$, can be generalized to the case that all edge costs
are at most a constant C (Exercise 8.4).

8.4.2   Line-Searching Algorithms

The prime disadvantage of maze-running algorithms is that they operate on
very large routing graphs. Since data must be kept for each vertex in the routing
graph, the memory requirements of maze-running algorithms are excessive.
This observation was made early on and, as a consequence, methods have been
investigated with which the path-finding process can be made more efficient.

The prime idea is effectively to eliminate a large number of vertices from the
routing graph by expanding the path search along only simple patterns that
avoid most vertices in the routing graph. Here we may or may not want to
insist on finding shortest paths. In a grid graph, the simplest search pattern we

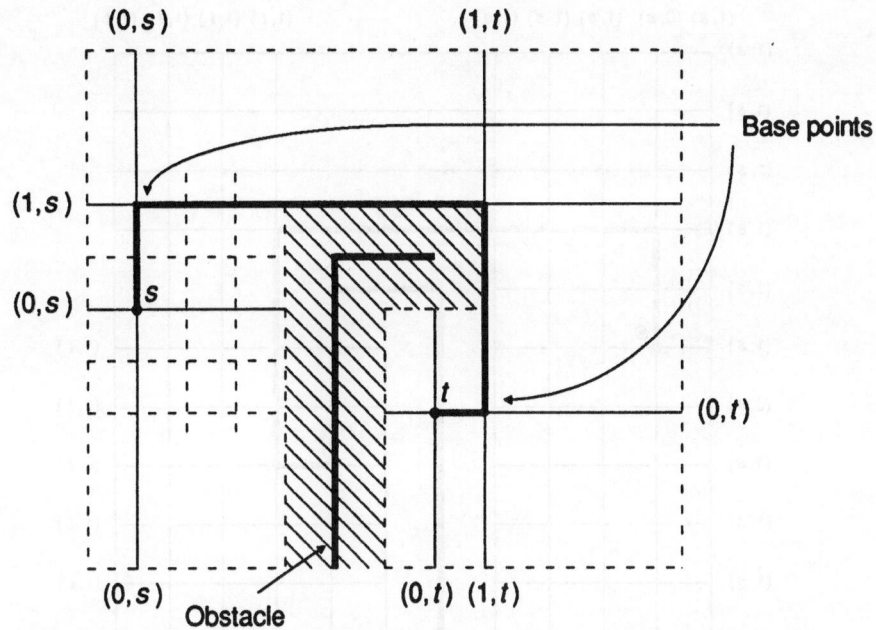

Figure 8.8: *The line-searching algorithm by Hightower. The points s and t are connected. Trial lines are labeled with* (level, source) *pairs. The constructed path is highlighted.*

can think of is a straight line. Therefore, the corresponding routing algorithms are also called *line-searching algorithms.*

8.4.2.1 Heuristic Techniques

The first line-searching algorithms were provided by Hightower [189] and Mikami and Tabuchi [322]. We describe the algorithm by Hightower (see Figure 8.8) intuitively.

The algorithm performs a bidirectional search. Straight search lines are emanated from both s and t, in all four directions. These search lines are called *level-0 trial lines.* If a search line emanated from t meets one emanated from s, then we have found a path with at most one bend. Otherwise, the trial lines hit obstacles. In this case, *level-1 trial lines* are emanated starting at *base points* on the level-0 trial lines. Hightower presents several heuristics for choosing the base points. The main goal is to escape from the obstacle that prevented the trial line on which the base point is located from continuing. We continue in this way until a trial line from s meets a trial line from t.

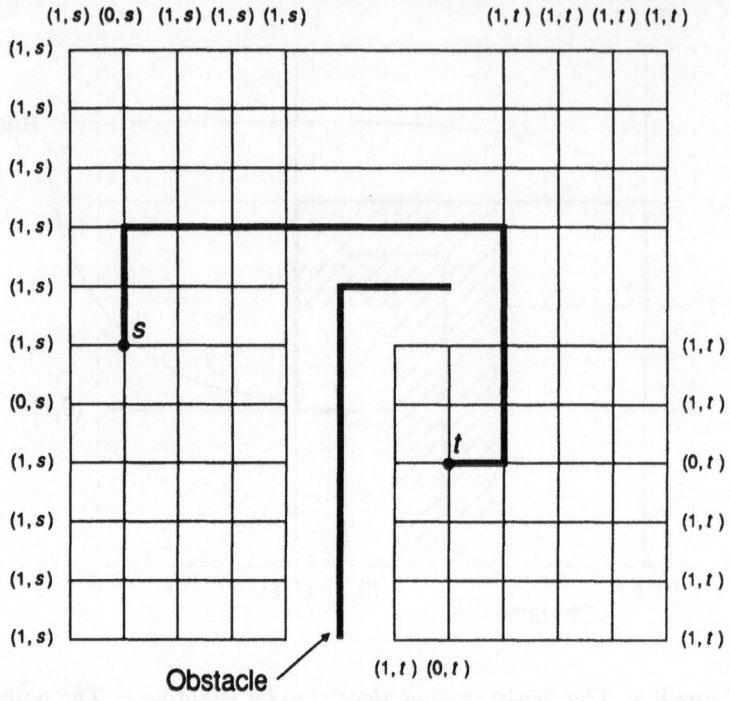

Figure 8.9: *The line-searching algorithm by Mikami and Tabuchi,
executed on the example of Figure 8.8.*

Hightower's algorithm does not necessarily find a route from s to t, even
if such a route exists. Mikami and Tabuchi [322] generate level-k trial lines
from all possible base points on level-$(k - 1)$ trial lines. In this way, they can
ensure that a path will be found if one exists. However, the path may not be
the shortest one. Figure 8.9 illustrates the algorithm by Mikami and Tabuchi,
executed on the example of Figure 8.8. We can see from the figure that the
algorithm by Mikami and Tabuchi tends to cover substantially more of the
routing graph than does Hightower's algorithm.

If the routing graph has many irregularly shaped obstacles, then we need
many levels of trial lines to connect to vertices. In this case, *pattern routers*
are of service. These routers do not just expand the search along straight
lines, but instead try different patterns—such as U- or Z-shaped routes. As
the routing graph becomes more irregular, maze running becomes preferable
to line searching.

Heyns et al. [188] present a heuristic method that combines maze-running
and line-searching techniques. They expand a trial line by generating wave-

fronts in both directions *orthogonal* to the trial line in a breadth-first fashion. If wavefronts originating from s and t meet, we have found a solution. Otherwise, the boundary points of the *expansion zone* covered by a wavefront are the base points of new wavefronts in the orthogonal direction. Figure 8.10 illustrates the algorithm. Special *stop lines* are created when two wavefronts originating from the same terminal cross, to prevent multiple exploration of the same region of the routing graph. Since the algorithm eventually explores the whole routing graph, it is sure to find a solution if one exists. Again, the algorithm may not find shortest paths (see Exercise 8.5).

8.4.2.2 An Algorithmic Framework for Line Searching

The methods presented in the previous section are highly heuristic. In the past decade, an algorithmic framework for line searching has been emerging that is based on computational geometry. To explain the basis for this framework, we shall review the problem we want to solve.

We want to find a path from s to t in the routing graph G. To do so, we have to explore some vertices in G. But not all vertices in G are necessary. Ideally, we would need to concentrate only on the vertices on a shortest path. Of course, this set of vertices is just what we want to find. Nevertheless, it may be possible to eliminate from consideration some vertices in G that we can show are not necessary for the creation of a shortest path. If we are interested only in finding some path, then we may even be able to eliminate more vertices. These observations lead us to the following definition.

Definition 8.1 (Connection Graph) *Let $G = (V, E)$ be a connected graph with edge length function $\lambda : E \to \mathbf{R}$; and let $V' \subseteq V$ be a subset of the vertices of G. Let G' be a graph to which a subgraph of G that spans V' is homeomorphic. (This means that edges in G' correspond to paths in G; see Definition 3.29. The lengths of the edges in G' must be the same as the lengths of the corresponding paths in G.)*

    1. G' is called a **connection graph** *for V' in G if all pairs of vertices in V' are connected in G'.*

    2. G' is called a **strong connection graph** *for V in G if, for all pairs of vertices in V', the lengths of shortest paths are the same in G and G'.*

If we are interested in finding shortest paths, we want to compute quickly a small strong connection graph for the set of vertices that we wish to connect. If we are interested only in finding some path, we are content with a small connection graph. The restriction of the path search on the connection graph is the formalization of the restricted line searches we did heuristically in the previous section. In fact, Hightower's algorithm is nothing but the attempt to search as small as possible a connection graph. Unfortunately, Hightower does not always explore a complete connection graph, such that the algorithm

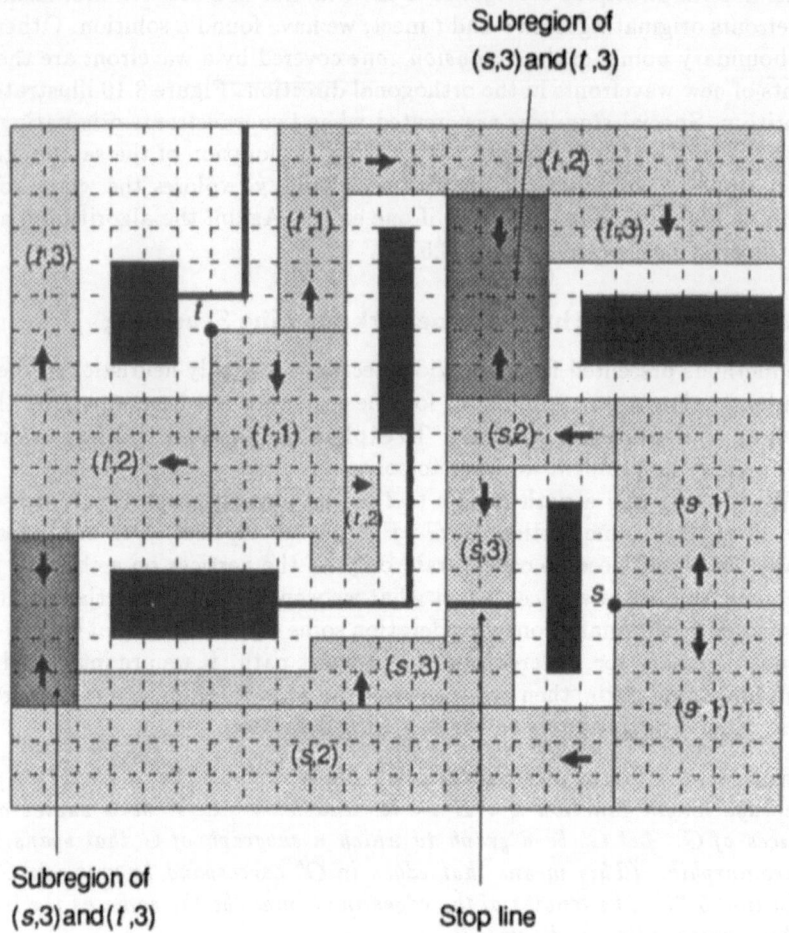

Figure 8.10: *The algorithm by Heyns et al.*

is not always successful. Mikami and Tabuchi as well as Heyns et al. search completely through connection graphs, but as Figures 8.9 and 8.10 illustrate, these connection graphs can be quite large.

The task of designing fast line-searching algorithms can be understood as the quest for finding small connection graphs. However, stated in terms of connection graphs, the line-searching paradigm extends beyond grid graphs. In fact, we can extend the paradigm even into the continuous realm, asking for connections in the plane instead of some graph [465]. We shall not detail these general results here, but shall instead confine ourselves to paths in grid graphs.

Usually, the global-routing problem entails a routing graph G that is a partial grid graph. We obtain G by eliminating some sets of vertices and edges from the grid that correspond to rectagonal cells or previously routed wires. We call such structures *obstacles* because they obstruct the straight path of a wire in the complete routing grid. We can identify the vertices of G with points Z^2. The vertices that we wish to connect are the pins of the cells. The set V' of pins is a prespecified vertex set of G that is determined by the instance of the global-routing problem.

Actually, the obstacles are sets of vertices and edges removed from the complete routing grid, but we can represent an obstacle more efficiently by listing the straight-line segments in the plane that make up the obstacle boundaries. Let the set of line segments delimiting obstacles be denoted with B. The sets V' and B, given as points and line segments in the plane, define G implicitly, and they do so in much less space than would an explicit description of G in terms of adjacency lists. Maze-running algorithms first generate all of G from this implicit definition, and then start the path search. In line searching, we do want to generate not G, but rather only a small (strong) connection graph for V' in G.

8.4.2.3 The Track Graph

Wu et al. [475] introduce a rather small connection graph—the *track graph* G_T—and show how it can be constructed. We describe the track graph informally. For an illustration, see Figure 8.11.

For the purpose of simplicity of the explanation, we restrict ourselves to sets B that describe b mutually disjoint *isoconvex* rectagonal polygons. We call each polygon an *obstacle*. These are polygons with horizontal and vertical boundary segments, such that, with the endpoints of a horizontal and vertical line segment, the whole line segment is contained in the cell (compare with Definition 9.3). The results we present can be generalized to arbitrary rectagonal polygons [475] (see Exercise 8.9).

A track graph is a graph that can be constructed as follows. Call the leftmost and rightmost vertical and the topmost and bottommost horizontal boundary line segments of an obstacle the *extreme boundary segments*. (An isoconvex obstacle has four extreme boundary segments.) We extend all extreme boundary segments of obstacles beyond both of their endpoints until they meet another cell boundary. The resulting lines are called *obstacle tracks*. The set C of corner points of the obstacle boundaries, as well as the set X of intersection points of obstacle tracks, are vertices of G_T. Furthermore, we extend tracks from each pin in all four directions, until they hit an already constructed track or an obstacle boundary. The set P of the respective points of intersection is also added to the vertex set of G_T. The edges of G_T are the vertical and horizontal segments of tracks and obstacle boundary segments connecting the vertices of G_T.

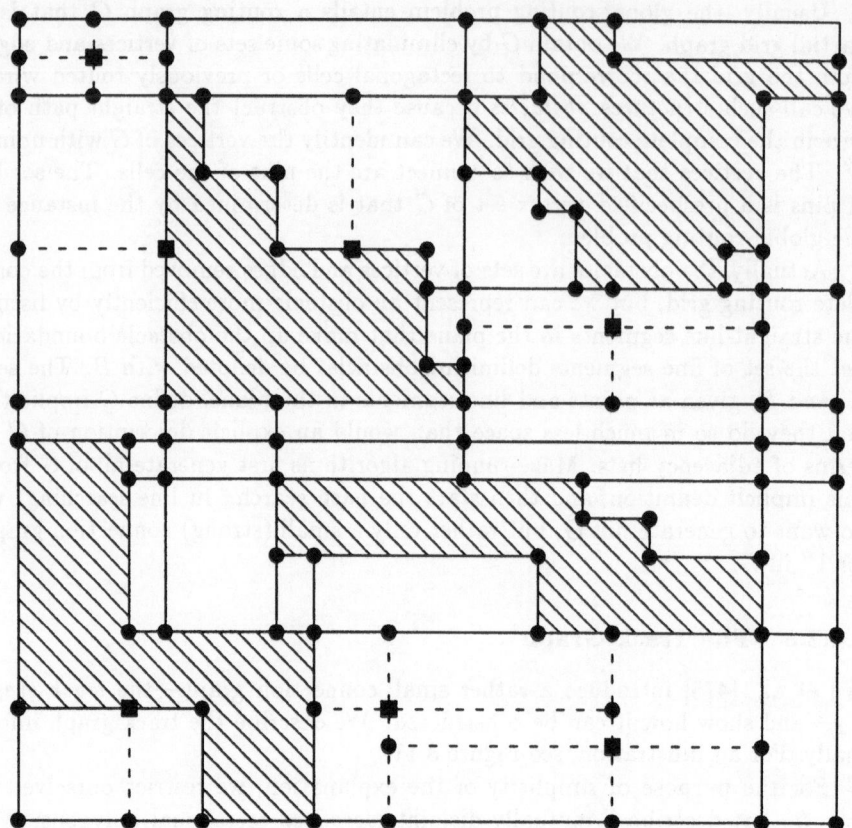

Figure 8.11: *The track graph. The interiors of obstacles are shaded.*
Obstacle tracks and obstacle boundaries are drawn solid. Pin tracks
are drawn dashed. Pins are denoted by squares.

A track graph G_T has $O(b^2 + |V'|)$ vertices and edges. It is not difficult to
see that two pins are connected in G_T if they are connected in G; thus, G_T
is a connection graph for V'. Wu et al. [475] even show that, under certain
conditions, the shortest path between two pins p_1 and p_2 in G_T is also a shortest
path in G. In particular, this situation occurs if some shortest path between
p_1 and p_2 in G crosses a horizontal and a vertical obstacle track. In general,
however, G_T is not a strong connection graph for V' (see Exercise 8.6).

There can be several track graphs for the same problem instance, because the length of the pin track depends on the order in which the pin tracks are generated. If we had run pin tracks up to the next obstacle track, then the track graph would have been unique, but its size could have grown with the square of the number of pins.

8.4.2.3.1 Construction of the Track Graph

We can construct a track graph using a straightforward version of the powerful plane-sweep technique from computational geometry. This technique is also applied in channel routing (Chapter 9) and compaction (Chapter 10). In the first phase, we construct the horizontal and vertical obstacle tracks. In the second phase, we generate the pin tracks. In the third phase, we build up the track graph from the information gained in the first two phases.

8.4.2.3.1.1 The First Phase

Since the first phase works the same for horizontal and vertical tracks, we restrict our attention to vertical tracks.

Intuitively, we sweep a vertical line from left to right across the set of obstacles. When the line crosses a track, we generate this track.

This intuition can be made precise as follows. First, we sort all vertical boundary segments according to increasing x coordinate. Then, we process the segments in this order. Processing a vertical boundary segment can be understood as handling the time instant that the sweep line crosses the boundary segment. Thus, we can associate with each boundary segment an *event*.

While processing the events, we maintain a data structure that indicates which part of the sweep line covers obstacles, and which part covers empty wiring space. This information can be stored in an ordered set D containing the y coordinates of horizontal obstacle boundaries crossed by the sweep line. Each value in D is labeled with information of whether it is the top or bottom boundary of an obstacle. If D is implemented as a balanced leaf-chained search tree, then we can insert and delete elements from D in time $O(\log |D|)$, and can find neighbors of elements in D in constant time. Initially, D is empty.

Case 1: If e is a left extreme edge, then we find the two points in D whose values are next above the top coordinate of e and next below the bottom coordinate of e, respectively. These two points must be neighbors in D if the obstacles do not overlap. The interval between these two points defines the track for e. Finally, we enter the y coordinates of both endpoints of e into D.

Case 2: If e is a right extreme edge, the corresponding track goes from the upper neighbor of the top endpoint of e in D to the lower neighbor of the bottom endpoint of e in D. We delete the y coordinates of both endpoints of e from D.

Case 3: If e is not an extreme edge, then the action depends on whether e forms a right or a left boundary of its obstacle. Which of these cases applies can be inferred from the specification of the obstacles. Say, e forms a left boundary. (The other case is analogous.) If the bottom endpoint of e is also the right endpoint of a horizontal boundary edge for the same obstacle, then we exchange the (y coordinate of) this endpoint with the (y coordinate of the) top endpoint of e in D. Otherwise, the top endpoint of e is also the right endpoint of a horizontal boundary edge for the same obstacle. In this case we exchange the (y coordinate of) this endpoint with the (y coordinate of the) bottom endpoint of e in D.

If there are several vertical boundary segments with the same x coordinate, then we have to modify the algorithm so as not to generate tracks multiply (see Exercise 8.7).

8.4.2.3.1.2 The Second Phase In the second phase, we compute the pin tracks. We can use the same algorithm as in Section 8.4.2.3.1.1. For the generation of the vertical pin tracks, we just add the horizontal obstacle tracks as additional obstacles. For the horizontal pin tracks, we proceed similarly.

8.4.2.3.1.3 The Third Phase After the second phase, we have a set of horizontal and a set of vertical line segments—namely, the cell and pin tracks—that we want to intersect. From these segments, we want to compute an adjacency-list description of the track graph. This task also can be accomplished with plane-sweep techniques; we perform a left-to-right plane sweep. We maintain the y coordinates of the horizontal segments crossing the sweep line in an ordered set D. We implement D with a balanced search tree. Events are now defined by the endpoints of horizontal segments and the x coordinates of vertical segments. An event corresponding to an endpoint of a horizontal segment triggers the insertion or deletion of an element in D. An event corresponding to the x coordinate of a vertical segment triggers a query of D that returns all elements whose values lie in the range between the y coordinates of the topmost and bottommost point in the segment. The implementation of these procedures is straightforward (see Exercise 8.8).

8.4.2.3.1.4 Complexity of the Construction We never have to sort more than $O(|B| + |V'|)$ elements. Also, the data structure D never contains more than $O(|B| + |V'|)$ elements. The handling of each event in any of the plane sweeps takes logarithmic time in the size of D. In the third phase, we generate up to $t = O(b^2 + |V'|)$ vertices. Thus, the run time is $O((|B| + |V'|) \log(|B| + |V'|) + t)$.

A tighter run-time analysis is contained in [475]. This reference also describes a method for finding *shortest* paths on the basis of the track graph. The approach here is not to enhance the track graph with edges so as to make it a strong connection graph—this modification would make the track graph too large. Rather, we use another plane sweep to find shortest paths between points for which no shortest path lies in the track graph. In effect, the track graph is used as a skeleton that divides the routing region into small subregions through which we can search for paths efficiently (see Exercise 8.6).

Several authors have investigated the efficient construction of small strong connection graphs. Assume that n is the number of obstacle corners plus the number of pins. De Rezende et al. [84] present a construction of a strong connection graph with $O(n \log n)$ edges in time $O(n(\log n))$, if all obstacles are rectangles. Their construction relies on the central fact that, in this case, there always is a shortest path between two pins that is monotonic either in the x or in the y dimension. Clarkson et al. [70] generalize this result to the case of rectagonal and, in fact, arbitrarily shaped obstacles. The run time increases to $O(n(\log n)^2)$. The central contribution of Clarkson et al. is to add judiciously to the graph points that can be used by shortest paths between many pairs of pins. Widmayer [465] modifies the construction of Clarkson et al., and manages to compute the graph in time $O(n \log n)$ in the general case.

8.4.2.4 Summary

Line searching provides an attractive framework for reducing the memory and run-time requirements of shortest-path computations in large routing graphs. The subject was initiated in the seventies, when it was approached highly heuristically. More recently, a graph-theoretic framework for line searching has emerged. Techniques for computational geometry provide valuable tools for the development of line-searching methods. In fact, these techniques are powerful enough to extend the line-searching paradigm to the continuous domain, where cells have arbitrary planar shapes. In this form, line searching is of great interest in other applications, such as robotics. On the other hand, the methods from computational geometry seem to rely on the planarity of the routing region. The algorithmic framework for line searching has yet to be made applicable to line searching for routing with many wiring layers.

8.4.3 Steiner Tree Heuristics in Sequential Routing

We can apply the Steiner tree heuristics given in Section 3.10 to sequential routing.

8.4.3.1 Steiner Tree Heuristics for Maze Running

Akers [9] suggests that we use the method described in Section 3.10.2 for the computation of Steiner trees. This method constructs a Steiner tree whose cost is within a factor of $2(1 - 1/\ell)$ of optimal in the worst case. It has a large run time on nets with many terminals, but such nets are rare in practical settings. Its attractiveness stems from the fact that it is a simple succession of ordinary executions of Lee's algorithm. Thus, it is quite robust, and tricks such as labeling schemes and goal-directed search can be employed.

Alternatively, we can use either of the methods described in Sections 3.10.3 and 3.10.4. These methods are asymptotically faster, but they are not amenable to incorporating goal-directed search techniques.

Soukup and Yoram [428] suggest a *tree-driven* area-routing algorithm that works in two phases. In the first phase, a *global route* is computed that disregards all obstacles in the routing region. In the second phase a *detailed route* is computed that tries to deviate from the global route as little as possible. In a sense, Soukup and Yoram apply the two-phase approach to area routing.

The global route is found quite simply. We just compute a Steiner tree for the net on the *complete k-grid*. This task is particularly simple because of a result by Hanan [170].

Lemma 8.2 *Let n be a net in a complete k-grid with unit length edges. Then, there is a minimum Steiner tree for n each of whose vertices can be connected with a straight path in one of the three grid dimensions to a terminal of n. (The path is not necessarily part of the minimum Steiner tree.)*

Hanan proved Lemma 8.2 for $k = 1$. His proof methods extend to $k > 1$; in fact, they extend to higher-dimensional grids (see Exercise 8.10). By Lemma 8.2, we can find the global route by solving a MINIMUM STEINER TREE problem on a graph with at most $(k|n|)^2$ vertices; namely, the k-grid that is created by rays from all terminals of n in all directions. Since the size of n is usually small, the corresponding instance of the MINIMUM STEINER TREE problem is small, as well. If it is small enough, we may solve the problem exactly. Alternatively, we can use any of the methods described in Section 3.10.1 to solve it. Note that, if the routing graph is a (complete) 1-grid, these approximation algorithms have an error bound of at most 3/2, by the result of Hwang [203].

A simple scaling argument shows that Lemma 8.2 and the bound by Hwang extend to the case where all edges *along the same dimension* have the same length. If the routing graph has a more uneven distribution of edge lengths, then Lemma 8.2 does not hold. However, we can still use the same method for computing Steiner trees; it may just yield worse results.

In a second pass, we find a Steiner tree on the actual routing graph (with obstacles) by successively adding paths to new terminals (Section 3.10.2). Here, a cost function measuring the deviation from the global route is put into the cost function for the search as a low-priority item.

This algorithm performs well if there are not too many obstacles. Otherwise, the global route, not taking the obstacles into account, does not provide a good guidance for the path-expansion process.

8.4.3.2 Steiner Tree Heuristics for Line Searching

In Section 8.4.2.2, we reduced the line-searching method to classical path finding on a sparse connection graph. Heuristically, we can therefore just apply the Steiner tree algorithms from Section 3.10 to this connection graph. However, we do not get analyzable results this way.

If we have a strong connection graph at hand, then we can use the Steiner tree heuristics described in Sections 3.10.1 to 3.10.4 and obtain generalized minimum spanning trees that meet the bound given in Theorem 3.2.

As we have mentioned in Section 8.4.2.3.1.4, the track graph is not a strong connection graph, but it can be used as a skeleton for efficient path search. Wu et al. [475] discuss methods for computing generalized minimum spanning trees of G on the basis of the track graph.

8.4.4 Incorporating Other Cost Measures

Shortest length is only one of the cost functions we want to optimize in area routing. Other possible cost functions include

- The number of bends of the path

- The number of layer changes

- The number of crossings over previously routed wires

- The number of segments that a wire runs parallel to another wire (on top or below) in a multilayer routing graph

- The distances to obstacles

The shortest-path framework we are using is general enough to accommodate some of these cost functions. Others lie outside the framework and must be attacked with different algorithmic methods.

8.4.4.1 Applications of Dijkstra's Algorithm

We can incorporate parallel wire segments, distances to obstacles, and layer changes by giving the respective edges appropriate lengths.

We can formulate the number of wire bends as an edge-cost assignment by expanding each vertex of a k-grid into the structure depicted in Figure 8.12. This construction can also be used to minimize the sum of the length of the path and the number of vertices on the path. To this end, the edges leading away from the construct receive appropriate length labels.

Each cost function in a set of cost functions can be minimized with a given priority by a vector of the cost functions being composed. As shown in Section 3.8.7, the shortest-path algorithms discussed in this book apply to such cost functions.

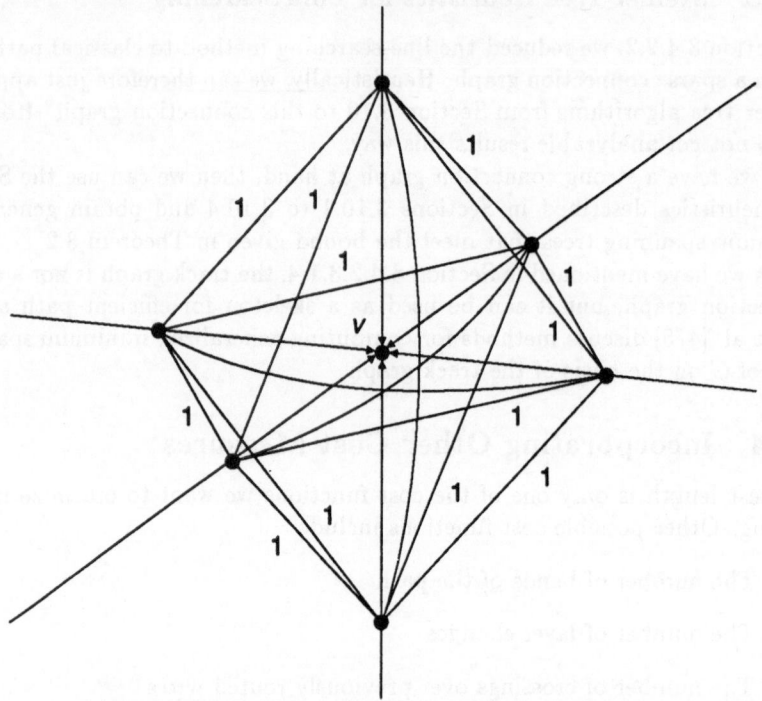

Figure 8.12: *We can optimize the number of wire bends by replacing each vertex in the k-grid with the constructs shown. All undirected edges represent pairs of antiparallel directed edges with equal length. Edges representing wire bends receive length 1. All other edges have length 0 (not shown in the figure).*

8.4.4.2* Beyond Dijkstra's Algorithm

Although this transformation does enable us to minimize wire bends, it is too expensive in most cases, since it increases the routing graph threefold. However, if we do not perform the transformation, but rather work on the original routing graph, then the number of wire bends cannot be captured with an edge-cost assignment. Figure 8.13 illustrates why this is the case.

Therefore, most approaches to minimizing the number of bends are heuristic. Several authors just try to advance without changing direction whenever possible. If the priority-queue implementation of Figure 3.13 is used, this goal can be achieved by arrangement of the *bucket* lists as stacks. When vertex v is scanned, the unique neighbor w of v such that going from v to w does not create a bend is pushed onto the stack *last*, such that it appears on top of the

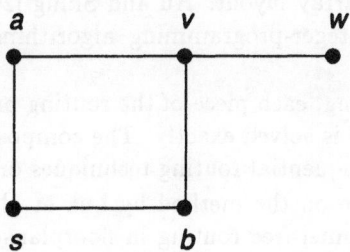

Figure 8.13: *Problems with shortest-path algorithms for minimizing the number of bends on a path. Assume that we want to minimize the sum of the edge lengths and the number of bends. All edges have length 1. Both paths from s to v are shortest path, but only the path via a extends to a shortest path to w. Thus, part 2 of Lemma 3.4 is violated, and the shortest-path algorithms we know do not work.*

stack after v has been scanned [193, 384, 426].

Hu and Shing [200] take a more thorough stab at this problem. They generalize shortest-path methods such that they can deal with cost values that incorporate the number of bends. Lengauer et al. [283] extend the method by Hu and Shing to a wide variety of cost functions that violate Lemma 3.4.

8.5 Hierarchical Global Routing Algorithms

Sequential routing algorithms disregard the fact that the routing of different nets is interdependent. This view enables us to consider one net at a time, and to use the wealth of approximation algorithms and heuristics that is known for shortest paths, minimum spanning trees, and Steiner trees. We pay for this flexibility by having only the (difficult) choice of the ordering in which the nets are routed as a tool for dealing with the interdependence between the routings of different nets.

On the other hand, the integer-programming formulation reflects all interdependencies between different nets. Unfortunately, this formulation is too complex to be used directly for solving the global-routing problem.

Hierarchical routing algorithms attempt to use a hierarchy on the routing graph to decompose the large routing problem into small, manageable pieces. The subproblems are solved in a way that deals explicitly with interdependencies between wires. The partial solutions are combined to create a solution of the original problem instance.

Hierarchical routing has first been proposed by Burstein and Pelavin [56] in the context of gate-array layout. Hu and Shing [201] and Luk et al. [299] present hierarchical integer-programming algorithms for global routing in general-cell layout.

In hierarchical routing, each piece of the routing problem is translated into an integer program and is solved exactly. The composition of the partial solution is achieved using sequential routing techniques or integer programming.

We shall concentrate on the method by Luk et al. [299], which applies to routing graphs for channel-free routing in floorplans. We shall also mention the other approaches.

Let I denote an instance of a global-routing problem (either the CON-STRAINED GLOBAL ROUTING problem or the UNCONSTRAINED GLOBAL ROUT-ING problem). Assume that $w \equiv 1$. (This assumption will help us to formulate *small* integer programs.) Let G be the related routing graph. We assume that G is the rectangular adjacency graph of a floorplan F. For purposes of illustration we use the floorplan of Figure 8.1 and its routing graph. Furthermore, let N be the set of nets. The length of an edge e in G is the Euclidean distance of the centers of the rooms in F that are connected by e. The capacity of e is taken to be the length of the dual wall in the floorplan F (see also Section 8.1.1).

8.5.1 Definition of the Routing Hierarchy

The most convenient way of imposing a hierarchy on G is by means of a cut tree T. The cut tree is a natural structure that is constructed during placement and floorplanning if partitioning methods are used. If the cut tree does not result from the preceding placement phase, it can be constructed specifically for the routing phase. Each interior node in the cut tree represents a primitive global-routing problem. Hierarchical routing algorithms solve these problems exactly using integer-programming methods.

A possible cut tree for the floorplan in Figure 8.1 is depicted in Figure 8.14. Each interior node z of the cut tree T is labeled with a sized floorplan pattern F_z that indicates how the subfloorplans corresponding to the children of the node are arranged. Each child of z is assigned a room in F_z. Such a cut tree can be generated with mincut or clustering techniques (see Sections 7.2.2). Note that here the degrees of the interior nodes are generally higher than 2. This is desirable, since higher-degree nodes lead to larger subproblems of the global-routing problem that are solved exactly. This provision increases the extent to which the interdependence between nets is taken into account. If the floorplanning phase is based on binary mincut, then the degrees of the interior nodes of the cut tree can be increased by merging of adjacent nodes. Usually, the maximum degree of an interior node of the cut tree that still allows the exact solution of the respective global-routing problem is 4 or 5. Hu and Shing [201] allow even higher degrees and resort to column-generating techniques for solving the respective primitive global-routing problems.

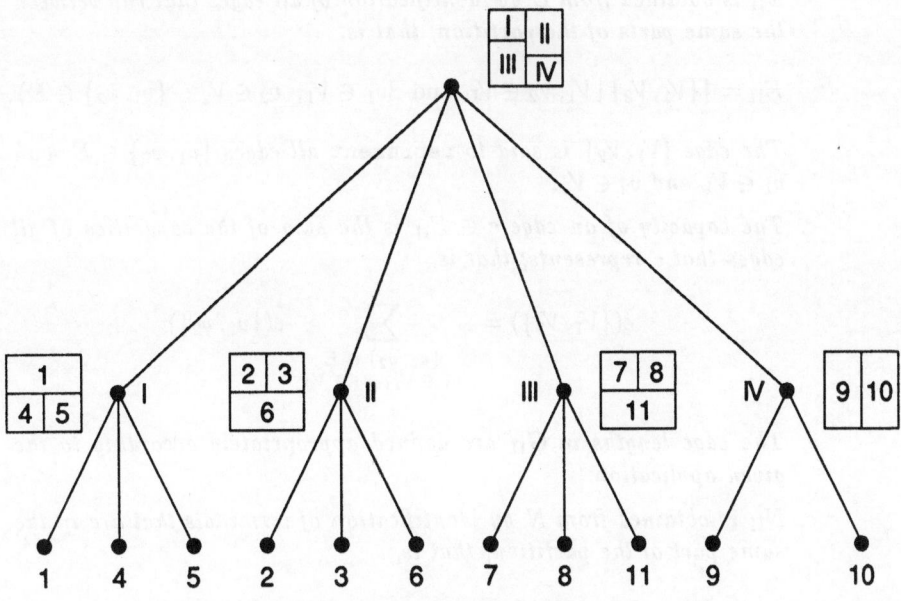

Figure 8.14: *A cut tree for the floorplan in Figure 8.1.*

8.5.2 The Primitive Global Routing Problems

Each interior node z of the cut tree is associated with a global-routing subproblem I_z. This problem instance is defined as follows.

Definition 8.2 (Hierarchical Routing) *Let I be an instance of a globalrouting problem. Let $G = (V, E)$ be the routing graph with edge lengths $\ell(e)$ and capacities $c(e)$. We assume $w \equiv 1$. Let N be the set of nets.*

1. *Let V' be a subset of V. The routing problem* induced *by V' has the graph G' induced on G by V' as a routing graph and the set N' as the set of nets, where N' is N restricted to G'; that is, all terminals of nets in n that do not belong to G' are deleted. Edge lengths and capacities remain unaltered.*

2. *Let Π be a partition of V such that the subgraphs induced by all parts of Π are connected. The* quotient routing problem *with respect to Π is defined via the following graph $G_\Pi = (V_\Pi, E_\Pi)$ and set of nets N_Π.*

    a. *V_Π is obtained from V by identification of all vertices that are in the same part of the partition; that is, $V_\Pi = \Pi$.*

b. E_Π is obtained from E by identification of all edges that run between the same parts of the partition; that is,

$$E_\Pi = \{\{V_1, V_2\} \mid V_1, V_2 \in V_\Pi \text{ and } \exists v_1 \in V_1, v_2 \in V_2 : \{v_1, v_2\} \in E\}$$

The edge $\{V_1, V_2\}$ is said to **represent** all edges $\{v_1, v_2\} \in E$ with $v_1 \in V_1$ and $v_2 \in V_2$.

c. The capacity of an edge $e \in E_\Pi$ is the sum of the capacities of all edges that e represents; that is,

$$c(\{V_1, V_2\}) = \sum_{\substack{\{v_1, v_2\} \in E \\ v_1 \in V_1, \ v_2 \in V_2}} c(\{v_1, v_2\})$$

d. The edge lengths in G_Π are defined appropriately according to the given application.

e. N_Π is obtained from N by identification of terminals that are in the same part of the partition; that is,

$$N_\Pi = \{n_\Pi \subseteq \Pi \mid \exists n \in N : \ n_\Pi = \{V' \in \Pi \mid V' \cap n \neq \emptyset\}\}$$

The routing problem induced from I by the vertex set V' is just the natural restriction of I to the vertices in V'. The quotient problem of I with respect to a partition Π of V results from shrinking all vertices in each part of V to a supervertex. The condition that each part of Π be connected is imposed because we will route nets that do not leave a part of Π completely inside the corresponding room.

Induced and quotient routing problems occur naturally in hierarchical routing. An induced routing problem can be associated with each interior node z in the hierarchy tree. The corresponding set V' is the set of all blocks that are descendants of z. A quotient routing problem represents a coarse view of the original routing problem. We can create it by deleting some subtrees of the hierarchy tree. This deletion results in a partial floorplan in which the internal structure of the subfloorplans represented by the deleted subtrees is disregarded. The corresponding partition groups together all blocks that are in the same deleted subtree. Usually, the edge lengths in the quotient problem are Euclidean distances between the centers of the subfloorplans represented by the roots of the subtrees. Putting together both concepts, we can associate a quotient problem of an induced routing problem with each interior node of T. The induced problem is obtained by restriction to the set of the blocks that are descendants of z. The quotient is based on the partition induced by the subtrees rooted at the children of z. The corresponding routing problem considers only how to route nets *between different rooms in the floorplan pattern* F_z *for node* z. This problem is called the *primitive routing problem for node*

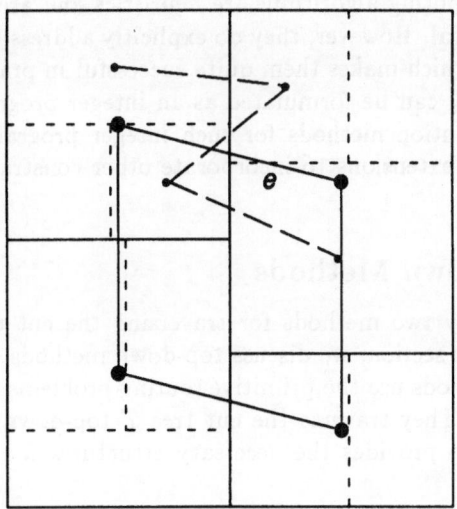

Figure 8.15: *The routing graph G_r associated with the root of the cut tree in Figure 8.14. The edges that edge e in G_r represents are dashed in the figure.*

z, and is denoted with I_z. (The components of this routing problem are also indexed with z.) The corresponding routing graph and set of nets are denoted with G_z and N_z, respectively. Figure 8.15 shows the primitive problem for the root r of the cut tree depicted in Figure 8.14.

Let I be an instance of the CONSTRAINED GLOBAL ROUTING problem, and let I_Π be a quotient problem for I based on some vertex partition Π. A routing R for I can be converted into a routing R_Π for I_Π. We do this conversion by routing every net $n \in N$ that connects blocks in different parts of Π across the edges e in G_Π that represent edges in G across which n is routed in R. Since the capacity of e is the sum of the capacities of all edges in G that are represented by e, the routing R_Π is legal if the routing R is legal. The converse does not have to hold, however, for two reasons. First, I_Π disregards all nets that are completely inside one part. Capacity violations could occur in I when such a net is routed. Second, even if there are no such nets, the refinement of a legal routing for I_Π to a legal routing for I may not be possible, because the distribution of nets across an edge e in G_Π to all edges in G represented by e may not be achievable within the capacities of edges in G that are not represented in G_Π (Exercise 8.11). Similarly, the costs of a routing R for I and its corresponding routing R_Π for I_Π are not formally correlated. Therefore, I_Π is a coarse version of I in an *informal* and *nonanalyzable* way. This fact implies that hierarchical routing algorithms are heuristics and are not approximation algorithms in general. However, they do explicitly address the interdependence of different nets, which makes them quite successful in practice.

The instance I_z can be formulated as an integer program, as discussed in Section 8.2.3. Solution methods for such integer programs are discussed in Section 8.2.5, and extensions to incorporate other constraints are the subject of Section 8.2.6.

8.5.3   Top-Down Methods

There are basically two methods for traversing the cut tree—top-down and bottom-up. In this section, we discuss top-down methods.

Top-down methods use the primitive routing problems that refine the routing step by step. They traverse the cut tree T top-down level by level. The following definition provides the necessary structures for describing the top-down methods.

Definition 8.3 (Abstraction) *Let T be a cut tree defining a floorplan F. Let T_i be the tree that results from T when all nodes at levels larger than i and the floorplan patterns from all nodes at level i are deleted. T_i defines another floorplan F_i called the* level-i abstraction of F *(with respect to T).*

Let I be a routing problem instance with a routing graph G that is the adjacency graph of the floorplan F. F_i induces a quotient routing problem I_i of I. The respective partition Π groups together all blocks that are in a common subtree rooted at level i. I_i is called the level-i abstraction of I *with respect to T. The components of I_i are also indexed with i.*

For instance, the part of the floorplan depicted in Figure 8.15 that is drawn with solid lines is the level-1 abstraction of the floorplan depicted in Figure 8.1 with respect to the cut tree depicted in Figure 8.14. Top-down methods compute solutions for the level-i abstractions I_i of the routing problem instance I for increasing i. They have the structure depicted in Figure 8.16. The height of T is denoted with d, and the root of T is denoted with r. To understand the algorithm, we have to note that $I_r = I_1$; that is, the primitive routing problem for the root r is the same as the level-1 abstraction of I. In line 5, a solution R_i of I_i is computed from a solution R_{i-1} of I_{i-1} and solutions R_z of the I_z for all nodes z at level $i-1$. This step, in effect, refines the routing R_i to cover one more level of the cut tree. Once i reaches the height d of T we are done, since $I_d = I$.

We compute the solutions R_z by solving the integer programs described in Section 8.2.3. The solution obtained by this method does not distinguish among different instances of the same net. However, in hierarchical routing, different instances of the same net on some level may actually be parts of instances of different nets in the original problem instance. Thus, the assignment of routes to nets is not arbitrary. Heuristic choices have to be made here. The literature

(1)   Compute a solution R_1 of the primitive routing problem I_r;

(2)   **for** i **from** 2 **to** d **do**
(3)   **for** z node at level $i - 1$ **do**
(4)   Compute the solution R_z of the primitive
                          routing problem I_z **od**;
(5)   Combine the R_z for all nodes z at level $i - 1$ and
                   the solution R_{i-1} to a solution R_i of I_i **od**;

Figure 8.16: *Top-down method for hierarchical global routing.*

does not elaborate on this point. The heuristics should be chosen, however, so as to assign short routes to subnets of a net that has other subnets that have been assigned long routes. In this way, the overall wire length per net can be evened out.

Neither is the solution R_d computed by this procedure optimal, in general, nor can we bound the error of this procedure, because the quality of the I_i as approximations for I cannot be provably bounded.

The crucial step in the whole procedure is the combination of the partial routings in line 5. There are several ways to implement this step. We shall now discuss them in detail.

8.5.3.1 Sequential Routing Refinement

One alternative is to use sequential routing algorithms in this step. This procedure introduces artificial orderings of the nets—just the thing we wanted to avoid by using hierarchical routing. However, the sequential procedure in line 5 is interleaved with the exact solution of the I_z in line 4. Thus the "communication" between different nets is enhanced significantly.

In sequential versions of line 5, we process each net separately. The order in which nets are processed is again chosen heuristically. For each net $n \in N_i$, we do the following. We first generate a special routing graph $G_{i,n}$ for computing the route for net n in R_i. $G_{i,n}$ is a subgraph of the routing graph G_i associated with I_i. It contains exactly the following edges (see Figure 8.17 for an example):

Case 1: All edges e of G_i that are also in the routing graph G_z for some node z at level $i - 1$ that is represented by a vertex in the route of net n inside R_{i-1}. (These edges are drawn solid in Figure 8.17.)

Case 2: All edges e of G_i that are represented (in the sense of Definition 8.2, part 2(b)) by edges e' in the route for n inside R_{i-1}. (Here, I_{i-1} is

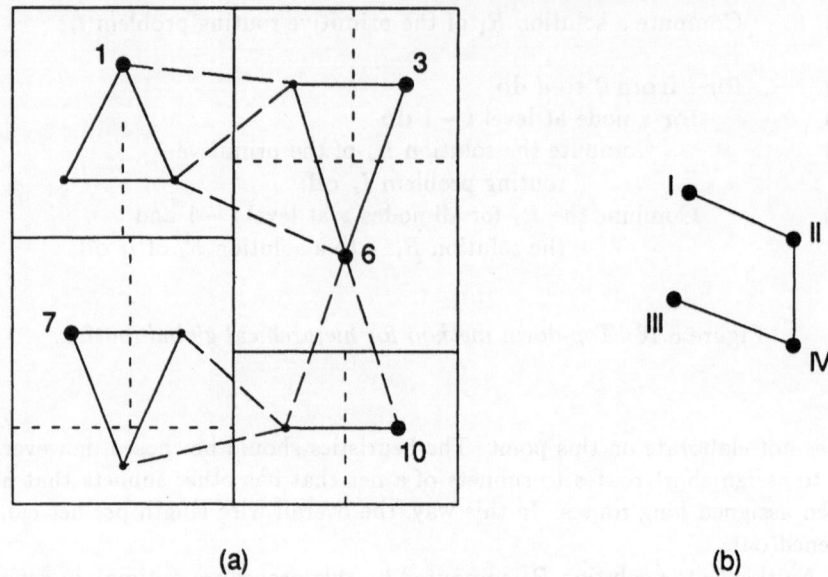

(a)   (b)

Figure 8.17: *The routing graph $G_{2,n}$ for a net connecting blocks 1, 3, 6, 7, and 10 in the floorplan depicted in Figure 8.1. (a) The routing graph $G_{2,n}$; (b) the route for net n inside R_1 that is used to generate $G_{2,n}$.*

considered as a quotient problem of I_i.) (These edges are dashed in Figure 8.17.)

On this routing graph, we compute a minimum Steiner tree for net n. This computation does not entail edge capacities. The edge lengths of $G_{i,n}$ for the computation of the minimum Steiner tree for the first net are assigned as follows:

Assignment 1: If Case 1 above applies to e,

    1. If e is in the route for net n in R_z, then its length is 0.

    2. Otherwise its length is the same as in G_z.

Assignment 2: If Case 2 above applies to e, then its length is the same as in G_i.

Assignment 1 incorporates the routings R_z into the Steiner tree. Assignment 2 accounts for the routing R_{i-1}.

After computing, the Steiner tree for the first net, we continue analogously with the subsequent nets. The routing graph changes with each net that is being routed. After their computation, all routes are considered as routes in G_i, not in the subgraph $G_{i,n}$ of G_i in which they were computed.

As net after net is routed, the loads on the edges in the graph G_i increase. Similar to sequential routing algorithms, the lengths of the edges in both classes are changed heuristically to make future nets avoid heavily congested regions.

So far, we have not specified whether we are solving the constrained or the unconstrained version of the routing problem. Note, however, that the framework outlined here works for both versions. We just have to choose the appropriate formulations of I_z, and to set the edge lengths for the MINIMUM STEINER TREE problems adequately.

Let us discuss the efficiency of this method. The run time is the sum of a linear term for solving the I_z plus a term amounting to $d-1$ times the time it takes to execute line 5 of Figure 8.16. The size of the latter term depends on the set of admissible floorplan patterns in T. It is *linear* in many cases, because the graphs $G_{i,n}$ have a very special structure.

Lemma 8.3 *If the number of floorplan patterns that can label the interior nodes of T is finite, then $G_{i,n}$ is a partial k-tree for some fixed k (see Section 4.7.5). A hierarchical definition for $G_{i,n}$ can be found in linear time in the size of $G_{i,n}$.*

Proof A hierarchical definition for $G_{i,n}$ can be derived from the route T_n of net n in the routing R_{i-1} as follows. T_n is a tree. Select a root inside T_n arbitrarily. Each vertex x in T_n represents a room in G_{i-1}. In G_i, this room corresponds to an interior node z that is labeled with F_z, one of the finitely many possible floorplan patterns. The corresponding routing graph G_z, together with the edges represented by the edge in T_n that connects vertex x with its parent in T_n, form a cell in the hierarchical definition. The cells for the different rooms are connected as prescribed by the tree T_n. Since there are only finitely many floorplan patterns, the size of all cells is bounded by a constant $s = k+1$. Figure 8.18 illustrates the hierarchical definition of $G_{i,n}$.

The construction of the hierarchical definition of $G_{i,n}$ takes linear time in the size of $G_{i,n}$ (see Exercise 8.12). □

Using dynamic-programming methods, the MINIMUM STEINER TREE problem can be solved in linear time on partial k-trees. The constant factor increases exponentially with k, but for small k very efficient Steiner tree algorithms result. Luk et al. [299] point out that, if all floorplan patterns come from the set depicted in Figure 8.19, then $G_{i,n}$ is a partial 2-tree (Exercise 8.12). Wald and Colbourn [454] have detailed the linear time algorithm for finding minimum Steiner trees in partial 2-trees. Dai [77] proved that $G_{i,n}$ is a partial 3-tree if each floorplan pattern has at most five rooms, and no four rooms touch the same side of the floorplan.

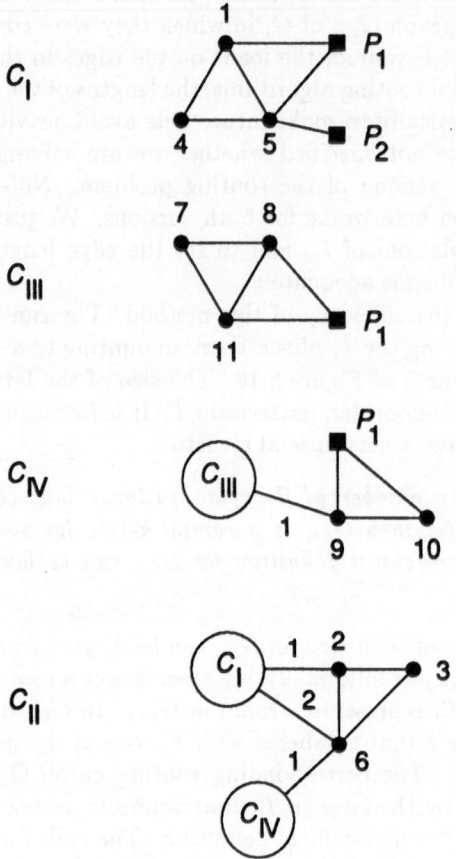

Figure 8.18: *Hierarchical definition of* $G_{2,n}$ *depicted in Figure 8.17(a).*
Node II has been chosen as the root of the tree depicted in Fig-
ure 8.17(b). The notation in the figure is similar to the one used in
Figure 3.27. Pins are denoted by squares; nonterminals are denoted
by large circles, inscribed with a symbol identifying their name and
type. The bijective correspondence between the neighbors of a non-
terminal and the pins of the cell that is its type is given by appropriate
edge labels.

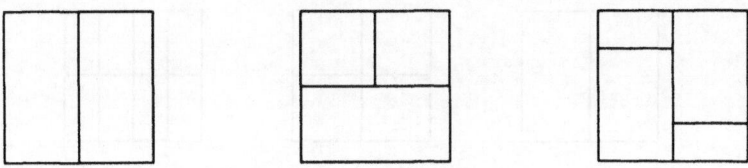

Figure 8.19: *A set of floorplan patterns; the set also includes all rotations and reflections of the patterns shown.*

The run time of the hierarchical routing algorithm is the sum of the time spent on solving the integer programs and the time spent on routing the Steiner trees. Assume T has bounded degree and the set of admissible floorplan patterns for interior nodes of T is finite. Then, the first term is linear in the size of T, and thus is linear in $|V|$, since the size of each integer program pertaining to a primitive routing problem is $O(1)$. Furthermore, at each level, constructing the Steiner tree for a net n takes linear time in the size of $G_{i,n}$. The overall run time of the hierarchical routing algorithm is $O(|N| \cdot |V| \cdot d)$. This is $O(|N| \cdot |V|^2)$ for general cut trees. If the cut tree is balanced, the run time reduces to $O(|N| \cdot |V| \log |V|)$. If, in addition, the levels of all leaves in the cut tree are the same or differ by only a constant, then the run time reduces to $O(|N| \cdot |V|)$. This is so because in this case the size of $G_{i,n}$ doubles when descending a level in T, except on the last few levels. Thus, the run time for computing all $G_{i,n}$ for a net n is $O(|V|)$.

8.5.3.2 Refinement with Integer Programming

Burstein and Pelavin [56] suggest that we refine R_{i-1} to R_i without explicitly computing the R_z. Their method is tailored to gate-array layout. The corresponding routing graphs are grid graphs. We generate the cut tree by doing bisections whose cut directions alternate. Depending on the orientation of the cut lines in level i of the cut tree the refinement from R_{i-1} to R_i is done by processing pairs of adjacent rows (if the hierarchy is refined by introducing horizontal cuts) or columns (see Figure 8.20).

Without loss of generality, consider a pair of adjacent rows. The routing R_{i-1} specifies what subnets of a net n have to be connected inside the two rows. We can break each net into its subnets accordingly, and modify the induced quotient routing problem on the two rows to realize these subnets instead of the net itself. The routing graph is a grid graph with two rows; thus, it is again a partial 2-tree, and we can run sequential routing algorithms on it that use linear-time minimum Steiner tree algorithms. Alternatively, we can solve this problem with a divide-and-conquer approach that amounts to applying to

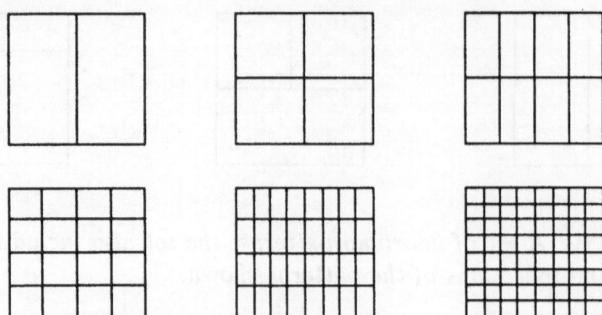

Figure 8.20: *A sequence of floorplan refinements in hierarchical global routing on gate arrays. The sequence alternates between doubling the numbers of columns and rows.*

the two rows the same refinement strategy that we just applied to the whole floorplan. This method is illustrated in Figure 8.21. It decomposes the routing problem into a sequence of primitive routing problems whose routing graph is a cycle of length 4. These problems are solved exactly using integer programming.

Note that, in this version, we do not use sequential routing algorithms at all. If the edge capacities in the grid graph are not uniform, it is advantageous to make heuristic adjustments to the edge capacities of the respective primitive routing problems [56].

8.5.3.3* A Method Based on Linear Assignment

Lauther [260] and Marek-Sadowska [309] present a top-down hierarchical global-routing method that is not derived directly from the integer-programming formulation, but rather is based on a heuristic use of the LINEAR ASSIGNMENT problem. The method can be adapted to channel-free routing or routing in channels; it can handle slicing or nonslicing floorplans. We describe the method as it applies to channel-free routing in slicing floorplans [260].

Assume that an oriented, ordered, and sized slicing tree T for the floorplan is given. We compose the global routing by processing T top down. We begin processing a node z in T by dividing the cutline represented by z into sections. Each section corresponds to a wall on the cutline. For each net n crossing the cutline and for each wall w on the cutline we determine a cost for running n across w. This cost is based on the half-perimeter of the bounding box for the net n that results from the terminals of the net that have to be connected on both sides of the cutline. This cost function is piecewise linear, as depicted in Figure 8.22, where the slope α of the inclines decreases with increasing

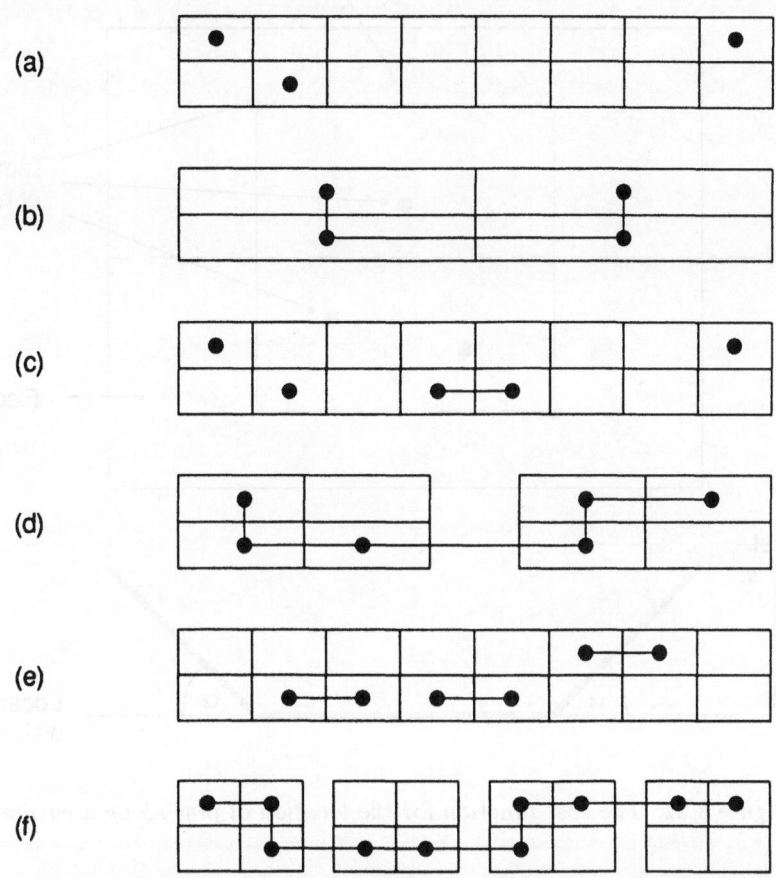

Figure 8.21: *Hierarchical solution of a 2 × n global-routing problem on gate arrays. (a) Original subproblem formulation with net terminals— this instance is derived from the routing R_{i-1}; (b) first refinement stage and the corresponding global route; (c) wire segment fixed after the first refinement stage; (d) subnets for second refinement stage and their global route; (e) wire segments determined after the second refinement stage; (f) final refinement stage.*

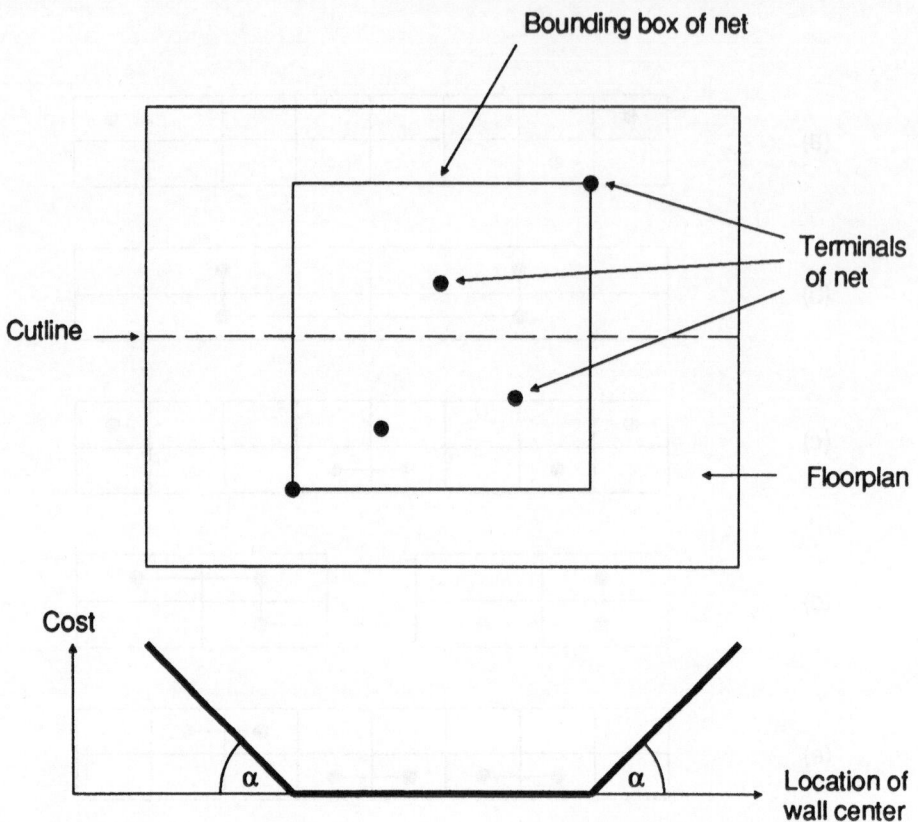

Figure 8.22: *The cost function for the location of pins along a cutline.*

distances from the cutline of the closest terminals on either side of the cutline. Now we are looking for an assignment of nets (crossing the cutline) to walls (on the cutline) that minimizes the sum of all cost contributions while satisfying the constraint that the number of nets routed across each wall not exceed the capacity of the wall. This problem is a generalized type of a linear-assignment problem. It can readily be translated into a LINEAR ASSIGNMENT problem if each wall is split into as many pieces as that wall's capacity indicates. If walls have large capacities, a better idea is to transform this problem into an instance of the MINCOST FLOW problem (Exercise 8.13). Because of the definition of the cost function, nets that are located "close" to the cutline receive optimal routes with preference.

This approach has the disadvantage that no net can cross the same cutline twice, even though such an arrangement may be highly preferable.

To overcome this disadvantage, we can combine the choice of global routes by linear assignment with other routing techniques. Lauther suggests that we find a minimum spanning tree for each net first, and use this tree to guide the generation of the global routes. Specifically, if a minimum spanning tree for some net crosses a cutline several times, all corresponding segments are routed across the cutline. To this end, we partition the net into subnets by deleting all wire segments in the tree that cross the cutline. Then the subnets connected with a wire segment across the cutline are submitted to the procedure for constructing the routing segments, one after another.

We can extend this method to nonslicing floorplans by observing that cutlines do not need to be straight; we can extend it to routing in channels by modifying the cost function [309].

8.5.4* Bottom-Up Methods

Hu and Shing [201] propose a bottom-up method for hierarchical routing on the basis of binary cut trees. In a first phase, the induced routing problem pertaining to a set of small subtrees of the cut tree is solved by integer programming. We then combine the resulting partial routings by processing internal tree nodes bottom-up. Each tree node corresponds to a slice through the floorplan. Thus, we have only to connect the trees for the two subnets of each net that runs across the cutline such as to maintain the capacity constraints and to minimize the cost of the resulting routing. Details are presented in [201].

Bottom-up methods have the disadvantage that they gain a global picture of the routing only late in the routing process, when the partial routings cannot be changed.

8.6   Routing by Relaxation

In this section, we discuss results by Raghavan and Thomborson (formerly Thompson) that apply randomized rounding techniques to the solution of the linear relaxation of the global-routing integer program [337, 369, 370, 371]. As we pointed out in Section 4.6.1, such techniques do not work on general integer programs, but they prove quite successful on versions of the UNCONSTRAINED GLOBAL ROUTING problem.

The basic idea of global routing by linear relaxation is simply to omit the integrality constraint in the integer program R formulating the global-routing problem, and to solve some linear relaxation L of R. If the resulting optimal solution $x = \alpha$ is integer, we have found an optimal routing. Otherwise, we have to transform α into an integer solution of R that is, hopefully, close to an optimal solution $x = z$ of R. In this section, we present randomized and

Step 1: Let L be the linear relaxation of R obtained by replacing the integrality constraint with the constraint $x \in [0,1]^n$. L is a linear program. Solve L. Let $x = \alpha$ be a corresponding optimal solution.

Step 2: Repeat Step 3 k times.

Step 3: For each variable x_i do the following. Interpret α_i as a probability. Toss a biased coin that yields a 1 with probability α_i, and a 0 with probability $1 - \alpha_i$. Assign the value of the outcome to x_i. The coin tosses for different variables are independent.

Step 4: Choose the best feasible solution among the k trials.

Figure 8.23: *The randomized rounding algorithm.*

deterministic techniques for doing this transformation. We show that the result is close to optimal with high probability for the randomized algorithms, and with certainty for the deterministic algorithms.

While we develop the method, we will keep track of which properties of the integer program R allow us to do the transformation. As a result, we will be able to make statements about the other types of integer programs to which this method of solution by linear relaxation applies.

The following subsections develop the method: In Section 8.6.1, we discuss the randomized method; in Section 8.6.2, we discuss the deterministic method.

8.6.1   Routing by Randomized Rounding

In this section, we discuss global-routing integer programs, for which rounding the optimal solution of the linear relaxation yields a high-quality feasible solution of the integer program.

We restrict our attention to 0, 1-integer programs. Let R be the integer program under consideration, and let $x = z$ be an optimal solution of R. Figure 8.23 presents the randomized rounding algorithm for solving the integer program R. Step 1 of the algorithm is deterministic—it just solves a linear relaxation of R; the corresponding optimal solution is α. Step 3 is randomized and can be modeled as a sequence of independent *Bernoulli trials*. A Bernoulli trial is the probabilistic model of a random coin toss, yielding 0 or 1. The coin toss for variable x_i in step 3 is modeled with a Bernoulli trial with probability α_i (that 1 comes up). Let X_i be the random variable denoting the outcome of the Bernoulli trial for variable x_i. The expected value of X_i is $E[X_i] = \alpha_i$, the value in the optimal solution of the linear relaxation.

Assume that the cost function to be optimized in R is

$$c(x) = \sum_{i=1}^{n} c_i x_i$$

where the c_i are real values in the interval $[0, 1]$. Let us consider the random variable Ψ that is the cost of the solution obtained by one execution of step 3 in the randomized rounding algorithm:

$$\Psi = \sum_{i=1}^{n} c_i X_i$$

Ψ has the expected value χ, where

$$\chi = E[\Psi] = E[\sum_{i=1}^{n} c_i X_i] = \sum_{i=1}^{n} c_i E[X_i] = \sum_{i=1}^{n} c_i \alpha_i$$

Thus, χ is just the optimal cost of the linear relaxation L.

The fortunate fact is that, with very high probability, the outcome of Ψ falls very close to χ (see Figure 8.24). Furthermore, as we shall see, we can derive strong bounds supporting this fact. The only other thing we need to ensure, to prove the effectiveness of the randomized rounding algorithm, is that a solution found by the randomized rounding algorithm is also feasible with high probability. The fact that $c(\alpha)$ is a lower bound on the optimal cost $c(z)$ of R then implies that the randomized rounding algorithm finds feasible solutions of R that are close to optimal with high probability.

By repeating step 3 k times, we can increase the probability of obtaining a high-quality feasible solution, at the expense of increased run time.

The randomized rounding algorithm of Figure 8.23 can be extended in two ways:

1. Other variants of the linear relaxation can be used. The only important requirement is that the optimal cost $c(\alpha)$ of the linear relaxation be a lower bound on the optimal cost $c(z)$ of R. The tightness of the bounds on the cost of the randomized solution decreases as $c(z) - c(\alpha)$ increases.

2. The randomization strategy may be modeled by sequences of Bernoulli trials in different ways. In the following sections, we shall see that this flexibility in providing a probabilistic model for the algorithm makes the method quite powerful and applicable in several different settings.

In the following sections, we shall apply randomized rounding to several global-routing problems. Before we do so, we cite the bounds on the deviation of Ψ from its expected value χ that enable us to provide a tight analysis of the randomized rounding algorithm. These bounds are called *Chernoff-type bounds* because this analysis was originated by Chernoff [64].

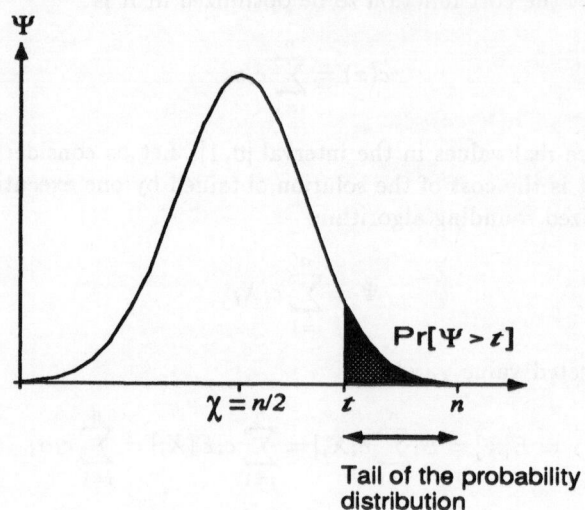

Figure 8.24: *The probability density of the sum of a sequence of* n
independent Bernoulli trials, each with probability $1/2$. *The proba-*
bility that $\Psi > t$ *holds decreases rapidly as* $t - \chi$ *increases above* 0.
An analogous phenomenon occurs for values of Ψ *that lie below* χ.

Lemma 8.4 (Raghavan, Spencer [370]) *Let the function* $\Psi = \sum_{i=1}^{n} c_i X_i$
be the weighted sum of n *independent Bernoulli trials. Here* $c_i \in [0, 1]$ *for all*
$i = 1, \ldots, n$. *Let* α_i *be the probability that trial* X_i *turns up 1. Assume that*
$\chi = E[\Psi] > 0$.

1. *Let* $\delta > 0$. *Then,*

$$\Pr[\Psi > (1 + \delta)\chi] < \left(\frac{e^\delta}{(1 + \delta)^{(1+\delta)}} \right)^\chi \quad (8.2)$$

2. *Let* $\gamma \in (0, 1]$. *Then,*

$$\Pr[\Psi - \chi < -\gamma\chi] < e^{-\gamma^2\chi/2} < \left(\frac{e^\gamma}{(1 + \gamma)^{(1+\gamma)}} \right)^\chi$$

Here $\Pr[$event$]$ *denotes the probability that the event* event *occurs.*

Proof

1.

$$\begin{aligned}
\Pr[\Psi > (1 + \delta)\chi] &= \Pr[e^{t\Psi} > e^{t(1+\delta)\chi}] \\
&< e^{-t(1+\delta)\chi} E[e^{t\Psi}]
\end{aligned} \quad (8.3)$$

The last inequality in (8.3) can be derived as follows. Let

$$\Psi' = \begin{cases} e^{t(1+\delta)\chi} & \text{if } \Psi > (1+\delta)\chi \\ 0 & \text{otherwise} \end{cases}$$

Then, $\Psi' \leq e^{t\Psi}$ and, since $\chi > 0$, $E[\Psi'] < E[e^{t\Psi}]$. We can now show the rightmost inequality in (8.3) by noting that

$$E[\Psi'] \geq e^{t(1+\delta)\chi} \Pr[\Psi > (1+\delta)\chi]$$

Now observe that the expected value of the product of independent random variables is the product of the expected values of each variable. Thus, the chain (8.3) of inequalities can be continued:

$$\leq e^{-t(1+\delta)\chi} \prod_{i=1}^{n} (\alpha_i \exp(tc_i) + 1 - \alpha_i) \tag{8.4}$$

$$\leq e^{-t(1+\delta)\chi} \prod_{i=1}^{n} \exp(\alpha_i(\exp(tc_i) - 1)) \tag{8.5}$$

In (8.5), we used the fact that $1 + x \leq e^x$. Substituting $t = \ln(1+\delta)$ into (8.5) yields

$$\Pr[\Psi > (1+\delta)\chi] < (1+\delta)^{-(1+\delta)\chi} \exp\left(\sum_{i=1}^{n} \alpha_i ((1+\delta)^{c_i} - 1)\right) \tag{8.6}$$

Now observe that

$$(1+\delta)^c \leq 1 + \delta c \quad \text{if } c \in [0, 1] \tag{8.7}$$

as we can see by differentiating $f(c) = (1+\delta)^c - 1 - \delta c$. Using (8.7), we can extend (8.6) to yield

$$\leq (1+\delta)^{-(1+\delta)\chi} \exp\left(\sum_{i=1}^{n} \delta c_i \alpha_i\right)$$

$$\leq \left(\frac{e^\delta}{(1+\delta)^{(1+\delta)}}\right)^\chi$$

2. The proof is similar.

□

Let us call the event, that the value of Ψ exceeds the expected value χ by more than a factor of $(1+\delta)$ a δ-*failure*. Part 1 of Lemma 8.4 states that the probability of a δ-failure decreases sharply as χ increases. Thus, the larger χ is, the better an approximation to the *real* value of Ψ is the *expected* value of Ψ. Part 2 of Lemma 8.4 makes an analogous statement for deviations of Ψ *below* its expected value. We shall only use part 1 of Lemma 8.4 in this book.

In Lemma 8.4, we are given a deviation and compute a probability. If we analyze approximation algorithms, this situation is reversed: We are given a (small) probability ε, and we want to find some (small) deviation $\delta = \delta(\chi, \varepsilon)$ such that the probability of δ-failure is smaller than ε. Because of the complicated form of the right-hand side of (8.2), we can give only an implicit definition of $\delta(\chi, \varepsilon)$. We denote the right-hand side of (8.2) with $B(\chi, \delta)$. Then, we define $\delta(\chi, \varepsilon)$ formally by the equation

$$B(\chi, \delta(\chi, \varepsilon)) = \varepsilon \tag{8.8}$$

Note that $\delta(\chi, \varepsilon)$ is monotonically decreasing in both arguments. The following bounds exist for $\delta(\chi, \varepsilon)$.

Lemma 8.5 (Raghavan [370])

1. *Assume that $\chi > \ln(1/\varepsilon)$. Then,*

$$\delta(\chi, \varepsilon) \leq (e - 1)\sqrt{\frac{\ln(1/\varepsilon)}{\chi}} \tag{8.9}$$

2. *Assume that $\chi \leq \ln(1/\varepsilon)$. Then,*

$$\delta(\chi, \varepsilon) \leq \frac{e \ln(1/\varepsilon)}{\chi \ln \frac{e \ln(1/\varepsilon)}{\chi}} \tag{8.10}$$

We will apply Lemmas 8.4 and 8.5 to prove that randomized routing algorithms yield good solutions of UNCONSTRAINED GLOBAL ROUTING problems with high probability. We start with a very restricted case.

8.6.1.1 Gate Arrays, Two-Terminal Nets, Few Routes per Net

In this section, we assume that all edges in the routing graph have the same capacity. (This situation mostly occurs in gate-array layout; hence, the title of this section.) Furthermore, we consider only two-terminal nets. As a third restriction, we assume that few enough admissible routes exist for each net such that each admissible route can be represented explicitly in the integer program by a variable.

We consider the UNCONSTRAINED GLOBAL ROUTING problem. The respective integer program is defined in Section 8.2.2. In our case, $w \equiv 1$. Since we can assume all edge capacities to be equal, we can set $c(e) = 1$ for all edges

e in the routing graph. (This convention amounts to minimizing the maximum number of wires along an edge, instead of the maximum load. It does not change the set of optimal solutions.) Not being concerned with total edge length, we modify the cost function by eliminating the second-priority term; that is, we discuss the cost function

$$c(x) = x_L$$

We now show that this integer program can be solved effectively by randomized rounding techniques.

We use a version of the randomized rounding algorithm that performs the following kind of random experiments. Recall that net (n, i) has the admissible routes T_n^j, $j = 1, \ldots, I_n$. For net (n, i), we throw an I_n-sided die whose faces have the probabilities $\alpha_{n,i,j}$, $j = 1, \ldots, I_n$. The face that comes up on top determines the unique variable among $x_{n,i,j}$, $j = 1, \ldots, I_n$, whose value is set to 1. All other variables of this sort are set to 0.

We do not throw a die for x_L. Instead, x_L is minimized, while all constraints are maintained, by a straightforward deterministic computation involving the other variables.

Theorem 8.1 *Let $\varepsilon > 0$. The preceding randomized rounding algorithm that repeats step 3 k times computes a solution of the routing problem whose cost does not exceed*

$$\alpha_L(1 + \delta(\alpha_L, \varepsilon^{1/k}/m)) \tag{8.11}$$

with probability larger than $1 - \varepsilon$. Here, $\delta(\alpha_L, \varepsilon^{1/k}/m)$ is defined implicitly by (8.8).

Proof Let α be the optimal solution of the linear relaxation L of the routing integer program that is obtained by requiring that $x_L \geq 0$ and that all other variables come from the interval $[0, 1]$. Consider the constraint for an edge e in the routing graph

$$\sum_{\{(n,j)\,|\,e \in T_n^j\}} \sum_{i=1}^{k_n} x_{n,i,j} \leq x_L \tag{8.12}$$

In step 3 of the randomized rounding algorithm, the left-hand side of (8.12)—that is, the traffic $U(e)$ on edge e—is the sum Ψ_e of the following set of independent Bernoulli trials. For each net (n, i), we toss a coin. The probability that this coin comes up with 1 is the sum of all $\alpha_{n,i,j}$ whose corresponding variables occur on the left-hand side of (8.12). Let χ_e be the expected value of Ψ_e. The fact that α_L meets the constraint (8.12) ensures that $\chi_e \leq \alpha_L$.

Note that this model does not identify single variables with Bernoulli trials. Rather, a set of variables that represent different routes for the same net corresponds to a Bernoulli trial. This model is accurate, since at most one variable out of such a set is assigned the value 1. The probability of the respective

Bernoulli trial is the *sum* of the probabilities that any of the involved variables are set to 1. This method of mapping a set of variables to a single Bernoulli trial is used frequently in the analysis of randomized rounding algorithms.

In the following, we denote the deviation claimed in the theorem with the shortened

$$\delta = \delta(\alpha_L, \varepsilon^{1/k}/m) \qquad (8.13)$$

Let us call the event that step 4 computes a value for x_L that is no greater than $(1 + \delta)\alpha_L$ a *success*, and the opposite event a *failure*. We will apply Lemma 8.4 to show that the probability of failure does not exceed ε. If we have shown this fact, we are done. To understand why, we note that x_L is computed deterministically to meet all edge constraints. Furthermore, the type of random experiments we are carrying out ensures that exactly one route is chosen for each net. Therefore, all random experiments yield feasible solutions.

The randomized rounding algorithm repeats step 3 k times. Thus, failure of step 4 means failure of *all* repetitions of step 3. The different executions of step 3 are independent from one another. The probability that all events in an independent set of events occur is the product of the probabilities of the events. Therefore, the probability of failure that should not be exceeded in each execution of step 3 is $\varepsilon^{1/k}$.

Failure of an execution of step 3 means failure at *at least one* edge. The sequences of Bernoulli trials for different edges are *not* independent from one another, because throwing the die for a net determines all variables representing routes for this net. However, we do not need independence here, because the probability of at least one of two *arbitrary* events occurring is at most the sum of the probabilities of either of the events occurring. Therefore, the probability of failure at each edge should not exceed $\varepsilon^{1/k}/m$.

Failure at edge e means that the outcome of the Bernoulli trials for this edge yields a sum that exceeds $(1 + \delta)\alpha_L$. If $\chi_e = \alpha_L$, then (8.8) and (8.13) immediately imply that failure happens with a probability smaller than $\varepsilon^{1/k}/m$. If $\chi_e < \alpha_L$, we can increase the probabilities of some of the Bernoulli trials to make $\chi_e = \alpha_L$. In the course of this modification, $\Pr[\Psi > (1 + \delta)\alpha_L]$ does not decrease. After the modification, we again apply (8.8) and (8.13). $\square$

Applying Lemma 8.5, we get the following bounds.

Corollary 8.2

1. *Let $\alpha_L > \ln(m/\varepsilon^{1/k})$. Then, the randomized rounding algorithm obtains a solution whose cost does not exceed*

$$\alpha_L + (e - 1)\sqrt{\alpha_L \ln(m/\varepsilon^{1/k})}$$

with probability greater than $1 - \varepsilon$.

2. Let $\alpha_L \leq \ln(m/\varepsilon^{1/k})$. Then, the randomized rounding algorithm obtains a solution whose cost does not exceed

$$\alpha_L + \frac{e\ln(m/\varepsilon^{1/k})}{\ln \frac{e\ln(m/\varepsilon^{1/k})}{\alpha_L}}$$

with probability greater than $1 - \varepsilon$.

Corollary 8.2 distinguishes between *dense* and *sparse* routing problems. Density is measured using the optimal cost of the relaxation L. If the problem is dense, then the edges in the routing graph are packed with wires, and part 1 of Corollary 8.2 applies. If the problem is sparse, then the routing edges do not have to carry many nets, and part 2 of Corollary 8.2 applies. In both cases, the corollary provides excellent bounds. In fact, for dense problems, we can with high probability achieve a routing whose cost is only a small additive term away from the optimum. Specifically, the additive term grows as the product of the *square root* of the optimum cost of R and the logarithm of the size of the routing graph. Note that this error is much smaller than, for instance, the multiplicative logarithmic factors achieved in Chapter 5. For sparse problems, the situation is even better. Here, the additive term depends only logarithmically on the size of the routing graph.

For fixed k, the randomized rounding algorithm can be made to run in polynomial time if appropriate methods are used for solving the linear relaxation of the global-routing integer program (see Section 4.5). In practice, solving the linear relaxations is quite efficient (see Section 8.6.1.5).

8.6.1.2 Gate Arrays, Two-Terminal Nets, Many Routes per Net

If the number of admissible routes per net is large, then we cannot represent each route by a variable in the integer program, because the integer program would become too large. In this case, we use the following technique. The relaxation L of the routing integer program R is a variant of the UNDIRECTED MULTICOMMODITY FLOW problem (see Section 3.11.3). The problem instance L is much smaller than the global-routing integer program. In fact, different routes are not represented explicitly in L. In L, a commodity corresponds to each net. We denote the commodity of net (n, i) also with (n, i). The flow graph of L is the routing graph. In step 1 of the randomized rounding algorithm, we solve L with linear-programming methods. We denote the resulting solution again with α. Then, for each execution of step 3 of the randomized rounding algorithm, we use the flow of the commodity (n, i) to determine randomly a route for (n, i). Here, we use random-walk techniques.

Let us first detail the relaxation L. The two terminals of a net (n, i) are the unique source and sink of the respective commodity. The demand and supply for each commodity are 1. The edges in the flow network have a uniform capacity that is represented by a new variable x_L. The cost function to be minimized is $c(x) = x_L$. That is, rather than minimizing the cost of the flow, we minimize the (uniform) capacities of the edges, subject to the constraints that the total flow along an edge should not exceed that edge's capacity. Flow values along edges are restricted to be in the interval $[0, 1]$ for each commodity.

It is straightforward to convert this variant of the UNDIRECTED MULTI-COMMODITY FLOW problem into a linear program (see also Section 3.11.3). Furthermore, any feasible solution of R also is a solution of L, because a route for a net is simply a unit flow along the path of the route. Therefore, the optimal cost α_L of L is again a lower bound on the optimal cost of R.

Now we explain how we convert the fractional flow for commodity (n, i) into a route for net (n, i) (step 3). For each net, we perform a random walk from its source to its sink. The random walks for different nets are probabilistically independent.

Consider net (n, i). We convert the routing graph into a directed graph as follows. We delete all edges that have zero flow for commodity (n, i). Then we direct all remaining edges in the direction of positive flow. On the resulting network we perform a random walk. We start at the source terminal of net (n, i). In general, if we are currently at vertex v, we choose the edge e out of v with probability

$$\frac{\alpha_{n,i}(e)}{\displaystyle\sum_{e' \in E_{v \to}} \alpha_{n,i}(e')}$$

Here, $\alpha_{n,i}(e)$ denotes the flow of commodity (n, i) along edge e in the optimal solution α of I. Each edge is processed only once, so each random walk takes at most m steps.

Note that the choices of edges along the same random walk are dependent. However, the proof of Theorem 8.1 requires only the Bernoulli trials contributing to the *same (edge) constraint* to be independent. This requirement is met here, because different random walks are independent.

Edge e is on the random walk for net (n, i) with a probability that does not exceed $\alpha_{n,i}(e)$. This fact can be shown by induction (see Exercise 8.14). Therefore, the expected value of the sum of all Bernoulli trials contributing to the same edge constraint never exceeds α_L and the proof of Theorem 8.1 carries over.

8.6.1.3 Multiterminal Nets

If multiterminal nets are involved, the relationship between multicommodity flows and wires for nets breaks down, because now each net corresponds to a Steiner tree. We know that the Steiner tree problem is difficult, in itself, because it is difficult to choose good Steiner vertices. However, if all nets have only few terminals, then this phenomenon does not dominate the complexity of the routing problem, and we can efficiently generalize the randomized rounding technique.

Let us assume for simplicity that each net has exactly three terminals. In this case, there is exactly one Steiner vertex to be chosen for each net. If we could determine the best Steiner vertices for each net, then we could use the approach of the previous section to route the nets. The Steiner vertex for net (n, i) becomes the unique sink of a commodity corresponding to the net. We denote this commodity also with (n, i). Since wires have to run from all three net terminals to the Steiner vertex, the demand of flow at the sink of commodity (n, i) is three units. Each terminal of net (n, i) is a source of commodity (n, i) with a supply of one unit of flow.

We have to incorporate the selection of the Steiner vertices into this framework. To this end, we augment the global-routing integer program with a new variable $x_{n,i,v}$ for each net (n, i) and each vertex v in the routing graph. The meaning of this variable is

$$x_{n,i,v} = \begin{cases} 1 & \text{if } v \text{ is the Steiner vertex for net } (n, i) \\ 0 & \text{otherwise} \end{cases}$$

The constraints

$$\sum_{v \in V \setminus n} x_{n,i,v} = 1 \quad \text{for all nets } (n, i)$$

$$x_{n,i,v} \in \{0, 1\} \quad \text{for all nets } (n, i) \text{ and vertices } v \in V$$

(8.14)

ensure this meaning. The constraints (8.14) are added to the integer program. We denote the resulting global-routing integer program with R.

The relaxation L of R is now defined to be a superposition of several multicommodity-flow problems of the form discussed in the previous section. First, we require that the $x_{n,i,v}$ come from the interval $[0, 1]$. We assign a commodity to each *triple* (n, i, v), where (n, i) denotes the net, and $v \in V \setminus n$ denotes a possible Steiner vertex. The flow of commodity (n, i, v) represents the wiring of net (n, i) in the case that v is chosen as the Steiner vertex for (n, i). Since the vertex v is chosen as the Steiner point for net (n, i) with probability $x_{n,i,v}$, we assign to this vertex a demand of $3x_{n,i,v}$ units of commodity (n, i, v). Each terminal t of net (n, i) receives a supply of $x_{n,i,v}$ units of commodity (n, i, v). Finally, the cost function to be minimized is again the uniform edge capacity x_L. L denotes the resulting variant of the UNDIRECTED MULTICOMMODITY FLOW problem.

Let α be an optimal solution of L. In particular, denote the optimal assignment to $x_{n,i,v}$ with $\alpha_{n,i,v}$ and the corresponding flow for commodity (n, i, v) along edge e with $\alpha_{n,i,v,e}$. Clearly, α_L is again a lower bound on the optimum cost of R, because each global routing is a solution of L.

The variant of the randomized rounding algorithm that we use in this context performs random experiments in two stages. First, for each net (n, i), we throw a biased die that selects a Steiner vertex for (n, i). Here, the vertex v is chosen with probability $\alpha_{n,i,v}$.

Having fixed the Steiner vertices, the second stage of the algorithm performs random walks similar to the previous section. For net (n, i), we consider only the flow of the commodity (n, i, v), where v is the vertex chosen as the Steiner vertex of net (n, i) in the first stage. We multiply all respective flow values with $1/\alpha_{n,i,v}$ to normalize the probabilities. (Remember that v has been chosen as the Steiner vertex for net (n, i) with probability $\alpha_{n,i,v}$.) On the basis of these probabilities, we perform three independent random walks, one from each terminal. Each of the random walks is carried out as described in the previous section. At the end, an edge receives a flow value of 1 if it does so in any of the three random walks. (If this process creates cycles, we can eliminate them in a postprocessing phase. This modification decreases the cost function, and therefore does not invalidate the analysis.)

We have to argue that this process can be mapped appropriately onto sequences of independent Bernoulli trials whose sums have expected values that do not exceed α_L. Then, the proof of Theorem 8.1 carries over. As in the previous section, the dependence that exists between different edges on the same random walk does not pose a problem.

Consider the capacity constraint for edge e. As in the proof of Theorem 8.1, we will attribute one Bernoulli trial to a *set* of variables. Specifically, all commodities pertaining to one net (n, i) contribute to the same Bernoulli trial. These commodities differ in only the Steiner vertex v, and the first stage of the randomized algorithm selects exactly *one* of these commodities. Therefore, at edge e, the Bernoulli trial corresponding to these commodities has a probability that is the sum of the probabilities of all commodities contributing to the trial; that is,

$$\sum_{v \in V \setminus n} p_{n,i,v,e} \alpha_{n,i,v} \tag{8.15}$$

Here, $p_{n,i,v,e}$ is the probability that the random walk for commodity (n, i, v) selects edge e. We cannot compute $p_{n,i,v,e}$, but this also will not be necessary for the analysis. As in the previous section, it follows from the definition of the edge probabilities for the random walk that $p_{n,i,v,e} \leq \alpha_{n,i,v,e}/\alpha_{n,i,v}$ (see Exercise 8.14). Substituting into (8.15) and summing over all nets (n, i) yields the following upper bound on the expected value χ_e of the sum of all Bernoulli trials for edge e:

$$\chi_e \leq \sum_{(n,i)} \sum_{v \in V \setminus n} p_{n,i,v,e} \alpha_{n,i,v} \leq \sum_{(n,i)} \sum_{v \in V \setminus n} \alpha_{n,i,v,e} \leq \alpha_L \tag{8.16}$$

The last inequality of (8.16) holds because α is a solution of L.

We established the appropriate upper bound for the sequences of Bernoulli trials modeling the randomized algorithm. With these observations, the proof of Theorem 8.1 now carries over.

If some of the nets have two and others three terminals, the methods of this and the previous section can be combined. If the number of terminals per net increases beyond three, the choice of the Steiner vertices becomes progressively more difficult. Furthermore, the decomposition of the Steiner tree into path segments allows for several topological variants. All these possibilities have to be explicitly represented in the integer program by variables. Then, the approach discussed in this section carries over. Because of the large number of resulting variables, this approach loses its attractiveness quickly, as the number of terminals per net increases.

8.6.1.4 Other Routing Variants

The randomized rounding algorithm is robust enough to be applicable to other variants of global-routing integer programs.

8.6.1.4.1 Nonuniform Edge Capacities If edge capacities are nonuniform, then the edge constraint does not take the form $U(e) \leq x_L$, but rather takes the form $U(e) \leq x_L c(e)$. (Remember that the left-hand side $U(e)$ of the edge constraint is the traffic across edge e.) Let $C = \max\{c(e) \mid e \in E\}$ and $c = \min\{c(e) \mid e \in E\}$. The cost of the global-routing integer program remains x_L. Propagating this update in Theorem 8.1 and its proof yields the following theorem.

Theorem 8.3 *Let $\varepsilon > 0$. The randomized rounding algorithm that repeats step 3 k times computes a solution of the routing problem in which the traffic $U(e)$ on edge e is bounded by*

$$U(e) \leq U_\alpha(e)(1 + \delta(U_\alpha(e), \varepsilon^{1/k}/m))$$

with probability larger than $1 - \varepsilon$. Here, $U_\alpha(e)$ is the traffic on edge e in the optimal solution α of L.

We can again apply Lemma 8.5 to get explicit bounds.

Corollary 8.4

1. *Let $\alpha_L C > \ln(m/\varepsilon^{1/k})$. Then, the randomized rounding algorithm obtains a solution whose cost does not exceed*

$$\alpha_L + (e - 1)\sqrt{\frac{\alpha_L}{c} \ln \frac{m}{\varepsilon^{1/k}}}$$

   with probability greater than $1 - \varepsilon$.

2. *Let $\alpha_L C \leq \ln(m/\varepsilon^{1/k})$. Then, the randomized rounding algorithm obtains a solution whose cost does not exceed*

$$\alpha_L + \frac{e \ln(m/\varepsilon^{1/k})}{c \ln \frac{e \ln(m/\varepsilon^{1/k})}{\alpha_L}}$$

   with probability greater than $1 - \varepsilon$.

If all edge capacities are uniform these bounds differ from the bounds in Corollary 8.2 by the factor $1/\sqrt{c}$ (in Case 1) and $1/c$ (in Case 2) in the additive term. These factors enter the expression, because here we are minimizing not the maximum number of wires along any edge, as in Theorem 8.1, but rather the load, which is a smaller value.

Similar variants of Theorem 8.1 can be found for other notions of load. For instance, Ng et al. [337] discuss a version of the routing problem that minimizes the value $U(e) - c(e)$—that is, the *excess capacity*, over all edges. The obtained routing solutions are feasible if they have nonpositive cost. It is straightforward to show that Theorem 8.1 carries over to this case as well (see Exercise 8.15).

8.6.1.4.2   Constrained Global Routing Problems   It is more difficult to adapt the method to formulations of the CONSTRAINED GLOBAL ROUTING problem, because in this case the feasibility of the rounded solution becomes an issue.   Raghavan and Thompson [371] offer general methods for attacking this problem.   Whether these methods apply to the CONSTRAINED GLOBAL ROUTING problem has yet to be investigated.

8.6.1.5   Experimental Results

Ng, Raghavan, and Thompson [337] have carried out experiments with the randomized rounding algorithm on grid graphs.   They allow for multiterminal nets.   The algorithm decomposes these nets into two-terminal nets using heuristics for rectilinear Steiner trees [170].   For each resulting two-terminal net, they allow a restricted number of routes.   The number of routes considered can be controlled by an input parameter.   The resulting global-routing problem is presented to the algorithm described in Section 8.6.1.1.

The experimental findings can be summarized as follows:

1. In general, many of the variables attain integer values already in the solution α.   Rounding can be restricted to the remaining few variables. Sometimes, α is all integer, making rounding completely unnecessary.

2. Run times are very small.   It takes substantially less than 1 CPU minute to solve a linear program for a problem instance with about 450 nets and a routing graph with about 800 edges.

3. The control parameter does have a significant influence on the quality of the resulting routing.   Although, in our problem formulation, all routings found are feasible, only the ones whose cost is smaller than c are realizable on a gate-array master with uniform edge capacity c.   As the control parameter is varied to allow for more admissible routes per net, the cost of the routing is more likely to decrease below this threshold.

In summary, the experiments show that randomized rounding is a useful routing technique in practical applications.

8.6.1.6 Prerequisites for the Applicability of Randomized Rounding

In this section, we summarize what properties of an integer program are required for randomized rounding.

1. The integer program must be $0, 1$. For some integer programs that do not fulfill this property, we are able to cut off the integral part of variables to turn the program into a $0, 1$-integer program.

2. The coefficients of the cost function must lie in the interval $[0, 1]$.

3. The constraints must be such that, after randomized rounding, a feasible solution is obtained with certainty or, at least, with high probability.

Given these properties, randomized rounding provides quite a flexible framework for the solution of the integer program. We have seen that there are a multitude of ways of modeling the randomization with sequences of Bernoulli trials. This flexibility can be exploited to fulfill the somewhat imprecisely stated requirement (3). Also, we worked with several relaxations of the original integer program, some being just straightforward linear relaxations, others being based on multicommodity-flow problems. The important property of the relaxation L is that the cost of its optimal solution α is a lower bound on the cost of an optimal solution z of the original integer program R.

Let us now discuss the quality of the solutions obtained by randomized rounding algorithms. The analysis of the quality of the solution is in terms of $c(\alpha)$. Therefore, if $c(\alpha)$ is substantially less than $c(z)$, then the bounds are not very tight. Thus, it is important to use a relaxation whose optimal cost is close to the optimal cost of R. The second parameter that affects the quality of the analysis is the expected value χ of the random experiments. By Lemma 8.4, the quality of the solution increases with χ. Specifically, the deviation $\delta(\chi, \varepsilon)$ is $O(1/\sqrt{\chi})$; see Lemma 8.5.

8.6.2 Making the Method Deterministic

Whereas the experiments show that, for practical applications, randomized rounding is a quite successful strategy, the property that good solutions are obtained only with high probability and not with certainty is undesirable from a theoretical point of view. Raghavan [370] has shown that the randomized rounding algorithm can be made deterministic without jeopardizing the quality of the solutions or excessively increasing the run time. In this section, we describe this method. We exemplify the method using the global-routing integer program discussed in Section 8.6.1.1. However, we shall not use any specific properties of this example. In fact, the method is applicable to all examples we discuss in this book and, more generally, to a wide class of integer programs susceptible to randomized rounding. Only problems with achieving feasibility in randomized rounding can interfere with the applicability of the deterministic method we discuss in this section.

First, we note that each of the versions of Theorem 8.1 also implies an existence proof for a routing with the predicted quality. In fact, setting $\varepsilon = 1$ and $k = 1$ results in a deviation from the optimum that can be achieved with probability larger than 0. Thus, a routing with the predicted deviation exists. We now describe a deterministic method for finding such a routing.

Let us call a routing *good* if its cost lies within the deviation $\delta(\alpha_L, 1/m)$ provided by Theorem 8.1 for $\varepsilon = 1$ and $k = 1$. The randomized rounding algorithm has a nonzero probability of finding a good routing in one execution of step 3. The execution of the randomized rounding algorithm can be modeled by a probabilistic decision tree T (see Figure 8.25).

The leaves of T correspond to routings. Each level ℓ of interior nodes of T corresponds to a net (n_ℓ, i_ℓ) that has fractional routes in the optimal solution α of L, and thus has to be rounded. The nodes in level ℓ correspond to random decisions that select a route for the net. Each node z has I_{n_ℓ} children, each one representing the choice of one route for the net. The edge from z to the child z_j representing the choice of $T_{n_\ell}^j$ for net (n_ℓ, i_ℓ) is labeled with the probability $\alpha_{n_\ell, i_\ell, j}$.

The randomized rounding algorithm starts by choosing a value for the net (n_0, i_0) corresponding to the root r of T. Depending on the outcome, we walk down the corresponding edge out of r and reside at a node at level 1. Here, we choose a value for the corresponding net (n_1, i_1), and so on (see Figure 8.25). All choices along a path in the tree are independent. Therefore, the probability of reaching a node z in T is the product of all probabilities of nodes along the path from z to the root of T.

We want to emulate this random choice deterministically, and to be sure to end up in a good leaf—that is, a leaf representing a good routing. Assume for a moment that, for all nodes z in T, we could compute the probability $p(z)$ that we reach a bad leaf, given that we start in node z and continue with the randomized rounding algorithm. (We cannot expect to be able to compute $p(z)$ exactly but, in several cases, we can compute tight estimates on $p(z)$.) How could we use $p(z)$ to find a good leaf ? Assume that z is on level ℓ of T. Of course, the most sensible choice is to go to the child z_j, $j = 1, \ldots, I_{n_\ell}$, of z that minimizes $p(z_j)$. In effect, this *optimistic strategy* eliminates the probabilities and moves to the more promising child deterministically, each time. Note that the optimistic strategy *deterministically* moves along the tree using the probabilities $p(z)$, whereas the probabilities $p(z)$ are determined on the basis of the probabilities used by the *randomized* algorithm.

An analysis of the values of the $p(z)$ shows that the optimistic strategy always ends up in a good leaf: The definition of the $p(z)$ implies that, for a

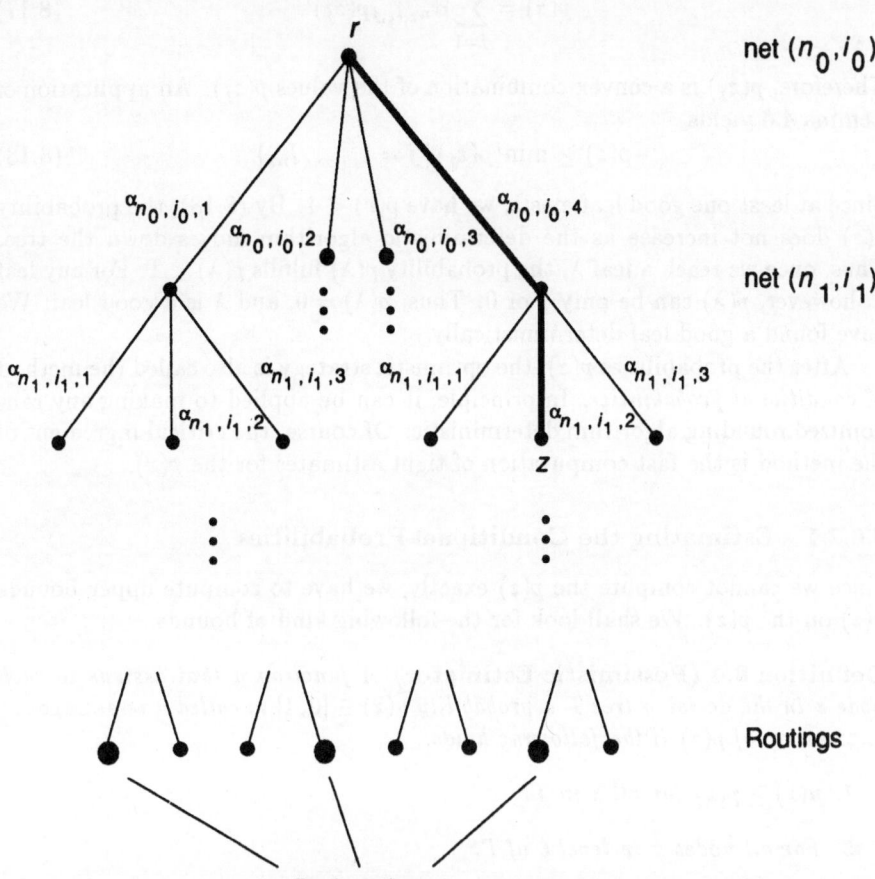

Figure 8.25: The decision tree T for the randomized rounding algorithm. Levels correspond to nets. Nodes of T correspond to partial outcomes of coin tosses. Node z represents the choice of route $T_{n_0}^4$ for net (n_0, i_0) and of route $T_{n_1}^2$ for net (n_1, i_1). The randomized rounding algorithm encounters node z with probability $\alpha_{n_0, i_0, 4} \cdot \alpha_{n_1, i_1, 2}$. The leaves of the tree represent total outcomes of the coin tosses—that is, routings. Bold leaves denote bad routings.

node z at level ℓ,

$$p(z) = \sum_{j=1}^{I_{n_\ell}} \alpha_{n_\ell, i_\ell, j} p(z_j) \qquad (8.17)$$

Therefore, $p(z_j)$ is a convex combination of the values $p(z_j)$. An application of Lemma 4.6 yields

$$p(z) \geq \min\{p(z_j) \mid j = 1, \ldots, I_{n_\ell}\} \qquad (8.18)$$

Since at least one good leaf exists, we have $p(r) < 1$. By (8.18), the probability $p(z)$ does not increase as the deterministic algorithm moves down the tree. Thus, once we reach a leaf λ, the probability $p(\lambda)$ fulfills $p(\lambda) < 1$. For any leaf λ, however, $p(\lambda)$ can be only 1 or 0. Thus, $p(\lambda) = 0$, and λ is a good leaf: We have found a good leaf deterministically.

After the probabilities $p(z)$, the optimistic strategy is also called the method of *conditional probabilities*. In principle, it can be applied to making any randomized rounding algorithm deterministic. Of course, the critical ingredient of the method is the fast computation of tight estimates for the $p(z)$.

8.6.2.1   Estimating the Conditional Probabilities

Since we cannot compute the $p(z)$ exactly, we have to compute upper bounds $u(z)$ on the $p(z)$. We shall look for the following kind of bounds.

Definition 8.4 (Pessimistic Estimator) *A function u that assigns to each node z in the decision tree T a probability $u(z) \in [0,1]$ is called a* **pessimistic estimator** *of $p(z)$ if the following holds:*

1. $u(z) \geq p(z)$ for all z in T.

2. *For all nodes z in level ℓ of T,*

$$u(z) \geq \min\{u(z_j) \mid j = 1, \ldots, I_{n_\ell}\}$$

3. $u(r) < 1$.

4. $u(z)$ can be computed efficiently.

Substituting a pessimistic estimator $u(z)$ for $p(z)$ in the deterministic algorithm will also lead to a good leaf, by an analogous argument. Furthermore, by (4), the algorithm will be efficient.

8.6.2.1.1   Defining the Pessimistic Estimator at the Root   We can find a pessimistic estimator by using the results of the analysis in the proof of Lemma 8.4. In effect this proof constructs easily computable upper bounds of the probability of failure, and proves that they are not too large.

Specifically, let $\varepsilon = 1$ and $k = 1$. Consider failure of the randomized rounding algorithm at edge $e \in E$. Considering the way the algorithm is modeled with Bernoulli trials—given in the proof of Theorem 8.1—and substituting the appropriate quantities into (8.4) yields the following upper bound on the probability of failure of the randomized rounding algorithm at edge e:

$$ e^{-t(1+\delta)\alpha_L} \prod_{(n,i) \in N} \sum_{\{j \,|\, e \in T_n^j\}} (\alpha_{n,i,j} \exp(t) + 1 - \alpha_{n,i,j}) \tag{8.19} $$

As in the proof of Lemma 8.4, we choose $t = \ln(1+\delta)$. As we have already seen in the proof of Theorem 8.1, the probability of failure of the whole algorithm is at most the sum of the probabilities of failure at each edge probability. Summing (8.19) over all edges, we get the following upper bound on the failure of the randomized rounding algorithm:

$$ \Pr[\Psi > (1+\delta)\alpha_L] \leq \tag{8.20} $$

$$ \sum_{e \in E} \left(e^{-t(1+\delta)\alpha_L} \prod_{(n,i) \in N} \sum_{\{j \,|\, e \in T_n^j\}} (\alpha_{n,i,j} \exp(t) + 1 - \alpha_{n,i,j}) \right) $$

In the proof of Theorem 8.1, we chose δ such that (8.20) is less than 1. Therefore, (8.20) is an appropriate choice for $u(r)$, such that requirement (3) of Definition 8.4 is met. Requirement (1) is met because we took $u(r)$ out of the analysis in the proof of Lemma 8.4.

8.6.2.1.2 Propagating the Estimate Through the Decision Tree

At level ℓ in T, we choose a route for net (n_ℓ, i_ℓ). This decision effectively sets

$$ \alpha_{n_\ell, i_\ell, j} = \begin{cases} 1 & \text{if } T_{n_\ell}^j \text{ is the route chosen for net } (n_\ell, i_\ell) \\ 0 & \text{otherwise} \end{cases} \tag{8.21} $$

Substituting (8.21) into (8.20) for the choices made on the path to a node z in T yields the formula for $u(z)$.

Again, because the definition of $u(z)$ is based on the analysis in Lemma 8.4, requirement (3) of Definition 8.4 is met. Exercise 8.16 shows that requirement (2) is met as well. Therefore, $u(z)$ is a pessimistic estimator of $p(z)$.

8.6.2.2 Run Time

To analyze the run time of the method of conditional probabilities, we observe that the optimistic strategy traverses one edge of the decision tree T for each Bernoulli trial executed by the randomized rounding algorithm. Traversing an edge out of the node z in T amounts to minimizing the conditional probabilities over all children of z, which, in turn, entails the computation of (8.20) for each child of z. Let $|x|$ denote the number of variables of R. The computation of

$u(r)$ takes time $O(m|x|)$. The update of $u(z)$, as we move from z to one of its children, can be done incrementally in time $O(m)$. Thus, traversing a path to a leaf can be done in time $O(m|x|)$. Therefore, the whole method takes time $O(m|x|)$, in addition to the time needed to solve L. This complexity contrasts with the time $O(k|x|)$ taken by steps 2 to 4 of the randomized rounding algorithm. As a result, the complexity of the method of conditional probabilities is essentially as large as the complexity of m repetitions of randomized rounding.

8.7*   Determining the Locations of the Pins

In global routing in floorplans, the Steiner tree $T_{n,i}$ for a net (n, i) in the routing graph does not provide enough information for the subsequent detailed-routing phase. $T_{n,i}$ specifies only *which* walls the wire for net (n, i) has to cross. For detailed routing, it is necessary to know the *exact location on a wall* where the wire crosses—that is, the exact locations of the pins for the wire. If the global routing does not compute this information, the latter has to be generated in a subsequent phase *before* detailed routing.

Most approaches to global routing that are in use today solve one of the routing problems defined in Section 8.1.1 first, and then assign the locations of pins heuristically.

If routing in channels is used, then the assignment of pin locations has to take into account the lengths of the wire segments through channels. Heuristically, we can formalize this aspect by having a pin experience a "pull" toward the end of the channel at which the pin is connected to the rest of the net. The size of the pull may depend on such variables as the location of the pin in the channel, the size of the net, and the load of the channel. (If the pin is connected to the rest of the route at both channel ends, then there is no pull to either side.) We can assign pin locations by processing one channel at a time and solving a LINEAR ASSIGNMENT problem (see Section 3.12). The corresponding problem instance attaches to each (net, pin location) pair a cost and assigns nets to pin locations such that the sum of all costs is minimized.

In the case of channel-free routing in floorplans we can use an approach similar to the one described in Section 8.5.3.3. After determining the global routing we process the walls of the floorplan in some order. When processing wall w, we assign a cost to each pair (n, p), where n is a net crossing w, and p is a pin location on w. These costs are defined on the basis of the net topologies in the neighborhood of the wall w, similarly to Section 8.5.3.3. Then we solve the corresponding LINEAR ASSIGNMENT problem.

8.8* Integration of Global Routing and Floorplanning

As we have mentioned several times already, floorplanning and routing have close interactions with each other. The routing phase determines the exact shape of the routing regions. If this shape deviates from the shape estimate computed during floorplanning, the floorplan topology may change. This change may make necessary at least a partial update and a renewed optimization of the floorplan, which in turn must be followed by a repeated routing phase. The interaction between floorplanning and routing is sufficiently complex that the dynamics of this loop are not well understood. In fact, it may be that the loop does not stabilize at all—that is, that a floorplan update implies a new routing that leads to the floorplan *before* the update. Therefore, we can hope for neither termination nor convergence. Thus, it is highly desirable to increase the interaction and communication between the floorplanning and routing phase in order to come up with a scheme that converges to a good solution in terms of both floorplanning and routing.

There are two ways of increasing communication. One is to devise an algorithmic scheme that intertwines floorplanning and routing sufficiently such that an agreement between the results of both phases can be achieved. We will call this method the *algorithmic approach* to integrating floorplanning and routing. The other approach is to provide a sufficiently powerful data structure that represents both floorplan and routing data and that allows us to carry out efficiently the updates of the floorplan implied by the routing, and vice versa. Such a data structure can be the basis of interactive or heuristic methods for integrated floorplanning and routing. We shall call this method the *data-structure approach* to integrating floorplanning and routing.

The data-structure approach naturally favors heuristics that *locally* change the floorplan and the global routing, respectively, whereas the algorithmic approach entails *global* interactions between floorplanning and global routing. On the other hand, the data-structure approach entails an extraordinary amount of flexibility for heuristic or manual intervention, which is not provided by the algorithmic approach. In the future, it may be possible to combine the advantages of both methods into one unified framework for integrated floorplanning and routing.

Ideally, the term *routing* should cover both global and detailed routing in this context, since the exact shape of the routing regions is determined in only detailed routing. However, the state of the art has advanced to only integrating floorplanning and *global routing*, so far.

We now sketch both approaches to integrating floorplanning and global routing.

8.8.1*   The Algorithmic Approach

All algorithmic approaches to integrating floorplanning and global routing are based on hierarchical methods. The basic idea is that, instead of generating a completely labeled cut tree first and then using it to compute a routing hierarchically, both phases can be merged.

In this section, we describe how to integrate the unoriented-mincut method with global routing [282]. Recall that, in the unoriented-mincut method, we first generate an unoriented slicing tree (Definition 7.8) top-down. This tree is sized and oriented in a subsequent bottom-up phase. During this phase, estimates of wiring area have to be used to approximate the total area needed inside a room. In Section 7.2.6, we mentioned statistical approaches for doing this estimation. Instead, we can also use hierarchical global-routing techniques to get more accurate estimates. This time, we are interested only in an area estimate, not actually in the routing itself. Therefore, the cost function to be optimized is some measure of the added area instead of the length of the routing. Furthermore, we do not attribute capacities to the edges in the routing graph corresponding to the floorplan pattern. This provision reflects the fact that we are still free to *add* area, if we need to. The resulting optimization problem can again be coached as an integer program. This integer program has a structure that is very similar to the integer program for global-routing problems. The techniques described in [181] are applicable to solving this program very quickly.

After sizing of the slicing tree, the tree has to be ordered. Section 7.2.1.3.2 discusses how this ordering can be done. There, we mention that, if the ordering is aimed at minimizing the total wire length (which is estimated by the sum of the half-perimeters of the bounding boxes of each net), then the ordering problem is NP-hard.

A heuristic method for accomplishing the ordering of the slicing tree is to let hierarchical global routing contribute the necessary data on wiring cost. This approach effectively merges hierarchical global routing with the ordering phase of unordered mincut floorplanning. Both procedures process the slicing tree top-down. Thus, they can be integrated easily.

First, we flatten the slicing tree by collapsing adjacent nodes to create floorplan patterns for internal nodes of the tree from the set that is depicted in Figure 8.19. Then, the resulting cut tree is processed top-down. The global routing is refined using the method by Luk et al. [299] that is described in Section 8.5.3. This time, however, the assignment of blocks to rooms is not completely specified yet, because the ordering of the cut tree has not been determined. Therefore, for each internal node of the cut tree, several floorplan patterns are still possible. We select one of these patterns on the basis of routing data. This selection also accomplishes the ordering of the cut tree. To evaluate the patterns, we perform a global routing on the pattern by solving an appropriate integer program (see Section 8.2.5), and then we route wires between the pattern and its neighborhood in the floorplan using heuristic techniques, as described in Section 8.5.3.1. Finally, we evaluate the cost of the resulting routing. The pattern that minimizes this cost is selected.

Since there are several patterns from which to choose for each node of the cut tree, we cannot try all combinations of patterns. Heuristic methods can be used to select patterns sequentially for each internal node of the cut tree. A popular technique is based on sweeping the floorplan diagonally—say, from the top-left to the bottom-right corner—and to select patterns as the corresponding subfloorplans are encountered during the sweep. For details, see [282].

Dai and Kuh [80] propose a scheme for integrating floorplanning and global routing that is based on a different floorplanning strategy. They generate a cut tree of the circuit bottom-up using clustering methods (see also Section 7.2.2). Then, they determine sized floorplan patterns for the interior nodes of the cut tree in a subsequent top-down pass. The selection of the floorplan patterns is integrated with a hierarchical top-down global-routing method, as described in Section 8.5.3.

Other algorithmic approaches to integrating floorplanning and global routing are described in [55, 422, 434, 435].

8.8.2* The Data-Structure Approach

The data-structure approach has been taken in the BEAR system developed at the University of California, Berkeley [81]. The goal of this approach is to break the loop between floorplanning and global routing in general-cell placement with channel routing. Here, the problem is that the channel widths are determined by the global routing, but they influence the floorplan topology. Rather than iterating the process of floorplanning and global routing several times, we can also choose a judicious ordering of the channels, route one channel at a time, and modify the floorplan directly according to the result. Then, we adapt the global routing, and route the next channel.

This approach to integrating floorplanning and global routing has the flavor of a local search strategy, but it is very flexible. The center of this approach is a dynamic data structure for representing rectangular subdivisions in the plane. Here, the rectangles represent both cells and routing regions. The data structure contains all information necessary to derive the corresponding floorplan. Efficient updating operations are developed for the data structure. These updating operations cover such instructions as inserting, deleting, or changing the size of a rectangle, and a compression-ridge approach to compaction (see Section 10.2.1). Floorplanning and global routing are done with algorithms that update the central data structure and perform channel routings and local optimizations (including compression-ridge compaction). Details can be found in [78, 79, 81].

8.9*   Simulated Annealing

Kirkpatrick et al. [234] apply simulated annealing to a version of the UNCON-
STRAINED GLOBAL ROUTING problem. The problem formulation is a variant
of the integer program discussed in Section 8.2.2. The routing graph is a grid,
all edges have capacity 1 only two-terminal nets are considered, and all routes
with at most one bend are admissible. We determine a configuration by assigning one of the at most two admissible routes to each net. Edges in the
configuration graph connect configurations that differ from each other in the
assignment of one route. The cost function sums up the squares of the loads
on all channels.

The various TimberWolf packages apply simulated annealing to both placement and global routing [406, 407, 408, 409]. They essentially solve the CON-
STRAINED GLOBAL ROUTING problem. We sketch the most recent algorithm
[406]. In a first phase, a set of admissible routes is generated for each net. How
many of them are generated can be set by a parameter M. For two-terminal
nets an algorithm by Lawler [261] is employed for finding the M shortest routes
between two points. For multiterminal nets, heuristic extensions are made to
this algorithm. Then, in a second phase, a route is chosen for each net by simulated annealing. The move from one configuration to another is performed
as follows. First, an edge in the routing graph is identified whose capacity
constraint is violated. (If no such edge exists, the algorithm halts and outputs
the solution.) A net that is running across the edge is chosen randomly, and
an alternative route is chosen randomly for this net.

The cost function minimizes the number $\sum_{e \in E} U(e) - c(e)$—that is, the
number of *excess tracks*—with first priority and the total length of the routing
with second priority.

Simulated annealing provides a robust framework for heuristically integrating placement, pin assignment, and global routing. However, tuning the algorithms and cooling schedules is an involved task.

8.10   Summary

Global routing is an integer-programming problem. Classically, the integer
program has been reduced to sequential Steiner tree computations—one for
each net—which, in turn, are reduced to shortest-path computations. This
approach is quite robust, since it is essentially based on Dijkstra's algorithm.
However, this approach does not take the interdependence between different
nets sufficiently into account.

Recently, more sophisticated integer-programming methods have been applied to global routing. One such method is hierarchical global routing. In
this method the global-routing integer program is decomposed into small pieces
that can be solved efficiently. Each piece represents a subproblem of the global-

routing problem. The decomposition is based upon a hierarchical decomposition of the routing graph that can be derived from the cut tree of a corresponding floorplan. The solutions of the integer program pieces are composed to an approximate solution of the original routing problem. During this composition either sequential Steiner tree methods are applied or integer programming is used again. Hierarchical methods of global routing are quite robust and yield good results in practice, but the quality of the resulting routing is not analyzable, in general.

The linear relaxation of the global-routing integer program can be used as the basis for random experiments that can be proven to yield good routings with high probability. In fact, the randomization can be taken out of this process to yield a deterministic algorithm for global routing with provable quality of the result. Hierarchy is no element of this routing method. However, currently this method can only be applied to versions of the UNCONSTRAINED GLOBAL ROUTING problem.

In conclusion, we can state that viewing global-routing problems as integer programs has lead to significant progress in global routing.

8.11 Exercises

8.1 Determine the integer program for the CONSTRAINED GLOBAL ROUTING problem with $w \equiv 1$, where the routing graph is the graph G depicted in Figure 8.4.

8.2 Even if an instance I of the CONSTRAINED GLOBAL ROUTING problem whose routing graph is the graph depicted in Figure 8.4 is solvable, the routing may not be achievable in the corresponding floorplan if only horizontal and vertical wire segments are allowed in the detailed-routing phase. Problems arise because the detailed routing of some global routes may use up tracks inside a room that have not been accounted for in the capacity constraints of I.

1. Construct an example illustrating this phenomenon.

2. Define additional constraints that ensure that enough tracks are available during detailed routing.

3. Show how to augment the integer program formulation to incorporate these constraints.

4. Calculate the size of the resulting integer program.

5. Generalize your solution to arbitrary floorplans and the corresponding routing graphs.

8.3 Let R be an integer program for the CONSTRAINED GLOBAL ROUTING problem, with a routing graph $G = (V, E)$ and edge lengths $\ell(e) > 0$. Assume that the edge lengths fulfill the *triangle inequality*—that is,

$$\ell(u, v) + \ell(v, w) \geq \ell(u, w) \quad \text{for all } u, v, w \in V \text{ and } (u, v), (v, w), (u, w) \in E$$

Call a two-terminal net (n, i) whose terminals are connected with an edge in the routing graph a *neighbor net*.

Show that in R we can preroute all neighbor nets across the edges connecting their terminals, without forfeiting an optimal solution of R.

8.4 In Section 8.4.1.3, a special implementation of the priority queue used in the algorithm depicted in Figure 8.6 is given for the case that all edge lengths are unity. This implementation allows for $O(1)$ time per operation. Generalize this implementation to the case that edge lengths are bounded by a constant C. Preserve the constant time per operation. How does the run time depend on C?

8.5 Provide an example for which the line-searching algorithms by Mikami and Tabuchi and by Heyns et al. do not find a shortest path.

8.6

1. Show that the track graph is a connection graph for the set V' of pins.

2. Show that the track graph is not always a strong connection graph.

3. Discuss how we can use the track graph to find a shortest path between any pair of pins.

   Hint: Let a *horizontal region* be the area delimited by two adjacent horizontal obstacle tracks on the top and bottom side and by two cell boundaries on the left and right side. Define a *vertical region* similarly. By the remarks in Section 8.4.2.3, if the track graph does not contain a shortest path between a pair of pins, this pair has to lie within the same horizontal or vertical region. In fact, the pin pair has to lie within a maximal isoconvex subregion of such a region. Apply plane-sweep techniques to find a shortest path for such a pair of pins.

8.7 Adjust the plane-sweep algorithm described in Section 8.4.2.3.1.1 to handle the existence of several vertical boundary segments with the same x coordinate.

8.8 Detail the third phase of the construction of the track graph described in Section 8.4.2.3.1.3.

1. Detail the plane-sweep procedure.

2. Take care of degenerate cases.

8.9 Generalize the definition of the track graph to sets of arbitrary disjoint rectagonal obstacles.

1. Show that the resulting track graph is a connection graph.

2. Detail the construction of the track graph.

Hint: Assign additional extreme edges to polygons that are not isoconvex.

8.10 Prove Lemma 8.2. To this end, show how any Steiner tree in the k-grid can be transformed into a Steiner tree that has the property in Lemma 8.2, without increasing the length.

Hint: Handle one grid dimension at a time so that your proof methods will extend to higher-dimensional grids.

8.11 Construct an instance I of the CONSTRAINED GLOBAL ROUTING problem and a corresponding quotient problem I_Π with respect to some vertex partition Π such that I_Π is solvable but I is not solvable. Use routing graphs such as occur in channel-free routing for general-cell layout.

8.12

1. Prove that, if all admissible floorplan patterns for internal nodes of the cut tree come from the set depicted in Figure 8.19, then the graph $G_{i,n}$ is a partial 2-tree.

2. Develop an algorithm for constructing the hierarchical definition of $G_{i,n}$ that runs in linear time in the size of $G_{i,n}$.

8.13 Transform the assignment problem in Section 8.5.3.3 to an instance of the MINCOST FLOW problem.

8.14 Show that the probability that edge e is contained in the random walk for net (n, i) described in Section 8.6.1.2 does not exceed $\alpha_{n,i}(e)$.

Hint: First, show that you can eliminate all cycles from the flow for commodity (n, i). Then, show by induction on the distance of vertex v from the source of the flow that the probability that vertex v is on the random walk does not exceed the flow into v. Use this fact to solve the exercise.

8.15 Tailor Theorem 8.1 to the global-routing integer program whose edge constraints have the form

$$U(e) - c(e) \leq x_L$$

and whose cost function is x_L.

8.16 This exercise proves that $u(z)$ as defined in Section 8.6.2.1 meets requirement (2) of Definition 8.4.

Let (n_0, i_0) be the net whose route is chosen at the root of T. Show that there is a choice of $j \in \{1, \ldots, I_{n_0}\}$ such that substituting 1 for $\alpha_{n_0,i_0,j}$ and substituting 0 for all $\alpha_{n_0,i_0,j'}$, $j' \neq j$ in (8.19) does not increase (8.19).

Extend this argument to the nets (n_ℓ, i_ℓ) for $\ell \geq 1$ whose routes are chosen in lower levels inside the tree T.

Hint: Use Lemma 4.6.

Chapter 9

Detailed Routing

The detailed-routing phase follows the global-routing phase in the two-phase approach to routing. Recall that the global-routing phase constructs routes that are paths in a routing graph whose edges represent routing regions. For the purpose of detailed routing, these edges have to be expanded to give an accurate graph representation of the respective routing region. The resulting *detailed-routing graph* looks very much like the routing grids used in area routing with the exception that it is (almost) always planar. However, in contrast to the area-routing problem, the input now contains not only a set of nets, but also a *global route* for each net. This route specifies how the net maneuvers around the obstacles in the routing region that are defined by the cells of the circuit. The detailed-routing problem is to find detailed routes through the routing graph that comply with the global routes and that obey a certain set of constraints. The constraint set is the *detailed-routing model*.

Figure 9.1(a) shows an instance of the detailed-routing problem and its solution. Note that the "holes" in the detailed-routing graph can be general rectagonal polygons. Figure 9.1(b) shows a solution of this instance of the detailed-routing problem. The detailed-routing model used here is the *knock-knee* model. It requires the routes for different nets to be edge-disjoint (see also Section 9.1.3). In the following section, we shall explain Figure 9.1 in detail, and shall develop the necessary formal definitions.

Before we do so, we make a few remarks on how we construct the input to the detailed-routing phase from the output of the global-routing phase. In gate-array layout, we get the detailed-routing graphs directly from the structure of the master. In general-cell layout, the detailed-routing graph naturally breaks into *channels*, each channel representing an edge in the routing graph. In general-cell layout, however, constructing the detailed-routing graph is quite difficult, because this process entails the exact placement of the cells inside their rooms and the allocation of routing area. Of course, we want to do the placement such that the minimum amount of routing area is needed.

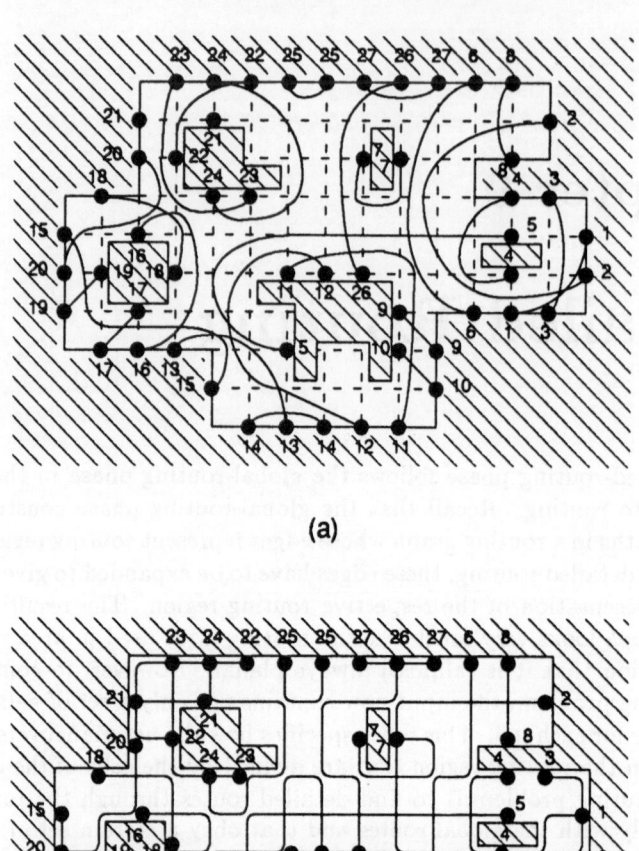

(a)

(b)

Figure 9.1: An instance of the detailed-routing problem and its solution. (a) The problem instance; (b) the solution. The detailed-routing model requires different wires to be edge-disjoint.

In practice, the detailed-routing problem is solved incrementally. The complete routing region is decomposed into *routing sections* that take the shape of channels (see Section 9.1.2.1) or switch boxes (see Section 9.1.2.2). The detailed-routing problem is then solved section by section in a prescribed order. When a routing section is routed, its exact shape is determined. On the way, the cells adjacent to the routing section are placed and the routing area inside the section is allocated. The major part of this chapter is concerned with the routing subproblems occurring in this decomposition approach and their solution. Note that the decomposition of the detailed-routing problem is a heuristic that yields nonanalyzable results, even though, in practice, it generally is quite effective.

Recently, significant mathematical progress has been made on solving the detailed-routing problem without decomposing the routing region. We introduce the respective mathematical framework in Section 9.2 and follow up on it throughout the remainder of the chapter. The application of these new mathematical results presupposes an appropriate expansion of the routing graph used in global routing. A method for doing this expansion is *topological compaction* (see Section 10.5).

This chapter will thoroughly discuss both approaches to detailed routing—the incremental approach and the newer direct approach. Detailed routing is quite a diverse combinatorial subject. Therefore, we have to defer an overview of the chapter until after we have introduced the relevant definitions.

9.1 Basic Definitions

To define the detailed-routing problem formally, we have to specify where the holes in the detailed-routing graph are located, since they define exactly in what way the global routes predetermine the detailed routes. To this end, we define the detailed-routing graph as an arbitrary plane graph—that is, as a planar graph with a planar embedding (see Section 3.14)—and we designate some faces as the holes. In Figure 9.1(a), the detailed-routing graph is a partial grid graph and the holes are the shaded faces. One hole is usually the exterior face. It represents the chip boundary. The global route for a net determines how this net maneuvers around the holes of the detailed-routing graph. The detailed-routing phase may change the course of the route as long as it complies with the global route. Formally, this means that the global and the local route have to be *homotopic*.

9.1.1 Homotopic Routing

Definition 9.1 (Homotopic Paths) *Let $G = (V, E)$ be a plane graph with a designated set H of faces called the* **holes**. *The pair (G, H) is called a* **detailed-routing graph**.

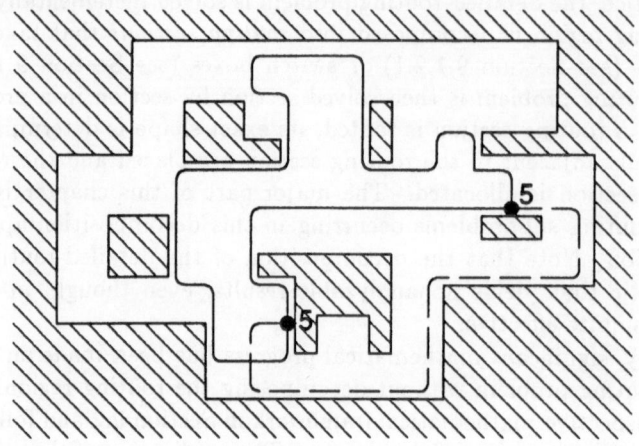

Figure 9.2: *The two solid paths are homotopic paths for net 5 in Figure 9.1. The dashed path is not homotopic to either solid path.*

Let p and p′ be two paths in G. p and p′ are called **homotopic** *if they have the same endpoints v and w, and there is a sequence $p = p_0, p_1, \ldots, p_k = p′$ of paths between v and w such that, for $i = 1, \ldots, k$, p_{i-1} can be transformed into p_i using the following kind of basic step: Choose a face F of G that is not in H and that contains some edges in p_{i-1}. Invert p_{i-1} on F; that is delete all edges in $F \cap p_{i-1}$ from p_{i-1} and add all edges in $F \setminus p_{i-1}$ to p_{i-1}.*

As an example, the two solid paths for net 5 in Figure 9.2 are homotopic, but the dashed path is not homotopic to either solid path.

Definition 9.1 is a graph-theoretic version of a well-known concept in topology [335]. Intuitively, each basic step implements a deformation of the path that does not pull the path across a hole. All such deformations are possible. It is easy to see that homotopy is an equivalence relation between paths in (G, H). The equivalence classes of this relation are called *homotopies*.

Definition 9.1 defines homotopy for two-terminal nets. For multiterminal nets, this definition can be extended in several ways, depending on how much of the tree topology specified by the global routing we want to preserve during the detailed routing. One definition that gives the detailed routing phase much freedom to change the tree topology is the following.

Definition 9.2 (Homotopic Trees) *Let (G, H) be a detailed-routing graph. Two trees T and T′ are* **homotopic** *if they have the same set of leaves and for all pairs v, w of leaves in T and T′ the paths between v and w in T and T′ are homotopic.*

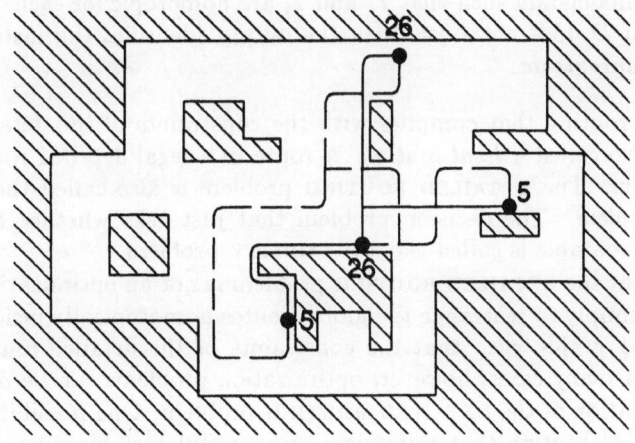

Figure 9.3: *Two trees for the multiterminal net we construct by merging nets 5 and 26 in Figure 9.1 that are homotopic by Definition 9.2. Note that the tree topologies differ. The solid tree has one interior nodes of degree 3, whereas the dashed tree has two such nodes.*

We shall discuss only detailed routing of multiterminal nets in detailed-routing graphs that have trivial homotopic structure, such as channels or switchboxes. In such graphs, all paths between the same pair of vertices are homotopic. In this context, Definition 9.2 implies that all trees with the same set of leaves are homotopic. If the detailed-routing graph has many holes, then we may want to preserve some of the tree topology of a route that has been computed in the global-routing phase. This is not done if we use Definition 9.2, as Figure 9.3 illustrates. We can require that aspects of the tree topology be preserved by suitably modifying the definition (Exercise 9.1). One problem with such definitions is that they have to allow for efficient algorithms for the detailed routing of multiterminal nets. No such algorithms are not known as of today.

We are now ready to formulate the detailed-routing problem formally.

DETAILED ROUTING

Instance: A detailed-routing graph (G, H), $G = (V, E)$, $|V| = n$, a set N of nets, such that all terminals of the nets are on the boundary of holes of G; a set R of global routes; one route T_i for each net $n_i \in N$. The route T_i is a Steiner tree of the terminals of net n_i in G.

Output: A set D of detailed routes, one route T_i' for each net $n_i \in N$, such that D complies with the constraints of the detailed-routing

model, and such that T_i' and T_i are homotopic for each $n_i \in N$.
If D does not exist, then the output identifies the instance as
unroutable.

A detailed routing that complies with the constraints of the detailed-routing
model is also called a *legal routing*. A route in a legal detailed routing is also
called a *wire*. The DETAILED ROUTING problem is also called the *homotopic
routing problem*. The decision problem that just asks whether the problem
instance is routable is called the ROUTABILITY problem.

Note that the DETAILED ROUTING problem is not an optimization problem.
Its goal is simply to rearrange the global routes homotopically inside the available routing graph, such that the constraints of the detailed-routing models
are met. In many cases, however, optimization problems can be derived from
this formulation of the DETAILED ROUTING problem. One possibility is to ask
for a detailed routing that minimizes some useful cost function, such as the
total wire length, the total number of jogs, or the total number of cross-overs.
Another, more popular, kind of optimization problem allows for considering a
class of instances with similar detailed-routing graphs and looks for the one
with the smallest detailed-routing graph that will accommodate the routes for
all nets. This kind of optimization problem has some of the flavor of layout
compaction, and therefore we shall discuss much of it in Chapter 10. But
we shall also encounter such optimization versions of the DETAILED ROUTING
problem in the succeeding sections of this chapter—specifically in the sections
that discuss channel routing (Sections 9.3.1, 9.4 and 9.6.1).

There are two independent taxonomies of versions of detailed routing. One
is via the detailed-routing model; the other is by the shape of the detailed-
routing graph. In Sections 9.3, 9.5, and 9.6, we choose to make the overall
classification of detailed-routing problems using the detailed-routing model.
For each model, we distinguish detailed-routing problems by the shape of the
routing graph. Sections 9.4, 9.7, and 9.8 transcend this classification. In these
sections, general routing schemes for detailed-routing problems are presented
that apply to several detailed-routing models.

We now introduce the detailed-routing models and the routing graphs that
we shall discuss in this chapter.

9.1.2   Detailed Routing Graphs

In this section, we classify detailed-routing problems by the shapes of the
detailed-routing graphs.

The detailed-routing problems on which the most research has been done so
far are the channel-routing problem and the switchbox-routing problem. Both
of them have the property that the exterior face is the only hole in the routing
graph; that is, all paths with the same endpoints are homotopic. Thus, global
routes can be disregarded, and the detailed-routing problem becomes a version

of the area-routing problem (in the respective detailed-routing model). Only recently has the theory advanced to dealing with detailed-routing graphs with nontrivial homotopic structure. Thus, the term *homotopic routing* is relatively new.

Most detailed-routing graphs are partial grid graphs. However, there also are some results on detailed routing in more general planar graphs.

9.1.2.1 Channel Routing

The CHANNEL ROUTING problem is the restriction of the DETAILED ROUTING problem that has the following properties:

1. G is a rectangular grid graph.

2. H contains only the exterior face of G.

3. All terminals of nets in N are on the top or bottom side of G. Each vertex at either of these sides can be the terminal of at most one net.

Since the exterior face is the only hole in G, the global routes can be disregarded. G is called the *channel*. The columns and rows of the channel are given integer coordinates. Columns are numbered from left to right, starting with 1. Rows are numbered from bottom to top, starting with 1. Each row in G is called a *track*. The number of tracks is called the *width* w of the channel. The number of columns of G is called the *spread* s of the channel. A column in G is called *empty* if there is no terminal on the top or bottom of this column. A net that has terminals on both sides of the channel goes *across* the channel. If all terminals of the net are on the same side of the channel the net is *one-sided*. A two-terminal net both of whose terminals are in the same column is a *trivial* net. The horizontal segments of a wire are called *trunks*; its vertical segments are called *branches*. Figure 9.4 illustrates these definitions.

There is a natural way to define an optimization version of the CHANNEL ROUTING problem. This problem's goal is to find the smallest channel width that accommodates the routing. Formally, we call two instances of the CHANNEL ROUTING problem *similar*, if they differ by only the channel width, and by different numbers of empty columns on the left and right side on the channel. The goal of the CHANNEL-WIDTH MINIMIZATION problem is, given an instance of the CHANNEL ROUTING problem, to determine a routable similar instance that has minimum width, and to construct the routing. Sometimes, among all such instances, an instance with minimum spread is required. Because the channel width is determined only in the course of the problem solution, an instance I of the CHANNEL-WIDTH MINIMIZATION problem is defined by a list of nets, each nets being a list of terminals, and each terminal being a pair of a column number and a channel side (top or bottom). If, for any channel width w and channel spread s, the corresponding problem instance $I_{w,s}$ of the CHANNEL

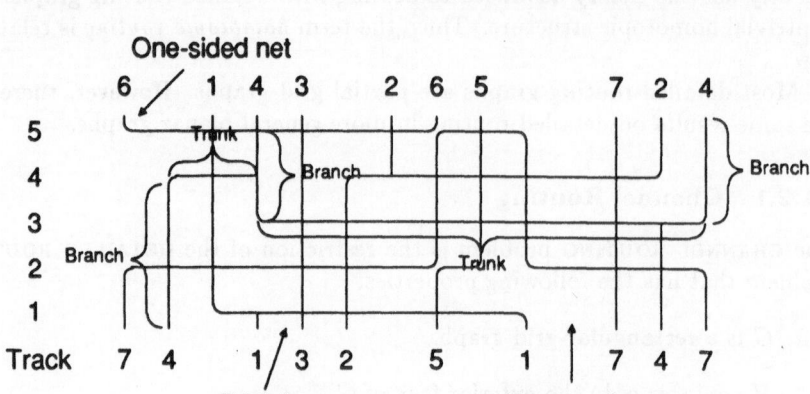

Figure 9.4: *An instance of the* CHANNEL ROUTING *problem. The routing channel has width 5 and spread 15. Terminals on the sides of the channel are denoted with numbers identifying the net to which they belong. A routing of this instance of the* CHANNEL ROUTING *problem in the knock-knee model—that is, with edge-disjoint routes— is given. Net 3 is a trivial net. Net 6 is a one-sided net. The trunks and branches of net 4 are indicated. Columns 11 and 15 are empty.*

ROUTING problem is unroutable, then I is *unsolvable*. Otherwise, the minimum channel width is denoted with $w_{min}(I)$.

The channel routing problem has received an very large amount of attention. One reason is that it is the easiest detailed-routing problem. The other is that it occurs naturally in several layout schemes. In standard-cell layout, the CHANNEL-WIDTH MINIMIZATION problem formalizes the detailed-routing problem between each pair of adjacent rows of standard cells. In an incremental approach to detailed routing in slicing floorplans, we perform the detailed routing by processing the slicing tree bottom up. Each internal tree node defines an instance of the CHANNEL-WIDTH MINIMIZATION problem (Figure 9.5). In this application, channels can have ragged boundaries. Thus, many of the newer channel routers relax property 1 of the CHANNEL ROUTING problem to allow for irregular top and bottom sides of the channel. The channel width is the number of partial rows in G. The corresponding problem is called the IRREGULAR CHANNEL ROUTING problem. Its minimization version is called the IRREGULAR CHANNEL-WIDTH MINIMIZATION problem. Often, an irregular channel is required to be vertically convex according to the following definition.

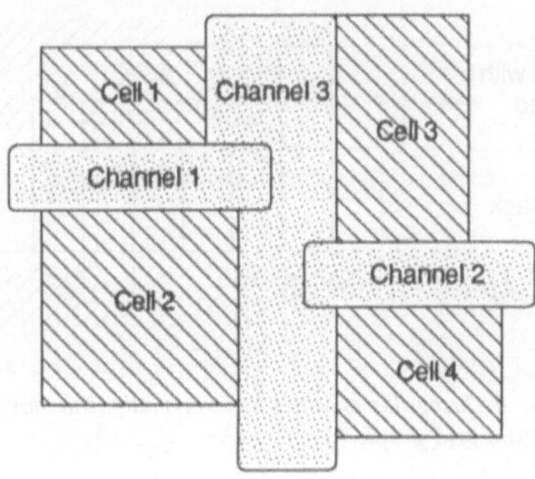

Figure 9.5: *Hierarchical detailed routing in slicing floorplans. This figure depicts the channel structure generated by hierarchical detailed routing in a slicing floorplan with a vertical and two horizontal cutlines (see Figure 8.4). The horizontal cutlines corresponding to nodes low in the slicing tree define instances of the* CHANNEL-WIDTH MINIMIZATION *problem that are solved first. As a result, blocks are formed that consist of cell 1, cell 2, and channel 1 on the one hand, and of cell 3, cell 4, and channel 2 on the other hand. These blocks define a third instance of the* CHANNEL-WIDTH MINIMIZATION *problem.*

Definition 9.3 (Isoconvex Channel) *A partial grid graph G is* **vertically (horizontally) convex**, *if each partial column (row) of G consists of one segment of edges. G is* **isoconvex** *if it is vertically and horizontally convex.*

For an illustration of Definition 9.3, see Figure 9.6.

The channel routing problem calls for the detailed routing of wires between two blocks of fixed terminals, one on either side of the channel. Often, we have the flexibility of offsetting the two sides of the channel laterally such as to improve the routing. This optimization is, for instance, applicable to all channels in Figure 9.5. Formally, we formulate this optimization problem by modifying the definition of similarity of problem instances to allow also for shifting all top against all bottom terminals horizontally by some arbitrary but uniform amount f. This flexibility to offset the top against the bottom side of the channel allows for the optimization of different cost criteria. The corresponding optimization problems are called *channel-offset problems:*

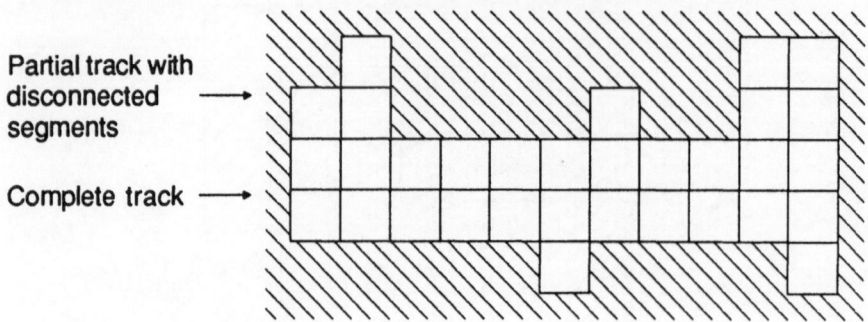

Figure 9.6: *An irregular channel is a vertically, but not horizontally, convex partial grid graph.*

CHANNEL OFFSET RANGE: Given a problem instance with width w compute the range of all offsets f that lead to routable problem instances.

OPTIMUM SPREAD CHANNEL OFFSET: Given a problem instance with channel width w choose the offset that minimizes the spread s.

OPTIMUM WIDTH CHANNEL OFFSET: Find the pair (w, f) such that the problem instance is routable with width w and offset f and w is minimum. Sometimes, f is required to minimize the spread among all offsets that are routable in width w.

OPTIMUM AREA CHANNEL OFFSET: Find the pair (w, f) such that the problem instance is routable with width w and offset f, and the channel area $w \cdot s$ is minimum.

In their original formulation, channel-offset problems call for rectangular channels. The application of channel routing in layout schemes based on slicing floorplans (see Figure 9.5) motivates us to generalize these problem to irregular channels. In the corresponding routing problems, which are identified by the prefix IRREGULAR in the problem name, the ragged boundaries on the top and bottom are allowed to be offset against each other. The terminals remain fixed on their respective channel boundaries.

The strongest optimization versions of the channel-routing problem assume the terminals to be grouped into *several* blocks on each side of the channel. All blocks are allowed to be offset against each other by a different amount. All offsets that can be achieved without swapping or overlapping different blocks are possible. This group of problems is called *channel-placement problems* (Figure 9.7). We discuss the following variants:

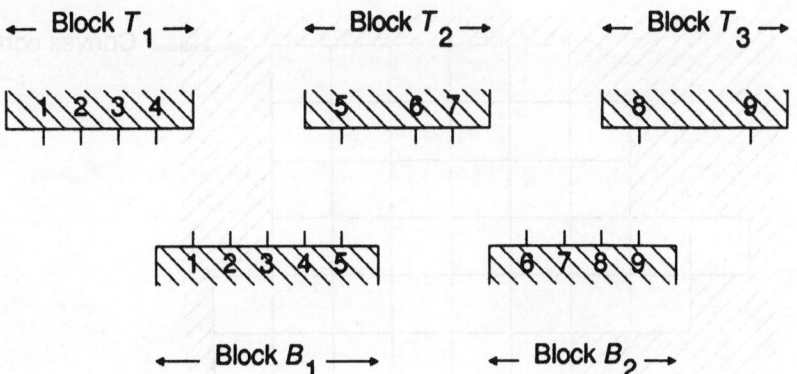

Figure 9.7: *An instance of the* OPTIMUM WIDTH CHANNEL PLACE-MENT *problem. The blocks can be offset horizontally by different amounts in order to minimize the channel width. Overlapping or swapping blocks is prohibited.*

OPTIMUM SPREAD CHANNEL PLACEMENT: Given a problem instance with width w, choose lateral offsets for the blocks that minimize the spread s.

OPTIMUM WIDTH CHANNEL PLACEMENT: Find offsets for the blocks such that the problem instance is routable with width w and w is minimum. Sometimes, the spread among all offsets that are routable in width w is required to be minimum, as well.

OPTIMUM AREA CHANNEL PLACEMENT: Find offsets for the blocks such that the problem instance is routable with width w, and the channel area $w \cdot s$ is minimum.

There are efficient algorithms for channel-placement problems in the planar routing model. Exercise 9.2 discusses applications of channel-placement problems.

In layout schemes that are based on floorplans that are no slicing floorplans, channels can have more general shapes—that is, L-shapes or T-shapes [78].

9.1.2.2 Switchbox Routing

If property 2 of the CHANNEL ROUTING problem is relaxed to allow terminals everywhere on the exterior face, then the CHANNEL ROUTING problem becomes the SWITCHBOX ROUTING problem. Irregular switchboxes are dealt with by the IRREGULAR SWITCHBOX ROUTING problem. Often, irregular switchboxes are

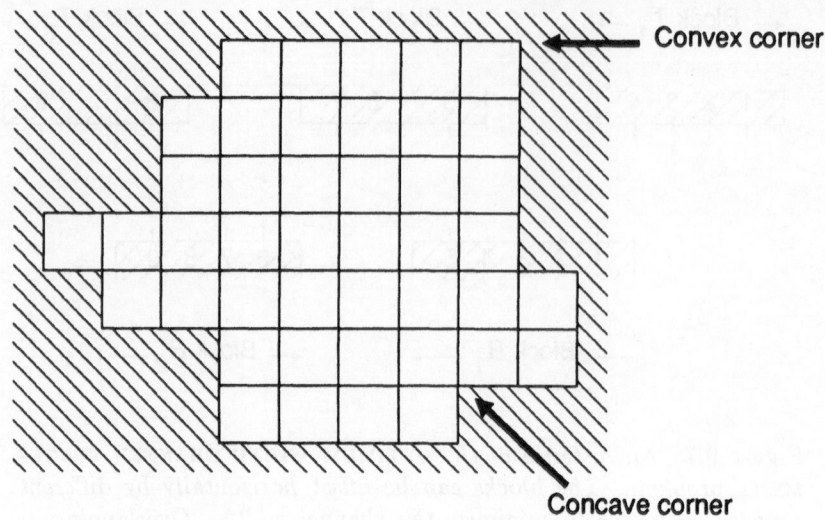

Figure 9.8: *An irregular but isoconvex switchbox.*

required to be isoconvex according to Definition 9.3. The following convention regulates the number of nets that may have terminals at a boundary vertex of the switchbox (see also Figure 9.8):

1. A convex corner of the switchbox can be the terminal of two nets.

2. A concave corner can be the terminal of no net.

3. Every other vertex can be the terminal of one net.

Since a switchbox is *closed* on all four sides, optimization versions of the SWITCHBOX ROUTING problem cannot optimize the shape of the switchbox, but rather need to optimize secondary criteria, such as the total wire length or total number of jogs.

9.1.2.3   Other Detailed-Routing Graphs

There are several results on other restricted sets of detailed-routing graphs. Among them a prominent role is played by graphs with few holes. The number of homotopies increases with the number of holes in the graph. Thus, graphs with few holes are more amenable to efficient routing algorithms. Sometimes, the fact that graphs with few holes have few homotopies even allows us to solve the AREA ROUTING problem on such graphs efficiently. (Remember that in the AREA ROUTING problem no global routes are given in the problem instance.)

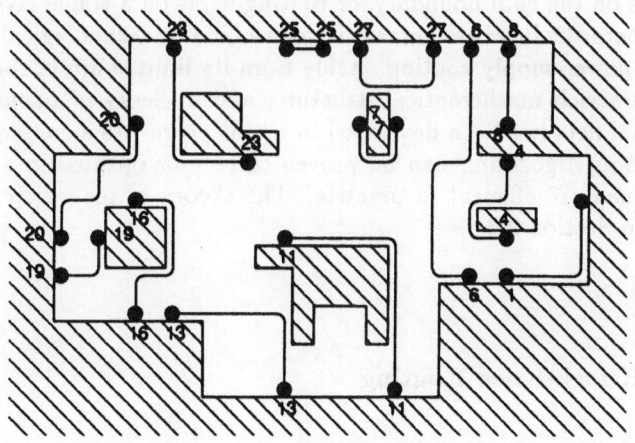

Figure 9.9: *A planar routing for the instance of the* DETAILED ROUT-
ING *problem given in Figure 9.1(a) with only nets 1, 4, 6, 7, 8, 11,
13, 16, 19, 20, 23, 25, 27 included. Adding any further net to the set
makes the problem instance unroutable in the planar routing model.*

In such cases, the global routing is determined together the detailed routing in
the problem solution. We shall specify the detailed-routing graphs for which
such results exist as we mention those results in the following sections.

9.1.3 Detailed Routing Models

We discuss five detailed-routing models. The first three of them—planar routing, knock-knee routing, and Manhattan routing—are the major graph-based
detailed-routing models. The forth model—gridless routing—is a geometric
detailed-routing model for which algorithms can be obtained by generalization
from the three major graph-based models. The fifth class of models—multilayer
models—is closely related to area routing, as discussed in Section 8.1.2.

9.1.3.1 Planar Routing

In planar routing, different routes are required to be vertex-disjoint (Figure 9.9). We know this restriction already from the AREA ROUTING problem,
except that there the routing graph is not planar. The planar routing model is
also called the *river routing* model, because rivers cannot cross without their
waters merging. This model is very restrictive; in fact, many instances of the
DETAILED ROUTING problem are not routable without crossing wires. The planar routing model is used for such tasks as routing the chip inputs and outputs

to the pads on the chip boundary, or routing wires on a single layer in routing schemes where the layer assignment is determined by technological constraints, such as in power-supply routing. Aside from its limited applicability in practice, it has a rich mathematical structure, and a theory of planar routing of two-terminal nets has been developed in recent years. As a consequence, most planar routing algorithms can be proven correct or optimal in a precise formal sense *and* are efficient in practice. The theory of planar routing will be discussed in Section 9.3.

9.1.3.2   Knock-Knee Routing

We have already used this model in our examples and figures. It is much more general than the planar routing model. Since wires can cross, many more instances of the DETAILED ROUTING problem are routable than in the planar routing model. Furthermore, much of the theory that exists for planar routing can be carried over to the knock-knee model. These results are discussed in Section 9.5. However, the theory of knock-knee routing is not as elegant as the theory of planar routing. Specifically, there is a nasty *evenness condition* that pervades the premises of correctness theorems. Furthermore, provably correct algorithms for knock-knee routing exist for only two-terminal nets.

Channel routing is especially easy in the knock-knee model. If only two-terminal nets are involved, a very simple algorithm solves the CHANNEL-WIDTH MINIMIZATION problem (Section 9.5.1). If multiterminal nets are involved the CHANNEL-WIDTH MINIMIZATION problem becomes NP-hard.

From a practical point of view, the knock-knee model has a strong disadvantage that has so far precluded it from entering the design automation scene: A detailed routing in the knock-knee model does not suggest an assignment of wiring layers to wire segments. Of course, in planar routing, such an assignment is trivial, since one layer suffices to route all wires. In knock-knee routing, wires can cross; thus, at least two layers are necessary. Unfortunately, two layers are not sufficient in some cases; sometimes, four layers are needed. Furthermore the assignment of layers is not trivial. In fact, the problem of deciding how many layers are necessary to realize a given detailed routing in the knock-knee model is NP-hard. Also, in many technologies, preferred wiring directions exist for different wiring layers. For instance, in a fabrication process with two wiring layers, one layer may favor vertical wires, whereas the other layer may favor horizontal wires. Such technological constraints do not fit well with the knock-knee model. For this reason, the following, slightly less general, routing model has been used in practice. The layer-assignment problem for detailed routings in the knock-knee model is discussed in more detail in Section 9.5.3.

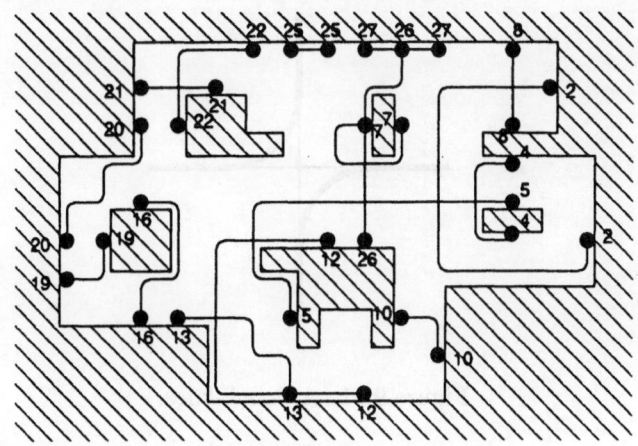

Figure 9.10: *A solution of the instance of the* DETAILED ROUTING
*problem depicted in Figure 9.1(a), with nets 1, 3, 6, 9, 11, 14, 15, 17,
18, 23, and 24 excluded in the Manhattan model.*

9.1.3.3 Manhattan Routing

In Manhattan routing, the detailed-routing graph is assumed to be a partial
grid graph. Different routes are required to be edge-disjoint. Furthermore two
routes may cross over at a grid vertex, but they may not bend away from each
other (Figure 9.10). Such *knock-knees* (Figure 9.11) are explicitly forbidden,
because they are the structure that introduces problems with layer assignment.
(The elimination of this restriction leads to the knock-knee model, and gives
that model its name.) In the Manhattan model, we can assign layers simply
by routing all vertical wire segments on one layer and all horizontal wire segments on the other layer. Contact cuts (*vias*) need to be placed only at grid
vertices, and the exclusion of knock-knees guarantees that there is no conflict
of two vias at any grid node. This ease of layer assignment makes the Manhattan model quite attractive for practical applications. Unfortunately, the
mathematical structure of the Manhattan model is much less conducive to the
development of a theory of Manhattan routing. In fact, many restrictions of the
DETAILED ROUTING problem that are in P for the planar and knock-knee model
are NP-hard for the Manhattan model (for instance, the CHANNEL ROUTING
problem involving only two-terminal nets). Thus, the Manhattan model also
does not lead to efficient solution algorithms. This fact has not been a problem in practice, because most instances of the DETAILED ROUTING problem in
the Manhattan model that occur in practice can be solved heuristically almost
optimally. Nevertheless, the fact remains that we do not understand the math-

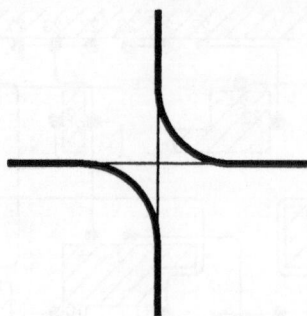

Figure 9.11: *A knock-knee.*

ematical structure of the Manhattan model very well, and that this model's exact solution is infeasible even for severely restricted problem instances. Analyzable algorithms only approximate the solution of the Manhattan routing problem, and so far the constant factors preclude the algorithms from being used in practice. Detailed routing in the Manhattan model is the subject of Section 9.6.1. .

9.1.3.4   Gridless Routing

The preceding three models are the main detailed-routing models investigated today. The fact that they are based on planar graphs has as a consequence that they model the real technological constraints only imprecisely. By moving away from the detailed-routing graph and using the continuous plane as the basic routing domain, and by incorporating the design rules of the fabrication process (see Section 1.2.3.1), we can model the real geometries more precisely, and can achieve tighter optimizations. Thus, particularly well-understood special cases of the detailed-routing problem, such as channel-routing problems, have recently been investigated in the context of gridless routing models. The respective results are summarized in Section 9.9.

9.1.3.5   Multilayer Models

In printed-circuit-board wiring, there are many more than two layers. For this application, multilayer models are needed that represent each wiring layer explicitly. Here, we determine a layer assignment for wiring segments at the time of detailed routing. The presence of many wiring layers usually enables routing across cells or components. Thus, the planar flavor of the detailed-routing problem is lost, homotopy is irrelevant, and routing in multilayer models becomes a version of area routing (see Chapter 8). Nevertheless, some of the results on routing in multilayer models, especially on channel routing, have a strong relationship with results in other detailed-routing models. These results will be discussed in this chapter, rather than in Chapter 8 (see Section 8.1.2). Several channel-routing algorithms for multilayer models result from a general channel-routing scheme that is described in Section 9.4.

The definition of the detailed-routing graph in area routing allows for *unrestricted overlap*. That is, wires can run on top of each other along arbitrary lengths. This assumption is suitable for a number of printed-circuit-board technologies, but long stretches of overlapping wires may introduce cross-talk problems in chip technologies. To account for this problem, researchers have introduced more restrictive models for multilayer wiring. These models have been studied especially in the context of channel routing. The Manhattan model qualifies as such a model, but we discuss it separately, because of its significance and the wealth of results on it. The knock-knee model is not a multilayer model, because a detailed routing in the knock-knee model does not entail a layer assignment. We now define several multilayer models with restricted overlap.

Two-Layer Knock-Knee Model: In this model, detailed-routing graphs are partial 2-grids. Wires have to be vertex-disjoint and no pair of distinct wires is allowed to be routed along a pair of edges in the detailed-routing graph that lie on top of each other. The two-layer knock-knee model is therefore an enhancement of the knock-knee model with a layer assignment using two wiring layers.

Three-Layer and Four-Layer Knock-Knee Model: We can increase the number of layers to get the three-layer or four-layer knock-knee model. (More than four layers are not interesting because every knock-knee routing can be done in four layers; see Section 9.5.3.) Again, the only two restrictions for a legal routing are vertex-disjointness in the k-grid and no overlap of different wires. Figure 9.12 shows two equivalent representations of a four-layer knock-knee routing, one based on the detailed-routing graph in the knock-knee model, and the other based on the 4-grid.

Unit-Vertical Overlap Model: Here, the overlapping wires are completely disallowed in the horizontal direction. In the vertical direction, wires may overlap, but *not for two adjacent vertical edges*.

k-Layer Model: Here, the detailed-routing graph is the k-layer grid graph and routes for different nets must be vertex-disjoint. No other restrictions are imposed.

(L Out of k)-Layer Model: If k layers are involved, $k \geq 3$, a reasonable restriction of overlap is to allow at most L wires to run on top of each other anywhere in the routing graph. Such a model is called a *sparse multilayer model* or, more precisely, the *(L out of k)-layer model*. Here, we

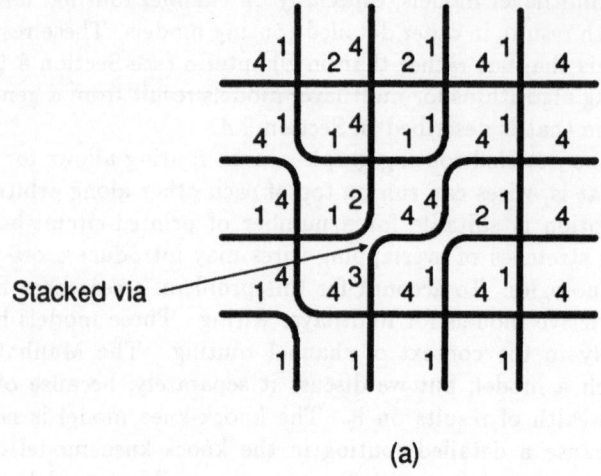

(a)

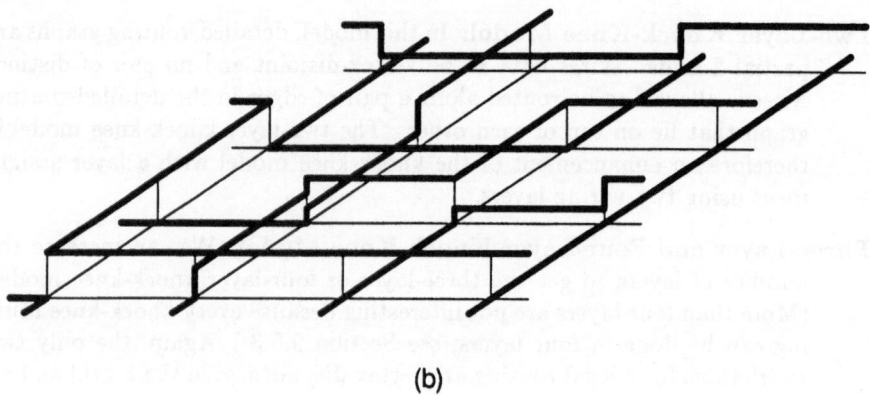

(b)

Figure 9.12: *Two equivalent representations of a layout in the four-layer knock-knee model. (a) Layer numbers label the edges; (b) three-dimensional representation of the layout in the four-layer grid graph.*

curb electrical instability by restricting the number of wires that overlap at any place. There are two restrictions of the $(L$ *out of* $k)$-*layer model* that are especially important. These restrictions are described now.

$(L|k)$-**Layer Model:** In this model (read "every Lth-in-k layer model"), overlapping wires have to be L layers apart. Thus, the $(L|k)$-layer model is a $(\lceil k/L \rceil$ out of $k)$-layer model. The rationale of this model is to reduce cross-talk by spacing wires far apart.

Directional Multilayer Models: These models attach preferred wiring directions, horizontal or vertical, to layers. A k-layer model can be specified by a string of letters from $\{$H, V$\}$ that define the wiring direction on layers 1, 2, and so on. For instance, the (HVHVH)-model is the five-layer model in which preferred wiring directions alternate, starting on the bottom layer with the horizontal direction. Formally, in a *directional multilayer model*, wires on a layer are allowed to run in *only* the direction specified for the layer. However, although preferred wiring directions are desired or demanded by many design and fabrication technologies, the strict adherence to the specified wiring direction on a layer is not mandatory, and it hampers both theoretical and practical algorithm design. (Evidence for this fact is given in Section 9.10.2.) Thus, strict directional models are not discussed much in theory or practice. Rather, algorithms for sparse multilayer models are developed that also implement preferred wiring directions (see Section 9.10.2).

For all multilayer models, there are two variants with respect to the location of terminals.

Surface Connection: Here, all terminals are located on the bottom layer.

Depth Connection: Here, terminals can be connected to on any layer. Formally, the location of a terminal is not fixed in the problem instance. Rather, any of a set of grid vertices lying on top of one another can be the location of the terminal.

As an example, the Manhattan model is a surface-connection model, whereas the multilayer models (with or without restrictions on overlap) are depth-connection models. Clearly a surface-connection model is more restrictive than is its corresponding depth-connection model. However, adding one track on either side of the channel allows for routing all terminal wires into the bottom layer. Thus, the minimum channel width in a surface-connection model is at most two more than in the corresponding depth-connection model.

Channel routing in multilayer models, both restricted and unrestricted, is discussed in Section 9.10.

9.2 Theories of Detailed Routing and Their Relevance

If only two-terminal nets are considered, the detailed-routing problem has a strong similarity with multicommodity flow problems. This fact is the key to the development of theories of detailed routing in several routing models. We have already alluded to it in Section 3.11.3, and we shall discuss it in more detail in this section.

We shall consider cuts through the detailed-routing graph that start and end at a hole. To define formally what we mean by a cut, we use a version of the planar dual of the detailed-routing graph that accounts for the holes.

Definition 9.4 (Multiple-Source Dual) *Let (G, H) be a detailed-routing graph. The* multiple-source *dual of (G, H) is a planar graph $D(G, H) = (V_D, E_D)$ that is constructed from the planar dual $\widetilde{G}$ of G (Definition 3.31) by splitting of each vertex v of degree k in $\widetilde{G}$ representing a hole in G with k edges into k vertices $v_1, \ldots, v_k$, each incident to one of the edges leaving v. The vertices thus created are called the* sources *of G.*

A cut *in (G, H) is the set of edges dual to a simple path between sources in $D(G, H)$*

Figure 9.13 illustrates Definition 9.4. There are two quantities that we can associate with a cut in (G, H): the *capacity* and the *density*.

Definition 9.5 (Cuts) *Let C be a cut in the detailed-routing graph (G, H) of an instance I of the* DETAILED ROUTING *problem that also has a set N of nets and a list $(T_i)_{n_i \in N}$ of global routes. The* capacity *$c_s(C)$ of C is the maximal number of wires that can cross C in any legal routing with respect to the given detailed-routing model. The subscript s denotes the detailed-routing model. We use the subscripts p (for planar routing), k (for knock-knee routing), and m (for Manhattan routing).*

For each net $n_i \in N$, the crossing number *$\chi(n_i, C)$ is*

$$\chi(n_i, C) := \min\{|T_i' \cap C| \mid T_i' \text{ homotopic to } T_i\}$$

that is, the minimum number of edges that cross C in any route for n_i that is homotopic to the global route for n_i. The density *$d(C)$ of C is*

$$d(C) := \sum_{n_i \in N} \chi(n_i, C)$$

The free capacity *of C is*

$$fc_s(C) := c_s(C) - d(C)$$

A cut C is called safe *if $fc_s(C) \geq 0$; otherwise, it is* unsafe. *C is* saturated *if $fc_s(C) = 0$.*

Note that the density of a cut does not depend on the routing model, whereas the capacity does. Specifically, in the knock-knee model, we have

$$c_k(C) = |C|$$

whereas in the planar model $c_p(C)$ is the size of the largest vertex-disjoint subset of edges in C.

Figure 9.13 illustrates these definitions.

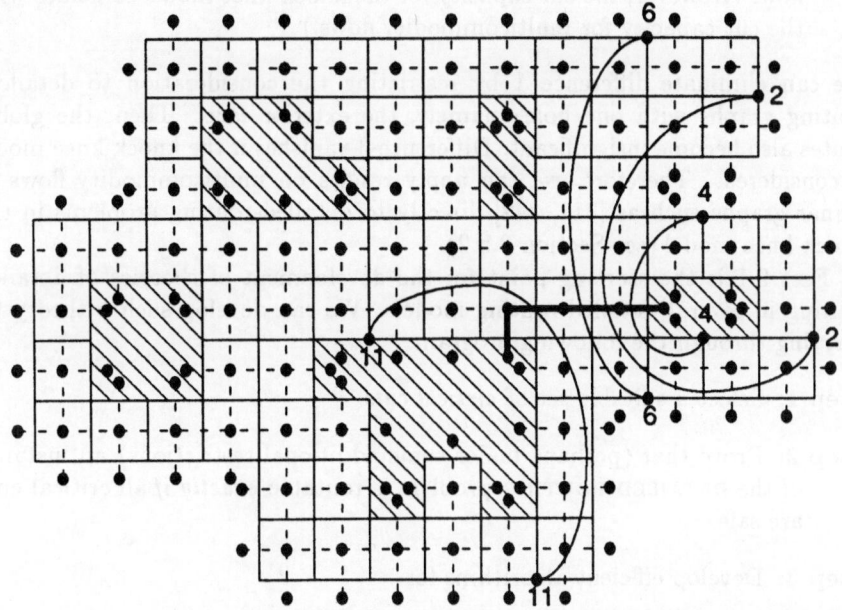

Figure 9.13: *The multiple-source dual of the detailed-routing graph of Figure 9.1(a). A cut is highlighted, and the homotopic routes of the nets 2, 4, 6, and 11 crossing the cut are given. The density of the cut is 4. The capacity of the cut is 3 in the planar routing model and 4 in the knock-knee model. Thus, the cut is unsafe in the planar routing model and is saturated in the knock-knee model.*

Fact 9.1 *The following holds trivially for any detailed-routing model: If an instance I of the* DETAILED ROUTING *problem has an unsafe cut, then I is unroutable.*

Fact 9.1 has a strong similarity with the unidirectional maxflow–mincut theorem for multicommodity flows (Lemma 3.13). Each net represents a commodity whose points of demand and supply are the two terminals of the net. The differences between detailed routing and multicommodity flow are that, in detailed routing,

1. A cut does not necessarily disconnect the graph.

2. Only zero units or one unit of flow can be carried across an edge.

3. In some detailed-routing models, such as the planar routing model, the definition of the cut capacity differs from the capacity in multicommodity

flow. (However, the cut capacity for the knock-knee model coincides with the cut capacity for multicommodity flows.)

We can eliminate difference 1 by restricting the consideration to detailed-routing graphs with one hole—namely, the exterior face. Then, the global routes also become insignificant. Difference 3 vanishes if the knock-knee model is considered. Therefore, we can apply results on multicommodity flows in planar graphs such as [311, 342] directly to detailed-routing problems in the knock-knee model (see Section 9.5.2).

Fact 9.1 is the starting point for the development of theories of detailed routing in several detailed-routing models. We can develop such a theory by carrying through the following program:

Step 1: Select a suitable set of *critical cuts*.

Step 2: Prove that (perhaps under some additional restrictions) an instance of the DETAILED ROUTING problem is routable *exactly if* all critical cuts are safe.

Step 3: Develop efficient algorithms for

1. Testing the safety of each critical cut.
2. Constructing a detailed routing if one exists.

All these steps can be carried through in the case of the planar routing model. In the case of the knock-knee model, there is an additional restriction in step 2, the *evenness condition*, that is a prerequisite for this theory to apply. In the case of the Manhattan model, this program cannot be carried through at all. Thus, there is no theory of Manhattan routing as of today.

The main drawback of all existing theories of detailed routing is that they cannot deal with multiterminal nets. The definition of density generalizes directly to multiterminal nets. However, the program indicated usually does not go through for multiterminal nets. A popular way out of this dilemma is to decompose multiterminal nets into sets of two-terminal nets. However, this decomposition may turn a routable problem instance into an unroutable one (see Figure 9.14).

There are two ways to ensure the routability of the problem instance I' that is generated from a problem instance I by decomposition of multiterminal nets into sets of two-terminal nets. The first is density-oriented. Specifically, we strengthen the requirement of the safety of all cuts in I. The decomposition of multiterminal nets into two-terminal nets at most doubles the density of a cut (Exercise 9.3). Thus, requiring that $c_s(C) - 2 \cdot d(C) \geq 0$ is sufficient to ensure that I' is routable. If step 2 introduces additional requirements for the problem instance, we also have to make sure that I' fulfills those requirements. This approach excludes some problem instances that may be routable.

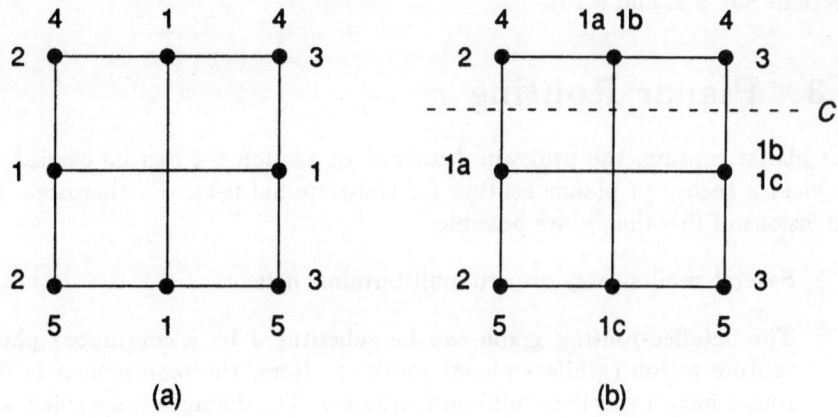

Figure 9.14: *Routing multiterminal nets by decomposition is suboptimal. (a) A routable instance of the* DETAILED ROUTING *problem. The routing model used is the knock-knee model. (b) The instance derived from (a) by decomposition of the four-terminal net 1 into three two-terminal nets. This instance is unroutable. Cut C is unsafe. There is no way to split net 1 into connected two-terminal nets such that a routable instance of the* DETAILED ROUTING *problem results.*

The second approach is capacity-oriented. Here, we enlarge the detailed-routing graph suitably by adding new vertices and edges such that the capacity of each cut is increased sufficiently such as to exceed the density of each cut in I' (which may be twice as large as in I). Again, additional restrictions may have to be met by I'. This approach routes every routable problem instance, but it uses a larger detailed-routing graph than is given in the instance. Both approaches are suboptimal, as they must be because several detailed-routing problems are NP-hard, even though their restrictions to two-terminal nets are in P. (An example is the channel-routing problem in the knock-knee model.)

Although they present significant insights, mathematical theories of detailed routing have not yet had much practical significance. This failure is mainly due to the restriction to two-terminal nets. Nevertheless, they are important contributions to the understanding of detailed-routing problems, and they may serve as the starting point to provably good heuristic algorithms that can be applied successfully in a practical setting.

Up to now, the design-automation community has mainly used robust heuristic Manhattan channel and switchbox routers to solve the detailed-routing problem hierarchically. Such routers handle multiterminal nets and can be extended to other models, such as gridless routing or multilayer models (see

Sections 9.6, 9.9, and 9.10).

9.3   Planar Routing

For planar routing, the program described in Section 9.2 can be carried out
to yield a theory of planar routing for two-terminal nets. Furthermore, two
extensions of this theory are possible:

1. Several results carry over to multiterminal nets.

2. The detailed-routing graph can be substituted by a continuous planar
   routing region (gridless planar routing). Here, the requirement is that
   routes have a specified minimum distance. The distance is specified with
   respect to a planar metric whose unit circle is a polygon or a circle. We
   shall touch on this geometric version of planar routing in Section 9.9.
   The reader is also referred to [95, 130, 305, 447]. Maley [305] gives an
   in-depth account of the state of the art of gridless planar routing.

9.3.1   Planar Channel Routing

Planar channel routing has the best developed theory of all detailed-routing
problems. Practically all optimization versions of the DETAILED ROUTING problem that involve two-terminal nets can be solved with efficient algorithms whose
run time does not exceed $O(n^2)$. The ideas underlying planar channel routing
are intuitively simple and appealing. Thus, planar channel routing is a good
starting point for a study of detailed-routing problems.

An instance of the CHANNEL ROUTING problem in the planar routing model
is depicted in Figure 9.15. The terminals on the bottom side of the channel
are denoted with $b_1, \ldots, b_{|N|}$. The terminals on the top side of the channel are
denoted with $t_1, \ldots, t_{|N|}$. For the instance to be solvable, the nets have to be
formed by terminal pairs (b_i, t_i), $i = 1, \ldots, |N|$, since wire crossings are not
allowed. We identify each terminal with the coordinate of its columns. We call
a net $n_i = (b_i, t_i)$ *rising* if $t_i \geq b_i$ and *falling* if $t_i < b_i$. Then we can partition
the set of nets into *chunks* by grouping together maximal sets of adjacent nets
that are all rising or all falling.

The following observation holds for the planar routing model, and also for
the knock-knee and Manhattan models.

Lemma 9.1 *Without loss of generality, we can assume that an instance of the*
CHANNEL ROUTING *or* CHANNEL-WIDTH MINIMIZATION *problem has no trivial*
nets.

Proof The following procedure solves an instance I of the CHANNEL ROUTING
or CHANNEL-WIDTH MINIMIZATION problem that has trivial nets.

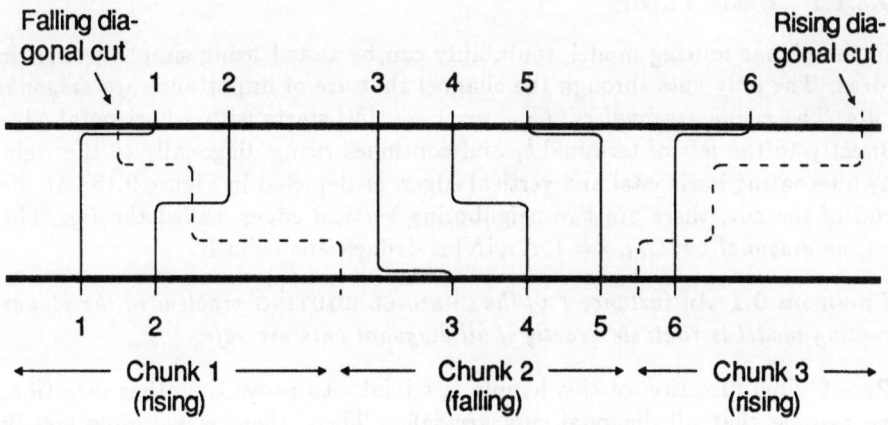

Figure 9.15: *An instance of the* CHANNEL ROUTING *problem in the planar routing model. Nets 1, 2, and 6 are rising nets. Nets 3, 4, and 5 are falling nets. The chunks are* {1,2} {3,4,5} *and* {6}. *The leftmost falling and rightmost rising diagonal cut are dashed in the figure. A routing resulting from the greedy strategy discussed in the proof of Theorem 9.1 is shown.*

Step 1: Delete all trivial nets and their columns from I.

Step 2: Solve the resulting problem instance I'.

Step 3: Introduce the trivial nets and their columns back into the routing, and route the trivial nets straight across the channel. If this routing is not possible, I is not solvable.

In the planar routing model, I is solvable only if nets do not cross. Thus, if I is solvable, no net crosses the column of a trivial net. Therefore, the trivial nets can be introduced in step 3. In the knock-knee and Manhattan models, nets may cross; thus, introducing the trivial nets in step 3 is always possible. Furthermore, step 3 does not increase the channel width. □

Note that Lemma 9.1 does not hold in some multilayer models (see Exercise 9.38). From now on, we shall assume that no trivial nets exist in instances of the CHANNEL ROUTING or CHANNEL-WIDTH MINIMIZATION problem, whenever we discuss the planar, knock-knee, or Manhattan models.

9.3.1.1 Basic Theory

In the planar routing model, routability can be tested using simple geometric ideas. The only cuts through the channel that are of importance are *diagonal cuts*. The *rising diagonal cut* $C_{b,i}$, $i = 1, \ldots, |N|$ starts with a horizontal edge directly to the left of terminal b_i and continues rising diagonally to the right by alternating horizontal and vertical edges, as depicted in Figure 9.15. At the end of the cut, there are two neighboring vertical edges, called the *leg*. The *falling diagonal cut* $C_{t,i}$, $i = 1, \ldots, |N|$ is defined analogously.

Theorem 9.1 *An instance I of the* CHANNEL ROUTING *problem in the planar routing model is routable exactly if all diagonal cuts are safe.*

Proof One direction of this lemma is trivial. To prove the other direction, we assume that all diagonal cuts are safe. Then, there is a simple greedy algorithm that routes the channel. The algorithm routes every chunk of nets independently. Without loss of generality, consider a chunk with rising nets. The nets are processed from left to right. Net $n_i = (b_i, t_i)$ is routed vertically, starting from b_i, until it hits an obstacle contour formed by the top side of the channel or previously routed nets. Then, the wire follows along the contour to terminal t_i. Figure 9.15 indicates the result of this process on an example. This algorithm can fail to route a net only if an obstacle contour extends to the bottom of the channel. To see that this cannot happen, we have to investigate the shape of the contours in more detail. The following statement can be shown by induction on j.

Fact 9.2 *The rightmost horizontal edge of each trunk of a wire is crossed by a cut $C_{t,j}$, if the wire does not lie on the bottommost track, otherwise the vertical segment following the trunk crosses the leg of a cut $C_{t,j}$. As the algorithm proceeds, the vertical edges in each cut $C_{t,j}$ are crossed by wires in order of increasing distance from t_j.*

By Fact 9.2, if all diagonal cuts are safe, no obstacle contour ever reaches the bottom side of the channel, and the routing algorithm succeeds. □

This routing algorithm routes each net greedily from its left to its right terminal. Thus, the total wire length of the routing is

$$|N|(w - 1) + \sum_{i=1}^{|N|} |t_i - b_i|$$

and the spread is

$$s := max\{b_{|N|}, t_{|N|}\} - \min\{b_1, t_1\}$$

Note that the algorithm constructs a routing that minimizes the length of each wire simultaneously.

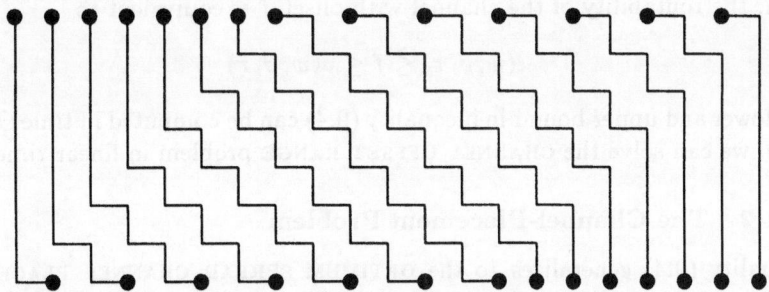

Figure 9.16: *A planar-channel routing problem that requires* $\Omega(|N|^2)$ *jogs.*

The requirement that cut $C_{t,j}$ be safe can be translated into the simple inequality

$$b_{j+w} \geq t_j + w, \quad j = 1, \ldots, |N| - w \qquad (9.1)$$

For $j > |N| - w$, $C_{t,j}$ is always safe, since there are more tracks in the channel than there are nets to the right of net j. Thus, no inequality is required for such cuts. The corresponding inequality for an rising cut $C_{b,j}$ is

$$t_{j+w} \geq b_j + w, \quad j = 1, \ldots, |N| - w \qquad (9.2)$$

Both inequalities can be tested in time $O(|N|)$. Furthermore, the greedy routing algorithm runs in time $O(|N| + g)$, where g is the number of jogs in the routing. (g can be as much as time $O(|N|^2)$, because the routing has many jogs, all of which have to be computed; see Figure 9.16. However, we can heuristically reduce the number of jogs to the minimum possible and speed up the routing algorithm accordingly; see Exercise 9.4.) The formulation of the routability criterion for channels in the planar routing model that is given by the inequalities (9.1) and (9.2) leads to a linear-time algorithm for solving the CHANNEL OFFSET RANGE problem. Set $\beta_i = b_i - i$ and $\tau_i = t_i - i$. Then, $\beta_1 \leq \beta_2 \leq \cdots \leq \beta_{|N|}$ and $\tau_1 \leq \tau_2 \leq \cdots \leq \tau_{|N|}$. Assume that the top terminals are offset by f columns against the bottom terminals. The routability of the respective routing problem is tantamount to the validity of the inequalities

$$\beta_i - \tau_{i+w} \leq f \leq \beta_{i+w} - \tau_i, \quad 1 \leq i \leq |N| - w \qquad (9.3)$$

Define

$$u(w, \beta, \tau) = \min\{\beta_{i+w} - \tau_i \mid 1 \leq i \leq |N| - w\}$$

and

$$\ell(w, \beta, \tau) = \max\{\beta_i - \tau_{i+w} \mid 1 \leq i \leq |N| - w\}$$

Then, the routability of the channel with offset f is equivalent to

$$\ell(w, \beta, \tau) \leq f \leq u(w, \beta, \tau) \tag{9.4}$$

The lower and upper bound in inequality (9.4) can be computed in time $O(|N|)$. Thus, we can solve the CHANNEL OFFSET RANGE problem in linear time.

9.3.1.2   The Channel-Placement Problem

Inequality (9.4) generalizes to the OPTIMUM SPREAD CHANNEL PLACEMENT problem. Assume that the terminals at the bottom of the channel are arranged in blocks $B_1, \ldots, B_{\ell_b}$. Block B_i consists of the terminals $b_{j_{i-1}+1}, \ldots, b_{j_i}$, where $j_0 = 0$ and $j_{\ell_b} = |N|$. The left corner of block B_i has the *offset* x coordinate $f_{b,i}$. On block B_i the terminals $b_{j_{i-1}+1}, \ldots, b_{j_i}$ have the x coordinates $y_{j_{i-1}+1}, \ldots, y_{j_i}$ with respect to the left corner of the block. The right corner of B_i has the x coordinate $r_{b,i}$ with respect to the left corner of the block. We assume that $y_{j_{i-1}+1} > 0$ and $y_{j_i} \leq r_{b,i}$; that is, no terminals are on the left corners of blocks. Similarly, the terminals at the top of the channel are arranged in blocks $T_1, \ldots, T_{\ell_t}$. The offset of block T_i is denoted with $f_{t,i}$. The x coordinate of terminal t_k on block T_i is denoted with x_k. For an assignment of offsets $(f_{t,i})_{i=1,\ldots,\ell_t}, (f_{b,i})_{i=1,\ldots,\ell_b}$ to be routable within width w, the following have to hold:

1. $f_{b,i} + r_{b,i} \overset{.}{\leq} f_{b,i+1}$ for all $i = 1, \ldots, \ell_b - 1$, and similarly for the top side of the channel. This constraint ensures that blocks do not overlap.

2. Assume that, for some $k \in \{1, \ldots, |N| - w\}$, t_k is in block T_j and b_{k+w} is in block B_i. Then,

$$f_{t,j} - f_{b,i} \leq y_{k+w} - (x_k + w)$$

These constraints follow directly from inequalities (9.1). Analogous constraints can be derived from inequalities (9.2). These constraints ensure that all cuts are safe.

9.3.1.2.1   A Solution via Shortest Paths

All the preceding constraints are of the form $v_j - v_i \leq a_{ij}$, where the v_i are the variables corresponding to the offsets to be chosen and the a_{ij} are constants derived from the problem instance. Such a system of inequalities can be solved with shortest-path methods. The reason is the triangle inequality (part 3 of Lemma 3.4), which allows us to reformulate the system of inequalities as the following *channel-placement graph* $G = (V, E)$. Choose a vertex for each block. (We denote the vertices for the top blocks with $v_{t,1}, \ldots, v_{t,\ell_t}$, and the vertices for the bottom blocks with $v_{b,1}, \ldots, v_{b,\ell_b}$.) Add an additional source vertex s that represents the right side of the layout. Create an edge of length r_{t,ℓ_t} from s to v_{t,ℓ_t} and an edge of length r_{b,ℓ_b} from s to v_{b,ℓ_b}. Create an edge from vertex v_i to vertex v_j if there

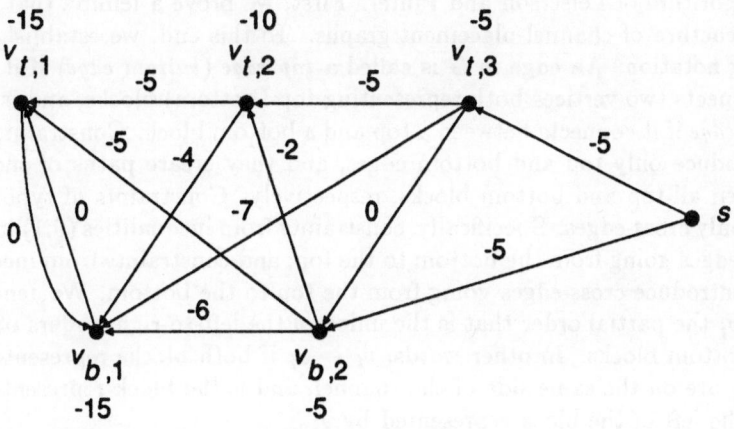

Figure 9.17: *The channel-placement graph for the problem instance depicted in Figure 9.7, with a channel width of 3. Optimal block positions label the respective vertices in the graph. The channel spread is 15.*

is a constraint $v_j - v_i \le a_{ij}$ of type 1 or 2 above. Choose the length of the edge (v_i, v_j) to be the minimum right-hand side in all constraints of the form $v_j - v_i \le a_{ij}$. Figure 9.17 shows the channel-placement graph for the problem instance given in Figure 9.7, with the channel-width set to 4.

By part 3 of Lemma 3.4, an assignment of shortest distances from v_0 to each vertex satisfies all constraints. By the definitions of the edges leaving the source s, the coordinate 0 is assigned to the right-hand side of the layout. By the constraints of type 1, all distances are negative. Furthermore, the shortest distance of any node is assigned to either $v_{t,1}$ or $v_{b,1}$, and it is the negative value of the channel spread.

The channel-placement graph does not need to be acyclic, as Figure 9.17 proves. Furthermore, it involves negative length edges. Thus, the only shortest-path algorithm we discussed in Chapter 3 that applies here is the Bellman–Ford algorithm. In the channel-placement graph, the numbers of vertices and of edges are both $O(|N|)$. Therefore, the run time of the Bellman–Ford algorithm is $O(|N|^2)$.

9.3.1.2.2 A Linear-Time Algorithm Leiserson and Pinter [273] have observed that, if G is a channel-placement graph with source s, then (G, s) is BF-orderable (Definition 3.15), and a BF-ordering of the edges can be found in linear time. This observation improves the time bound for the shortest-path search on the channel-placement graph to $O(|N|)$. In this section, we discuss the algorithm of Leiserson and Pinter. First, we prove a lemma that exhibits the structure of channel-placement graphs. To this end, we establish the following notation. An edge in G is called a *top edge* (*bottom edge*) if it leaves s or connects two vertices both representing top (bottom) blocks, and is called a *cross-edge* if it connects between a top and a bottom block. Constraints of type 1 introduce only top and bottom edges, and they create paths of such edges between all top and bottom blocks, respectively. Constraints of type 2 introduce only cross-edges. Specifically, constraints from inequalities (9.1) introduce cross-edges going from the bottom to the top, and constraints from inequalities (9.2) introduce cross-edges going from the top to the bottom. We denote with $v_i \leftarrow v_j$ the partial order that is the union of the left-to-right orders of the top and bottom blocks. In other words, $v_i \leftarrow v_j$ if both blocks represented by v_i and v_j are on the same side of the channel, and is the block represented by v_i is to the left of the block represented by v_j.

Lemma 9.2 *In a placement graph G, the following three statements hold:*

1. *There are no cross-edges (v_i, v_j) and (v_k, v_ℓ) such that $v_k \leftarrow v_i$ and $v_j \leftarrow v_\ell$.*

2. *There are no cross-edges (v_i, v_j) and (v_k, v_ℓ) such that $v_k \leftarrow v_j$ and $v_i \leftarrow v_\ell$.*

3. *All circuits in G have two consecutive cross-edges (v_i, v_j) and (v_j, v_k) such that $v_i \leftarrow v_k$ or $v_i = v_k$.*

Proof Figure 9.18 illustrates the proof.

1. Consider Figure 9.18(a). Assume, without loss of generality, that the two cross-edges e_1 and e_2 are going from the bottom to the top, as depicted in the figure. The two cross-edges are caused by two constraints and $b_{i_1+w} \geq t_{i_1} + w$ and $b_{i_2+w} \geq t_{i_2} + w$ from inequalities (9.1). If the edges cross as in the figure, then $i_1 + w < i_2 + w$; and, at the same time, $i_2 < i_1$. This is a contradiction.

2. Here, e_1 arises from a constraint of the form $t_{i_1+w} \geq b_{i_1} + w$ from inequalities (9.2), and e_2 arises from a constraint of the form $b_{i_2+w} \geq t_{i_2} + w$ from inequalities (9.1). If e_1 and e_2 cross, then $i_1 + w < i_2$ and $i_2 + w < i_1$. Since $w > 0$, we again get a contradiction.

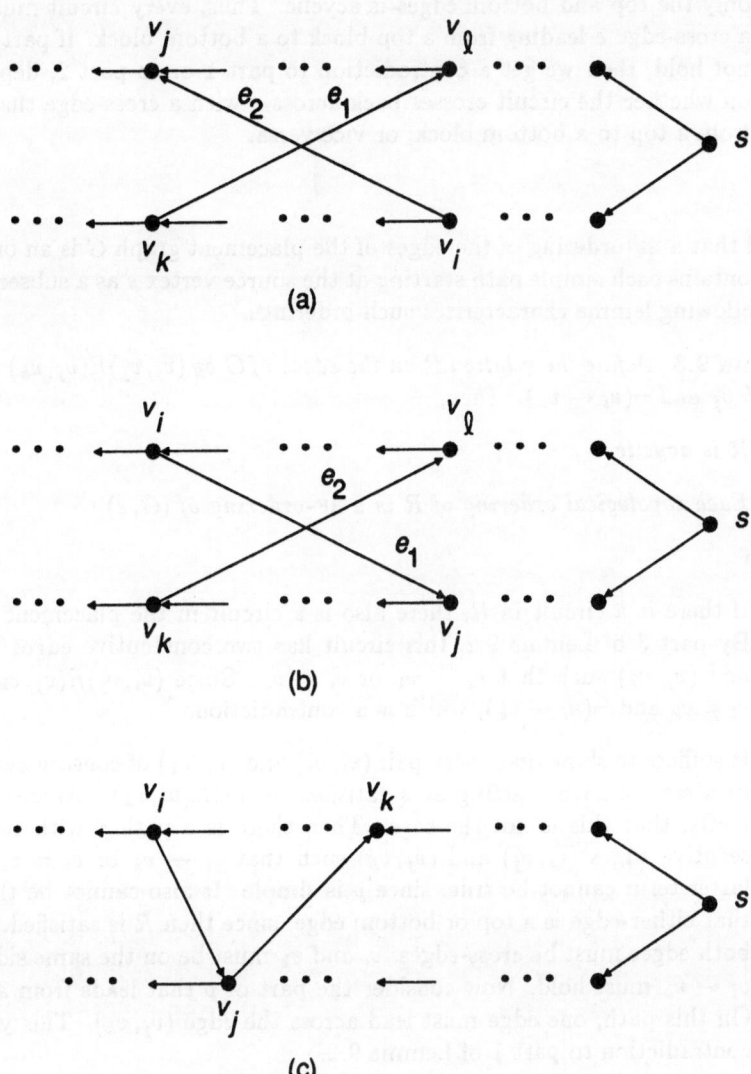

Figure 9.18: *The structure imposed on placement graphs by Lemma 9.2. (a) Structure excluded by part 1; (b) structure excluded by part 2; (c) structure imposed on each simple circuit by part 3.*

3. It suffices to consider simple circuits. The subgraph of G consisting of only the top and bottom edges is acyclic. Thus, every circuit must have a cross-edge e leading from a top block to a bottom block. If part 3 does not hold, then we get a contradiction to part 1 or to part 2, depending on whether the circuit crosses back across e with a cross-edge that leads from a top to a bottom block, or vice versa.

□

Recall that a BF-ordering of the edges of the placement graph G is an ordering that contains each simple path starting at the source vertex s as a subsequence. The following lemma characterizes such orderings.

Lemma 9.3 *Define the relation R on the edges of G by $(v_i, v_j)R(v_j, v_k)$ exactly if $v_i \neq v_j$ and $\neg(v_i \leftarrow v_k)$. Then,*

1. *R is acyclic.*

2. *Each topological ordering of R is a BF-ordering of (G, s).*

Proof

1. If there is a circuit in R, there also is a circuit in the placement graph. By part 3 of Lemma 9.2, this circuit has two consecutive edges (v_i, v_j) and (v_j, v_k) such that $v_i \leftarrow v_k$ or $v_i = v_k$. Since $(v_i, v_j)R(v_j, v_k)$, also $v_i \neq v_k$ and $\neg(v_i \leftarrow v_k)$, which is a contradiction.

2. It suffices to show that every pair (v_i, v_j) and (v_j, v_k) of consecutive edges in a simple path starting at s satisfies $(v_i, v_j)R(v_j, v_k)$. Assume, indirectly, that this is not the case. Then there is a path p with two consecutive edges (v_i, v_j) and (v_j, v_k) such that $v_i \leftarrow v_k$ or $v_i = v_k$. The latter term cannot be true, since p is simple. It also cannot be the case that either edge is a top or bottom edge, since then R is satisfied. Thus, both edges must be cross-edges, v_i and v_k must be on the same side, and $v_i \leftarrow v_k$ must hold. Now consider the part of p that leads from s to v_i. On this path, one edge must lead across the edge (v_j, v_k). This yields a contradiction to part 1 of Lemma 9.2.

□

Part 2 of Lemma 9.3 proves the BF-orderability of placement graphs. We now describe an algorithm that computes a topological ordering of R.

The algorithm constructs a list Λ of edges representing the BF-ordering. We successively attach edges to the end of Λ. We also refer to attaching an edge to Λ as *labeling* the edge. Initially, Λ contains the two edges leaving s in arbitrary order. Λ is extended by scanning across the top and bottom vertices of the placement graph from right to left. Let T and B be the index of the vertex on the top and bottom side, respectively, that is processed currently. Initially $T = \ell_t$ and $B = \ell_b$.

In the general step, we first label all unlabeled (cross-) edges that enter $v_{t,T}$ from vertices $v_{b,i}$ with $i \geq B$—that is, vertices that have been scanned already. Then we label all unlabeled (cross-) edges entering $v_{b,B}$ from vertices $v_{t,j}$ with $j \geq T$. Finally, if all edges entering $v_{t,T}$ are labeled, we label the top edge leaving $v_{t,T}$, and decrease T by 1. Then, we process $v_{b,B}$ analogously.

When both T and B are set to 0, we are finished.

The proof that this algorithm is correct, and an efficient implementation, are the subject of Exercise 9.5 (see also [273]).

9.3.1.2.3* Minimizing Total Wire Length

The approach taken in the previous section that solves the OPTIMUM SPREAD CHANNEL PLACEMENT problem using shortest-path methods suffers from a disadvantage that is common to many solutions of inequality systems by shortest-path methods. In the problem statement, we are interested only in minimizing the *minimum over the distances of all vertices from s*, since this value is the negative of the spread. However, the shortest-path algorithm minimizes *the distance of each vertex individually*. Clearly, minimizing the spread requires minimizing the distance of each vertex on a shortest path starting at s. But many vertices are not located on such a path. They can assume other offset values without invalidating the inequality system or increasing the spread. Choosing other offset values for these vertices may optimize other cost measures, such as the total wire length or total number of jogs in the routing. The shortest-path algorithm intuitively pushes all blocks as much to the right as possible. However, a solution that is optimal with respect to wire length or other natural cost measures intuitively is one that distributes blocks that are off the shortest path evenly throughout the channel.

The relocation of vertices that are off the shortest path to optimize other cost measures can be done in a subsequent pass following the solution of the shortest-path problem. Kolla [238, 239] describes an algorithm that minimizes the spread and the total wire length simultaneously in linear time.

The problem that shortest-path methods lead to unbalanced solutions that do not optimize secondary cost measures occurs in other layout subproblems as well—above all, in compaction. In the case of compaction, the graphs on which the shortest paths are searched have a different shape, and the cost functions differ, as well. As a consequence, the balancing problem is much more difficult (see Section 7.1.4).

9.3.1.3 Channel-Width Minimization

In this section, we discuss problem versions that minimize channel width—that is, the CHANNEL-WIDTH MINIMIZATION problem, the OPTIMUM WIDTH CHANNEL OFFSET problem, and the OPTIMUM WIDTH CHANNEL PLACEMENT

(1) w : **integer**;

(2) $w := 1$;
(3)   **for** $i := 1$ **to** $|N| - w$ **do**.
(4)   **if** $\beta_{i+w} - \tau_i < 0$ **or** $0 < \beta_i - \tau_{i+w}$ **then**
(5) $w := w + 1$ **fi**;
(6)   **exit od**;

Figure 9.19: *Computation of the minimum channel width.*

problem. The first method for solving these problems that comes to mind is based on binary search. We start with $w = 1$ and double w until the routing problem becomes feasible. Let this happen for $w = w_{\max}$. Then, we use binary search to determine $w_{\min}$ in the interval $[0, w_{\max}]$. This algorithm takes time $O(|N| \log w_{\min})$ to find $w_{\min}$, and the additional time discussed in Section 9.3.1.1 to route the channel. The same method also applies to the other optimization problems in planar channel routing, and there it has the same asymptotic run time.

We now develop more clever linear-time solutions for these problems. Let us first discuss the CHANNEL-WIDTH MINIMIZATION problem.

Theorem 9.2 *The algorithm depicted in Figure 9.19 computes the minimum channel width.*

Proof We use induction on $|N|$. For $|N| = 1$, the algorithm is correct by inspection. Now consider $|N| > 1$. By the induction hypothesis, after the loop is executed with $i = |N| - 1$, w is the minimum channel width for the problem instance consisting of the first $|N| - 1$ nets. There are only two possibilities. Either the $|N|$th net can also be accommodated in w tracks, or the $|N|$th net requires one extra track. (Because of the properties of the greedy routing algorithm, we can always route the $|N|$th net with one extra track using one vertical and one horizontal segment.) By inequality (9.3), line 4 tests which of these two is the case. $\square$

Clearly, the algorithm runs in time $O(|N|)$.

Now let us discuss the OPTIMUM WIDTH CHANNEL PLACEMENT problem. Our discussion will also cover the OPTIMUM WIDTH CHANNEL OFFSET problem, since the latter is a restriction of the former.

Mirzaian [328] suggests the use of a *halving technique* for solving this problem. (Actually, Mirzaian considers only the OPTIMUM WIDTH CHANNEL OFFSET problem, but the technique generalizes to the OPTIMUM WIDTH CHANNEL PLACEMENT problem.) To explain the halving technique, let us consider an instance I of the CHANNEL-WIDTH MINIMIZATION problem. We transform I into a smaller instance I^* by deleting every other net—specifically, every *even-numbered net*—in I, and scooting the terminals of each of the remaining nets n_{2i-1}, $i = 1, \ldots, \lceil n/2 \rceil$ to the left by i units. This procedure effectively eliminates the sites of the terminals of the deleted nets. The result is a reduced instance I^* of the CHANNEL-WIDTH MINIMIZATION problem. We show in Theorem 9.3 that the minimum channel width $w^*_{\min}$ of I^* is very close to one-half of the minimum channel width $w_{\min}$ of I. Thus, $2w^*_{\min}$ serves as an appropriate starting point for a local search for $w_{\min}$. The minimum channel width for I^* is computed recursively to determine $w^*_{\min}$.

Theorem 9.3 *Let I be an instance of the* CHANNEL-WIDTH MINIMIZATION *problem, and let I^* be defined as before. Then,*

$$2w^*_{\min} - 1 \le w_{\min} \le 2w^*_{\min} + 1 \tag{9.5}$$

Proof We prove each inequality separately.

For proving the left inequality in (9.5), we use Theorem 9.1 to infer the safety of all diagonal cuts in instance I with channel width $w_{\min}$. We now consider how the cuts for the odd-numbered nets that stay in the channel change during the transformation from I (with channel width $w_{\min}$) to I^* (with channel width $\lceil w_{\min}/2 \rceil$). Let C be an arbitrary diagonal cut associated with an odd-numbered net (see Figure 9.20). Clearly, the capacity of C changes from $w_{\min}$ to $\lceil w_{\min}/2 \rceil$. The density $d(C)$ changes as follows: The deletion of all even-numbered nets reduces the density of C to at most $\lceil d(C)/2 \rceil$. Scooting the remaining nets over to the left may increase the density of C again by as much as $\lfloor w_{\min}/2 \rfloor$. However, the new nets entering C can advance only $\lfloor w_{\min}/2 \rfloor$ units into C. Thus, none of the new nets across C crosses the initial segment of C that is relevant when the number of tracks is reduced to $\lceil w_{\min}/2 \rceil$. Therefore, after the transformation from I to I^*, C is still safe, and I^* is routable in $\lceil w_{\min}/2 \rceil$ tracks. As a consequence, $w^*_{\min} \le \lceil w_{\min}/2 \rceil \le (w_{\min} + 1)/2$, which is a restatement of the left inequality in (9.5).

For the right inequality in (9.5), we consider a routing of I^* with channel width $w^*_{\min}$. To transform this routing into a routing of I, we expand the vertical segments of the routes for all nets n_{2i-1} by one unit in the horizontal dimension (see Figure 9.21). Then, we add new tracks to the channel, one between each pair of adjacent tracks and one at the bottom and the top of the channel. This transformation separates the routes of each pair of adjacent nets n_{2i-1} and n_{2i+1} in I^* by two units, both horizontally and vertically. Thus, there is enough space to route the additional even-numbered net n_{2i} between n_{2i-1} and n_{2i+1}. This observation proves that I is routable with $2w^*_{\min} + 1$ tracks. $\square$

Theorem 9.3 can also be applied to the OPTIMUM WIDTH CHANNEL PLACEMENT problem. Let I_{pl} be an instance of this problem. We generate the corresponding

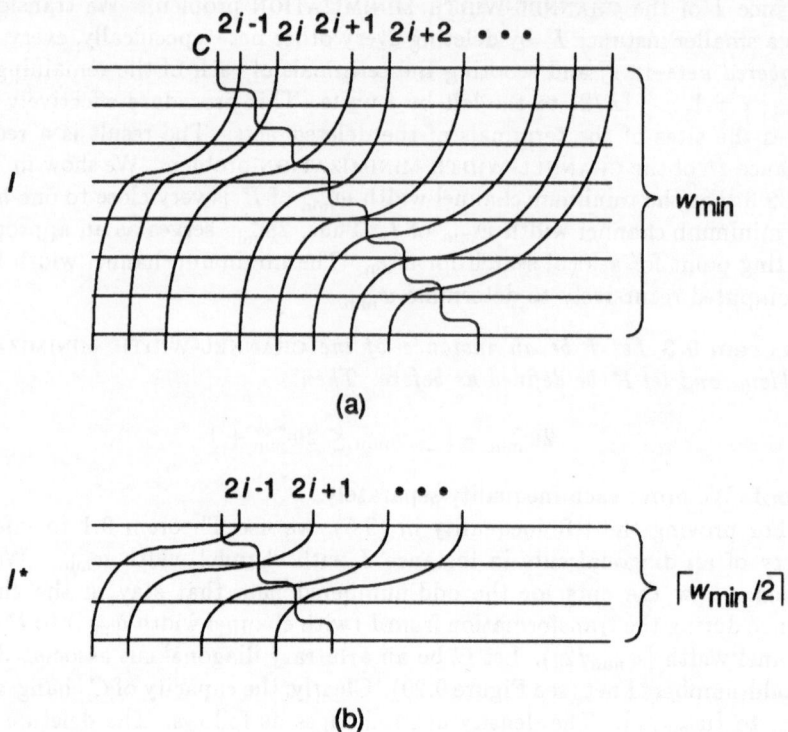

Figure 9.20: *Transformation of a cut; (a) the cut C in I. The cut has capacity $w_{\min}$. (b) The cut in I^*; net $2i+k$ has been scooted over by $k/2$ units with respect to cut C. Only the nets originally crossing C in I also cross C in I^*.*

reduced instance I^*_{pl} from I by first deleting all even-numbered nets. Then, the remaining nets are scooted over inside each block in turn. Let B_i be a block. Each terminal b_k inside B_i is scooted over to the left by one unit for each terminal b_j, $j < k$ that has been deleted from B_i. Also, $r_{b,i}$ is reduced by as many units as there are terminals deleted from B_i. An application of Theorem 9.3 shows that, if $w_{\min,pl}$ is the minimum channel width for I_{pl} and $w^*_{\min,pl}$ is the minimum channel width for I^*_{pl}, then

$$2w^*_{\min,pl} - 1 \le w_{\min,pl} \le 2w^*_{\min,pl} + 1$$

The algorithm depicted in Figure 9.22 solves the OPTIMUM WIDTH CHANNEL PLACEMENT problem. (Here, β and τ pertain to the instance I_{pl}, whereas β^* and τ^* pertain to the instance I^*_{pl}.) The correctness of the algorithm fol-

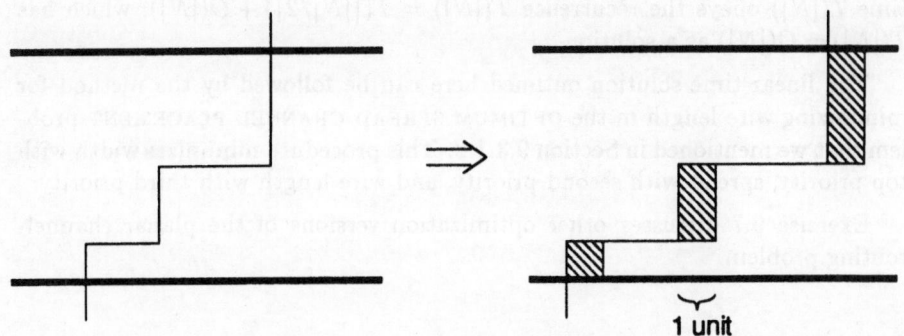

Figure 9.21: *Expanding the channel along the route for a net.*

```
(1)        recursive function minwidth
(2)                (n : integer; β, τ : array [1 : n] of integer) : integer;
(3)        w : integer;

(4)        begin
(5)        if n ≤ 1 then
(6)            return 0;
(7)        else
(8)            w := minwidth(⌈n/2⌉, β*, τ*);
(9)            if ℓ(2w, β, τ) > u(2w, β, τ) then
(10)               return 2w + 1
(11)           else if ℓ(2w − 1, β, τ) > u(2w − 1, β, τ) then
(12)               return 2w
(13)           else return 2w − 1 fi
(14)       end;

(15)       minwidth(|N|, β, τ);
```

Figure 9.22: *An algorithm for solving the* OPTIMUM WIDTH CHANNEL
PLACEMENT *problem.*

lows from Theorem 9.3 and inequality (9.4) by induction on $|N|$. The run time $T(|N|)$ obeys the recurrence $T(|N|) = T(\lceil |N|/2 \rceil) + O(|N|)$, which has $T(|N|) = O(|N|)$ as a solution.

The linear-time solution outlined here can be followed by the method for minimizing wire length in the OPTIMUM SPREAD CHANNEL PLACEMENT problem that we mentioned in Section 9.3.1.2. This procedure minimizes width with top priority, spread with second priority, and wire length with third priority.

Exercise 9.7 discusses other optimization versions of the planar channel-routing problem.

9.3.1.4* Irregular Channels and Multiterminal Nets

The case that both terminals of a net are on the same side can be handled as follows. We route the terminals that are on the same side of the channel first. We do this again by moving away from the channel boundary as little as possible. Afterwards, the rising and falling nets are routed chunkwise as before. This channel is then routed with the method outlined in Section 9.3.1.1. The correctness of this procedure can be established by a suitable extension of Theorem 9.1 (Exercise 9.9).

This approach also generalizes to vertically convex irregular channels whose terminals are on only those sites at the channel boundaries that are on horizontal boundary segments and are not on corners. This time, the set of critical cuts contains all diagonal cuts that extend from terminals diagonally to the right or from convex corners of the channel boundaries into the diagonal direction that bisects the corner. The correctness of this procedure is established with an application of Theorem 9.1. The irregular channel boundaries are here modeled as contours of one-sided nets (see Exercise 9.10).

Furthermore, we can incorporate nets with sliding terminals. Assume a regular channel, where the terminals of a net are specified by nonoverlapping intervals on the channel boundaries. The terminal site can be located anywhere within its corresponding interval. Inequalities (9.1) and (9.2) imply that it is always best to choose terminal sites that are as close to each other as possible. This observation reduces the problem of sliding terminals to the problem of fixed terminals for both regular and irregular channels.

Finally, we can incorporate multiterminal nets by putting together all these ideas. First, the terminals of each multiterminal net that are located on either side of the channel are wired up separately. The resulting routing defines an irregular channel with sliding terminals. This channel is then routed to make the final connections (Exercise 9.13). Note that, in the planar routing model, only severely restricted problem instances with multiterminal nets are routable.

9.3.1.5* Stacks of Channels

Heng and LaPaugh [186] present a polynomial-time algorithm solving a generalization of the OPTIMUM AREA CHANNEL PLACEMENT problem that deals with $k + 1$ rows of blocks stacked on top of one another and separated by k channels. Each row can have at most two blocks. There are n two-terminal nets in the problem instance. The algorithm runs in time $O(kn^3)$ and produces a sizing function (Definition 7.2) for the layout trading off width (spread) against height (sum of the widths of all channels). This algorithm can be combined with the floorplanning sizing method discussed in Section 7.2.1.2 to optimize the area of slicing floorplans composed of rows of components if only one wiring layer is available. If a row can have an arbitrary number of blocks, the channel-placement problem with stacked channels becomes NP-hard [355].

9.3.2 General Planar Routing

In this section, we survey results on planar routing in detailed-routing graphs that are general partial grid graphs with several holes. We again restrict ourselves to two-terminal nets. Multiterminal nets can be incorporated by similar methods, as described in Section 9.3.1.4.

Pinter [355] discusses the problem of deciding whether a problem instance of the DETAILED ROUTING problem is solvable given that the routing graph is expanded to make enough room for the wires. This question is closely related to planarity testing. In fact, if the cells that are represented by the holes in the routing graph were flippable, then we could answer the question as follows. Construct a graph in which each cell is represented by a wagon wheel with as many spokes as there are terminals in the cell. Connect the vertices on the periphery of the wagon wheels as the nets of the problem instance of the DETAILED ROUTING problem prescribe. Test the resulting graph for planarity. If the cells are not flippable, then the planarity test can be modified. Pinter [355] gives a construction that still runs in linear time $O(|N|)$. Syed and El Gamal [433] discuss other versions of the planar ROUTABILITY problem.

If we are not allowed to increase the routing area, the problem becomes more difficult. In the remainder of this section, we discuss efficient algorithms for solving this problem on partial grid graphs, and we examine the theory behind them. A first version of these results was developed by Cole and Siegel [72]. The theory was made more precise by Leiserson and Maley [272]. Maley extended the results from partial grid graphs to gridless routing (see also Section 9.9.1). His doctoral dissertation [305] is an excellent treatise of the mathematical theory and algorithms for planar detailed routing. Independently, Gao et al. [130] presented similar results on homotopic routing in the gridless planar model.

At the basis of all algorithms is a cut theorem as described in Section 9.2. From now on, we use the notation set down in this section. The set of critical cuts is defined as follows.

Definition 9.6 (Critical Cut) *A* critical cut *is a path between two distinct sources in the multiple-source dual that can be constructed as follows. Choose two vertices of the detailed-routing graph. Either vertex must be either a terminal of a net or a corner of a hole. Draw a straight line that starts just to the left of the leftmost vertex and ends just to the right of the rightmost vertex. (We can always choose such a line that does not cross any grid vertex.) The cut consists of all edges in the multiple-source dual whose dual edges are crossed by the straight line, plus two edges connecting to the nearest sources (see Figure 9.23).*

The capacity *of a critical cut the the maximum of its number of horizontal and vertical grid edges.*

Theorem 9.4 (Cole, Siegel [72]) *An instance of the* DETAILED ROUTING *problem with two-terminal nets and a detailed-routing graph that is a partial grid graph is routable in the planar routing model exactly if all critical cuts are safe.*

The proof of the Theorem 9.4 is quite involved. If the proof is done in sufficient generality to apply also to gridless models, it requires advanced methods from topology. A detailed discussion of the general form of Theorem 9.4 is contained in [305].

On the basis of Theorem 9.4, there are several ways to develop efficient routing algorithms. Perhaps the most natural way is to implement a constructive version of the theorem. This approach is pursued by Leiserson and Maley [272], Maley [305], and Gao et al. [130].

In this approach, we first compute a normal form of the input that is suitable for computing the density of the critical cuts. In the input, we are given global routes of the nets. However, the global routes may cross the same cut several times, even though there is a homotopic route that does not cross the cut at all (see Figure 9.24). Therefore, the input is not a suitable representation for computing the density of the cuts. The normal form that is suitable for this computation is the *rubberband equivalent*. In the rubberband equivalent, every route is pulled taught, so as to maneuver around the holes of the detailed-routing graph in the shortest possible way. Therefore, situations such as that in Figure 9.24 cannot occur. The rubberband equivalent can be computed efficiently using plane-sweep techniques [272].

On the basis of the rubberband equivalent, we can compute the densities and can test whether all critical cuts are safe. (Capacities are easy to compute.) If they are, we can again use plane-sweep techniques to compute a legal planar routing. The methods used here differ from those in [272, 305] and [130].

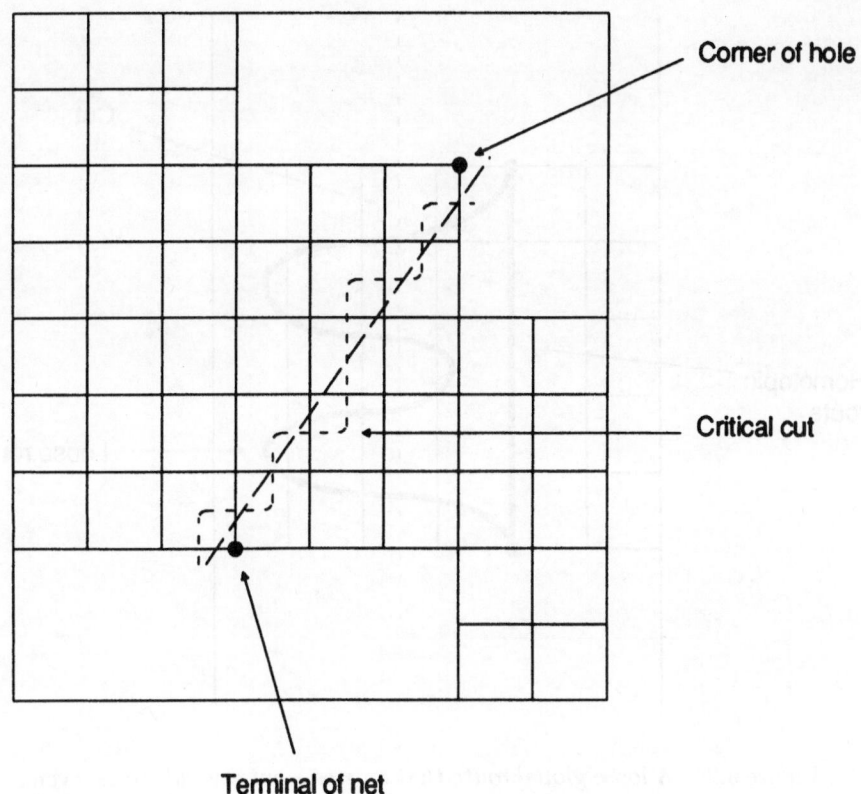

Corner of hole

Critical cut

Terminal of net

Figure 9.23: *A critical cut.*

The resulting run time is slightly greater than quadratic in the number of nets and corners on holes, if partial grids are used. Run times of $O(n^3)$ and $O(n^4)$ plus logarithmic factors are reported in the gridless case.

Schrijver [401] generalizes these results to arbitrary detailed-routing graphs. In this case, it is not sufficient to require all critical cuts to be safe. Another technical criterion involving cycles in the multiple-source dual has to be fulfilled.

Schrijver proves the necessity and sufficiency of these conditions for routability. Furthermore, he gives an efficient routing algorithm that reduces the problem to a feasible special case of integer programming. His algorithm applies to general detailed-routing graphs.

The algorithm consists of four steps.

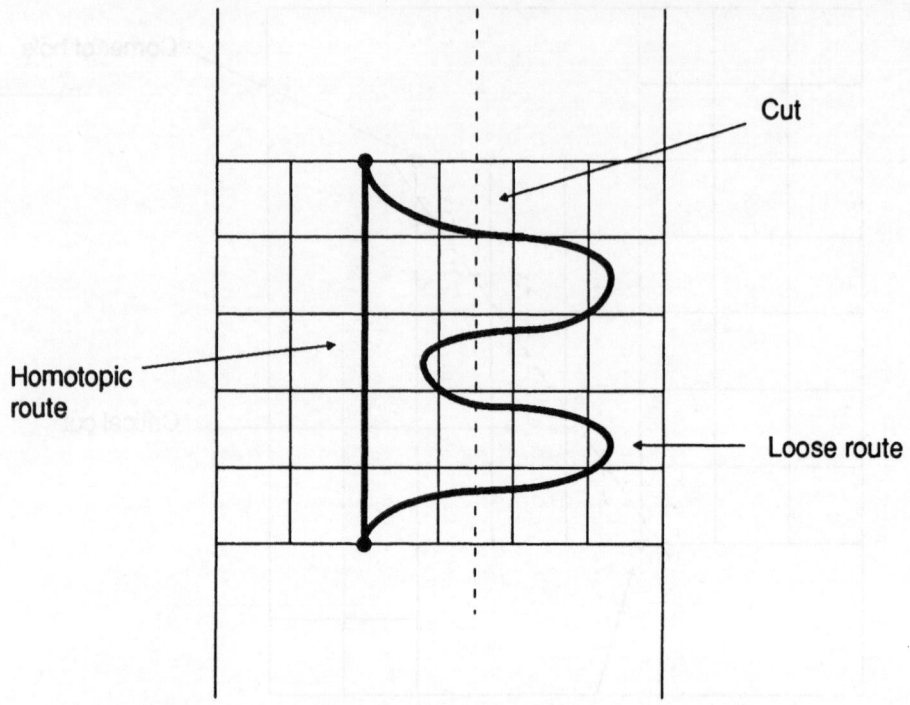

Figure 9.24: *A loose global route that crosses a cut several times, even though there is a homotopic route that does not cross the cut at all.*

Step 1: Modify the global routes such that they do not cross or share edges.

Step 2: Construct a system $Ax \leq b$ of linear inequalities with integer variables.

Step 3: Solve the system.

Step 4: Shift the global routes homotopically as the solution of the system prescribes.

Schrijver also outlines a method for solving the inequality system generated in step 2 in polynomial (cubic) time.

9.4 A Provably Good Scheme for Nonplanar Channel Routing

Different nonplanar detailed-routing models—for instance, the knock-knee, Manhattan, and multilayer models—have widely differing properties. Nevertheless, there are certain similarities between the channel-routing problems in these models. Specifically, there is a general strategy for routing two-terminal nets across a channel that is close to optimal in many nonplanar models for detailed routing. This strategy yields the minimum channel width in the knock-knee model and comes provably close to the minimum channel width in the Manhattan model and in multilayer models. It was first developed by Baker et al. [18] for the Manhattan model, and was subsequently applied by Berger et al. [26] to other nonplanar models of detailed routing. Due to several disadvantages, the method is not directly usable in practice. We discuss it here because it presents useful insight into the structure common to channel-routing problems in a wide variety of nonplanar models for detailed routing. Furthermore, there is hope that the inclusion of practical heuristics will make the strategy efficient enough on typical problem instances to be practically applicable.

We start with an instance of the CHANNEL-WIDTH MINIMIZATION problem that contains $|N|$ two-terminal nets $n_i = (\ell_i, r_i)$, $i = 1, \ldots, |N|$. Here, ℓ_i (r_i) is the pair of column number and channel side of the leftmost (rightmost) terminal of net n_i. We write $\ell_i \leq r_i$ to denote that the column of ℓ_i is not to the right of the column of r_i, and we assume that this is always the case. Note that both ℓ_i and r_i can be on the same side of the channel.

The method breaks up the channel as depicted in Figure 9.25. Vertically, the channel is partitioned into *blocks*. Horizontally, the channel is divided into *sections*. In a first phase, the middle sections of the blocks of the channel are routed separately in left-to-right order. In a second phase, the top and bottom sections of the channel are routed. We now discuss the first phase of the routing process in more detail. To this end, we establish the following notation.

Definition 9.7 (Net Classification) *Let B_j be the jth block (from the left). Let $n_i = (\ell_i, r_i)$ be a net. n_i is called*

1. vertical *if ℓ_i and r_i are both in block B_j and on opposite sides of the channel*

2. local *if ℓ_i and r_i are both in block B_j and on the same side of the channel*

3. starting *if ℓ_i is in block B_j and r_i is to the right of block B_j; in this case, we call net n_i* low *(*high*)* starting *if r_i is on the bottom (top) side (the attributes "low" and "high" denote the side of the channel through which the net is exiting the channel to the right of block B_j)*

4. ending *if r_i is in block B_j and ℓ_i is to the left of block B_j; in this case, we call net n_i* low *(*high*)* ending *if r_i is on the bottom (top) side*

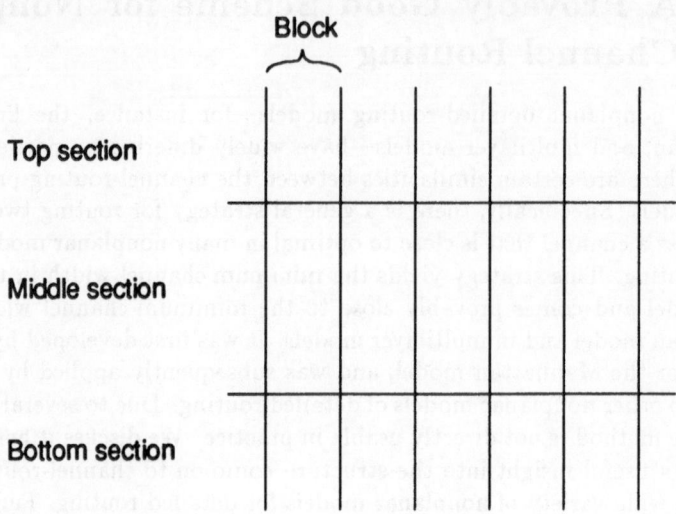

Figure 9.25: *The decomposition of the channel.*

5. continuing *if ℓ_i is to the left of block B_j and r_i is to the right of block B_j; in this case, we call net n_i* low (high) *continuing if r_i is on the bottom (top) side*

The local nets are not routed through to the middle section of the channel. Thus, the first phase of the routing method can disregard the local nets. The routing method maintains the following invariant as blocks are routed.

Invariant: At the left side of a block, the high (low) continuing and ending nets occupy the topmost (bottommost) tracks of the middle section of the channel (in arbitrary order). Between the occupied tracks at both sides of the middle section of the channel there is a *pyramid* of free edges in the channel, (A pyramid is a number of adjacent free tracks such that, for all i, the ith tracks from the top and bottom are connected with a vertical path; see Figure 9.26.) Note that the vertical edges in the pyramid do not need to be in adjacent columns. Each starting (ending) net leaves the middle section at the corresponding side of the channel. The column through which the net leaves the middle section of the channel is arbitrary.

When block B_j is routed, all nets ending in B_j are routed to their relevant sides of the middle section in the channel. Similarly, the starting nets are routed from their relevant sides of the middle section. This routing is done such that the variant is restored for block B_{j+1}. The exact columns at which the nets cross

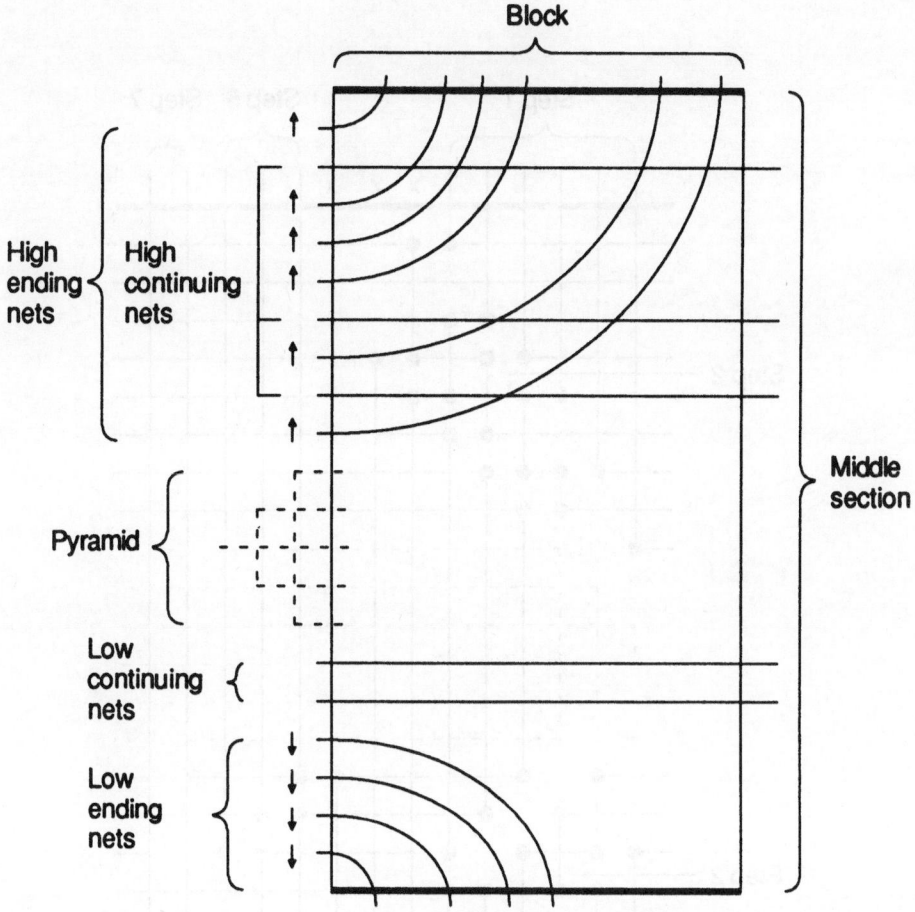

Figure 9.26: *The invariant of the routing method.*

the top and bottom side of the middle section of the channel is irrelevant—
as the last sentence of the invariant formulates—because the top and bottom
sections of the channel are large enough to route the nets to their respective
terminals in the block. This final routing is done in the second phase.

We now explain in more detail how the middle section of a block is routed.
Figure 9.27 illustrates this process using the Manhattan model. The dotted
lines show the empty pyramid in the middle. By using staircase patterns, we
make sure that, at each time in the routing process, the pyramid includes all
empty tracks in the middle of the channel. The initial situation is depicted in
Figure 9.26.

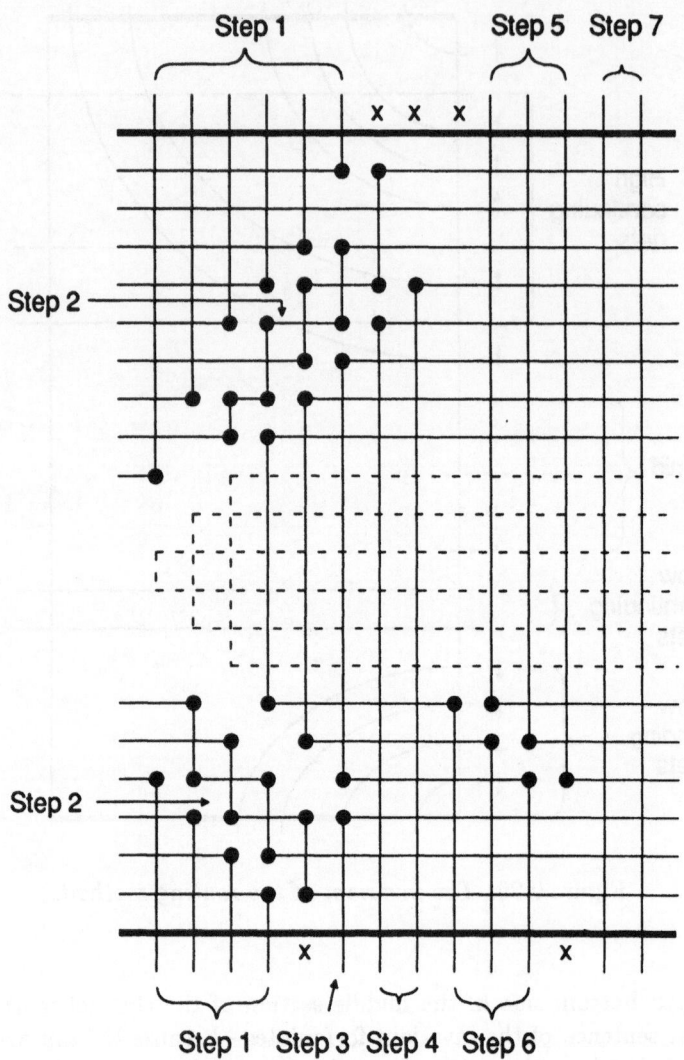

Figure 9.27: *Routing the middle section of a block.*

Step 1: Use *staircase patterns* to route the low ending and high ending nets to the bottom and top side of the middle section of the channel, respectively, starting with the nets that occupy the innermost tracks and continuing to the outside.

Step 2: Route the continuing nets to the boundaries of the middle sections in staircase patterns, again starting with the innermost nets and continuing to the outside.

Step 3: Route the one-sided starting nets on both sides of the channel.

Step 4: If more columns are used up on the top side of the channel than on the bottom side (as in Figure 9.27) route high starting nets to the top side until the used up columns on the top side *balance* the used up columns on the bottom side of the channel.

If after step 3 more columns are used up on the bottom side of the channel than on the top side, proceed symmetrically.

Step 5: If more low starting nets still have to be routed than high starting nets (as in Figure 9.27), then route all low starting nets to the bottom side of the channel.

If after step 4 more high starting nets still have to be routed than low starting nets, proceed symmetrically.

Step 6: Route the remaining starting nets by backtracking them through the empty pyramid.

Step 7: Route all vertical nets straight across the channel.

This general channel-routing method has to be tailored to the specific detailed-routing model. This process entails choosing a suitable size of the blocks and the sections of the channel and making the above steps of the first phase as well as the channel-routing algorithm used in the second phase more precise. In general, a small number of tracks suffices for the top and bottom sections of the channel, routed in the second phase, because the respective routing problems decompose into independent small subproblems, one for each block. That the appropriate choice of the detailed algorithm implies the provable quality of the routing must be established for each routing model separately. In Sections 9.5.1, 9.6.1.2, and 9.10.1, we carry out the adjustments of the method to the different detailed-routing models.

The method can be extended to multiterminal nets. The basic idea is to run two wires for each net, one on either side of the channel that connects to all terminals on that side. Definition 9.7 and the routing process must be modified accordingly:

Definition 9.8 (Classification of Multiterminal Nets) *Let B_j be the jth block (from the left). Let $n_i = (t_{i,1}, \ldots, t_{i,k_i})$ be a multiterminal net whose terminals are listed from left to right. n_i is called*

1. **vertical** *if all terminals of n_i are in block B_j and there are terminals on both sides of the channel*

2. **local** *if all terminals of n_i are in block B_j and on the same side of the channel*

3. **starting** *if $t_{i,1}$ is in block B_j and t_{i,k_i} is to the right of block B_j; in this case, we call net n_i **low** (high) **starting** if some t_h, $h > 1$ to the right of block B_j is on the bottom (top) side*

4. **ending** *if t_{i,k_i} is in block B_j and $t_{i,1}$ is to the left of block B_j; in this case, we call net n_i **low** (high) **ending** if r_i is on the bottom (top) side*

5. **continuing** *if $t_{i,1}$ is to the left of block B_j and t_{i,k_i} is to the right of block B_j; in this case, we call net n_i **low** (high) **continuing** if some t_h, $h > 1$ in or to the right of block B_j is on the bottom (top) side*

If Definition 9.8 is used in the place of Definition 9.7, then the invariant carries over directly. The routing process carries over with the following modifications:

1. All terminals of a wire within a block and at the top side of the channel are concentrated to a single terminal that is located at an arbitrary column on the bottom side of the top section of the channel. An analogous statement holds for the bottom side of the channel. Thus, in the middle section, each multiterminal wire has only one terminal per block on each side of the channel.

2. Middle terminals of a net n_i are connected with a straight vertical wire segment to the wire for n_i in the middle section of the channel that is closest to the terminal.

3. A starting wire may be both high and low continuing. In this case, it generates two horizontal wires, one at the top side and the other at the bottom side of the middle section of the channel.

If the sizes of the blocks and sections in the channel can be computed in time $O(|N|)$, the rest of the method takes time $O(|N| + g)$, where g is the number of jogs.

We conclude this section with an appraisal of the method from a practical point of view. The foremost advantage of the method is that the channel width it computes is provably close to the minimum in several detailed-routing models. The following properties of the method stand in the way of its immediate practical application.

1. Usually, the sizes of the top and bottom region of the channel are larger than necessary. They contribute only a small term to the asymptotic channel width, but this term adds possibly unnecessary and costly tracks in practice.

2. The staircase patterns introduce possibly unnecessary jogs, which are undesirable for practical purposes.

3. Some of the nets—namely, the ones routed in step 6—can become quite long. Multiterminal nets can be backtracked through several blocks. The algorithm does not try to keep critical nets short.

4. The algorithm may use several empty columns at the left side of the channel, but it uses no empty column at its right side. This asymmetry creates lopsided channels, which may be undesirable in practice.

5. When the method is extended to multiterminal nets, the channel width doubles. This increase is much greater than will result when heuristics are used on practical problem instances.

Nevertheless, the method is an important stepping stone toward the development of channel-routing algorithms that have provable performance and are good in practice. Further work on this subject is presented in [467].

Gao and Kaufmann [131] have addressed specifically the fifth point in the list. They can reduce the factor of 2 by which the channel width must be multiplied when incorporating multiterminal nets to 3/2. Their result solves a long standing open problem. Earlier, Mehlhorn et al. [319] have given a simpler algorithm that achieves this bound if all multiterminal nets have three terminals.

La Paugh and Pinter [254] present an up-to-date survey on channel routing in a variety of detailed-routing models.

9.5 Detailed Routing in the Knock-Knee Model

In this section, we discuss the knock-knee model. This model allows quite a surprising number of detailed-routing problems to be solved exactly.

9.5.1 Channel Routing in the Knock-Knee Model

The most trivial application of the general channel-routing method described in Section 9.5 is to the knock-knee model. Here, we can choose every block to have just one column. This convention simplifies the algorithm considerably. First, the top and bottom sections of the channel that are used for locally relocating terminals of wires become unnecessary. Second, the phase of the

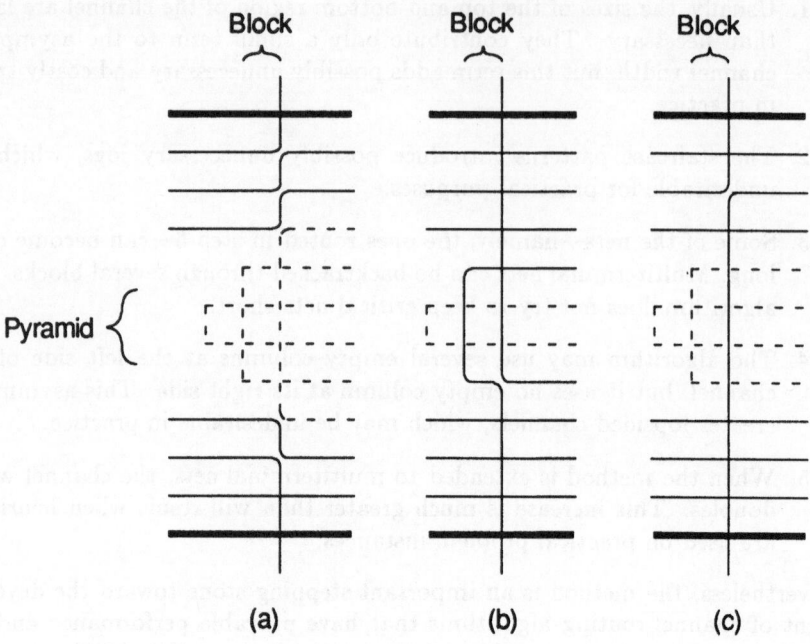

Figure 9.28: *Channel routing in the knock-knee model. (a) Routing two ending nets; (b) routing two starting nets; (c) routing an ending and a starting net.*

algorithm that routes the middle section of the channel becomes very simple. The more difficult cases for two-terminal nets are summarized in Figure 9.28. The staircase patterns collapse horizontally, because knock-knees are allowed. A proof by induction on j shows that the number of tracks needed for routing block j is the following important parameter.

Definition 9.9 (Open Density) *The* **open density** $d_o(I)$ *of an instance I of the* CHANNEL-WIDTH MINIMIZATION *problem is the maximum density $d(C)$ of any vertical cut C through the channel.*

In the context of knock-knee routing, we shall omit the word *open*, and shall just speak of *density* unless this use is ambiguous. In the knock-knee model, the capacity of a cut is the number of edges that cut contains. Thus, Fact 9.1 implies that, in the case of two-terminal nets, the preceding channel-routing algorithm uses minimum channel width. If multiterminal nets are involved, the routing algorithm resulting from the general routing scheme may use as many as $2d_o(I)$ tracks. Thus, it may use twice as many tracks as required.

Actually, we can save one track by slightly modifying the algorithm (part 1 of Exercise 9.14 [319]). Gao and Kaufmann [131] can reduce this value to $3d_o(I)/2 + O(\sqrt{d_o(I)} \log d_o(I))$. Preparata and Sarrafzadeh have presented an algorithm that is simpler than the algorithm by Gao and Kaufmann and that routes problem instances with two- and three-terminal nets in channel width $3d_o(I)/2$. This upper bound is matched with a lower bound of $d_o(I) + \Omega(1)$ (see part 3 of Exercise 9.14 [26, 319, 364].). Sarrafzadeh [388] has shown that it is NP-complete to decide whether a problem instance I involving multiterminal nets with at least six terminals can be laid out in channel width $d_o(I)$. It is an open problem to increase the lower bound on channel width above $d_o(I) + 1$ or decrease the upper bound below $3d_o/2 + O(\sqrt{d_o(I)} \log d_o(I))$.

Kaufmann and Tollis [226] and Formann et al. [118] present fast algorithms that route two-point nets across a channel with minimum width and a total wire length that is close to minimum in the case of [226] and exactly minimum in the case of [118].

9.5.2 Other Detailed-Routing Graphs

In the past few years, numerous results have appeared on the knock-knee model. They show that a theory of knock-knee routing exists for problem instances with only two-terminal nets that fulfill a technical *evenness condition*. We shall detail one of the simpler results in this area, and shall survey what other results there are.

9.5.2.1 The General Method

In this section, we describe the method that is common to most algorithms in knock-knee routing. This method has the advantage of routing every routable (and even) problem instance. But it has some disadvantages that stand in the way of its practical application.

Let us first state the evenness property that is a prerequisite for many results on knock-knee routing.

Definition 9.10 (Even Problem) *A problem instance I of the* DETAILED ROUTING *problem is* **even** *if $fc_k(C)$ is even for all cuts through the routing graph.*

We can explain the significance of the evenness condition by resorting to multicommodity flow. As we pointed out in Section 9.2, there is a strong relationship between knock-knee routing and multicommodity flow. An integer multicommodity flow in a planar flow graph in which all edge capacities are 1 can be regarded as a legal knock-knee routing. Thus, finding knock-knee routings amounts to finding integer multicommodity flows in planar graphs. Algorithms for finding *arbitrary* multicommodity flows in planar graphs are described by [177, 311, 342]. The corresponding flows can be guaranteed to be integer—and thus to correspond to routings—only if the evenness condition is fulfilled, however.

Evenness is quite an involved property, and it is not easy to see how to test for evenness efficiently. In fact, it is possible to check evenness along the way when we test for routability. In many cases, however, a less restrictive condition that is also easier to test can replace the evenness condition.

Definition 9.11 (Locally Even Problem) *The* **extended degree** *of a vertex* $v \in V$ *in the detailed-routing graph is the sum of the degree of the vertex and the number of terminals at this vertex.*

A problem instance I *of the* DETAILED ROUTING *problem is* **locally even** *if the extended degree of all vertices in the routing graph is even.*

Theorem 9.5

1. *Every even problem instance is locally even.*

2. *If there is only one hole in the routing graph, then every locally even problem instance is even.*

Proof

1. Assume that the problem instance is even, and consider an arbitrary vertex $v \in V$. Recall that cuts are paths in the multiple-source dual $D(G, H)$ of G. Vertex v corresponds to a face F_v in $D(G, H)$. Consider a cut C that passes along F_v, and the cut C' that we obtain from C by going around F_v the other way. Both of these cuts are even. We make a case distinction:

    Case 1: The degree of v is odd. Then, the capacities of C and C' must have different parity. Therefore, their densities must have different parity, as well. This situation is possible only if there is an odd number of terminals at v. Thus, the extended degree of v is even.

    Case 2: The degree of v is even. Then, the capacities of C and C' must have the same parity. Therefore, their densities must have the same parity, as well. This situation is possible only if there is an even number of terminals at v. Thus, the extended degree of v is again even.

2. If there is only one hole in G (the exterior face), then cuts actually disconnect G; that is, there are two sides X_C and $Y_C = V \setminus X_C$ to each cut. We write $C = (X_C, Y_C)$, just as in Chapter 3. In this case, we can modify the notation as follows.

Definition 9.12 (Notation) *Let G be a detailed-routing graph with one hole. Let C be a cut in G and $X, Y \subseteq V$. Then,*

$$
\begin{aligned}
c_k(X, Y) &:= \{e = \{x, y\} \in E \mid x \in X \text{ and } y \in Y\} \\
c_k(C) &:= c_k(C, V \setminus C) \\
d(X, Y) &:= \{n_i = \{s_i, t_i\} \in N \mid s_i \in X \text{ and } t_i \in Y\} \\
d(C) &:= d(X_C, V \setminus X_C) \\
fc_k(C) &:= c_k(C) - d(C)
\end{aligned}
$$

$c_k(X, Y)$ counts the edges running between X and Y. $d(X, Y)$ counts the nets running from X to Y. The theorem now follows from the equation

$$
\begin{aligned}
fc_k(C) &= c_k(C) - d(C) \\
&= \sum_{v \in X_C} c_k(\{v\}) - 2c_k(X_C, X_C) - \left(\sum_{v \in X_C} d(\{v\}) - 2d(X_C, X_C) \right)
\end{aligned}
$$

and from the observation that the extended degree of v is $c_k(\{v\}) + d(\{v\})$.

$\square$

Exercise 9.15 shows that the reverse of part 1 of Theorem 9.5 does not hold, in general.

As a rule, results on the DETAILED ROUTING problem in the knock-knee model assume that the problem instance is even. In some cases, such as in the example we discuss over the next few sections, the evenness condition can be relaxed; in other cases, it can be omitted altogether.

Many detailed-routing algorithms in the knock-knee model follow the scheme depicted in Figure 9.29. We execute line 7 by selecting an appropriate net $n_i = \{s_i, t_i\}$ that goes across C_j, creating the new terminal t_i' located at the endpoint of e that is on the same side of C_j as is s_i, creating the new terminal s_i', located at the endpoint of e that is on the side of C_j as t_i, deleting net n_i, adding nets $n_{i,1} = \{s_i, t_i'\}$ and $n_{i,2} = \{s_i', t_i\}$ to N, and deleting e.

The correctness of the algorithm has to be proved separately for every application of this scheme. The proof hinges on the correct choice of n_i and establishes a theorem that states that, if all critical cuts are safe, then the problem instance is routable—that is, the reverse of Fact 9.1. Which cuts are critical has to be defined depending on the application.

The efficiency of the scheme is based on

1. An efficient way of testing for unsafe cuts in line 4 and saturated cuts in line 5

2. An efficient way of finding the cut C_j and the net n_i in line 7

(1) e : edge;

(2)   while $E \neq 0$ do
(3) $e :=$ an arbitrary edge on a hole of G;
(4)   if there is an unsafe critical cut containing e then
(5)   fail
(6)   else if there is a saturated cut containing e then
(7)   route an appropriate net n_i that goes across an
                            appropriate saturated cut C_j through e
                            and delete e
(8)   else
(9)   add a net whose terminals are the endpoints of e;
(10)   delete e fi od;

Figure 9.29: *A method for knock-knee routing in arbitrary graphs.*

Although this routing scheme finds a detailed routing whenever there is one, the detailed routing it constructs may be undesirable, because in line 9 nets are added that do not belong to the original problem instance, and edges are discarded unused. Thus, "easy" problem instances that have no saturated cuts are first made "harder" until they are barely routable; that is, until there is a saturated cut. Only then a net is routed. In the process of making a problem instance harder, a lot of the routing region is thrown away unused. This procedure leads to an uneven distribution of wires through the routing region, and it stands in the way of minimizing the length of wires, especially in easy problem instances where short wire lengths can be obtained.

Thus, the algorithmic scheme has to be enhanced with other methods to shorten wire lengths, to minimize jogs, and so on.

9.5.2.2   An Example: Routing in Switch Graphs

We detail a result presented by Becker and Mehlhorn [24]. There are three reasons for this choice.

1. The result is one of the few that concerns detailed-routing graphs other than grid graphs.

2. The result illustrates the central issues of knock-knee routing such as cut arguments, the evenness condition and possibilities of going around it, and issues of handling multiterminal nets.

3. The proofs are short enough to be discussed in a textbook.

The result by Becker and Mehlhorn concerns routing two-terminal nets in the knock-knee model inside so-called switch graphs. It is an algorithm that is custom-tailored to the knock-knee model. The algorithm does not generalize to noninteger multicommodity flow, in contrast to [311], for instance. However, the algorithm is quite simple and easy to implement.

Definition 9.13 (Switch Graph) *A* switch *graph is a plane graph whose only hole is the exterior face. Terminals may appear only on the hole of the switch graph.*

The homotopy structure of switch graphs is trivial. A switch graph is a generalization of a switchbox in which the internal graph structure can be arbitrary instead of being gridlike. In particular, irregular switchboxes with obstacles in them are switch graphs. By the same token, switchboxes with obstacles (such as previously routed wires) are switch graphs.

Let us now tailor the algorithmic scheme of Figure 9.29 to switch graphs. In this case, all possible cuts are critical. (Remember that a cut corresponds to a simple path in the multiple-source dual of G that starts and ends in a source.) We use the notation of Definition 9.12.

We first describe how the cut C_j and the net n_i are to be picked in line 7 of Figure 9.29. Let $e = \{v, w\}$. Number the edges around the exterior face clockwise $e = e_0, e_1, \ldots, e_{k-1}$. We called these edges *boundary edges*. Denote with C_j the cut starting at edge e and ending at edge e_j that has the smallest capacity. (Note that all cuts starting at edge e and ending at edge e_j have the same density. Therefore, if C_j is safe for $j = 1, \ldots, k - 1$, then all cuts starting at e are safe; and if there are any saturated cuts, then one of the C_j is saturated.) In line 7, we choose j minimum such that C_j is saturated. The corresponding cut C_j is the one we select. We choose a net n_i across C_j whose left terminal s_i is closest to edge e (see Figure 9.30). The proof that this algorithm is correct uses the following simple lemma.

Lemma 9.4 (Okamura, Seymour [342]) *For all sets $X, Y \subseteq V$*

$$
\begin{aligned}
d(X) + d(Y) &= d(X \cup Y) + d(X \cap Y) + 2d(X \setminus Y, Y \setminus X) \\
c_k(X) + c_k(Y) &= c_k(X \cup Y) + c_k(X \cap Y) + 2c_k(X \setminus Y, Y \setminus X)
\end{aligned}
$$

Proof Exercise 9.16. □

We now state the theorem establishing the correctness of the algorithm.

Theorem 9.6 *Let I be an even instance of the* DETAILED ROUTING *problem for switch graphs in the knock-knee model, and let I have only two-terminal nets. If all critical cuts in I are safe, then the above algorithm constructs a routing.*

Proof Call the instance I' that is obtained by going through the loop of lines 2 to 10 once the *reduced problem instance*. $c'_k(C)$ denotes the capacity of cut

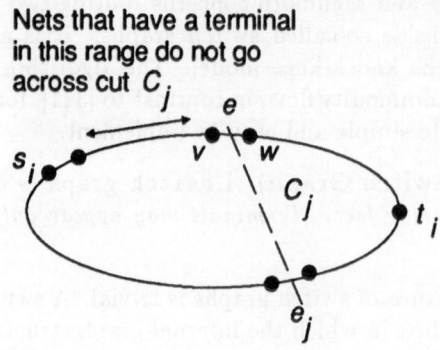

Figure 9.30: *Illustration of the knock-knee routing algorithm for switch graphs.*

C in I'. Similarly, we denote the density and the free capacity of C in I'. We have to show only that, if I is even and all cuts in I are safe, the same holds true for I'. The rest follows by induction.

Assume now that I is even. To show that I' is even, we observe that, for all $v \in V$, the parities of $fc_k(\{v\})$ and $fc'_k(\{v\})$ are the same.

We still have to show that all cuts in I' are safe. Assume that C' is a cut in I'. Either $C = C'$ is a cut in I, as well, or $C = C' \cup \{e\}$ is. Let us analyze the free capacities of C and C'. A quick inspection shows that the free capacities of C and C' can differ only if one of the three cases depicted in Figure 9.31 happens. In all cases,

$$fc'_k(C') \geq fc_k(C) - 2$$

Thus, C' can be unsafe only if C is saturated and

$$fc'_k(C') = fc_k(C) - 2$$

in one of the cases. Let us assume this happens.

We can eliminate case 3 from our discussion. If, in case 3, C is saturated, this cut would have been chosen in line 7 instead of C_j. Thus, we are left with cases 1 and 2. We can make the following observation for both cases. Let X_C be the side of the cut C that contains v and let X_{C_j} be the side of the cut C_j that contains w.

Fact 9.3

$$d(X_C \setminus X_{C_j}, X_{C_j} \setminus X_C) = 0$$

Fact 9.3 can be inferred from Figure 9.31, since no nets that cross cut C_j terminate between v and s_i. The sets $X_C \setminus X_{C_j}$ and $X_{C_j} \setminus X_C$ are shaded in the figure.

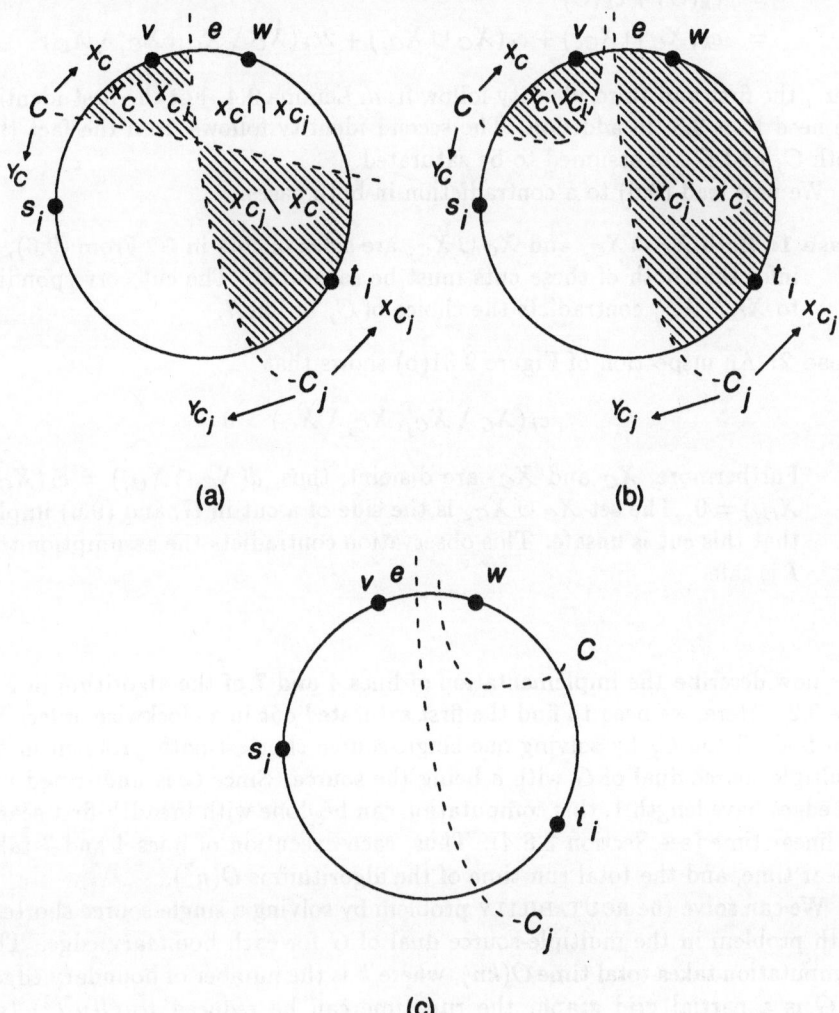

Figure 9.31: *Three cases for Theorem 9.6. (a) Case 1; (b) case 2; (c) case 3.*

With this observation, we can apply Lemma 9.4. We obtain in both cases

$$
\begin{aligned}
d(X_C \cap X_{C_j}) + d(X_C \cup X_{C_j}) &= d(X_C) + d(X_{C_j}) \\
&= c_k(C) + c_k(C_j) \\
&= c_k(X_C \cap X_{C_j}) + c_k(X_C \cup X_{C_j}) + 2c_k(X_C \setminus X_{C_j}, X_{C_j} \setminus X_C)
\end{aligned}
\tag{9.6}
$$

Here, the first and third identity follow from Lemma 9.4. For the first identity, we need Fact 9.3, in addition. The second identity follows from the fact that both C_j and C are assumed to be saturated.

We now lead (9.6) to a contradiction in both cases.

Case 1: Both $X_C \cap X_{C_j}$ and $X_C \cup X_{C_j}$ are sides of cuts in G. From (9.6), we infer that both of these cuts must be saturated. The cut corresponding to $X_C \cap X_{C_j}$ contradicts the choice of C_j in line 7.

Case 2: An inspection of Figure 9.31(b) shows that

$$
c_k(X_C \setminus X_{C_j}, X_{C_j} \setminus X_C) > 0
$$

Furthermore, X_C and X_{C_j} are disjoint; thus, $d(X_C \cap X_{C_j}) = c_k(X_C \cap X_{C_j}) = 0$. The set $X_C \cup X_{C_j}$ is the side of a cut in G, and (9.6) implies that this cut is unsafe. This observation contradicts the assumption that I is safe.

□

We now describe the implementation of lines 4 and 7 of the algorithm in Figure 9.29. Here, we need to find the first saturated cut in a clockwise order. We can find all the C_j by solving one single-source shortest-path problem in the multiple-source dual of G with e being the source. Since G is undirected and all edges have length 1, this computation can be done with breadth-first search in linear time (see Section 3.8.4). Thus, each execution of lines 4 and 7 takes linear time, and the total run time of the algorithm is $O(n^2)$.

We can solve the ROUTABILITY problem by solving a single-source shortest-path problem in the multiple-source dual of G for each boundary edge. This computation takes total time $O(kn)$, where k is the number of boundary edges. If G is a partial grid graph, the run time can be reduced to $O(n^{3/2})$ (see Exercise 9.17 [24]).

9.5.2.2.1   Relaxing the Evenness Condition   In this section, we show that, in the case of switch graphs, it suffices to require the problem instance to be half-even, as defined next.

Definition 9.14 (Half-Even Problem) *An instance I of the* DETAILED ROUTING *problem is* **half-even** *if all vertices that are not on a hole have even extended degree.*

In half-even problem instances, the local evenness property is demanded of only those vertices that are not on holes. Since the degree of a vertex that is not on a hole is also the extended degree of that vertex , in a half-even problem instance all vertices that are not on a hole have even degree.

We route a half-even problem instance as follows:

Step 1: Extend the half-even problem instance I to an even problem instance I' by adding more two-terminal nets. I' must be routable whenever I is.

Step 2: Solve I'.

Step 3: Delete the additional nets.

We now explain the intuition of how to extend I suitably. Since I is half-even, there are some vertices on the exterior face that have odd free capacity. Let the set of these vertices be denoted by U. To make I even, we have to have a new net start at each of these vertices. We want to add as few nets as possible, and we do not want to change the parity of the free capacity of any other vertex. Furthermore, we can add only two-terminal nets. Thus, we have to pair up the vertices in U. Each pair represents the endpoints of one new net. The pairing must be done such that I' is routable whenever I is. That this pairing is possible is not difficult to see. In fact, let us consider a detailed routing D of I. After deleting all routes for nets in I from G, we obtain a detailed-routing graph in which the vertices of odd degree are exactly the vertices in U. Using Euler path methods, these vertices can be paired up through a set of paths to yield a legal routing D' of I' (see Section 4.7.5).

Unfortunately, this method of finding a pairing presupposes the knowledge of a detailed routing. But this routing is just what we want to compute! Thus, we need a suitable way of finding a pairing without resorting to a detailed routing. The following lemma is the basis for an efficient algorithm.

Lemma 9.5 *Let I be a routable instance of the* DETAILED ROUTING *problem in the knock-knee model such that G is a switch graph and only two-terminal nets are involved. Let C be a saturated cut. Then, $|U \cap X_C|$ is even.*

Proof Consider a detailed routing D' of a suitable problem instance I'. As we have seen, D' and I' exist. If $|U \cap X_C|$ were odd, then one of the new nets in D' would have to cross C. Thus, C would be unsafe in D'. This consequence contradicts the existence of D'. □

The algorithm for computing I' from I is based on the following notion.

Definition 9.15 (Minimal Cut) *Let $C = (X_C, V \setminus X_C)$ be a saturated cut and let $X_C \cap U = \{u_0, \ldots, u_{2t-1}\}$. (By Lemma 9.5, this set has an even number of elements.) If there is no saturated cut $C' = (X_{C'}, V \setminus X_{C'})$ such that $X_{C'} \cap U \subset X_C \cap U$, then C is called* minimal.

If there is no saturated cut, then $C = (V, \emptyset)$ is the only minimal cut.

(1) $U := \{v \in V \mid fc_k(\{v\})$ is odd $\}$;
(2)   **while** $U \neq \emptyset$ **do**
(3)   **if** there is an unsafe cut **then fail**
(4)   **else**
(5) $C :=$ a minimal cut;
(6)   pair the vertices in $X_C \cap U = \{u_i, \ldots, u_j\}$ into
                        new nets (u_i, u_{i+1}), (u_{i+2}, u_{i+3}), and so on;
(7) $U := U \setminus X_C$ **fi od**;

Figure 9.32: *Transformation of half-even routing problems into even ones.*

The algorithm that computes I' from I is depicted in Figure 9.32.

Lemma 9.6 *The algorithm in Figure 9.32 computes an even problem instance I' that is routable whenever I is routable.*

Proof By Lemma 9.5, line 6 always pairs up an even number of vertices. Since a minimal cut exists whenever $U \neq \emptyset$, the algorithm eventually pairs up all vertices in U. Thus, I' is even.

Now assume that I is routable. We show by induction on the number of executions of the while-loop that the modified problem instance I_k obtained after k executions of lines 2 to 7 is a half-even problem instance that is routable.

Since $I_0 = I$, this statement is trivial for $k = 0$. Now assume $k > 0$. First assume that there is no saturated cut in I_{k-1}. Then we pair up all vertices in U by adding new nets. Furthermore, each cut is crossed by at most two new nets. Therefore, the free capacities of all cuts are at least -1, and since I_k is even all cuts are safe. Therefore, I_k is routable.

Now assume that I_{k-1} has a saturated cut and let C be a minimal cut. We make a case distinction according to requirements that a cut C' has to fulfill in order to become unsafe in I_k. In each case we show that no cut fulfilling the requirements does, in fact, become unsafe in I_k.

For a cut C' to become unsafe in I_k one of the following two cases must happen.

Case 1: C' is saturated in I_{k-1} and crosses C. If C' is saturated in I_{k-1} and is unsafe in I_k, then at least one boundary edge of C' must lie on the side $V \setminus C$ of C; otherwise, C would not be minimal. But the other boundary edge of C' must then lie on the side X_C of C, since otherwise no new net would cross C', and C' would not be unsafe in I_k. Thus, C' must cross C.

We will now show that, in fact, C' is safe in I_k. To this end, we note that, as a routable half-even problem instance, I_{k-1} has an even routable extension $\widehat{I_{k-1}}$ that can, for instance, be obtained with Euler methods. In $\widehat{I_{k-1}}$, no new nets cross C, since C is saturated. Thus, we can add the new nets in $\widehat{I_{k-1}}$ that pair up vertices in $(V \setminus X_C) \cap U$ to I_k to get an even extension $\widehat{I_k}$ of I_k. Since $\widehat{I_{k-1}}$ is routable, a cut C' fulfilling the requirement considered in this case is safe in $\widehat{I_{k-1}}$. Since at most one new net crosses C' in I_k, the free capacity of C' in $\widehat{I_k}$ is at least -1. Since $\widehat{I_k}$ is even, the free capacity of C' in $\widehat{I_k}$ is at least 0; that is, C' is safe in $\widehat{I_k}$. Since $\widehat{I_k}$ is an extension of I_k, C' is also safe in I_k.

Case 2: C' is unsaturated in I_{k-1}, and both boundary edges of C' lie on the side X_C of C. If C' is unsaturated, then two new nets have to cross C' if C' is to be unsafe in I_k. This can happen only if both boundary edges of C' lie on the side X_C of C. Since, even in this case, *at most two* new nets cross C', the free capacity of C' is at least -1. The even extension $\widehat{I_k}$ of I_k does not change the free capacity of C'. Since $\widehat{I_k}$ is even, C' is safe both in $\widehat{I_k}$ and in I_k.

Thus, all cuts in I' are safe, and I' is routable. $\square$

To make the algorithm efficient, we have to provide a data structure for finding minimal cuts. We do not have to look at all cuts in this context. Rather, we can group the cuts into classes. Class S_{ij} contains all cuts C such that $X_C \cap U = \{u_i, u_{i+1}, \ldots, u_j\}$. Out of class S_{ij}, we need to consider only one cut with the smallest free capacity. Denote this cut with C_{ij}. If there is a saturated cut, then there always is a minimal cut among the C_{ij}, $0 \le i < j \le 2\ell - 1$.

Exercise 9.18 provides a data structure for maintaining the C_{ij}. Plugging this data structure into the algorithm yields a run time of $O(kn)$.

9.5.2.2.2 Multiterminal Nets

The extension to multiterminal nets is density-oriented. We strengthen the routability criterion that all cuts be safe to the criterion that

$$c_k(C) - 2d(C) \ge 0$$

for all cuts. This criterion is sufficient but not necessary.

Theorem 9.7 *Let I be a half-even instance of the* DETAILED ROUTING *problem, with G being a switch graph. Assume that $c_k(C) - 2d(C) \ge 0$ for all cuts. Then, I is routable and a routing can be generated in time $O(n^2)$.*

Proof Consider any multiterminal net with k terminals $t_1, \ldots, t_k$, where the terminals are numbered in clockwise order around the exterior face. This net is split up into $k - 1$ two-terminal nets $(t_1, t_2), (t_2, t_3), \ldots, (t_{k-1}, t_k)$. Splitting up all multiterminal nets increases the density of each cut by at most a factor of 2. The resulting two-terminal problem instance is again half-even; and, by the condition on the cuts, it is routable. $\square$

9.5.2.3*   Other Results on Knock-Knee Routing

In this section, we survey other results on knock-knee routing. For a more detailed survey, see [225].

9.5.2.3.1*   **Switchboxes**   Mehlhorn and Preparata [318] present an algorithm that routes two-terminal nets in a regular switchbox. The algorithm does not presuppose the evenness condition, and it runs in time $O(k + |N| \log |N|)$, where k is the number of boundary edges. The ROUTABILITY problem can be solved in time $O(k)$. This is the only algorithm, besides channel routing, where the run time may be asymptotically smaller than the size of the detailed-routing graph. The algorithm also limits the number of knock-knees to $O(N)$. However, there may be nets with many knock-knees.

Kaufmann and Mehlhorn [224] introduce an algorithm for routing not-necessarily-even problem instances in not-necessarily-convex switchboxes. The algorithm runs in time $O(n(\log n)^2)$ if the problem instance is even, and in time $O(n(\log n)^2 + b^2)$ if the problem instance is not even. Here, b is the number of vertices $v \in V$ such that $fc_k(\{v\})$ is odd. Kaufmann [219] presents a much simpler linear-time algorithm for routing even problem instances with two-terminal nets in a convex irregular switchbox. He needs the evenness condition. Nishizeki et al. [340] present another linear-time algorithm for routing two-terminal nets in convex switchboxes. Their notion of convexity is more restrictive: They call an irregular switchbox convex if each pair of boundary vertices can be joined with a straight or L-shaped path. They claim not to need the evenness condition. However, there is a mistake in their proof of this fact. This mistake is identified and corrected by Lai and Sprague [256].

9.5.2.3.2*   **Switch Graphs**   Matsumoto et al. [311] present an alternative construction of a quadratic algorithm for routing two-terminal nets in switch graphs. Their algorithm actually solves the UNIDRECTED MULTICOMMODITY FLOW problem in the case that each commodity has to be routed from a single source to a single sink. (This problem is more general than is the DETAILED ROUTING problem, since partial flow can take different paths from source to sink.) The algorithm yields integer flows whenever the problem instance is even. The run time of the algorithm as reported in [311] is $O(n^{5/2} \log n)$, but can be improved to $O(n^2)$ [219].

9.5.2.3.3*   **Homotopic Routing in General Detailed-Routing Graphs**
Kaufmann and Mehlhorn [223] solve the general case of the DETAILED ROUTING

problem for two-terminal nets in partial grid graphs. They need the evenness condition. The run time of the algorithm is $O(n^2)$. Kaufmann [220] improves the run time to $O(n)$.

Kaufmann [220] extends the results of [223] to partial grid graphs that also contain the main or the minor diagonals of the grid squares. In this case, the run time for routing is $O(n^2)$. Schrijver [399, 400] extends these results two more general classes of planar graphs.

9.5.2.3.4* **The Evenness Condition** Kaufmann and Maley [222] have proved that the evenness condition is a necessary ingredient for homotopic routing in the knock-knee model. They consider only partial grid graphs, and they prove the following result: The DETAILED ROUTING problem in the knock-knee model is NP-complete if the problem instances may be uneven. This result contrasts with a linear-time algorithm for detailed routing of even instances in the knock-knee model.

Therefore, if the placement of cells is fixed during detailed routing, then the evenness condition presents an unattractive but unavoidable technical restriction for routing in the knock-knee model. In partial grids, however, we can relax the evenness condition if we are allowed to move the cells slightly. In this case, we need to assert only the local evenness of the problem instance. More precisely, given a routable locally even but not even problem instance, we can slide each cell by at most one unit to the left ot right, such that the problem instance becomes even while staying routable. Such movements of cells are considered in topological compaction (see Section 10.5). Unfortunately, this method does not apply if the instance is not locally even. In fact, even if we may slide cells such as to make a problem instance even, the DETAILED ROUTING problem stays NP-hard if half-even—that is, not locally even—instances are allowed as input. Therefore, testing for necessarily local evenness cannot be avoided in efficient detailed routing in the knock-knee model [222].

9.5.3 Layer Assignment

In Section 9.1.3.2, we mentioned that one disadvantage of the knock-knee model is that it requires an assignment of wiring layers *after* routing. This characteristic of the knock-knee model is unattractive both because the layer assignment is computationally difficult and because assigning layers *during* routing conceivably and actually leads to more efficient layouts (see Sections 9.10.1.1 and 9.10.1.2).

Formally, an assignment of k wiring layers in the knock-knee model converts a knock-knee routing into the k-layer knock-knee model. Such a routing is a vertex-disjoint routing in the k-grid that, when viewed from the top, is identical with the knock-knee routing. Figure 9.12 depicts the two equivalent representations of a routing in the four-layer knock-knee model.

Note that *vias*—that is, layer changes—occur exclusively at grid nodes,
and if there are more than two layers, then we can route wires across vias
without causing electrical contact. If there are four layers, then we can even
stack vias on top of one another. Some fabrication technologies do not allow
stacked vias. In such technologies, a via connects all wiring layers, instead of
just two adjacent layers. The k-layer knock-knee model allows for stacked vias;
therefore, all results we present in this section incorporate stacked vias, and
most of them *need* stacked vias.

We restrict our attention to detailed-routing graphs that are partial grids.
Brady and Brown [46] were the first researchers to present an algorithm that
assigns layers to any knock-knee routing in a partial grid. They need only four
layers; thus, k-layer knock-knee models with $k > 4$ are uninteresting. The layer
assignment can be found in linear time in the size of the detailed-routing graph.
Of course, some knock-knee routings require fewer layers. As we shall see, it
is easy to decide whether a knock-knee routing needs only one or two layers.
On the other hand, Lipski [289] proves that the problem of deciding whether
a knock-knee routing needs only three layers is NP-complete. Note that most
layer-assignment algorithms apply to multiterminal nets as well as to two-
terminal nets. This observation contrasts with the difficulty of incorporating
multiterminal nets into the actual routing process.

Lipski and Preparata [290] present a unified theory of layer assignment in
the knock-knee model. The theory is based on a two-coloring paradigm and a
classification of routing patterns in the neighborhood of a grid node. The results
are applied to square grids, but the framework encompasses other regular grids,
such as hexagonal grids. Tollis [443, 444] actually develops layer-assignment
algorithms for routing in uniform grids other than the square grid, using this
framework.

Here, we shall describe another layer-assignment algorithm by Tollis [443,
445]. The algorithm applies to partial (square) grids. It has the advantage that
it uses layers greedily. Specifically, each routing that can be wired in two layers
is wired in two layers. Of course, we expect that some layouts that actually
need only three layers are wired in four layers; otherwise, the algorithm could
not be feasible unless $P = NP$.

9.5.3.1   The Layer-Assignment Algorithm by Tollis

The algorithm by Tollis is based on a graph called the *layer graph* that is
obtained from a given knock-knee routing. The layer graph formalizes the
interdependencies that exist between the assignments of layers to different wire
segments. It is similar in flavor to the constraint graphs used in Section 9.6.1.3
to describe and solve Manhattan channel-routing problems.

Before we discuss the details of the algorithm, we introduce some notation.
In this section, a knock-knee routing will be called a *(knock-knee) layout*. We
allow multiterminal nets, and assume that each multiterminal net is routed

with a Steiner tree. We make the restriction that each node on a hole of the detailed-routing graph may be the terminal of *at most two* nets. This restriction is not critical, and is also assumed in many results on homotopic knock-knee routing.

Definition 9.16 (Layer Assignment) *A* knock-knee layout *is a partial mapping* $\Lambda : E \rightarrow N$ *from the set of edges of the detailed-routing graph into the set N of nets. $\Lambda(e)$ is the net that is routed across edge e. The set of edges that compose the routing of a net is called a* wire.

We assume the set $L_1, \ldots, L_k$ of wiring layers *to be ordered from bottom to top; that is , layer L_1 is the bottommost layer and layer L_k is the topmost layer.*

A layer assignment *$W : E \rightarrow L$ is a partial mapping that assigns to each edge e in the domain of the layout Λ a wiring layer $W(e)$. The layer assignment is* legal *if, whenever $p \in V$ is a node in the detailed-routing graph and e_1 and e_2 are two (not necessarily distinct) edges incident to p such that $\Lambda(e_1) = \Lambda(e_2)$ and $W(e_1) \leq W(e_2)$, then there is no edge e_3 incident to p such that $\Lambda(e_3)$ is defined, $\Lambda(e_3) \neq \Lambda(e_1)$, and $W(e_1) \leq W(e_3) \leq W(e_2)$.*

The definition of the legality of a layer assignment formalizes the concept of a *stacked via*. It asserts that, in a legal layer assignment, if two layers L_i and L_j are occupied by a wire in a node p, then no layer between L_i and L_j is occupied by a different wire in p.

We now discuss how to find legal layer assignments with few wiring layers.

9.5.3.2 Net Orderings and Layer Graphs

The first step of the development of an algorithm for layer assignment is to represent a legal layer assignment by a labeling the nodes in the detailed-routing graph, creating a *net ordering*. We then will represent the net ordering in the form of a graph, the *layer graph*. The net ordering and the layer graph formalize the dependencies that exist between different wires with respect to a given layer assignment. These notions are similar in flavor to the constraint graphs that are used in jog-free Manhattan channel routing (see Section 9.6.1.3), but they differ in the sense that they depend on a *specific* layer assignment, and not just on the problem instance. Therefore, layer assignment in terms layer graphs involves not just the *analysis* of the layer graph, but rather the *construction* of a layer graph representing an efficient layer assignment.

Definition 9.17 (Net Ordering) *Let p be a node of the routing graph that has two different incident nets n_1 and n_2. Such a node is called* nontrivial. *The set of nontrivial nodes is denoted by V'. (Note that no more than two nets can be incident to any node of V.)*

A net ordering *is a mapping $w : N \times V' \rightarrow \{\bot, x, \top\}$, such that, for each trivial node p and each net n routed across p, we have $w(n, p) = x$, and* *for each nontrivial node p and the two nets n_1 and n_2 routed across p, we have $w(\{n_1, n_2\}, p) = \{\perp, \top\}$. The value $\perp$ denotes a* **bottom** *net; the value $\top$ denotes a* **top** *net; and the value x denotes an unspecified net ordering at a trivial node.*

The net ordering w **induced** *by the legal layer assignment W is defined as follows. If p is a nontrivial node with incident nets n_1 and n_2, then, since W is legal, all edges of n_1 are on layers, say, below all edges of n_2. We set $w(n_1, p) = \perp$ and $w(n_2, p) = \top$.*

Definition 9.18 (Layer Graph) *Let $e = \{p_1, p_2\}$ be an edge in V. We define the edge $\vec{e} = (p_1, p_2)$ to be* **layer-increasing** *if for the net n routed across e we have $w(n, p_1) = \top$ and $w(n, p_2) = \perp$.*

The **layer graph** *D_w induced by the net ordering w has the nodes of the detailed-routing graph. The edge set of the layer graph D_w is the set of layer-increasing edges.*

Note that not every directed graph on the set of vertices in the knock-knee layout that, in addition, contains only directed versions of edges in the knock-knee layout is a layer graph. Rather, whether such a graph is a layer graph is determined by several structural properties that ensure that the graph is induced by a suitable net ordering. The details of these properties are not of interest to us. We will perform layer assignment by constructing net orderings. The quality of the layer assignment will be determined by a structural parameter of the induced layer graph—namely, by its depth.

Theorem 9.8 *Let Λ be a knock-knee layout, and let w be a net ordering for Λ. Let D_w be the layer graph induced by w.*

1. *There is a legal layer assignment W that induces w exactly if D_w is acyclic.*

2. *If D_w is acyclic, the legal layer assignment W that induces w and has the fewest number of layers has $depth(D_w) + 2$ layers.*

Proof Clearly, a cycle in D_w induces a cyclic ordering of the layers among the set of nets routed across the edges contained in the cycle. Such an ordering is impossible.

Assume, on the other hand, that D_w is acyclic. Any path of length ℓ in D_w induces a total ordering on the set of layers assigned to all nets incident to the nodes on the path in any legal layer assignment inducing w. Since all of these nodes are nontrivial, there are $\ell + 2$ such nets. Therefore, at least $depth(D_w) + 2$ layers are needed in any legal layer assignment inducing w.

To complete the proof of the theorem, we will construct a legal layer assignment W with $depth(D_w) + 2$ layers that induces w.

The layer assignment assigns to each edge e in Λ that has a directed version $\vec{e}$ in D_w the layer number $\ell(e) + 1$, where $\ell(e)$ is the length of the longest directed path in D_w that ends in $\vec{e}$. For any other edge $e = \{p_1, p_2\}$ in Λ carrying net n, the set $\{w(n, p_1), w(n, p_2)\}$ contains at most one of the values $\perp$ and $\top$. If this value is $\perp$, the layer number assigned to edge e is 1—that is, the bottom layer; if the value is $\top$, the layer number assigned to e is $depth(D_w) + 2$—that is, the top layer. Otherwise, the layer number can be set arbitrarily.

We can see that this layer assignment is legal by checking each node p individually. The time for constructing the layer assignment is linear in the size of the layout. $\square$

The layer-assignment procedure described in the proof of Theorem 9.8 does not minimize the number of vias. Specifically, the assignment of the bottom and top layers to edges that have no directed versions in D_w may introduce unnecessary vias and stacked vias. We discuss via minimization in Section 9.5.3.4. However, there are layouts for which we cannot avoid running wires across vias if we want to use only four layers (see Exercise 9.19).

9.5.3.3 Constructing Layer Graphs

In this section, we construct, given a layout Λ, a net ordering w whose layer graph D_w has a depth of at most 2. Therefore, with Theorem 9.8, we can get by with four layers for wiring the layout. We also show that, if the layout can be wired in two layers, then our construction does not generate any layer-increasing edge, such that our algorithm uses only two layers.

We process the layout by visiting the nodes in the detailed-routing graph rowwise from left to right, starting with the bottommost row and proceedings upward. When we visit a node, we generate the net orderings pertaining to that node. Here, the goal is to minimize the depth of D_w.

For the purpose of the exposition, we use the following notation. Nodes in the routing grid are denoted with pairs of positive integers, signifying their x and y coordinates in the grid. The neighborhood around the node p that is currently processed is depicted in Figure 9.33. When we process node p, net orderings have already been assigned at nodes p_1 and p_2. We call p a *conflict node* if p is not trivial and one of the following two facts holds:

1. Edges e_1 and e_2 carry the same wire n and $w(n, p_1) \neq w(n, p_2)$.

2. Edges e_1 and e_2 carry different wires n_1 and n_2, respectively, and $w(n_1, p_1) = w(n_2, p_2)$.

If p is not a conflict node, then it is straightforward to assign the net-ordering labels pertaining to p such that no layer-increasing edge is introduced at e_1 and e_2. We simply propagate the net-ordering labels from p_1 across edge e_1 to p without change, and do the same for p_2. Then we extend this net ordering to all nets incident to p.

If p is a conflict node, then we must introduce layer-increasing edges. However, we need to introduce only *one* such edge, and we can freely choose whether

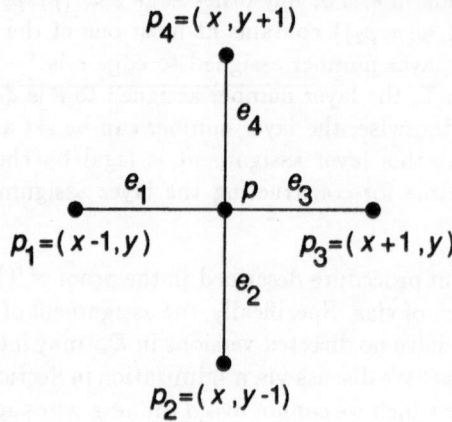

Figure 9.33: *The neighborhood of the currently processed node.*

we do so at edge e_1 or at edge e_2. To prove this observation, we consider the following two cases.

Case 1: e_1 and e_2 carry the same net n and $w(n, p_1) \neq w(n, p_2)$. Setting $w(n, p) = w(n, p_1)$ introduces a layer-increasing edge at e_2. Setting $w(n, p) = w(n, p_2)$ introduces a layer-increasing edge at e_1. In each case, the other edge is not layer increasing.

Case 2: e_1 and e_2 carry different nets n_1 and n_2, respectively, and $w(n_1, p_1) = w(n_2, p_2)$. Setting $w(n_1, p) = w(n_1, p_1)$ and extending this labeling appropriately to n_2 introduces a layer-increasing edge at e_2. Setting $w(n_2, p) = w(n_2, p_2)$ and extending this labeling appropriately to n_1 introduces a layer-increasing edge at e_1. In each case, the other edge is not layer increasing.

In both cases, the layer-increasing edge could be directed either way.

We now detail how we make the decision whether we introduce a layer-increasing edge at e_1 or e_2 if p is a conflict point. Our objective is to keep the depth of D_w low.

A very important observation is that we can ensure the node p_3 to be conflict-free by choosing an appropriate layer-increasing edge among e_1 and e_2. Indeed, let n_1 be the net routed across the edge between p_3 and its lower neighbor $p_{3,2}$, and let n_2 be the net routed across e_3. The value of $w(n_1, p_{3,2})$ was already determined when the previous row was processed. This value uniquely determines a value for $w(n_2, p)$ such that p_3 is conflict-free. All other values of net-ordering labels at p, and therefore also the direction of the layer-increasing edge, are implied by this choice.

(1)   **for** $y = 1, \ldots, n$ **do**
(2)   **for** $x = 1, \ldots, n$ **do**
(3) $p := (x, y)$;
(4)   **if** p is a conflict node **then**
(5)   **if** $\ell(p_2) = 2$ **then**
(6)   introduce a layer-increasing edge at $e_1 = \{p_1, p\}$
(7)   **else if** $\ell(p_1) = 2$ **then**
(8)   introduce a layer-increasing edge at $e_2 = \{p_2, p\}$
(9)   **else comment** $\ell(p_1), \ell(p_2) \le 1$;
(10)   let $p^* = (x^*, y^*)$ be the most recently processed
                                 node with $y^* \ge y - 1$ and $\ell(x^*, y^*) = 2$;
(11)   **if** p^* exists and $|x - x^*| + |y - y^*|$ is odd **then**
(12)   introduce a layer-increasing edge at $e_1 = \{p_1, p\}$
(13)   **else comment** p^* does not exist or
                                   $|x - x^*| + |y - y^*|$ is even;
(14)   introduce a layer-increasing edge such that
                                       p_3 becomes conflict-free **fi fi**;
(15)   update the ℓ values of the nodes that lie on the path
                             in D_w that got extended by the introduced
                             layer-increasing edge
(16)   **else comment** p is conflict-free;
(17)   propagate the net ordering to p **fi od od**;

Figure 9.34: *The layer-assignment algorithm.*

The idea of the layer-assignment algorithm is to limit the depth of D_w to at most 2. To achieve this goal, we have to keep track, for each node p in the row currently processed and the row below that, of the length $\ell(p)$ of the longest undirected path in the part of D_w constructed so far that has p as an endpoint. The algorithm is depicted in Figure 9.34. In line 12 of the algorithm, we choose the layer-increasing edge for p as described previously, such that the right neighbor p_3 of p becomes conflict-free. In line 9, the parity criterion that $|x - x^*| + |y - y^*|$ be odd causes horizontal layer-increasing edges to be introduced only *every other time*. This choice effectively prohibits long paths in the layer graph.

We now prove the correctness of the algorithm.

Theorem 9.9 *The algorithm in Figure 9.34 computes a net ordering w such that the depth of D_w does not exceed 2.*

•**Proof** We prove the following two invariants of the algorithm. Let $p = (x, y)$ be the node that the algorithm has just finished processing, and let

$\tilde{p} = (\tilde{x}, \tilde{y})$ be any node processed up to p. We shall index the values involved with superscripts to denote the time instant at which we look at them. The superscript ^{p} will denote values at the time just after p was processed.

I1 Let $p' = (x', y')$ be a node processed before $\tilde{p}$ such that $y' \geq \tilde{y} - 1$; that is, p' is a node in the row of $\tilde{p}$ or in the row below. Then, $\ell^{\tilde{p}}(x', y') \leq 2$ and, if $|\tilde{x} - x'| + |\tilde{y} - y'|$ is odd and $\ell^{\tilde{p}}(\tilde{x}, \tilde{y}) = 2$, then $\ell^{\tilde{p}}(x', y') < 2$.

I2 If the node $(x + 1, y)$ is a conflict node, then $\ell^p(x, y) + \ell^p(x + 1, y - 1) \leq 2$.

Invariant I1 gains its significance by the following important conclusion. Call two nodes p and p' such that $|x - x'| + |y - y'|$ is odd *at odd distance*.

Lemma 9.7 *If Invariant I1 holds after processing node p, then no two nodes p' and p'' that have been processed up to p, are at odd distance, and lie in the same or adjacent rows can both have $\ell^p(p') = \ell^p(p'') = 2$.*

Proof The proof is indirect. Assume that, after processing of p, there are two paths of length 2, one having p' and one having p'' as an endpoint. Let r be the endpoint of these two paths that is processed last. Note that $\ell^r(q) = 2$ for any endpoint q of the two paths. Applying Invariant I1 with $\tilde{p} = r$ and p' being the endpoint of the other path that is processed last yields the contradiction, because any pair of endpoints, one on each path, is at odd distance. □

By Lemma 9.7, Invariant I1 intuitively means that nodes that are ends of long undirected paths in D_w are only sparsely dispersed in the layout. Only every other node can be such a node. This invariant implies that the paths ending in such nodes cannot combine to form long paths. This statement is necessary to maintain the invariant. Invariant I2 intuitively means that only one of the left and bottom neighbors of any node can be critical in the sense that it is on a long path already.

On the assumption that the invariants hold just after processing p_1, we now prove that they also hold after processing p. We make a case distinction. Invariant I2 allows us to limit the case distinction to the following four cases.

Case 1: p is the leftmost node in a row or p is in the bottom row. Then, p is not a conflict node, and both invariants are maintained.

Case 2: $\ell^{p_1}(x, y-1) = 2$ and $\ell^{p_1}(x-1, y) = 0$. Then, the algorithm introduces a layer-increasing edge between p and p_1 in line 6. Afterward, it sets $\ell^p(x - 1, y) = \ell^p(x, y) = 1$ in line 14. The node $p_3 = (x + 1, y)$ may be a conflict node. But an application of Lemma 9.7 with $p' = (x, y - 1)$ and $p'' = (x + 1, y - 1)$ shows that $\ell^p(x + 1, y - 1) = \ell^{p_1}(x + 1, y - 1) < 2$. Therefore, $\ell^p(x, y) + \ell^p(x + 1, y - 1) \leq 2$ and Invariant I2 is maintained. Invariant I1 is trivially maintained, since $\ell^p(x, y) = 1$.

Case 3: $\ell^{p_1}(x, y - 1) = 0$ and $\ell^{p_1}(x - 1, y) = 2$. Analogous to case 2.

Case 4: $\ell^{p_1}(x-1,y), \ell^{p_1}(x,y-1) \leq 1$. Before processing p, let $p^* = (x^*, y^*)$ be the most recently processed node with $\ell^{p_1}(p^*) = 2$ and $y^* \geq y - 1$.

> **Case 4.1:** p^* exists and $|x - x^*| + |y - y^*|$ is odd. Applying Lemma 9.7 with $p' = (x+1, y-1)$ and $p'' = p^*$, we infer $\ell^p(x+1, y-1) = \ell^{p_1}(x+1, y-1) < 2$, because $|x - 1 - x^*| + |y - 1 - y^*|$ is odd.
>
> We now claim that $\ell^{p_1}(x-1, y) = 0$. To prove this claim, we note that $|x + 1 - x^*| + |y - y^*|$ is even. If $\ell^{p_1}(x-1, y)$ were 1, then (x, y) should be conflict-free by line 14 of the algorithm. This is not the case; therefore, $\ell^{p_1}(x-1, y) = 0$. Thus, again, after processing p, we have $\ell^p(x-1, y) = \ell^p(x, y) = 1$. Putting everything together, we get $\ell^p(x, y) + \ell^p(x+1, y-1) \leq 2$, and Invariant I2 is maintained. Invariant I1 is trivially maintained, since $\ell^p(x, y) = 1$.
>
> **Case 4.2:** p^* does not exist or $|x - x^*| + |y - y^*|$ is even. The algorithm executes line 14. Afterward, $\ell^p(x, y) \leq 2$, and $(x+1, y)$ is conflict-free. Thus, Invariant I2 is maintained.
>
> Executing line 14 keeps $\ell^p(x-1, y) = \ell^{p_1}(x-1, y) \leq 1$ and $\ell^p(x, y-1) = \ell^{p_1}(x, y-1) \leq 1$. Therefore, the algorithm alters the ℓ values of only those nodes that are at distance 1 from $(x-1, y)$ or $(x, y-1)$. These are exclusively nodes p' such that $|x - x'| + |y - y'| = 2$.
>
> To show that Invariant I1 is maintained, we assume that $\ell^p(x, y) = 2$. Let p' be a node visited before p such that $|\tilde{x} - x'| + |\tilde{y} - y'|$ is odd and $\ell^p(p') = 2$. This assumption implies, in particular, that p^* exists. By the remark in the preceding paragraph, $\ell^{p_1}(p') = \ell^p(p') = 2$.
>
> By definition of p^*, $|x' - x^*| + |y' - y^*|$ must be odd. But this is a contradiction to Lemma 9.7, with $p'' = p^*$.

□

As a consequence, we can wire each knock-knee layout in four layers. The construction takes linear time in the size of the layout.

Figure 9.35 applies the algorithm to the knock-knee layout depicted in Figure 9.12. The resulting layer assignment also is depicted in Figure 9.12. Although the algorithm uses four layers on this layout, three layers would be sufficient, proving that at times the algorithm does not minimize the number of layers. Tollis and Vaguine [446] present heuristics that reduce the number of layers from four to three in many cases. However, if the layout can be wired in two layers, the layer-assignment algorithm in Figure 9.34 does so.

Theorem 9.10 *If the layout can be wired in two layers, the layer-assignment algorithm does not find a conflict point.*

Proof We prove this theorem indirectly. Assume that $p = (x, y)$ is the first conflict point encountered by the algorithm. We make a case distinction; see Figure 9.33.

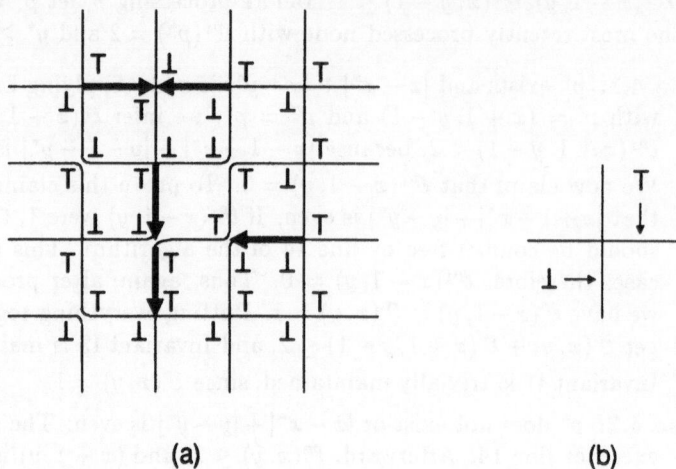

(a)   (b)

Figure 9.35: *Application of the layer-assignment algorithm to the knock-knee layout in Figure 9.12(a). Layer-increasing edges are denoted by bold arrows. (a) Layer graph and net ordering; (b) convention for representing the net ordering at crossings. The resulting layer assignment is shown in Figure 9.12.*

Case 1: Edges e_1 and e_2 carry the same wire n and $w(n, p_1) \neq w(n, p_2)$. Let n_1 be the other wire at p_1 and n_2 the other wire at p_2. Two nets from the set $\{n, n_1, n_2\}$ are incident to node $(x - 1, y - 1) = p_{1,2} = p_{2,1}$. We make another case distinction that can be checked by inspection of the neighborhood of p in the layout.

> **Case 1.1:** If n_1 and n_2 are incident to $(x - 1, y - 1)$, then the layout needs three wiring layers.

> **Case 1.2:** If, without loss of generality, n and n_1 are incident to $(x - 1, y - 1)$, then p cannot be a conflict point.

Case 2: Let n_1 be the net routed across edge e_1 and n_2 be the net routed across edge e_2. Let n_3 be the other net at p_1, and n_4 be the other net at p_2. This time, there are four subcases.

> **Case 2.1:** If n_1 and n_2 are incident to $(x - 1, y - 1)$, then either p_1 or p_2 must have been a conflict point, in contradiction to the assumption that p is the first conflict point encountered.

> **Case 2.2:** If n_1 and n_4 are incident to $(x - 1, y - 1)$, then wiring the layout needs at least three layers.

Case 2.3: If n_2 and n_3 are incident to $(x - 1, y - 1)$, then wiring the layout needs at least three layers.

Case 2.4: If n_3 and n_4 are incident to $(x - 1, y - 1)$, then either p_1 or p_2 must have been a conflict point, in contradiction to the assumption that p is the first conflict point encountered.

□

9.5.3.4 Via Minimization

The algorithm by Tollis uses at most four layers on any knock-knee layout. However, it does not always use the minimum number of layers, even though, if only one or two layers are needed, the algorithm uses only that many layers. Furthermore, the algorithm does not minimize the number of vias.

Via minimization is an issue, because vias introduce electrical instabilities and the potential of defects during the fabrication of the circuit. It turns out that via minimization is feasible if the number of wiring layers is two [357], but the problem becomes NP-hard if the number of wiring layers is larger than two [330].

9.5.3.4.1 Via Minimization with Two Wiring Layers

The via-minimization problem with two wiring layers can be reduced to the problem of finding a maximum vertex cut in a planar graph. Let Λ be a knock-knee layout. We restrict our attention to layouts with two-terminal nets first. Figure 9.36(a) depicts an example layout.

Starting with Λ, we construct a *layout graph* that represents the layer conflicts between segments belonging to different nets. To this end, we break up each wire into two kinds of segments. *Critical segments* are maximal segments of wires that connect midpoints of edges in the detailed-routing graph and along which we cannot place a via. (Note that stacked vias are prohibited, because we have only two wiring layers.) *Free segments* are segments that represent portions of the wire connecting critical segments. Figure 9.36(b) shows the decomposition of the layout in Figure 9.36(a) into segments. The critical segments are numbered.

Crossing or touching critical segments have to be wired on distinct layers. Therefore, each pair of critical segments making up a crossing or a knock-knee represents a layer conflict. The following graph represents all such conflicts.

Definition 9.19 (Layout Graph) *The* layout graph $L_\Lambda = (S, A \cup B)$ *for a layout Λ is an undirected graph whose vertex set is the set S of critical segments. There are two kinds of edges in L_Λ. A* conflict edge *connects two vertices representing critical segments that cross or touch. A* continuation

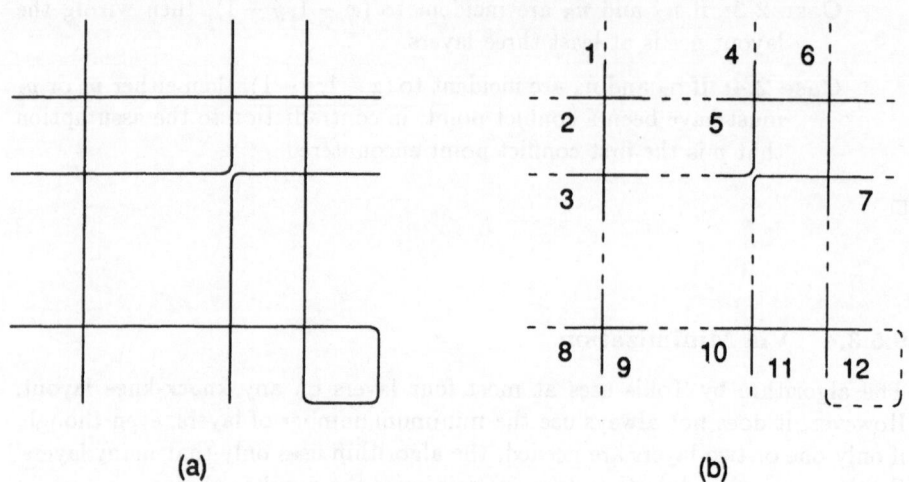

(a)   (b)

Figure 9.36: *A knock-knee layout with two-terminal nets. No node of the detailed-routing graph is incident to four edges carrying the same net. (a) The knock-knee layout; (b) a decomposition of the layout into segments. The critical segments are drawn with solid lines; the free segments are drawn with dashed lines. The critical segments are numbered.*

edge *connects two vertices that represent critical segments that are connected by a free segment. The set of conflict edges is denoted with* A. *The set of continuation edges is denoted with* B.

Figure 9.37 shows the layout graph of the layout in Figure 9.36. Whether the layout is wirable in two layers can be seen from the layout graph. Indeed, we can interpret each wiring layer as a color that labels the vertices of the conflict graph. Since each critical segment must be assigned a unique layer—layer changes along a critical segment are not possible—the corresponding vertex in the conflict graph can be labeled with the corresponding color. In this way, a legal layer assignment with two layers translates into a (vertex) coloring with two colors of the portion of the layout graph that contains only the conflict edges. (For the definition of a coloring, see Definition 3.3). Vice versa, each such coloring can be translated into a layer assignment with two wiring layers, because we can place a via on every free wire segment. Thus, deciding whether the layout is wirable in two layers amounts to showing that the *conflict subgraph* $C_\Lambda = (S, A)$ is two-colorable. By Exercise 3.1, this decision can be made in linear time. Therefore, we have found an alternative to Tollis' algorithm for deciding in linear time whether a knock-knee layout can be wired in two layers.

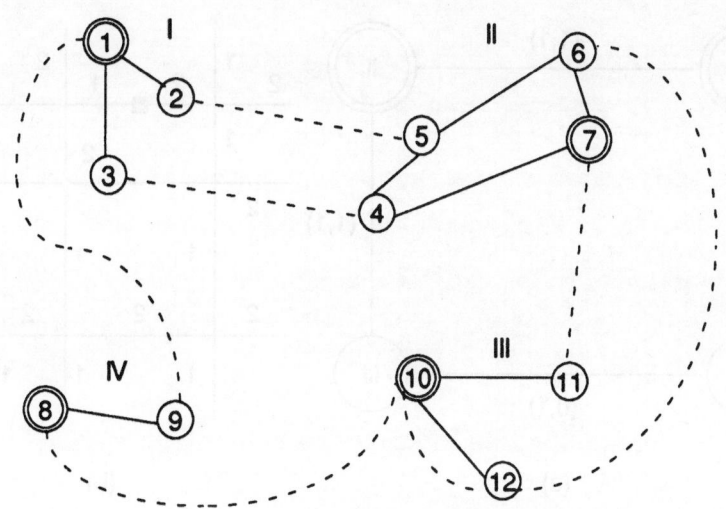

Figure 9.37: *The layout graph for the layout in Figure 9.36; the conflict edges are drawn with solid lines; the continuation edges are drawn with dotted lines. Representative vertices for the connected components are double-circled. The conflict subgraph is two-colorable. Thus, the knock-knee layout in Figure 9.36 is wirable using two layers.*

Assume now that the conflict subgraph C_Λ is two-colorable. In this case, two-colorings of the conflict subgraph correspond bijectively to layer assignments with two wiring layers. The number of vias can be read off the layout graph. A via is necessary on each free segments that corresponds to a continuation edge that connects two differently colored vertices in the conflict graph. (Obviously, there is no need to place more than one via on any free segment.) Therefore, we need to find a two-coloring of the conflict subgraph that minimizes the number of continuation edges that connect differently colored vertices in the layout graph. We will convert this problem into a graph problem that can be solved efficiently.

In general, the conflict subgraph C_Λ is disconnected. Consider a connected component S' of C_Λ. We choose some vertex $v_{S'} \in S'$ as the *representative* of S'. Coloring $v_{S'}$ determines the color of all other vertices in S'. Furthermore, for a continuation edge that connects two vertices v and w in S', the colors of v and w are identical or different *independently of how S' is colored*. Thus, we can disregard such continuation edges during the via minimization.

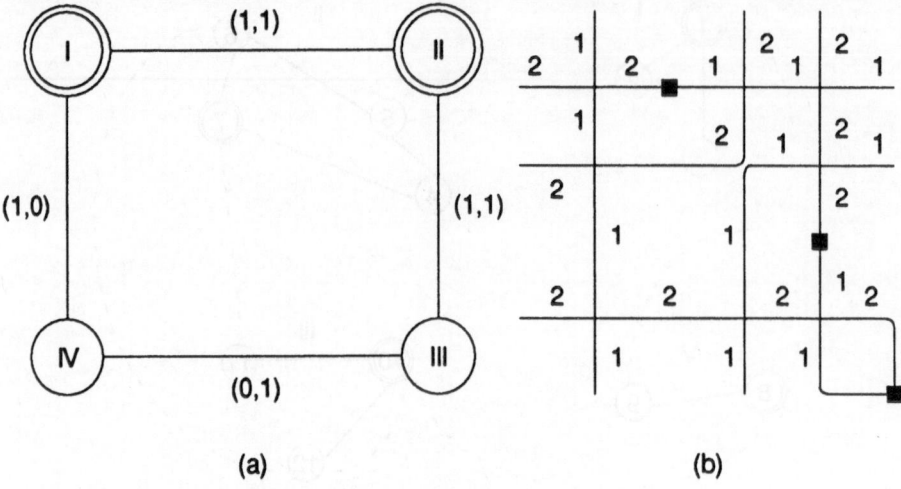

(a)   (b)

Figure 9.38: (a) The residue graph of the layout in Figure 9.36. The minimum cut is highlighted in the figure. It results an a layer assignment with three vias, as depicted in (b).

The following graph is a succinct representation of the information necessary for via minimization.

Definition 9.20 (Residue Graph) *The residue graph $R_\Lambda = (X, T)$ results from the layout graph by collapsing all connected components of the conflict subgraph to single vertices. Specifically, the vertex set X of R_Λ is the set of connected components of the conflict subgraph C_Λ. T contains the edge $e = \{S', S''\}$ between two connected components S' and S'' exactly if there are one or more continuation edges in L_Λ, connecting vertices in S' and S''. Each edge $e = \{S', S''\}$ in T is labeled with a pair $(\mu_=(e), \mu_{\neq}(e))$ of integers. The integer $\mu_=(e)$ denotes the number of continuation edges connecting differently colored vertices in S' and S'' if the representative vertices $v_{S'}$ and $v_{S''}$ of S' and S'' receive the same color. Analogously, $\mu_{\neq}(e)$ denotes the number of continuation edges connecting differently colored vertices in S' and S'' if the representative vertices $v_{S'}$ and $v_{S''}$ of S' and S'' receive different colors.*

Figure 9.38 depicts the residue graph of the layout in Figure 9.36. The residue graph can be computed in linear time in the size of the layout. In terms of the residue graph, the via-minimization problem takes the following form.

VIA MINIMIZATION

Instance: A residue graph $R = (X, T)$

Configurations: All two-colorings $\lambda : X \to \{0, 1\}$ of the vertices of R

Solutions: All configurations

Minimize: $c(R) = \sum_{e \in T} \rho(e)$, where, for $e = \{S', S''\}$,

$$\rho(e) = \begin{cases} \mu_=(e) & \lambda(S') = \lambda(S''), \\ \mu_{\neq}(e) & \lambda(S') \neq \lambda(S'') \end{cases}$$

Clearly, a two-coloring of a graph divides the vertex set of the graph into two parts—one for each color. Thus, such a coloring can be viewed as a *bipartition* in the sense of Chapter 6. In other words, we are looking for a bipartition of the residue graph that minimizes (or maximizes) some cost. The natural cost of a bipartition is its cutsize in terms of some edge weighting—that is, the sum of the weights of all edges across the cut of the bipartition. As we defined the VIA MINIMIZATION problem, the cost function does not have this structure, but we can reformulate the edge labels so as to give the cost function this structure.

To this end, we subtract the value $\sum_{e \in T} \mu_=(e)$ from the cost function. This change does not alter the set of minima. Afterward, the contribution of edge $e = \{S', S''\}$ is $\mu_{\neq}(e) - \mu_=(e)$ if $\lambda(S') \neq \lambda(S'')$, and is 0 otherwise. Therefore, labeling the edge e with weight $\mu_{\neq}(e) - \mu_=(e)$ makes the cost of the labeling exactly the size of the cut.

With this reformulation, the VIA MINIMIZATION problem becomes the problem of finding a minimum cut in a graph that is edge weighted with integer numbers. Adding the same constant to all edge weights such as to make all edge weights negative, and then multiplying all edge weights with -1, converts the problem into the classical MAXCUT problem—that is, the problem of finding a maximum-size cut in a positively edge-weighted graph (see Exercise 9.21). This problem is NP-hard [133] on general graphs.

It is, however, an immediate consequence of the definition of the layout graph that the residue graph is always *planar* (see Exercise 9.20). On planar graphs the MAXCUT problem can be solved in polynomial time [162]. The solution reduces the MAXCUT problem to the MINIMUM WEIGHT MATCHING problem on the planar dual of the graph (see Section 3.12 and Exercise 9.21). Thus, it takes time $O(n^3)$, where n is the number of vertices in the graph.

9.5.3.4.2 Incorporation of Multiterminal Nets

We cannot extend this approach to via minimization to arbitrary knock-knee layouts with multiterminal nets. Instead, we have to make the restriction that, in the knock-knee layout Λ, there is no node in the detailed-routing graph to which the *same* net is incident from all four sides (*four-way split*). This restriction is rather

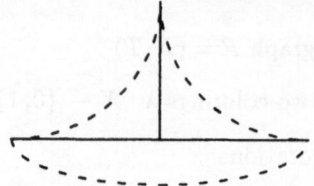

Figure 9.39: *Modeling of free segments with three terminals in the layout graph. The solid lines depict a free segment with three terminals. The dotted lines depict the continuation edges added to the layout graph to represent the free segment.*

technical. We shall soon see where we need it. Assume now that a layout is given that meets the restriction.

The decomposition of the layout into free and critical segments is modified as follows. Note that each critical segment has two terminals, whereas there can be free segments with three or more terminals. We introduce additional critical segments of length 0 in the middle of edges of the detailed-routing graph appropriately such that no free segment has more than three terminals. This modification increases the size of the layout graph but does not restrict the set of possible layer assignments. After this modification, we have two kinds of free segments: those with two terminals and those with three terminals— that is, those with exactly one three-way split. We model each two-terminal free segment by one continuation edge in the layout graph, as before. For each three-terminal free segment we introduce three continuation edges, one connecting each pair of adjacent critical segments as we pass around the free segment in clockwise order (see Figure 9.39).

The central observation now is that, if the critical segments adjacent to a three-terminal free segment are wired on the same layer, then no continuation edge for the free segments connects differently colored vertices. However, if the three adjacent critical segments to a three-terminal frees segment are colored with different colors, then *exactly two* continuation edges connect differently colored vertices. At the same time, we can realize the corresponding layer assignment by placing *one* via, say, on the central node of the three-way split. Therefore, all we need to do to be able to extend the reduction to the MAXCUT problem is to give the continuation edges for three-terminal free segments the weight of 1/2, while the continuation edges for two-terminal free segments receive a weight of 1. The values $\mu_=(e)$ and $\mu_{\neq}(e)$ are then based on these edge weights.

This approach does not extend to more general shapes of free segments, because we cannot maintain the property that the total weight of the edges connecting differently colored critical segments adjacent to the free segment is *independent* of the coloring. (We note the affinity of this observation to the observations made in Section 6.1.5.)

9.5.3.4.3* Other Results Molitor [330] presents an extension to the algorithm for via minimization discussed here. He points out that minimizing the total number of vias may cause a single wire to receive many vias. Then, he presents a modification of the via-minimization algorithm that allows for balancing the number of vias on different wires, at the cost of increasing the total number of vias. Molitor also defines via-minimization problems involving several layers, and shows their NP-hardness. His multilayer models do not coincide with the model we discussed in this section. In fact, they are not very practical, since they disallow putting vias on nontrivial grid nodes, which is necessary in some cases (see Exercise 9.19). Nevertheless, his NP-hardness results carry over to via minimization on more than two layers in our model.

Barahona et al. [20] incorporate the possibility of preassigning layers to wire segments and of defining a preferred layer for a wire segment in the problem formulation. Wiring as much as possible of the layout in the preferred layer is a goal that conflicts with the goal of via minimization. Barahona et al. modify the VIA MINIMIZATION problem such that a combination of the two conflicting goals is optimized. The relative priority of each goal can be set by parameter. The corresponding layout graph now becomes nonplanar; thus, the MAXCUT problem becomes difficult. Barahona et al. apply cutting-plane techniques to solving this problem. Xiong and Kuh [478] present similar extensions of the VIA MINIMIZATION problem and solve the corresponding MAXCUT problem heuristically.

Several authors discuss algorithms that aim at minimizing the number of vias in a *topological* variant of the two-layer knock-knee model [198, 308, 389, 390, 477]. Here, we do not assume the knock-knee layout to be given; rather, we want to find a knock-knee layout that requires few vias during layer assignment. Thus, in some sense, these algorithms are *routing* algorithms. However, they perform only *topological* routing, in the sense that they do not presuppose a certain detailed-routing graph with specific overlap rules for wires, but rather perform routing in the continuous plane. The only structure in which these algorithms are interested is the topological arrangement of wire crossings. Vias may be placed on any wire segment between two crossings. The algorithms aim at arranging crossings such that a minimum number of vias is necessary to wire the routing in two layers. The algorithms leave open the problem of appropriately placing the vias on nodes of the detailed-routing graph. Especially in a dense layout, this procedure may require additional area for placing vias, and therefore the topological routing algorithms do not optimize the area needed for the routing. Nevertheless, the algorithms are of interest in applications where sparse layouts occur, such as in PRINTED CIRCUIT BOARD routing. The general problem is NP-hard [390]. Heuristics are presented in [198, 308, 477], and special cases are solved in [389, 390].

9.5.4 Summary

Layer assignment of knock-knee layouts in four layers is always possible, and can be done in linear time in the size of the layout. Deciding whether one or two layers suffice is easy. However, deciding whether three layers suffice is NP-hard. Furthermore, via minimization also can be done for two wiring layers (with certain restrictions for layouts with multiterminal nets), but is NP-hard for three or more wiring layers.

From a practical point of view, layer assignment in the knock-knee model poses several problems:

1. A routing in the four-layer knock-knee model is quite a sparse routing. We would probably prefer to introduce some overlap, in order to be able to make the routing denser.

2. Often, fabrication technologies require that we assign preferred directions to wiring layers. In contrast, the layer assignments in the four-layer knock-knee model are quite irregular.

For these reasons the layer-assignment procedures for routings in the knock-knee model are not attractive for practical applications, and this fact also reduces the attractiveness of the knock-knee model itself for practical applications. Nevertheless, the model is quite attractive from a theoretical point of view, because it is based on the notion of edge-disjoint paths and has a strong affinity to multicommodity flow problems. Also, the methods for via minimization in two wiring layers are used in practical applications.

9.6   Detailed Routing in the Manhattan Model

In this section, we discuss the Manhattan model. Most of the results are on channel routing, but there are efficient algorithms for a few other detailed-routing problems.

9.6.1 Channel Routing in the Manhattan Model

Most of the work in all of detailed routing has been done on the problem of channel routing. Practical approaches use a slightly more restrictive model, the *jog-free model*. The corresponding results are discussed in Section 9.6.1.3. Recently, new lower bounds on channel routing in the Manhattan model have been discovered; these are described in Section 9.6.1.1. With their use, an application of the general channel-routing scheme introduced in Section 9.4

can be proved to come within a constant factor of minimum density. This algorithm is discussed in Section 9.6.1.2.

The channel-routing algorithm for the knock-knee model that is described in Section 9.5.1 scans through the channel from left to right column by column and extends the routing appropriately. This method actually is quite a good heuristic even for other detailed-routing models. For instance, Rivest and Fiduccia [378] present a greedy channel router for the Manhattan model using this heuristic. The router has quite a good performance in practice. Furthermore, it is robust in the following sense:

1. It can handle multiterminal nets.

2. It has parameters whose adjustment controls the number of jogs in the routing.

3. The greedy heuristics can be adapted to incorporate special technological rules.

However, the algorithm performs very badly on some problem instances. To evaluate Manhattan channel routers such as the greedy channel router, we must investigate lower bounds on the channel width in the Manhattan model.

9.6.1.1 Lower Bounds on the Channel Width

Since the Manhattan model is a restriction of the knock-knee model, the open density is a lower bound in the Manhattan model, as well. In fact, this bound can be strengthened slightly. An alternative definition of open density is the following. Assume the channel to be embedded in the plane. The columns of the channel have integer abscissae. The open density is the maximum over all *noninteger* abscissae x of the number of nets crossing x. A net crosses x if one of its terminals is in a column less than x and another one is in a column greater than x. Here, we specifically leave the integers x out of the maximization. This is the reason why this kind of density is called *open* density. To define the *closed* density, we include integer abscissae into the maximization.

Definition 9.21 (Closed Density) *Let I be an instance of the* CHANNEL ROUTING *problem. The* **closed density** $d_c(I)$ *of I is defined as*

$$d_c(I) := \max_x |N_x|$$

where x is an arbitrary real number and N_x is the set of (nontrivial) nets that have one terminal in a column $\leq x$ and another in a column $\geq x$. (Compare with Definition 9.9.)

Here, we need to maximize over only *integer abscissae*. Figure 9.40 illustrates an example of a problem instance whose open and closed density are different. This is the *shift-right-1 problem* $SR1(|N|)$, which consists of the nets

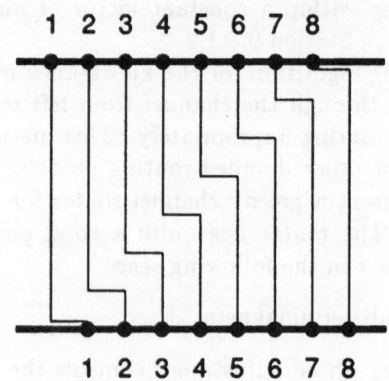

Figure 9.40: *The problem instance SR1(8) and its solution.*

$n_1, \ldots, n_{|N|}$ with $n_i = ((i+1, b), (i, t))$. However, the closed density is never more than one greater than the open density (see Exercise 9.22). It is an important observation that the closed density is not necessarily the density of any straight vertical cut through the channel.

Since knock-knees are disallowed in the Manhattan model, the following fact is obvious.

Fact 9.4 $d_c(I)$ *is a lower bound on the channel width in the Manhattan model.*

In the context of Manhattan routing, we shall omit the word "closed" and shall just speak of "density" unless this notation is ambiguous.

The greedy channel router routes most practical problem instances *in line-break density*—that is, it uses $d_c(I)$ tracks—and it constructs a routing fast. However, there are problem instances on which the algorithm performs very poorly, in terms of either its running time or the resulting channel width. An example for an instance on which a large channel width results is the shift-right-1 problem (Figure 9.40). Here, the greedy channel router tends to use $\Omega(|N|)$ tracks.

More fundamentally, in the case of Manhattan channel routing, the density is not even a good lower bound on the minimum channel width, in general. As an example, $d_c(SR1(|N|)) = 2$, but we are quickly convinced that we need many more tracks to route the instance. What makes $SR1(|N|)$ such a difficult problem instance is that there are few free terminal sites at the sides of the channel. Thus, there is not enough room inside the channel to allow sufficient wire jogging. Brown and Rivest [51] prove that $\Omega(\sqrt{n})$ tracks are necessary to route $SR1(|N|)$. Part 1 of Exercise 9.23 shows that $O(\sqrt{n})$ tracks suffice for this problem. Baker et al. [18] generalize the proof method of Brown and

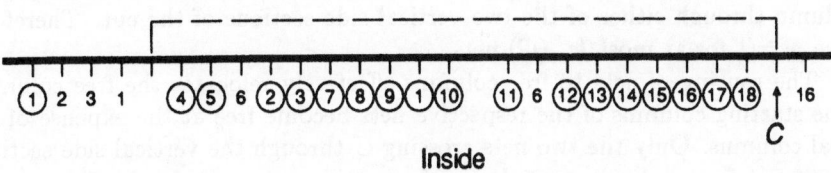

Figure 9.41: *A flux cut; the subinstance I' consists of the circled terminals.*

Rivest to provide a second parameter that is a lower bound on the channel width. This parameter is called *flux*.

The flux is related to the free capacities of certain cuts. This time, the cuts have a long horizontal segment to be able to measure the density of terminals along the sides of the channel.

Definition 9.22 (Flux Cut) *Let I be an instance of the* CHANNEL-WIDTH MINIMIZATION *problem. A cut that starts and ends at the same side of the channel, and consists of two horizontal edges at the ends at all vertical edges in the middle, is called a* flux cut *(Figure 9.41). The two sides of the cut are called the* inside *and the* outside.

The following lower bound on the minimum channel width of a problem instance depends on the capacities and densities of flux cuts.

Theorem 9.11 *Let I be an instance of the* CHANNEL-WIDTH MINIMIZATION *problem, and let C be a flux cut. Then,*

$$fc_m(C)\, w + w(w - 1) \geq d(C) \tag{9.7}$$

Proof Let I be an instance of the CHANNEL-WIDTH MINIMIZATION problem. Let C be a flux cut, and assume, without loss of generality, that C is on the bottom side of the channel. Construct a subinstance I' of I by including just two terminals of each net split by C, one on each side of C. There are $d(C)$ such nets.

Assume a routing of I' with w tracks. We think of the routes for all nets as being directed from the inside to the outside terminal. Call the column of the inside and outside terminal of a net the *start* and *goal column*, respectively. Each route consists of alternating vertical and horizontal segments. Since no net is trivial (Lemma 9.1), each net has to change columns sometime. Let p_i be the number of the track at which the route for net n_i reaches the goal column of n_i for the first time. Call a column *free* if it is on the inside of the cut and does not have a terminal. There are $fc_m(C) - 2$ free columns. Thus, only $fc_m(C)$ nets can change their columns in the first track: At most $fc_m(C) - 2$ nets can be routed to a free column, and one net can be routed to its new column through either of the two vertical side sections of the cut. Therefore, also $p_i = 1$ for at most $fc_m(C)$ nets.

The routing of nets to free columns effectively relocates the free columns. The starting columns of the respective nets become free at the expense of the goal columns. Only the two nets crossing C through the vertical side sections of the cut free up two more columns for routing nets on track 2. Thus, $p_i = 2$ can be the case for at most $fc_m(C) + 2$ nets. In general, $p_i = j$ can be the case for at most $fc_m(C) + 2(j - 1)$ nets. Since the routing uses only w tracks, we have

$$\sum_{j=1}^{w} (fc_m(C) + 2(j - 1)) \geq d(C)$$

Substituting the closed form of the sum yields the formula in the theorem. $\square$

Theorem 9.11 leads to the following definition of flux.

Definition 9.23 (Flux) *The* **flux** $f(I)$ *of an instance* I *of the* CHANNEL-WIDTH MINIMIZATION *problem is defined as the minimum value of* w *such that inequality (9.7) holds for all flux cuts.*

Obviously, the flux is a lower bound on the minimum channel width. (We can even tighten this lower bound a little more, see Exercise 9.24 or [254].) Computing the flux entails finding the largest root of the polynomial

$$fc_m(C)(w + 1) + w(w - 1) - c_m(C)$$

with variable w for each flux cut, and maximizing over the obtained values. We can restrict the maximization to flux cuts whose rightmost and leftmost inside column are occupied. This observation leads to an efficient algorithm for computing the flux (see Exercise 9.25). The flux can never exceed $\lceil \sqrt{|N|} \rceil$. We can see this fact by inspecting inequality (9.7) and noting that $d(C) \leq |N|$ and $fc_m(C) \geq 2$. There are problem instances whose flux is $\lceil \sqrt{n} \rceil$—for instance, the shift-right-1 problem (part 2 of Exercise 9.23). However, the flux is not always a tight lower bound, even if it exceeds the density (part 3 of Exercise 9.23).

The definition of flux given here differs from that in the literature [18, 26, 131]. There, the flux $f'(I)$ is defined to be the maximum integer f such that there is a flux cut C with capacity $c_m(C) = 2f^2 + 2$ and density $d(C) \geq 2f^2 - f$. Our definition of flux provides a tighter lower bound by a constant factor (see Exercise 9.26).

The discovery of flux as a lower bound on the channel width is quite a recent development. The flux has evaded discovery for a long time, because in practice channel-routing problems have small flux. (For instance, we avoid the large flux of the shift-right-1 problem (Figure 9.40) by shifting one side of the channel one unit over to the left.) Thus, in practice people try to route *in density*, and they come fairly close to achieving this goal.

9.6.1.2 Application of the General Channel-Routing Method

The general scheme for nonplanar channel routing from Section 9.4 can be applied to the Manhattan model such that the routing uses $d_c(I) + O(f(I))$ tracks on instances involving only two-terminal nets. We shall now discuss how this is done. For the purpose of simplicity, we consider only instances with two-terminal nets and without one-sided nets. One-sided nets and multiterminal nets can be incorporated as discussed in Section 9.4.

When applied to Manhattan channel routing, the general channel-routing method may require up to three free terminal sites on each side of the channel inside a block. In Figure 9.27, these empty terminal sites are marked with a × symbol. To provide these empty terminal sites, we have to distribute free columns evenly throughout the channel. This distribution is done in a pre-processing phase that actually generates one more section on either side of the channel, called the boundary sections. Figure 9.42 illustrates the decomposition of the channel. The sizes of the respective regions are given in the figure. They are stated in terms of a parameter k. k is the smallest integer for which the following holds: The channel can be broken up into *superblocks* of k^2 columns each such that each superblock contains at least $3k$ empty terminal sites on either side of the channel.

Let us discuss how large k has to be. Assume that a certain value of k does not fulfill this requirement. Then, there is a section of k^2 columns such that on at least one side of the channel there are less than $3k$ free columns. Thus, there is a flux cut C with $c_m(C) = k^2 + 2$ and $fc_m(C) \leq 3k + 1$; that is, $d(C) \geq k^2 - 3k + 1$. We want to determine a value of k such that no such cut exists. Since all flux cuts have to fulfill inequality (9.7), we need to substitute these values into the negation of inequality (9.7) and set $w = f(I)$. This yields $k = 3.302... f(I) + O(1)$ (see Exercise 9.27). We can find the minimum value of k by testing the requirement on increasing k until a suitable k is found.

The empty terminal sites may be distributed unevenly in a superblock. The preprocessing phase has the purpose of assigning three empty terminal sites on each side of the channel to each block. We achieve this goal by simply freeing up

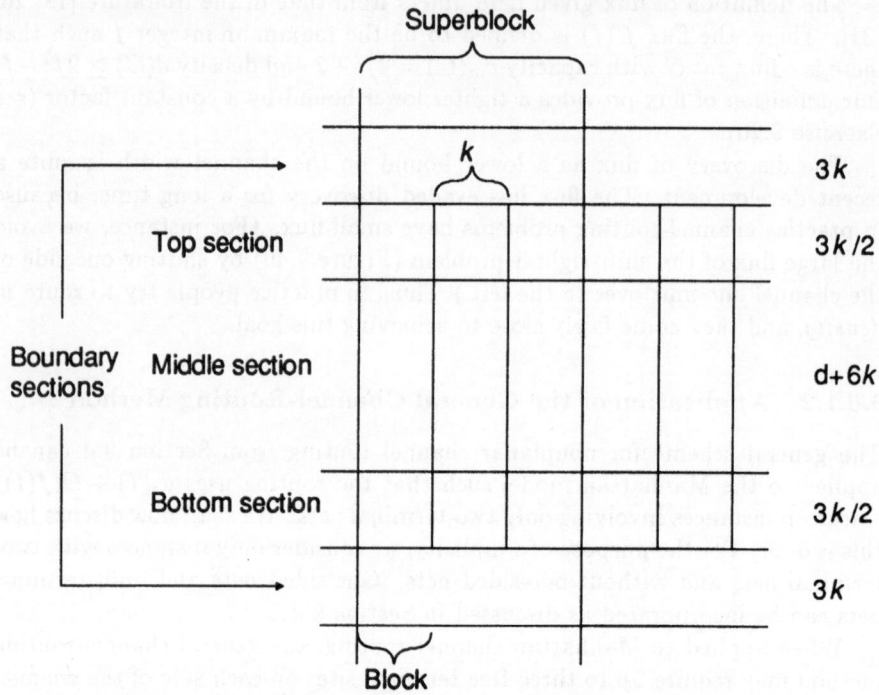

Figure 9.42: *Channel decomposition for Manhattan channel routing.*

three columns in each block by rerouting their nets to unnecessary free columns. In each superblock, only $3k$ nets are rerouted; thus, $3k$ tracks suffice in each boundary region. Furthermore, the density of the channel-routing problem that is induced on the three sections in the middle of the channel may be affected by the rerouted nets. But it can increase to only at most $d_c(I) + 6k$. Thus, the middle section can be routed in $d_c(I) + 6k$ tracks.

The routing of the middle section is illustrated in Figure 9.27. Note that the reordering of the nets inside a block does not increase the density of the respective routing problem, since ending nets are moved to the left and starting nets are moved to the right. Thus, each block can be routed in $d_c(I) + 6k$ tracks. Finally, the top and bottom sections are routed. Here, we can apply a reasonably good heuristic that uses at most $3k/2$ tracks (Exercise 9.29 [227]).

In total, the algorithm uses $d_c(I) + 15k = d_c(I) + 49.5f(I) + O(1)$ tracks. The constant factor in the flux term is still quite large, and this fact makes the method unattractive in practice. However, the method can be enhanced with heuristics to render it more efficient in practice [467].

The running time of the algorithm is a small polynomial (the exact asymptotic running time is asked for in Exercise 9.28). The algorithm can be sped up to run in linear time in the smallest area of a minimum width channel.

It has been proved by Szymanski [436] that the CHANNEL-WIDTH MINIMIZATION problem in the Manhattan model is NP-hard, even if only two-terminal nets are allowed.

9.6.1.3 Routing Based on the Jog-Free Model

Historically, the first work on Manhattan channel routing considered a restricted version of the Manhattan model, in which no wire jogs are allowed in the channel—that is, each wire could occupy only one track. This model is called the *jog-free model*. The knowledge about the jog-free model is summarized in this section. Jogs can be incorporated into jog-free algorithms to decrease the channel width of the routing. Section 9.6.1.4 discusses how this extension is done.

We again assume that there are no trivial nets. In the jog-free model, each wire has only one trunk. Thus, each net n_i is assigned a unique track τ_i in the channel, and the sequence $(\tau_1, \ldots, \tau_{|N|})$ uniquely determines the routing.

To characterize legal routings in the jog-free model, we make the following definitions.

Definition 9.24 (Constraints) *Let ℓ_i and r_i be the leftmost and rightmost terminal of net n_i, respectively. Let (ℓ_i, r_i) be the open interval of columns between the outside terminals of net n_i (excluding the columns of the outside terminals). $[l_i, r_i]$ denotes the corresponding closed interval (including the columns of the outside terminals).*

1. *There is a* **horizontal constraint** *between two nets n_i and n_j if $[\ell_i, r_i] \cap [\ell_j, r_j] \neq \emptyset$. In this case, we also say that n_i and n_j* **overlap**.

2. *There is a* **vertical constraint** *from net n_i to net n_j if two outside terminals, one on the bottom side of the channel from n_i and one on the top side of the channel from n_j, share a column. In this case, we also say that n_i is* **below** *n_j or n_j is* **above** *n_i.*

Fact 9.5 *A routing $(\tau_1, \ldots, \tau_n)$ of an instance of the CHANNEL ROUTING problem in the jog-free Manhattan model is legal exactly if the following two conditions hold:*

1. *If n_i and n_j overlap, then $\tau_i \neq \tau_j$.*

2. *If n_i is below n_j, then $\tau_i < \tau_j$.*

Using the Definition 9.24, we can define the following graphs.

Definition 9.25 (Constraint Graphs) *Let I be an instance of the* CHANNEL-WIDTH MINIMIZATION *problem in the jog-free Manhattan model.*

1. *The* **horizontal constraint graph** $G_h(I) = (V_h, E_h)$ *of I is an undirected graph such that V_u is the set of nets of I and there is an edge $\{n_i, n_j\}$ in E_h exactly if n_i and n_j overlap.*

2. *The* **vertical constraint graph** $G_v(I) = (V_v, E_v)$ *is a directed graph such that V_v is the set of nets of I and there is an edge (n_i, n_j) exactly if $n_i \neq n_j$ and n_i is below n_j. (Note that $G_v(I)$ is a directed version of a subgraph of $G_h(I)$.)*

3. *The* **total constraint graph** $G_t(I) = (V_t, E_t)$ *of I is a partially directed graph that is obtained by addition to $G_v(I)$ of all undirected edges in $G_h(I)$ that connect vertices that are not adjacent in $G_v(I)$. Informally speaking, $G_t(I)$ is results from merging $G_v(I)$ and $G_h(I)$.*

Figure 9.43 illustrates Definition 9.25. The horizontal constraint graph $G_h(I)$ has a very special structure.

Definition 9.26 (Interval Graph) *A graph G is called an* **interval graph** *if the vertices of G can be represented by nonempty intervals on the real axis such that edges exist exactly between pairs of vertices whose intervals intersect. Some of the intervals may be closed, others may be open or half-open.*

$G_h(I)$ is an interval graph. Specifically, the interval $[\ell_i, r_i]$ represents net n_i. Interval graphs have a lot of structure. Especially the *clique structure* of an interval graph is striking. Recall from Definition 3.3 that a clique of a graph G is a complete subgraph of G. The following lemma characterizes interval graphs.

Lemma 9.8

1. *For each abscissa x on the real axis, the set of vertices in an interval graph G that are represented by intervals that contain x is a clique.*

2. *For each clique C in an interval graph G, there is an abscissa x on the real axis such that all intervals representing vertices in C contain x.*

3. *G is an interval graph exactly if there is an ordering of the maximal cliques of G such that, for all vertices v in G, the cliques that contain v occur consecutively.*

Proof

1. Obvious.

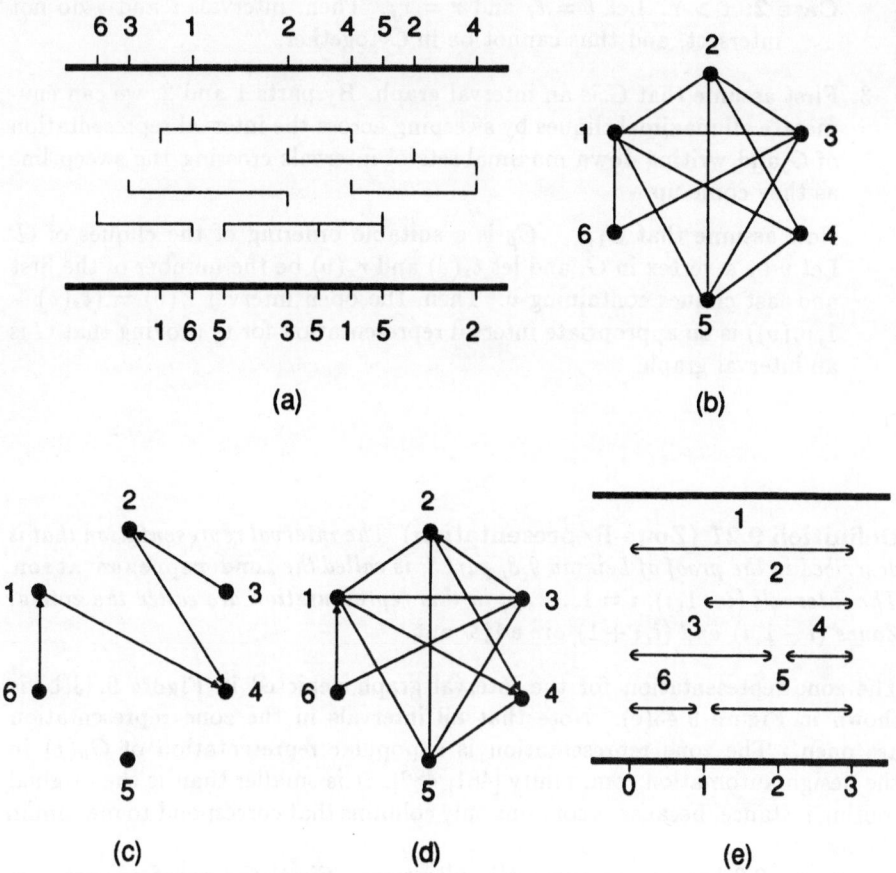

Figure 9.43: *The constraint graphs of a channel-routing problem in the jog-free model. (a) An appropriate representation of the problem instance; (b) the horizontal constraint graph; (c) the vertical constraint graph; (d) the total constraint graph; (e) the zone representation for the horizontal constraint graph.*

2. Let the intervals for the vertices in C be $[\ell_1, r_1], \ldots, [\ell_k, r_k]$. (If some of the intervals are open or half-open, we can close all intervals by reducing them at their open sides by a suitable small amount $\varepsilon > 0$. If ε is small enough, this transformation does not change the interval graph.) Let $\ell = \max\{\ell_i \,|\, i = 1, \ldots, n\}$ and $r = \min\{r_i \,|\, i = 1, \ldots, n\}$. We make a case distinction.

Case 1: $\ell \leq r$. In this case, $x = \ell$ is an appropriate choice.

Case 2: $\ell > r$. Let $\ell = \ell_i$ and $r = r_j$. Then, intervals i and j do not intersect, and thus cannot be in C together.

3. First assume that G is an interval graph. By parts 1 and 2, we can enumerate all maximal cliques by sweeping across the interval representation of G and writing down maximal sets of intervals crossing the sweep line as they come up.

   Now assume that $C_1, \ldots, C_p$ is a suitable ordering of the cliques of G. Let v be a vertex in G, and let $\ell_z(v)$ and $r_z(v)$ be the number of the first and last cliques containing v. Then, the open interval $Z(v) = (\ell_z(v) - 1, r_z(v))$ is an appropriate interval representation for v, proving that G is an interval graph.

□

Definition 9.27 (Zone·Representation) *The interval representation that is described in the proof of Lemma 9.8, part 3, is called the* **zone representation**. *The intervals* $(i-1, i)$, $i = 1, \ldots, p$, *in this representation are called the* **zones**. *Zones* $(i-1, i)$ *and* $(i, i+1)$ *are* **adjacent**.

The zone representation for the interval graph depicted in Figure 9.43(b) is shown in Figure 9.43(e). Note that all intervals in the zone representation are open. The zone representation is a popular representation of $G_h(I)$ in the design-automation community [481, 482]. It is smaller than is the original routing instance, because it contains only columns that correspond to maximum cliques.

Lemma 9.8 has two important implications. First, the proof of part 3 of Lemma 9.8 also provides a linear-time algorithm for finding a maximum clique in an interval graph, given by its interval representation. (This problem is NP-hard on general graphs.) This fact exemplifies the phenomenon that many problems that are difficult on general graphs can be solved efficiently on interval graphs. Second, the characterization of the cliques of an interval graph given by parts 1 and 2 of Lemma 9.8 proves that the size of the maximum clique in $G_h(I)$ is the density $d_c(I)$.

The constraint graphs can be used to characterize channel-routing problems in the jog-free Manhattan model. By part 1 of Fact 9.5, an assignment of tracks to nets that represents a legal routing is, in particular, a coloring of $G_h(I)$ (see Definition 3.3). The colors are track numbers and the vertices represent nets.

Can we extend the coloring paradigm to include the requirements of part 2 of Fact 9.5? Indeed we can. To this end, we define a *rainbow coloring* of a partially directed graph to be an assignment of colors to the vertices. Here, we assume colors to be ordered (according to their wavelength; thus, the name *rainbow coloring*). We require vertices that are joined by an undirected edge to have different colors. For each directed edge, the head of the edge has to receive a *smaller* color than does the tail. With this definition, minimizing the channel width of an instance I of the CHANNEL-WIDTH MINIMIZATION problem is tantamount to finding a rainbow coloring of $G_t(I)$ with as few colors as possible. This important observation allows us to connect results on jog-free Manhattan channel routing with other application areas, such as scheduling and decision theory.

Definition 9.28 (Rainbow Coloring) *A partially directed graph G is called a* **partially directed interval graph** *if the underlying undirected graph G_h—that is, the graph that is obtained when the directions on all edges are disregarded—is an interval graph. The subgraph of G consisting of all vertices and only the directed edges is denoted with G_v. The* RAINBOW COLORING *problem is, given an partially directed interval graph, to construct a rainbow coloring with the smallest number of colors. This number is called the* **chromatic number** *of G.*

Lemma 9.9 *Each instance G of the* RAINBOW COLORING *problem can be transformed into an instance I of the* CHANNEL-WIDTH MINIMIZATION *problem in the jog-free Manhattan model.*

Proof Let G be a partially directed interval graph that is an instance of the RAINBOW COLORING problem. We can construct the corresponding instance I of the CHANNEL-WIDTH MINIMIZATION problem by transforming the zone representation of G as follows:

1. Add columns to the channel and modify the interval lengths such that the interval graph G_h does not change, but all intervals are closed and no two interval boundaries share a coordinate on the x axis (see also part 2 of the proof of Lemma 9.8).

2. Arbitrarily assign terminals to the boundaries of the closed intervals. This step converts the zone representation into an instance of the CHANNEL-WIDTH MINIMIZATION problem with two-terminal nets in the jog-free Manhattan model.

3. For each edge $e = (v, w)$ in G_v, add a column to the channel that intersects both the interval for v and w, and place a terminal of v on the bottom and a terminal of w on the top side of the channel in this column. This procedure generates the desired instance I of the CHANNEL-WIDTH MINIMIZATION problem with multiterminal nets in the jog-free Manhattan model.

□

Partially directed interval graphs are an important class of graphs even in other application areas, such as decision theory and resource allocation. Because of Lemma 9.9, results on channel routing in the jog-free Manhattan model can be translated directly to these application areas, and vice versa.

We shall now discuss the solution of the CHANNEL-WIDTH MINIMIZATION problem in the jog-free Manhattan model. Because of the equivalence of this problem with the RAINBOW COLORING problem on partially directed interval graphs, we shall use notation from both problem areas. Thus, we shall sometimes refer to the assignment of a track to an interval as *coloring* the interval. Similarly, we shall use the notions *vertex* and *interval* interchangeably with those of *track* and *color*.

Theorem 9.12 *Let I be an instance of the* CHANNEL-WIDTH MINIMIZATION *problem in the jog-free Manhattan model.*

1. *The* CHANNEL-WIDTH MINIMIZATION *problem is solvable for I exactly if $G_v(I)$ is a dag. This condition can be tested in time $O(|N|)$. If a legal routing exists, then*

$$w_{\min}(I) \geq \max\{d_c(I), \operatorname{depth}(G_v(I)) + 1\}$$

   where $\operatorname{depth}(G_v(I))$ is the depth of $G_v(I)$ (see Definition 3.5).

2. *If $E_v = \emptyset$, then $w_{\min}(I) = d_c(I)$. A legal routing minimizing the number of tracks can be found in time $O(|N|)$.*

3. *There is an $O(|N| \log |N|)$–time algorithm that constructs a legal routing if one exists. It does not use a minimum number of tracks, however.*

Proof

1. The solvability criterion is obvious. The lower bound on $w_{\min}(I)$ follows from part 2 of Lemma 9.8. Testing for circuits in a directed graph can be done in linear time with the strong components algorithm (Section 3.4).

2. We present the *left-edge algorithm* by Hashimoto and Stevens [176]. We base the description on the zone representation of the horizontal constraint graph.

   The algorithm tries to pack tracks as tightly as possible. The algorithm processes one track after the other. For each track, it scans the channel from left to right and colors any possible interval that it encounters and that is not colored yet. Specifically, the uncolored interval (ℓ_i, r_i) with the smallest ℓ_i is colored first. The uncolored interval (ℓ_j, r_j) with the smallest ℓ_j such that $\ell_j \geq r_i$ is colored next, and so on.

   To prove that the algorithm uses a minimum number of colors, we consider the leftmost interval (ℓ_k, r_k) that is assigned the highest color $w_{\max}$. For each color $w \leq w_{\max}$, there must be an interval (l_j, r_j) that is colored with w such that $\ell_j \leq \ell_k < r_j$. Otherwise, let w' be the smallest color for which this is not the case. Then, the interval (ℓ_k, r_k) should have been colored with w'. Therefore, the abscissa $\ell_k + 1/2$ witnesses to a density of at least $w_{\max}$. This observation implies that $w_{\max} \leq d_c(I) \leq w_{\min}(I)$, where the last inequality follows from part 1.

Let us now discuss how to implement this algorithm efficiently. We use a balanced search tree to store the intervals, sorted by their left boundary [314, 411]. Such a tree supports the operations *insert*, *delete*, and *find the smallest element larger than a given value* in $O(\log |N|)$ time per operation. Thus, the overall algorithm has run time $O(|N| \log |N|)$.

Exercise 9.30 discusses a linear-time implementation of the left-edge algorithm; see also [158]. The two implementations are amenable to incorporating different heuristics dealing with vertical constraints.

3. It is not difficult to extend the left-edge algorithm to incorporate vertical constraints. We have only to make sure that, when processing color w, we color only intervals from the set of *sources* of the subgraph $G_{v,w}$ of $G_v(I)$ that is induced by the intervals that have not received a color smaller than w. Since $G_v(I)$ is a dag, this set is always nonempty.

With an appropriate implementation, the run time of this algorithm is again $O(|N| \log |N|)$.

□

On bad examples, the left-edge algorithm can use many more colors than is necessary. La Paugh [252] proved that at most $O(w_{\min}(I)\sqrt{n})$ colors are used, and that there are examples on which the left-edge algorithm uses that many colors (see Exercise 9.31). The same thesis also shows that the CHANNEL-WIDTH MINIMIZATION problem in the jog-free Manhattan model is NP-hard. Kernighan et al. [230] present a branch-and-bound algorithm for solving this problem. The worst-case run time of the algorithm is exponential. It is an open problem to find efficient approximation algorithms for this problem that come closer to the minimum number of colors (see Exercise 9.32). For the purposes of channel routing, the practical relevance of this problem is questionable, because we should use jogs anyway to reduce the channel width. But, in light of the fact that this problem has many applications in other areas, this question is quite interesting.

Let us now discuss heuristic ways of improving the left-edge algorithm.

1. First, we can, of course, assign color w to the interval starting from the right, instead of from the left. This choice may be advantageous for some problem instances. For each color, we can also choose whether to color sources or sinks of $G_v(I)$. Since both choices can be made independently for each column, quite a variety of combinations is possible here. We can try a few, and select the best result [88] (see also part 4 of Exercise 9.30).

2. It is a good idea to construct the zone representation at the beginning and to do all other processing on that representation. We have already explained the left-edge algorithm as it operates on the zone representation. The zone representation can be found in time $O(|N|)$ using the same data structures as in Exercise 9.30. In the zone representation, several intervals have the same left-corner coordinates. This fact introduces nondeterminism into the left-edge algorithm, which can be used to decrease the number of colors. Specifically, assume that at some time one of the intervals $Z(n_{i_1}), \ldots, Z(n_{i_k})$ can be colored. Let d_j, $j = 1, \ldots, k$ be the length of the longest path in the graph $G_v(I)$ that starts at n_{i_j}. Then, at least $\max\{d_j \mid j = 1, \ldots, k\}$ more colors have to be used. Thus, it is preferable to color the interval $Z(n_{i_j})$ that maximizes d_j, in order to shorten long paths in $G_v(I)$.

Shortening paths in $G_v(I)$ could even take priority over the goal of packing tracks in the channel tightly, which is pursued by the left-edge algorithm. When modified in this way, the routing algorithm always chooses the uncolored interval $Z(n_j)$ that can be colored with the present color and that maximizes d_j. Again, we can resolve nondeterminism by choosing the leftmost interval that is eligible.

Rather than having one of the two goals take priority over the other, we can merge them, and possibly include other cost functions. Yoshimura [481] presents a general way of doing this generalization. He proposes that we assign a weight $w(n_j)$ to each uncolored net $n_j \in N$. This weight can be a function of such parameters as d_j, the length of $Z(n_j)$, and the degree of n_j in $G_v(I)$. Then, the algorithm processes a color w by choosing among the set of sources in $G_{v,w}$ a set N' of nonoverlapping intervals that maximizes $w(N') = \sum_{n_j \in N'} w(n_j)$. We can find N' by solving a single-source shortest-path problem on a dag G_z that is derived from the zone representation of I. G_z contains a vertex v_i for each integer abscissa $i = 0, \ldots, p$ in the zone representation. There are directed edges (v_i, v_{i+1}), with weight 0, for all i. Furthermore, the interval $Z(n_j) = (\ell_z(n_j), r_z(n_j))$ for the uncolored net n_j is translated into an edge $(v_{\ell_z(n_j)}, v_{r_z(n_j)})$ with a weight of $-w(n_j)$. Clearly, a shortest path through G_z starting at v_0 traverses the edges corresponding to a suitable set N'.

Determination of N' can be done in time $O(p)$, where p is the number of zones (see Definition 9.27). Thus, this modification of the left-edge algorithm runs in time $O(wp + |N| \log d)$, where w is the number of colors used. This amount is at most $O(|N|^2)$. For experimental results on the effect of different choices for the weights, we refer the reader to [481].

3. Another heuristic is based on the *merging* of nets. Merging two nets means deciding that they share a track. At the time of merging, it is not decided which track this is. In terms of operations on the constraint

graphs, nets n_i and n_j can be merged if there are no constraints between them. Merging amounts to

a. Substituting the intervals $Z(n_i)$ and $Z(n_j)$ in the zone representation with the smallest interval containing $Z(n_i) \cup Z(n_j)$

b. Merging the vertices for n_i and n_j in $G_v(I)$

The merge cannot introduce a circuit into $G_v(I)$, but it may increase the depth of $G_v(I)$. Since the depth of $G_v(I)$ is a lower bound on w_{min}, we want it to increase as little as possible. Merging nets is continued until no more merges are possible. Then, the normal left-edge algorithm is used.

Yoshimura and Kuh [482] discuss heuristics based on merging nets. Experiments show that the channel widths achieved by the heuristics of Yoshimura [481] and Yoshimura and Kuh [482] are comparable; both of them route almost in density in most cases. But the former heuristic is quite a bit faster than is the latter one.

9.6.1.4 Routing with Jogs and Other Generalizations

The left-edge algorithm is quite amenable to incorporating jogs for a net n_i in columns in which n_i has a terminal [88]. We can do this jog introduction by decomposing net n_i into subnets, each spanning the region between two neighboring terminal columns of n_i. Formally, the subnets of n_i are represented by *open* intervals (in the horizontal constraint graph) between the respective terminal columns. The intervals must be open to avoid horizontal constraints between subnets of the same net. Then, vertical constraints are added appropriately. Breaking up nets into subnets can break circuits in $G_v(I)$ (see Figure 9.44). Introducing jogs only in terminal columns both reduces the size of the configuration space and ensures that no additional contacts are used. But there are examples of channel-routing problems in which adding jogs in terminal columns is not enough (see Figure 9.45). In this case, we have to place a jog in a column that does not contain terminals. Only heuristics for choosing such columns are known today [259, 352, 376]. After deciding on a column, we split the net as discussed previously. Sometimes, such as in Figure 9.45, there is no suitable column such that a new column has to be created on one of the channel sides.

The left-edge algorithm can also be extended to handle irregular channels. To this end, we model straight segments of the sides of the channel as one-sided two-terminal nets. (A similar idea has been discussed in Section 9.3.1.4.) We enforce the accurate placement of the different segments on a channel side with respect to each other by adding vertical constraints between them. The corresponding edges in $G_v(I)$ are given a length that is the vertical distance of the corresponding segments. All other edges have unit length. Although the left-edge algorithm may not place the sections of the channel sides correctly at first, the placement can always be modified such that it is correct.

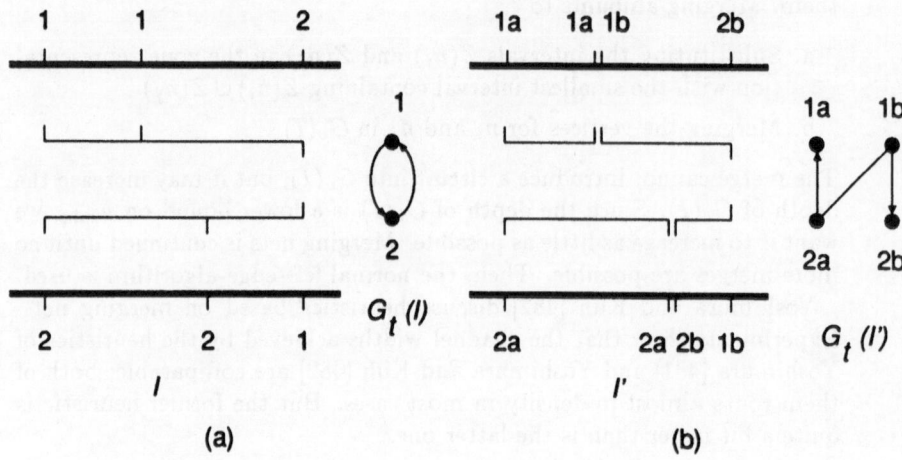

Figure 9.44: *Breaking a circuit of vertical constraints. (a) Original problem instance I, $G_t(I)$ has a circuit; (b) modified problem instance (I'), $G_t(I')$ has no circuit.*

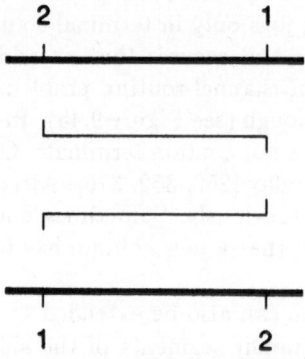

Figure 9.45: *A problem with a vertical constraint circuit that cannot be broken by introducing jogs in terminal columns.*

The channel router YACR2 [376] contains one of the most refined sets of heuristics based on the left-edge algorithm. The main heuristic ideas employed in YACR2 are these:

1. Reserve $d_c(I)$ tracks at first. Create new tracks later if you need them.

2. Use the sweep technique discussed in Exercise 9.30 for implementing the left-edge algorithm. Start the sweep at a column of maximum density and extend it first to one side and then to the other (see part 4 of Exercise 9.30).

3. Use a judicious scheme of selecting the track for each net, as it is colored. The goals here are to minimize vertical constraint violations, and to achieve a routing in density. However, vertical constraint violations are allowed.

4. Try to eliminate vertical constraint violations with maze-running techniques. Only if this attempt fails, add a new track. During this phase, YACR2 allows horizontal jogs on the vertical layer. Thus, it is not a Manhattan router, but rather uses a two-layer model. This means that the density $d_c(I)$ is no longer a lower bound on the channel width! However, by design, YACR2 never uses less than $d_c(I)$ tracks.

In summary, the left-edge algorithm is a powerful method that is the basis of a large number of successful heuristics in Manhattan channel routing. In addition, the relationship between the jog-free Manhattan model and partially directed interval graphs makes routing algorithms in this model also applicable to other problem areas, such as resource allocation and decision theory.

9.6.2* Manhattan Routing in Other Detailed-Routing Graphs

The SWITCHBOX ROUTING problem and the DETAILED ROUTING problem on general detailed-routing graphs are both extensions of the CHANNEL ROUTING problem, and thus they are NP-hard in the Manhattan model. Heuristic switchbox routers are presented in [71, 167, 213, 298]. These routers incorporate features that go beyond the Manhattan model, such as nontrivial layer assignment of multiple layers.

9.7 Channel-Offset and Channel-Placement Problems in Nonplanar Routing Models

In nonplanar routing models (the Manhattan and the knock-knee model), channel-offset and channel-placement problems are more difficult to solve than in the planar routing model.

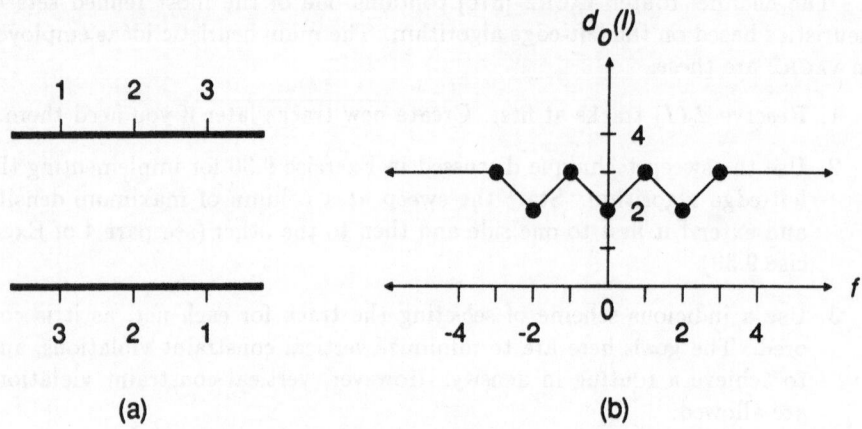

Figure 9.46: *Channel offset in nonplanar routing models. (a) A problem instance; (b) the open density as a function of the offset. The closed density and the minimum channel width in the Manhattan model, as well as in the jog-free model, are all the same as the open density for this problem instance.*

9.7.1   Channel-Offset Problems

In the knock-knee model, the open density is a tight lower bound on the channel width if only two-terminal nets are involved. Therefore, solving the OPTIMUM WIDTH CHANNEL OFFSET problem amounts to choosing an offset that minimizes the open density.

It follows from inequality (9.4) that, in the planar routing model, the minimum channel width as a function of the offset decreases to the left of an offset f, yielding minimum channel width, and increases to the right of f. This *bitonicity* of the minimum channel width in the planar routing model is critical for the development of efficient algorithms for many problem in the area of channel routing as well as in that of compaction (see Section 10.5). Unfortunately, this property is not shared by the open density of a nonplanar instance of the MINIMUM WIDTH CHANNEL OFFSET problem. Figure 9.46 depicts a problem instance for which the open density has several local minima and maxima. As a consequence, no linear-time algorithms for solving the MINIMUM WIDTH CHANNEL OFFSET or even the CHANNEL OFFSET RANGE problem in the Manhattan or knock-knee model are known as of today. The simplest idea for solving the OPTIMUM WIDTH CHANNEL OFFSET problem is to go through all relevant choices of the offset. Such an algorithm can be made to run in time $O(|N|b)$, where b is the sum of the differences between the column numbers of the leftmost and rightmost terminals on either channel side (Exercise 9.33). La Paugh and Pinter [253] present the idea of an algorithm that runs in time $O(|N|^2 \log |N|)$. This run time is preferable to $O(|N|b)$ if $b \geq |N| \log |N|$.

In the Manhattan model, the CHANNEL ROUTING problem is NP-hard even for two-terminal nets. But the closed density is a good lower bound in most practical cases. Thus, it makes sense to approximate a solution of the OPTIMUM WIDTH CHANNEL OFFSET problem by choosing the offset that minimizes the closed density. Both the closed density and the minimum channel width in the Manhattan model as functions of the offset can oscillate. Indeed, since in Figure 9.46 for all offsets there are minimum-width routings without knock-knees, the open density, the closed density, and the minimum channel width in the Manhattan model, as well as in the jog-free model, are all the same function in this case.

The algorithms for minimizing the open density that we have discussed here can be modified to minimize the closed density.

9.7.2 Channel-Placement Problems

Gopal et al. [152] consider channel-placement problems in the special case that each block contains just one terminal. In this case, the OPTIMUM WIDTH CHANNEL PLACEMENT problem can be solved in time $O(|N|^2)$. Note that, since each block has only one terminal and the spread of the channel is not an issue, vertical constraints can always be avoided. Thus, the OPTIMUM WIDTH CHANNEL PLACEMENT problem both in the knock-knee and in the Manhattan model amounts to finding x coordinates for the corners of the intervals in $G_h(I)$ such that

1. No vertical constraints exist

2. The ordering of the terminals on the top and bottom side of the channel is not changed

3. The (closed or open) density is minimized

Gopal et al. scan the channel from left to right and solve this problem using dynamic programming. Define $S_{i,j}$ to be the set of the leftmost i top and j bottom terminals in the channel. Define $N_{i,j}$ to be the set of nets that have terminals in $S_{i,j}$. Let $w(i,j)$ be the minimum channel width of all placements of the restriction of the problem instance to the nets in $N_{i,j}$. Furthermore, let $d(i,j)$ be the number of nets that have a terminal both in $S_{i,j}$ and its complement. $w(i,j)$ obeys the following recurrence:

$$
\begin{aligned}
w(0,0) &= 0 \\
w(0,j) &= j, \; j = 1, \ldots, \ell_b \\
w(i,0) &= i, \; j = i, \ldots, \ell_t \\
w(i,j) &= \max\{d(i,j), \min\{w(i-1,j), w(i,j-1), w(i-1,j-1)\delta(i,j)\}\}
\end{aligned}
$$

Here, $\delta(i,j) = 1$ exactly if the ith top and jth bottom terminal from the left belong to the same net; otherwise, $\delta(i,j) = \infty$. The first two terms in the minimum inside the recurrence cover the cases that the rightmost terminal in an optimal placement is on the bottom and top of the channel, respectively. The third term covers the case that the rightmost terminals are in the same column. The incorporation of $\delta(i,j)$ ensures that vertical constraints are eliminated. $w(\ell_t, \ell_b)$ is the minimum channel width of any placement. A corresponding placement can be recovered using the methods discussed in Section 4.7 (see Exercise 9.34).

In the Manhattan model, the algorithm does not generalize to blocks with several terminals, because then vertical constraints cannot be eliminated. However, in the knock-knee model with restriction to two-terminal nets only, the algorithm does generalize. The reason is that, in this model, we can allow terminals of different nets to be in the same column and, as long as only two terminals are involved, open density is a tight lower bound on channel width (see Exercise 9.35).

The OPTIMUM AREA CHANNEL PLACEMENT and the OPTIMUM SPREAD CHANNEL PLACEMENT problem are NP-hard in the Manhattan model, even if each block has only one terminal [152]. The complexity of these problems for the knock-knee model is open.

9.8* Detailed Routing Without Homotopies

There are, in fact, routing graphs with nontrivial homotopies on which the DETAILED ROUTING problem can be solved in polynomial time without taking the homotopies into account. In other words, no global routes are given in the problem instance. Since the homotopy is nontrivial, the routing algorithm has to choose a global route *and* detail that route's course. The problem therefore has the flavor of a global-routing problem. Results in this area apply to both the Manhattan and the knock-knee model.

The first family of detailed-routing graphs that was studied in this respect is the *ring* depicted in Figure 9.47. Terminals are allowed at the boundary of only the inside hole. La Paugh [252] develops an $O(|N|^3)$ algorithm that minimizes the size of the routing graph while preserving routability. (The size can be varied by the width of the ring on each of the four sides being changed.) This algorithm deals with only two-terminal nets. Gonzalez and Lee [150] improve the running time to $O(|N| \log |N|)$. Sarrafzadeh and Preparata [392] present a simple $O(|N| \log |N|)$–time algorithm for this problem. The solution in the Manhattan model is the same as the solution in the knock-knee model. This problem is not very interesting from a practical point of view. It gains its significance from the desire to find the hardest problems that are still in P for the Manhattan model.

Kaufmann and Klär [221] extend these results to the case that terminals

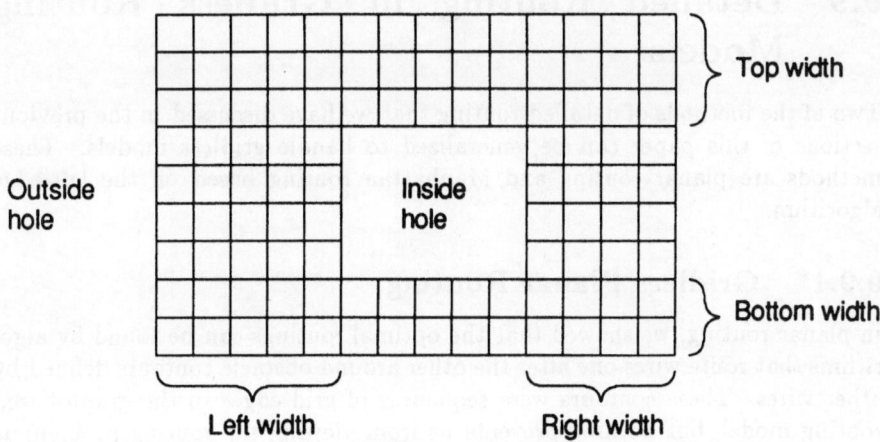

Figure 9.47: *A ring graph.*

are also allowed on the exterior hole, and the knock-knee model is used. As a generalization of the CHANNEL ROUTING problem, this problem is NP-hard in the Manhattan model. This problem occurs in wiring circuitry to peripheral pads.

Baker [17] presents an $O(|N|\log|N|)$-time approximation algorithm for a special version of an NP-hard two-cell routing problem in the Manhattan model. The algorithm optimizes the perimeter of the wired layout and comes to within a factor of 19/10 of optimal. Sarrafzadeh and Preparata [392] solve a variation of this problem in the knock-knee model. Their algorithm takes time $O(|N|\log|N|)$ and minimizes the dimensions of the wired layout separately with certain priorities.

9.9   Detailed Routing in Gridless Routing Models

Two of the methods of detailed routing that we have discussed in the previous sections of this paper can be generalized to handle gridless models. These methods are planar routing and Manhattan routing based on the left-edge algorithm.

9.9.1*   Gridless Planar Routing

In planar routing, we showed that the optimal routings can be found by algorithms that route wires one after the other around obstacle contours defined by other wires. These contours were sequences of grid edges in the graph-based routing model, but nothing prevents us from viewing the routing problem as a purely geometric task and defining the obstacle contours as curves in the plane. The constraint that different wires be vertex-disjoint translates into a minimum-distance requirement. The unit circle of metric used as a distance measure defines the shape of the contours. (The unit circle is the set of all points at unit distance from the origin.) The graph-based model is obtained if a minimum distance of 1 is required in the Manhattan (L_1-) metric. Figure 9.48(a) shows a routing that obeys the Euclidean (L_2-) metric. Several fabrication technologies prohibit the use of circular arcs in wires but allow, say, 45° angles. In this case, another "polygonal" metric is in order whose unit circle is a regular octagon (see Figure 9.48(b)). Inequalities (9.1) and (9.2) in the routability criterion must be adapted to the metric used. Let us first assume that the same distance has to be maintained between any pair of different wires (uniform distance requirement). Then, the inequalities take the following form.

For the 45° norm:

$$b_{j+r} \geq t_j + r\sqrt{2} - w, \ w/\sqrt{2} \leq r \leq w, \ j = 1, \ldots, |N| - r \quad \text{and}$$
$$t_{j+r} \geq b_j + r\sqrt{2} - w, \ w/\sqrt{2} \leq r \leq w, \ j = 1, \ldots, |N| - r$$

Here w is the geometric width of the channel.

For the Euclidean norm:

$$b_{j+r} \geq t_j + \sqrt{r^2 - w^2}, \ w \leq r \leq n, \ j = 1, \ldots, |N| - r \quad \text{and}$$
$$t_{j+r} \geq b_j + \sqrt{r^2 - w^2}, \ w \leq r \leq n, \ j = 1, \ldots, |N| - r$$

All parts of Section 9.3.1 can be generalized to the gridless model with the Euclidean metric or metrics with polygons as unit circles. The corresponding algorithms become more complex.

The greedy routing algorithm that follows contours may not minimize the lengths of wires if metrics other than the Manhattan metric are used. In the

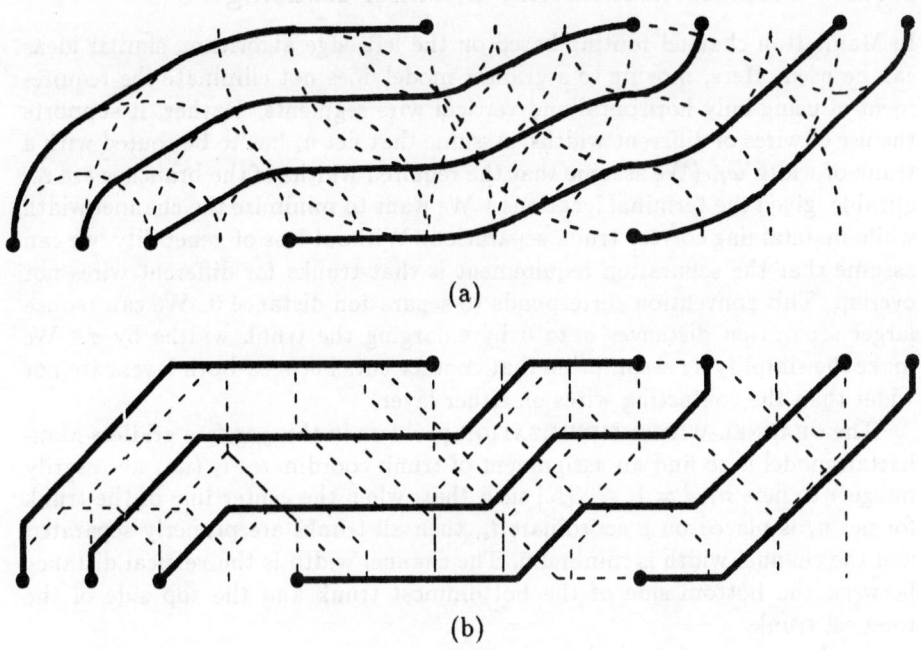

Figure 9.48: *Routing in the continuous plane. (a) A minimum-width routing with circular arcs (Euclidean distance metric); (b) a minimum-width routing with 45° angles. The obstacle contours are indicated in the figure. Both routings minimize the length of all wires simultaneously.*

case of the Euclidean norm the simultaneous minimization of all wire lengths is still possible [447]. We can gain an intuition for why this is true by thinking of each wire as being a rubberband that runs between the contours defined by the minimum distances to all other wires. If these contours do not leave a corridor from one terminal to the other for a wire, then the routing problem is not solvable. However, if such a corridor exists for all wires, then the corridors are disjoint and we can minimize the length of each wire independently by stretching the wire as much as possible inside its corridor. This intuition also guides an extension of the results in Section 9.3.2 to gridless models [130, 305]. In fact, this intuition supports the idea that even nonuniform minimum distances between different pairs of wires can be supported. This feature is important when wires of different widths have to be routed in a planar manner. The details of these generalizations are involved, however. We refer the reader to the original references [130, 305].

9.9.2   Gridless Manhattan Channel Routing

In Manhattan channel routing based on the left-edge algorithm, similar ideas can be used. Here, moving to a gridless model does not eliminate the requirement of using only horizontal and vertical wire segments. Rather, it supports the use of wires of different widths. Assume that net n_i has to be routed with a trunk of width w_i. (We assume that the required widths of the branches are realizable, given the terminal locations.) We want to minimize the channel width while maintaining correct trunk separation. Without loss of generality, we can assume that the separation requirement is that trunks for different wires not overlap. This convention corresponds to separation distance 0. We can reduce larger separation distances σ to 0 by enlarging the trunk widths by σ. We make the simplifying assumption that contact cuts between both layers are not wider than the contacting wires on either layer.

The CHANNEL-WIDTH MINIMIZATION problem in this jog-free gridless Manhattan model is to find an assignment of trunk coordinates t_i (not necessarily integer) to nets n_i, $i = 1, \ldots, |N|$ such that, when the center line of the trunk for net n_i is placed on y coordinate t_i, then all trunks are properly separated and the channel width is minimum. The channel width is the vertical distance between the bottom side of the bottommost trunk and the top side of the topmost trunk.

The key to extending the left-edge algorithm to solving the CHANNEL-WIDTH MINIMIZATION problem in the jog-free gridless Manhattan model is a weighted version of the definition of closed density, and the introduction of an edge labeling of the vertical constraint graph. The weighted version of $d_c(I)$ is defined as follows (compare with Definition 9.21):

$$d_{c,gl}(I) := \max_x \sum_{n_i \in N_x} w(n_i)$$

The edge weighting of $G_t(I)$ is defined as follows. The edge between nets n_i and n_j in $G_t(I)$ receives a weight that amounts to the minimum distance between the center lines of the trunk for both nets. This weight is

$$w(n_i, n_j) := \frac{1}{2}\left(w(n_i) + w(n_j)\right)$$

In addition, we must account for the bottom half-widths of the trunks on the bottommost track and the top half-widths of the trunks on the topmost tracks. To this end, a source s and a sink t are adjoined to $G_t(I)$ and, for all i, s and t are connected to net n_i with a directed edge of length $w(n_i)/2$. All edges are directed away from the source and toward the sink. Figure 9.49 shows an example. With these modifications, part 1 of Theorem 9.12 generalizes directly to the jog-free gridless Manhattan model. The left-edge algorithm also generalizes, even in the presence of vertical constraints. But the algorithm does not necessarily find a routing *with minimum channel width*, even if no

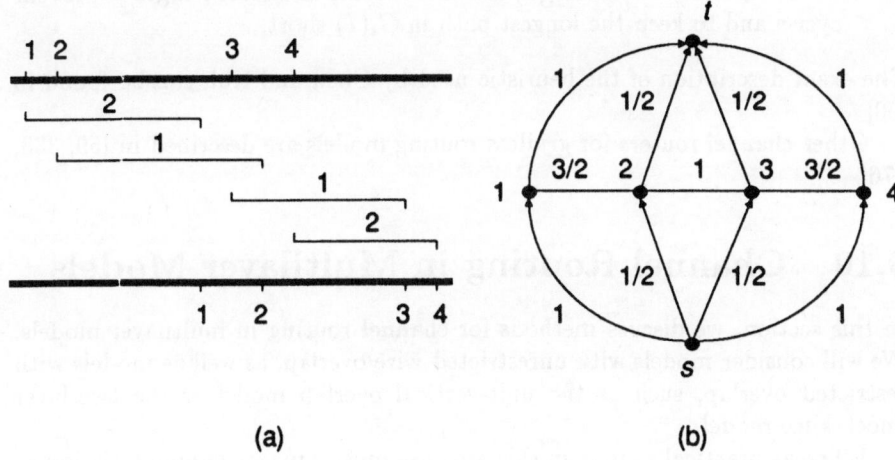

Figure 9.49: *Channel routing in the gridless Manhattan model. (a)
The problem instance I, with nets labeled with their wire widths; (b)
the total constraint graph $G_t(I)$.*

vertical constraints are present. The CHANNEL-WIDTH MINIMIZATION problem
in the jog-free gridless Manhattan model is the same as a special case of the
1D-COMPACTION problem that is NP-hard (see Section 10.1.2 [94]). Therefore,
$d_{c,gl}(I)$ is not necessarily equal to the minimum channel width, even if no
vertical constraints are present (see Exercise 9.36).

Chen and Kuh [60] propose a heuristic algorithm for solving the CHANNEL-
WIDTH MINIMIZATION problem in the jog-free gridless Manhattan model. The
algorithm alternates between assigning trunk coordinates to nets and assigning
directions to undirected edges in $G_t(I)$. It incorporates the following ideas:

1. Work from both sides of the channel inward. In other words, the nets eligible for assignment of a trunk coordinate are those that have a minimum
   distance from the source of $G_t(I)$ (bottom side) and sink of $G_t(I)$ (top
   side). The trunk coordinates are assigned relative to the bottom side of
   the channel in the first and to the top side of the channel in the second
   case.

2. At the bottom side, give priority to the nets that have a large distance to
   the sink of $G_t(I)$. Analogously, at the top side, give priority to the nets
   that have a large distance to the source of $G_t(I)$.

3. As nets are assigned trunk coordinates, formerly undirected edges in
   $G_t(I)$ become directed. This process changes the path structure in $G_t(I)$.

Thus, regularly inspect $G_t(I)$ and direct some undirected edges to prevent cycles and to keep the longest path in $G_t(I)$ short.

The exact description of the heuristic used by Chen and Kuh can be found in [60]

Other channel routers for gridless routing models are described in [59, 338, 376].

9.10   Channel Routing in Multilayer Models

In this section, we discuss methods for channel routing in multilayer models. We will consider models with unrestricted wire overlap, as well as models with restricted overlap, such as the unit-vertical overlap model or the two-layer knock-knee model.

Whereas practical routers in this area use mostly maze-running techniques, in recent years there has been a number of theoretical investigations aimed at minimizing the channel width in the presence of several layers.

9.10.1   Derivatives of the Knock-Knee Model

9.10.1.1   The Two-Layer Knock-Knee Model

The general channel-routing method for nonplanar detailed-routing models that is discussed in Section 9.4 can also be applied to overlap models. In the two-layer knock-knee model, we can use the channel-routing algorithm described in Section 9.5.1. Then, we add an auxiliary track between any two tracks in the channel. The auxiliary tracks are used to place contacts for switching layers. We can construct the layer assignment by processing the tracks in the channel bottom up. Each track is processed from left to right, and layers are assigned such that wires crossing or meeting at knock-knees lie on different layers. If only two-terminal nets are involved, the knock-knee router uses $d_o(I)$ tracks; thus, the two-layer knock-knee routing can be done in $2d_o(I) - 1$ tracks. The layer-assignment process works independently of the knock-knee router that is used and applies also to instances involving multiterminal nets, where it uses at most $4d_o(I) - 1$ tracks [377]. Thus, the upper bounds on the channel width for multiterminal nets that are described in Section 9.4 and in [131] carry over. The bound of $2d_o(I) - 1$ tracks is tight for some problem instances involving only two-terminal nets [267].

9.10.1.2*   The Three-Layer Knock-Knee Model

Preparata and Lipski [363] were the first investigators to show that a channel with only two-terminal nets can be routed in (open) density in the three-layer knock-knee model. Their method first computes a minimum-width knock-knee routing of the channel that fulfills some property. Then, the algorithm assigns layers. The details of the algorithm are quite involved.

Gonzalez and Zheng [151] present a scheme for layer assignment of knock-knee channel routings that applies to *conservative* channel-routing algorithms. A conservative algorithm routes the channel greedily from left to right. In other words, after column c is processed, none of the columns 1 to c receive more wire segments. Furthermore, the use of knock-knees is somewhat restricted. The channel-routing algorithm discussed in Section 9.5.1 does not have this property, but it is not difficult to develop conservative channel-routing algorithms that are optimal for two-terminal nets and that come within a factor of 2 of optimal for multiterminal nets [319].

Gonzalez and Zheng provide a simple greedy method for modifying conservative channel-routing algorithms and computing a layer assignment in three layers, simultaneously.

Kuchem et al. [250] present an algorithm that routes channels in three layers with minimum channel width and minimum spread. Other channel-routing algorithm for the three-layer knock-knee model are presented in [391]. Lipski [289] proves that it is NP-complete to decide whether a given knock-knee routing with only two-terminal nets can be wired in three layers.

A characteristic of the algorithms in the three-layer knock-knee model is that their structure is dominated by technical detail. Furthermore, the model is not very interesting from a practical point of view, because

1. Not all knock-knee routings are wirable in three layers [289].

2. In practice, if three layers are given, some overlap can be allowed.

Therefore, we do not present the routing algorithms for the three-layer knock-knee model in detail, here.

9.10.1.3 The Unit-Vertical Overlap Model

In the unit-vertical overlap model, $d_o(I)$ is still a lower bound on the channel width, since overlap is allowed in only the vertical direction. Berger et al. [26] strengthen this lower bound to $d_o(I) + \Omega(\log d_o(I))$. This bound holds even if arbitrary vertical overlap is allowed, but horizontal overlap is prohibited. Berger et al. also present an application of the general routing method that uses $d_o(I)+O(\sqrt{d_o(I)})$ tracks for two-terminal nets. This result supersedes constructions by Gao [128] and Gao and Hambrusch [129] that use $3d_o(I)/2+O(1)$ and $d_o(I)+O(d_o(I)^{2/3})$ tracks, respectively. The main idea of the construction is to create a Manhattan routing, but, instead of relying on free columns to separate the staircases, using additional free horizontal *tracks* that expand the pyramid structure in the middle of the channel. These tracks are called *good* tracks. After routing of a block, the five free tracks are dispersed in channel as *bad* tracks. It turns out that five additional free tracks are sufficient (see

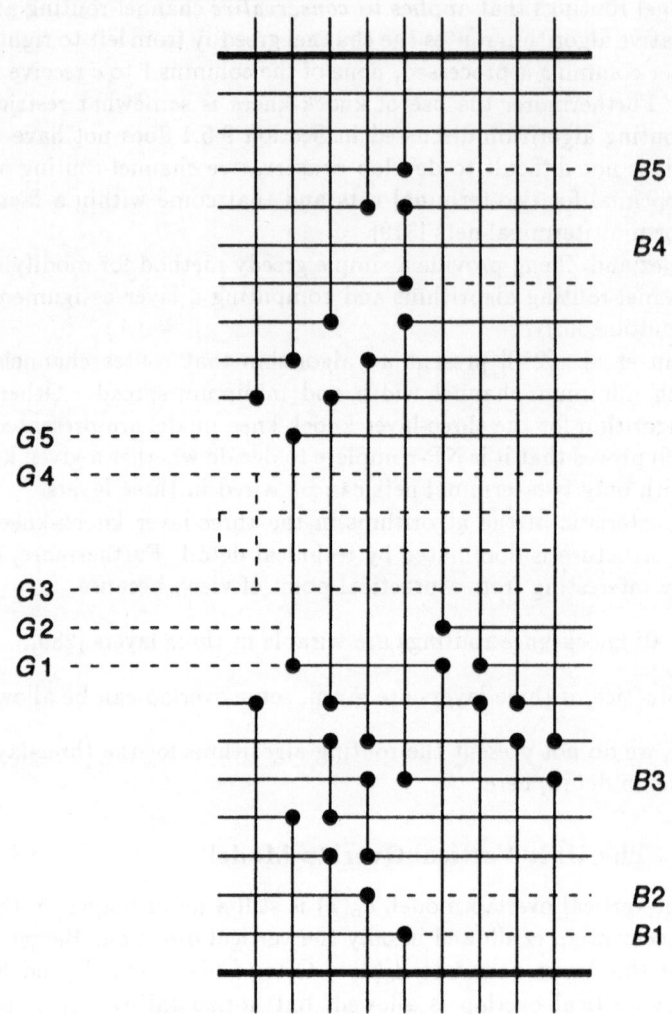

Figure 9.50: *The nonplanar channel-routing scheme for the unit-vertical overlap model.*

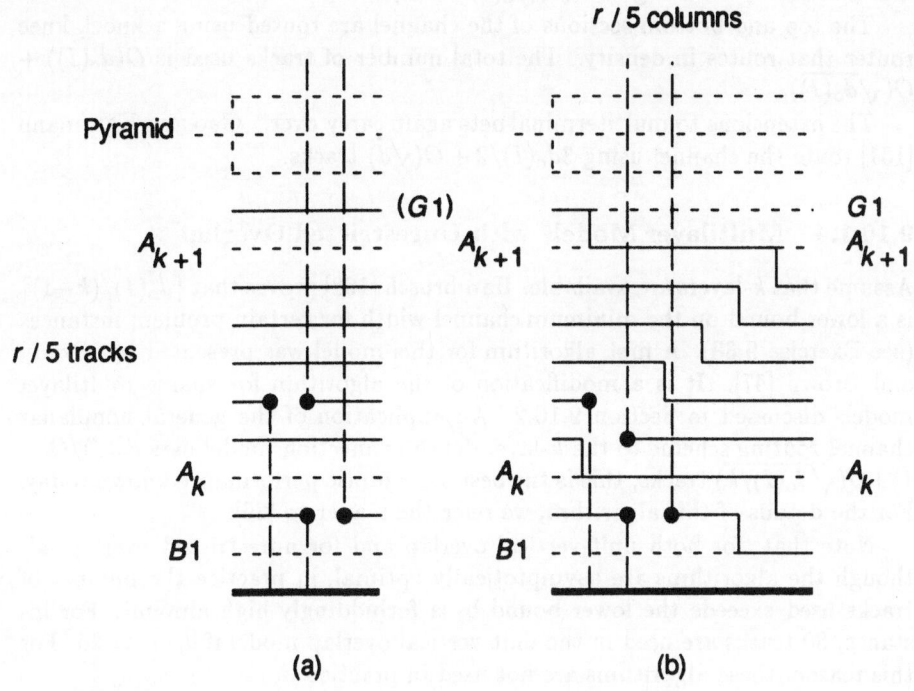

Figure 9.51: *Relocating free tracks. (a) Before relocation; (b) after relocation.*

Figure 9.50). Using a good track to separate staircases may make it bad, as indicated in Figure 9.50. We can convert a bad track into a good track by shifting the free track into the right position in the channel. We achieve this shift by using two sets of five free tracks each, one for the odd-numbered blocks, and one for the even-numbered blocks. Besides routing the nets through B_i, block B_i also serves to convert the bad tracks left by block B_{i-1} to good tracks for block B_{i+1}. This relocation of free tracks is done *after* the middle section in the block is routed. It introduces unit-length vertical jogs into all tracks and consequently also into the horizontal wire segments on them (without changing layers; see Figure 9.51). Here, we use more auxiliary tracks, denoted by A_k and A_{k+1} in Figure 9.51. If the block size is r, we need an auxiliary track for every $r/5$ tracks to be able to convert a bad track within $r/5$ columns into a good track; that is, to convert all five bad tracks to good tracks with the horizontal extent of a block. The conversion of bad tracks to good tracks introduces unit-vertical overlap. In total, we use $(d_o(I) + 10)(1 + 5/r)$ tracks. We minimize this number by choosing $r = \sqrt{5(d_o(I) + 10)}/2$.

The top and bottom sections of the channel are routed using a knock-knee router that routes in density. The total number of tracks used is $O(d_o(I)) + O(\sqrt{d_o(I)})$.

The extensions to multiterminal nets again carry over. Gao and Kaufmann [131] route the channel using $3d_o(I)/2 + O(\sqrt{d})$ tracks.

9.10.1.4 Multilayer Models with Unrestricted Overlap

Assume that k-layers are available. Hambrusch [168] proves that $\lceil d_o(I)/(k-1) \rceil$ is a lower bound on the minimum channel width for certain problem instances (see Exercise 9.38). A first algorithm for this model was presented by Brady and Brown [47]. It is a modification of the algorithm for sparse multilayer models discussed in Section 9.10.2. An application of the general nonplanar channel-routing scheme to the k-layer depth-connection model uses $d_o(I)/(k-1) + O(\sqrt{d_o(I)/k})$ tracks; this is the best asymptotic performance known today. For the details of this algorithm, we refer the reader to [26].

Note that, for both unit-vertical overlap and for unrestricted overlap, although the algorithms are asymptotically optimal, in practice the number of tracks used exceeds the lower bound by a forbiddingly high amount. For instance, 80 tracks are used in the unit-vertical overlap model if $d_o(I) = 20$. For this reason, these algorithms are not used in practice.

9.10.2 Channel Routing in Sparse Multilayer Models

Hambrusch [169] presents the first analysis of channel-routing algorithms in a sparse multilayer model. Hambrusch discusses the (L out of k)-layer model. Brady and Brown [47] present channel-routing algorithms for ($L|k$)-layer models and directional multilayer models whose performance is comparable with the one Hambrusch achieves. Thus, we shall limit our discussion here to the algorithms by Brady and Brown.

In the ($L|k$)-layer model, $\lceil \frac{d_o(I)}{\lceil k/L \rceil} \rceil$ is a trivial lower bound on the channel width. Brady and Brown [47] present an algorithm for the ($L|k$)-layer model with $k \geq 4$, $L \geq 2$ that uses $\lceil \frac{d_o(I)+1}{\lceil k/L \rceil} \rceil + 2$ tracks. This figure is at most three tracks more than optimal.

The development of the algorithm by Brady and Brown starts with a very simple routing algorithm that applies to the (depth-connection) (HVHV...)-model with at least three layers. This model is a restriction of the ($2|k$)-layer model. (Later, we shall generalize the algorithm to other sparse multilayer models.) The algorithm has been independently presented by [47, 103]. Chen and Liu [61] describe the algorithm for the (VHV)-model. Even this first simple algorithm serves to illustrate the two main phenomena occurring in sparse multilayer models:

1. We can come quite close to—that is, to within a small multiplicative factor of—the lower bound on channel width.

2. Routing multiterminal nets does not require many more tracks than does routing two-terminal nets.

Both of these phenomena are related to the fact that, if more than two layers are available, vertical constraints do not exist in the channel, since branches of different nets in the same column can be routed on different layers. Thus, we can essentially do a jog-free Manhattan routing using the left-edge algorithm, and can connect the trunks to the terminals on appropriate layers. We now give the details of this construction.

Assume that the number k of layers is odd. (If k is even, we do not use the top layer.) Designate the odd layers for vertical and the even layers for horizontal wire segments. Therefore, the odd layers are also called *vertical* and the even layers are also called *horizontal* layers. We call a *lane* a pair (i, t), where i is the number of a horizontal layer and t is a track number. The assignment of horizontal wire segments is an execution of the left-edge algorithm where the set of colors is the set of lanes on horizontal layers; that is,

$$\{(i, t) \mid 2 \leq i \leq k, \ i \text{ even}, \ 1 \leq t \leq w\}$$

Here,

$$w = \lceil \frac{d_c(I)}{\lceil k/2 \rceil - 1} \rceil$$

is the channel width. As we showed in part 2 of Theorem 9.12, the assignment is always successful. Exercise 9.30 shows that the assignment can be done in linear time in the total number of terminals.

The assignment of vertical wire segments is not difficult. Each horizontal wiring layer has two adjacent vertical wiring layers. We can route at most two terminals in each column in those two wiring layers. The run time of the algorithm is linear in the number of terminals.

For $k = 3$, the channel width w is about twice the lower bound, but as k increases the channel width gets closer and closer to optimal. Furthermore, the algorithm has the following features:

1. It does not introduce jogs in wires.

2. It routes multiterminal nets without a sacrifice in terms of the channel width.

3. Its does not extend the channel on the left or right side.

All these points are not addressed by any known approximation algorithms in the knock-knee, Manhattan, and unit-vertical overlap models that have bounded error.

We shall now discuss how to improve this very simple algorithm to yield a channel width that is only a bounded *additive* term away from the lower bound. To this end, we leave directional models and turn to the $(2|k)$-layer model. We will need the additional freedom this model gives us for saving tracks. However, we will still mainly route in the preferred direction for a layer. The key idea is to reverse the role of the horizontal and vertical layers. Now, the *odd* layers are the horizontal layers and the *even* layers are the vertical layers.

Now, a problem is posed by columns that have two terminals, one for each of two nets n_i and n_j such that the trunks of n_i and n_j run on, say, the bottom layer, and the terminal on the bottom side of the channel has to connect to the topmost trunk of n_i and n_j. Such a column requires two vertical layers adjacent to the bottom layer, but only one layer exists. To eliminate such situations, we do two things. First, we modify the instance of the routing problem such that such difficult columns cannot appear. Second, we solve the modified problem instance.

Here is a description of both steps of the algorithm:

Step 1: The idea is to pair up adjacent columns i and $i + 1$ and to scoot over terminals between the two columns, such that no column has terminals on both sides of the channel. This transformation is done at the expense of sometimes having two terminals in one column *on the same side* of the channel. This situation poses no problem, however, since several vertical layers exist, such that we can route more than one terminal in a column on each side of the channel.

The following lemma can be shown by a straightforward case distinction.

Lemma 9.10 *We can always scoot over terminals between matched columns such that*

1. *No column has terminals on both sides of the channel.*

2. *The closed density of the problem instance does not increase.*

Step 1 of the algorithm modifies the problem instance I as Lemma 9.10 prescribes. Thus, a problem instance I' is obtained.

Step 2: In the second step, the trunks of I' are placed on tracks and layers. This procedure decomposes into three substeps.

> **Step 2.1:** The trunks for I' are placed using the left-edge algorithm. By Lemma 9.10, only $\lceil \frac{d_c(I)}{\lceil k/2 \rceil} \rceil$ tracks are needed.

> **Step 2.2:** This step routes the vertical connections in problem instance I'. In almost all columns, branches can be routed directly to the terminals on appropriate vertical layers. The only difficult case occurs if there are two terminals for nets n_i and n_j in a column and the

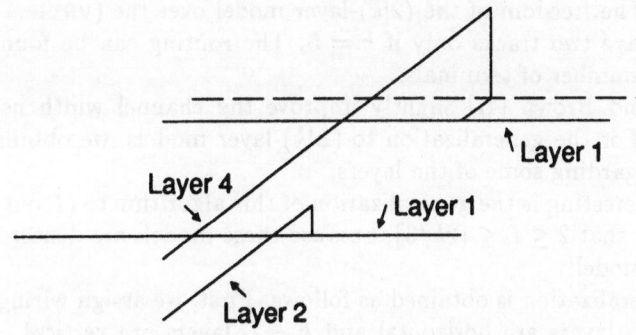

Figure 9.52: *Routing two terminals in a column.*

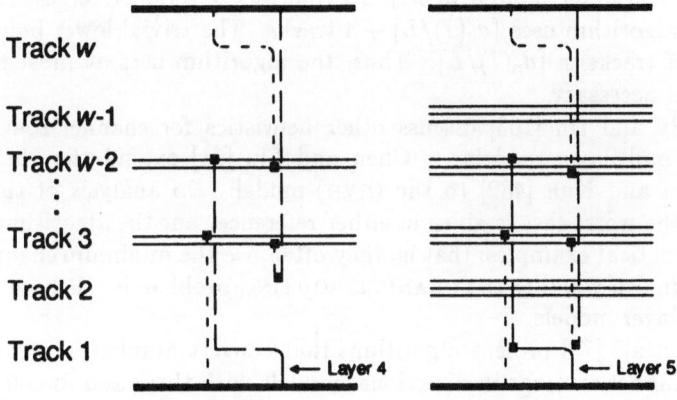

Figure 9.53: *Scooting back terminals.*

trunks of both nets are on the bottom layer (or analogously on the top layer). This case is handled as shown in Figure 9.52. We need two additional tracks, one on either side of the channel, to perform this step.

Step 2.3: Finally, the problem instance I has to be restored from I'. We do this transformation by processing columns pairwise. Figure 9.53 shows the resulting routing patterns. If $k = 4$, one more track on either side of the channel may be necessary here.

The algorithm uses $\lceil \frac{d_c(I)}{\lceil k/2 \rceil} \rceil + 2$ tracks if $k \geq 5$, and $\lceil \frac{d_c(I)}{2} \rceil + 4$ tracks if $k = 4$. This figure is no more than five tracks (three tracks if $k \geq 5$) more than

necessary. The freedom of the $(2|k)$-layer model over the (VHVH ...)-model is
needed to save two tracks only if $k = 5$, The routing can be found in linear
time in the number of terminals.

Brady and Brown [47] slightly improve the channel width used by this
algorithm. For the generalization to $(L|k)$-layer models, we obtain $L \geq 3$ by
simple disregarding some of the layers.

More interesting is the generalization of this algorithm to (L out of k)-layer
models such that $2 \leq L \leq \lceil 2k/3 \rceil$, because these models are denser than is the
$(2|k)$-layer model.

The generalization is obtained as follows. First, we assign wiring directions
to layers. L layers are horizontal and $k - L$ layers are vertical. We assign
directions such that each horizontal layer is next to a vertical layer. This
assignment is always possible, as long as $L \leq \lceil 2k/3 \rceil$. Then, the algorithm is
run as before, but the special column routing has to be done for each horizontal
layer that is adjacent to only one vertical layer. Also, in this case, we need the
two extra tracks introduced in step 2.2, unless $k \geq 6$ (see Exercise 9.39).

They algorithm uses $\lceil d_c(I)/L \rceil + 4$ tracks. The trivial lower bound on the
number of tracks is $\lceil d_o(I)/L \rceil$. Thus, the algorithm uses at most five tracks
more than necessary.

Enbody and Du [103] discuss other heuristics for channel routing in directional multilayer models.   Chen and Liu [61] extend the heuristic by
Yoshimura and Kuh [482] to the (HVH)-model.   No analysis of the channel
width in the worst case is given in either reference, but the algorithms perform
well on practical examples; that is, they often use the minimum channel width.
It is not known whether the CHANNEL ROUTING problem is NP-hard for any of
the multilayer models.

Cong et al. [73] present algorithms that convert Manhattan channel routings to channel routings in directional models with three and four layers.

The reason why it is relatively simple to come close to the optimum channel
width in sparse multilayer models is that the problem instances are inherently
sparse. In fact, the detailed-routing model itself dictates that only a small
fraction (less than one-half) of the edges of the detailed-routing graph can
contain wires. If we eliminate the restrictions on overlap, this is no longer the
case, and it becomes more difficult to come close to the lower bound on the
channel width.

9.10.3   Practical Multilayer Channel Routing

In a setting in which overlap is not restricted, the algorithms presented in
Section 9.10.1 are not practical because they still use too many tracks. However,
the algorithms from Section 9.10.2, especially the algorithm for the (L out of
k)-layer model with $L \leq \lceil 2k/3 \rceil$ also have practical significance. This algorithm
runs in linear time and comes to within five tracks of the minimum channel
width. Furthermore, it has many attractive features, such as the following:

1. It handles multiterminal nets.

2. Each net has only one trunk.

3. The channel is not extended much on the left or right side. In fact, only one additional column on one side is needed.

4. Branches never deviate from their column by more than 1.

It is definitely worthwhile to investigate whether this router can be extended to become even more robust in practical situations that involve additional complexities, such as gridless models.

Braun et al. [48] present a heuristic multilayer channel router that has been used on many practical problems. It is based on a directional model in which each horizontal layer is next to a vertical layer. In practice, of course, this model is not strictly adhered to; local changes of direction are allowed.

Preferred wire directions are maintained even if they are not imposed by the technology, because we want to be able to use some version of the left-edge algorithm. For the same reason, a horizontal layer has to be adjacent to at least one vertical layer. With this restriction, the (HVHHVH)-model and its relative are the densest multilayer models available.

Braun et al. use a combination of the following ideas:

1. Divide up the set of layers into small sets of two or three layers on which the routing proceeds independently.

2. Use a combination of maze-running techniques and a derivative of the left-edge algorithm.

9.11 Summary

Detailed routing is a diverse field. There are many different combinatorial problems tailored to special cases arising in different design and fabrication technologies. As a rule, the heuristics perform quite well, compared to the heuristics for global routing and placement. In Manhattan channel routing, for instance, improving the behavior of the heuristics on typical cases seems to be pushing the limits of the technology. Nevertheless, many detailed-routing problems are not well understood, combinatorially. On the other hand, there are signs of emerging mathematical theories for several detailed-routing models.

However, there is still a gap between theoretical and practical research in the area. The recent mathematical theories of detailed routing have not yet gained much significance with respect to practice. The reason is not so much that the algorithms are not efficient enough, but rather that the models are too unrealistic and lack the required amount of robustness. However, the mathematical progress in the area of detailed routing has widened research attention to considering detailed-routing graphs with nontrivial homotopies. This development is relatively new; as recently as ten years ago, the channel and the switchbox were essentially the only detailed-routing graphs considered.

In the future, we can expect that progress in the mathematical part of the subject may make dealing with nontrivial homotopies feasible in practice. This development would greatly change the perception of the process of detailed routing and its relationship to the preceding and subsequent layout phases.

Another phenomenon is expected to influence the area of detailed routing: With the advance of the fabrication technology, detailed routing in several layers (more than two or three) is becoming increasingly important. Special methods for doing this kind of routing have yet to be developed. So far, planarity has been an important aspect of detailed routing. As the number of layers increases, the planarity of the problem is lost. There are some results on multilayer channel routing, but the concept of a channel itself loses much of its significance in multilayer routing. The only other methods to which we can resort in this case are area-routing techniques based on maze running or line searching. We need to do more work in the area of multilayer routing.

9.12 Exercises

9.1 Discuss different modifications of Definition 9.2 that require that homotopic trees have the same or similar tree topologies.

9.2 Devise a hierarchical layout scheme that calls for the solution of channel-placement problems during the detailed-routing phase. (See also [207, 423].)

9.3 Prove that the multiterminal nets in a problem instance can be decomposed into two-terminal nets such that the density of each cut at most doubles.

9.4 Modify the greedy algorithm for channel routing in the planar routing model such that the number g of jogs is kept low. (Pinter [356] presents an algorithm and claims that it minimizes the number of jogs for each wire individually. Kolla et al. [239] have given a counterexample to this claim.)

9.5

1. Consider the algorithm for topologically sorting the relation R defined in Lemma 9.3 that is discussed in Section 9.3.1.2.2.

    a. Prove that the algorithm produces a topological ordering of R.

    b. Describe an implementation of the algorithm that runs in linear time $O(|N|)$.

2. Show that computing R explicitly and topologically sorting the edges afterward costs $\Omega(|N|^2)$ time in some cases.

9.6 Prove that, in the planar routing model, the minimum channel width that solves the OPTIMUM WIDTH CHANNEL OFFSET problem does not exceed $\lfloor n/2 \rfloor$, and that this bound is tight in some cases.

9.7 Develop an efficient algorithm for solving the OPTIMUM AREA CHANNEL PLACEMENT problem in the planar routing model. (Quadratic running time is not difficult to achieve.)

Discuss how you can extend the algorithm to optimize other cost measures, such as

1. Minimize the length of the longest wire

2. Minimize total wire length

All optimizations should be done *without minimizing channel width first*. See also [328].

9.8 Prove that the bounds given in Theorem 9.3 are tight.

9.9 Incorporate nets that have both terminals on the same boundary of the channel into planar channel routing.

1. Devise a routing algorithm and a correctness criterion of the form of Theorem 9.1 (see also Section 9.3.1.4).

2. Extend the greedy routing algorithm.

9.10 Extend Theorem 9.1 to vertically convex irregular channels with terminals on the horizontal boundary segments and not on corners.

Hint: Model the irregular channel boundaries as contours of additional one-sided nets. Then use the results of Exercise 9.9.

9.11 Extend Exercise 9.4 to irregular channels.

9.12 Develop solution algorithms for channel-offset problems with irregular channels.

9.13 Incorporate multiterminal nets into planar channel routing (see also Section 9.3.1.4).

1. Develop an efficient algorithm for testing whether a problem instance involving multiterminal nets is solvable if the channel width is chosen large enough.

2. Solve the ROUTABILITY problem.

3. Extend the greedy routing algorithm.

9.14 This exercise discusses the CHANNEL-WIDTH MINIMIZATION problem in the knock-knee model.

1. Modify the application of the general scheme for nonplanar channel routing to the knock-knee model such that, on an instance I with multiterminal nets and open density $d_o(I)$, at most $2d_o(I) - 1$ (instead of $2d_o(I)$) tracks are used. Construct an example on which this bound is attained.

   Hint: Whenever the algorithm scans across a cut with maximum density, the multiterminal net in the middle of the channel needs only one track.

2. Modify the application of the general scheme for nonplanar channel routing to the knock-knee model such that, on an instance I with two- and three-terminal nets and open density $d_o(I)$, at most $3d_o(I)/2$ tracks are used. Construct an example on which this bound is attained.

   Hint: As you process the channel, pair up nets such that one net in a pair uses only one track, the other two tracks.

3. Construct a problem instance that shows that a lower bound for routing multiterminal (even two- and three-terminal) nets in the knock-knee model is $d_o(I) + 1$.

9.15 Give an example of an instance of the DETAILED ROUTING problem in the knock-knee model that is locally even but is not even.

9.16 Prove Lemma 9.4 by a case distinction.

9.17 Reduce the run time of the algorithm for knock-knee routing in switch graphs (Section 9.5.2) to $O(n^{3/2})$ in the case that G is a partial grid graph.

   Hint: Use the planar separator algorithm described in Section 6.4. You need to use an additional property of the separators constructed there.

9.18

1. Devise a data structure that maintains the set C_{ij}, $0 \leq i < j \leq 2\ell - 1$ of cuts. Specifically, the following operations must be supported:

   Init: Initialize the data structure to contain all cuts C_{ij}, $0 \leq i < j \leq 2\ell - 1$.

   Select: Return a minimal cut.

   Update: Update the data structure after adding a new net to the problem instance.

   The initialization should take time $O(kn)$; the select and update operations should each take time $O(|U|)$.

   Hint: Use linked lists with appropriate additional pointers.

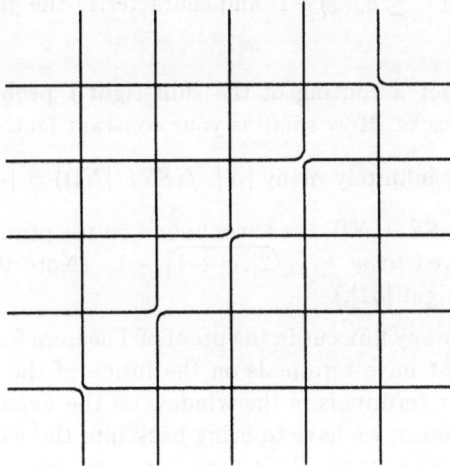

Figure 9.54: *A knock-knee layout.*

2. Use the data structure in the algorithm in Figure 9.32, and prove a worst-case run time of $O(kn)$.

9.19

1. Prove that, in Figure 9.54, we need to run a wire across a via if the layout is to be wired in four layers.

2. In the layout in Figure 9.54, can you avoid stacking two vias on top of each other if the layout is to be wired in four layers? Explain your answer.

9.20 Show that the residue graph of a knock-knee layout is always planar.

9.21 Reduce the MAXCUT problem on a positively edge-weighted planar graph $G = (V, E)$ to a MAXIMUM WEIGHT MATCHING problem on the planar dual of G.

Hint: Call an edge set E' of a graph $G = (V, E)$ an *odd-cycle cover* exactly if each cycle in G with an odd number of edges contains an edge in E'. Prove that an edge set E' of a graph $G = (V, E)$ is a maximum cut exactly if the complement of E' is an odd-cycle cover of minimum cost. Then, prove that, if E' is an odd-cycle cover in G, then the set of edges that are dual to E' corresponds to a set of paths whose endpoints are the vertices of odd degree in the dual of G. Use these results to complete the exercise.

9.22 Show that $d_c(I) \leq d_o(I) + 1$, and characterize the problem instances for which $d_c(I) = d_o(I)$.

9.23   1. Construct a routing of the shift-right-1 problem $SR1(|N|)$ that uses $O(\sqrt{n})$ tracks. How small is your constant factor?

2. Show that, for infinitely many $|N|$, $f(SR1(|N|)) = \lceil \sqrt{n} \rceil$.

3. Show that, for $SR1(|N|)$, the lower bound on the minimum channel width can be improved to $w \geq \lceil \sqrt{2|N| + 1} \rceil - 1$. (Note that even this lower bound is not tight [51].)

   Hint: Consider any flux cut in the proof of Theorem 9.11. Call the interval of columns that have terminals on the inside of the cut the *window*. If there are many terminals in the window on the *outside* of the cut, then, to free up columns, we have to bring back into the window wires that we routed out of the window. This operation costs free columns. Modify the argument in the proof of Theorem 9.11 by accounting for this fact.

9.24 The lower bound on the minimum channel width in the Manhattan model that is given by the flux can be tightened a little more. The idea is to observe that one-sided nets all of whose terminals are on the inside of the flux cut also have to leave their start column at some point. In Figure 9.41 net 6 is such a net. Use this idea to extend Theorem 9.11 to include such nets.

9.25 Let I be an instance of the CHANNEL-WIDTH MINIMIZATION problem.

1. Prove that, for computing the flux, we need to consider only flux cuts whose leftmost and rightmost inside column are occupied.

2. Devise an efficient algorithm for computing the flux. How fast is your algorithm?

9.26 Let $f'(I)$ be the maximum integer f such that there is a flux cut C with capacity $c_m(C) = 2f^2 + 2$ and density $d(C) \geq 2f^2 - f$. Prove the following:

1. $f'(I) \leq f(I)$.

2. For each $\varepsilon > 0$, there is a problem instance $I(\varepsilon)$ such that $f(I(\varepsilon)) \geq f'(I(\varepsilon)) \cdot (\sqrt{2} - \varepsilon)$.

9.27 Show that the value of k in the Manhattan channel-routing algorithm of Section 9.6.1.2 can be bounded from above by

$$k = \left(\frac{3}{2} + \sqrt{\frac{13}{4}} \right) f(I) + O(1)$$

The constant factor evaluates to about 3.302.

9.28 1. Determine the asymptotic run time of the algorithm for Manhattan channel routing that was described in Section 9.6.1.2.

2. Speed up the algorithm to run in linear time in the smallest area of a minimum width channel.

   Hint: Allow for different sizes of different superblocks. Find the suitable size of each superblock by scanning the channel from left to right.

9.29 Devise a linear-time algorithm that routes a channel using at most $3|N|/2$ tracks in the Manhattan model.

Hint: Process nets one by one. Use one track per net, and two tracks for each last-processed net in a circuit inside the vertical constraint graph (Definition 9.25).

9.30 Discuss the following variation of the left-edge algorithm for problem instances I without vertical constraints. The algorithm uses $d_c(I)$ colors. A vertical sweep line is moved across the channel from left to right. A dynamic data structure (balanced tree) is used to maintain the set of intervals that cross the sweep line and the intervals' colors. Whenever an interval enters the sweep line, it receives an arbitrary unused color.

1. Prove that, given the appropriate nondeterministic choices, the algorithm produces the same coloring as does the left-edge algorithm if the smallest unused color is assigned every time.

2. Detail the data structures of the algorithm such that the latter runs in time $O(|N|)$ if we can assign an arbitrary unused color, and in time $O(|N| \log d_c(I))$ if we have to assign the smallest unused color.

3. Incorporate directed constraints into this implementation of the left-edge algorithm. Prove an upper bound on the number of colors it uses. Is your bound tight? (See Exercise 9.31.)

4. Discuss the following modification implemented in YACR2 [376]. Here, the sweep across the channel extends to either side, starting from a column of maximum density. Why is this modification advantageous?

Note that this implementation of the left-edge algorithms follows the sweep paradigm that has been used successfully by the greedy channel router and that is also the basis for the general scheme for channel routing in nonplanar detailed-routing models.

9.31 This exercise discusses an upper bound on the number of tracks used by the left-edge algorithm. Let I be a solvable instance of the CHANNEL-WIDTH MINIMIZATION problem in the jog-free Manhattan model. The *level* of a net $n_i \in N$ is the length of the shortest path from a source of $G_v(I)$ to n_i. That is, the sources of $G_v(I)$ have level 0, and the largest level of any vertex in $G_v(I)$ is $depth G_v(I)$. Let G_i be the subgraph of $G_t(I)$ that is induced by the vertices at level i. G_i is an interval graph; that is, it has no directed edges. Let $d_i = d_c(G_i)$ be the chromatic number of G_i. Let $w_{le}(I)$ be the number of tracks used by the left-edge algorithm on I.

1. Prove both of the following inequalities:

$$w_{le}(I) \leq d_c(I) \cdot \min\left\{ depth(G_v(I)) + 1, \frac{\sum_{i=0}^{depth(G_v(I))} d_i}{depth(G_v(I)) + 1} \right\}$$

$$\leq w_{\min}(I)\sqrt{n}$$

The second term in the minimum is the average density over all i of the subproblems consisting of all intervals at level i in $G_v(I)$.

Hint: Prove first that $w_{le}(I) \leq \sum_{i=0}^{depth(G_v(I))} d_i$.

2. Prove that there are problem instances I such that

$$w_{le}(G_v(I)) = \Omega(w_{\min}(I) \cdot \sqrt{n})$$

9.32 * Develop efficient approximation algorithms for the CHANNEL-WIDTH MINIMIZATION problem in the jog-free Manhattan model that use fewer colors than does the left-edge algorithm in the worst case.

9.33 Devise an algorithm for solving the OPTIMUM WIDTH CHANNEL OFFSET problem in the Manhattan model that runs in time $O(|N|b)$, where b is the sum of the differences between the columns of the leftmost and rightmost terminals on either channel side.

9.34 Detail the dynamic-programming algorithms for computing an optimum-width channel placement in the Manhattan model (all blocks have one terminal) and in the knock-knee model.

9.35 Generalize the dynamic-programming algorithm for computing an optimum-width channel placement in the knock-knee model to the case that blocks have several terminals.

9.36 Prove that the problem instance depicted in Figure 9.55 has a minimum channel width of 7.
   Hint: Scanning the channel from left to right leads to a contradiction to the assumption that the minimum channel width is 6.

9.37 * The following questions are open:

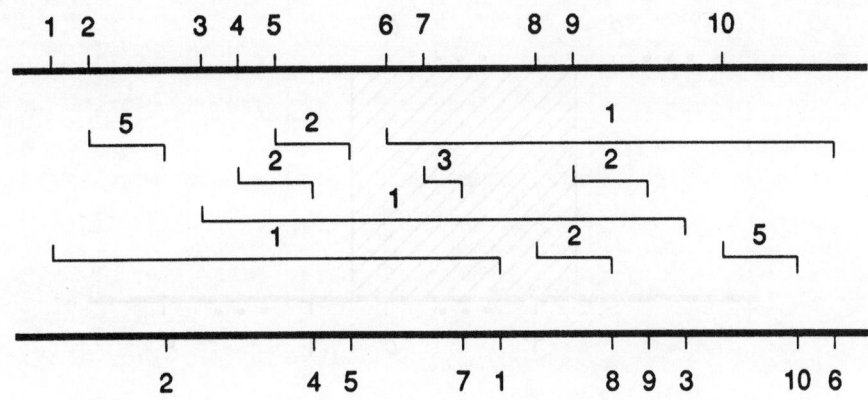

Figure 9.55: *A problem instance I of the channel-routing problem in the gridless Manhattan model. Nets are labeled with their wire width,* $d_{c,gl}(I) = 6$ *but* $w_{min} = 7$.

1. Is $d_{c,gl}(I)$ an asymptotically tight lower bound on the minimum channel width for the CHANNEL-WIDTH MINIMIZATION problem without vertical constraints in the jog-free gridless Manhattan model?

2. Is the CHANNEL-WIDTH MINIMIZATION problem without vertical constraints in the jog-free gridless Manhattan model still NP-hard if trunk widths are from the set $1, \ldots, k$ for some small k? (If, in addition, $d_{c,gl}(I) \leq B$, then a trivial $O(|N|^{kB})$–time algorithm exists.)

9.38 Consider the instance I of the CHANNEL-WIDTH MINIMIZATION problem that is depicted in Figure 9.56. In this instance, there are d falling nets crossing over c trivial nets. We have $d_o(I) = d$. Assume that we are wiring in k layers with depth connection, and there are no restrictions on overlap. Show that, if c is chosen large enough, the minimum channel width is at least $\lceil d/(k-1) \rceil$. Choosing $c > d/(k-1)$ suffices. Note that I can be routed in a channel of width $\lceil d/k \rceil$ if the trivial nets are omitted. Therefore, Lemma 9.1 does not hold for the k-layer model.

Hint: Make the following case distinction. Case 1: At least one trivial net is routed completely within the region R. Case 2: Otherwise.

9.39 Prove that, in the algorithm for channel routing in the $(L$ out of $k)$-layer model with $2 \leq L \leq \lceil 2k/3 \rceil$ that is presented in Section 9.10.2, the two extra tracks introduced in step 2.2 are necessary only if $k < 6$.

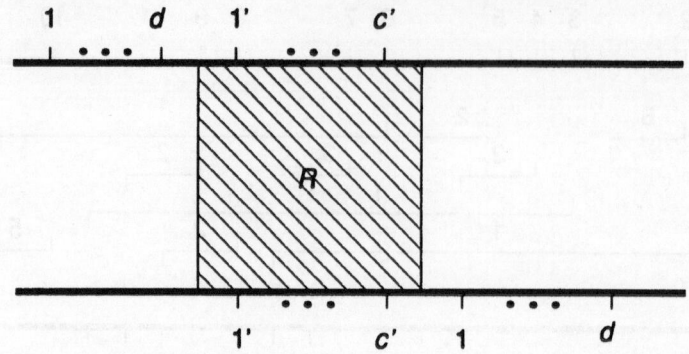

Figure 9.56: *A channel-routing problem requiring* $\lceil d/(k-1) \rceil$ *tracks in k layers.*

Chapter 10

Compaction

Except for the gridless routing models, detailed routing still occurs in a discrete graph setting. Its output is an embedding of a graph representing the circuit into a mostly very regular routing graph. This representation of a layout is also called *symbolic*, because transistors and wires are represented by symbols, such as vertices and edges labeled in appropriate ways. Labels can, for instance, determine the type and strength of a transistor, or the width of a wire. These labels can be attached to the circuit description in a variety of stages of the circuit or layout design process. One possibility is to attach the labels to the netlist before the layout design is started. In this case, the appropriate labels are derived from a simulation or timing analysis of the netlist. Another possibility is to run a performance-optimization program after the detailed-routing phase. Such optimizers are described in [179, 310]. Attaching labels after the detailed-routing phase allows for incorporating layout aspects into the performance optimization of the circuit. For instance, the delay of propagating a wire depends on the length of the wire. If the layout is not known at the time that the labels are computed, some nominal value has to be used here. This approach is viable only if the dependence of the delay on the wire length is a second-order phenomenon. As the complexity of chips increases, this is no longer the case, and performance optimizers that make use of layout data become increasingly important. We do not discuss this issue further here, but refer the reader to the literature [179, 310, 385].

Historically, symbolic layout not only has been produced by detailed routers, but it has been the domain of hand layout by the designer. To this end, special languages for describing symbolic layout [75, 101, 278], and special graphic editors for symbolic layouts [190, 191, 199], have been developed. To compute an actual mask layout, we have to translate the symbolic layout into geometric mask features on the silicon. The basic features out of which a mask is composed can, in principle, have any shape. But most often they are polygons—in fact, they are usually rectangles. We shall discuss only rectagonal features in this chapter. These are features whose contours have only horizontal and vertical edges. As explained in Section 1.2.3.1, the features have to obey certain design rules that fall into the following categories:

Spacing Rules: Different features on the same or on different masks that belong to different circuit elements have to have a minimum separation distance that depends on the masks involved (see Figure 1.6).

Minimum Size Rules: All features must have a minimum size, depending on the mask and the feature.

Shape Rules: Different features belonging to the same circuit element must have a prescribed shape and be located in a prescribed position with respect to one another (see Figure 1.5(a) and (b)).

The reader is encouraged to read through Section 1.2.3.1 again at this point, to refamiliarize her- or himself with this material.

The translation of the output of the detailed-routing phase into mask data has to convert the circuit elements into the appropriate mask features. At the same time, it should ensure that all design rules are met while minimizing the layout area. This last goal gives this process the name *compaction*. We should note, however, that compaction is more than just an optimization problem. It performs a translation from the graph domain into mask geometries.

If we are careful about the data representation, the compacted mask layout can still be interpreted as a symbolic layout. To do this interpretation, we have to preserve the notions of symbolic circuit elements and their relationships (electrical connectivity, and so on) during compaction. This provision makes it possible to submit a *mask layout* to repeated compactions to improve the layout area. In fact, according to this view, the symbolic layout is just a different view of the data describing an annotated mask layout. All interactive symbolic layout systems that incorporate compacters take this approach. With symbolic layout languages, it is more difficult to achieve this equivalence between symbolic and mask layout, because the symbolic layout specifications written in the layout language can be parameterized cell generators, whereas the compaction applies to only a single circuit at a time.

Compaction is quite a difficult problem not only from a combinatorial, but also from a systems, point of view. The task of compaction is closely related to intricate details of the fabrication technology. Many subtasks in compaction, such as special-case handling of certain circuit elements (for instance, the placement of substrate contacts in CMOS bulk processes) or smoothing out mask geometries (that is, eliminating spurious corners and ragged edges in wells in CMOS bulk processes) to avoid undesirable electrical instabilities cannot be easily cast into a purely combinatorial framework. Furthermore, combinatorial and technological issues mix. For instance, in many technologies, the design rules that apply in the vicinity of a transistor gate are not easily cast in combinatorial terms. Advanced compacters aim at separating combinatorial and technological issues as much as possible. They provide an interface to the technology in terms of a design-rule database that comprises all important facts about the technology that are needed by the compacter. The compaction algorithm uses the information in the design-rule database to compute the mask features. Nevertheless, the kind of information that the compaction algorithm has to extract from the design rule database can be quite complicated: It involves not only simple numerical parameters, such as minimum distances between features on different masks, but also connectivity information (that is, the location at which wires can connect to a circuit element on different mask layers) and electrical information. For instance, whether two mask features have to satisfy a minimum distance requirement sometimes depends on the electrical signal they carry.

The interplay between the technological and the combinatorial aspect of compaction is not discussed in detail in this book. We refer to the excellent survey by Wolf and Dunlop [471] for this purpose. Here, we shall emphasize the combinatorial aspect of compaction. To this end, in Section 10.1, we describe and justify the abstraction from the real problems to the combinatorial problems we discuss in this chapter. The subsequent sections describe algorithms that solve these combinatorial problems and handle generalizations.

We distinguish between *one-dimensional* and *two-dimensional* compaction. In one-dimensional compaction only the, say, x coordinates of the mask features are changed. This kind of compaction is also called x *compaction*; y *compaction* is defined analogously. The goal is to minimize the width of the layout, while preserving the design rules and not altering the function of the circuit. In two-dimensional compaction, both x and y coordinates can be changed simultaneously in order to minimize area. In a nutshell, most versions of one-dimensional compaction can be done efficiently; in fact, they can be done with almost-linear-time algorithms. A few versions of one-dimensional compaction, as well as most versions of two-dimensional compaction, are NP-hard. The difficulty of two-dimensional compaction lies in determining how the two dimensions of the layout must interact to minimize the area. To circumvent the intrinsic complexity of this question, some heuristics that decide locally how the interaction is to take place. These heuristics are sometimes referred to as $1\frac{1}{2}$-dimensional compaction. We discuss all three types of compaction in separate sections.

Several approaches to compaction yield to the intricate interplay between the combinatorial and technological aspects of compaction by using compaction algorithms that substantially simplify the combinatorics such that we can concentrate on the technological issues. One such class of algorithms is *virtual-grid compaction*. Here, the coordinates of mask features are restricted to be aligned with a square grid that is much coarser than the minimum resolution of the fabrication process. The mesh size of the grid is closer to the size of a complete circuit element than to the size of the smallest geometric feature. (There can be a factor of 50 difference between these two parameters.) The compaction operation compresses the grid along with all features placed on it, but it keeps gridlines straight along the way. The advantage of virtual-grid compaction is that the algorithms are simple and easily implemented, and they are usually very fast. Furthermore, adhering to the virtual grid can simplify the handling of hierarchy during compaction. The disadvantage is that the restriction to the virtual grid leads to increased layout area. We will not discuss virtual grid compaction in detail, because its difficulties do lie not on the combinatorial, but rather on the technological, side. We refer the reader to [43, 44, 104, 438, 471].

Most of the emphasis of this chapter is on the compaction of mask layouts. This application is also where most of the developmental work on compaction has been invested. In recent years, however, applications of compaction during earlier stages in layout design have been introduced. For instance, cells are compacted after placement and global routing, but *before* the final detailed routing is done. We call this type of compaction *topological compaction*. Topological compaction is surveyed in Sections 10.1.4 and 10.5.

10.1   Basic Definitions

In this section, we introduce the relevant formal definitions and illustrate their relationship to the actual compaction problems occurring in practice.

10.1.1   Geometric Layouts

We shall view compaction as a problem of arranging rectagonal—in fact, mostly rectangular—features in the plane.

Definition 10.1 (Geometric Layout) *A* feature F_i *consists of a* contour C_i *and a* mask layer m_i. *The contour is a rectagonal closed curve in the plane that we can think of as being described by a listing of its straight-line segments, say, in clockwise sequence. In most cases, we assume that F_i is a rectangle. In this case, we also describe F_i in the form $R_i = (w_i, h_i, x_i, y_i, m_i)$, where w_i is the width, h_i is the height, and (x_i, y_i) are the Cartesian coordinates of the lower-left corner of the rectangle. The mask layer m_i is a small integer $1 \leq m_i \leq M$.*

A collection of features on a set of mask layers is called a geometric layout—*specifically, an M-layer layout.*

The most straightforward representation of mask data by a geometric layout breaks up the mask data into rectangular features. Each feature belongs to a circuit element or to a wire. The number m_i identifies the mask of each feature. We allow for nonrectangular features, in order to be able to do hierarchical compaction. In hierarchical compaction, a feature can represent a whole subcell, instead of just a single circuit element. As the shape of subcells can be complicated, it is necessary to allow for general rectagonal features.

Some compacters are based not on a rectagonal representation of the mask data, but rather on an edge representation. Each edge of a mask feature is represented individually. From the systems point of view, this edge representation has advantages and disadvantages. An advantage is that, with an edge representation, we can handle different edges of the same feature differently, and this feature is required in some cases. Disadvantages are a larger memory requirement, and the fact that, when different features overlap, some edges may be temporarily hidden from the view. From a combinatorial point of view, there is not much difference between both representations, because an edge is a rectangle whose extent in one dimension is zero. In this chapter, we shall stick to the rectagonal representation.

To define the legality of a layout with respect to a set of design rules, we must introduce a few more concepts. First, a minimum distance requirement (*spacing constraint*) must be imposed on pairs of distinct features. We will assume that there is a number $d_{ij} \geq 0$ specifying the separation between any feature on mask i and any feature on mask j. Here, the *separation* is the minimum distance between any pair of points, one in each of the two features. The distance is measured in the L_∞-*metric*; that is, the distance of two points (x_1, y_1) and (x_2, y_2) is defined by

$$d_\infty(x_1, y_1, x_2, y_2) = \max\{|x_1 - x_2|, |y_1 - y_2|\}$$

Many fabrication processes support a concept of minimum distance that is based on the *Euclidean metric*. However, the Euclidean metric is much more difficult to handle combinatorially. Therefore, we will here stick to the L_∞-metric; see Figure 10.1.

The number d_{ii} specifies the minimum separation between different features on mask i. Often, we will consider 1-layer layouts with $d_{11} = 0$. In this case, features are not allowed to overlap, but they may touch. Note that, if we restrict ourselves to one-layer layouts, then we can reduce any positive value of d_{11} to $d_{11} = 0$ by expanding each feature of the layout by $d_{11}/2$ on each of its four sides. However, with multilayer layouts for which the different d_{ij} have different values, positive minimum distances have to be considered explicitly.

Second, we have to give a formal account of the fact that not all features need to be separated. Some features form groups, such as the features that make up the mask data for a circuit element. We will bind together such features with *grouping constraints*. Grouping constraints can be of two types. Assume that two features F_1 and F_2 contribute to the same circuit element. If F_1 and F_2 are bound *tightly*, their location with respect to each other is specified uniquely. This is the case for all circuit elements shown in Figure 1.5. In this case, the grouping constraint is a set of two equations of the form

$$\begin{aligned} x_1 + c &= x_2 \\ y_1 + d &= y_2 \end{aligned} \tag{10.1}$$

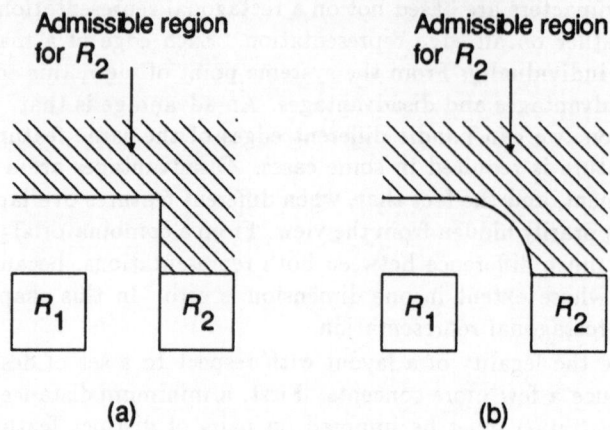

Figure 10.1: *The separation of two features in two metrics. R_2 has to be completely contained in the shaded region. (a) L_∞-metric; (b) Euclidean metric.*

where (x_1, y_1) and (x_2, y_2) are two distinguished points of F_1 and F_2. (If F_1 and F_2 are rectangles, we will choose the lower-left corners as distinguished points. If F_1 is the rectangle on the cut mask and F_2 is a rectangle on one of the contacted masks in Figure 1.5(a), then $c = d = 1$ is the appropriate choice.) In some cases, there is some freedom as to how to locate two features with respect to each other. Consider the case of a wire attaching to a contact between two layers. Since the contact is larger than the wire, there is a variety of valid attachments (see Figure 10.2). This arrangement is sometimes called a *sliding port*. The corresponding set of constraints is the following:

$$
\begin{aligned}
x_1 - x_2 &\leq c_1 \\
x_2 - x_1 &\leq c_2 \\
y_1 + d &= y_2
\end{aligned}
\tag{10.2}
$$

If F_1 is the rectangle representing the bottom vertical wire and F_2 is the rectangle representing the contact in Figure 10.2, then $c_1 = 0$ and $c_2 = 1$ are appropriate choices.

Often, other constraints, in addition to the grouping constraints discussed so far, have to be obeyed during compaction. A popular kind of additional constraints is maximum length requirements on horizontal wire segments. Such a requirement translates into a linear inequality involving the coordinates of the two circuit elements connected by the wire. We call such additional constraints *diverse constraints*. We will allow any set of diverse constraints that can be couched in terms of inequalities of the form $x_j - x_i \leq a_{ij}$. Usually, there

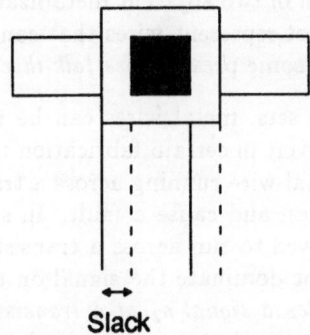

Figure 10.2: *A sliding port. The bottom vertical wire can be shifted horizontally by one unit.*

are comparatively few diverse constraints, but some of them may be essential during compaction.

Some features may be completely immune to spacing constraints, even though they are on interfering layers. For instance, two segments of the same wire can overlap even if they are on the same layer. We will use a *compatibility relation* π to specify what pairs of features are immune to the spacing constraints. Formally, the relation π is a binary relation on the features that is reflexive (that is, $F\pi F$ for all features) and symmetric (that is, $F_1\pi F_2 \Leftrightarrow F_2\pi F_1$), but is not necessarily transitive. We call two features F_1 and F_2 *compatible* if $F_1\pi F_2$. Examples for the interpretation of the compatibility of two features are the following:

1. Whole mask layers may be compatible. For instance, different metallization layers usually are compatible; that is, the location of no feature on one layer is restricted by any feature on the other layer.

2. Two features may be compatible if they are on the same layer and carry the same electrical signal.

A compatibility relation that contains only the types of feature pairs mentioned is also transitive; that is, it is an equivalence relation. Special compaction algorithms apply if the compatibility relation is an equivalence relation. But there are other notions of compatibility that do not lead to an equivalence relation:

3. We may want to eliminate overlap of long wires on different metallization layers so as to keep the danger of cross-talk low. In this case, the notion of compatibility defined in case 1 is replaced by the following notion: Two

features, one on each of two adjacent metallization layers, are compatible exactly if they do not represent wires that can be made to overlap for a length that exceeds some preset *cross-talk threshold*.

4. In most design-rule sets, metal wires can be run across transistors un-conditionally. However, in certain fabrication technologies, there may be a danger that a metal wire running across a transistor gate can turn the transistor partially on and cause a fault. In such technologies, a metal wire should be allowed to run across a transistor gate only if the signal on the wire does not dominate the signal on the transistor gate. Here, a signal s_1 *dominates a signal* s_2 *at a transistor t* if there is a possible state of the circuit such that signal s_1, if placed on the gate of t, turns on t while signal s_2 does not.

The last two notions of compatibility are substantially more intricate than are the first two. Deciding whether two features are compatible with respect to case 3 involves conducting a geometric analysis of the features *and* depends on the power of the compaction operation we want to perform. For instance, if one-dimensional compaction is performed, we have to determine the extent of overlap of the two features in the dimension orthogonal to the dimension in which the compaction takes place. If two-dimensional compaction is performed, then the maximum extent of both features has to be analyzed in each dimension. Deciding whether two features are compatible in case 4 is even more difficult; it may involve an analysis of the functional behavior of the circuit, and an analysis can be NP-hard. Thus, in case 4, a compatibility can be incorporated efficiently only if sufficient information about the functional behavior of the circuit is provided as part of the input.

To be suitable for compaction, a compatibility relation must be easy to compute. We provide the compatibility relation not by listing all pairs of com-patible or incompatible features, but rather by giving a *constant-time* algorithm that, given two features, decides whether they are compatible. (Such an algo-rithm can be provided easily in cases 1 to 3.) We can think of the algorithm for computing the compatibility relation as being part of the design-rule database.

10.1.2   One-Dimensional Mask-Layout Compaction

In this section, we formally define one-dimensional compaction problems. Fur-thermore, we review their complexity and illustrate the process of converting a symbolic (or mask) layout into an instance of a one-dimensional compaction problem. We concentrate on rectangular features for the purpose of discussing compaction algorithms. However, we mention extensions to general rectagonal features.

Without loss of generality, we assume that one-dimensional compaction takes place along the x axis (x *compaction*). The following concept is central to one-dimensional compaction.

Definition 10.2 (Overlapping Rectangles) *Let R_i and R_j be two rectangular features on mask layers m_i and m_j. R_j overlaps horizontally with R_i (written $R_j \prec R_i$) if and only if*

$$(y_j < y_i + h_i + d_{m_i m_j}) \wedge (y_i < y_j + h_j + d_{m_i m_j}) \wedge (x_j < x_i \vee (x_i = x_j \wedge i < j))$$

Intuitively, two rectangles overlap horizontally if there is a way of violating the spacing constraint between them by changing only their x coordinates. It is important to note that the relation $\prec$ is a partial ordering. If $R_j \prec R_i$, then R_j is always to the *left* of R_i. Ties $x_i = x_j$ are broken using the indices i and j.

We now define the one-dimensional compaction problem formally.

1D-COMPACTION

Instance: A geometric layout $L = \{R_1, \ldots, R_n\}$ with rectangular features on M mask layers; a matrix $D = ((d_{ij}))_{1 \le i, j \le M}$ of minimum separations for pairs of layers, such that $d_{ij} = d_{ji}$ for $1 \le i, j \le M$, $1 \le i \le j \le M$; a set I comprising all diverse and grouping constraints on L; and a compatibility relation π on L

Configurations: All geometric layouts $L[\alpha] = (R_1[\alpha], \ldots, R_n[\alpha])$, where $\alpha = (\alpha_1, \ldots, \alpha_n)$ is a replacement vector. The feature $R_i[\alpha]$ is obtained from the feature $R_i = (w_i, h_i, x_i, y_i, m_i)$ by a replacement along the x axis; that is, $R_i[\alpha] = (w_i, h_i, \alpha_i, y_i, m_i)$

Solutions: All *legal* geometric layouts; that is, all layouts $L[\alpha]$ such that all pairs of features in $L[\alpha]$ satisfy the grouping constraints and the spacing constraints unless they are compatible. The spacing constraints apply to only the x dimension; the y coordinates of the layout are unchanged.

Minimize: The width of $L[\alpha]$; that is, the difference between the maximum and minimum x coordinate of any point inside any rectangle of $L[\alpha]$

The size of an instance of the 1D-COMPACTION problem is the number n of features. The compatibility relation does not enter into the problem size because it is defined by a constant-time algorithm.

The 1D-COMPACTION problem is NP-hard, even with the restrictions that $M = 1$, $d_{11} = 0$, $\pi = $ id where id is the identity relation, and no diverse or grouping constraints exist [94]. This restriction will be used to exemplify most of the compaction algorithms discussed in this chapter. For this reason, it receives the special name 1D-PACKING. The intuitive reason why 1D-PACKING is hard is that it is hard to decide for two rectangles R_1 and R_2 that overlap horizontally, whether R_1 should be located to the right of R_2 or vice versa. Its complexity makes the 1D-COMPACTION problem unattractive in practice.

There is another reason why the 1D-COMPACTION problem is not used in practice. To understand this reason, we have to discuss how a symbolic layout together with a design-rule database is translated into an instance of the 1D-COMPACTION problem. The design-rule database details how the circuit elements represented by symbols in the symbolic layout are made up of mask features. It also contains the grouping constraints for the elements making up each circuit element, as well as the grouping constraints between wires contacting a circuit element and the respective mask features of the circuit element. Furthermore, it contains the algorithm computing the compatibility relation. The instance of the 1D-COMPACTION problem corresponding to a symbolic layout is defined as follows. The features of the layout are placed on the vertices and edges of a grid graph (see Figure 10.3). The circuit elements are placed on vertices, the wires are placed along edges. The mesh size of the grid is chosen large enough, initially, such that all circuit elements can be placed and connected with wires while preserving all spacing constraints. The mask features for each circuit element are placed on the location of the corresponding grid vertex such that all grouping constraints pertaining to the circuit element are fulfilled. For x compaction, the *vertical* wires are translated into rectangles on the corresponding mask layers (see Figure 10.4). The heights of these rectangles are chosen such that they fulfill all grouping constraints (with adjacent circuit elements) involving y coordinates. The rectangles are centered on the corresponding grid edges. This placement may violate some grouping constraints involving the x coordinates of a circuit element and one of its connecting wires. This phenomenon is not of concern to us, since the ensuing compaction will aim at fulfilling *all* grouping constraints.

There is a subtle point here. We mentioned that two-dimensional compaction is NP-hard. It turns out that what makes this problem NP-hard is not the area minimization. In fact, Rülling [386] has shown that, given a symbolic layout, it is NP-hard in general just to find *any* geometric layout that satisfies all grouping constraints between circuit elements and wires. This means that polynomial-time algorithms sometimes may not be able to satisfy all grouping constraints and to compute a legal layout, even if such a layout exists. Fortunately, in most technologies, there are strong restrictions on the grouping constraints between circuit elements and wires. For instance, in many technologies, only one wire can connect to a circuit element on each of its four sides. In this case, satisfying all grouping constraints is easy and can be done in linear time as part of the translation from the symbolic to the mask domain (see Exercise 10.1). This is also the reason why the NP-hardness of computing legal geometric layouts has been observed only recently. However, as we move to hierarchical compaction, circuit elements are replaced by more complicated representations of whole subcells. There can be many connections to these subcells on each side. In this case, the NP-hardness of computing legal geometrical layout can become a problem.

For x compaction, horizontal wires represent *stretchable* mask features.

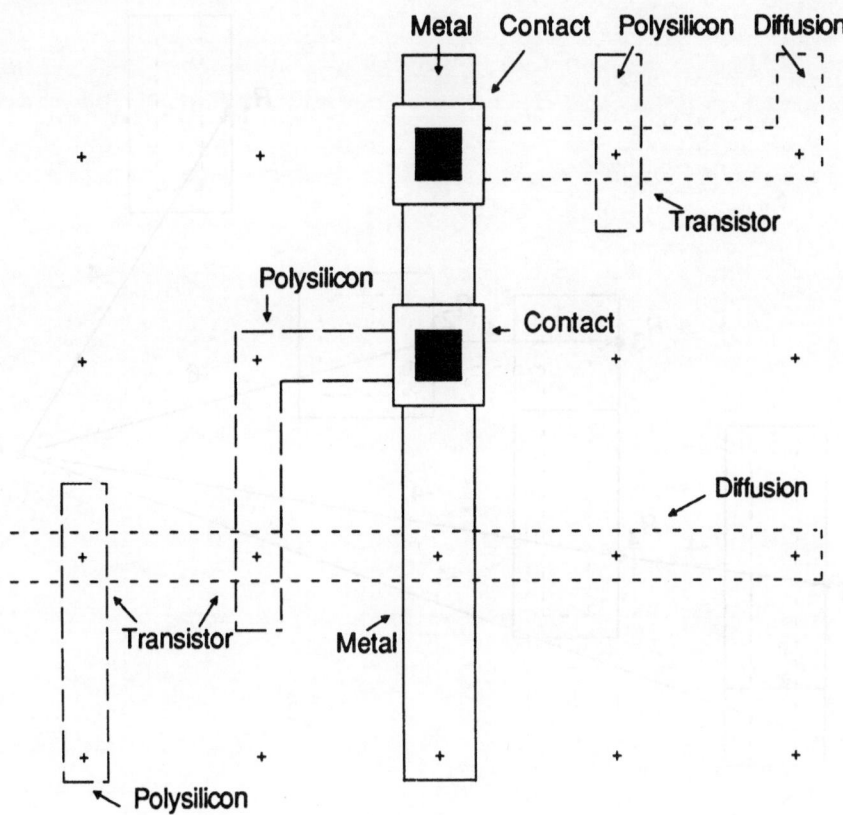

Figure 10.3: *A "fleshed-out" version of the symbolic layout before compaction. The nodes of the underlying grid are depicted with "+" signs.*

They do not generate any rectangles in the geometric layout. Rather, they translate into grouping constraints *involving y coordinates* of the circuit elements they connect. (We assume here that wires are never wider than the mask features to which they connect on the same layer. This assumption is technologically reasonable. Furthermore, it can be eliminated, if necessary; see Exercise 10.2.)

Now we can understand why, in one-dimensional compaction, we cannot make use of the freedom of choosing which of two horizontally overlapping features comes left and which right. If the compaction did switch a pair of horizontally overlapping features that are connected with a wire, then the wire would have to be rerouted in a subsequent pass. This modification would

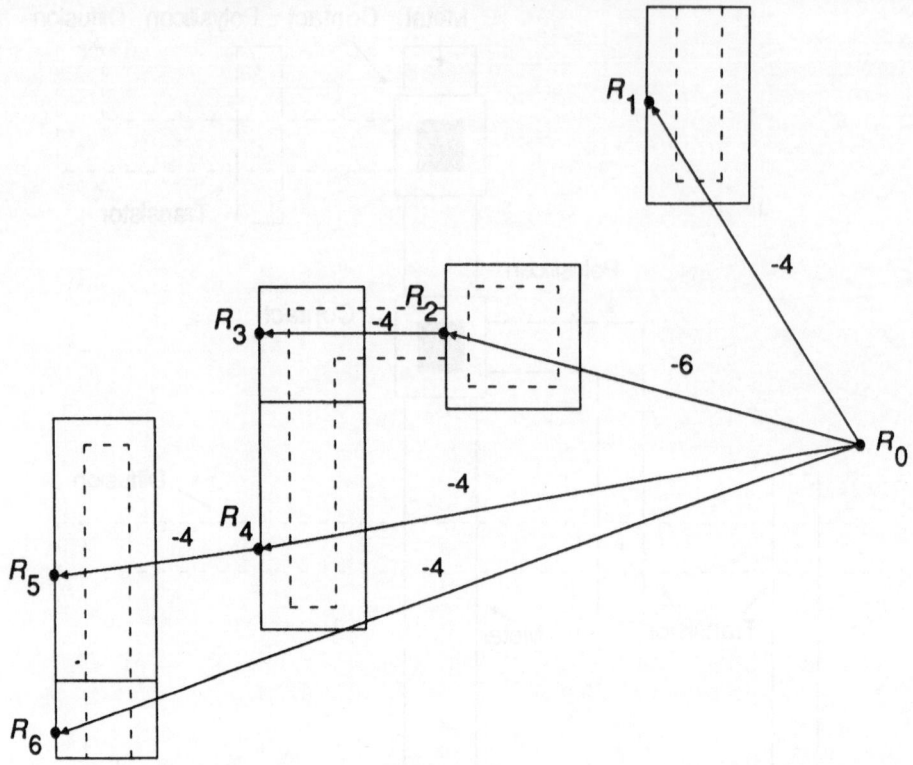

Figure 10.4: *The geometric layout for x compaction corresponding to only the features on the polysilicon layer of the layout in Figure 10.3. The rectangles are enlarged versions of the vertical features, in order to ensure minimum spacing. The layout graph is also shown.*

change the layout substantially, and would require yet another compaction.

For this reason, the following modification of the 1D-COMPACTION problem is used as the basis for practically all one-dimensional compacters.

ORDERED 1D-COMPACTION

Instance: Same as in the 1D-COMPACTION problem

Configurations: All geometric layouts $L[\alpha]$ such that, for all incompatible rectangle pairs (R_i, R_j) with $R_j \prec R_i$, we have $\alpha_j \leq \alpha_i$

Solutions: All legal geometric layouts $L[\alpha]$ that are configurations

Minimize: The width of $L[\alpha]$

In the ORDERED 1D-COMPACTION problem, we are not allowed to switch the relative order of two incompatible horizontally overlapping features. The compaction satisfies all constraints involving x coordinates if possible, and minimizes the width of the layout. A subsequent one-dimensional y compaction satisfies the constraints involving y coordinates, and minimizes the height of the layout *subject to the x placement obtained in the x-compaction phase.*

As before, the restriction that $M = 1$, $d_{11} = 0$, $\pi = $ id, and no diverse or grouping constraints exist is called ORDERED 1D-PACKING.

10.1.3 Two-Dimensional Compaction

It is quite simple to generalize the 1D-PACKING problem to two dimensions. We just need to allow that both the x and the y coordinates of the rectangle locations be variable. The cost function to be minimized is some function of the width and height of the smallest rectangle enclosing the total layout. Area and half-perimeter are examples of suitable cost functions. The corresponding problem is called the 2D-PACKING problem. The 2D-PACKING problem is not directly suitable for application in mask layout compaction. However, it is of combinatorial interest, and solution algorithms for this problem extend to two-dimensional mask layout compaction. Solution algorithms for the 2D-PACKING problem and their modifications for the application to two-dimensional mask layout compaction are discussed in Section 10.3.

10.1.4 Topological Compaction

So far, we have discussed compaction as it applies to the translation of symbolic layouts into mask data. However, variants of compaction are also of interest in general-cell placement, especially if the placement and routing phase are integrated.

Consider the general-cell placement depicted in Figure 10.5. The cells have rectagonal shapes and are separated by wiring channels. The width of a wiring channel can be estimated using the closed density of the channel implied by a global routing. The width of some channel in the placement may exceed this estimate; that is, there is *empty space* in the cell placement. Removing this empty space can be formulated as an instance of the ORDERED 1D-COMPACTION problem. In this instance, we have $M = 1$, $d_{11} = 0$, the features are derived by expanding the cells so as to account for the estimated channel widths, $\pi = $ id, and no diverse or grouping constraints are present. Dai and Kuh [79] discuss this version of one-dimensional topological compaction in an incremental designer-guided approach to layout design [81]. Note that the

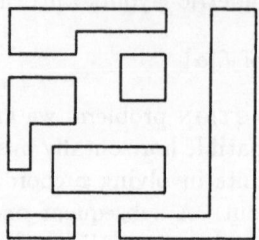

Figure 10.5: *A block placement with empty space.*

derived instance is a simple special case of the ORDERED 1D-COMPACTION problem.

This version of one-dimensional topological compaction does not take into account that moving the cells also changes the densities of some channels. Thus, after a compaction, new channel width estimates have to be computed. This computation changes the cell placement once more and, in fact, it may increase the height of the placement. So that the new cell placement can be evaluated, some global routes may have to be changed—and this again changes the estimated channel widths. It becomes clear that there is heavy interdependence among the individual subtasks in this incremental approach to layout design. So far, no scheme for systematically organizing the information flow between the subtasks has been developed.

A step toward a more systematic linkage among placement, routing, and topological compaction has been taken in the case that the detailed-routing model is the planar routing model and only two-terminal nets are involved [303, 317]. In this model, it is possible to perform a one-dimensional compaction of the cells that minimizes layout width while preserving routability. Note that neither horizontal nor vertical wire segments are represented explicitly during topological compaction. The wires are represented only implicitly by their contribution to the densities of critical cuts. Thus, the compaction takes into account all possible detailed routes of wires. However, the compaction does not alter the global route of any wire. Therefore, a subsequent iteration of the global-routing phase followed by a topological compaction phase may still lead to layout improvement.

Two properties of the planar routing model make this kind of topological compaction possible:

1. The planar routing model allows for a theory of detailed routing that reduces routability of a problem instance to the safety of all critical cuts (see Sections 9.2 and 9.3.2).

2. As cells move against each other, the density and capacity of each cut change in well-behaved ways.

However, because the spacing constraints between cells change with the densities of critical cuts during the compaction, topological compaction does not reduce to the ORDERED 1D-COMPACTION problem. Instead, a more difficult combinatorial problem has to be solved.

All attempts to lift the methods for *two-dimensional* mask layout compaction to the topological domain have been unsuccessful, as of today.

We discuss topological compaction in more detail in Section 10.5.

10.2 One-Dimensional Compaction Algorithms

There are basically two approaches to one-dimensional compaction: compression ridges and graph-based compaction. The compression-ridge approach was historically the first, but it has been largely abandoned for some time, and has only recently been revived. Most one-dimensional compacters for mask layout that are in use today use the graph-based approach.

10.2.1 Compression-Ridge Method

The compression-ridge approach was pioneered by Akers et al. [11]. Akers et al. applied the approach in a virtual-grid setting. Later Dunlop [99] applied it to the ORDERED 1D-COMPACTION problem.

In the compression-ridge approach as described by Akers et al. [11], the geometric layout presented as an input to the ORDERED 1D-COMPACTION problem is required to be legal. We illustrate the method in the single-layer case with $d_{11} = 0$, with $\pi = $ id, and without diverse or grouping constraints.

In the compression-ridge approach, a region of empty space is identified in the geometric layout that separates the layout into a left and a right part. Such a region is subsequently removed by translation of the right part of the layout to the left by as much as the spacing constraints will allow (see Figure 10.6). This step is repeated until no more compression ridges are found.

Originally, this approach was quickly abandoned because of the following disadvantages:

1. Compression ridges were grown from the top to the bottom of the layout. This growing process amounts to a search through the layout that involves backtracking (if a compression ridge cannot be grown further) and thus is computationally complex. Furthermore, it finds only horizontally convex compression ridges in the sense of Definition 9.3. However, as Figure 10.6 (b) and (c) proves some layouts are compactible even though they do not have horizontally convex compression ridges.

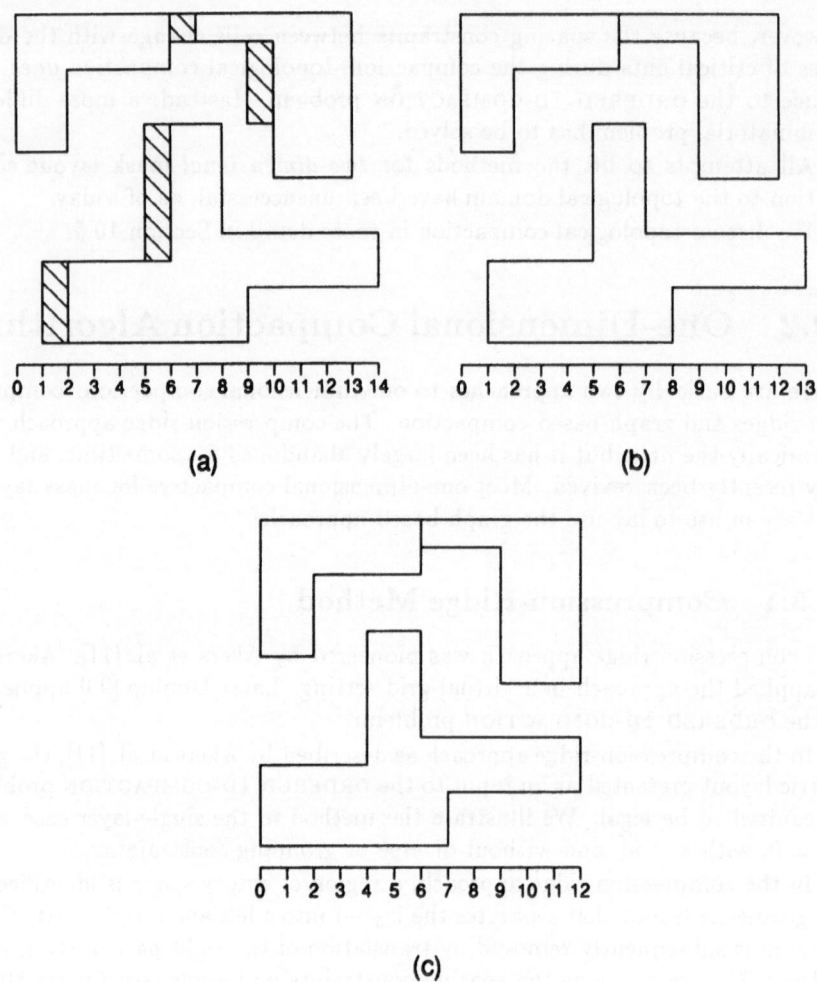

(a)

(b)

(c)

Figure 10.6: *Compaction of a one-layer layout with rectagonal features*
and $d_{11} = 0$ along a compression ridge. (a) Layout before compaction;
the compression ridge is depicted. (b) Layout after compaction; no
more horizontal compression ridge exists. Nevertheless, the layout
can be compacted further to yield (c).

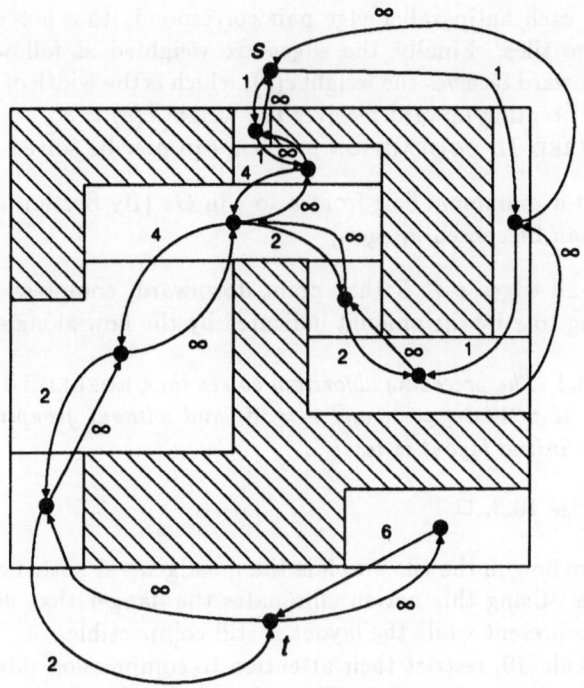

Figure 10.7: *The tile graph. The amount of the maximum flow is 2.*
The maximum flow results in the compacted layout of Figure 10.6(c).

2. Incorporating additional designer constraints on mask data made it difficult to satisfy the requirement that the initial layout be legal.

Recently, Dai and Kuh [79] have revived the compression-ridge approach. They overcome the first disadvantage by introducing a more efficient and powerful algorithm for finding compression ridges. The second disadvantage is overcome by restriction of the application of the compaction algorithm to compacting cell placements. As we noted in Section 10.1.4, this restriction leads to simple instances of the ORDERED 1D-COMPACTION problem.

The key to finding powerful compression ridges is the representation of the empty space in the layout by a directed edge-weighted graph, the *tile graph*. The tile graph is constructed as follows (see Figure 10.7). First, the horizontal segments of the contours of the features are extended through the adjacent regions of empty space. This process divides the empty space between the features into rectangular tiles. The tile graph $G = (V, E)$ has a vertex for each tile. Special vertices s and t represent the top and bottom side of the layout.

Vertices representing vertically adjacent tiles are connected with antiparallel edges. Thus, each antiparallel edge pair corresponds to a horizontal segment separating two tiles. Finally, the edges are weighted as follows: An edge e pointing downward receives the weight $c(e)$, which is the width of corresponding tile. An edge pointing upward receives the weight $c(e) = \infty$.

The ORDERED 1D-COMPACTION problem can now be solved as follows:

Step 1: Find a maximum flow from s to t in G. (By Section 3.11.1, all flow values can be chosen integer.)

Step 2: For all edges $e \in E$ that point downward, compress the tile corresponding to e by the amount indicated by the flow along e.

Theorem 10.1 *The preceding algorithm solves the* ORDERED 1D-COMPACTION *problem with $M = 1$, $d_{11} = 0$, and $\pi =$ id, and without grouping constraints, as long as the initial layout is legal.*

Proof Exercise 10.3. □

The maximum flow in the tile graph is the most general notion of compression ridge possible. Using this notion eliminates the danger that no compression ridges may be present while the layout is still compactible.

Dai and Kuh [79] restrict their attention to compression ridges that correspond to paths in the tile graph. They give an $O(n \log n)$–time algorithm for finding such a ridge that allows for the largest decrease in layout width. They propose alternation between finding compression ridges for x and y compaction as a way of dealing with the interdependence between the two dimensions. Note that finding a pathlike compression ridge takes only $O(n \log n)$ time, which is quite a bit shorter than the time required for finding maximum flows in a planar graph. However, before the compaction is complete, many compression ridges may have to be found.

The advantages of the compression ridge method are that the compaction can be broken up into small steps and that it is possible to interleave compactions in both dimensions. This feature is especially important in an environment in which compaction is done interactively, step by step. For solving the ORDERED 1D-COMPACTION problem in a batch environment in which compaction is done without designer interaction, the graph-based method described in Section 10.2.2 is more powerful and more efficient.

10.2.2 Graph-Based Compaction

The graph-based compaction method turns an instance of the ORDERED 1D-COMPACTION problem into a system of linear inequalities that is subsequently solved using shortest-path methods. We discussed similar systems of linear inequalities in Section 9.3.1.2.

We describe the graph-based method as it applies to rectangular features. Let (L, D, I, π) be an instance of the ORDERED 1D-COMPACTION problem. We have already discussed how grouping constraints translate into linear inequalities of the type

$$x_j - x_i \leq a_{ij} \tag{10.3}$$

Spacing constraints translate into inequalities of the same type where, in addition, $a_{ij} \leq 0$. For instance, the spacing constraint between two rectangles $R_i = (w_i, h_i, x_i, y_i, m_i)$ and $R_j = (w_j, h_j, x_j, y_j, m_j)$ such that $R_j \prec R_i$ amounts to the inequality

$$x_j - x_i \leq -w_j - d_{ij}$$

We can augment the system of linear inequalities that results from all grouping and spacing constraints by adding one more variable x_0 standing for the right boundary of the layout, together with the inequalities

$$x_i - x_0 \leq -w_i$$

for all rectangles R_i. The inequalities are again of the type (10.3). The variable x_0 can be thought of representing a new rectangle R_r with width zero that horizontally overlaps with all other features in the layout and is located on the right side of the layout. The separation between R_r and any other rectangle is zero. We use the notation $\overline{L} := L \cup \{R_r\}$.

The resulting system of inequalities can be solved using shortest-path methods. We represent each variable x_i by a vertex, which we also denote with R_i. Each inequality $x_j - x_i \leq a_{ij}$ is converted into a directed edge $e = (R_i, R_j)$ with length $\lambda(e) = a_{ij}$. The resulting graph $G(L, D, I, \pi) = (\overline{L}, E(L))$ is called the *layout graph*. (In the literature, $G(L, D, I, \pi)$ is also often called the *constraint graph*). Figure 10.8(a) depicts a one-layer layout without any grouping constraints and with $\pi = \mathrm{id}$, and the associated layout graph. Another example of a layout graph is given in Figure 10.2(b).

By part 3 of Lemma 3.4, an assignment of shortest distances $d_{R_r}(R_i)$ from R_r to each vertex R_i satisfies all inequalities. Furthermore, $\min_{1 \leq i \leq n} d_{R_r}(R_i)$ is the negative of the width of the resulting layout L_0. Since in L_0 all inequalities corresponding to edges on a shortest path are fulfilled with equality, L_0 has minimum width among all legal layouts. Thus, it solves the ORDERED 1D-COMPACTION problem. Figure 10.8(b) shows the solution of the instance depicted in Figure 10.8(a).

The complexity of this solution method is $O(n^3)$: The layout graph G can be generated in time $O(n^2)$. It has $n + 2$ vertices and $O(n^2)$ edges. Solving the shortest path problem on G takes time $O(n^3)$ if the Bellmann–Ford algorithm is used (Section 3.8.3).

Given the large number of features that may constitute a geometric layout, we need to improve on the run time of the algorithm. There are two ways of doing so:

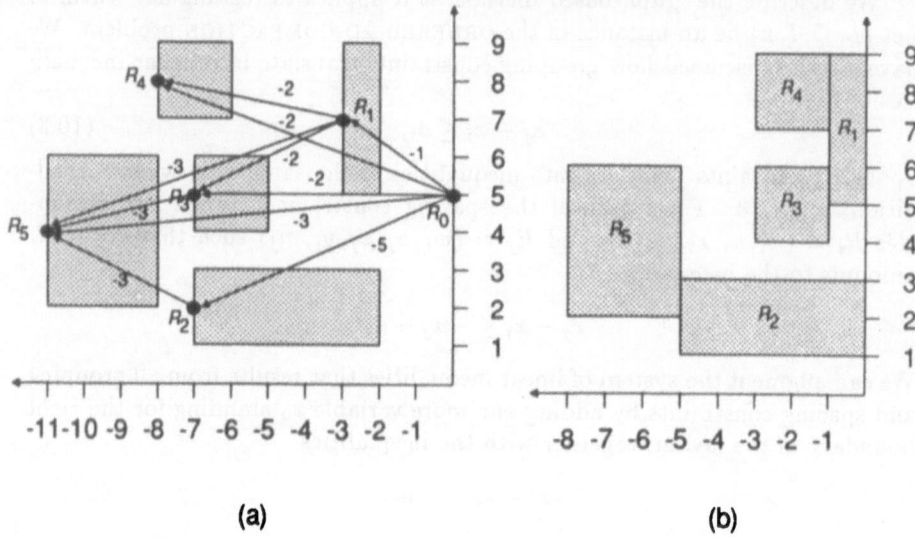

Figure 10.8: *A one-layer layout and its associated layout graph,* $d_{11} =$ 0. *(a) The original layout L and its layout graph; (b) the compacted layout L_0.*

1. Generate small variants of the layout graph quickly (Section 10.2.2.1)

2. Speed up the solution of the shortest-path problem (Section 10.2.2.2)

Beside performance issues, there is another problem with the graph-based method as outlined here. The shortest-path algorithm tends to push all features to the right as much as possible. Features that lie on a shortest path have no freedom to move, but features that do not lie on a shortest path can be placed in a variety of possible locations. This additional freedom can be used to perform further optimizations on the layout. One popular such optimization is *wire balancing*. Here, the features are placed such that the total length of the horizontal wires is minimized. (Note that horizontal wires are not explicitly represented in the geometric layout.) Technological aspects can enter this optimization. For instance, wires that are run in a medium with high resistivity can be made shorter than wires that consist of a good conductor. The wire-balancing problem is discussed in Section 10.2.2.3.

10.2.2.1   Generating Small Layout Graphs

The material in this section is taken from [94]. We reduce the size of layout graphs, and we do so by deleting spacing constraints only. Grouping constraints and diverse constraints are comparatively small in number and will not be considered in the reduction of the size of the layout graph. For this reason, we discuss layout graphs $G(L)$ for only layouts L without grouping or diverse constraints. The edges representing grouping or diverse constraints are simply added to the layout graph after it is generated by the methods discussed in this section.

10.2.2.1.1 The One-Layer Case In this section, we restrict ourselves to one-layer layouts. Section 10.2.2.1.2, we shall discuss how to handle multiple layers.

We assume again that $d_{11} = 0$. Note that the layout graphs discussed in this section are dags.

As an example, consider the layout graph in Figure 10.8(a). A quick glance at this graph shows that several edges are superfluous. Specifically, an edge between two horizontally overlapping features R_i and R_k does not need to be present if the rectangles are separated by a third rectangle R_j between them that horizontally overlaps with both R_i and R_k. For instance, the edge (R_r, R_5) in Figure 10.8(a) ensuring that R_5 comes to lie to the left of the right layout margin is superfluous. This constraint is implicitly ensured by the edges (R_r, R_3) and (R_3, R_5). By the same token, the edge (R_r, R_3) can be omitted, because of the presence of the edges (R_r, R_1) and (R_1, R_3). In general, an edge $e = (R_i, R_j)$ in the layout graph can be omitted if R_j can still be reached from R_i after deletion of e—that is, if e is a shortcut. We shall prove this fact formally now.

Definition 10.3 (Redundant Edge) *Let L be a one-layer layout, and let R_i and R_k be rectangles in $\overline{L}$, such that $R_k \prec R_i$.*

 1. *The set of rectangles between R_k and R_i is defined as follows:*

$$\beta(R_k, R_i) = \{R_j \in \overline{L} \mid R_k \prec R_j \prec R_i\}$$

 2. *An edge (R_i, R_j) in $G(L)$ is redundant if $\beta(R_j, R_i) \neq \emptyset$.*

The following lemmas show that the redundant edges are exactly the shortcuts in the layout graph.

Lemma 10.1 *If $R_k \prec R_i$, p is a path from R_i to R_k in $G(L)$, and the length of p is at least 2, then p contains a rectangle in $\beta(R_k, R_i)$.*

Proof Let $p = (R_i, \ldots, R_k)$ be a path from R_i to R_k in $G(L)$ with length at least 2. Assume indirectly that no rectangle on p is in $\beta(R_k, R_i)$. The rectangle R_j following R_i on p overlaps horizontally with R_i, and thus $\neg R_k \prec R_j$. Therefore, we have

$$y_k + h_k \;\leq\; y_j \tag{10.4}$$

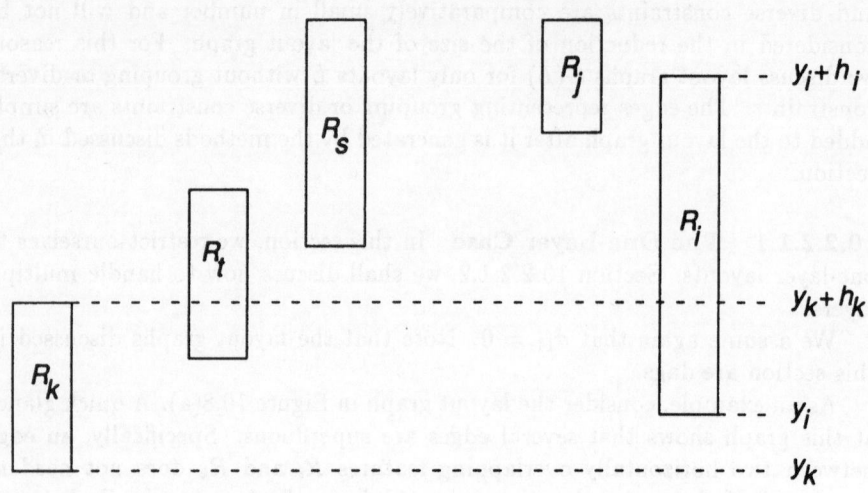

Figure 10.9: *Illustration of the proof of Lemma 10.1.*

or

$$y_j + h_j \quad \leq \quad y_k \tag{10.5}$$

We discuss only case (10.4) (see Figure 10.9); case (10.5) can be handled analogously.

From $R_j \prec R_i$ and (10.4), it follows that $y_k + h_k \leq y_j < y_i + h_i$. Let R_s be the last rectangle on p such that $y_k + h_k \leq y_s$. Then, $\neg R_k \prec R_s$. Thus, the rectangle R_t following R_s on p cannot be R_k. Putting everything together, we get

$$
\begin{array}{llll}
y_i & < & y_k + h_k & \text{since } R_k \prec R_i \\
    & \leq & y_s & \text{by definition of } s \\
    & < & y_t + h_t & \text{since } R_t \prec R_s
\end{array}
\tag{10.6}
$$

and

$$
\begin{array}{llll}
y_t & < & y_k + h_k & \text{by definition of } R_s \text{ and } R_t \\
    & \leq & y_j & \text{case (10.4)} \\
    & < & y_i + h_i & \text{since } R_j \prec R_i
\end{array}
\tag{10.7}
$$

From (10.6) and (10.7), we can infer that $R_k \prec R_t \prec R_i$; that is, $R_t \in \beta(R_k, R_i)$. This is a contradiction to our assumption that $\beta(R_k, R_i) = \emptyset$.
□

Corollary 10.2 *An edge in $G(L)$ is redundant exactly if it is a shortcut.*

Proof Lemma 10.1 proves that each shortcut is a redundant edge. On the other hand, if (R_i, R_k) is redundant, then there is a rectangle $R_j \in \beta(R_k, R_i)$. In this case, (R_i, R_j, R_k) is a path in $G(L)$; thus, (R_i, R_k) is a shortcut. $\square$

We denote the *transitive reduction* of $G(L)$—that is, the graph resulting from $G(L)$ by the deletion of all shortcut edges (Definition 3.5)—with $G^r(L) = (\overline{L}, E^r(L))$. The following lemma proves that omitting shortcuts does not change the lengths of shortest paths in $G(L)$.

Lemma 10.2 *Let $G'(L) = (\overline{L}, E'(L))$ such that $E'(L) \subseteq E(L)$. The length of the shortest path from R_i to R_j is the same in $G'(L)$ and $G(L)$ for all pairs of rectangles $R_i, R_j \in \overline{L}$ exactly if $E^r(L) \subseteq E'(L)$.*

Proof

" $\Leftarrow$ ": No removal of a redundant edge changes the length of any shortest path, because all rectangle widths are nonnegative and $d_{11} = 0$.

" $\Rightarrow$ ": If we remove a nonredundant edge (R_i, R_j), then R_i can no longer be reached from R_j. Thus, the shortest path length from R_i to R_j increases to ∞.

$\square$

By Lemma 10.2, it suffices to generate the edges in $E^r(L)$ for one-dimensional compaction. However, if we omit nonredundant constraints, we may miss a critical constraint for compaction. Thus, the goal for efficient one-dimensional compaction is to generate all nonredundant constraints but as few redundant constraints as possible.

Figure 10.10 shows the transitive reduction of the layout graph depicted in Figure 10.8(a). $G^r(L)$ may have many fewer edges than $G(L)$ does, but it is not always computationally efficient to generate $G^r(L)$. In general, computing the transitive reduction takes as much time as does computing the transitive closure. This operation, in turn, takes as long as multiplying two $n \times n$ matrices. The best asymptotic value for this quantity is $O(n^{2.38})$ [74]. The fastest *practical* algorithms for computing the transitive reduction take time $O(mn)$. This is more than the size of the original layout graph, and thus is prohibitively large for our application. What we need is a way of computing the transitive reduction or only a slightly larger graph in almost linear time. Fortunately, if $\pi = \mathrm{id}$, the transitive reduction of a one-layer layout graph can be computed in time $O(n \log n)$. In the following section, we discuss how this is done.

10.2.2.1.1.1 Trivial Compatibility In this section, we consider the case $\pi = \mathrm{id}$; that is, we discuss the ORDERED 1D-PACKING problem. The critical observation is that the corresponding layout graphs are interval dags.

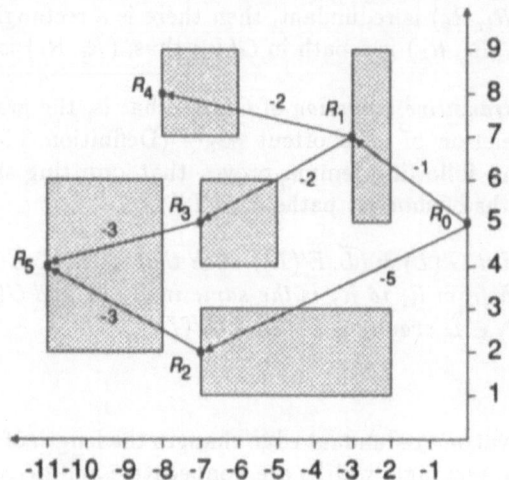

Figure 10.10: *The transitive reduction of the layout graph in Figure 10.8(a).*

Definition 10.4 (Interval Dag) *An* **interval dag** *is a dag whose underlying undirected graph is an interval graph (see Definition 9.26).*

In fact, we obtain an interval representation of the layout graph by representing the vertex R_i by the open interval $(y_i, y_i + h_i)$. Since $\pi = \mathrm{id}$, all pairs of horizontally overlapping intervals define constraints. On the other hand, each interval dag with a unique source whose underlying undirected graph is an interval graph is the layout graph of a one-layer layout. In the respective layout, the vertex v_i represents a rectangle $R_i = (w_i, h_i, x_i, y_i)$ such that $-x_i$ is the number of v_i in a topological ordering of the dag, y_i is the left endpoint of an open interval representing v_i in an interval representation, h_i is the length of the corresponding interval, and $w_i = 0$. Thus, we are faced with the task of finding the transitive closure of interval dags. This goal can be achieved using a plane-sweep method.

We have already encountered plane-sweep algorithms in the construction of the track graph for line-searching algorithms in global routing (Section 8.4.2.3) and in channel routing in the jog-free Manhattan model (Exercise 9.30). We also mentioned that plane-sweep methods are an essential tool in homotopic planar routing (Section 9.3.2).

Conceptually, a horizontal scan line is swept vertically across the layout, say, from bottom to top. During the sweep, events of two kinds can happen:

1. $low(R_j)$: The sweep line crosses the bottom side of rectangle R_j.

2. $high(R_j)$: The sweep line crosses the top side of rectangle R_j.

Each of these events occurs at some specific time during the sweep. This instant is called the *time stamp* of the event. Also, each event leads to some algorithmic action. In general, the plane-sweep paradigm uses two data structures:

The event queue Q: This data structure is a sorted list of the events, in the order of increasing time stamp. If several events have the same time stamp, a tie-breaking procedure is invoked that depends on the application.

The sweep data structure D: This dynamic data structure contains the information that has been accumulated during the sweep so far and that is relevant for future events. We distinguish between two kinds of sweep.

> **Memoryless sweep:** Here the sweep data structure D contains only information about the features currently crossing the sweep line.

> **Sweep with memory:** Here the sweep data structure D also contains information about features that the sweep has passed over completely.

In our present application of the plane-sweep paradigm, the event $low(R_j)$ has the time stamp y_j, and the event $high(R_j)$ has the time stamp $y_j + h_j$. Among several events occurring at the same time, *high* events precede *low* events in the event queue.

The sweep data structure D implements a memoryless sweep; that is, we keep only information that pertains to rectangles currently crossing the sweep line. Specifically, D maintains a set of intervals I_j. The interval $I_j = (x_j, x_j + w_j)$ is in D for as long as the rectangle $R_j = (w_j, h_j, x_j, y_j)$ crosses the sweep line. Thus, the algorithmic action at the event $low(R_j)$ and $high(R_j)$ includes inserting R_j in D and removing R_j from D, respectively. In addition, the edges of the transitive reduction have to be generated.

So that this can be done, the intervals I_j are kept in sorted order with respect to the $\prec$ relation. Furthermore, in addition to insertion and deletion of elements, D supports the following operations:

$\ell(R_j)$: Returns the left neighbor of R_j in D with respect to $\prec$

$r(R_j)$: Returns the right neighbor of R_j in D with respect to $\prec$

In addition to maintaining intervals in D, we keep a set of candidate edges for inclusion in the transitive reduction. These edges are stored in an array *cand* that is indexed with rectangles from $\overline{L}$. The fact that $cand(R_i) = R_j$ is to be interpreted as "(R_i, R_j) is a candidate edge for the transitive reduction".

At the event $low(R_j)$, the edges $(R_j, \ell(R_j))$ and $(r(R_j), R_j)$ in $G(L)$ become candidate edges for the transitive reduction. On the other hand, the edge

```
(1)         procedure Insert(Rⱼ : rectangle);
(2)         begin
(3)             insert Iⱼ into D;
(4)             cand(Rⱼ) := ℓ(Rⱼ);
(5)             cand(r(Rⱼ)) := Rⱼ
(6)         end;

(7)         procedure Delete(Rⱼ : rectangle);
(8)         begin
(9)             if ℓ(Rⱼ) ≠ nil and ℓ(Rⱼ) = cand(Rⱼ) then
(10)                output (Rⱼ, ℓ(Rⱼ)) fi;
(11)            if cand(r(Rⱼ)) = Rⱼ then
(12)                output (r(Rⱼ), Rⱼ) fi;
(13)            delete Iⱼ from D
(14)        end;
```

Figure 10.11: *The event actions during the plane sweep.*

$(r(R_j), \ell(R_j))$ ceases to be a candidate edge, because it has been established that $R_j \in \beta(\ell(R_j), r(R_j))$. If at the event $high(R_j)$ the edges $(R_j, \ell(R_j))$ and $(r(R_j), R_j)$ are still candidate edges, then they belong to the transitive reduction, because now it has been established that $\beta(\ell(R_j), R_j) = \emptyset$ and $\beta(R_j, r(R_j)) = \emptyset$, respectively.

These observations lead to the algorithms Insert(R_j) and Delete(R_j) in Figure 10.11 that are executed at the event $low(R_j)$ and $high(R_j)$, respectively. Initially all elements of the *cand* array are **nil**, and $D = \emptyset$. We denote the resulting plane-sweep algorithm with Gen_0. In the following theorem, we formally prove the correctness of Gen_0.

Theorem 10.3 *The algorithm Gen_0 outputs exactly the edges in $E^r(L)$.*

Proof We use induction on the time during the plane sweep. Without loss of generality, let us assume that no two events have the same time stamp. (Otherwise, we modify the layout L by slightly changing the time stamps. This assumption can be made to hold without affecting $G(L)$ or $G^r(L)$.) Let ev_i denote the ith event in the event queue, and let $t(ev_i)$ denote its time stamp. Note that $ev_0 = low(R_0)$ and $ev_{2n+1} = high(R_0)$ by definition of R_0. We choose snapshot times t_ℓ, $-1 \leq \ell \leq 2n+1$ between events; that is,

$$t_{-1} < t(ev_0), \qquad t_{2n+1} > t(ev_{2n+1}),$$
$$t(ev_\ell) < t_\ell < t(ev_{\ell+1}), \quad 0 \leq \ell \leq 2n$$

We then prove the following statement by induction on ℓ.

Inductive hypothesis: Let L_{t_ℓ} be the portion of the layout $\overline{L}$ over which the sweep line has crossed up to time t_ℓ. An edge $e = (R_i, R_j)$ is contained in $E^r(L_{t_\ell})$ exactly if

$$e \text{ has been output before time } t_\ell \tag{10.8}$$

or

$$cand(R_i) = R_j \text{ and } I_i, I_j \in D \text{ at time } t_\ell \tag{10.9}$$

Furthermore, if (10.8) holds, then $e \in E^r(L_{t_s})$ for all $s \geq \ell$.

If applied to time t_{2n+1}, the inductive hypothesis implies the theorem. Indeed, at time t_{2n+1}, $D = \emptyset$, and $L_{t_{2n+1}} = \overline{L}$. Thus, all edges in $E^r(L)$ have been output.

For $\ell = -1$, $D = \emptyset$, and no edge fulfills (10.8) or (10.9). Furthermore, $L_{t_{-1}} = \emptyset$. Thus, the inductive hypothesis holds.

Now assume $\ell \geq 0$ and the inductive hypothesis holds for values less than ℓ. By definition of the procedures Insert and Delete, if $cand(R_j) \in D$, then $cand(R_j) = \ell(R_j)$. We make the following case distinction.

Case 1: $ev_\ell = low(R_j)$. The new candidate edges $(R_j, \ell(R_j))$ and $(r(R_j), R_j)$ are accounted for in lines 4 and 5 of the algorithm in Figure 10.11. The removal of the edge $(r(R_j), \ell(R_j))$ from $E^r(L_{t_{\ell-1}})$ is accounted for by resetting of $cand(r(R_j))$ to a value different from R_j in line 5.

Case 2: $ev_\ell = high(R_j)$. By induction, if $cand(R_j) = \ell(R_j)$, then $(R_j, \ell(R_j)) \in E^r(L_{t_{\ell-1}})$. Since, in the event ev_ℓ, rectangle R_j has left the sweep line, $\beta(\ell(R_j), R_j) = \emptyset$ has been established. Thus, $(R_j, \ell(R_j)) \in E^r(L_{t_s})$ for all $s \geq \ell$. An analogous statement holds for the edge $(r(R_j), R_j)$. Thus, these two edges can now be output (lines 10 and 12 of Figure 10.11).

□

The data structure D can be implemented using a leaf-chained balanced search tree. This implementation leads to the following execution times for the operations to be performed on D:

Insert I_j into D	$O(\log|D|)$
Delete I_j from D	$O(\log|D|)$
$\ell(R_j)$	$O(1)$
$r(R_j)$	$O(1)$
Initialization	$O(n)$

Exactly $n + 1$ events of each kind are executed. Thus, the total time for the algorithm is $O(n \log n)$. This value is also a lower bound on the time needed to compute the transitive reduction of an interval dag (see Exercise 10.4).

We can also make a strong statement on the number of generated constraints.

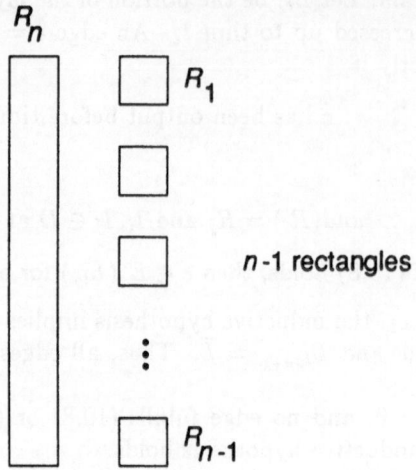

Figure 10.12: *A lower-bound example for Lemma 10.3.*

Lemma 10.3

$$|E^r(L)| \le \begin{cases} 1 & n = 1 \\ 2n - 2 & n > 1 \end{cases}$$

Proof $G^r(L)$ is a triangle-free graph with $n + 1$ vertices. (A graph is called *triangle-free* if it has no cycle of length 3.) For such graphs, the preceding bound is well known [174]. Alternatively, the bound can be derived by a direct inspection of the algorithm: $E^r(L_{-1}) = \emptyset$. If ev_ℓ is a *high* event, then $E^r(L_{t_{\ell-1}}) = E^r(L_{t_\ell})$. If ev_ℓ is the first or second *low* event, then $E^r(L_{t_\ell})$ has at most one edge more than $E^r(L_{t_{\ell-1}})$. Otherwise, $E^r(L_{t_\ell})$ has at most two edges more than $E^r(L_{t_{\ell-1}})$. □

Thus, the number of edges generated by Gen_0 is linear in the number of rectangles. In contrast, if we assume roughly uniformly distributed rectangles in an approximately square layout, then the number of edges in $G(L)$ can be expected to be $O(n^{1.5})$. This difference indicates how much superior generating the transitive reduction of a layout graph can be. Figure 10.12 shows that the bound shown in Lemma 10.3 is tight.

Exercise 10.6 discusses an alternative method for constructing sparse layout graphs, which has been used in compacters [14].

10.2.2.1.1.2 Nontrivial Compatibilities The algorithm Gen_0 for generating the layout graph assumes that any pair of rectangles is incompatible.

Two compatible rectangles can overlap in the compacted layout without violating any constraints. Such overlap is prohibited by the layout graph produced by Gen_0.

We can increase the mobility of compatible rectangles with respect to each other by modifying the edge lengths in the layout graph. Assume that R_i and R_j are two rectangles such that $R_j \prec R_i$. Thus, there is an edge $e = (R_i, R_j)$ in $G(L)$. In general, the larger the length of e, the more freedom we have in the movement of R_i with respect to R_j. If R_i and R_j are compatible (that is, $R_i \pi R_j$), then the edge length of e can be increased from $\lambda(e) = -w_j$ to $\lambda(e) = \min\{w_i - w_j, 0\}$. This value is the largest possible, because

- If $\lambda(e) > 0$, then the left boundary of R_j can move to the right of the left boundary of R_i. Thus, some shortcut edge diverging from R_i may become significant for the shortest-path computation.

- If $\lambda(e) > w_i - w_j$, then the right boundary of R_j can move to the right of the right boundary of R_i. Thus, some redundant edge converging in R_j may become significant for the shortest-path computation.

In both cases, Lemma 10.2 would be invalidated. Consequently, the layout graph $G^r(L)$ does not allow a swap in the positions of two compatible but horizontally overlapping rectangles.

As an example, we return to Figure 10.3. Rectangles R_2 and R_3 in Figure 10.3(b) are compatible, because they represent different segments of the same wire. Figure 10.13(a) shows the result of one-dimensional compaction using $G(L)$ or equivalently $G^r(L)$ with adjusted edge lengths, taking into account the compatibility of R_2 and R_3. However, the mask layout depicted in Figure 10.13(b) is also valid. This layout can be obtained only by swapping of R_2 with R_3. To do this swap, we must delete an edge between a pair of compatible rectangles in $G(L)$—namely, between R_2 and R_3.

If we generalize this observation to arbitrary layouts, we see that the layout graph we want to use for compaction is the graph $G_\pi(L) = (\overline{L}, E_\pi(L))$, where

$$E_\pi(L) = \{(R_i, R_j) \in E(L) \,|\, \neg R_i \pi R_j\}$$

Although $G_\pi(L)$ may be a small subgraph of $G(L)$, the transitive reduction $G_\pi^r(L)$ of $G_\pi(L)$ may be much larger than $G^r(L)$. As an example, consider the layout depicted in Figure 10.14(a). The transitive reduction $G_\pi^r(L)$ of the corresponding layout graph is depicted in Figure 10.14(b). It has $\Omega(n^2)$ edges. Therefore, in general, computing the transitive reduction of $G_\pi(L)$ may not lead to layout graphs small enough for efficient compaction. Besides $G_\pi^r(L)$ potentially being large, it also may be difficult to compute. The reason for this difficulty is that $G_\pi(L)$ is not an interval dag any more.

There are several routes out of this dilemma. One is that an example such as depicted in Figure 10.14(a) does not naturally occur in layout applications.

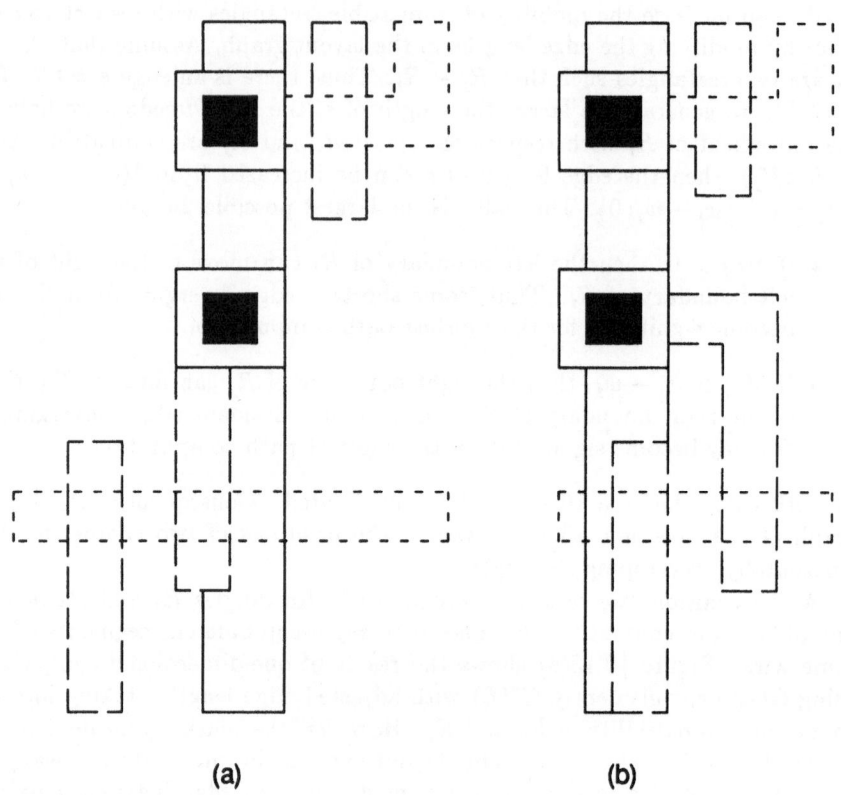

(a)   (b)

Figure 10.13: *Swapping compatible rectangles. (a) The result of compacting the layout in Figure 10.3(a) using the layout graph depicted in Figure 10.3(b) for the polysilicon layer, and setting the length of the edge (R_2, R_3) to 0 to account for the compatibility of R_2 and R_3. (b) Another compacted version of the layout in Figure 10.3(a). In this version, the polysilicon features represented by the rectangles R_2 and R_3 have been swapped. We obtain the corresponding layout graph from Figure 10.3(b) by deleting the edge (R_2, R_3). The width of the layout is six units smaller than in (a).*

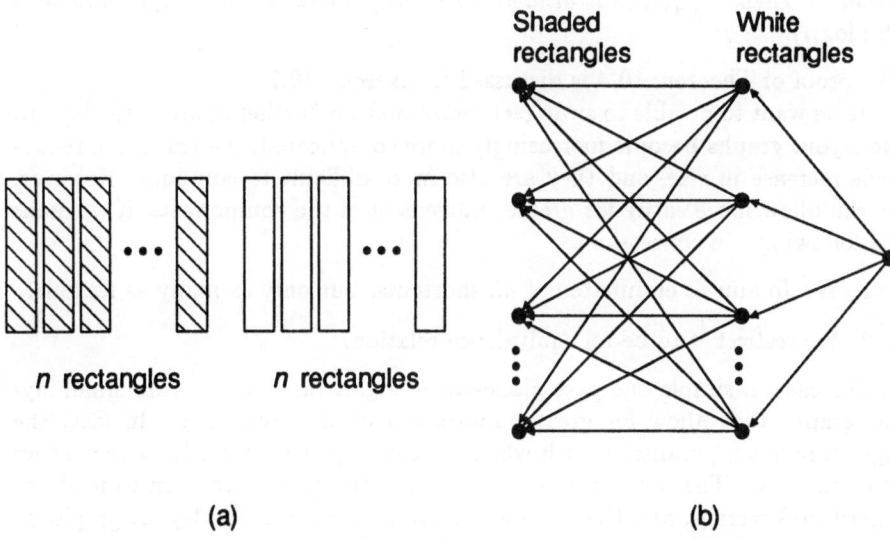

Figure 10.14: *A layout whose layout graph has many nonredundant edges. (a) The layout L; the shaded rectangles are pairwise compatible as are the white rectangles. No shaded rectangle is compatible with a white rectangle. (b) The layout graph $G_\pi(L)$.*

Specifically, it is seldom the case that, on the same layer, a rectangle can move across many compatible rectangles. In fact, most often a rectangle can move across only *one* compatible rectangle next to it before it encounters an incompatible rectangle. This is the case, for instance, in Figure 10.13(b). In a few situations, a rectangle may be able to move across a few more compatible rectangles, but the number is never very large. We can exploit this observation for modifying the layout graph considered during compaction.

If we can be certain that only pairs of rectangles can be swapped that are next to each other in the layout, we can use the layout graph $G_{1,\pi}(L) = (\overline{L}, E_{1,\pi}(L))$, where

$$E_{1,\pi}(L) = \{(R_i, R_j) \in E(L) \mid \beta(R_j, R_i) \neq \emptyset \text{ or } \neg R_i \pi R_j\}$$

In $G_{1,\pi}$, only the compatibilities of *neighboring* rectangles R_i and R_j—that is, rectangles such that $\beta(R_j, R_i) = \emptyset$—are taken into account. It turns out that the transitive reduction $G_{1,\pi}^r(L)$ of $G_{1,\pi}(L)$ is always small and can be computed quickly.

Theorem 10.4 (Doenhardt, Lengauer [94]) *Let L be a one-layer layout without diverse or grouping constraints and with an arbitrary compatibility re-*

lation π. Then, $G^r_{1,\pi}(L)$ *has at most* $4n-6$ *edges and can be computed in time* $O(n \log n)$.

The proof of Theorem 10.4 is discussed in Exercise 10.7.

If we want to be able to swap rectangles that are farther apart in the layout, the layout graphs become increasingly more complicated, the transitive reductions increase in size, and they are also more difficult to compute. However, we can allow more easily for greater movement of the components, if we make the following two concessions:

1. We do aim to eliminate not *all* shortcuts, but only as many as feasible.

2. We restrict π to be an equivalence relation.

In this case, a simple one-pass plane-sweep algorithm can generate small layout graphs that allow for greater movement of the rectangles. In fact, the algorithm has a parameter with which we can adjust how much movement we want to allow. The more movement we want, the greater the run time of the algorithm becomes, and the more edges are generated in the layout graph. In the remainder of this section, we explain this algorithm.

The parameter that can be used to tune the power of the algorithm is denoted with k, and the corresponding algorithm is called $Gen(k)$. Roughly, $Gen(k)$ allows us to move a rectangle across at most k compatible rectangles to its left or right. $Gen(k)$ has a run time of $O(n(k + \log n))$ and generates at most $2(k + 1)n$ constraints. We first describe the algorithm $Gen(0)$. Then, we extend the discussion to general $k \geq 0$.

Instead of checking for two rectangles R_i and R_j with $R_j \prec R_i$ whether $\beta(R_j, R_i) = \emptyset$ for the *complete* time interval that R_i and R_j cross the sweep line, the algorithm $Gen(k)$ guesses whether $\beta(R_j, R_i) = \emptyset$ by observing R_i and R_j at only *one* time instant. This instant is called the *test time* for R_i and R_j, and it is the time instant at which the last rectangle of the two is encountered by the sweep; that is, it is the time $\max\{y_i, y_j\}$. If at this time there is no rectangle R crossing the sweep line such that $R_j \prec R \prec R_i$, then the edge (R_i, R_j) is added to the layout graph. The corresponding event action procedures are depicted in Figure 10.15. The graph constructed by the algorithm $Gen(0)$ is denoted by $G(0, L, \pi) = (\overline{L}, E(0, L, \pi))$. Of course, $E(0, L, \pi)$ may contain redundant edges. For instance, in Figure 10.8(a), the redundant edge (R_0, R_3) is contained in $G(0, L, \pi)$. However, the total number of edges in $E(0, L, \pi)$ is still at most $2n$. This is the case because edges are added in only Insert actions, and each such action adds at most two edges. Of course, compaction using $G(0, L, \pi)$ has the same result as compaction using $G(L)$, because $E^r(L) \subseteq E(0, L, \pi) \subseteq E(L)$.

We now generalize the algorithm $Gen(0)$ to allow for swapping rectangles. To do so, we define a layout graph $G(k, L, \pi) = (\overline{L}, E(k, L, \pi))$ and modify $Gen(0)$ to compute $G(k, L, \pi)$. To this end, we keep the test time for the edge (R_i, R_j), but we modify the test criterion to allow for more rectangle movement. The generalized test criterion depends on k and is defined as follows.

```
(1)        procedure Insert(R_j : rectangle);
(2)        begin
(3)            insert I_j into D;
(4)            if ℓ(R_j) ≠ nil then
(5)            output (R_j, ℓ(R_j)) fi
(6)        output (r(R_j), R_j)
(7)        end;

(8)        procedure Delete(R_j : rectangle);
(9)        delete I_j from D;
```

Figure 10.15: *Fast generation of a layout graph.*

Definition 10.5 (Test Criterion) *Let L be a layout with a compatibility relation π that is an equivalence relation, and let $R_i, R_j \in \overline{L}$ such that $R_j \prec R_i$. Then, $(R_i, R_j) \in E(k, L, \pi)$ exactly if the following condition holds. If $R'_1, \ldots, R'_\ell$ are the rectangles between R_j and R_i crossing the sweep line at the test time for (R_i, R_j)—that is, $R_j \prec R'_\ell \prec R'_{\ell-1} \cdots \prec R'_2 \prec R'_1 \prec R_i$—then either of the following is true:*

1. *$\neg R_i \pi R_j$, all $R'_1, \ldots, R'_h$ are compatible with R_i, and $R'_{h+1}, \ldots, R'_\ell$ are all compatible with R_j, for some h, $0 \leq h, \ell - h \leq k$.*

2. *$R_i \pi R_j$, all $R'_1, \ldots, R'_\ell$ are compatible with R_i and R_j, and $k \leq \ell \leq 2k$.*

Intuitively, the test for (R_i, R_j) is positive if, at the test time, we can make R_j and R_i neighbors by moving each of them across at most k compatible rectangles. Definition 10.5 is adjusted carefully to implement this idea and to make $E(k, L, \pi)$ quickly computable.

In the following, we prove the relevant structural properties of $E(k, L, \pi)$.

Lemma 10.4

$$E^r_\pi(L) \subseteq E(n, L, \pi) \subseteq E_\pi(L)$$

Proof Obviously, $E(n, L, \pi) \subseteq E_\pi(L)$. Assume now indirectly that $(R_i, R_j) \in E^r_\pi(L) \setminus E(n, L, \pi)$. Then, $R_j \prec R_i$, $\neg R_i \pi R_j$, and, at the test time of (R_i, R_j), the test criterion of Definition 10.5 does not hold. We distinguish two cases:

Case 1: Using the notation of option 1 in Definition 10.5, there is a rectangle R'_m such that $\neg R_i \pi R'_m$ and $\neg R_j \pi R'_m$. Then, $(R_i, R'_m), (R'_m, R_j) \in E_\pi(L)$, and thus $(R_i, R_j) \notin E^r_\pi(L)$.

Case 2: There are two rectangles R'_s and R'_t, $s < t$, such that $R_i \pi R_t$ and $R_j \pi R_s$. Then, since π is an equivalence relation, $\neg R_s \pi R_t$ holds. Thus, $(R_i, R_s), (R_s, R_t), (R_t, R_j) \in E_\pi(L)$, and again $(R_i, R_j) \notin E_\pi^\tau(L)$.

Since π is an equivalence relation, no other cases are possible. $\square$

Lemma 10.4 proves that compaction on the basis of $G(n, L, \pi)$ yields the same result as does compaction on the basis of $G_\pi(L)$—that is, compaction with completely unrestricted rectangle movement. We have already seen that compaction on the basis of $G(0, L, \pi)$ is the same as compaction on the basis of $G(L)$. Thus, we now have established that the cases $k = 0$ and $k = n$ span the whole range between completely free movement and excluding swapping of any rectangle pair.

The following lemma shows that, as k increases from 0 to n, we have a smooth transition between these two extremes. Specifically, we will show that, as k increases, the flexibility of movement among the rectangles does so as well. Let $G^* = (V, E^*)$ denote the transitive closure of $G = (V, E)$ (see Definition 3.5). By Lemma 10.2, compaction on the basis of G^* yields the same result as does compaction on the basis of G.

Lemma 10.5

$$E^*(k, L, \pi) \subseteq E^*(k - 1, L, \pi)$$

Proof We prove the lemma by induction on decreasing k. We show the general step. The initial step $k = n$ is subsumed in the proof. Our argument proceeds by showing that, if for some k, $(R_i, R_j) \in E(k, L, \pi)$, then R_j is also reachable from R_i in $E(k - 1, L, \pi)$. We again make a case distinction:

Case 1: $\neg R_i \pi R_j$. If $(R_i, R_j) \in E(k, L, \pi)$, then the test criterion of option 1 in Definition 10.5 holds. Let ℓ, h be as in option 1 in Definition 10.5.

> **Case 1.1:** $h, \ell - h \le k - 1$. In this case, $(R_i, R_j) \in E(k - 1, L, \pi)$ trivially by option 1 in Definition 10.5.

> **Case 1.2:** $h = k$ and $\ell - h \le k - 1$. By option 2 in Definition 10.5, $(R_i, R'_h) \in E(k - 1, L, \pi)$; and, by option 1 in Definition 10.5, $(R'_h, R_j) \in E(k - 1, L, \pi)$.

> **Case 1.3:** $h \le k - 1$ and $\ell - h = k$. Analogous to case 1.2.

> **Case 1.4:** $h, \ell - h = k$. In this case, $(R_i, R'_h), (R'_{h+1}, R_j) \in E(k-1, L, \pi)$ by option 2 in Definition 10.5. Furthermore, $\neg R'_h \pi R'_{h+1}$, which implies

$$
\begin{aligned}
(R'_h, R'_{h+1}) \quad &\in \quad E_\pi(L) \\
&\subseteq \quad E_\pi^*(L) \\
&= \quad E^*(n, L, \pi) \quad \text{by Lemma 10.4} \\
&\subseteq \quad E^*(k, L, \pi) \quad \text{by induction.}
\end{aligned}
$$

The last inequality is trivial for the initial step $k = n$.

Case 2: $R_i \pi R_j$. Then, option 2 in Definition 10.5 holds. Let ℓ be as in this option.

  Case 2.1: $\ell \leq 2k - 2$. Directly by option 2 in Definition 10.5, we have $(R_i, R_j) \in E(k - 1, L, \pi)$.

  Case 2.2: $\ell = 2k - 1$. Again by option 2 in Definition 10.5, we have $(R_i, R'_k), (R'_k, R_j) \in E(k - 1, L, \pi)$.

  Case 2.3: $\ell = 2k$. By option 2 in Definition 10.5, $(R_i, R'_k) \in E(k - 1, L, \pi)$. If $k > 1$, then $k < 2k - 2$ and we get $(R'_k, R_j) \in E(k - 1, L, \pi)$, as in case 2.1. If $k = 1$, then (R_i, R'_1, R'_2, R_j) is a path in $E(0, L, \pi)$.

□

Exercise 10.8 discusses the extension of algorithm $Gen(0)$ to the efficient algorithm $Gen(k)$ computing $G(k, L, \pi)$. The general idea of the algorithm is to augment the Insert procedure in Figure 10.15 such that, after entering R_j, a scan is started toward the left of R_j to test the criteria of Definition 10.5. This scan is then repeated toward the right. In total, $2k + 1$ rectangles must be checked on either side. Thus, the run time per call of Insert is $O(k + \log |D|)$. At most $2k + 1$ edges are generated per call to Insert. Thus, the total number of edges is at most $2(k + 1)n$. Exercise 10.9 discusses the quality of this upper bound.

10.2.2.1.2 Multiple Layers The algorithms in Section 10.2.2.1.1 consider only one-layer layouts. In principle, we can handle multilayer layouts by analyzing one layer at a time, generating the corresponding edges, and merging the edges for all layers in a single layout graph. In this approach, interfering layers have to be mapped onto sets of mutually compatible layers. We can achieve this goal by defining one "virtual" layer for each pair of interfering layers. The virtual layer contains all rectangles on the two interfering layers. Two rectangles that are on the same layer are compatible on the virtual layer.

10.2.2.2 Finding Shortest Paths Quickly

In Section 10.2.2.1, we discussed how to generate small layout graphs. We were quite successful in that we could generate graphs with $O(n)$ edges in time $O(n \log n)$. In this section, we want to take advantage of these results by complementing them with specially tailored efficient shortest-path algorithms.

To devise efficient shortest-path algorithms, we need to understand the structure of layout graphs. We have seen that the spacing constraints lead to a dag. This dag is then augmented with edges representing grouping and diverse constraints. We now discuss the influence of these additional constraints on the structure of the layout graph.

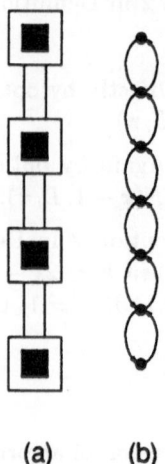

Figure 10.16: *(a) A chain of sliding ports; (b) its representation in the layout graph.*

10.2.2.2.1 Grouping Constraints

Grouping constraints either are *tight*—that is, equality constraints, see (10.1)—or they come from sliding ports, see (10.2). In the first case, only one of the rectangles R_i and R_j that are tied together by the grouping constraint—say, R_i—need to be represented as a vertex in the layout graph. The other rectangle R_j can be omitted, and edges incident in R_j can be converted to edges incident in R_i, where the edge length is suitably adjusted. A sliding port leads to a pair of antiparallel edges in the layout graph. In layouts, sliding ports occur in chains. For instance, for x compaction, grouping constraints that represent sliding ports tie together chains of rectangles that are arranged vertically in the layout; see Figure 10.16(a). Different such chains are disjoint, and two rectangles inside a single chain do not observe any spacing constraints. Therefore, the layout graph has the structure illustrated in Figure 10.16(b). Specifically, the layout graph has the property that each strong component is a chain of antiparallel edges. Furthermore, it has a unique source—namely, R_r. We call such a graph a *chain dag*. The following lemma is the basis for a linear-time shortest-path algorithm on chain dags.

Lemma 10.6 (Lengauer, Mehlhorn [279]) *Let G be a chain dag with a single source s. Then, (G, s) is BF-orderable (Definition 3.15), and a BF-ordering can be computed in time $O(m + n)$.*

(1)   Compute a list $(C_1, \ldots, C_k)$ of the strong components of G
                in topological order;
(2) $\sigma := \emptyset$;

(3)   **for** $i = 1$ **to** k **do**
(4)   attach all edges converging in G_i to σ;
(5)   let σ_i be the list of edges that is obtained by walking around
                      the chain C_i starting at a vertex at one of its two ends;
(6)   attach σ_i to σ **od**;

Figure 10.17: *An edge-test sequence for chain dags.*

Proof The BF-ordering σ is computed by the procedure depicted in Figure 10.17: Clearly, the procedure takes linear time and generates an ordering σ of the edges of G.

To prove that σ is a BF-ordering, we consider a simple path p in G starting at s. The path p consists of segments inside strong components and edges connecting between strong components. The strong components are traversed by p in topological order. Thus, all edges connecting between components occur in the same order in p and in σ. Because C_i is a chain of antiparallel edges, the segment of p that goes through C_i must be a subsegment of σ_i. In σ, the segment σ_i follows all edges entering C_i and precedes all edges leaving C_i. Thus, p is a subsequence of σ. $\square$

By Lemma 10.6, the single-source shortest-path problem on a chain dag G with source s can be solved in time $O(m + n)$. Note that sliding-port constraints do not create negative circuits. Therefore, in Ford's method, we need only to go through the edge tests in σ. The final check whether the triangle inequality holds at each edge is superfluous.

10.2.2.2.2 Diverse Constraints Finally, we have to discuss the effects of diverse constraints on the layout graph. These constraints can produce circuits. In fact, the circuits may be *global*; that is, they may contain rectangles that are far apart in the layout. This happens if maximum distance constraints exist between rectangles that are far apart from each other. Sometimes, a maximum distance constraint cannot be met. In this case, there is a negative circuit in the layout graph. In Section 10.2.2.2.3, we discuss how to deal with negative circuits in layout graphs.

Although diverse constraints can destroy the structure we have exhibited in layout graphs, there are not very many such constraints. This observation can be exploited to develop efficient shortest-path algorithms.

Let us group diverse constraints into two groups:

Group I: Edges (R_i, R_j) such that $x_j < x_i$ or $(x_j = x_i$ and $j > i)$ and R_i and R_j are in different chains

Group II: All other diverse constraints

Constraints in group I do not create circuits in the layout graph. If only such constraints exist, then the shortest-path algorithm discussed in the previous section can be used. Constraints in group II may create new circuits. Let $G = (V, E)$ be the total layout graph. Denote the set of diverse constraints in group II with C and assume that $|C| = k$, where k is a small number compared to n. We can now devise the following edge sequence to be used in Ford's method. First, we test all edges in $E \setminus C$ in the sequence σ described in Section 10.2.2.2.1. Then, we test all constraints in C. This total sequence of edge tests is repeated $k + 1$ times.

Lemma 10.7 *Consider the test sequence $\sigma' = (\sigma \oplus C)^{k+1}$—that is, the $(k+1)$-fold repetition of the concatenation of σ with any ordering of C. If the last test of any edge in C reveals that the triangle inequality is not met, then G has a negative circuit. Otherwise, Ford's method with the use of σ' as a test sequence computes the shortest distances for all vertices in G from R_r.*

Proof Each simple path in G starting at R_r contains each edge in C at most once. Therefore, such a path is a subsequence of the prefix $\tilde{\sigma} = (\sigma \oplus C)^k \oplus \sigma$ of σ'. Thus, if there are no negative circuits, then even $\tilde{\sigma}$ yields the shortest distances of all vertices in G from R_r. If there are negative circuits then, after testing of all edges in $\tilde{\sigma}$, the triangle inequality is still violated at some edge. Since $E \setminus C$ does not have negative circuits, the triangle inequality must be violated for some edge in C. This condition is tested in the last k tests of σ'. □

Clearly, σ' can be computed in time $O(m + n)$. Thus, the run time of the resulting shortest-path algorithm is $O(nk)$. This adjustment of the shortest-path algorithm to layout graphs was suggested by [288].

For practical purposes, we can further improve the algorithm by not testing all edges in $E \setminus C$ every time (see Exercise 10.10).

10.2.2.2.3 Negative Circuits If there is a negative circuit in the layout graph, the compaction problem is unsolvable. In this case, we must either generate a suitable error display or perform some error recovery.

The simplest kind of error message is to display one or all violated constraints. Such constraints are edges in C that do not meet the triangle inequality. To point out such edges, we could, for instance, output the rectangle locations after testing $\tilde{\sigma}$, and highlight the respective constraints in this picture.

A more instructive error message is the display of a complete negative circuit. Let $e = (R_j, R_i)$ be an edge in C for which the triangle inequality is not met after $\tilde{\sigma}$. We can find such a circuit by moving backward along the p-pointers starting at an edge that does not fulfill the triangle inequality (Section 3.8.3).

What should be done if a negative circuit is found? Three approaches to error recovery have been suggested in the literature:

Freezing [93]: Fix the locations of rectangles on a negative circuit as they are in the original layout. Then compact in the orthogonal direction and, perhaps, try the original compaction once more.

Constraint Relaxation [233]: Select some constraints in the negative circuit and relax them. *Relaxing* means increasing the edge length such that the triangle inequality is satisfied. However, it is difficult to guess which of the constraints in C are the best to choose for relaxation.

Jog Introduction: Insert jogs into vertical wires to break the negative cycle (see Section 10.3.3), and repeat the compaction.

Schiele [395] presents an algorithm that detects negative circuits in the layout graph and removes them one by one, using constraint relaxation and jog introduction.

10.2.2.2.4   Use of Dijkstra's Algorithm   We can exploit an observation made in Section 3.8.6 to see that, if the initial layout satisfies all constraints, then we can use Dijkstra's algorithm instead of the algorithm by Bellman and Ford to solve the single-source shortest-path problem [304]. Let $b(v)$ be the initial location of the feature represented by vertex v in the layout graph. We simply modify the edge lengths of the graph as follows:

$$\lambda'(e) = \lambda(e) + b(v) - b(w)$$

Since the initial layout is assumed to satisfy all constraints, the resulting edge lengths are all nonnegative. Furthermore, the lengths of all paths between the same two vertices v and w change by the same amount $b(v) - b(w)$. Thus, shortest paths remain shortest paths after the modification of the edge lengths, and we can run Dijkstra's algorithm on the modified graph to carry out the compaction. In many cases, we initially start with a legal layout and can apply this observation.

10.2.2.3   Wire Balancing

Graph-based x compaction with the shortest-path algorithm pushes each rectangle as far as possible to the right. The resulting layouts are unpleasing because they distribute rectangles unevenly. More important, such layouts often lead to technologically inferior mask layouts. The reason is that horizontal

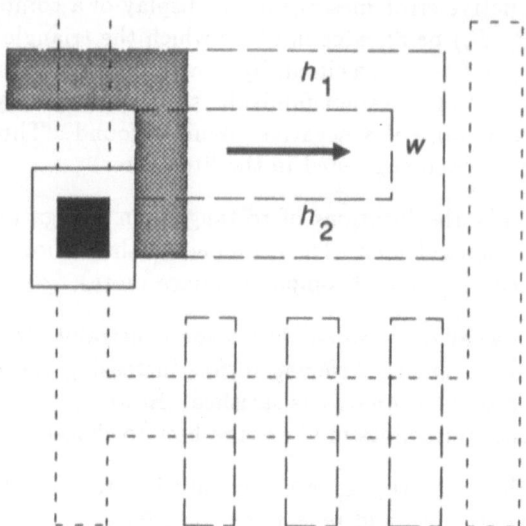

Figure 10.18: *A compacted layout with long wires. The horizontal
wire segments h_1 and h_2 are not represented by rectangles during
x compaction. The shortest-path algorithm pushes the vertical wire
segment w as far as possible over to the right side of the layout. We
would prefer a layout that has w on the left side, thus keeping the
wire segments h_1 and h_2 short; see shaded alternative.*

wires are not represented explicitly during x compaction, and an uneven layout
can result in long wires. Figure 10.18 shows an example of this phenomenon.

Thus, in a second pass after compaction, the layout features should be
distributed evenly without the width of the layout being increased. The corresponding layout optimization is commonly referred to as *wire balancing*. It
changes the x coordinates of rectangles that do not lie on a shortest path from
R_r to the leftmost rectangle. Such rectangles can move without the width of the
layout being changed. (For simplicity, we will assume throughout this section
that the leftmost rectangle R_ℓ horizontally overlaps with all other rectangles
in the layout. Thus, any rectangle that is not on a shortest path from R_r to
R_ℓ can be moved in the wire-balancing step.) The function to be optimized by
relocation of the rectangles should reflect the cost of the wires. We determine
the cost $c(w)$ of a wire w by multiplying its length with a weight factor. This
weight factor usually is assigned according to the electric resistivity of the wire.
Wires with high resistivity receive large weights because they have to be kept
short with priority. The goal of the optimization is then to move the components that are not on a shortest path between R_r and R_ℓ such that, on the one

hand, the width of the layout is not increased, and, on the other hand, the sum of the wire costs is minimized.

The corresponding optimization problem is quite similar to the RELAXED PLACEMENT problem described in Section 7.1.4.1. We can assign to each horizontal wire w variables $z_{\ell,w}$ and $z_{r,w}$ denoting the x coordinates of the rectangles at the left and right end of the wire. The restriction that each rectangle R_i connected to by the wire should be located within the left and right limits of the wire can then be formulated as a set of inequalities of the type (10.3) in the variables $z_{\ell,w}$, $z_{r,w}$, and x_i. The spacing, grouping, and diverse constraints are added to this constraint set. In addition, there is a constraint

$$x_r - x_\ell = d$$

where d is the width of the layout determined by the compaction algorithm. This constraint ensures that the layout width does not increase. The cost function to be minimized is

$$\sum_w c(w)(z_{r,w} - z_{\ell,w})$$

The result of this formulation is a linear program that is the dual of a MINCOST FLOW problem (Section 3.11.2, Exercise 10.11).

Schiele [394] has proposed an iterative heuristic method for doing wire balancing. He identifies groups of rectangles that can be moved as a unit such that the total edge cost is reduced.

10.2.3* Incremental Compaction

In graphical layout editors, it is often convenient to have an operation that does not compact the whole layout, but rather compacts (or stretches) only small portions of the layout (in one dimension). The designer specifies the area that she or he wants to compact—say, by defining a layout window—and the system responds with the partially changed layout. In this setting, it is important to be able to maintain compaction constraints dynamically. Often, interactive layout changes cause only small modifications in the set of constraints for compaction. The data structure must support efficiently the update of the constraint set. Furthermore, the data structure must allow for fast access to the usually small subset of constraints that is needed for a partial compaction. Dynamic data structures from algorithmic geometry, such as the *corner-stitching* data structure [349], are useful for this purpose. We do not go into detail here, but instead refer the reader to the literature [57, 405].

10.3 Two-Dimensional Compaction

In this section, we shall solve the 2D-PACKING problem and discuss how to apply the problem to the two-dimensional mask layouts.

10.3.1   Two-Dimensional Packing

Recall that an instance of the 2D-PACKING problem is a layout L of iso-oriented rectangles in the plane. Since we are discussing the *packing* problem, all rectangles are on the same layer and are mutually incompatible. We are looking for an arrangement of the rectangles in which no two rectangles overlap and some cost function of the width and height of the smallest enclosing rectangle, such as are or perimeter, is minimized.

The 2D-PACKING problem is strongly NP-hard for many cost functions; easy polynomial-time transformations from the 3-PARTITION problem exist (see Exercise 10.12).

The 2D-PACKING problem can be written as a mixed $0, 1$-integer program. We associate with each ordered pair (R_i, R_j) $i < j$ of rectangles in L two binary decision variables s_{ij} and t_{ij}. The variable s_{ij} is called the *orientation variable*. If $s_{ij} = 0$ (1), then R_i and R_j are supposed to be separated in the horizontal (vertical) dimension. The variable t_{ij} is called the *ordering variable*. It signifies in which order R_i and R_j are arranged. Specifically, the different combinations of values of s_{ij} and t_{ij} have the following meaning:

$$
\begin{aligned}
s_{ij} = 0, t_{ij} = 0 \quad &\text{if } x_i + w_i \le x_j; \\
&\text{that is, } R_i \text{ is to the left of } R_j \\
s_{ij} = 0, t_{ij} = 1 \quad &\text{if } x_j + w_j \le x_i; \\
&\text{that is, } R_i \text{ is to the right of } R_j \\
s_{ij} = 1, t_{ij} = 0 \quad &\text{if } y_i + h_i \le y_j; \\
&\text{that is, } R_i \text{ is below } R_j \\
s_{ij} = 1, t_{ij} = 1 \quad &\text{if } y_j + h_j \le y_i; \\
&\text{that is, } R_i \text{ is above } R_j
\end{aligned}
\tag{10.10}
$$

In addition to the constraint that $s_{ij}, t_{ij} \in \{0, 1\}$, the integer program has constraints that ensure the meaning of the s_{ij} and t_{ij}. Specifically, the constraints pertaining to the pair (R_i, R_j) are

$$
\begin{aligned}
(x_i - x_j)(1 - s_{ij})(1 - t_{ij}) &\le -w_i(1 - s_{ij})(1 - t_{ij}) \\
(x_j - x_i)(1 - s_{ij})t_{ij} &\le -w_j(1 - s_{ij})t_{ij} \\
(y_i - y_j)s_{ij}(1 - t_{ij}) &\le -h_i s_{ij}(1 - t_{ij}) \\
(y_j - y_i)s_{ij}t_{ij} &\le -h_j s_{ij}t_{ij}
\end{aligned}
$$

Note that, for each assignment to the two binary variables s_{ij} and t_{ij}, exactly one of these four constraints is nontrivial. For each such assignment, the set of nontrivial constraints makes up two independent systems of inequalities of the type (10.3), one exclusively with variables that represent x coordinates in the layout, the other exclusively with variables that represent y coordinates in the layout. Both systems can be solved independently with the shortest-path methods for one-dimensional compaction.

The difficult part of 2D-PACKING is finding an optimal assignment to the s_{ij} and t_{ij}. Here, we can use the branch-and-bound technique for linear programming described in Section 4.6.2.1. At each interior node z of the branch-and-bound tree, we decide on the values for one pair s_{ij} and t_{ij} of decision variables; that is, we fix the relative position of one pair (R_i, R_j) of rectangles. Therefore, each interior node has four children, one for each combination of binary values for s_{ij} and t_{ij}. We can obtain the lower bound at the node by minimizing both width and height of the layout incorporating *only* the constraints that are determined by decision variables that have already being fixed on the path from the root to node z. This minimization again amounts to a short-path computation on two independent layout graphs. We denote the corresponding layout graph for the x dimension with $G_{z,x}$, and the layout graph for the y dimension with $G_{z,y}$. The graph $G_{z,x}$ contains an edge for each constraint that has already been selected for a pair (R_i, R_j) with $s_{ij} = 0$. Analogously, $G_{z,y}$ handles the pairs with $s_{ij} = 1$. We call $G_{z,x}$ and $G_{z,y}$ the *lower-bound graphs at z*.

The cost function chosen to be minimized determines the cost of the inspected layouts and the size of the computed lower bounds as a function of the width and height of the layout. In this way, the cost function controls the traversal of the branch-and-bound tree. Any cost function of the width and height of a layout can be chosen, since, during the computation of upper *and* lower bounds, the width and height of the layout are handled independently.

Due to the large size of the branch-and-bound tree, depth-first traversals of the tree are used with preference.

The resulting algorithm for 2D-PACKING is conceptually simple. It shares its advantages and disadvantages with the general branch-and-bound method for 0, 1-integer programs. In general, the optimal solution is found quickly, but verifying optimality takes a long time. However, the computation can be stopped early. In this case, the preempted branch-and-bound algorithm delivers a compacted layout together with a lower bound on the cost of any legal layout.

Two-dimensional compaction algorithms based on this branch-and-bound paradigm were introduced by Kedem and Watanabe [228] and Schlag et al. [397]. To date, these authors have presented the only systematic algorithm approach to optimal two-dimensional compaction.

Several observations can be made to speed up the algorithm:

Observation 1: As soon as one of the two graphs $G_{z,x}$ and $G_{z,y}$ for computing the lower bound at an interior node z has a negative circuit, the corresponding interior node is fathomed, because none of its leaves can yield a legal layout. This observation eliminates a large number of nodes.

Observation 2: If none of the graphs $G_{z,x}$ and $G_{z,y}$ contains a circuit, then the solution of the shortest-path problems on $G_{z,x}$ and $G_{z,y}$ yields a new layout. If no rectangles overlap in this layout, the node z is also fathomed, and the obtained layout is a suitable sample solution s_z for z, with $c(s_z) \leq L_z$. (Here, we use the notation of Section 4.6.2).

Observation 3: If all rectangles have positive width, then all edge lengths in the layout graphs constructed during the branch-and-bound process are negative. Therefore, the existence of a circuit is equivalent to the existence of a negative circuit, and we can solve the shortest path problem on each layout graph in linear time by topologically sorting the layout graph (Section 3.6). If the sort detects a circuit, the shortest-path problem is unsolvable.

Because of the large flexibility of movement, there is no apparent method for reducing the size of the layout graphs constructed during two-dimensional packing. Ideas based on the redundancy of edges do not carry over. Also, the layout graphs are not interval dags any more.

10.3.2   Extension of the Method

In this section, we extend the branch-and-bound method for 2D-PACKING to the two-dimensional compaction problem.

First, we note that, as in one-dimensional compaction, we have to restrict the movement of the rectangles if we do not want to create major problems with the routing of wires. In one-dimensional compaction, the way out of this problem was to solve the ORDERED 1D-COMPACTION problem, which prohibits swapping incompatible rectangles. We can do the same here. This restriction effectively eliminates the decision variables t_{ij}. The constraints selected by each value of s_{ij} are the ones from (10.10) that do *not* cause rectangles to be swapped. This modification decreases the size of the branch-and-bound tree significantly. We now make several observations that tailor the algorithm for 2D-PACKING to two-dimensional compaction, and make it efficient in practice.

For the discussion of two-dimensional compaction, we have to leave the domain of layouts composed of fixed-size rectangles. The reason is that, say, horizontal wire segments stretch or compress during x compaction, but they nevertheless have to be explicitly represented in the layout graph, because they pose obstacles for the y compaction. Therefore, wire segments have to be represented explicitly in the layout.

From now on, a *layout* will be composed of fixed-size rectangles called *components* and of stretchable rectangles called *wires*. Each wire connects two components either horizontally (*horizontal wire*) or vertically (*vertical wire*). An example is given in Figure 10.19(a). During the compaction, all rectangles can change their position in both dimensions. In addition, the horizontal (vertical) wires can change their width (height). Connections between wires and components have to be maintained.

For many pairs of rectangles (R_i, R_j), the dimension in which the rectangles have to be separated can be deduced from the initial layout. For instance,

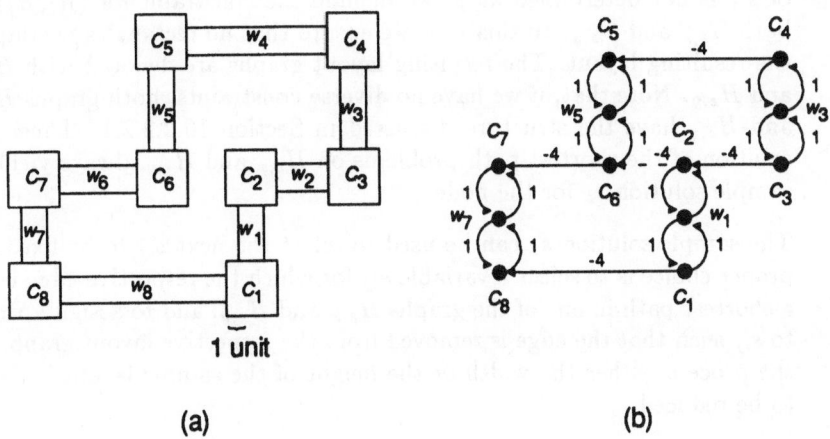

Figure 10.19: *A layout that is composed of fixed-size components and stretchable wires. (a) The layout; (b) the layout graph for the x dimension. The layout graph for the y dimension is isomorphic up to vertex labelings. The dotted edge represents an alternative. Only one of the two dotted edges in the two layout graphs has to be included. Which edge is included is determined by the value of the decision variable associated with the component pair (C_2, C_6). The edge lengths are assigned such that different components do not overlap and wires connect fully to their adjacent components.*

two rectangles that are connected by a horizontal wire *have to be* separated in the x dimension. In fact, their y coordinates have to be related by a grouping constraint. Therefore, in this case, the variable s_{ij} can be eliminated, and the suitable constraints must be added to all layout graphs. This modification reduces the number of decision variables dramatically. For instance, in Figure 10.19, only one decision variable is necessary for two-dimensional compaction.

The reduction of the number of decision variables dramatically increases the efficiency of the two-dimensional compaction. But we can exploit the structure of the problem further to gain even more efficiency. We make several observations to this end:

Observation 4: The algorithm for 2D-PACKING computes sample solutions only for leaves and fathomed nodes in the branch-and-bound tree. In contrast, with the reduced flexibility of movement in two-dimensional compaction, there is a natural way of computing a sample solution for each interior node z: For each rectangle pair (R_i, R_j) such that the value of s_{ij} is not determined at z, we include the constraint for (R_i, R_j) in both $G_{z,x}$ and $G_{z,y}$. In this way, we ensure that no rectangles overlap in the resulting layout. The resulting layout graphs are denoted with $H_{z,x}$ and $H_{z,y}$. Note that, if we have no diverse constraints, both graphs $H_{z,x}$ and $H_{z,y}$ have the structure discussed in Section 10.2.2.2.1. Thus, the solution of the shortest-path problems on $H_{z,x}$ and $H_{z,y}$ always yields a sample solution s_z for the node z.

The sample solution s_z can be used to select the next s_{ij} to be fixed. A proper choice is to select a variable s_{ij} for which the respective edge is on a shortest path in one of the graphs $H_{z,x}$ and $H_{z,y}$, and to assign a value to s_{ij} such that the edge is removed from the respective layout graph. In the process, either the width or the height of the sample layout is likely to be reduced.

In this variant, the two-dimensional compaction algorithm can be viewed as a process of deleting dotted edges from the two layout graphs $H_{z,x}$ and $H_{z,y}$ and inserting them into the graphs $G_{z,x}$ and $G_{z,y}$ (as we move down the branch-and-bound tree), and solving shortest-path problems on the resulting graphs on the way to compute upper and lower bounds. When backing up the tree, we have to undo the modifications of the layout graphs.

Observation 5: The reduced flexibility of movement in two-dimensional compaction enables us to eliminate redundant edges from the layout graphs considered during the two-dimensional compaction. As an example, consider Figure 10.19(a). Assume that this ring of features is part of a larger layout that is composed of additional elements inside the ring and other elements outside the ring. Then, no constraints have to be generated between any element inside the ring and any element outside the ring, because the ring always separates such a pair of elements. We can use plane-sweep methods to determine which constraints have to be generated, and we generate only those constraints (see Exercise 10.13). This computation is done in a preprocessing step that computes $H_{r,x}$ and $H_{r,y}$, where r is the root of the branch-and-bound tree.

Watanabe [456] discusses other heuristics that can be used to speed up the compaction. Most notably, Watanabe begins at a leaf of the branch-and-bound tree that represents an initial estimate of the optimal layout. Special heuristics are developed to find an initial estimate with small cost. Then, the branch-and-bound algorithm proceeds by backing up from this leaf. The lower the cost of the initial estimate, the more efficient the computation.

No methods for incorporating nontrivial compatibility relations into two-dimensional compaction have been reported.

Mosteller et al. [333] have applied simulated annealing to two-dimensional compaction. Because of the universality of the method, they can handle very general layouts involving nonrectangular—even nonpolygonal—features.

10.3.3* $1\frac{1}{2}$-Dimensional Compaction

Since two-dimensional compaction is a time-consuming process, people have investigated heuristic methods for making heuristic decisions relating the two dimensions of the compaction. Since such heuristics do not solve the two-dimensional compaction problem optimally, but still interrelate the two dimensions, they are referred to as $1\frac{1}{2}$-dimensional compaction.

Wolf et al. [472] and Shin and Lo [419] present methods of compaction in which the layout is essentially compacted in one dimension—the *preferred dimension*. The goal is to decrease the extent of the layout in the preferred dimension. However, on the way, the compaction can also cause coordinate changes in the other dimension—the *shear dimension*. Figure 10.20 exhibits two situations that are resolved by different methods. Each local change is called a *reorganization*. Wolf et al. [472] present a heuristic framework called *supercompaction* for carrying out feasible reorganizations such that the length of the shortest path in the layout graph for the preferred dimension is decreased.

An alternative method, called *zone refinement*, was proposed by Shin et al. [420]. This method is reminiscent of a physical process with the same name that is used to purify crystal ingots in semiconductor manufacturing. Figure 10.21 shows the essential idea of zone refining, as it applies to y compaction. The layout is processed from bottom to top. At each time, the features in the layout are partitioned into a floor and a ceiling part. These two parts are separated by a *zone*. A step of the zone-refining method consists of moving a rectangle from the ceiling across the zone to the floor. Here, the rectangle can also move in the horizontal direction. The place of the rectangle in the floor part is selected such as to decrease the height of the floor part of the layout.

Although heuristic methods for $1\frac{1}{2}$-dimensional compaction may yield good results in practice, their performance is not *provably* good in the worst case or even in typical cases.

10.4 Hierarchical Compaction

The methods for hierarchical circuit design discussed in Section 1.2.2.1 naturally lead to hierarchical descriptions of geometric layouts. In a hierarchical geometric layout, a set of rectangles that describes a cell of the circuit is specified independently and is referred to by name in the geometric layouts of other subcircuits using the cell. Figure 10.22 shows such a hierarchical layout. The rectangles pertaining to each subcell are enclosed in a rectangular *box*. The goal

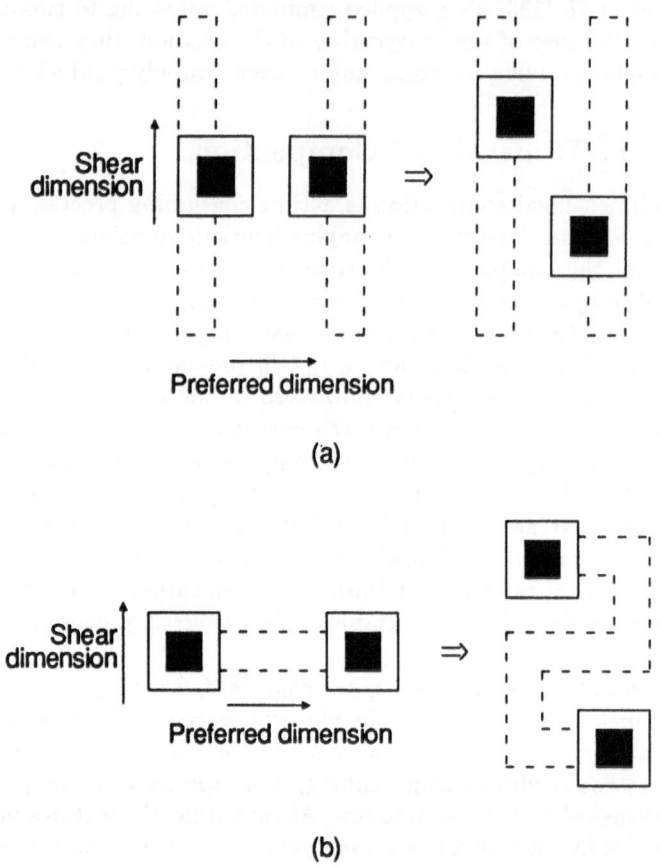

Figure 10.20: *Two heuristics for $1\frac{1}{2}$-dimensional compaction.   (a) Compaction by shearing; (b) compaction by jog introduction.*

of this convention is to ensure that the compaction leaves subcells rectangular. We can discuss, controversially, whether this goal is reasonable, and we shall do so in Section 10.4.4.1. For now, we take the goal as given.

10.4.1   Hierarchical One-Dimensional Packing

We now discuss how to solve the ORDERED 1D-PACKING PROBLEM on a hierarchical geometric layout. To this end, we suggest that the reader familiarize her- or himself with the material in Sections 3.13.1 and 3.13.2.

For the purposes of one-dimensional x compaction, we can abstract the box

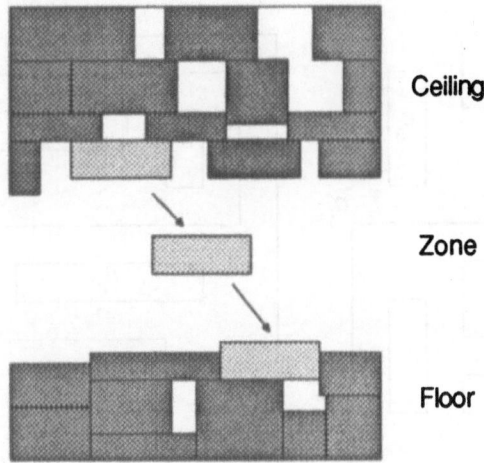

Figure 10.21: *Zone refining.*

of each cell to a set of two rectangles with zero width, one representing the left
and the other representing the right side of the box. The top and bottom side
of the box can stretch during the x compaction and are therefore handled as
horizontal wires; that is, they are not represented explicitly in the geometric
layout.

A hierarchical geometric layout gives rise in a natural way to a hierarchical
layout graph in the sense of Definition 3.21. The hierarchical layout graph for
the layout in Figure 10.22 is depicted in Figure 10.23. It is obvious that solving
the single-source shortest-path problem on the expansion of the hierarchical
layout graph amounts to a one-dimensional compaction that maintains the
rectangular shape of all cells and minimizes the width of the layout under this
constraint. We can apply the bottom-up method to solving this problem. The
appropriate burner solves the single-source shortest-path problem whose source
and sink are the left and right boundary of the cell, respectively. This process
takes linear time in the size of the graph. Figure 10.24 shows the bottom-up
table for the example in Figure 10.23. The shortest path with source R_r in the
graph $\widetilde{G}_4$ has length -10. Therefore, the layout can be compressed to width
10. Note that the burner is optimal in the sense of Lemma 3.15. Therefore, we
can determine the optimal width of the layout in linear time *in the length of
the hierarchical description of the layout graph.*

Instead of constructing the full layout graph for each cell, it suffices to
construct the transitive reduction (see Section 10.2.2.1.1.1).

We can obtain a layout with optimum width by laying out each cell indepen-

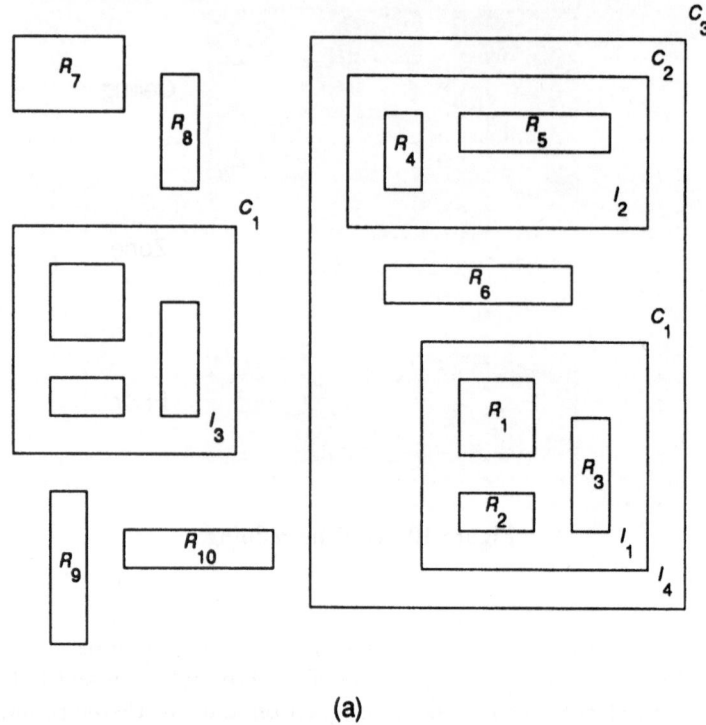

(a)

Figure 10.22: *A hierarchical geometric layout. Cells of the layout are denoted with C_1, C_2, and C_3. Instances of cells are denoted with I_1 to I_4. Rectangles are denoted with R_1 to R_{10}. Cell C_1 has two instances. All other cells have one instance.*

dently in the empty neighborhood—that is, just the cell itself—and substituting this layout for each instance of the cell. Therefore, the compacted layout can be described hierarchically again.

We can eliminate the requirement that cells in the layout remain rectangular. To this end, we delete the rectangles $C_{i,\ell}$ and $C_{i,r}$ representing the left and right boundary of a cell C_i in the hierarchical layout from the cell G_i representing C_i in the hierarchical layout graph. Then, we make all rectangles that were visible from $C_{i,\ell}$ or $C_{i,r}$ in G_i pins of G_i. Call the pins that were visible from $C_{i,\ell}$ *left pins*, and the pins that were visible from $C_{i,r}$ *right pins*. The burner now solves a single-source shortest-path problem from each right pin to each left pin, and constructs a bipartite graph with edges directed from the right to the left pins and labeled with the lengths of the corresponding shortest

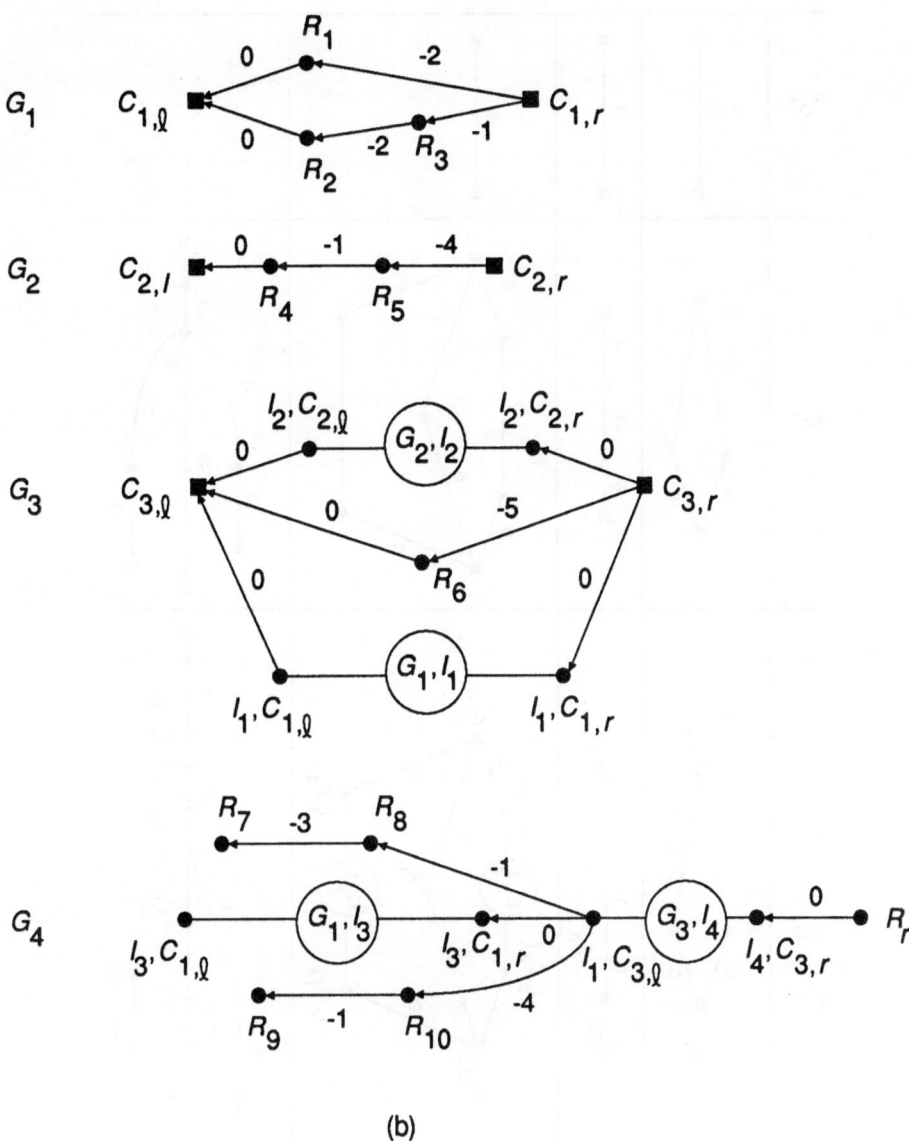

(b)

Figure 10.23: *The hierarchical layout graph of the layout in Figure 10.22 for the* ORDERED 1D-PACKING *problem. Cell G_i in the layout graph represents cell C_i in the layout. The rectangles with zero width representing the left and right boundary of a cell in the layout are also the two terminals of the corresponding cell in the layout graph.*

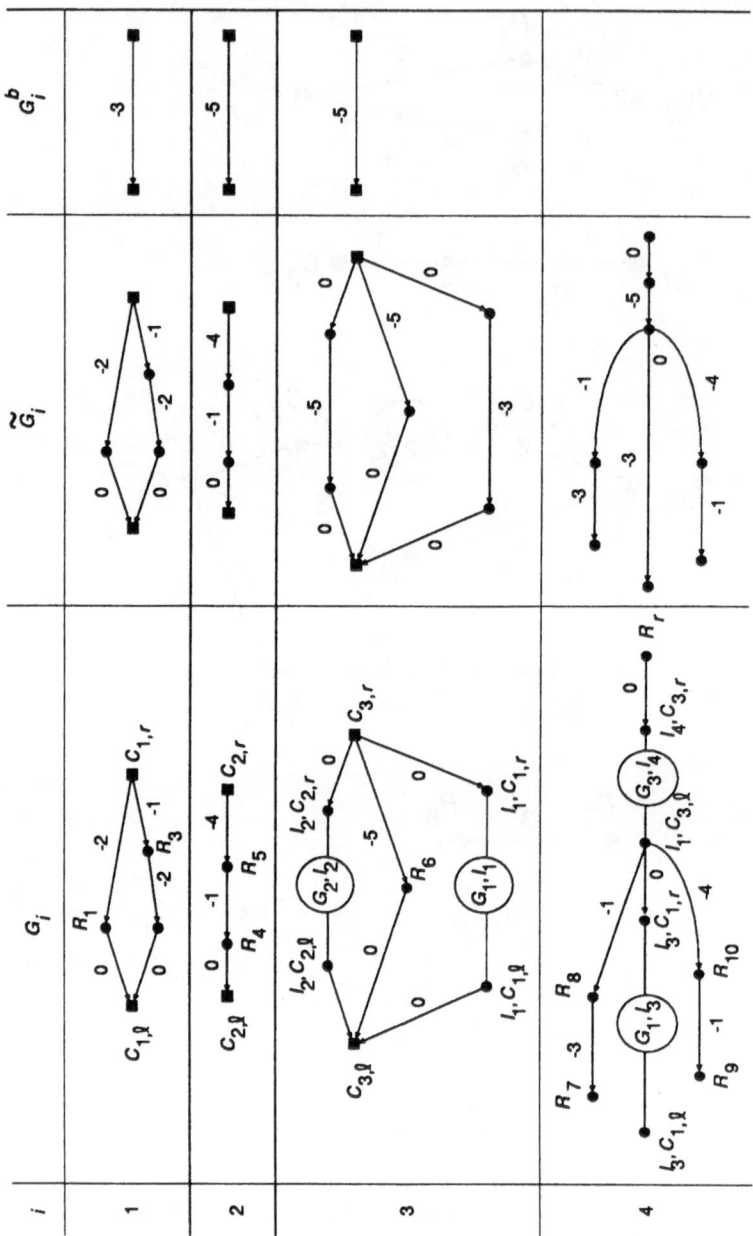

Figure 10.24: *The* BU-*table for the example in Figure 10.23.*

paths. This burner is no longer optimal. It may take quadratic time in the size of the graph, because it constructs a dense burnt graph. Also, the compacted layout cannot be described hierarchically, in general, because there are cases in which a cell has many instances, and each instance requires a different layout to achieve the minimum width for the total expanded layout. Lengauer [275] provides an example of this phenomenon (see also Exercise 10.14).

Nevertheless, we can compute the layout efficiently. To this end, we first determine the width of the layout with the bottom-up method. As a result, we obtain burnt graphs for all cells in the hierarchical layout graph. In a second pass that processes the hierarchy tree top-down, we compute the locations of the rectangles in an optimally compacted layout. We start by solving a single-source shortest-path problem on the graph $\widetilde{G_i}$ pertaining to the root of the hierarchy tree. This computation determines the locations of all rectangles belonging to the root of the hierarchy tree. Then, we descend in the hierarchy tree in a depth-first fashion. When processing an interior node z that represents a cell instance of type G_i in the hierarchy tree, we have already determined the locations for the pins of the instance. These locations now become additional constraints for a shortest-path computation on the burnt graph $\widetilde{G_i}$ that computes the locations of all rectangles in the cell instance represented by node z. The resulting algorithm may take time that is of the order of the size of the *expanded* layout graph; that is, the algorithm may be quite slow compared to the bottom-up method. At all times, however, the algorithm is working on a number of data that is of the order of the size of a single cell. Therefore, the required working storage is quite small. Since the algorithm does not have to retrieve large numbers of data computed earlier, expensive secondary-storage management becomes unnecessary. It is the swapping overhead arising through secondary-storage management that presents the performance bottleneck in the computation on large sets of data, and this overhead is eliminated completely. Therefore, for practical purposes, hierarchical compaction is a big win, even if the resulting layout cannot be described hierarchically.

10.4.2 Hierarchical One-Dimensional Compaction

In one-dimensional compaction, grouping and diverse constraints must be added to the problem instance. Grouping constraints may connect pins on the top and bottom side of a cell C_i in the hierarchical circuit to the neighborhood of the cell. Such pins translate into new pins of the cell G_i representing C_i in the hierarchical layout graph. For this reason, the burnt graph is also called the *port abstraction* in this application. (A *port* is the same as a pin in the layout.) Now, the burner must solve a single-source shortest-path problem *for each ordered pair of pins*. Since the layout graphs are no longer acyclic, computing the shortest paths can no longer be done in linear time. Diverse constraints may even introduce negative circuits into the expansion of the hierarchical layout graph. Thus, in general, burning may take up to cubic time in the size of the graph (see Section 3.8.6). However, the methods described in Section 10.2.2.2 carry over.

Compacters based on this hierarchical method are described by Kingsley [233] and Eichenberger [101].

10.4.3   Maintaining the Layout Hierarchy

We mentioned in Section 10.4.1 that the compacted version of a hierarchical layout may not be describable hierarchically in small space. This phenomenon can already occur in one-dimensional packing, if subcells are not required to remain rectangular (see Exercise 10.14), or it can be introduced in one-dimensional compaction by pitch matching of pins on the top or bottom sides of different cells. In the latter case, the neighborhood can require the positions of pins on the top and bottom side of a cell, respectively, to be in different locations for each instance of the cell. As a result, a hierarchical layout may become quite irregular during x compaction. If we then want to y compact, we cannot resort to hierarchical techniques.

This phenomenon is undesirable, because it dramatically increases the time for processing the layout, and the resulting layout is quite irregular and is difficult for the designer to inspect.

One way out of this problem is to augment the hierarchical layout graph for the x compaction with additional constraint edges that ensure that, during x compaction, the layout does not become too irregular for a subsequent hierarchical y compaction. We explain the method as it applies to ORDERED 1D-PACKING. Let C_i be a cell of the hierarchical layout, and let $G_{i,x}$ $(G_{i,y})$ be the cell in the hierarchical layout graph for the x- (y-) compaction representing C_i. Eichenberger [101] suggests that we add constraints between pairs (R_1, R_2) of rectangles in a cell C_i to $G_{i,x}$, if R_1 and R_2 are not separated in the y dimension (that is, there is no directed path between R_1 and R_2 in $G_{i,y}$) and R_1 and R_2 can overlap *vertically* after the x compaction of the cell in some neighborhood (that is, there is no directed path between R_1 and R_2 in $G_{i,x}$). We can test this criterion by solving two all-pairs shortest-path problems, one on $G_{i,x}$ and one on $G_{i,y}$. The directions of the added constraint edges can be chosen arbitrarily. Choosing optimum directions such as to minimize the length of the shortest path in $G_{i,x}$ is closely related to the RAINBOW COLORING problem (Section 9.6.1.3), and is probably NP-hard. Therefore, we have to resort to heuristics here. One heuristic is to choose the constraint that maintains the relative position of the involved two rectangles in the original layout. After adding the respective constraints to $G_{i,x}$, we do an x compaction. The result is a compacted layout whose layout graph for y compaction is a subgraph of $G_{i,y}$. Therefore, $G_{i,y}$ is a valid layout graph for y compaction, even though it imposes some unnecessary constraints.

This method has two disadvantages. First, it is inefficient because the resulting layout graphs become large and unstructured and the additional constraints are difficult to compute. (It takes cubic time to compute them.) Second, the freedom of movement is severely restricted by the additional constraints, and therefore the resulting layout width is likely to be increased appreciably. Also, there are superfluous constraints for the y compaction, so the height of the y-compacted layout is increased, as well.

Note that, although the same layout graph can be used on the y compaction of all instances of a cell, different instances of the same cell may still look different after x compaction.

Eichenberger [101, 471] also describes another method for interrelating the two dimensions during hierarchical compaction that is faster but less robust. There is more work to be done in this area.

10.4.4 Hierarchical Compaction with Identical Cells

So far, we have assumed that, during compaction, a cell yields to the constraints imposed by its neighborhood. Since different instances of the cell are located in different neighborhoods, after compaction they will have different layouts. In this section, we consider the case that we require all instances of a cell to look the same. This additional restriction is even stronger than the one made in the previous section. It ensures that the layout hierarchy is maintained, and it does so in a very restricted but quite elegant way.

Of course the method of compaction maintaining this restriction that first comes to mind is to process the hierarchical definition bottom up and to compact each cell by itself in the empty environment, using the already-calculated compacted layouts for the subcells. This process presupposes that the empty environment is a good environment in which to compact a cell. This assumption is not necessarily true, in general, however. In fact, the neighborhoods of all instances of a cell may be identical but may force the cell into a shape that differs greatly from the shape the cell takes after the compaction in the empty environment.

10.4.4.1 Toroidal Compaction of Two-Dimensional Cell Arrays

Consider Figure 10.25(a). It depicts the x- and y-compacted layout of a cell in an empty environment. If we create an array of identical cells of the form given in Figure 10.25(a) we obtain the layout in Figure 10.25(b). A much more efficient layout is the layout presented in Figure 10.25(d). Here, the cells are all identical again, but their layout is as in Figure 10.25(c).

Two-dimensional arrays of identical cells occur frequently in layout design, and they deserve special attention. To be independent from the size of the array, a model in which we think of one cell placed on a torus is appropriate. A torus comes about if we identify the two lateral side and the top and bottom side of a rectangle. Thus, one cell, placed on the torus, models an infinite array of identical cells. The "fatness" of the torus measures the efficiency of the layout

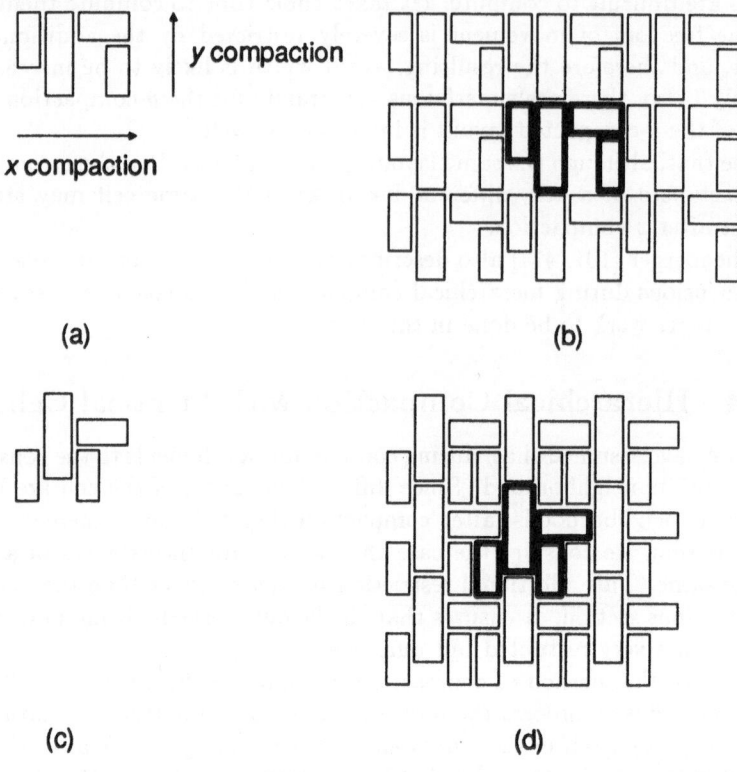

Figure 10.25: *Compaction of cell arrays. (a) x- and y-compacted layout of a cell in the empty environment. (b) An array of identical cells as laid out in (a). (c) A different layout of the cell in (a). An array of identical cells as laid out in (c). As the number of cells increases, the area of the layout in (b) tends to increase to 4/3 times the area of the layout in (c).*

in in the y dimension. The circumference of the torus represents measures the efficiency of the layout in the x dimension (see Figure 10.26) Therefore, during x compaction, we minimize the circumference of the torus; in y compaction, we minimize its fatness. Both problems can be stated as linear programs in a straightforward manner. To see how, we consider x compaction. Note that minimizing the width of the layout is a shortest-path problem—that is, it is a special linear program. We can minimize the circumference of the torus as follows. First, we generate the usual constraints for the x compaction of cell C. Then, we add constraints to make the left and right side of cell C fit. To this

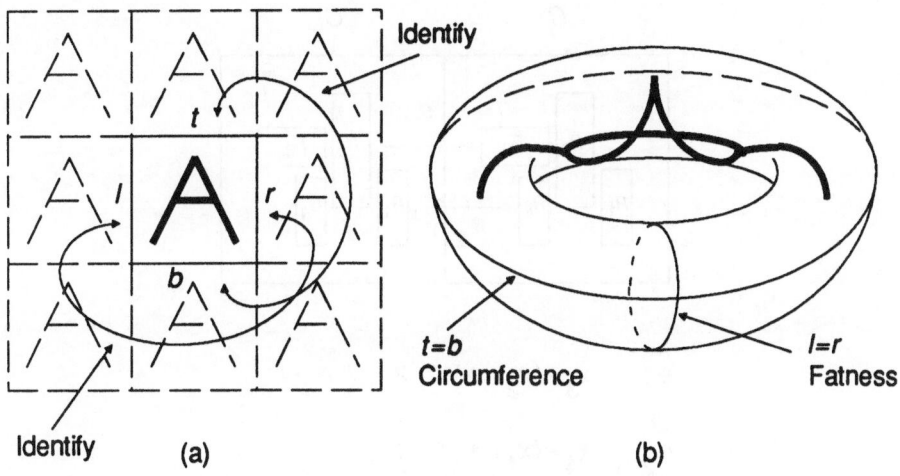

Figure 10.26: *Toroidal layout. (a) An infinite array of rectangular cells. The A symbol represents the layout features inside each cell. (b) The infinite array folded into a torus containing a single cell instance.*

end, we think of a replication C' of cell C attached on the right side of C. If we were to generate a layout graph for the respective layout, we would create constraints between some rectangles in C and some rectangles in C'. Let R_i be a rectangle in C that would have a constraint with a replica of a rectangle R_j in C'. In this situation, we create the constraint

$$x_i - (x_j + \Delta) \le -w_i \qquad (10.11)$$

Here, Δ is a new variable denoting the circumference of the layout. The term $x_j + \Delta$ denotes the location of R_j in the cell instance C', because the term x_j denotes the location in the left cell instance C; see Figure 10.27

We want to minimize Δ with respect to all constraints. The resulting optimization problem is a linear program, but it is not as simple as a shortest-path problem, because the constraints (10.11) involve three variables.

Even if all values in the problem instance are integer, the minimum value of Δ may not be integer (see Exercise 10.15). If we are required to place the layout features on integer coordinates, we have to solve the corresponding integer program.

Note that the branch-and-bound method for two-dimensional compaction presented in Section 10.3 can be directly extended to toroidal compaction. We just replace the shortest-path problems with the respective one-dimensional toroidal-compaction problems.

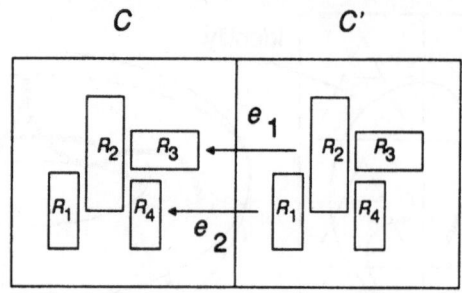

$$\theta_1: \quad x_3 - (x_2 + \Delta) \leq -2$$

$$\theta_2: \quad x_4 - (x_1 + \Delta) \leq -1$$

Figure 10.27: *An illustration of the constraints that have to be added for torus compaction.*

Eichenberger and Horowitz [102] introduce toroidal compaction and present heuristic methods for speeding up the solution of the linear program. Mehlhorn and Rülling [320] present graph-based methods for toroidal compaction that also apply to topological compaction (see Section 10.5). Their algorithm runs in time $O(m^2 \log m)$, where m is the number of constraints. Schlag et al. [396] present a method based on shortest-path algorithms that applies just to compaction of mask layouts on a single layer but runs in time $O(m \log m)$.

Toroidal compaction maximizes the efficiency of the layout of an infinite array of cells. In real applications, we always have cell arrays of finite size, and such an array is placed into some layout neighborhood that is composed of different cells. The effect of the neighborhood decreases as the size of the array grows. Thus, with large arrays, a suitable approach to compaction is compact the array toroidally first, and then to compact the neighborhood, given the side constraints placed on it by the precompacted array. If the array is small, the effect of the neighborhood on the shape of the compacted array should be taken into account explicitly. The framework of toroidal compaction can be generalized to solving this problem efficiently using graph-based methods [180].

10.4.4.2   The General Case

In the previous section, we compacted two-dimensional arrays of identical cells with the restriction that all cell instances be identical. The case of two-dimensional arrays is a very important one, and it deserves the special attention that we have awarded it. However, the basic method for toroidal compaction discussed in the previous section extends to the general case.

In this section, we consider arbitrary hierarchical layouts. We maintain the restriction that all instances of a cell be compacted identically, and we want to minimize the layout width under this restriction.

To x compact the layout, we first generate the hierarchical layout graph. Then, we introduce additional constraints that ensure that the pins of a cell occur in the same position in all instances of the cell. We achieve this goal easily by providing a new variable for each cell pin and adding constraints that equate the position of this pin in all cell instances with the new variable. Figure 10.28 illustrates this process, incorporating the additional restriction that all cells remain rectangular.

The resulting hierarchical system of linear constraints decomposes into portions that pertain to the different cells in the layout graph. The variables p_k provide the cross-links between different parts of the hierarchical linear-constraint system that ensure that all instances of the same cell are identical.

In total, we get a complex hierarchically described linear-programming problem. We solve this problem hierarchically as follows:

Step 1: Processing cells independently, and generate a modified port abstraction for each cell (see Section 10.4.2). The port abstraction of cell C_i solves a single-source shortest-path problem for each vertex in G_i that either is a pin of C_i (such as x_k in Figure 10.28(a)) or represents a pin of a subcell of C_i (such as $x_{k,j}$ in Figure 10.28(b)). Add to the constraints of the port abstraction the constraints depicted in Figure 10.28. The cells can be processed independently in this step, because there are no more nonterminals after the introduction of the new variables.

Step 2: Solve the linear program whose constraints are the constraints of the port abstractions for *all* cells and that minimizes the width of the largest cell. If the linear program is unsolvable, there is no layout such that all instances of the same cell are identical. Otherwise, the optimum solution of the linear program determines the relative positions of the pins in all cells.

Step 3: Process the cells independently. Augment the layout graph of each cell with the constraints implied by the positions of the pins found in step 2, and solve the corresponding shortest-path problem.

Because we use port abstractions only to define the linear program solved in step 2, the size of this linear program is of the order of the length of the hierarchical layout description, and is not as large a the expanded layout description. Therefore, the hierarchical compaction process is efficient.

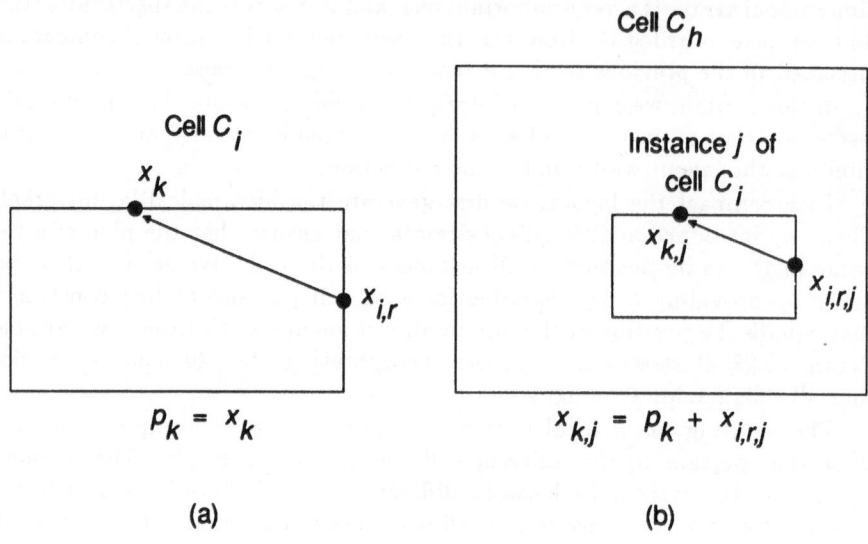

Figure 10.28: *The method by Rülling and Schilz.* (a) *The definition of a cell in a hierarchical layout. Pin p is represented by the vertex* x_k *in the layout graph for cell* C_i. *A new variable* p_k, *representing the position of p in* C_i, *is added to the whole constraint system. The constraint* $p_k = x_k$ *is added, as well. There is a path to* x_k *from the right boundary* $x_{i,r}$ *of the cell in the layout graph.* (b) *In an instance j of cell* C_i *inside a cell* C_h, *the right boundary of* C_i *is represented by a vertex* $x_{i,r,j}$, *and pin p is represented by a vertex* $x_{k,j}$. *We add to the layout graph of cell* C_h *the constraint* $x_{k,j} = p_k + x_{i,r,j}$. *After we have done this for all pins of* C_i, *we eliminate the nonterminal for cell* C_i *from the layout graph for cell* C_h.

This hierarchical compaction method was introduced by Rülling and Schilz [387]. Note that the method can be applied with or without the additional restriction that all cells stay rectangular. In the latter case, the port abstractions have to formulate the exact shape of the cell boundaries, and they become correspondingly larger.

This hierarchical compaction method can be combined with the method discussed in Section 10.4.1 that allows for different instances of the same cell. We have to make up our minds beforehand which cells may appear in a variety of versions—that is, which cells are *flexible* and which cells always look the same, or are *fixed*. We add constraints pertaining to the fixed cells, as in Figure 10.28. This process eliminates all nonterminals representing fixed cells in the hierarchical layout graph. Then, we use the bottom-up method to compute the port abstractions of the flexible cells. The constraint systems of the port abstractions also contain the new variables p_k. In the next step, we solve the linear program consisting of all port abstractions. This computation determines the locations of the pins in the fixed cells. Then, we compact hierarchically by traversing the hierarchy tree top-down, as described in Section 10.4.1. At the end, we solve the layout graphs for the fixed cells and fill in the corresponding layouts.

At the end of this section, we note that hierarchical compaction with fixed cells requires the solution of general linear programs. No special techniques that eliminate the need for a general algorithm for linear programming, such as the simplex algorithm, are known.

10.5* Topological Compaction

Mask layout compaction, as it is described in this chapter, is quite a restrictive kind of layout optimization. The topology of the layout is kept invariant in the following sense:

1. The relative positions of the circuit components are maintained.

2. The course of the wires is kept invariant.

Intuitively, mask layout compaction squeezes the air out of the mask layout while maintaining the correctness with respect to the design rules. Here, the invariant 2 is maintained in a very strong sense. Not only do we stick to the homotopy of the path taken by the wire, but the decomposition of the wire into a sequence of horizontal and vertical segments remains unaltered, as well. The reason is that a wire segment poses a rigid obstacle for the compaction in the direction orthogonal to the course of the segment. Therefore, a modification of the wire such as depicted in Figure 10.20(b) cannot be achieved by graph-based compaction, even though it may lead to a decrease in layout area. $1\frac{1}{2}$-dimensional compaction as described by Wolf et al. [472] and Shin and Lo [419] enhances graph-based compaction with specific heuristics allowing for topological changes of this sort. These heuristics effectively introduce jogs in wires. Therefore, the process is called *jog introduction*. Heuristics for jog introduction can be easily combined with one-dimensional compaction. Consider x compaction. Any vertical wire segment that is represented by a vertex on the shortest path in the layout graph is a candidate for jog introduction. After the jog has been introduced into the wire segment, that segment is divided into two parts. Thus, the shortest path is split as well, and it is likely that the length of the shortest path in the layout is increased, decreasing the width of the layout. However, this improvement is made at the expense of possibly increasing the height of the layout in the subsequent y compaction. Since this kind of jog introduction is heuristic, there is no guarantee that the introduction of any specific jog will yield an improvement in the layout size. Furthermore, each jog is introduced individually. There is no overall strategy ensuring that a given sequence of jog introductions will lead to the optimum layout size possible by the combination of x compaction and jog introduction.

The recent advances in theories of detailed routing afford us the possibility of basing compaction with jog introduction on a much more general foundation. In Section 9.2, we discussed theories of detailed routing that reduce the routability of a problem instance to the *safety* of all of a set of critical cuts. Safety of a cut means that the number of nets that cross the cut (the *density*) is no greater than the number of wires that can cross the cut in the given detailed-routing model (the *capacity*). The safety of each critical cut amounts to a numerical constraint. The problem instance is routable exactly if all such constraints are satisfied.

The basic idea of topological compaction is to parameterize this constraint system such that the changes in the density and capacity of the safety constraint for each critical cut are taken into account, as the x compaction of the problem instance proceeds. In other words, we want to compress the cells and change the wires homotopically along the way such that the system of safety constraints is met at all times.

To help us to think of x compaction as a continuous operation in time, we will view detailed routing as a geometric rather than graph-based process. We mentioned in Section 9.9.1 that theories of detailed routing can be coached in geometric terms.

We assume that we start with a problem instance that is routable. During x compaction, we modify only the x coordinates of the cells continuously. At all times during the compaction, we require the problem instance to stay routable. This requirement extends the requirement from mask layout compaction that different (incompatible) cells not be moved across each other. As the locations of the cells change, wires change their course. However, we assume that wires are not moved across cells, such that the homotopies of the wires are maintained. We do not represent wires explicitly during compaction. Their presence is incorporated only implicitly, because the homotopies of the wires determine the densities of the cuts.

We consider only straight-line cuts. (It turns out that this restriction is not critical for planar and knock-knee routing—that is, the two detailed-routing models for which cut conditions are known.) Throughout the compaction, we identify a cut with its two endpoints (on two cells). As the cells on which the two endpoints are located move, the cut changes its shape. Sometimes, during the compaction, the cut can actually move across one or more cells. At such times, the safety of the cut does not have to be maintained. During the compaction, both the capacity and the density of the cut can change (see Figure 10.29). Whereas the capacity usually is a simple function of the geometric shape of the cut, the density is a more complicated function. The density changes

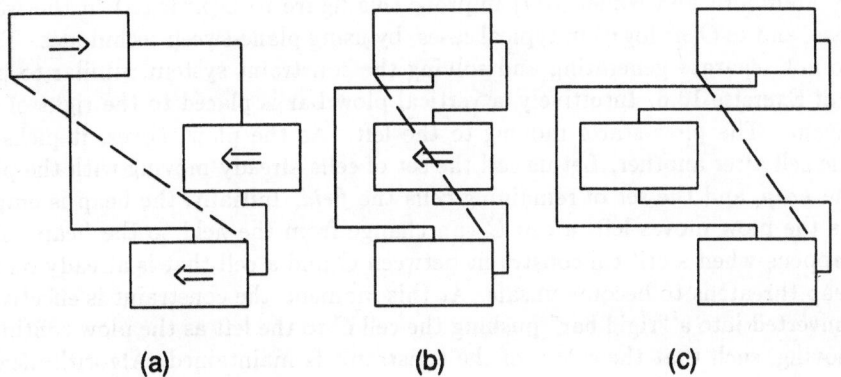

Figure 10.29: *Topological compaction in the gridless version of the planar routing model, in which wires can run horizontally and vertically and different wires must be separated by at least one unit. In this model, the capacity of a cut connecting two points (x_1, y_1) and (x_2, y_2) is $\lfloor \max\{|x_1 - x_2|, |y_1 - y_2|\} \rfloor$. (a) Initial layout with global wire routings. A cut is depicted by the dotted line. The cut has density 1. (b) A modified layout during x compaction. The cut now passes through a cell. (c) The cut is recreated, but now with density 3.*

whenever a cell moves through the cut, and it is not obvious how the change of the density can be taken into account efficiently. Nevertheless, we want to formulate the topological compaction problem as a problem of minimizing a simple cost function (the layout width) subject to the maintenance of a simple constraint system (the safety constraints for the critical cuts).

Maley [303] showed that such a constraint system can be generated efficiently in the case of the planar routing model. Furthermore, he proved that the corresponding constraint system amounts to a shortest-path problem. However, because of the problems with cuts changing density during the compaction, not all constraints can be generated at once. Rather, Maley introduces an iterative process in which the generation of additional constraints alternates with the solution of the partial constraint systems obtained so far. His algorithm runs in time $O(n^4)$, where n is the size of the problem instance (the sum of the number of cell boundary segments and the number of wires). In [306], Maley presents a general account of the topological compaction algorithm. He lists the requirements a detailed-routing model has to fulfill in order for topological compaction to be applicable. Then, he details the compaction algorithm, and exemplifies it with the planar routing model.

Mehlhorn and Näher [317] improve this figure to $O(n^3 \log n)$ in the worst case, and to $O(n^2 \log n)$ in typical cases, by using plane-sweep techniques. They do not separate generating and solving the constraint system, similar to part 4 of Exercise 10.6. Intuitively, a vertical plow bar is placed to the right of the layout. The plow starts moving to the left. As the plow moves, it picks up one cell after another. Let us call the set of cells already moving with the plow the *heap*, and the set of remaining cells the *field*. Initially, the heap is empty. As the plow moves left, a cell C can change from the field to the heap. This happens when a critical constraint between C and a cell that is already on the heap threatens to become unsafe. At this moment, the constraint is effectively converted into a "rigid bar" pushing the cell C to the left as the plow continues moving, such that the safety of the constraint is maintained. Algorithmically, the plane-sweep deals with two kinds of events. One kind represents the time instants at which a cut threatens to become unsafe. The second kind of event occurs when a cut starts or finishes moving across a cell. Mehlhorn and Näher provide the data structures maintained during the plane-sweep to identify and handle both kinds of events. Unfortunately, the algorithm by Mehlhorn and Näher cannot be applied to the knock-knee model, since crossing wires cannot be handled.

Gao et al. [132] present the fastest algorithm for topological compaction known to date. Its run time is $O(n^2 \log n)$ in the worst case, and the algorithm is likely to be much faster in typical cases. It can handle both the knock-knee model and the planar routing model. In connection with the results on the evenness condition for knock-knee routing, we can use homotopic compaction to compact *locally even* problem instances in which all critical cuts are safe. The result is always an even problem instance [132, 222]. Another advantage of this algorithm over the algorithm by Mehlhorn and Näher is that it separates generating and solving the constraint system. Therefore, it has the flexibility of modifying the constraint system by adding additional diverse constraints before it is solved.

The theory of topological compaction is probably one of the deepest mathematical facets of advanced circuit layout. The theory is still in an infant state, and it is not ready to be applied in practice. First, it presupposes the existence of theories of detailed routing that reduce routability to the maintenance of simple cut conditions. Such theories exist only for the planar and knock-knee model, and in the case of the knock-knee model they cover only two-terminal nets. (In the planar routing model, multiterminal nets can be handled; see Schrijver [401].) Second, no way of extending the two-dimensional compaction algorithm discussed in Section 10.3 to topological compaction is known. The problem is that, in topological compaction, we cannot trade off the two dimensions simply by choosing an alternative with respect to which dimension should separate a pair of cells. Rather, each maintained distance in the y dimension implies its specific set of constraints for the x dimension.

The topological compaction operation discussed here is probably more powerful than what is needed in practice. Generally, floorplanning can be expected to yield quite an accurate position for each cell, such that large movements of cells during compaction would not occur. Rather, in compacting floorplans or general-cell layouts, we are interested in making small adjustments to the positions of cells in both dimensions such as to decrease the size of the layout. Here, handling both dimensions simultaneously is more important than is being able to handle widely ranging movements in one dimension. In other words, cells can be moved in both dimensions, but they are *sticky*; that is, they resist moving large distances. Ciesielski and Kinnen [69] make a first step at addressing compaction in this context. The algorithm for generalized channel placement by Heng and LaPaugh [186] can be viewed as a simplified variant of two-dimensional topological compaction of slicing floorplans (see also Section 9.3.1.5). However, no systematic combinatorial methods have been developed to solve the two-dimensional topological compaction problem with sticky cells. In fact, it seems to be difficult even to formulate the problem in a precise way.

10.6 Summary

Compaction is the last layout subproblem in the phase approach to layout. The (mask layout) compaction operation turns symbolic layouts into mask data and optimizes the area of the layout on the way while maintaining the correctness with respect to the design rules. The main purpose of compaction is to achieve independence from the specific fabrication technology. The designer is freed from having to know the design rules, and updating a layout after a (small) change of the design rules can be done simply by recompaction of the symbolic layout with the new design rules. However, the technology independence does not go very far. The introduction of a new wiring layer, for instance, requires changes in the preceding layout phases such as detailed routing and layer assignment. If new devices become available, we may even have to adapt the netlist of the circuit.

Today, the most attractive approach to mask layout compaction is graph-based compaction. This framework provides a robust basis for one- and two-dimensional compaction. Heuristics exist for interrelating both dimensions efficiently during the compaction without solving the two-dimensional compaction problem optimally. Also, wire balancing can be incorporated into this framework. Wire balancing provides a tool for optimizing cost measures other than area, such as delay or power consumption, with second priority.

Graph-based compaction handles only rectagonal layout features. For methods incorporating more general layout features, see [333, 457]. Mosteller [333] uses the expressive power of the simulated annealing approach to formulate and solve compaction problems with general geometries. Sun [432] and Waterkamp et al. [457] use plane-sweep methods for the ordered one-dimensional compaction of octagonal and general polygonal layout features, respectively.

A large portion of the compaction problem deals with intricate sets of design rules. Since we are interested in only the combinatorial structure of the problem, we abstracted from these technological details. On the way, we made several strong restrictions. One is that all design rules have the form of minimum or maximum distance requirements. In practice, geometric rules with a different structure come into play, for instance, during the generation of wells in CMOS layouts. Lo [293] addresses this issue. Fabrication technology may dictate many complicated design rules—for instance, if subtle electrical phenomena have to be taken care of. Then, the compaction problem quickly loses much of its combinatorial structure, and we have to resort to general techniques, such as simulated annealing or rule-based systems.

10.7   Exercises

10.1 Explain how a symbolic layout can be translated into a legal geometric layout in linear time if each circuit element can attach to only one wire on each of its four sides.

Relax the condition on the circuit elements as far as possible such that you can still quickly translate a symbolic into a legal geometric layout.

10.2 Explain how to deal with horizontal wires that are wider than the mask features to which they connect on the same layer when translating a symbolic layout into an instance of the ORDERED 1D-COMPACTION problem.

10.3 Prove Theorem 10.1. To this end, show that the layout computed by the algorithm is legal and has minimal width.

Hint: For the first part, prove that each flow f corresponds to a compacted layout, in which each horizontal segment that corresponds to a downward edge e with flow $f(e)$ decreases in length by $f(e)$ units, and each horizontal segment that corresponds to an upward edge e with flow $f(e)$ increases in length by $f(e)$ units. For the second part, observe that all legal layouts for an instance of the ORDERED 1D-COMPACTION problem differ by only the widths of tiles. Lead the assumption that the obtained layout does not have minimum width to a contradiction to the premise that the flow in G that was computed is maximum.

10.4 Show that, in the worst case, computing the transitive reduction of an interval dag with n vertices takes time $\Omega(n \log n)$.

Hint: Reduce the problem of sorting n numbers to the problem of finding the transitive reduction of a interval dag.

10.5 Whereas all layout graphs have a unique source, the algorithm Gen_0 can be extended to computing the transitive reduction of interval dags with multiple sources. Indicate the necessary changes in Figure 10.11.

10.6 Develop an alternative algorithm for generating a sparse layout graph for an instance of the ORDERED 1D-PACKING problem. Use a right-to-left plane sweep. Generate edges between pairs of rectangles that are (partially) visible from each other. (These are rectangles R_i and R_j, for which there is a horizontal line segment connecting R_i with R_j without touching other rectangles.) This method is called *shadowing*.

1. What is the run time of your algorithm?

2. How many edges are generated in the worst case? Compare with algorithm Gen_0.

3. Discuss disadvantages and advantages of shadowing with respect to the algorithm Gen_0.

4. Modify the algorithm such that the pass that generates the layout graph and the pass that solves the shortest-path problem are merged. In this modification, no layout graph is explicitly constructed. Rather, the plane sweep pushes a rectangle as far as possible, to the right, as it is encountered during the sweep.

10.7 This exercise discusses the proof of Theorem 10.4. For two rectangles $R_i, R_j \in \overline{L}$ with $R_j \prec R_i$, define

$$R_i \pi_0 R_j :\Leftrightarrow R_i \pi R_j \text{ and } \beta(R_j, R_i) = \emptyset$$

The relation π_0 is the part of the compatibility relation π taken into account by $G_{1,\pi}(L)$. The following lemma is at the heart of the proof of the theorem.

Lemma: *For all $R_i, R_j \in \overline{L}$ with $R_j \prec R_i$, we have*

$$(R_i, R_j) \in E^r_{1,\pi}(L) \Leftrightarrow (\neg R_i \pi_0 R_j \text{ and}$$
$$\forall R_k \in \beta(R_j, R_i) : R_i \pi_0 R_k \text{ or } R_j \pi_0 R_k)$$

1. Prove this lemma.

2. Use the lemma to modify the algorithm Gen_0 such as to compute $E^r_{1,\pi}(L)$.

   Hint: Use a two-pass algorithm. In the first pass, run Gen_0. In the second pass, perform a top-to-bottom sweep, check the criterion stated in the lemma, and output edges accordingly.

3. Prove the following statements, and use them to make the modified algorithm efficient.

   Let L_t be the layout portion that the sweep line has swept over up to time t in the top-to-bottom sweep.

a. If $(R_i, R_j) \in E_{1,\pi}^r(L_t)$, then, for at most two rectangles $R \in \overline{L_t}$, we have

$$R_j \prec R \prec R_i, \text{ and } R \text{ intersects the sweep line at time } t.$$

b. The maximum indegree and outdegree of any vertex in $G_{1,\pi}(L_t)$ is 2.

4. Give an example of a layout that illustrates that a memoryless one-pass sweep algorithm is not sufficient for the computation of $G_{1,\pi}^r(L)$.

5. Prove that $|E_{1,\pi}^r(L)| \leq 4n - 6$ if $|L| = n$.

   Hint: Use a method analogous to the proof of Lemma 10.3

10.8 Generalize the Insert procedure of $Gen(0)$ depicted in Figure 10.15 to test the criterion of Definition 10.5. Prove the correctness of this modification. Analyze the run time of the algorithm.

10.9 Construct an example for which $Gen(k)$ computes many edges. How many edges does $Gen(k)$ compute?

10.10 Improve on the shortest-path algorithm presented in Section 10.2.2.2.2 by eliminating superfluous edge tests.

10.11 Prove that the wire-balancing problem in one-dimensional compaction is the dual of a MINCOST FLOW problem.

10.12 Transform the NP-complete 3-PARTITION problem [133] in polynomial time to the following decision problem versions of the 2D-PACKING problem:

1. Given a layout and two values w and h, can the rectangles in the layout be packed into a rectangle with width w and height h?

2. Given a layout and a value A, can the rectangles in the layout be packed into a rectangle with area A?

3. Given a layout and a value p, can the rectangles in the layout be packed into a rectangle with perimeter p?

10.13 Use plane-sweep methods to generate sparse versions of $H_{r,x}$ and $H_{r,y}$ that exclude redundant edges.

10.14 Construct an example of a hierarchical layout with 2^n instances of the same cell, such that, in an optimally x-compacted version of the layout, each instance of the cell has to be laid out differently. The length of the hierarchical layout description must be $O(n)$.

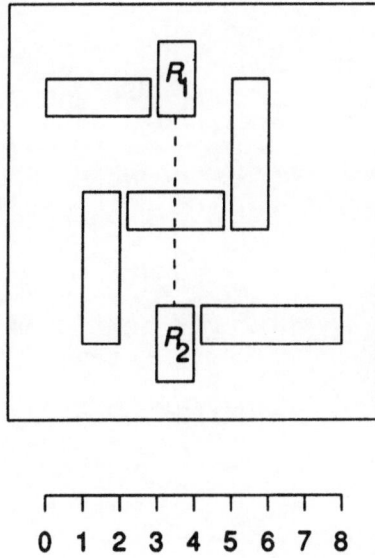

Figure 10.30: *A cell layout that, after toroidal compaction, yields a fractional circumference Δ. The rectangles R_1 and R_2 are assumed to be connected with a grouping constraint of the form $x_1 = x_2$, indicated by the dotted line. Such a constraint can come about, for instance, through a vertical wire connecting R_1 and R_2 on a different layer.*

10.15

1. Consider the cell in Figure 10.30. Prove that the minimum circumference Δ of a torus accommodating this cell is not integer. Calculate the minimum value of Δ.

2. Give general criteria for determining when the minimum value of Δ is integer.

Bibliography

[1] E. H. L. Aarts, F. M. J. de Bont, J. H. A. Habers, and P. J. M. van Laarhoven. Parallel implementations of the statistical cooling algorithm. *Integration*, 4(3):209–238, 1986.

[2] E. H. L. Aarts and P. J. M. van Laarhoven. Statistical cooling: A general approach to combinatorial optimization problems. *Philips Journal of Research*, 40(4):193–226, 1985.

[3] B. D. Ackland. Knowledge-based physical design automation. In B. T. Preas and M. J. Lorenzetti, editors, *Physical Design Automation of VLSI Systems*, Chapter 9, pages 409–460. Benjamin Cummings, Menlo Park, CA, 1988.

[4] D. Adolphson and T. C. Hu. Optimal linear ordering. *SIAM Journal of Applied Mathematics*, 25(3):403–423, 1973.

[5] A. V. Aho, J. E. Hopcroft, and J. D. Ullman. *The Design and Analysis of Computer Algorithms*. Addison-Wesley Series in Computer Science and Engineering. Addison-Wesley, Reading, MA, 1974.

[6] R. K. Ahuja, K. Mehlhorn, J. B. Orlin, and R. E. Tarjan. Faster algorithms for the shortest path problem. Technical Report 193, Operations Research Center, Mass. Inst. of Technology, Cambridge, MA, 1988.

[7] R. K. Ahuja, J. B. Orlin, and R. E. Tarjan. Improved time bounds for the maximum flow problem. Technical Report 1966-87, Sloan School of Management, Mass. Inst. of Technology, Cambridge, MA, 1987.

[8] S. B. Akers. A modification of Lee's path connection algorithm. *IEEE Transactions on Electronic Computers*, EC-16(2):97–98, 1967.

[9] S. B. Akers. Routing. In M. A. Breuer, editor, *Design Automation of Digital Systems, Volume 1: Theory and Techniques*, pages 283–333. Prentice Hall, Inc., Englewood Cliffs, NJ, 1972.

[10] S. B. Akers. On the use of the linear assignment algorithm in module placement. In *Proceedings of the 18th Design Automation Conference*, pages 137–144. ACM/IEEE, 1981.

[11] S. B. Akers, M. E. Geyer, and D. L. Roberts. IC mask layout with a single conductor layer. In *Proceedings of the 7th Design Automation Workshop*, pages 7–16. ACM/IEEE, 1970.

[12] T. M. Apostol. *Calculus, Volume 2: Multi-Variable Calculus and Linear Algebra, with Applications to Differential Equations and Probability.* Xerox College Publishing, Waltham, MA, 1969.

[13] S. Arnborg, D. Corneil, and A. Proskurowski. Complexity of finding embeddings in a *k*-tree. *SIAM Journal on Algebraic and Discrete Methods,* 8(2):277–284, 1987.

[14] R. A. Auerbach, B. W. Lin, and E. A. Elsayed. Layouts for the design of VLSI circuits. *Computer Aided Design,* 13(5):271–276, 1981.

[15] D. Avis. A survey of heuristics for the weighted matching problem. *Networks,* 13:475–493, 1983.

[16] U. G. Baitinger and R. Schmid. The role of floor-plan tools in the VLSI design process. *Elektronische Rechenanlagen,* 26(6):290–297, 1984.

[17] B. S. Baker. A provably good algorithm for the two module routing problem. *SIAM Journal on Computing,* 15(1):162–189, 1986.

[18] B. S. Baker, S. N. Bhatt, and F. T. Leighton. An approximation algorithm for Manhattan routing. In F. P. Preparata, editor, *Advances in Computing Research, Volume 2: VLSI Theory,* pages 205–229. JAI Press, Greenwich, CT, 1984.

[19] P. Banerjee and M. Jones. A parallel simulated annealing algorithm for standard cell placement on a hypercube computer. In *Proceedings of the International Conference on Computer-Aided Design,* pages 34–37. IEEE, 1986.

[20] F. Barahona, M. Grötschel, M. Jünger, and G. Reinelt. An application of combinatorial optimization to statistical physics and circuit layout design. *Operations Research,* 36(3):493–513, 1988.

[21] E. R. Barnes. An algorithm for partitioning the nodes of a graph. *SIAM Journal of Algebraic and Discrete Methods,* 3(4):541–550, 1982.

[22] E. R. Barnes, A. Vannelli, and J. Q. Walker. A new heuristic for partitioning the nodes of a graph. *SIAM Journal of Discrete Mathematics,* 1(3):299–305, 1988.

[23] J. E. Beasley. An algorithm for the Steiner problem in graphs. *Networks,* 14:147–159, 1984.

[24] M. Becker and K. Mehlhorn. Algorithms for routing in planar graphs. *Acta Informatica,* 23(2):163–176, 1986.

[25] R. E. Bellman. On a routing problem. *Quarterly Applied Mathematics,* 16:87–90, 1958.

[26] B. Berger, M. L. Brady, D. J. Brown, and F. T. Leighton. Nearly optimal algorithms and bounds for multilayer channel routing. *Journal of the Association for Computing Machinery.* In press.

[27] M. R. C. M. Berkelaar and J. A. G. Jess. Technology mapping for standard-cell generators. In *Proceedings of the International Conference on Computer-Aided Design,* pages 470–473. IEEE, 1988.

[28] M. Bern. Faster exact algorithms for Steiner trees in planar networks. *Networks.* To appear.

[29] M. Bern, E. L. Lawler, and A. L. Wong. Linear-time computation of optimal subgraphs of decomposable graphs. *Journal of Algorithms*, 8:216–235, 1987.

[30] P. Bertolazzi, M. Lucertini, and A. Marchetti Spaccamela. Analysis of a class of graph partitioning problems. *R.A.I.R.O. Theoretical Informatics*, 16(3):255–261, 1982.

[31] D. P. Bertsekas. *Constrained Optimization and Lagrange Multiplier Methods.* Computer Science and Applied Mathematics. Academic Press, New York, 1982.

[32] J. Bhasker and S. Sahni. A linear algorithm to check for the existence of a rectangular dual of a planar triangulated graph. *Networks*, 17:307–317, 1987.

[33] J. Bhasker and S. Sahni. A linear algorithm to find a rectangular dual of a planar triangulated graph. *Algorithmica*, 3:247–278, 1988.

[34] S. N. Bhatt and F. T. Leighton. A framework for solving VLSI graph layout problems. *Journal of Computer and System Sciences*, 28(2):300–343, 1984.

[35] K. Binder. *Monte Carlo Methods in Statistical Physics.* Springer Verlag, New York, 1986.

[36] J. P. Blanks. Near optimal placement using a quadratic objective function. In *Proceedings of the 22nd Design Automation Conference*, pages 609–615. ACM/IEEE, 1985.

[37] J. P. Blanks. *Use of a Quadratic Objective Function for the Placement Problem in VLSI Design.* PhD thesis, University of Texas, Austin, TX, 1985.

[38] H. Bollinger. A mature DA system for PC layout. In *Proceedings of the First International Printed Circuit Board Conference*, pages 85–99, 1979.

[39] B. Bollobas. *Random Graphs.* Academic Press, New York, 1985.

[40] K. S. Booth and G. S. Lueker. Testing for the consecutive ones property, intervals graphs, and graph planarity using *pq*-trees. *Journal of Computer and System Sciences*, 13(3):335–379, 1976.

[41] R. B. Boppana. Eigenvalues and graph bisection: An average case analysis. In *Proceedings of the 28th Annual Symposium on Foundations in Computer Science*, pages 280–285. IEEE, 1987.

[42] R. B. Boppana, 1989. Personal Communication.

[43] D. G. Boyer. Virtual grid compaction using the most recent layers algorithm. In *Proceedings of the International Conference on Computer-Aided Design*, pages 92–93. IEEE, 1986.

[44] D. G. Boyer. Split grid compaction for a virtual grid symbolic design system. In *Proceedings of the International Conference on Computer-Aided Design*, pages 134–137. IEEE, 1987.

[45] H. N. Brady. An approach to topological pin assignment. *IEEE Transactions on Computer-Aided Design of Integrated Circuits and Systems*, CAD-3(3):250–255, 1984.

[46] M. L. Brady and D. J. Brown. VLSI routing: Four layers suffice. In F. P. Preparata, editor, *Advances in Computing Research, Volume 2: VLSI Theory*, pages 245–258. JAI Press, Greenwich, CT, 1984.

[47] M. L. Brady and D. J. Brown. Optimal multi-layer channel routing with over-lap. In C. E. Leiserson, editor, *Proceedings of the Fourth MIT Conference on Advanced Research in VLSI*, pages 281–296. MIT Press, Cambridge, MA, 1986.

[48] D. Braun, J. L. Burns, F. Romeo, A. L. Sangiovanni-Vincentelli, K. Mayaram, S. Devadas, and H.-K. T. Ma. Techniques for multi-layer channel routing. *IEEE Transactions on Computer-Aided Design of Integrated Circuits and Systems*, CAD-7(6):698–712, 1988.

[49] M. A. Breuer. Min-cut placement. *Design Automation and Fault Tolerant Computing*, 1(4):343–362, 1977.

[50] M. A. Breuer and A. Kumar. A methodology for custom VLSI layout. *IEEE Transactions on Circuits and Systems*, CAS-30(6):358–364, 1983.

[51] D. J. Brown and R. R. Rivest. New lower bounds for channel width. In H. T. Kung, B. Sproull, and G. Steele, editors, *VLSI Systems and Computations*, pages 178–185. Computer Science Press, Rockville, MD, 1981.

[52] T. N. Bui, S. Chauduri, F. T. Leighton, and M. Sipser. Graph bisection algo-rithms with good average case behavior. *Combinatorica*, 7(2):171–191, 1987.

[53] T. N. Bui and C. Jones. Sequential and parallel algorithms for partitioning simple classes of graphs. Technical Report CS-89-45, Department of Computer Science, The Pennsylvania State University, University Park, PA, 1989.

[54] R. E. Burkard. Quadratic assignment problems. *European Journal of Opera-tional Research*, 15:283–289, 1984.

[55] M. Burstein and S. J. Hong. Hierarchical VLSI layout: Simultaneous wiring and placement. In F. Anceau and E. J. Aas, editors, *Proceedings of VLSI'83*, pages 45–60. Elsevier Science Publishers B.V., Amsterdam, The Netherlands, 1983.

[56] M. Burstein and R. Pelavin. Hierarchical wire routing. *IEEE Transactions on Computer-Aided Design of Integrated Circuits and Systems*, CAD-2(4):223–234, 1983.

[57] C. W. Carpenter and M. Horowitz. Generating incremental VLSI compaction spacing constraints. In *Proceedings of the 24th Design Automation Conference*, pages 291–297. ACM/IEEE, 1987.

[58] A. Casotto, F. Romeo, and A. L. Sangiovanni-Vincentelli. A parallel simulated annealing algorithms for the placement of macro-cells. In *Proceedings of the International Conference on Computer-Aided Design*, pages 30–33. IEEE, 1986.

[59] H. H. Chen. Trigger: A three-layer gridless channel router. In *Proceedings of the International Conference on Computer-Aided Design*, pages 196–169. IEEE, 1986.

[60] H. H. Chen and E. S. Kuh. Glitter: A gridless variable-width channel router. *IEEE Transactions on Computer-Aided Design of Integrated Circuits and Sys-tems*, CAD-5(4):459–465, 1986.

[61] Y. K. Chen and M. L. Liu. Three-layer channel routing. *IEEE Transactions on Computer-Aided Design of Integrated Circuits and Systems*, CAD-3(2):156–163, 1984.

[62] C.-K. Cheng. Linear placement algorithms and applications to VLSI design. *Networks*, 17:439–464, 1987.

[63] D. Cheriton and R. E. Tarjan. Finding minimum spanning trees. *SIAM Journal on Computing*, 5(4):724–742, 1977.

[64] H. Chernoff. A measure of asymptotic efficiency for tests of a hypothesis based on the sum of observations. *Annals of Mathematical Statistics*, 23:493–507, 1952.

[65] S. Chowdury. Analytical approaches to the combinatorial optimization in linear placement problems. *IEEE Transactions on Computer-Aided Design of Integrated Circuits and Systems*, 8(6):630–639, 1989.

[66] N. Christofides. *Graph Theory: An Algorithmic Approach*. Academic Press, New York, 1975.

[67] F. R. K. Chung. On optimal linear arrangements of trees. *Computers and Mathematics with Applications*, 10(1):43–60, 1984.

[68] M. J. Chung and K. K. Rao. Parallel simulated annealing for partitioning and routing. In *Proceedings of the International Conference on Computer Design: VLSI in Computers*, pages 238–242. IEEE, 1986.

[69] M. J. Ciesielski and E. Kinnen. An analytical method for compacting routing area in integrated circuits. In *Proceedings of the 19th Design Automation Conference*, pages 30–37. ACM/IEEE, 1982.

[70] K. L. Clarkson, S. Kapoor, and P. M. Vaidya. Rectilinear shortest paths through polygonal obstacles in $O(n(\log n)^2)$ time. In *Proceedings of the Third Annual Conference on Computational Geometry*, pages 251–257. ACM, 1987.

[71] J. P. Cohoon and P. L. Heck. Beaver: A computational-geometry–based tool for switchbox routing. *IEEE Transactions on Computer-Aided Design of Integrated Circuits and Systems*, 7(6):684–697, 1988.

[72] R. Cole and A. R. Siegel. River routing every which way but loose. In *Proceedings of the 25th Annual IEEE Symposium on Foundations of Computer Science*, pages 65–73. IEEE, 1984.

[73] J. Cong, D. F. Wong, and C. L. Liu. A new approach to three- or four-layer channel routing. *IEEE Transactions on Computer-Aided Design of Integrated Circuits and Systems*, 7(10):1094–1104, 1988.

[74] D. Coppersmith and S. Winograd. Matrix multiplication via arithmetic progressions. In *Proceedings of the 19th Annual ACM Symposium on Theory of Computing*, pages 1–6. ACM, 1987.

[75] J. M. da Mata. Allende: A procedural language for the hierarchical specification of VLSI layout. In *Proceedings of the 22nd Design Automation Conference*, pages 183–189. ACM/IEEE, 1985.

[76] E. Dahlhaus, D. S. Johnson, C. H. Papadimitriou, P. Seymour, and M. Yannakakis. The complexity of multiway cuts. Typescript, 1983.

[77] W. W.-M. Dai, 1988. Personal Communication.

[78] W. W.-M. Dai, T. Asano, and E. S. Kuh. Routing region definition and order-
     ing scheme for building block layout. *IEEE Transactions on Computer-Aided
     Design of Integrated Circuits and Systems*, CAD-4(3):189–197, 1985.

[79] W. W.-M. Dai and E. S. Kuh. Global spacing of building-block layout. In
     C. H. Sequin, editor, *VLSI'87*, pages 193–205. Elsevier Science Publishers B.V.,
     Amsterdam, The Netherlands, 1987.

[80] W. W.-M. Dai and E. S. Kuh. Simultaneous floorplanning and global routing
     for hierarchical building block layout. *IEEE Transactions on Computer-Aided
     Design of Integrated Circuits and Systems*, CAD-6(5):828–837, 1987.

[81] W. W.-M. Dai, M. Sato, and K. E. S. A dynamic and efficient representa-
     tion of building-block layout. In *Proceedings of the 24th Design Automation
     Conference*, pages 376–384. ACM/IEEE, 1987.

[82] S. Daijavad, E. Polak, and R.-S. Tsay. A combined deterministic and random
     optimization algorithm for the placement of macro cells. Department of Electri-
     cal Engineering and Computer Science, University of California Berkeley, CA,
     1988.

[83] F. Darema, S. Kirkpatrick, and V. A. Norton. Parallel algorithms for chip
     placement by simulated annealing. *IBM Journal of Research and Development*,
     31(3):391–402, 1987.

[84] P. J. De Rezende, D. T. Lee, and Y.-F. Wu. Rectilinear shortest paths with
     rectangular barriers. In *Proceedings of the Second Annual Conference on Com-
     putational Geometry*, pages 204–213. ACM, 1985.

[85] W. A. Dees, Jr. and P. G. Karger. Automated rip-up and reroute techniques.
     In *Proceedings of the 19th Design Automation Conference*, pages 432–439.
     ACM/IEEE, 1982.

[86] W. A. Dees, Jr. and R. J. Smith II. Performance of interconnection rip-up and
     reroute strategies. In *Proceedings of the 18th Design Automation Conference*,
     pages 382–390. ACM/IEEE, 1981.

[87] E. Detjens, G. Gannot, R. Rudell, A. S. Sangiovanni-Vincentelli, and A. Wang.
     Technology mapping in MIS. In *Proceedings of the International Conference on
     Computer-Aided Design*, pages 116–119. IEEE, 1987.

[88] D. N. Deutsch. A 'dogleg' channel router. In *Proceedings of the 13th Design
     Automation Conference*, pages 425–433. ACM/IEEE, 1976.

[89] R. B. Dial. Shortest path forest with topological ordering. *Communications
     of the Association for Computing Machinery*, 12(11):632–633, 1969. Algorithm
     360.

[90] E. W. Dijkstra. A note on two problems in connexion with graphs. *Numerische
     Mathematik*, 1:269–271, 1959.

[91] E. A. Dinic. Algorithm for solution of a problem of maximum flow in a network
     with power estimation. *Soviet Math. Doklady*, 11:1277–1280, 1970.

[92] H. N. Djidjev. On the problem of partitioning planar graphs. *SIAM Journal
     on Algebraic and Discrete Methods*, 3(2):229–240, 1982.

[93] J. Do and W. M. Dawson. SPACER II: A well-behaved IC layout compactor. In *Proceedings of VLSI 85*, pages 277–285. Elsevier Science Publishers B.V., Amsterdam, The Netherlands, 1985.

[94] J. Doenhardt and T. Lengauer. Algorithmic aspects of one-dimensional layout compaction. *IEEE Transactions on Computer-Aided Design of Integrated Circuits and Systems*, CAD-6(5):863–879, 1987.

[95] D. Dolev and A. Siegel. The separation for general single-layer wiring barriers. In H. T. Kung, B. Sproull, and G. Steele, editors, *CMU Conference on VLSI Systems and Computations*, pages 143–152. Computer Science Press, Rockville, MD, 1981.

[96] W. E. Donath. Logic partitioning. In B. T. Preas and M. J. Lorenzetti, editors, *Physical Design Automation of VLSI Systems*, Chapter 3, pages 65–86. Benjamin Cummings, Menlo Park, CA, 1988.

[97] W. E. Donath and A. J. Hoffman. Lower bounds for the partitioning of graphs. *IBM Journal of Research and Development*, 17(9):420–425, 1973.

[98] S. E. Dreyfus and R. A. Wagner. The Steiner problem in graphs. *Networks*, 1:195–208, 1972.

[99] A. E. Dunlop. SLIP: Symbolic layout of integrated circuits with compaction. *Computer-Aided Design*, 10(6):387–391, 1978.

[100] A. E. Dunlop and B. W. Kernighan. A procedure for the placement of standard-cell VLSI circuits. *IEEE Transactions on Computer-Aided Design of Integrated Circuits and Systems*, CAD-4(1):92–98, 1985.

[101] P. A. Eichenberger. *Fast Symbolic Layout Translation for Custom VLSI Integrated Circuits*. PhD thesis, Computer Systems Laboratory Technical Report No. 86-295, Department of Electrical Engineering, Stanford University, Stanford, CA, 1986.

[102] P. A. Eichenberger and M. Horowitz. Toroidal compaction of symbolic layouts for regular structures. In *Proceedings of the International Conference on Computer Design: VLSI in Computers and Processors*, pages 142–145. IEEE, 1987.

[103] R. J. Enbody and H. C. Du. Near-optimal n-layer channel routing. In *Proceedings of the 23rd Design Automation Conference*, pages 708–714. ACM/IEEE, 1986.

[104] G. Entenman and S. W. Daniel. A fully automatic hierarchical compacter. In *Proceedings of the 22nd Design Automation Conference*, pages 69–75. ACM/IEEE, 1985.

[105] B. Eschermann, W. W.-M. Dai, E. S. Kuh, and M. Pedram. Hierarchical placement and floorplanning in BEAR. *IEEE Transactions on Computer-Aided Design of Integrated Circuits and Systems*, CAD-8(12):1335–1349, 1989.

[106] S. Even. *Graph Algorithms*. Computer Science Press, Potomac, MD, 1979.

[107] S. Even, A. Itai, and A. Shamir. On the complexity of timetable and multi-commodity flow problems. *SIAM Journal on Computing*, 5(4):691–703, 1976.

[108] U. Faigle and W. Kern. Note on the convergence of simulated annealing algorithms. Technical Report 774, Faculty of Applied Mathematics, University of Twente, Enschede, The Netherlands, 1989.

[109] U. Faigle and R. Schrader. On the convergence of stationary distributions in simulated annealing algorithms. *Information Processing Letters*, 27(4):189–194, 1988.

[110] W. Feller. *An Introduction to Probability Theory and Its Applications*, Volume 1. John Wiley & Sons, New York, 1968.

[111] M. R. Fellows and M. A. Langston. Layout permutation problems and well-partially ordered sets. In J. Allen and F. T. Leighton, editors, *Proceedings of the Fifth MIT Conference on Advanced Research in VLSI*, pages 315–330. MIT Press, Cambridge, MA, 1988.

[112] M. R. Fellows and M. A. Langston. Nonconstructive tools for proving polynomial-time decidability. *Journal of the Association for Computing Machinery*, 35(3):727–739, 1988.

[113] T. A. Feo and M. Khellaf. A class of bounded approximation algorithms for graph partitioning. typescript, 1987.

[114] C. M. Fiduccia and R. M. Mattheyses. A linear time heuristic for improving network partitions. In *Proceedings of the 19th Design Automation Conference*, pages 175–181. ACM/IEEE, 1982.

[115] R. Fiebrich and C. Wang. Circuit placement based on simulated annealing on a massively parallel computer. In *Proceedings of the International Conference on Computer Design: VLSI in Computers and Processors*, pages 78–82. IEEE, 1987.

[116] G. Finke, R. E. Burkard, and F. Rendl. Quadratic assignment problems. *Annals of Discrete Mathematics*, 31:61–82, 1987.

[117] L. R. Ford. Network flow theory. Technical Report P-923, RAND Corp., Santa Monica, CA, 1956.

[118] M. Formann, D. Wagner, and F. Wagner. Routing through a dense channel with minimum total wire length. Technical Report B-89-13, Department of Mathematics, Free University of Berlin, Berlin, West Germany, 1989.

[119] M. Formann and F. Wagner. The VLSI layout problem in various embedding models. Technical Report B-89-16, Department of Mathematics, Free University of Berlin, Berlin, West Germany, 1989.

[120] J. A. Frankle. *Circuit Placement Methods Using Multiple Eigenvectors and Linear Probe Techniques*. PhD thesis, Department of Electrical Engineering and Computer Science, University of California, Berkeley, CA, 1987.

[121] J. A. Frankle and R. M. Karp. Circuit placements and cost bounds by eigenvector decompositions. In *Proceedings of the International Conference on Computer-Aided Design*, pages 414–417. IEEE, 1986.

[122] M. R. Fredman and R. E. Tarjan. Fibonacci heaps and their uses in improved network optimization algorithms. *Journal of the Association for Computing Machinery*, 34(3):596–615, 1987.

[123] H. N. Gabow. Data structures for weighted matching and nearest common ancestors with linking. In *Proceedings of the First ACM-SIAM Symposium on Discrete Algorithms*, pages 434–443. ACM/SIAM, 1990.

[124] H. N. Gabow, Z. Galil, and T. H. Spencer. Efficient implementation of graph algorithms using contraction. *Journal of the Association for Computing Machinery*, 36(3):540–572, 1989.

[125] H. N. Gabow, Z. Galil, T. H. Spencer, and R. E. Tarjan. Efficient algorithms for finding minimum spanning trees in undirected and directed graphs. *Combinatorica*, 6(2):109–122, 1986.

[126] Z. Galil. Efficient algorithms for finding maximum matching in graphs. *ACM Computing Surveys*, 18(1):23–38, 1986.

[127] Z. Galil, S. Micali, and H. N. Gabow. An $O(EV \log V)$ algorithm for finding a maximal weighted matching in general graphs. *SIAM Journal on Computing*, 15(1):120–130, 1986.

[128] S. Gao. An algorithm for two-layer channel routing. In K. Mehlhorn, editor, *Proceedings of the Second Annual Symposium on Theoretical Aspects of Computer Science*, pages 151–160. Springer Lecture Notes in Computer Science, No. 182, Springer Verlag, New York, 1985.

[129] S. Gao and S. Hambrusch. Two-layer channel routing with vertical unit-length overlap. *Algorithmica*, 1:223–232, 1986.

[130] S. Gao, M. Jerrum, M. Kaufmann, K. Mehlhorn, W. Rülling, and C. Storb. On continuous homotopic one layer routing. In *Proceedings of the Fourth Annual Symposium on Computational Geometry*, pages 392–402. ACM, 1988.

[131] S. Gao and M. Kaufmann. Channel routing of multiterminal nets. In *Proceedings of the 28th Annual IEEE Symposium on Foundations in Computer Science*, pages 316–325. IEEE, 1987.

[132] S. Gao, M. Kaufmann, and F. M. Maley. Advances in homotopic layout compaction. In *Proceedings of the 1989 Symposium on Parallel Algorithms and Architectures*, pages 273–282. ACM, 1989.

[133] M. R. Garey and D. S. Johnson. *Computers and Intractability: A Guide to the Theory of NP-Completeness*. Freeman, San Francisco, CA, 1979.

[134] S. B. Gelfand and S. K. Mitter. Analysis of simulated annealing for optimization. In *Proceedings of the 24th Conference on Decision and Control*, pages 779–786. IEEE, 1985.

[135] S. Geman and D. Geman. Stochastic relaxation, Gibbs distributions, and the Bayesian restoration of images. *IEEE Transactions on Pattern Analysis and Machine Intelligence*, PAMI-6(6):721–741, 1984.

[136] J. M. Geyer. Connection routing algorithm for printed circuit boards. *IEEE Transactions on Circuit Theory*, CT-18(1):95–100, 1971.

[137] B. Gidas. Nonstationary markov chains and the convergence of the annealing algorithm. *Journal of Statistical Physics*, 39(1):73–131, 1985.

[138] P. E. Gill, W. Murray, and M. H. Wright. *Practical Optimization*. Academic Press, New York, 1981.

[139] L. P. P. P. v. Ginneken and R. H. J. M. Otten. An inner loop criterion for annealing. Technical Report RC-12980, IBM Thomas J. Watson Research Center, Yorktown Heights, NY, 1987.

[140] L. A. Glasser and D. W. Dobberpuhl. *The Design and Analysis of VLSI Circuits*. Addison-Wesley, Reading, MA, 1985.

[141] A. W. Goldberg and R. E. Tarjan. A new approach to the maximum flow problem. In *Proceedings of the 18th Annual ACM Symposium on Theory of Computing*, pages 136–146. ACM, 1986.

[142] A. W. Goldberg and R. E. Tarjan. Solving minimum-cost flow problems by successive approximation. In *Proceedings of the 19th Annual ACM Symposium on Theory of Computing*, pages 7–18. ACM, 1987.

[143] A. W. Goldberg and R. E. Tarjan. Finding minimum-cost circulations by canceling negative cycles. In *Proceedings of the 20th Annual ACM Symposium on Theory of Computing*, pages 388–397. ACM, 1988.

[144] M. K. Goldberg and M. Burstein. Heuristic improvement technique for bisection of VLSI networks. In *Proceedings of the International Conference on Computer Design: VLSI in Computers*, pages 122–125. IEEE, 1983.

[145] M. K. Goldberg and Z. Miller. A parallel algorithm for bisection width in trees. *Computers and Mathematics with Applications*, 15(4):259–266, 1988.

[146] O. Goldschmidt and D. S. Hochbaum. A polynomial algorithm for the k-cut problem. Technical report, Department of Industrial Engineering and Operations Research and School of Business Administration, University of California, Berkeley, CA, 1987.

[147] G. H. Golub and C. F. van Loan. *Matrix Computations*. The Johns Hopkins University Press, Baltimore, MD, second edition, 1989.

[148] R. E. Gomory. Outline of an algorithm for integer solution to linear programs. *Bulletin of the American Mathematical Society*, 64(5):275–278, 1958.

[149] R. E. Gomory. An algorithm for integer solution to linear programs. In R. L. Graves and P. Wolfe, editors, *Recent Advances in Mathematical Programming*, pages 269–302. McGraw-Hill, New York, 1963.

[150] T. F. Gonzalez and S.-L. Lee. An optimal algorithm for optimal routing around a rectangle. In *Proceedings of the 20th Annual Allerton Conference on Communication, Control, and Computing*, pages 636–645, 1982.

[151] T. F. Gonzalez and S.-Q. Zheng. Simple three-layer channel routing algorithms. In J. Reif, editor, *Proceedings of Third International Workshop on Parallel Computation and VLSI Theory*, pages 237–246. Springer Lecture Notes in Computer Science, No. 319, Springer Verlag, New York, 1988.

[152] I. S. Gopal, D. Coppersmith, and C. K. Wong. Optimal wiring of movable terminals. *IEEE Transactions on Computers*, C-32(9):845–858, 1983.

[153] S. Goto and T. Matsuda. Partitioning, assignment and placement. In T. Ohtsuki, editor, *Advances in CAD for VLSI, Volume 4: Layout Design and Verification*, Chapter 2, pages 55–98. North-Holland, New York, 1986.

[154] R. I. Greenberg and C. E. Leiserson. A compact layout for the three-dimensional tree of meshes. *Applied Mathematics Letters*, 1(2):171–176, 1988.

[155] R. I. Greenberg and C. E. Leiserson. Randomized routing on fat-trees. In S. Micali, editor, *Advances in Computing Research, Volume 5: Randomness and Computation*. JAI Press, Greenwich, CT, 1989.

[156] J. W. Greene and K. J. Supowit. Simulated annealing without rejected moves. In *Proceedings of the International Conference on Computer Design: VLSI in Computers*, pages 658–663. IEEE, 1984.

[157] B. Grünbaum. *Convex Polytopes*, Volume XVI of *Pure and Applied Mathematics*. Wiley-Interscience, New York, 1967.

[158] U. I. Gupta, D. T. Lee, and J. Y.-T. Leung. An optimal solution for the channel assignment problem. *IEEE Transactions on Computers*, C-28(11):807–810, 1979.

[159] A. Habel and H. J. Kreowski. May we introduce to you: Hypergraph languages generated by hyperedge replacement. In H. Ehrig, A. Rosenfeld, and G. Rozenberg, editors, *Graph Grammars and Their Application to Computer Science*, pages 15–26. Springer Lecture Notes in Computer Science, No. 291, Springer Verlag, New York, 1987.

[160] R. W. Haddad and A. A. Schaeffer. Recognizing Bellman-Ford-orderable graphs. *SIAM Journal on Discrete Mathematics*, 1(4):447–471, 1988.

[161] F. O. Hadlock. Minimum spanning forests of bounded trees. In *Proceedings of the Fifth Southeastern Conference on Combinatorics, Graph Theory and Computing (Congressus Numerantium X)*, pages 7–18, 1974.

[162] F. O. Hadlock. Finding a maximum cut of a planar graph in polynomial time. *SIAM Journal on Computing*, 4(3):221–225, 1975.

[163] F. O. Hadlock. A shortest path algorithm for grid graphs. *Networks*, 7:323–334, 1977.

[164] B. Hajek. Cooling schedules for optimal annealing. *Mathematics of Operations Research*, 13(2):311–329, 1988.

[165] D. W. Hall and G. Spencer. *Elementary Topology*. John Wiley & Sons, New York, 1955.

[166] K. M. Hall. An r-dimensional quadratic placement algorithm. *Management Science*, 17:219–229, 1970.

[167] G. T. Hamachi and J. K. Ousterhout. A switchbox router with obstacle avoidance. In *Proceedings of the 21st Design Automation Conference*, pages 173–179. ACM/IEEE, 1984.

[168] S. E. Hambrusch. Using overlap and minimizing contact points in channel routing. In *Proceedings of the 21st Allerton Conference on Communication, Control, and Computing*, pages 256–257, 1983.

[169] S. E. Hambrusch. Channel routing in overlap models. *IEEE Transactions on Computer-Aided Design of Integrated Circuits and Systems*, CAD-4(1):23–30, 1985.

[170] M. Hanan. On Steiner's problem with rectilinear distance. *SIAM Journal on Applied Mathematics*, 14(2):255–265, 1966.

[171] M. Hanan and J. M. Kurtzberg. A review of the placement and quadratic assignment problems. *SIAM Review*, 14(2):324–342, 1972.

[172] M. Hanan, P. K. Wolff Sr., and B. J. Agule. A study of placement techniques. *Design Automation and Fault-Tolerant Computing*, 2:28–61, 1978.

[173] D. W. Hanson. Interconnection analysis. In B. Preas and M. Lorenzetti, editors, *Physical Design Automation of VLSI Systems*, Chapter 2, pages 31–64. Benjamin Cummings, Menlo Park, CA, 1988.

[174] F. Harary. *Graph Theory*. Addison-Wesley, Reading, MA, 1969.

[175] P. Hart, N. Nilsson, and B. Raphael. A formal basis for the heuristic determination of minimum cost paths. *IEEE Transactions on Systems, Science and Cybernetics*, SCC-4(2):100–107, 1968.

[176] A. Hashimoto and J. Stevens. Wire routing by optimizing channel assignment with large apertures. In *Proceedings of the Eighth Design Automation Workshop*, pages 155–169. ACM/IEEE, 1971.

[177] R. Hassin. On multicommodity flows in planar graphs. *Networks*, 14:225–235, 1984.

[178] P. S. Hauge, R. Nair, and E. J. Yoffa. Circuit placement for predictable performance. In *Proceedings of the International Conference on Computer-Aided Design*, pages 88–91. IEEE, 1987.

[179] K. S. Hedlund. Aesop, a tool for automatic transistor sizing. In *Proceedings of the 24th Design Automation Conference*, pages 114–120. ACM/IEEE, 1987.

[180] J. Heistermann and T. Lengauer. Context-sensitive tile compaction of finite two-dimensional cell arrays. In preparation, Department of Mathematics and Computer Science, University of Paderborn, Paderborn, West Germany.

[181] J. Heistermann and T. Lengauer. The efficient solution of integer programs for hierarchical global routing. Manuscript, Department of Mathematics and Computer Science, University of Paderborn, Paderborn, West Germany, 1989.

[182] M. Held and R. M. Karp. The traveling-salesman problem and minimum spanning trees. *Operations Research*, 18(6):1138–1162, 1970.

[183] M. Held and R. M. Karp. The traveling-salesman problem and minimum spanning trees: Part II. *Mathematical Programming*, 1:6–25, 1971.

[184] M. Held, P. Wolfe, and H. P. Crowder. Validation of subgradient optimization. *Mathematical Programming*, 6:62–88, 1974.

[185] W. R. Heller, G. Sorkin, and K. Maling. The planar package planner for system designers. In *Proceedings of the 19th Design Automation Conference*, pages 253–260. ACM/IEEE, 1982.

[186] F. L. Heng and A. S. LaPaugh. Optimal compaction of multiple two component channels under river routing. Manuscript, Department of Computer Science, Princeton University, Princeton, NJ, 1988.

[187] A. Herrigel and W. Fichtner. An analytic optimization technique for placement of macro-cells. In *Proceedings of the 26th Design Automation Conference*, pages 376–381. ACM/IEEE, 1989.

[188] W. Heyns, W. Sansen, and H. Beke. A line-expansion algorithm for the general routing problem with a guaranteed solution. In *Proceedings of the 17th Design Automation Conference*, pages 243–249. ACM/IEEE, 1980.

[189] D. W. Hightower. A solution to line routing problems on the continuous plane. In *Proceedings of the Sixth Design Automation Workshop*, pages 1–24. IEEE, 1969.

[190] D. D. Hill. ICON: A tool for design at schematic, virtual-grid and layout levels. *IEEE Design and Test*, 1(4):53–61, 1984.

[191] D. D. Hill, K. Keutzer, and W. H. Wolf. Overview of the IDA system: A toolset for VLSI layout and synthesis. In W. Fichtner and M. Morf, editors, *VLSI CAD Tools and Applications*, Chapter 8, pages 233–263. Kluwer Academic, Boston, MA, 1987.

[192] C. H. A. Hoare. Quicksort. *Computer Journal*, 5(1):10–15, 1962.

[193] J. H. Hoel. Some variations of Lee's algorithm. *IEEE Transactions on Computers*, C-25(1):19–24, 1976.

[194] A. J. Hoffman and H. W. Wielandt. The variation of the spectrum of a normal matrix. *Duke Mathematical Journal*, 20(1):37–39, 1953.

[195] J. E. Hopcroft and R. E. Tarjan. Dividing a graph into triconnected components. *SIAM Journal on Computing*, 2(3):145–169, 1973.

[196] J. E. Hopcroft and R. E. Tarjan. Efficient planarity testing. *Journal of the Association for Computing Machinery*, 21(4):549–568, 1974.

[197] T. M. Hsieh, H. W. Leong, and C. L. Liu. Two-dimensional compaction by simulated annealing. Manuscript.

[198] C. P. Hsu. Minimum via topological routing. *IEEE Transactions on Computer-Aided Design of Integrated Circuits and Systems*, CAD-2(4):235–246, 1983.

[199] M. Y. Hsueh. Symbolic layout and compaction of integrated circuits. Technical Report UCB/ERL M79/80, Electronics Research Laboratory, University of California, Berkeley, CA, 1979.

[200] T. C. Hu and M. T. Shing. The alpha-beta routing. In T. C. Hu and E. S. Kuh, editors, *VLSI Layout: Theory and Design*, pages 139–143. IEEE Press, New York, 1985.

[201] T. C. Hu and M. T. Shing. A decomposition algorithm for circuit routing. In T. C. Hu and E. S. Kuh, editors, *VLSI Layout: Theory and Design*, pages 144–152. IEEE Press, New York, 1985.

[202] M. D. Huang, F. Romeo, and A. L. Sangiovanni-Vincentelli. An efficient general cooling schedule for simulated annealing. In *Proceedings of the International Conference on Computer-Aided Design*, pages 381–384. IEEE, 1986.

[203] F. K. Hwang. On Steiner minimal trees with rectilinear distance. *SIAM Journal on Applied Mathematics*, 30(1):104–114, 1976.

[204] L. Hyafil and R. L. Rivest. Graph partitioning and constructing optimal decision trees are polynomially complete problems. Technical report, IRIA-Laboria, Roquencourt, France, 1973.

[205] M. A. B. Jackson and E. S. Kuh. Performance-driven placement of cell based IC's. In *Proceedings of the 26th Design Automation Conference*, pages 370–375. ACM/IEEE, 1989.

[206] D. W. Jepsen and C. D. Gelatt Jr. Macro placement by Monte Carlo annealing. In *Proceedings of the International Conference on Computer Design: VLSI in Computers*, pages 495–498. IEEE, 1983.

[207] D. L. Johannsen. Bristle blocks: A silicon compiler. In *Proceedings of the 16th Design Automation Conference*, pages 310–313. ACM/IEEE, 1979.

[208] D. B. Johnson. Efficient algorithms for shortest paths in sparse networks. *JACM*, 24(1):1–13, 1977.

[209] D. S. Johnson. The NP-completeness column, an ongoing guide. *Journal of Algorithms*, 5:147–160, 1984. 10th edition.

[210] D. S. Johnson. The NP-completeness column, an ongoing guide. *Journal of Algorithms*, 6:434–451, 1985. 16th edition.

[211] D. S. Johnson, C. R. Aragon, L. A. McGeoch, and C. Schevon. Optimization by simulated annealing: An experimental evaluation (part I). Preprint, AT&T Bell Laboratories, Murray Hill, NJ, 1985.

[212] C. Jones. Improving the performance of the Kernighan-Lin and simulated annealing graph bisection algorithms. In *Proceedings of the 26th Design Automation Conference*. ACM/IEEE, 1989.

[213] R. Joobani and D. P. Siewiorek. Weaver: A knowledge-based routing expert. *IEEE Design and Test*, 3(1):12–33, 1986.

[214] A. B. Kahng. Fast hypergraph partition. In *Proceedings of the 26th Design Automation Conference*, pages 762–766. ACM/IEEE, 1989.

[215] N. Karmarkar. A new polynomial algorithm for linear programming. *Combinatorica*, 4(4):373–395, 1984.

[216] R. M. Karp, F. T. Leighton, R. L. Rivest, C. D. Thompson, U. Vazirani, and V. Vazirani. Global wire routing in two-dimensional arrays. *Algorithmica*, 2:113–129, 1987.

[217] A. V. Karzanov. Determining the maximal flow in a network by the method of preflows. *Soviet Math. Doklady*, 15:434–437, 1974.

[218] L. Kaufman and F. Broeckx. An algorithm for the quadratic assignment problem using Bender's decomposition. *European Journal of Operational Research*, 2:204–211, 1978.

[219] M. Kaufmann. A linear-time algorithm for routing in a convex grid. Technical Report SFB 124-14/87, Department of Computer Science, University of the Saarland, Saarbrücken, West Germany, 1987.

[220] M. Kaufmann. *Über Lokales Verdrahten von Zwei-Punkt-Netzen*. PhD thesis, Department of Computer Science, University of the Saarland, Saarbrücken, West Germany, 1987.

[221] M. Kaufmann and G. Klär. Routing around a rectangle—the general case. Technical Report SFB 124-07/89, Department of Computer Science, University of the Saarland, Saarbrücken, West Germany, 1989.

[222] M. Kaufmann and F. M. Maley. Parity conditions in homotopic knock-knee routing. Manuscript, Department of Computer Science, University of the Saarland, Saarbrücken, West Germany, 1988.

[223] M. Kaufmann and K. Mehlhorn. On local routing of two-terminal nets. Technical Report SFB 124-03/86, Department of Computer Science, University of the Saarland, Saarbrücken, West Germany, 1986.

[224] M. Kaufmann and K. Mehlhorn. Routing through a generalized switchbox. *Journal of Algorithms*, 7:510–531, 1986.

[225] M. Kaufmann and K. Mehlhorn. Routing problems in grid graphs. Technical Report SFB 124-05/89, Department of Computer Science, University of the Saarland, Saarbrücken, West Germany, 1989.

[226] M. Kaufmann and I. G. Tollis. Channel routing with short wires. In J. Reif, editor, *AWOC'88: VLSI Algorithms and Architectures*, pages 226–236. Springer Lecture Notes in Computer Science, No. 319, Springer Verlag, New York, 1988.

[227] T. Kawamoto and Y. Kajitani. The minimum width routing of a 2-row 2-layer polycell-layout. In *Proceedings of the 16th Design Automation Conference*, pages 290–296. ACM/IEEE, 1979.

[228] G. Kedem and H. Watanabe. Graph-optimization techniques for IC-layout and compaction. *IEEE Transactions on Computer-Aided Design of Integrated Circuits and Systems*, CAD-3(1):12–20, 1984.

[229] B. W. Kernighan and S. Lin. An efficient heuristic procedure for partitioning graphs. *Bell Systems Technical Journal*, 49(2):291–307, 1970.

[230] B. W. Kernighan, D. G. Schweikert, and G. Persky. An optimum channel routing algorithm for polycell layouts of integrated circuits. In *Proceedings of the 10th Design Automation Workshop*, pages 50–59. ACM/IEEE, 1973.

[231] K. Keutzer. DAGON: Technology binding and local optimization by DAG matching. In *Proceedings of the 24th Design Automation Conference*, pages 341–347. ACM/IEEE, 1987.

[232] L. G. Khachian. A polynomial algorithm for linear programming. *Doklady Akad. Nauk. USSR*, 244(5):1093–1096, 1979. Translated in *Soviet Math. Doklady* 20:191–194, 1979.

[233] C. Kingsley. A hierarchical, error-tolerant compactor. In *Proceedings of the 21st Design Automation Conference*, pages 126–132. ACM/IEEE, 1984.

[234] S. Kirkpatrick, C. D. Gelatt Jr., and M. P. Vecchi. Optimization by simulated annealing. *Science*, 220(4598):671–680, 1983.

[235] J. M. Kleinhans, G. Sigl, and F. M. Johannes. GORDIAN: A new global optimization/rectangle dissection method for cell placement. In *Proceedings of the International Conference on Computer-Aided Design*, pages 506–509. IEEE, 1988.

[236] J. M. Kleinhans, G. Sigl, and F. M. Johannes. Sea-of-gates placement by simultaneous quadratic programming combined with improved partitioning. In *Proceedings of VLSI'89*. Elsevier Science Publishers B.V., Amsterdam, The Netherlands, 1989.

[237] D. E. Knuth. Big omicron and big omega and big theta. *SIGACT News*, pages 18–24, 1976.

[238] R. Kolla. *Spezifikation und Expansion logisch-topologischer Netze*. PhD thesis, Department of Computer Science, University of the Saarland, Saarbrücken, West Germany, 1986.

[239] R. Kolla, P. Molitor, and H. G. Osthof. *Einführung in den VLSI Entwurf.* Teubner Verlag, Stuttgart, West Germany, 1989.

[240] N. L. Koren. Pin assignment in automated printed circuit board design. In *Proceedings of the Ninth Design Automation Workshop*, pages 72–79. IEEE, 1972.

[241] L. T. Kou and K. Makki. An even faster approximation algorithm for the Steiner tree problem in graphs. *Congressus Numerantium*, 59:147–154, 1987.

[242] L. T. Kou, G. Markowsky, and L. Berman. A fast algorithm for Steiner trees. *Acta Informatica*, 15:141–145, 1981.

[243] K. A. Kozminski and E. Kinnen. Rectangular duals of planar graphs. *Networks*, 15:145–157, 1985.

[244] K. A. Kozminski and E. Kinnen. Rectangular dualization and rectangular dissections. *IEEE Transactions on Circuits and Systems*, CAS-35(11):1401–1416, 1988.

[245] M. R. Kramer and J. van Leeuwen. The complexity of wire routing and finding minimum area layouts for arbitrary VLSI circuits. In F. P. Preparata, editor, *Advances in Computing Research, Volume 2: VLSI Theory*, pages 129–146. JAI Press, Reading, MA, 1984.

[246] S. A. Kravitz and R. R. Rutenbar. Placement by simulated annealing on a multiprocessor. *IEEE Transactions on Computer-Aided Design of Integrated Circuits and Systems*, CAD-6(4):534–549, 1987.

[247] B. Krishnamurthy. An improved min-cut algorithm for partitioning VLSI networks. *IEEE Transactions on Computers*, C-33(5):438–446, 1984.

[248] B. Krishnamurthy and P. Mellema. On the evaluation of mincut partitioning algorithms for VLSI networks. In *Proceedings of the International Symposium on Circuits and Systems*, pages 12–15. IEEE, 1983.

[249] J. B. Kruskal. On the shortest spanning subtree of a graph and the traveling salesman problem. *Proceedings of the American Mathematical Society*, 7(1):48–50, 1956.

[250] R. Kuchem, D. Wagner, and F. Wagner. Area-optimal three-layer channel routing. In *Proceedings of the 30th Annual Symposium on Foundations in Computer Science*. IEEE, 1989.

[251] K. Kuratowski. Sur le probléme des courbes gauche en topologie. *Fundamenta Mathematicae*, 15:271–283, 1930.

[252] A. S. La Paugh. *Algorithms for Integrated Circuit Layout: An Analytic Approach.* PhD thesis, Departyment of Electrical Engineering and Computer Science, Mass. Inst. of Technology, Cambridge, MA, 1980.

[253] A. S. La Paugh and R. Y. Pinter. On minimizing channel density by lateral shifting. In *Proceedings of the International Conference on Computer Design: VLSI in Computers,* pages 121–122. IEEE, 1983.

[254] A. S. La Paugh and R. Y. Pinter. Channel routing for integrated circuits (a survey). Technical Report YALEU/DCS/TR-713, Department of Computer Science, Yale University, New Haven, CT, 1989.

[255] D. P. La Potin and S. W. Director. Mason: A global floorplanning approach for VLSI design. *IEEE Transactions on CAD of Integrated Circuits and Systems,* CAD-5(4):477–489, 1986.

[256] T.-H. Lai and A. Sprague. On the routability of a convex grid. *Journal of Algorithms,* 8:372–384, 1987.

[257] J. Lam and J.-M. Delosme. Performance of a new annealing schedule. In *Proceedings of the 25th Design Automation Conference,* pages 306–311. ACM/IEEE, 1988.

[258] U. Lauther. A min-cut based algorithm for general cell assemblies based on a graph representation. *Journal of Digital Systems,* 4(1):21–34, 1980.

[259] U. Lauther. Channel routing in a general cell environment. In E. Hörbst, editor, *Proceedings of VLSI'85,* pages 389–401. Elsevier Science Publishers B.V., Amsterdam, The Netherlands, 1985.

[260] U. Lauther. Top down hierarchical global routing for channelless gate arrays based on linear assignment. In C. H. Sequin, editor, *Proceedings of VLSI'87,* pages 141–151. Elsevier Science Publishers B.V., Amsterdam, The Netherlands, 1987.

[261] E. L. Lawler. *Combinatorial Optimization: Networks and Matroids.* Holt, Rinehart and Winston, New York, 1976.

[262] E. L. Lawler. Sequencing jobs to minimize total weight completion time subject to precedence constraints. *Annals of Discrete Mathematics,* 2:75–90, 1978.

[263] E. L. Lawler, J. K. Lenstra, A. H. G. Rinnooy Kan, and D. B. Shmoys. *The Traveling Salesman Problem: A Guided Tour of Combinatorial Optimization.* Wiley-Interscience Series in Discrete Mathematics. John Wiley & Sons, New York, 1985.

[264] E. L. Lawler, M. G. Luby, and B. Parker. Finding shortest paths in very large networks. In M. Nagl and J. Perl, editors, *Proceedings of the Ninth International Workshop on Graphtheoretic Concepts in Computer Science,* pages 184–199. Trauner Verlag, Linz, Austria, 1983.

[265] C. Y. Lee. An algorithm for path connection and its applications. *IRE Transactions on Electronic Computers,* EC-10(3):346–365, 1961.

[266] D. T. Lee, S. J. Hong, and C. K. Wong. Number of vias: A control parameter for global wiring of high-density chips. *IBM Journal of Research and Development,* 25(4):261–271, 1981.

[267] F. T. Leighton. New lower bounds for channel routing. Technical Report VLSI Memo No. 82-71, Department of Electrical Engineering and Computer Science, Mass. Inst. of Technology, Cambridge, MA, 1981.

[268] F. T. Leighton, C. E. Leiserson, B. Maggs, S. Plotkin, and J. Wein. Theory of parallel and VLSI computation. Technical Report RSS-1, Laboratory for Computer Science, Mass. Inst. of Technology, Cambridge, MA, 1988.

[269] F. T. Leighton and S. Rao. An approximate max-flow min-cut theorem for uniform multicommodity flow problems with applications to approximation algorithms. In *Proceedings of the 28th Annual Symposium on Foundations of Computing*, pages 422–431. IEEE, 1988.

[270] F. T. Leighton and A. L. Rosenberg. Three-dimensional circuit layouts. *SIAM Journal on Computing*, 15(3):793–813, 1986.

[271] C. E. Leiserson. Fat trees: Universal networks for hardware-efficient supercomputing. *IEEE Transactions on Computers*, C-34(10):892–901, 1985.

[272] C. E. Leiserson and F. M. Maley. Algorithms for routing and testing routability of planar VLSI layouts. In *Proceedings of the 17th Annual ACM Symposium on Theory of Computing*, pages 69–78. ACM, 1985.

[273] C. E. Leiserson and R. Y. Pinter. Optimal placement for river routing. *SIAM Journal on Computing*, 12(3):447–462, 1983.

[274] T. Lengauer. Upper and lower bounds on the complexity of the min-cut linear arrangement problem on trees. *SIAM Journal on Algebraic and Discrete Methods*, 3(1):99–113, 1981.

[275] T. Lengauer. The complexity of compacting hierarchically specified layouts of integrated circuits. In *Proceedings of the 23rd Annual Symposium on Foundations in Computer Science*, pages 358–368. IEEE, 1982.

[276] T. Lengauer. Efficient algorithms for finding minimum spanning forests of hierarchically defined graphs. *Journal of Algorithms*, 8:260–284, 1987.

[277] T. Lengauer. Hierarchical planarity testing algorithms. *JACM*, 36(3):474–509, 1989.

[278] T. Lengauer and K. Mehlhorn. The HILL system: A design environment for the hierarchical specification, compaction, and simulation of integrated circuit layouts. In P. Penfield Jr., editor, *Proceedings of the Second MIT Conference on Advanced Research in VLSI*, pages 139–149. Artech House, Dedham, MA, 1984.

[279] T. Lengauer and K. Mehlhorn. VLSI complexity, efficient VLSI algorithms, and the HILL design system. In C. Trullemans, editor, *Algorithmics for VLSI*, pages 33–89. Academic Press, New York, 1986.

[280] T. Lengauer and R. Müller. The complexity of floorplanning based on binary circuit partitions. Technical Report 46, Department of Mathematics and Computer Science, University of Paderborn, Paderborn, West Germany, 1988.

[281] T. Lengauer and R. Müller. Linear arrangement problems on recursively partitioned graphs. *Zeitschrift für Operations Research*, 32(3):213–230, 1988.

[282] T. Lengauer and R. Müller. A robust framework for hierarchical floorplanning with integrated global routing. Technical Report 70, Department of Mathematics and Computer Science, University of Paderborn, Paderborn, West Germany, 1990.

[283] T. Lengauer, D. Theune, and A. Feldmann. Path problems with general cost criteria and their application in wire routing. Manuscript, Department of Mathematics and Computer Science, University of Paderborn, Paderborn, West Germany, 1990.

[284] T. Lengauer and K. W. Wagner. The correlation between the complexities of the non-hierarchical and hierarchical versions of graph problems. In F. J. Brandenburg, G. Vidal-Naquet, and M. Wirsing, editors, *Proceedings of the Fourth Annual Symposium on Theoretical Aspects of Computer Science*, pages 100–113. Springer Lecture Notes in Computer Science, No. 247, Springer Verlag, New York, 1987.

[285] T. Lengauer and E. Wanke. Efficient analysis of graph properties on cellular graph languages. In T. Lepistö and A. Salomaa, editors, *15th International Symposium on Automata, Languages, and Programming*, pages 379–393. Springer Lecture Notes in Computer Science, No. 317 Springer Verlag, New York, 1988.

[286] T. Lengauer and E. Wanke. Efficient solution of connectivity problems on hierarchically defined graphs. *SIAM Journal on Computing*, 17(6):1063–1081, 1988.

[287] T. Lengauer and C. Wieners. Efficient solutions of hierarchical systems of linear equations. *Computing*, 39:111–132, 1987.

[288] Y.-Z. Liao and C. K. Wong. An algorithm to compact a VLSI symbolic layout with mixed constraints. *IEEE Transactions on Computer-Aided Design of Integrated Circuits and Systems*, CAD-2(2):62–69, 1983.

[289] W. Lipski, Jr. On the structure of three-layer wireable layouts. In F. P. Preparata, editor, *Advances in Computing Research, Volume 2: VLSI Theory*, pages 231–244. JAI Press, London, England, 1984.

[290] W. Lipski, Jr. and F. P. Preparata. A unified approach to layout wirability. *Mathematical Systems Theory*, 19:189–203, 1987.

[291] R. J. Lipton and R. E. Tarjan. A separator theorem for planar graphs. *SIAM Journal on Applied Mathematics*, 36(2):177–189, 1979.

[292] R. Lisanke, F. Brglez, and G. Kedem. McMAP: A fast technology mapping procedure for multi-level logic synthesis. In *Proceedings of the International Conference on Computer Design: VLSI in Computers and Processors*, pages 252–256. IEEE, 1988.

[293] C.-Y. Lo. Automatic tub generation for symbolic layout compaction. In *Proceedings of the 26th Design Automation Conference*, pages 302–306. ACM/IEEE, 1989.

[294] B. Lokanathan and E. Kinnen. Performance optimized floor planning by graph planarization. In *Proceedings of the 26th Design Automation Conference*, pages 116–121. ACM/IEEE, 1989.

[295] L. Lovász. *An Algorithmic Theory of Numbers, Graphs and Convexity.* CBMS-NSF Regional Conference Series in Applied Mathematics No. 50. Society for Industrial and Applied Mathematics, Philadelphia, PA, 1986.

[296] M. G. Luby and P. Ragde. A bidirectional shortest-path algorithm with good average case behavior. *Algorithmica*, 4:551–567, 1989.

[297] F. Luebbert and M. Ulrey. Gate assignment and pack placement: Two approaches compared. In *Proceedings of the 17th Design Automation Conference*, pages 472–482. ACM/IEEE, 1980.

[298] W. K. Luk. A greedy switchbox router. *Integration*, 3(2):129–149, 1985.

[299] W. K. Luk, P. Sipila, M. Tamminen, D. Tang, L. S. Woo, and C. K. Wong. A hierarchical global wiring algorithm for custom chip design. *IEEE Transactions on Computer-Aided Design of Integrated Circuits and Systems*, CAD-6(4):518–533, 1987.

[300] J. A. Lukes. Efficient algorithm for the partitioning of trees. *IBM Journal of Research and Development*, 18(5):217–224, 1974.

[301] M. Lundy and A. Mees. Convergence of an annealing algorithm. *Mathematical Programming*, 34:111–124, 1986.

[302] R. M. MacGregor. *On Partitioning a Graph: A Theoretical and Empirial Study.* PhD thesis, Electronics Research Laboratory, University of California, Berkeley, CA, 1988.

[303] F. M. Maley. Compaction with automatic jog introduction. In H. Fuchs, editor, *Proceedings of the 1985 Chapel Hill Conference on VLSI*, pages 261–284, Rockville, MD, 1985. Computer Science Press.

[304] F. M. Maley. An observation concerning constraint-based compaction. *Information Processing Letters*, 25(2):119–122, 1987.

[305] F. M. Maley. *Single-Layer Wire Routing.* PhD thesis, Department of Electrical Engineering and Computer Science, Mass. Inst. of Technology, Cambridge, MA, 1987.

[306] F. M. Maley. A generic algorithm for one-dimensional homotopic compaction. Manuscript, Department of Computer Science, Princeton University, Princeton, NJ, 1988.

[307] S. Mallela and L. K. Grover. Clustering based simulated annealing for standard cell placement. In *Proceedings of the 25th Design Automation Conference*, pages 312–317. ACM/IEEE, 1988.

[308] M. Marek-Sadowska. An unconstrained topological via minimization. *IEEE Transactions on Computer-Aided Design of Integrated Circuits and Systems*, CAD-3(3):184–190, 1984.

[309] M. Marek-Sadowska. Route planner for custom chip design. In *Proceedings of the International Conference on Computer-Aided Design*, pages 246–249. IEEE, 1986.

[310] M. D. Matson. Optimization of digital MOS VLSI circuits. In H. Fuchs, editor, *Proceedings of the 1985 Chapel Hill Conference on Very Large Scale Integration*, pages 109–126. Computer Science Press, Rockville, MD, 1985.

[311] K. Matsumoto, T. Nishizeki, and N. Saito. Planar multicommodity flows, maximum matchings and negative cycles. *SIAM Journal on Computing*, 15(2):495–510, 1986.

[312] D. W. Matula and F. Shahrokhi. Graph partitioning by sparse cuts and maximum concurrent flow. Technical Report 86-CSE-6, Department of Computer Science and Engineering, Southern Methodist University, Dallas, TX, 1986.

[313] C. Mead and L. Conway. *Introduction to VLSI Systems*. Addison-Wesley Series in Computer Science. Addison-Wesley, Reading, MA, 1980.

[314] K. Mehlhorn. *Data Structures and Algorithms I: Sorting and Searching*, Volume 1 of *EATCS Monographs in Theoretical Computer Science*. Springer Verlag, New York, 1984.

[315] K. Mehlhorn. *Data Structures and Algorithms II: Graph Algorithms and NP-completeness*, Volume 2 of *EATCS Monographs on Theoretical Computer Science*. Springer Verlag, New York, 1984.

[316] K. Mehlhorn. A faster approximation algorithm for the Steiner problem in graphs. *IPL*, 27(3):125–128, 1988.

[317] K. Mehlhorn and S. Näher. A faster compaction algorithm with automatic jog insertion. In J. Allen and F. T. Leighton, editors, *Proceedings of the Fifth MIT Conference on Advanced Research in VLSI*, pages 297–314. The MIT Press, Cambridge, MA, 1988.

[318] K. Mehlhorn and F. P. Preparata. Routing through a rectangle. *Journal of the Association for Computing Machinery*, 33(1):60–85, 1986.

[319] K. Mehlhorn, F. P. Preparata, and M. Sarrafzadeh. Channel routing in knock-knee mode: Simplified algorithms and proofs. *Algorithmica*, 1:213–221, 1986.

[320] K. Mehlhorn and W. Rülling. Compaction on the torus. In J. Reif, editor, *AWOC'88: VLSI Algorithms and Architectures*, pages 212–225. Springer Lecture Notes in Computer Science, No. 319, Springer Verlag, New York, 1988.

[321] K. Mehlhorn and B. H. Schmidt. On BF-orderable graphs. *Discrete Applied Mathematics*, 15:315–327, 1986.

[322] K. Mikami and K. Tabuchi. A computer program for optimal routing of printed circuit connectors. *IFIPS Proceedings*, H47:1475–1478, 1968.

[323] C. E. Miller, A. W. Tucker, and R. A. Zemlin. Integer programming formulations and traveling salesman problems. *Journal of the Association for Computing Machinery*, 7:326–329, 1960.

[324] G. L. Miller. Finding small simple cycle separators for 2-connected planar graphs. In *Proceedings of the 16th Annual ACM Symposium on Theory of Computing*, pages 376–382. ACM, 1984.

[325] G. L. Miller and J. Naor. Flow in planar graphs with multiple sources and sinks. Manuscript, 1988.

[326] G. L. Miller and V. Ramachandran. A new triconnectivity algorithm and its parallelization. In *Proceedings of the 19th Annual ACM Symposium on Theory of Computing*, pages 335–344. ACM, 1987.

[327] Z. Miller and I. H. Sudborough. A polynomial algorithm for recognizing small cutwidth in hypergraphs. In F. Makedon, K. Mehlhorn, T. Papatheodorou, and P. Spirakis, editors, *AWOC'86: VLSI Algorithms and Architectures*, pages 252–260. Springer Lecture Notes in Computer Science, No. 227, Springer Verlag, New York, 1986.

[328] A. Mirzaian. River routing in VLSI. *Journal of Computer and System Sciences*, 31(1):43–54, 1987.

[329] D. Mitra, F. Romeo, and A. L. Sangiovanni-Vincentelli. Convergence and finite-time behavior of simulated annealing. *Advances in Applied Probability*, 18(3):747–771, 1986.

[330] P. Molitor. On the contact minimization problem. In F. J. Brandenburg, G. Vidal-Naquet, and M. Wirsing, editors, *Proceedings of the Fourth Annual Symposium on Theoretical Aspects of Computer Science*, pages 420–431. Springer Lecture Notes in Computer Science, No. 247, Springer Verlag, New York, 1987.

[331] B. Monien and I. H. Sudborough. Min-cut is NP-complete for edge weighted trees. In L. Kott, editor, *Proceedings of 13th International Colloquium on Automata, Languages, and Programming*, pages 265–273. Springer Lecture Notes in Computer Science, No. 226, Springer Verlag, New York, 1986.

[332] E. F. Moore. Shortest path through a maze. In *Proceedings of the International Symposium on Switching Circuits*, pages 285–292. Harvard University Press, Cambridge, MA, 1959. (in Annals of the Harvard Computing Laboratory, Volume 30, Part II).

[333] R. C. Mosteller, A. H. Frey, and R. Suaya. 2-D compaction: A Monte Carlo method. In P. Losleben, editor, *Proceedings of the 1987 Stanford Conference on Advanced Research in VLSI*, pages 173–197. MIT Press, Cambridge, MA, 1987.

[334] A. Mukherjee. *Introduction to NMOS and CMOS VLSI Systems Design*. Prentice Hall, Englewood Cliffs, N. J., 1986.

[335] J. R. Munkres. *Topology, A First Course*. Prentice-Hall, Englewood Cliffs, NJ, 1975.

[336] G. L. Nemhauser and L. A. Wolsey. *Integer and Combinatorial Optimization*. John Wiley & Sons, New York, 1988.

[337] A. P.-C. Ng, P. Raghavan, and C. D. Thompson. Experimental results for a linear program global router. *Computers and Artificial Intelligence*, 6(3):229–242, 1987.

[338] C. H. Ng. A 'gridless' variable-width channel router for macro cell design. In *Proceedings of the 24th Design Automation Conference*, pages 633–636. ACM/IEEE, 1987.

[339] I. Nishioka, T. Kurimoto, S. Yamamoto, I. Shirakawa, and H. Ozaki. An approach to gate assignment and module placement for printed wiring boards. In *Proceedings of the 15th Design Automation Conference*, pages 60–69. ACM/IEEE, 1978.

[340] T. Nishizeki, N. Saito, and K. Suzuki. A linear-time routing algorithm for convex grids. *IEEE Transactions on Computer-Aided Design of Integrated Circuits and Systems*, CAD-4(1):68–75, 1985.

[341] T. Ohtsuki. Maze-running and line-search algorithms. In T. Ohtsuki, editor, *Advances in CAD for VLSI, Volume 4: Layout Design and Verification*, pages 99–131. North-Holland, New York, 1986.

[342] K. Okamura and P. D. Seymour. Multicommodity flows in planar graphs. *Journal of Combinatorial Theory, Series B*, 31(1):75–81, 1981.

[343] J. B. Orlin. A faster strongly polynomial minimum cost flow algorithm. In *Proceedings of the 20th Annual ACM Symposium on Theory of Computing*, pages 377–387. ACM, 1988.

[344] J. B. Orlin and R. K. Ahuja. New distance-directed algorithms for maximum flow and parametric maximum flow problems. Technical Report 192, Operations Research Center, Mass. Inst. of Technology, Cambridge, MA, 1988.

[345] R. H. J. M. Otten. Automatic floorplan design. In *Proceedings of the 19th Design Automation Conference*, pages 261–267. ACM/IEEE, 1982.

[346] R. H. J. M. Otten. Efficient floorplan optimization. In *Proceedings of the International Conference on Computer Design: VLSI in Computers*, pages 499–502. IEEE, 1983.

[347] R. H. J. M. Otten and L. P. P. P. van Ginneken. Floorplan design using simulated annealing. In *Proceedings of the International Conference on Computer-Aided Design*, pages 96–98. IEEE, 1984.

[348] R. H. J. M. Otten and L. P. P. P. van Ginneken. Stop criteria in simulated annealing. In *Proceedings of the International Conference on Computer Design: VLSI in Computers and Processors*, pages 549–553. IEEE, 1988.

[349] J. K. Ousterhout. Corner stitching: A data-structuring technique for VLSI layout tools. *IEEE Transactions on Computer-Aided Design of Integrated Circuits and Systems*, CAD-3(1):87–100, 1984.

[350] C. H. Papadimitriou and K. Steiglitz. *Combinatorial Optimization: Algorithms and Complexity*. Prentice Hall, Inc., Englewood Cliffs, NJ, 1982.

[351] A. M. Patel. A wirability placement algorithm for hierarchical VLSI layout. In *Proceedings of the International Conference on Computer Design: VLSI in Computers*, pages 344–350. IEEE, 1984.

[352] G. Persky, D. N. Deutsch, and D. G. Schweikert. LTX—a minicomputer-based system for automated LSI layout. *Journal of Design Automation and Fault Tolerant Computing*, 1:217–255, 1977.

[353] R. Philipp and E.-J. Prauss. Separators in planar graphs. *Acta Informatica*, 14:87–106, 1980.

[354] L. T. Pillage and R. A. Rohrer. A quadratic metric with a simple solution scheme for initial placement. In *Proceedings of the 25th Design Automation Conference*, pages 324–329. ACM/IEEE, 1988.

[355] R. Y. Pinter. *The Impact of Layer Assignment Methods on Layout Algorithms for Integrated Circuits*. PhD thesis, Department of Electrical Engineering and Computer Science, Mass. Inst. of Technology, Cambridge, MA, 1982.

[356] R. Y. Pinter. On routing two-point nets across a channel. In *Proceedings of the 19th Design Automation Conference*, pages 894–902. ACM/IEEE, 1982.

[357] R. Y. Pinter. Optimal layer assignment for interconnect. *Journal of VLSI and Computer Systems*, 1:123–137, 1984.

[358] J. Plesnik. A bound for the Steiner tree problem in graphs. *Mathematica Slovaca*, 31(2):155–163, 1981.

[359] I. Pohl. Bi-directional search. In B. Meltzer and D. Mitchie, editors, *Machine Intelligence*, Volume 6, pages 127–140. Edinburgh University Press, Edinburgh, Scotland, 1971.

[360] S. Prasitjutrakul and W. J. Kubitz. Path-delay constrained floorplanning: A mathematical programming approach for initial placement. In *Proceedings of the 26th Design Automation Conference*, pages 364–369. ACM/IEEE, 1989.

[361] B. T. Preas and C.-S. Chow. Placement and routing algorithms for topological integrated circuit layout. In *Proceedings of the International Symposium on Circuits and Systems*, pages 17–20. IEEE, 1985.

[362] B. T. Preas and P. T. Karger. Placement, assignment and floorplanning. In B. T. Preas and M. J. Lorenzetti, editors, *Physical Design Automation of VLSI Systems*, Chapter 4, pages 87–155. Benjamin Cummings, Menlo Park, CA, 1988.

[363] F. P. Preparata and W. Lipski, Jr. Optimal three-layer channel routing. *IEEE Transactions on Computers*, C-33(5):427–437, 1984.

[364] F. P. Preparata and M. Sarrafzadeh. Channel routing of nets of bounded degree. In P. Bertolazzi and F. Luccio, editors, *Proceedings of VLSI: Algorithms and Architectures*, pages 189–203. North-Holland, New York, 1984.

[365] F. P. Preparata and J. Vuillemin. The cube-connected cycles: A versatile network for parallel computation. *Communications of the Association for Computing Machinery*, 24(5):300–309, 1981.

[366] R. C. Prim. Shortest connection networks and some generalizations. *Bell System Technical Journal*, 36(6):1389–1401, 1957.

[367] W. R. Pulleyblank. Polyhedral combinatorics. In G. L. Nemhauser, A. H. G. Rinnooy Kan, and M. J. Todd, editors, *Handbooks in Operations Research and Management Science, Volume 1*, pages 371–446. North-Holland, New York, 1989.

[368] N. R. Quinn. The placement problem as viewed from the physics of classical mechanics. In *Proceedings of the 12th Design Automation Conference*, pages 173–178. ACM/IEEE, 1975.

[369] P. Raghavan. Integer programming in VLSI design. *Discrete Applied Mathematics*. In press.

[370] P. Raghavan. Probabilistic construction of deterministic algorithms: Approximating packing integer programs. *Journal of Computer and System Sciences*, 37(2):130–143, 1988.

[371] P. Raghavan and C. D. Thompson. Randomized rounding: A technique for provably good algorithms and algorithmic proofs. *Combinatorica*, 7(4):365–374, 1987.

[372] P. Rao, R. Ramnarayan, and G. Zimmermann. SPIDER: A chip planner for ISL technology. In *Proceedings of the 21st Design Automation Conference*, pages 665–666. ACM/IEEE, 1984.

[373] S. Rao. Finding near optimal separators in planar graphs. In *Proceedings of the 28th Annual IEEE Symposium on Foundations of Computer Science*, pages 225–237. IEEE, 1987.

[374] C. P. RaviKumar and L. M. Patnaik. Parallel placement based on simulated annealing. In *Proceedings of the International Conference on Computer Design: VLSI in Computers and Processors*, pages 91–94. IEEE, 1987.

[375] V. J. Rayward-Smith and A. Clare. On finding Steiner vertices. Technical report, School of Comp. Studies and Accountancy, University of East Anglia, 1984.

[376] J. Reed, A. L. Sangiovanni-Vincentelli, and M. Santomauro. A new symbolic channel router, YACR2. *IEEE Transactions on Computer-Aided Design of Integrated Circuits and Systems*, CAD-4(3):208–219, 1985.

[377] R. L. Rivest, A. E. Baratz, and G. L. Miller. Provably good channel routing algorithms. In H. T. Kung, B. Sproull, and G. Steele, editors, *VLSI Systems and Computations*, pages 153–159. Computer Science Press, Rockville, MD, 1981.

[378] R. L. Rivest and C. M. Fiduccia. A "greedy" channel router. In *Proceedings of the 19th Design Automation Conference*, pages 418–424. ACM/IEEE, 1982.

[379] N. Robertson and P. D. Seymour. Graph minors XIII: The disjoint paths problem. Manuscript. Bell Communications Research, Morristown, NJ, 1986.

[380] N. Robertson and P. D. Seymour. Graph minors XV: Wagner's conjecture. Manuscript. Bell Communications Research, Morristown, NJ, 1988.

[381] D. J. Rose. On simple characterizations of k-trees. *Discrete Mathematics*, 7:317–322, 1974.

[382] J. Rose, W. Klebsch, and J. Wolf. Equilibrium detection and temperature measurement of simulated annealing placements. In *Proceedings of the International Conference on Computer-Aided Design*, pages 514–517. IEEE, 1988.

[383] E. Rosenberg. Optimal module sizing in VLSI floorplanning by nonlinear programming. AT&T Bell Laboratories, Holmdel, NJ, 1988.

[384] F. Rubin. The Lee path connection algorithm. *IEEE Transactions on Computers*, C-23(9):907–914, 1974.

[385] A. E. Ruehli and D. L. Ostapko. VLSI circuit analysis, timing verification and performance optimization. In W. Fichtner and M. Morf, editors, *VLSI CAD Tools and Applications*, Chapter 5, pages 129–146. Kluwer Academic, Norwell, MA, 1987.

[386] W. Rülling. Legalization of mask layouts. Manuscript, Department of Computer Science, University of the Saarland, Saarbrücken, West Germany, 1987.

[387] W. Rülling and T. Schilz. A new method for hierarchical compaction. Manuscript, Department of Computer Science, University of the Saarland, Saarbrücken, West Germany, 1988.

[388] M. Sarrafzadeh. Channel-routing problem in the knock-knee mode is NP-complete. *IEEE Transactions on Computer-Aided Design of Integrated Circuits and Systems*, CAD-6(4):503–506, 1987.

[389] M. Sarrafzadeh and D. T. Lee. Topological via minimization revisited. Technical Report CIMS-88-01, Department of Electrical Engineering and Computer Science, Northwestern University, Evanston, IL, 1988.

[390] M. Sarrafzadeh and D. T. Lee. A new approach to topological via minimization. *IEEE Transactions on Computer-Aided Design of Integrated Circuits and Systems*, CAD-8(8):890–900, 1989.

[391] M. Sarrafzadeh and F. P. Preparata. Compact channel routing of multiterminal nets. *Annals of Discrete Mathematics*, 25:255–280, 1985.

[392] M. Sarrafzadeh and F. P. Preparata. A bottom-up layout technique based on two-rectangle routing. *Integration*, 5(3,4):231–246, 1987.

[393] G. H. Sasaki and B. Hajek. The time complexity of maximum matching by simulated annealing. *Journal of the Association for Computing Machinery*, 35(2):387–403, 1988.

[394] W. Schiele. Improved compaction with minimized length of wires. In *Proceedings of the 20th Design Automation Conference*, pages 121–127. ACM/IEEE, 1983.

[395] W. L. Schiele. Compaction with incremental over-constraint resolution. In *Proceedings of the 25th Design Automation Conference*, pages 390–395. ACM/IEEE, 1988.

[396] M. D. F. Schlag, R. Anderson, and S. Kahan. An $O(n \log n)$ algorithm for 1D-tile compaction. Technical Report UCSC-CRL-89-09, Computer Research Laboratory, University of California, Santa Cruz, CA, 1989.

[397] M. D. F. Schlag, Y.-Z. Liao, and C. K. Wong. An algorithm for optimal two-dimensional compaction of VLSI layouts. *Integration*, 1(2,3):179–209, 1983.

[398] M. D. F. Schlag, L. S. Woo, and C. K. Wong. Maximizing pin alignment by pin permutations. *Integration*, 2(4):279–307, 1984.

[399] A. Schrijver. Decomposition of graphs on surfaces and a homotopic circulation theorem. Technical Report OS-R8719, Centre for Mathematics and Computer Science, Amsterdam, The Netherlands, 1986.

[400] A. Schrijver. Edge-disjoint homotopic paths in straight-line planar graphs. Manuscript, Centre for Mathematics and Computer Science, Amsterdam, The Netherlands, 1987.

[401] A. Schrijver. Disjoint homotopic paths and trees in a planar graph. Part I: Description of the method. Part II: Correctness of the method. Part III: Disjoint trees. Manuscript, Department of Econometrics, Tilburg University, Tilburg, The Netherlands, 1988.

[402] B. Schürmann. Hierarchisches top-down chip planning. *Informatik Spektrum*, 11(2):57–71, 1988.

[403] D. G. Schweikert. A two-dimensional placement algorithm for the layout of electrical circuits. In *Proceedings of the 13th Design Automation Conference*, pages 408–416. ACM/IEEE, 1976.

[404] D. G. Schweikert and B. W. Kernighan. A proper model for the partitioning of electrical circuits. In *Proceedings of the Ninth Design Automation Workshop*, pages 57–62. ACM/IEEE, 1972.

[405] W. S. Scott and J. K. Ousterhout. Plowing: Interactive stretching and compaction in Magic. In *Proceedings of the 21st Design Automation Conference*, pages 166–172. ACM/IEEE, 1984.

[406] C. Sechen. Chip-planning, placement, and global routing of macro/custom cell integrated circuits using simulated annealing. In *Proceedings of the 25th Design Automation Conference*, pages 73–80. ACM/IEEE, 1988.

[407] C. Sechen. *VLSI Placement and Routing Using Simulated Annealing*. Kluwer Academic, Boston, MA, 1988.

[408] C. Sechen and A. L. Sangiovanni-Vincentelli. The TimberWolf placement and routing package. *IEEE Journal of Solid-State Circuits*, SC-20(2):510–522, 1985.

[409] C. Sechen and A. L. Sangiovanni-Vincentelli. TimberWolf 3.2: A new standard cell placement and global routing package. In *Proceedings of the 23rd Design Automation Conference*, pages 602–608. ACM/IEEE, 1986.

[410] R. Sedgewick. The analysis of quicksort programs. *Acta Informatica*, 7:327–355, 1977.

[411] R. Sedgewick. *Algorithms*. Addison-Wesley, Reading, MA, second edition, 1988.

[412] R. Sedgewick and J. S. Vitter. Shortest paths in euclidean graphs. In *Proceedings of the 25th Annual IEEE Symposium on Foundations of Computer Science*, pages 417–424. IEEE, 1984.

[413] E. Seneta. *Non-negative Matrices and Markov Chains*. Springer Verlag, New York, 1981.

[414] P. D. Seymour. On odd cuts and planar multicommodity flows. *Proceedings of the London Mathematical Society*, 42(3):178–192, 1981.

[415] L. Sha and R. W. Dutton. An analytical algorithm for placement of arbitrarily sized rectangular blocks. In *Proceedings of the 22nd Design Automation Conference*, pages 602–608. ACM/IEEE, 1985.

[416] C. E. Shannon. The synthesis of two-terminal switching circuits. *Bell System Technical Journal*, 28(1):45–98, 1949.

[417] J. F. Shapiro. A survey of Lagrangian techniques for discrete optimization. *Annals of Discrete Mathematics*, 5:113–138, 1979.

[418] D. D. Sherlekar and J. JáJá. Input sensitive VLSI layouts for graphs of arbitrary degree. In J. Reif, editor, *AWOC'88: VLSI Algorithms and Architectures*, pages 268–277. Springer Lecture Notes in Computer Science, No. 319, Springer Verlag, New York, 1988.

[419] H. Shin and L. Chi-Yuan. An efficient two-dimensional layout compaction algorithm. In *Proceedings of the 26th Design Automation Conference*, pages 290–295. ACM/IEEE, 1989.

[420] H. Shin, A. L. Sangiovanni-Vincentelli, and C. H. Sequin. Two-dimensional compaction by "zone refining". In *Proceedings of the 23rd Design Automation Conference*, pages 115–122. ACM/IEEE, 1986.

[421] T. Shiple, P. Kollaritsch, D. Smith, and A. Jonathan. Area evaluation metrics for transistor placement. In *Proceedings of the International Conference on Computer Design: VLSI in Computers and Processors*, pages 428–433. IEEE, 1988.

[422] E. Shragowitz, J. Lee, and S. Sahni. Placer–router for "sea of gates" design style. In *Proceedings of the International Conference on Computer Design: VLSI in Computers and Processors*, pages 330–335. IEEE, 1987.

[423] H. E. Shrobe. The data path generator. In P. Penfield Jr., editor, *Proceedings of the MIT Conference on Advanced Research in VLSI*, pages 175–181, Dedham, MA, 1982. Artech House.

[424] D. D. Sleator and R. E. Tarjan. An $O(mn \log n)$ algorithm for maximum network flow. Technical Report STAN-CS-80-381, Computer Science Department, Stanford University, Stanford, CA, 1980.

[425] D. D. Sleator and R. E. Tarjan. A data structure for dynamic trees. *Journal of Computer and System Sciences*, 24:362–391, 1983.

[426] J. Soukup. Fast maze router. In *Proceedings of the 15th Design Automation Conference*, pages 100–102. ACM/IEEE, 1978.

[427] J. Soukup. Global router. *Journal of Digital Systems*, 4(1):59–69, 1980.

[428] J. Soukup and Y. Lapid. A tree driven maze router. Manuscript, CADENCE Corp., Ontario, Canada, 1988.

[429] L. Stockmeyer. Optimal orientations of cells in slicing floorplan design. *Information and Control*, 57:91–101, 1983.

[430] P. R. Suaris and G. Kedem. A quadrisection-based combined place and route scheme for standard cells. *IEEE Transactions on Computer-Aided Design of Integrated Circuits and Systems*, CAD-8(3):234–244, 1989.

[431] G. F. Sullivan. Approximation algorithms for Steiner tree problems. Technical Report 249, Department of Computer Science, Yale University, New Haven, CT, 1982.

[432] P. K. Sun. An octagonal geometry compactor. In *Proceedings of the International Conference on Computer Design: VLSI in Computers and Processors*, pages 190–193. IEEE, 1988.

[433] Z. A. Syed and A. El Gamal. Single layer routing of power and ground networks in integrated circuits. *Journal of Digital Systems*, 6(1):53–63, 1982.

[434] A. A. Szepieniec. *HECTIC—Highly Efferent Construction of Topology of Integrated Circuits*. PhD thesis, Eindhoven University of Technology, Eindhoven, The Netherlands, 1986.

[435] A. A. Szepieniec. Integrated placement/routing in sliced layouts. In *Proceedings of the 23rd Design Automation Conference*, pages 300–307. ACM/IEEE, 1986.

[436] T. G. Szymanski. Dogleg channel routing is NP-complete. *IEEE Transactions on Computer-Aided Design of Integrated Circuits and Systems*, CAD-4(1):31–41, 1985.

[437] H. Takahashi and A. Matsuyama. An approximate solution for the Steiner problem in graphs. *Mathematica Japonica*, 24(6):573–577, 1980.

[438] D. Tan and N. Weste. Virtual grid symbolic layout 1987. In *Proceedings of the International Conference on Computer Design: VLSI in Computers and Processors*, pages 192–196. IEEE, 1987.

[439] E. Tardos. A strongly polynomial minimum cost circulation algorithm. *Combinatorica*, 5:247–255, 1985.

[440] R. E. Tarjan. Depth first search and linear graph algorithms. *SIAM Journal on Computing*, 1(2):146–160, 1972.

[441] R. E. Tarjan. *Data Structures and Network Algorithms*. CBMS-NSF Regional Conference Series in Applied Mathematics No. 44. Society for Industrial and Applied Mathematics, Philadelphia, PA, 1983.

[442] C. D. Thompson. *A Complexity Theory for VLSI*. PhD thesis, Department of Computer Science, Carnegie-Mellon University, Pittsburgh, PA, 1980.

[443] I. G. Tollis. *Algorithms for VLSI Layouts*. PhD thesis, Technical Report No. UILU-ENG-87-2268, Coordinated Science Laboratory, University of Illinois at Urbana-Champaign, Urbana, IL, 1987.

[444] I. G. Tollis. Wiring in uniform grids and two-colorable maps. Technical Report UTDCS 18-88, Department of Computer Science, The University of Texas at Dallas, Dallas, TX, 1987.

[445] I. G. Tollis. A new algorithm for wiring layouts. In J. Reif, editor, *AWOC'88: VLSI Algorithms and Architectures*, pages 257–267. Springer Lecture Notes in Computer Science, No. 319, Springer Verlag, New York, 1988.

[446] I. G. Tollis and A. V. Vaguine. Improved techniques for wiring layouts in the square grid. In *Proceedings of the International Symposium on Circuits and Systems*, pages 1875–1878. IEEE, 1989.

[447] M. Tompa. An optimal solution to a wire-routing problems. *Journal of Computer and System Sciences*, 23(2):127–150, 1981.

[448] R.-S. Tsay, E. S. Kuh, and C.-P. Hsu. PROUD: A fast sea-of-gates placement algorithm. In *Proceedings of the 25th Design Automation Conference*, pages 318–323. ACM/IEEE, 1988.

[449] S. Tsukiyama, M. Fukui, and I. Shirakawa. A heuristic algorithm for a pin assignment problem of gate array LSI's. In *Proceedings of the International Symposium on Circuits and Systems*, pages 465–469. IEEE, 1984.

[450] J. Valdes. *Parsing Flowcharts and Series-Parallel Graphs*. PhD thesis, Computer Science Department, Stanford University, Stanford, CA, 1979.

[451] R. Varadarajan. Algorithms for circuit layout compaction of building blocks. Master's thesis, Texas Tech University, 1985.

[452] M. P. Vecchi and S. D. Kirkpatrick. Global wiring by simulated annealing. *IEEE Transactions on Computer-Aided Design of Integrated Circuits and Systems*, CAD-2(4):215–222, 1983.

[453] A. F. Veinott Jr. and G. B. Dantzig. Integral extreme points. *SIAM Review*, 10(3):371–372, 1968.

[454] J. A. Wald and C. J. Colbourn. Steiner trees, partial 2-trees, and minimum ifi networks. *Networks*, 13:159–167, 1983.

[455] E. Wanke. Algorithms for graph problems on BNLC structured graphs. Technical Report Theoretische Informatik No. 58, Department of Mathematics and Computer Science, University of Paderborn, Paderborn, West Germany, 1989.

[456] H. Watanabe. *IC Layout Generation and Compaction Using Mathematical Optimization*. PhD thesis, Department of Computer Science, University of Rochester, Rochester, NY, 1984.

[457] J. Waterkamp, R. Wicke, R. Brück, M. Reinhardt, and G. Schrammeck. Technology tracking of non Manhattan VLSI layout. In *Proceedings of the 26th Design Automation Conference*, pages 296–301. ACM/IEEE, 1989.

[458] B. X. Weis. *Globale Plazierung und Anordnung von Gate-Arrays*. PhD thesis, Department of Electrical Engineering, University of Karlsruhe, Karlsruhe, West Germany, 1987.

[459] B. X. Weis and D. A. Mlynski. A graphtheoretic approach to the relative placement problem. *IEEE Transactions on Circuits and Systems*, CAS-35(3):286–293, 1988.

[460] N. H. E. Weste and K. Eshragian. *Principles of CMOS VLSI Design: A Systems Perspective*. The VLSI Systems Series. Addison-Wesley, Reading, MA, 1985.

[461] H. Whitney. Congruent graphs and the connectivity of graphs. *American Journal of Mathematics*, 54:150–168, 1932.

[462] H. Whitney. Non-separable and planar graphs. *Transactions of the American Mathematical Society*, 34:339–362, 1932.

[463] H. Whitney. Planar graphs. *Fund. Math.*, 21:73–84, 1933.

[464] P. Widmayer. Fast approximation algorithms for Steiner's problem in graphs. Habilitation, University of Karlsruhe, West Germany, 1987.

[465] P. Widmayer. Network design issues in VLSI. Manuscript, Department of Computer Science, University of Freiburg, Freiburg, West Germany, 1989.

[466] P. Widmayer and C. K. Wong. An optimal algorithm for the maximum alignment of terminals. *Information Processing Letters*, 20(2):75–82, 1985.

[467] C. Wieners-Lummer. A Manhattan channel router with good theoretical and practical performance. In *Proceedings of the First ACM-SIAM Symposium on Discrete Algorithms*, pages 465–474. ACM/SIAM, 1990.

[468] S. Wimer, I. Koren, and I. Cederbaum. Optimal aspect ratios of building blocks in VLSI. *IEEE Transactions on Computer-Aided Design of Integrated Circuits and Systems*, CAD-8(2):139–145, 1989.

[469] P. Winter. Steiner problem in networks: A survey. *Networks*, 17:129–167, 1987.

[470] G. J. Wipfler, M. Wiesel, and D. A. Mlynski. A combined force and cut algorithm for hierarchical VLSI layout. In *Proceedings of the 19th Design Automation Conference*, pages 671–677. ACM/IEEE, 1982.

[471] W. H. Wolf and A. E. Dunlop. Symbolic layout and compaction. In B. T. Preas and M. J. Lorenzetti, editors, *Physical Design Automation of VLSI Systems*, Chapter 6, pages 211–281. Benjamin Cummings, Menlo Park, CA, 1988.

[472] W. H. Wolf, R. G. Mathews, J. A. Newkirk, and R. W. Dutton. Algorithms for optimizing, two-dimensional symbolic layout compaction. *IEEE Transactions on Computer-Aided Design of Integrated Circuits and Systems*, CAD-7(4):451–466, 1988.

[473] D. F. Wong, H. W. Leong, and C. L. Liu. *Simulated Annealing for VLSI Design*. Kluwer Academic, Boston, MA, 1988.

[474] D. F. Wong and P. S. Sakhamuri. Efficient floorplan area optimization. In *Proceedings of the 26th Design Automation Conference*, pages 586–589. ACM/IEEE, 1989.

[475] Y.-F. Wu, P. Widmayer, M. D. F. Schlag, and C. K. Wong. Rectilinear shortest paths and minimum spanning trees in the presence of rectilinear obstacles. *IEEE Transactions on Computers*, C-36(3):321–331, 1987.

[476] Y.-F. Wu, P. Widmayer, and C. K. Wong. A faster approximation algorithm for the Steiner problem in graphs. *Acta Informatica*, 23:223–229, 1986.

[477] X.-M. Xiong. A new algorithm for topological routing and via minimization. In *Proceedings of the International Conference on Computer-Aided Design*, pages 410–413. IEEE, 1988.

[478] X.-M. Xiong and E. S. Kuh. The constrained via minimization problems for PCB and VLSI design. In *Proceedings of the 25th Design Automation Conference*, pages 573–578. ACM/IEEE, 1988.

[479] M. Yannakakis. A polynomial algorithm for the min-cut linear arrangement of trees. *Journal of the Association for Computing Machinery*, 32(4):950–988, 1985.

[480] X. Yao, M. Yamada, and C. L. Liu. A new approach to the pin assignment problem. In *Proceedings of the 25th Design Automation Conference*, pages 566–572. ACM/IEEE, 1988.

[481] T. Yoshimura. An efficient channel router. In *Proceedings of the 21st Design Automation Conference*, pages 38–44. ACM/IEEE, 1984.

[482] T. Yoshimura and E. S. Kuh. Efficient algorithms for channel routing. *IEEE Transactions on Computer-Aided Design of Integrated Circuit and Systems*, CAD-1(1):25–35, 1982.

[483] X. Zhang, L. T. Pillage, and R. A. Rohrer. Efficient final placement based on nets-as-points. In *Proceedings of the 26th Design Automation Conference*, pages 578–581. ACM/IEEE, 1989.

[484] G. Zimmermann. A new area and shape function estimation technique for VLSI layouts. In *Proceedings of the 25th Design Automation Conference*, pages 60–65. ACM/IEEE, 1988.

Author Index

Subject Index